MICROBIAL PHYSIOLOGY

Fourth Edition

Albert G. Moat
John W. Foster
Michael P. Spector

WILEY-LISS

A JOHN WILEY & SONS, INC., PUBLICATION

Cover images (clockwise from upper-left corner): 1) Coralline shape of the bacterial nucleoid. Reprinted with permission from Bohrmann, B. et al. 1991. *J. Bacteriol.* 173:3149–3158 (see Chapter 7), 2) Cell envelope-free nucleoid from *E. coli* (left frame RNAase treated, right frame washed). Reprinted with permission from Kavenoff, R. and B.C. Bowen, 1976. *Chromosome* 59:89–101 (see Chapter 7), 3) Pili (fimbriae) and flagella of *Proteus mirabilis*. Reprinted with permission from Hoeniger, J.F.M. 1965. *J. Gen. Microbiol.* 40:29. 4) Cell envelope-free nucleoid from *E. coli*. Reprinted with permission from Kavenoff, R. and B.C. Bowen, 1976. *Chromosome* 59:89–101 (see Chapter 7).

Library of Congress Cataloging-in-Publication Data:

Moat, Albert G.
    Microbial physiology / Albert G. Moat, John W. Foster, Michael P. Spector.–4th ed.
        p.cm.
    Includes bibliographical references and index.
    ISBN 0-471-39483-1 (paper: alk. paper)
    1. Microorganisms—Physiology. I. Foster, John Watkins. II. Spector, Michael P.
III. Title.

QR84.M64 2002
571.29–dc21   2002071331

CONTENTS

5 REGULATION OF PROKARYOTIC GENE EXPRESSION   194

8 CENTRAL PATHWAYS OF CARBOHYDRATE METABOLISM 350

PREFACE

The field of microbial physiology has expanded at an incredibly rapid pace since the last edition of this text. The development and implementation of new, highly sophisticated, techniques to study the molecular genetics and physiology of an ever broadening range of microbes has prompted us to write a fourth edition to this book. To give full measure to the extraordinary advances made in microbial physiology we have found it necessary to reorder, separate, and add new material. However, in doing so we have attempted to remain true to the goal of the first edition of "Moat's Notes" and each subsequent edition by targeting discussions to undergraduate and beginning graduate students while providing sufficient detail useful to established microbial physiologists. This new edition continues the tradition of addressing the physiology of a variety of microbes and not just *Escherichia coli*. We have updated chapters on bacterial structures, intermediary metabolism, genetics and growth; and added chapters discussing the genomic and proteomic methodologies employed by the new breed of microbial physiologist. We have reorganized, updated and expanded chapters on microbial stress responses and bacterial differentiation and have added a chapter on host-parasite interactions that correlates microbial physiology with microbial pathogenesis. We hope that the reader, be they an advanced undergraduate entering the field or a professor who has been in the field for forty years, will come to better appreciate the elegant simplicities and the intricate complexities of microbial physiology, while at the same time realizing that there is still much to be learned.

The authors would like to thank the many students, colleagues, and family who provided help and encouragement as we compiled this new edition. We are particularly thankful to those who granted us permission to use figures or illustrations and/or provided us with original materials for this purpose.

ALBERT G. MOAT
JOHN W. FOSTER
MICHAEL P. SPECTOR

Huntington, West Virginia
Mobile, Alabama

CHAPTER 1

INTRODUCTION TO MICROBIAL PHYSIOLOGY

THE *ESCHERICHIA COLI* PARADIGM

Microbial physiology is an enormous discipline encompassing the study of thousands of different microorganisms. It is, of course, foolhardy to try to convey all that is known on this topic within the confines of one book. However, a solid foundation can be built using a limited number of organisms to illustrate key concepts of the field. This text helps set the foundation for further inquiry into microbial physiology and genetics. The gram-negative organism *Escherichia coli* is used as the paradigm. Other organisms that provide significant counterexamples to the paradigm or alternative strategies to accomplish a similar biochemical goal are also included. In this chapter we paint a broad portrait of the microbial cell with special focus on *E. coli*. Our objective here is to offer a point of confluence where the student can return periodically to view how one aspect of physiology might relate to another. Detailed treatment of each topic is provided in later chapters.

CELL STRUCTURE

As any beginning student of microbiology knows, bacteria come in three basic models: spherical (coccus), rod (bacillus), and spiral (spirillum). They do not possess a membrane-bound nucleus as do eukaryotic microorganisms; therefore, they are prokaryotic. In addition to these basic types of bacteria, there are other more specialized forms described as budding, sheathed, and mycelial. Figure 1-1 presents a schematic representation of a typical (meaning *E. coli*) bacterial cell.

The Cell Surface

The interface between the microbial cell and its external environment is the cell surface. It protects the cell interior from external hazards and maintains the integrity of the cell

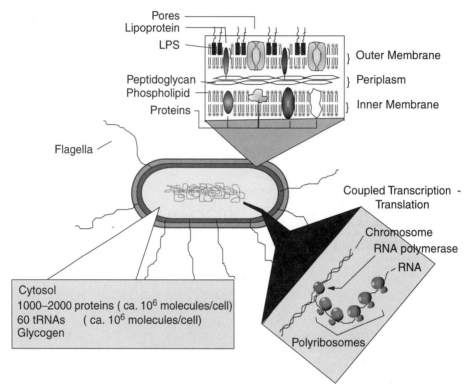

Fig. 1-1. Diagrammatic representation of a "typical" bacterial cell (*Escherichia coli*). Portions of the cell are enlarged to show further details.

as a discrete entity. Although it must be steadfast in fulfilling these functions, it must also enable transport of large molecules into and out of the cell. These large molecules include carbohydrates (e.g., glucose), vitamins (e.g., vitamin B_{12}), amino acids, and nucleosides, as well as proteins exported to the exterior of the cell. The structure and composition of different cell surfaces can vary considerably depending on the organism.

Cell Wall. In 1884, the Danish investigator Christian Gram devised a differential stain based on the ability of certain bacterial cells to retain the dye crystal violet after decoloration with 95% ethanol. Cells that retained the stain were called gram positive. Subsequent studies have shown that this fortuitous discovery distinguished two fundamentally different types of bacterial cells. The surface of gram-negative cells is much more complex than that of gram-positive cells. As shown in the schematic drawings in Figure 1-2, the gram-positive cell surface has two major structures: the cell wall and the cell membrane. The cell wall of gram-positive cells is composed of multiple layers of peptidoglycan, which is a linear polymer of alternating units of N-acetylglucosamine (NAG) and N-acetylmuramic acid (NAM). A short peptide chain is attached to muramic acid. A common feature in bacterial cell walls is cross-bridging between the peptide chains. In a gram-positive organism such as *Staphylococcus aureus*, the cross-bridging between adjacent peptides may be close to 100%. By contrast, the frequency of cross-bridging in *Escherichia coli* (a gram-negative organism) may be as low as 30% (Fig. 1-3). Other components—for example, lipoteichoic acid (only

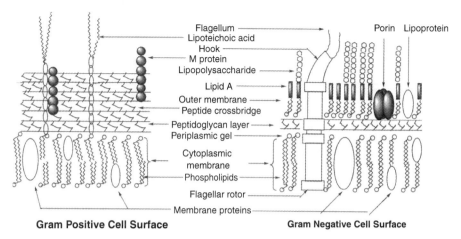

Gram Positive Cell Surface   Gram Negative Cell Surface

Fig. 1-2. Composition of the cell surfaces of gram-positive and gram-negative bacteria. Not all structures shown are found in all organisms. For example, M protein is only used to describe a structure in some of the streptococci. Also, not all organisms have flagella.

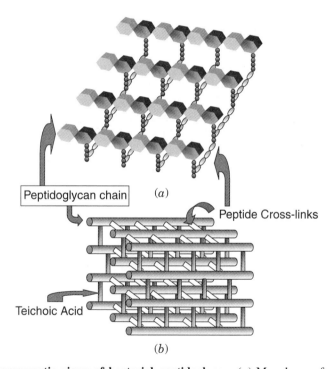

Peptidoglycan chain   (a)   Peptide Cross-links

Teichoic Acid

(b)

Fig. 1-3. Diagrammatic views of bacterial peptidoglycan. (a) Monolayer of peptidoglycan. Lightly shaded hexagons represent N-acetylglucosamine; darkly shaded hexagons represent N-acetylmuramic acid; vertically arranged spheres represent the peptide side chains; horizontal ovals represent the amino acid cross-bridges between peptide chains. (b) Diagrammatic representation of the multilayered peptidoglycan in the gram-positive cell wall. Long horizontal bars denote the chains of N-acetylglucosamine and N-acetylmuramic acid. Short horizontal bars indicate peptide cross-bridges and vertical bars represent teichoic acid.

present in gram-positive organisms) — are synthesized at the membrane surface and may extend through the peptidoglycan layer to the outer surface.

The peptidoglycan layer of a gram-negative cell is generally a single monolayer. An outer membrane surrounding the gram-negative cell is composed of phospholipids, lipopolysaccharides, enzymes, and other proteins, including lipoproteins. The space between this outer membrane and the inner membrane is referred to as the **periplasmic space**. It may be traversed at several points by various enzymes and other proteins (Fig. 1-2).

Membranes. The cytoplasmic membrane of both gram-positive and gram-negative cells is a lipid bilayer composed of phospholipids, glycolipids, and a variety of proteins. The proteins in the cytoplasmic membrane may extend through its entire thickness. Some of these proteins provide structural support to the membrane while others function in the transport of sugars, amino acids, and other metabolites.

The outer membrane of gram-negative cells contains a relatively high content of **lipopolysaccharides**. These lipid-containing components represent one of the most important identifying features of gram-negative cells: the **O antigens,** which are formed by the external polysaccharide chains of the lipopolysaccharide. This lipid-containing component also displays **endotoxin** activity — that is, it is responsible for the shock observed in severe infections caused by gram-negative organisms. Bacterial cell surfaces also contain specific carbohydrate or protein receptor sites for the attachment of **bacteriophages**, which are viruses that infect bacteria. Once attached to these receptor sites, the bacteriophage can initiate invasion of the cell.

Gram-positive and gram-negative cells have somewhat different strategies for transporting materials across the membrane and into the cell. The cytoplasmic membrane of gram-positive organisms has immediate access to media components. However, chemicals and nutrients must first traverse the outer membrane of gram-negative organisms before encountering the cytoplasmic membrane. Gram-negative cells have **pores** formed by protein triplets in their outer membrane that will permit passage of fairly large molecules into the periplasmic space. Subsequent transport across the inner or cytoplasmic membrane is similar in both gram-positive and gram-negative cells.

Capsules. Some bacterial cells produce a capsule or a **slime layer** (Fig. 1-4) of material external to the cell. Capsules are composed of either polysaccharides (high-molecular-weight polymers of carbohydrates) or polymers of amino acids called polypeptides (often formed from the D- rather than the L-isomer of an amino acid). The capsule of *Streptococcus pneumoniae* type III is composed of glucose and glucuronic acid in alternating β-1, 3- and β-1, 4- linkages:

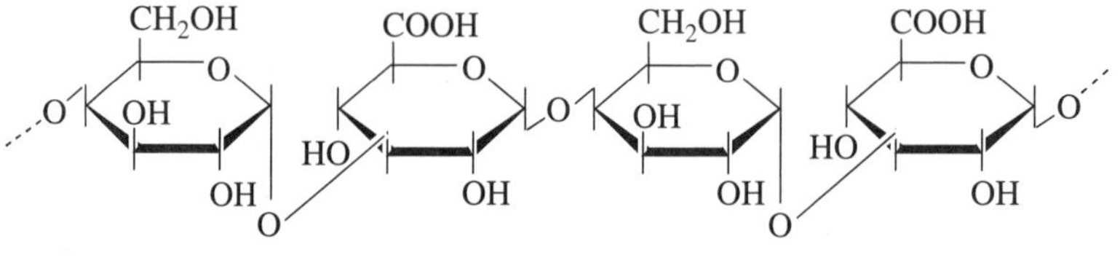

This capsular polysaccharide, sometimes referred to as pneumococcal polysaccharide, is responsible for the virulence of the pneumococcus. *Bacillus anthracis*, the anthrax

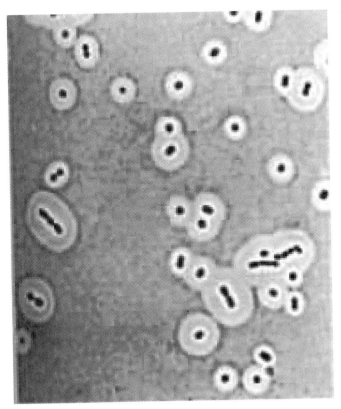

Fig. 1-4. Capsules of *Streptococcus pneumoniae*.

bacillus, produces a polypeptide capsule composed of D-glutamic acid subunits, which is a virulence factor for this organism.

Organs of Locomotion. Many microorganisms are motile — that is, able to move from place to place in a concerted manner — especially in an aqueous environment. In the case of bacteria, this motility is accomplished by means of simple strands of protein (flagellin) woven into helical organelles called flagella. The bacterial flagellum is attached at the cell surface by means of a basal body (Fig. 1-5a). The basal body contains a motor that turns the flagellum, which propels the organism through the liquid environment.

Pili or Fimbriae. Many bacteria possess external structures that are shorter and more rigid than flagella. These structures have been termed pili (from Latin meaning "hair") or fimbriae (from Latin meaning "fringe"). These appendages also appear to arise from a basal body or granule located either within the cytoplasmic membrane or in the cytoplasm immediately beneath the membrane (Fig. 1-5b). Generalized or common pili play a role in cellular adhesion to surfaces or to host cells.

Ribosomes. The cytoplasm of all cells has a fine granular appearance observed in many electron micrographs. Tiny particles called ribosomes are responsible for this look. Ribosomes contain approximately 65% RNA and 35% protein (see Fig. 1-1).

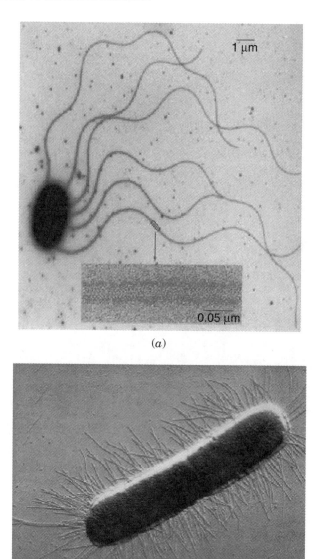

Fig. 1-5. Microbial appendages. (*a*) Flagella of *Salmonella typhimurium.* (*b*) Pili of *Escherichia coli.* (*Source*: Pili Image courtesy Indigo Instruments. Visit http://www.indigo.com.) Reprint permission is granted with this footer included.

The ribosome orchestrates the polymerization of amino acids into proteins (i.e., protein synthesis). At higher magnification under the electron microscope the ribosome particles are spherical. In properly prepared specimens the ribosomes are observed as collections or chains held together on a single messenger RNA (mRNA) molecule and are referred to as **polyribosomes** or simply polysomes.

The more or less spherical ribosome particle, when examined by sucrose gradient sedimentation, has been found to have a svedberg coefficient of 70S. (A svedberg unit denotes the rate of sedimentation of a macromolecule in a centrifugal field and is related to the molecular size of that macromolecule.) The prokaryotic ribosome may be separated into two lower-molecular-weight components: one of 50S and another of 30S. Only the complete 70S particle functions in polypeptide synthesis. By comparison, the ribosomes of eukaryotic cells are associated with the endoplasmic reticulum, are larger (80S), and are composed of 40S and 60S subunits. The function of both 70S and 80S ribosomes in protein synthesis is identical. Curiously, eukaryotic mitochondria characteristically display 70S ribosomes—not the 80s particles that you would expect—because mitochondria probably evolved from endosymbiotic prokaryotic cells, a hypothesis supported by extensive analyses comparing bacterial and mitochondrial genomes.

SYNTHESIS OF DNA, RNA, AND PROTEIN

The chromosome of *E. coli* is a single, circular, double-stranded DNA molecule whose nucleotide sequence encodes all the information required for cell growth and structure. The major molecular events required for propagating the species start with the chromosome and include DNA replication, transcription, and translation. In bacteria, replication involves the accurate duplication of chromosomal DNA and the formation of two daughter cells by **binary fission**. In binary fission the cell grows until a certain mass-to-DNA ratio is achieved, at which point new DNA is synthesized and a centrally located cross-wall is constructed that will ultimately separate the two daughter cells.

A simplified view of **DNA replication** in *E. coli* is shown in the diagram in Figure 1-6. The double-stranded DNA molecule unwinds from a specific starting point (**origin**). The new DNA is synthesized opposite each strand. The enzyme involved in

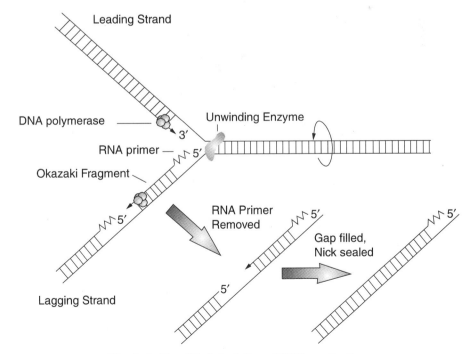

Fig. 1-6. Simplified depiction of DNA replication.

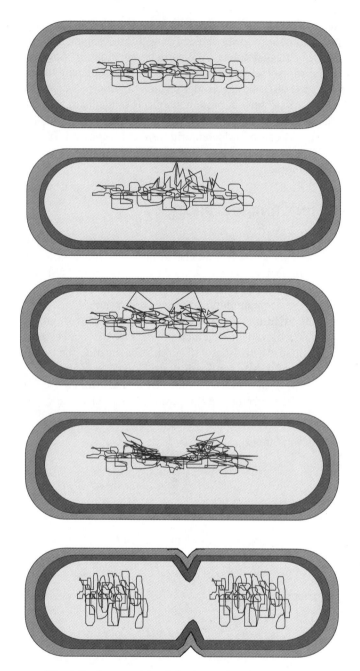

Fig. 1-7. Segregation of the bacterial chromosome.

replication (**DNA polymerase**) uses a parent strand as a template, placing adenine residues opposite thymine, and cytosine residues opposite guanine. New DNA is synthesized in both directions from the origin and continues until both replication forks meet at the **terminus** 180° from the origin. At this point, cell division proceeds with cross-wall formation occurring between the two newly synthesized chromosomes

(Fig. 1-7). Note that the chromosome appears attached to the cell membrane as the daughter chromosomes begin to separate. At some point about midway to the ends of the cell, the nascent chromosomes separate from the membrane but continue to move toward the cell poles by a still undefined mechanism. Chromosomal segregation into pre-daughter cells must occur before the cell completes construction of the cross-wall that will separate the two offspring (see "Termination of DNA Replication and Chromosome Partitioning" in Chapter 2).

The genetic information contained within DNA is processed in two steps to produce various proteins. Protein synthesis (**translation**) is depicted in Figure 1-8. The enzyme **RNA polymerase** (DNA-dependent RNA polymerase) first locates the beginning of a gene (**promoter**). This area of the chromosome then undergoes a localized unwinding, allowing RNA polymerase to transcribe RNA from the DNA template. Before the RNA — called messenger RNA (mRNA) — is completely transcribed, a ribosome will attach to the beginning of the message.

As already noted, the ribosome contains of two subunits, 30S and 50S, each composed of special ribosomal proteins and **ribosomal ribonucleic acids (rRNA)**. rRNA molecules do not, by themselves, code for any protein but form the architectural scaffolding that directs assembly of the proteins to form a ribosome. The ribosome translates mRNA into protein by reading three nucleotides (known as a **triplet codon**) as a specific amino acid. Each amino acid used by the ribosome must first be attached to an adaptor or **transfer RNA (tRNA)** molecule specific for that amino acid. tRNA containing an attached amino acid is referred to as a **charged tRNA molecule**. A part of the tRNA molecule called the **anticodon** will base-pair with the codon in mRNA.

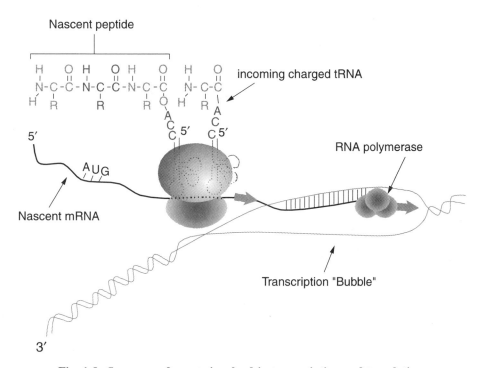

Fig. 1-8. Sequence of events involved in transcription and translation.

When two such charged tRNA molecules simultaneously occupy adjacent sites on the ribosome, the ribosome catalyzes the formation of a peptide bond between the two amino acids.

At this point, the two amino acids are attached to one tRNA while the other tRNA is uncharged and eventually released from the ribosome. The ribosome is then free to move along the message to the next codon. The process continues until the ribosome reaches the end of the message, at which point a complete protein has been formed. Notice that synthesis of the protein begins with the N-terminal amino acid and finishes with the C-terminal amino acid. Also note that the ribosome begins translating at the 5′ end of the mRNA while the DNA strand encoding the mRNA is read by RNA polymerase starting at the 3′ end. Although the beginning of a gene is usually called the 5′ end, this doesn't refer to the strand that is actually serving as a template for RNA polymerase. It refers to the complementary DNA strand whose sequence is the same as the mRNA (except for containing T instead of U). The details of replication, transcription, and translation are discussed in Chapter 2.

Metabolic and Genetic Regulation

For a cell to grow efficiently, all the basic building blocks and all the macromolecules derived from them have to be produced in the correct proportions. With complex metabolic pathways, it is important to understand the manner by which a microbial cell regulates the production and concentration of each product. Two common mechanisms of metabolic and genetic regulation are

1. **Feedback inhibition** of enzyme activity (**metabolic regulation**)
2. **Repression** or **induction** of enzyme synthesis (**genetic regulation**)

In feedback inhibition, the *activity* of an enzyme already present in the cell is inhibited by the end product of the reaction. In genetic repression, the *synthesis* of an enzyme (see previous discussion of transcription and translation) is inhibited by the end product of the reaction. Induction is similar except the substrate of a pathway stimulates synthesis of the enzyme. Hypothetical pathways illustrating these concepts are presented in Figure 1-9. In Figure 1-9a, excessive production of intermediate B results in the inhibition of enzyme 1 activity, a phenomenon known as feedback or end-product inhibition. Likewise, an excess of end-product C may inhibit the activity of enzyme 1 by feedback inhibition.

In contrast to feedback inhibition, an excess intracellular concentration of end-product C may cause the cell to stop synthesizing enzyme 1, usually by inhibiting transcription of the genes encoding the biosynthetic enzymes (Fig. 1-9b). This action is referred to as **genetic repression**. The logic of this control is apparent when considering amino acid biosynthesis. If the cell has more than enough of a given amino acid, that amino acid will activate a repressor protein, which then blocks any further transcription of the biosynthesis genes. In contrast, substrates such as carbohydrates can stimulate the transcription of genes whose protein products consume that carbohydrate. This genetic process is called **induction** (Fig. 1-9c). Different organisms may employ quite different combinations of feedback inhibition, repression, and induction to regulate a metabolic pathway. In Chapter 5, these and other regulatory mechanisms are discussed in greater detail.

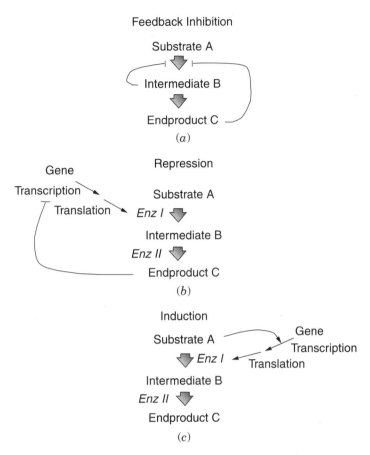

Fig. 1-9. Diagrammatic presentation of feedback inhibition of enzyme activity and end-product repression of enzyme synthesis. *a*, *b* and *c* are chemical intermediates in the hypothetical pathway. Arrow indicates activation, line with cross indicates inhibition.

MICROBIAL GENETICS

Having just outlined the processes of transcription, translation, and replication, it is now possible to define several genetic terms. The gene may be defined as a heritable unit of function composed of a specific sequence of purine and pyrimidine bases, which in turn determines the base sequence in an RNA molecule, and, of course, the sequence of bases in an RNA molecule specifies the sequence of amino acids incorporated into a polypeptide chain. The **genotype** of an organism is the sum total of all of the hereditary units of genes. The observed expression of the genetic determinants — that is, the structural appearance and physiological properties of an organism — is referred to as its **phenotype**.

An individual gene can exist in different forms as a result of nucleotide sequence changes. These alternative gene forms are referred to as **alleles**. Genetic material is not absolutely stable but can change or mutate. The process of change is known as **mutagenesis**. Altered genes are referred to as mutant alleles in contrast to the normal or **wild-type** alleles. **Spontaneous mutations** are thought to arise during replication, repair, and recombination of DNA as a result of errors made by the enzymes involved in DNA metabolism. Mutations may be increased by the activity of a number of environmental influences. Radiation in the form of X rays, ultraviolet (UV) rays, or cosmic rays may affect the chemical structure of the gene. A variety of chemicals may also give rise to mutations. Physical, chemical, or physicochemical agents capable of increasing the frequency with which mutations occur are referred to as **mutagens**. The resulting alterations are **induced mutations** in contrast to *spontaneous mutations*, which appear to occur at some constant frequency in the absence of intentionally applied external influences. Since bacterial cells are haploid, mutants are usually easier to recognize because the altered character is more likely to be expressed, particularly if the environment is favorable to mutant development.

The use of mutants has been a tremendous tool in the study of most, if not all, biochemical processes. Genes are usually designated by a three-letter code based on their function. For example, genes involved in the biosynthesis of the amino acid arginine are called *arg* followed by an uppercase letter to indicate different *arg* genes (e.g., *argA, argB*). A gene is always indicated by lowercase italic letters (e.g., *arg*), whereas an uppercase letter in the first position (e.g., ArgA) indicates the gene product. At this point, we need to expose a common mistake made by many aspiring microbial geneticists concerning the interpretation of mutant phenotypes. Organisms such as *E. coli* can grow on basic minimal media containing only salts, ammonia as a nitrogen source, and a carbon source such as glucose or lactose because they can use the carbon skeleton of glucose to synthesize all the building blocks necessary for macromolecular synthesis. The building blocks include amino acids, purines, pyrimidines, cofactors, and so forth. A mutant defective in one of the genes necessary to synthesize a building block will require that building block as a supplement in the minimal medium (e.g., an *arg* mutant will require arginine in order to grow). Microorganisms also have an amazing capacity to catabolically use many different compounds as carbon sources. However, a mutation in a carbon source utilization gene (e.g., *lac*) does not mean it requires that carbon source. It means the mutant will *not grow* on medium containing that carbon source if it is the only carbon source available (e.g., a *lac* mutant will not grow on lactose).

The chromosome of our reference cell, *E. coli*, is 4,639,221 base pairs long. Gene positions on this map can be given in base pairs starting from the gene *thrL*, or in minutes based on the period of time required to transfer the chromosome from one cell to another by conjugation (100-minute map with *thrL* at 0).

CHEMICAL SYNTHESIS

Chemical Composition

Our paradigm cell (the gram-negative cell *E. coli*) can reproduce in a minimal glucose medium once every 40 minutes. As we proceed through a detailed examination of all the processes involved, the amazing nature of this feat will become increasingly obvious. It is useful to discuss the basic chemical composition of our model cell. The total weight of an average cell is 9.5×10^{-13} g, with water (at 70% of the cell) contributing 6.7×10^{-13} g. The total dry weight is thus 2.8×10^{-13} g. The components that form the dry weight include protein (55%), ribosomal RNA (16.7%), transfer RNA (3%), messenger RNA (0.8%), DNA (3.1%), lipids (9.1%), lipopolysaccharides (3.4%), peptidoglycans (2.5%), building block metabolites, vitamins (2.9%), and inorganic ions (1.0%). It is interesting to note that the periplasmic space forms a full 30% of the cell volume, with total cell volume being approximately 9×10^{-13} ml (0.9 femtoliters). An appreciation for the dimensions of the cell follows this simple example. One teaspoon of packed *E. coli* weighs approximately 1 gram (wet weight). This comprises about one trillion cells — more than 100 times the human population of the planet. When calculating the concentration of a compound within the cell, it is useful to remember that there are 3 to 4 microliters of water per 1 milligram of dry weight. Our reference cell, although considered haploid, will contain two copies of the chromosome when growing rapidly. It will also contain 18,700 ribosomes and a little over 2 million total molecules of protein, of which there are between 1000 and 2000 different varieties. As you might gather from these figures, the bacterial cell is extremely complex. However, the cell has developed an elegant strategy for molecular economy that we still struggle to understand. Some of what we have learned is discussed throughout the remaining chapters.

In just 40 minutes an *E. coli* cell can make a perfect copy of itself growing on nothing more than glucose, ammonia, and some salts. How this is accomplished seems almost miraculous! All of the biochemical pathways needed to copy a cell originate from just 13 precursor metabolites. To understand microbial physiology, you must first discover what the 13 metabolites are and where they come from.

The metabolites come from glucose or some other carbohydrate. The catabolic dissimilation of glucose not only produces them but also generates the energy needed for all the work carried out by the cell. This work includes biosynthetic reactions as well as movement, transport, and so on. Figure 1-10 is a composite diagram of major pathways for carbohydrate metabolism with the 13 metabolites highlighted. Most of them are produced by the Embden-Meyerhof route and the tricarboxylic acid cycle (see Fig. 1-10). Three are produced by the pentose phosphate pathway. Figure 1-11 illustrates how these compounds are siphoned off from the catabolic pathways and used as starting material for the many amino acids, nucleic acid bases, and cofactors that must be produced. Subsequent chapters deal with the specifics of each pathway, but this figure presents an integrated picture of cell metabolism.

Energy

Another mission of carbohydrate metabolism is the production of energy. The most universal energy transfer compound found in living cells is adenosine triphosphate (ATP) (Fig. 1-12). The cell can generate ATP in two ways: (1) by **substrate-level phosphorylation** in which a high-energy phosphate is transferred from a chemical compound (e.g., phosphoenol pyruvate) to adenosine diphosphate (ADP) during the course of carbohydrate catabolism; or (2) by **oxidative phosphorylation** in which the energy from an electrical and chemical gradient formed across the cell membrane is used to drive a membrane-bound ATP hydrolase complex to produce ATP from ADP and inorganic phosphate.

The generation of an electrical and chemical gradient (collectively called the **proton motive force**) across the cell membrane requires a complex set of reactions in which H^+ and e^- are transferred from chemical intermediates of the Embden-Meyerhof and TCA cycles to a series of membrane-associated proteins called **cytochromes**. As the e^- is passed from one member of the cytochrome chain to another, the energy released

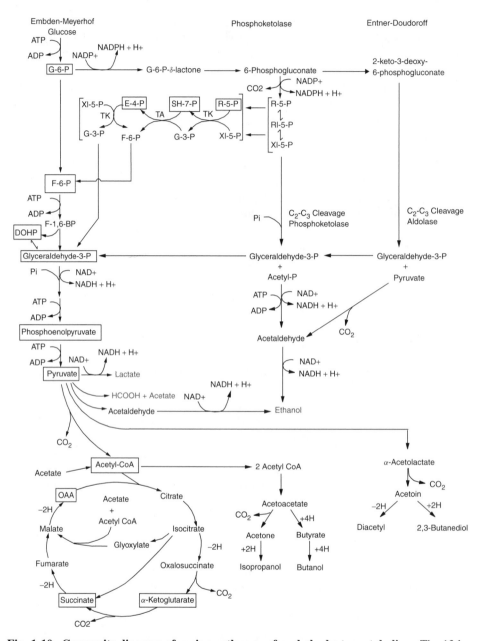

Fig. 1-10. Composite diagram of major pathways of carbohydrate metabolism. The 13 key metabolites are boxed. G, glucose; E, erythrose; R, ribose; Xl, xylulose; SH, sedoheptulose; DOHP, dihydroxyacetone phosphate; OAA, oxaloacetate; TK, transketolase; TA, transaldolase.

is used to pump H^+ out of the cell. The resulting difference between the inside and outside of the cell in terms of charge (**electrical potential**) and pH (**chemical potential**) can be harnessed by the cell to generate ATP. Of course, in order for the cytochrome system to work, there must be a terminal electron acceptor molecule. Under aerobic conditions, oxygen will serve that function, but under anaerobic conditions, *E. coli*

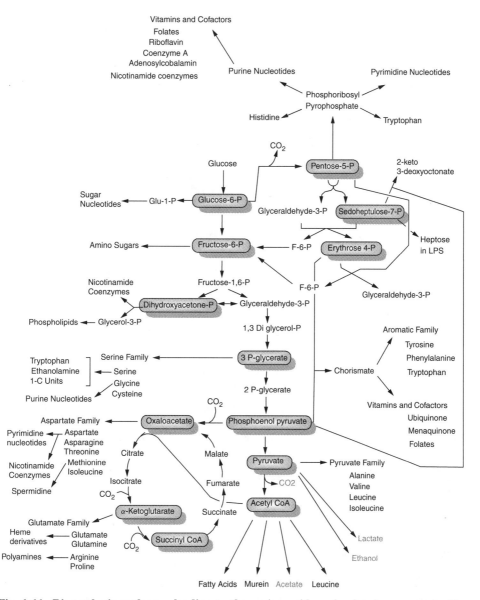

Fig. 1-11. Biosynthetic pathways leading to the amino acids and related compounds. The oblong-circled intermediates are the 13 key compounds that serve as biosynthetic precursors for a variety of essential end products.

has a menu of alternate electron acceptors from which it can choose depending on availability (e.g., nitrate). A more detailed accounting of this process is discussed in Chapter 9.

Oxidation–Reduction Versus Fermentation

Carbohydrate metabolism is the progressive oxidation of a sugar in which hydrogens are transferred from intermediates in the pathway to hydrogen-accepting molecules.

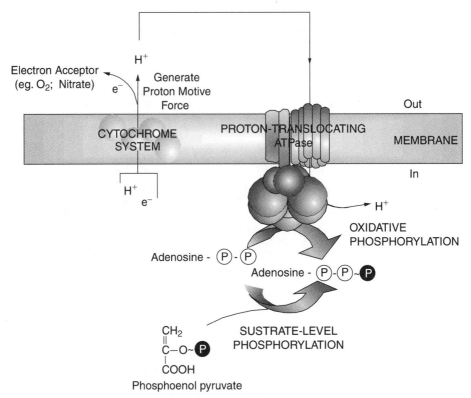

Fig. 1-12. Reactions essential to energy production. Oxidative phosphorylation. The energy that comprises the proton motive force can be harnessed and used to generate ATP when protons from outside the cell pass through the membrane-associated proton-translocating ATPase. The energy released will run the ATPase in reverse. It is estimated that passage of three H^+ through the ATPase is required to generate one ATP. Substrate-level phosphorylation. Energy contained within high-energy phosphate bonds of certain glycolytic intermediates can be transferred to ADP, forming ATP. The example shows phosphoenolpyruvate.

The most commonly used hydrogen acceptor compound is **nicotinamide adenine dinucleotide (NAD)** (Fig. 1-13). It is the reduced form of NAD (NADH) that passes the H^+ and e^- to the cytochrome system. However, a problem can develop when a cell is forced to grow in an anaerobic environment without any alternate electron acceptors. This situation could lead to a complete depletion of NAD^+, with all of the NAD pool converted to NADH. NADH, produced during the early part of glycolysis, would not be able to pass its H along and therefore the cell could not regenerate NAD^+. If this situation were allowed to develop, the cell would stop growing because there would be no NAD^+ to continue glycolysis! To avoid this problem, many microorganisms, including *E. coli*, can regenerate NAD^+ by allowing NADH to transfer H to what would otherwise be dead-end intermediates in the glycolytic pathway (e.g., pyruvate or acetyl CoA). The process, known as **fermentation**, produces lactic acid, isopropanol, butanol, ethanol, and so on, depending on the organism. *E. coli* does not perform all of these fermentation reactions. It is limited to lactate, acetate, formate, ethanol, CO_2, and H_2 production (Fig. 1-10). Table 1-1 lists the fermentation patterns for some other common organisms.

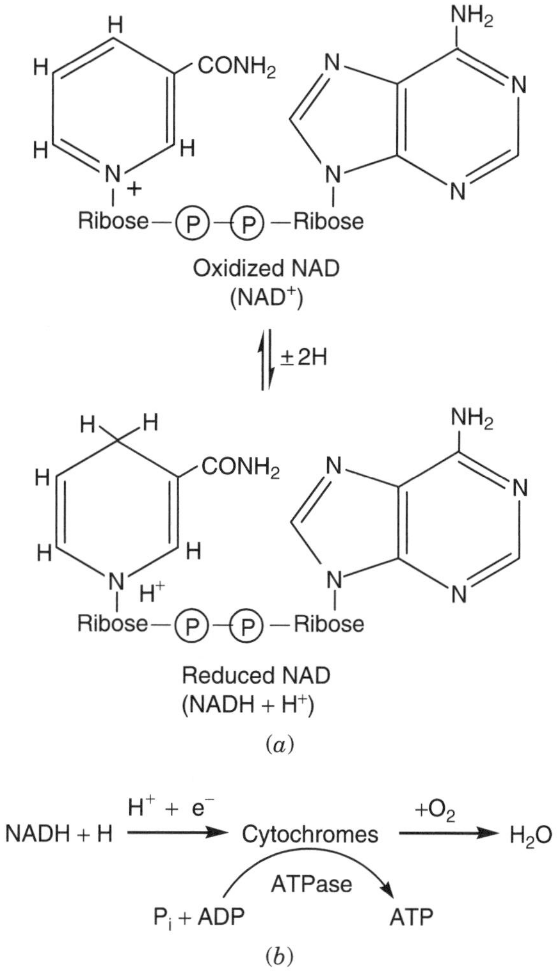

Fig. 1-13. Nicotinamide adenine dinucleotide (NAD). Function of NAD in oxidation–reduction reactions. (*a*) Hydrogen atoms removed from a hydrogen donor are transferred to the nicotinamide portion of NAD. (*b*) The hydrogen atoms can be transferred from NAD to an acceptor such as cytochrome pigments.

TABLE 1-1. Variation in Fermentation Products Formed from Pyruvate

Organism	Product(s)
Saccharomyces (yeast)	Carbon dioxide, ethanol
Streptococcus (bacteria)	Lactic acid
Lactobacillus (bacteria)	Lactic acid
Clostridium (bacteria)	Acetone, butyric acid, isopropanol, butanol
Enterobacter (bacteria)	Acetone, carbon dioxide, ethanol, lactic acid
E. coli (bacteria)	Lactic acid, acetic acid, H_2, ethanol, formic acid, carbon dioxide

The cell does not only catabolize glucose via glycolysis. There are alternate metabolic routes available for the dissimilation of glucose. One use for alternate pathways of carbohydrate metabolism (e.g., the phosphoketolase pathway; see Fig. 1-10) is the generation of biosynthetic **reducing power**. The cofactor NAD is actually divided functionally into two separate pools. NAD(H) is used primarily for catabolic reactions, whereas a derivative, NAD phosphate (NADP), and its reduced form, NADPH, are involved in biosynthetic (anabolic) reactions. The phosphoketolase pathway is necessary for the generation of the NADPH that is essential for biosynthetic reactions.

Nitrogen Assimilation

A major omission in our discussion to this point involves the considerable amount of nitrogen (N) needed by microorganisms. Every amino acid, purine, pyrimidine, and many other chemicals in the cell include nitrogen in their structures. Since glucose does not contain any nitrogen, how do cells acquire it? Some microorganisms can fix atmospheric nitrogen via nitrogenase to form ammonia (NH_4^+) and then assimilate the ammonia into amino acids (e.g., Rhizobium). Other organisms such as *E. coli* must start with NH_4^+. The assimilation of N involves the amidation of one of the 13 key metabolites, α-ketoglutarate, to form glutamic acid (Fig. 1-14). After assimilation into glutamate, the amino nitrogen is passed on to other compounds by transamination reactions. For example, glutamate can pass its amino group to oxaloacetate to form aspartate. From Figure 1-11, it can be seen that aspartate, like glutamate, is a precursor for several other amino acids. The subject of nitrogen assimilation is covered in depth in Chapter 14.

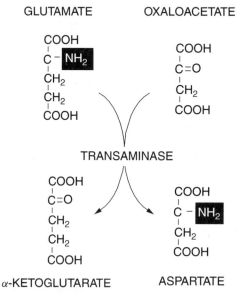

Fig. 1-14. Transamination. In this example, the amine group from glutamic acid is transferred to oxaloacetate, forming aspartic acid.

SPECIAL TOPICS

Endospores

A few bacteria such as *Bacillus* and *Clostridium* produce specialized structures called endospores. Endospores are bodies that do not stain with ordinary dyes and appear as unstained highly refractile areas when seen under the light microscope. They provide resistance to heat, desiccation, radiation, and other environmental factors that may threaten the existence of the organism. Endospores also provide a selective advantage for survival and dissemination of the species that produce them. Under the electron microscope, spores show a well-defined multilayered exosporium, an electron-dense outer coat observed as a much darker area, and a thick inner coat. In the spore interior, the darkly stained ribosomes and the nuclear material may also be visible (Fig. 1-15).

Growth

Growth of a cell is the culmination of an ordered interplay among all of the physiological activities of the cell. It is a complex process involving

1. Entrance of basic nutrients into the cell
2. Conversion of these compounds into energy and vital cell constituents
3. Replication of the chromosome
4. Increase in size and mass of the cell
5. Division of the cell into two daughter cells, each containing a copy of the genome and other vital components

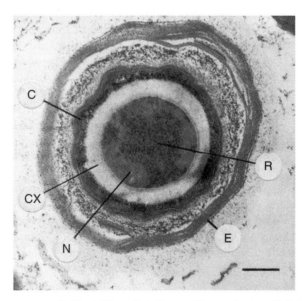

Fig. 1-15. Mature spore of *Clostridium botulinum*. Shown is a well-defined, multilayered exosporium (E), an electron-dense outer coat layer, a thick inner coat (C) and a less dense cortex (CX). The darkly stained ribosomes (R) and nucleoid areas (N) are clearly differentiated in the spore interior. Bar equals 0.2 μm. (*Source*: From Stevenson, K. E., R. H. Vaughn, and E. V. Crisan, 1972. *J. Bacteriol.* **109:**1295.)

Microbiologists usually consider the phenomenon of growth from the viewpoint of population increase, since most current techniques do not allow the detailed study of individual cells. A study of the increase in population implies that each cell, as it is produced, is capable of producing new progeny.

Growth Cycle. Under ideal circumstances in which cell division commences immediately and proceeds in unhampered fashion for a protracted period of time, prokaryotic cell division follows a geometric progression:

$$2^0 \longrightarrow 2^1 \longrightarrow 2^2 \longrightarrow 2^3 \longrightarrow 2^4 \longrightarrow 2^5 \longrightarrow 2^6 \longrightarrow 2^7$$
$$\longrightarrow 2^8 \longrightarrow 2^9 \longrightarrow \text{etc.}$$

This progression may be expressed as a function of 2 as shown in the line above. The number of cells (b) present at a given time may be expressed as

$$b = 1 \times 2^n$$

The total number of cells (b) is dependent on the number of generations (n = number of divisions) occurring during a given time period. Starting with an inoculum containing more than one cell, the number of cells in the population can be expressed as

$$b = a \times 2^n$$

where a is the number of organisms present in the original inoculum. Since the number of organisms present in the population (b) is a function of the number 2, it becomes convenient to plot the logarithmic values rather than the actual numbers. Plotting the number of organisms present as a function of time generates a curvilinear function. Plotting the logarithm of the number, a linear function is obtained as shown in Figure 1-16. For convenience, logarithms to the base 10 are used. This is possible because the logarithm to the base 10 of a number is equal to 0.3010 times the logarithm to the base 2 of a number.

Up to this point it has been assumed that the individual generation time (i.e., the time required for a single cell to divide) is the same for all cells in the population. However, in a given population, the generation times for individual cells vary, so the term **doubling time** encompasses the doubling time for the total population. As shown in Figure 1-16, the cells initially experience a period of adjustment to the new environment, and there is a lag in the time required for all of the cells to divide. Actually, some of the cells in the initial inoculum may not survive this **lag phase** and there may be a drop in the number of viable cells. The surviving cells eventually adjust to the new environment and begin to divide at a more rapid rate. This rate will remain constant until conditions in the medium begin to deteriorate (e.g., nutrients are exhausted). Since plotting the cell number logarithm during this period results in a linear function, this phase of growth is referred to as the **logarithmic (log) phase** or, more correctly, the **exponential phase**.

All cultures of microorganisms eventually reach a maximum population density in the **stationary phase**. Entry into this phase can result from several events. Exhaustion of essential nutrients, accumulation of toxic waste products, depletion of oxygen, or development of an unfavorable pH are the factors responsible for the decline in the

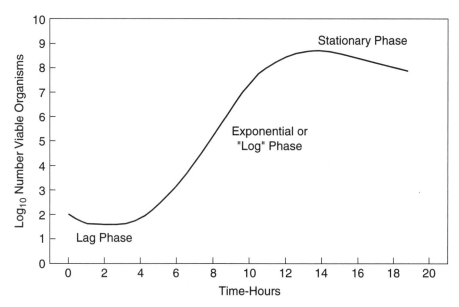

Fig. 1-16. A typical growth curve for a bacterial culture.

growth rate. Although cell division continues during the stationary phase, the number of cells that are able to divide (viable cells) are approximately equal to the number that are unable to divide (nonviable cells). Thus, the stationary phase represents an equilibrium between the number of cells able to divide and the number that are unable to divide.

Eventually, the death of organisms in the population results in a decline in the viable population and the **death phase** ensues. The exact shape of the curve during the death phase will depend on the nature of the organism under observation and the many factors that contribute to cell death. The death phase may assume a linear function such as during heat-induced death where viable cell numbers decline logarithmically.

Some additional considerations of the growth curve are important in assessing the effect of various internal as well as external factors on growth. Since the number of cells in a population (b) is equal to the number of cells in the initial inoculum (a) $\times 2^n$,

$$b = a \times 2^n$$

Then

$$\log_2 b = \log_2 a + n$$
$$\log_{10} b = \log_{10} a + n \log_{10} 2$$
$$\log_{10} b = \log_{10} a + (n \times 0.3010)$$

Solving the equation for n, the number of generations that occurred between the time of inoculation and the time of sampling is

$$n = \frac{\log_{10} b - \log_{10} a}{0.3010}$$

The generation time (t_g) or doubling time may be determined by dividing the time elapsed (t) by the number of generations (n):

$$t_g = t/n$$

Continuous Culture

Usually bacteria are grown in "batch" culture in which a flask containing media is inoculated and growth is allowed to occur. This is a closed system where it is actually very difficult to manipulate growth rate. In batch cultures, growth rate is determined internally by properties of the bacteria themselves. A batch culture can be used to grow bacteria at different rates as long as the nutrient added is at a concentration that does not support maximal growth. But, to accomplish this, the cell density, and thus the cell number, will be too low for certain analyses. To grow bacteria at slow growth rates and at high cell density, a chemostat is used. In this apparatus, fresh medium containing a limiting nutrient is added from a reservoir to the culture vessel at a set rate. The volume in the culture vessel is kept constant by an overflow device that removes medium and cells at the same rate as fresh medium is added. In a chemostat, growth rate is determined externally by altering the rate-limiting nutrient added to the culture vessel. The faster the limiting nutrient is added, the faster the growth rate.

FACTORS AFFECTING GROWTH

Nutrition

All living organisms have certain basic nutritional requirements: sources of carbon, nitrogen, energy, and essential growth factors (minerals and vitamins) are needed to support growth. Microorganisms vary widely in their nutritional requirements. Two main groups of organisms are classified on the basis of their ability to gain energy from certain sources and the manner in which they satisfy their carbon and nitrogen requirements for growth:

1. **Lithotrophs** utilize carbon dioxide as the sole source of carbon and gain energy through the oxidation of inorganic compounds (**chemolithotrophs** or "rock eaters") or light (**photolithotrophs**). Inorganic nitrogen is utilized for the synthesis or organic compounds.

2. **Organotrophs** generally prefer organic substrates as a source of energy and carbon. **Photoorganotrophs** utilize light as a source of energy for the assimilation of carbon dioxide as well as organic compounds. **Chemoorganotrophs** utilize organic compounds for growth.

Although their nutritional requirements are remarkably simple, chemolithotrophic bacteria must be metabolically complex since they synthesize all of their cellular components and provide the energy for this activity through the oxidation of inorganic compounds. One fundamental characteristic of strict chemolithotrophs is that they are unable to grow on or assimilate exogenous organic compounds. Facultative chemolithotrophs can utilize exogenous organic carbon sources. Chemolithotrophs possess unique mechanisms for carbon dioxide fixation such as the ribulose bisphosphate (Calvin-Benson) cycle and the reductive carboxylic acid (Campbell-Evans) cycle (see Chapter 9).

Some organotrophic organisms utilize carbon dioxide as a source of carbon, but most prefer organic carbon sources and generally cannot subsist on carbon dioxide as the sole carbon source. Organotrophs may use inorganic nitrogen, but most members of the group grow better when supplied with organic nitrogen compounds. For example, *E. coli, Enterobacter aerogenes*, yeasts, and molds grow luxuriantly on glucose as the only organic nutrient. Other organotrophs such as streptococci and staphylococci also exhibit specific requirements for one or more nitrogen sources as amino acids, purines, or pyrimidines (see Table 1-2).

Fatty acids are required by some organisms, particularly in the absence of certain B vitamins. Replacement of a growth factor requirement by the addition of the end product of a biosynthetic pathway in which the vitamin normally functions is referred to as a **sparing action**. This type of activity has been reported for many growth factors, including amino acids, purines, pyrimidines, and other organic constituents. If a vitamin can completely replace a particular organic nutrient in a defined medium, that nutrient cannot be regarded as a true growth requirement since it can be synthesized in the presence of the requisite vitamin.

TABLE 1-2. Nutritional Requirements of Some Organotrophs

	Escherichia coli	*Salmonella typhi*	*Staphylococcus aureus*[a]	*Leuconostoc paramesenteroides*[b]
Basic Nutrients				
Glucose				
NH_4^+				
Mn^{2+}				
Mg^{2+}				
Fe^{2+}	Required by all for maximum growth in defined medium			
K^+				
Cl^-				
SO_4^{2-}				
PO_4^{3-}				
Additional Requirements				
	None	Tryptophan	Nicotinic acid	Nicotinic acid
			Thiamine	Thiamine
			10 amino acids	Pantothenate
				Pyridoxal
				Riboflavin
				Cobalamin
				Biotin
				p-Aminobenzoate
				Folate
				Guanine
				Uracil
				16 Amino acids
				Sodium acetate
				Tween 80

[a]From Gladstone, G. P. 1937. *Br. J. Exp. Pathol.* **18**:322.
[b]From Garvie, E. I. 1967. *J. Gen. Microbiol.* **48**:429.

Although most bacterial membranes do not contain sterols, they are required in the membranes of some members of the Mycoplasmataceae. (These organisms do not possess a cell wall.) *Mycoplasma* require sterols for growth. *Acholeplasma* do not require sterols; however, they produce terpenoid compounds that function in the same capacity as sterols. Fungi (yeasts and molds) contain sterols in their cell membranes but in most cases appear to be capable of synthesizing them.

Oxygen

Microorganisms that require oxygen for their energy-yielding metabolic processes are called **aerobes**, while those that cannot utilize oxygen for this purpose are called **anaerobes**. **Facultative organisms** are capable of using either respiratory or fermentation processes, depending on the availability of oxygen in the cultural environment. Aerobic organisms possess cytochromes and cytochrome oxidase, which are involved in the process of oxidative phosphorylation. Oxygen serves as the terminal electron acceptor in the sequence and water is one of the resultant products of respiration. Some of the oxidation–reduction enzymes interact with molecular oxygen to give rise to superoxide ($^{\bullet}O_2^{-}$), hydroxyl radicals ($OH^{\bullet}$), and hydrogen peroxide (H_2O_2), all of which are extremely toxic:

$$O_2 + e^{-} \xrightarrow{\text{oxidative enzyme}} O_2^{-}$$

$$O_2 + H_2O_2 \xrightarrow{\text{nonenzymatic}} O_2 + OH^{\bullet} + OH^{-}$$

The enzyme **superoxide dismutase** dissipates superoxide:

$$2O_2^{-} + 2H^{+} \longrightarrow H_2O_2 + O_2$$

Superoxide dismutase is present in aerobic organisms and those that are aerotolerant, but not in strict anaerobes. Many, but not all, aerobes also produce catalase, which can eliminate the hydrogen peroxide formed:

$$2H_2O_2 \longrightarrow 2H_2O + O_2$$

Aerotolerant organisms generally do not produce catalase. Hence, growth of these organisms is frequently enhanced by culture on media containing blood or other natural materials that contain catalase or peroxidase activity. Organisms that do not utilize oxygen may tolerate it because they do not interact in any way with molecular oxygen and do not generate superoxide or peroxide.

Anaerobic bacteria from a variety of genera are present in the normal flora of the animal and human body as well as in a number of natural habitats such as the soil, marshes, and deep lakes. A number of the more widely known genera of anaerobic organisms are listed in Table 1-3.

Carbon Dioxide

Many organisms are dependent on the fixation of carbon dioxide. Certain organisms thrive better if they are grown in an atmosphere containing increased carbon dioxide.

TABLE 1-3. Genera of Anaerobic Bacteria

Bacilli		Cocci	
Gram Positive	Gram Negative	Gram Positive	Gram Negative
Clostridium	*Bacteroides*	*Peptococcus*	*Veillonella*
Actinomyces	*Fusobacterium*	*Peptostreptococcus*	
Bifidobacterium	*Vibrio*	*Ruminococcus*	
Eubacterium	*Desulfovibrio*		
Lactobacillus			
Propionibacterium			

Haemophilus, Neisseria, Brucella, Campylobacter, and many other bacteria require at least 5 to 10% carbon dioxide in the atmosphere to initiate growth, particularly on solid media. Even organisms such as *E. coli* use carbon dioxide to replenish intermediates in the TCA (tricarboxylic acid) cycle that have been siphoned off as precursors for amino acid synthesis. These anapleurotic reactions include pyruvate carboxylase, phosphoenolpyruvate carboxylase, or malic enzyme (see Chapter 8).

Extremophiles

Microorganisms vary widely in their ability to initiate growth over certain ranges of temperature (Table 1-4), hydrogen ion concentration (Table 1-5), and salt concentration. Organisms that function best under extreme environmental conditions are called **extremophiles**. Examples include bacteria found in hot springs and in the thermal vents on the ocean floor. These organisms prefer to grow at extremely high temperatures. Some microorganisms prefer to live in an acidic environment (**acidophilic** organisms) while others prefer an alkaline pH (**alkaliphilic** organisms). *E. coli* prefers a neutral pH environment and thus is classified as **neutralophilic**. (The older term, neutrophilic,

TABLE 1-4. Temperature Ranges of Bacterial Growth

Type of Organism	Growth Temperature (°C)		
	Minimum	Optimum	Maximum
Psychrophilic	$-5-0$	5–15	15–20
Mesophilic	10–20	20–40	40–45
Thermophilic	25–45	45–60	>80

TABLE 1-5. pH Limits for Growth of Various Microorganisms

Organism	Minimum	Optimum	Maximum
Bacteria	2–5	6.5–7.5	8–11
Yeasts	2–3	4.5–5.5	7–8
Molds	1–2	4.5–5.5	7–8

is not consistent with the nomenclature of the other two groups and can be confused with neutrophiles, a form of white blood cell, and thus should not be used.) The ability of certain organisms to grow in extreme environments can be linked to the possession of unique membrane compositions and/or enzymes with unusual temperature or pH optima that are more suitable to their environment.

Microbial Stress Responses

For **normalophiles**, meaning organisms that prefer to grow under conditions of 37 °C, pH 7, and 0.9% saline, variations in pH and temperature have a marked impact on enzyme activity and, ultimately, viability. Outside their optimal parameters, enzymes function poorly or not at all, membranes become leaky, and the cell produces compounds (e.g., superoxides) that damage DNA and other macromolecular structures. All of these factors contribute to cell death when the cell is exposed to suboptimal environments. However, many, if not all, microorganisms have built-in stress response systems that sense when their environment is deteriorating, such as when medium acidifies to dangerous levels. At this point, signal transduction systems perceive the stress and transmit instructions to the transcription/translation machineries to increase expression of specific proteins whose job is to protect the cell from stress. The various genetic regulatory systems and protection strategies used by the cell to survive stress are discussed in Chapters 5 and 18.

SUMMARY

This chapter is a highly condensed version of the remainder of this book, provided to build a coherent picture of microbial physiology from the start. Too often textbooks present a student with excruciatingly detailed treatments of one specific topic after another without ever conveying the "big picture." As a result, the information overload is so great that the student, lost in the details, never develops an integrated view of the cell and what makes it work. Our hope is that the framework in this chapter will be used to build a detailed understanding of microbial physiology and an appreciation of its future promise.

CHAPTER 2

MACROMOLECULAR SYNTHESIS AND PROCESSING: DNA, RNA, AND PROTEIN SYNTHESIS

The way nature has designed and interwoven thousands of biochemical reactions to form the cell is still not fully understood. Yet, what we have managed to decipher is nothing short of amazing. Our discussion starts with the molecular epicenter of the cell — DNA — and builds layer upon layer, illustrating what is known about the biochemical basis of life and how perturbing just one small aspect of a cell's physiology can have a massive ripple effect on hundreds of seemingly unrelated biochemical processes.

This chapter focuses on macromolecules including peptidoglycan, the rigid, cross-bridged molecule that provides cell shape and structural integrity; proteins that carry out biochemical activities; and nucleic acids (DNA and RNA), which contain the information necessary to coordinate cell activities into a balanced system of orderly growth and development. Our discussion of the bacterial cell begins with what are arguably the core processes of all life: nucleic acid and protein syntheses. Peptidoglycan synthesis is described in Chapter 7.

During growth, the continuous synthesis of various building blocks and their proper assembly into the structures of the cell are dependent on the fidelity with which DNA is replicated. The chain of events emanating from DNA follows the general scheme:

$$DNA \xleftarrow{\text{replication}} DNA \xrightarrow{\text{transcription}} \begin{bmatrix} tRNA \\ mRNA \\ rRNA \end{bmatrix} \xrightarrow{\text{translation}} Protein$$

Knowledge of these processes is fundamental to understanding the biological processes of life including energy production, building block biosynthesis, growth and its regulation. Throughout the remainder of the book we refer to numerous genetic loci from *Escherichia coli* and *Salmonella enterica*. More detail on each gene can be found by referring to [Berlyn, 1998] or on the Internet at www.ucalgary.ca/%7Ekesander/

(*Salmonella*), cgsc.biology.yale.edu/(*Escherichia*), or genolist.pasteur.fr/ (*E. coli, Mycobacterium, Bacillus, Helicobacter*).

STRUCTURE OF DNA

In order to appreciate the processes of replication, transcription, and translation of the information contained within a DNA molecule, it is necessary to describe the structure of DNA and to illustrate its ability to direct the events occurring in the cell. DNA is composed of the four deoxyribonucleosides [purine (adenine or guanine) and pyrimidine (cytosine or thymine) bases + deoxyribose] arranged in a chain with phosphodiester bonds connecting the 5′-carbon on the deoxyribose of one nucleoside to the 3′-carbon of the deoxyribose of the adjacent nucleoside as shown in Figure 2-1. The property of polarity (illustrated by the lower part of Fig. 2-1) is maintained by the orientation of the phosphodiester bonds on the 5′ and 3′ positions of the deoxyribose in the DNA strand. In the bacterial chromosome, DNA is composed of two such strands arranged in apposition; that is, it is double stranded (dsDNA). In dsDNA, the two strands have opposite **polarities** as shown in Figure 2-2 and are described as being antiparallel. One strand will end with a 3′-hydroxyl group (the 3′ terminus), while the other will end with a 5′-phosphate (the 5′ terminus). Notice, too, that DNA is a negatively charged molecule under physiological conditions.

The purine and pyrimidine bases are planar structures that pair with one another through the forces of hydrogen bonding, usually depicted as dashed or dotted lines in diagrammatic presentations. These base pairs are stacked parallel to one another and lie perpendicular to the phosphodiester backbone of the structure. As a result of covalent

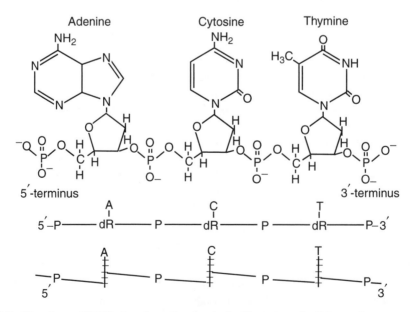

Fig. 2-1. Structure of DNA showing alignment of adjacent nucleotides and some common methods of depicting polarity in abbreviated structures. The crossbars on the vertical lines represent the carbon atoms in deoxyribose.

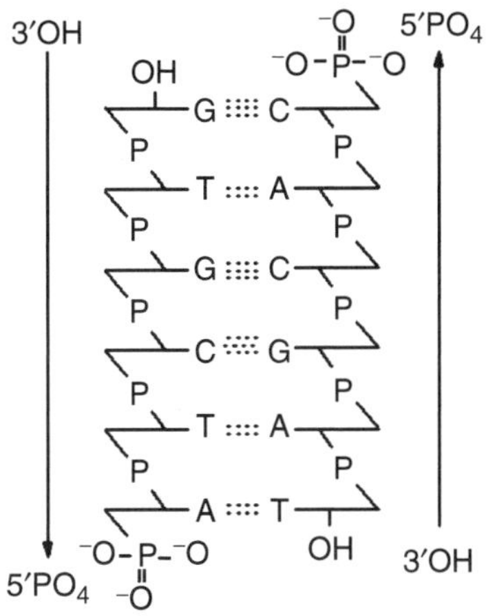

Fig. 2-2. Polarity of the double-stranded DNA molecule. This abbreviated diagrammatic view uses the conventions described in Figure 2-1.

bond angles and noncovalent forces, native DNA assumes a helical structure that exists in different forms depending on the composition of the DNA and the environment. In an aqueous environment, DNA exists primarily in a stable conformation (B form) in which there are 3.4 Å between the stacked bases and 34 Å per right-handed helical turn. Therefore, there are 10 base pairs per turn, and the bases will be arranged so that they are perpendicular to the helical axis as shown in Figure 2-3.

Under certain conditions, DNA can shift into either the **A** conformation (11 rather than 10 base pairs per turn and tilted bases) or become **Z-DNA** (left-handed rather than right-handed helix having 12 base pairs per turn). Native DNA (in vivo) very likely exists in a variety of conformations. Different areas of the bacterial chromosome may well differ topologically as a result of variations in the local environment. Structural differences of this kind may be essential to certain functions of DNA, particularly with regard to the binding of regulatory proteins to specific areas of the chromosome.

Z-DNA is most often formed as a result of extended runs of deoxyguanosine-deoxycytosine [poly(dG-dC)]. In vitro this also requires a fairly high salt environment because poly(dG-dC) will assume a B structure in low salt. Over the years there has been some controversy regarding the in vivo existence and significance of Z-DNA conformations. Proof that Z-DNA can exist in vivo was provided by R. D. Wells and colleagues by placing segments of poly(dG-dC) immediately next to a sequence of DNA that is specifically methylated only when in the B form. This sequence was dramatically undermethylated in vivo when the Z-DNA sequence was included. Evidence that Z-DNA naturally exists in vivo was based on the finding that Z-DNA, unlike B-form DNA, is a reasonable antigen. Antibodies to Z-DNA have been used to demonstrate that the *E. coli* genome has an average of one Z-DNA segment for every 18 kilobars of DNA.

Under certain circumstances, regions of double-stranded DNA may be converted from the double-stranded helix into a nonbase-paired random coil bubble. Both of these states may exist in different regions of the chromosome at the same time, and the process of conversion may be freely reversible. This helical-random coil transition may be essential for the function of DNA in replication and transcription.

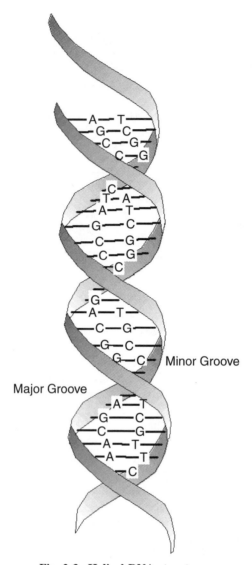

Fig. 2-3. Helical DNA structures.

An interesting biological question concerns how regulatory DNA-binding proteins recognize their proper DNA-binding sequences, since DNA duplexes present a fairly uniform surface to potential regulatory proteins (i.e., a uniform phosphodiester backbone that faces outward). There are two potential mechanisms for nucleic acid–protein interaction. The first model predicts that the specificity of interaction relies on groups present in the major or minor grooves of dsDNA (see Fig. 2-3). Thymine and adenine have groups that will protrude into the grooves. Proteins interact with these groups to bind to specific regions of DNA. The second model is that direct access to bases may occur by a protein interacting with DNA in a randomly melted region. DNA-binding proteins can have one of several characteristic structures that will aid in sequence-specific binding. They are discussed in Chapter 5.

Bacterial Nucleoids

In all living cells, DNA takes on a highly compacted tertiary structure. The first stage of compaction for eukaryotic cells involves the winding of DNA around an octomeric assembly of histone proteins to form the nucleosome structure (see Chapter 7). At this stage, a linear stretch of DNA takes on a bead-like appearance. The second stage is thought to be the solenoidal organization of these nucleosomes into a helical network.

The packaging of DNA in prokaryotic cells is understood less completely, but the problem is obvious. The following example illustrates the extent to which the chromosome of *E. coli* must be compacted to fit in the cell. The *E. coli* genome is about 1.5 mm long (4.2×10^{-15} g). In exponentially growing cells, the DNA comprises 3 to 4% of the cellular dry mass, or about $12-15 \times 10^{-15}$ g per cell. Thus, there are around three genomes in the average growing cell. If chromosomal DNA were homogeneously distributed over the entire cell volume (1.4×10^{-12} ml), the packing density would be 10 mg/ml. However, DNA is confined to ribosome-free areas of the cell, bringing its packing density to 15–30 mg/ml. The inevitable conclusion is that the chromosome is packaged into what is loosely referred to as a "compactosome." One component in this process involves supercoiling the DNA molecule. The bacterial chromosome is a covalently closed, circular, double-stranded DNA molecule that is actually *underwound* in the cell, resulting in a negatively supercoiled structure. DNA supercoiling occurs when a circular dsDNA molecule becomes twisted so that the axis of the helix is itself helical (Fig. 2-4). If DNA is overwound (tighter coils made by twisting DNA opposite to the direction of the helix) or underwound (twisted in the same direction as the helix), a significant amount of torsional stress is introduced into the molecule. (Try this with a coiled telephone cord.) In order to relieve this stress, the DNA double helix will twist upon itself. Underwound DNA will form negative supercoils, whereas overwinding results in positive supercoils.

Topoisomerases. Topoisomerases are enzymes that alter the topological form (supercoiling) of a circular DNA molecule (Table 2-1). There are two classic types — I and II — named for whether they cleave one or both strands of DNA, respectively. Type I topoisomerases are further subdivided as type IA (e.g., *E. coli* TopA), which nicks one strand of DNA and passes the other strand through the break; and type IB (e.g., *E. coli* Top III) enzymes, which relax negatively supercoiled DNA by breaking one of the phosphodiester bonds in dsDNA, subsequently allowing controlled rotational diffusion ("swivel") of the 3′-hydroxyl end around the 5′-phosphoryl end. Both subtypes then reseal the "nicked" phosphodiester backbone (Fig. 2-5).

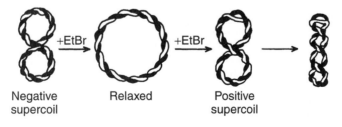

Negative supercoil   Relaxed   Positive supercoil

Ethidium bromide concentration ⟶

Fig. 2-4. Superhelicity of a closed circular DNA molecule changing from negative supercoil to positive supercoil with increasing concentrations of ethidium bromide.

TABLE 2-1. DNA Topoisomerases

Enzyme	Type	Relaxation of (−) or (+) Twists	Introduction of (−) Twists	Linking of Duplex DNA Circles	ATP Requirement
E. coli topoisomerase I (*topA*, (omega protein)	IA	(−)	No	Yes	No
Phage λ int	?	(−), (+)	No	Yes	No
Bacterial DNA gyrase (Top II)	II	(−)	Yes	Yes	Yes
E. coli topoisomerase III (*topB*)	IB	(−), (+)	No	Yes	No
Phage T4 topoisomerase	II	(−), (+)	No	Yes	Yes

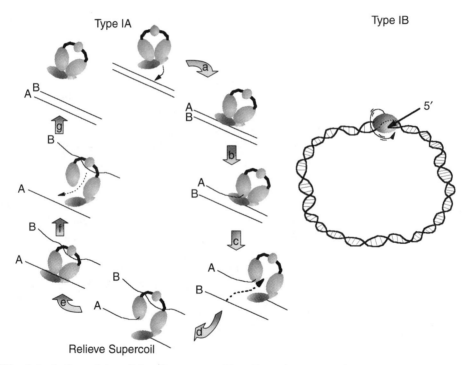

Fig. 2-5. Action of type I topoisomerases. Type I topoisomerases (e.g., *E. coli* TopA) relax negatively supercoiled DNA apparently by one of two mechanisms. Both types IA and IB break one of the phosphodiester bonds in one strand of a DNA molecule. Type IA topoisomerases catalyze a cleavage and transport cycle to relieve supercoils. Domain III of TopA binds, cleaves, and opens an ssDNA segment. Separation is achieved by domain III lifting away from domains I and IV while remaining tethered to the rest of the protein through the domain II arch. After the other strand passes through the break, domain II reassociates with domains I and IV and religates the cleaved strand. Following religation, domain III again lifts away from the rest of the protein, allowing exit of the segment. Notice strands A and B have switched places. Type IB enzymes such as TopB also cleave a single strand but then allow the protein-bound 3′-hydroxyl end of the DNA to swivel around the 5′-phosphoryl end. The topoisomerase then reseals the nicked phosphodiester backbone and the DNA is partially unwound.

Type II topoisomerases, in contrast, require energy to underwind DNA molecules and introduce negative supercoils. Type II topoisomerases, such as DNA gyrase (GyrA and GyrB, called topoisomerase II) and topoisomerase IV (ParC, E), can also relax negative supercoils in the absence of energy. These enzymes utilize the mechanism described in Figure 2-6. The model requires the passage of one dsDNA molecule through a second molecule, which has a double-stranded break in its phosphodiester

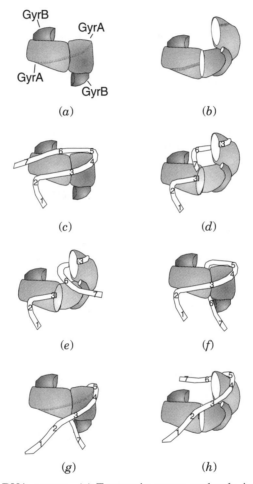

Fig. 2-6. Model for DNA gyrase. (*a*) Tetrameric gyrase molecule in closed conformation. (*b*) Gyrase molecule in open conformation. Illustrations *c* through *h* represent steps in a scheme for negative supercoiling by inversion of a right-handed DNA loop (*c*) into a left-handed one (*f*). (*c*) A small section of a circular double-stranded DNA molecule is represented as a tube (numbered). The DNA is bound to gyrase in a right-handed coil. Three is the point of DNA cleavage and covalent attachment of subunit A polypeptides. Sections 1–3 and 3–5 are wrapped around the left and right A subunits, respectively, while DNA at 6 contacts a B subunit. (*d*) Gyrase cleaves DNA at 3 and opens. (*e*) DNA (point 6) is transferred from the left to the right B subunit through the cleavage point. (*f*) Gyrase closes and reseals DNA at point 3. The DNA between points 3 and 7 now describes a left-handed coil. (*g* and *h*) DNA partially unwraps from gyrase. The gyrase opens, releasing the DNA, and then recloses. (Modified From Wang et al., in Alberts, Mechanistic Studies of DNA replication and genetic recombination 1980.)

backbone. The steady-state level of DNA supercoiling in vivo appears to involve interplay between Top I and Top III to relax negative supercoils and DNA gyrase, which introduces negative supercoils. As is discussed later, this supercoiling is required for efficient replication and transcription of procaryotic DNA.

Nucleoid domains and Nucleoid-Associated proteins. As already noted, the bacterial nucleoid is highly compacted, but supercoiling alone cannot account for the level of compaction. A cell 1 μm in length must accommodate a chromosome 1500 μm in length, and it must do this in a way that does not entangle the molecule so that transcription and replication are not impeded! The solution is that the bacterial chromosome actually contains between 30 and 200 negatively supercoiled loops or domains (Fig. 2-7). Each domain represents a separate topological unit, the boundaries of which may be defined by sites on DNA (see "REP Elements" in this chapter) that bind anchoring proteins. Gathering the DNA loops at their bases would compact the chromosome to a radius of less that 1 μm — small enough to fit in the cell. These loops are supercoiled, compacted, and centrally located within the cell. A single-stranded nick in one loop leads to its relaxation without disturbing the other supercoiled loops.

Four different histone-like proteins isolated from *E. coli* contribute to compactosome structure (Table 2-2). They are all basic, heat-stable, DNA-binding proteins present in high proportions on the *E. coli* nucleoid. Their similarity to eukaryotic histones was used to extrapolate a role in forming prokaryotic nucleosomes. However, there is some controversy as to the existence of nucleosomes in *E. coli*. For example, the level of histone-like proteins does not appear sufficient to generate a nucleosome structure. The most abundant histone-like protein, HU, is not even associated with the bulk of the DNA. Instead, it is situated where transcription and translation occur and may play a more direct role in those processes. However, the proteins that associate with the nucleoid are distributed equally over the DNA.

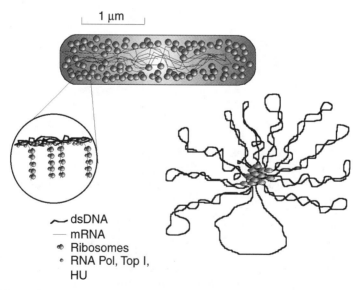

Fig. 2-7. Schematic representation of *E. coli* nucleoids in situ and after isolation and spreading. The insert illustrates an area active in transcription and translation.

TABLE 2-2. Histone-like Proteins of *Escherichia coli*

Name	Subunits	Molecular Weight	Function
Hu (*hupA*)	Hu(α)[HLPII$_a$]	9000	Can form nucleosome-like
(*hupB*)	Hu(β)[HLPII$_b$] (associated as heterologous dimers)	9000	structures similar to eukaryotic histone H2B
H-NS (H1, *hns*)		15,000	May selectively modulate in vivo transcription, nucleoid compaction
StpA		15,000	H-NS analog

The most abundant proteins associated with the nucleoid include HU, H-NS, Fis, and RNA polymerase. Recently it has been shown that H-NS, which forms dimers or tetramers, can actually compact DNA in vitro. Two H-NS dimers bound to different regions of a DNA molecule will bind to each other and condense DNA. The tethering points for each domain are suspected of being DNA gyrase, as is the case in eukaryotes, and perhaps HU. In contrast to eukaryotic nucleosomes, the prokaryotic DNA would be in continuous movement such that transcriptionally active segments would be located at the ribosome–nucleoid interface. The inactive parts would be more centrally located. An alternative hypothesis is that the coupled transcription/translation/secretion of exported proteins serves to tether regions of the chromosome to the membrane, thereby defining different chromosomal domains (see Protein Trafficking).

Another role of histones in eukaryotes is charge neutralization of DNA. Could this also be the case in *E. coli*? The data suggest that this is not the situation. Because there is so little histone-like protein in *E. coli*, neutralization of DNA charge must be accomplished by other molecules such as polyamines and Mg^{2+}.

What role might DNA supercoiling have besides nucleoid compaction? There is mounting evidence that the expression of many genes can vary with the extent of negative supercoiling. The cell may even regulate certain sets of genes by differentially modulating domain supercoiling. Evidence suggests that this may occur in response to environmental stresses such as an increase in osmolarity or change in pH. The subsequent, coordinated increase in expression of some genes and decreased expression of others would enable the cell to adapt and survive the stress. Further detail on nucleoid structure of both prokaryotic and eukaryotic microbes is presented in Chapter 7.

REP Elements

One sequence element present in prokaryotic genomes that may play a role in the organization of the nucleoid is called the repetitive extragenic palindrome (REP) or palindromic unit (PU) sequence. It is a 38 base pair palindromic consensus sequence capable of producing a stem-loop structure. A consensus sequence is defined by comparing several like sequences and determining which bases are conserved among them. A palindromic sequence is one in which the 5'-3' sequence of the top strand is the same as the 5'-3' sequence read on the bottom strand. As a result, intrastrand base pairing can occur, resulting in hairpin structures. The consensus REP sequence for *E. coli* is 5'A(AT)TGCC(TG)GATGCG(GA)CG(CT)NNNN(AG)CG(TC)CTTATC

(AC)GGCCTAC(AGX). N indicates any base can be found at that position. Two bases within parentheses indicate one or the other base can be found at that position. The imperfect palindromic region can be observed by starting at the NNNN region and reading in both directions, noting that the bases are complementary (see underlined regions). The REP elements are located in different orientations and arrays between genes within an operon or at the end of an operon. Although there are between 500 and 1000 REP copies, they are always found outside of structural genes. The suggestion that REP or a complex of REP sequences called BIMEs (bacterial-interspersed mosaic elements) play a part in nucleoid structure comes from in vitro studies showing specific interactions between REPs, DNA polymerase I, and DNA gyrase. However, the actual function for these elements is not known.

DNA REPLICATION

DNA Replication is Bidirectional and Semiconservative

For all organisms, the production of viable progeny depends on the faithful replication of DNA by DNA polymerase. DNA replication proceeds in a semiconservative manner where one strand of the parental molecule is contributed to each new (daughter) dsDNA molecule as shown in Figure 2-8. During the first round of DNA replication, each strand is duplicated, so at the end of this step each of the resulting DNA molecules will contain one old and one new strand. Only after the second round of replication will two completely new strands form. Autoradiographic studies provide evidence that the bacterial chromosome remains circular throughout replication and that the ring is doubled between the point of initiation and the growing point (replication fork). If the replication process goes to completion, the entire cyclic chromosome will be duplicated and the two new daughter chromosomes will separate.

Enzymes called DNA polymerases catalyze the polymerization of deoxyribonuleotides in an ordered sequence dictated by a preexisting DNA template. *E. coli* contains five DNA polymerase activities: Pol I (*polA*), Pol II (*dinA*), Pol III (*dnaE*), Pol IV (*dinB*), and Pol V (*umuDC*). We focus on Pol I and III in this section. All DNA polymerases synthesize DNA in the $5' \rightarrow 3'$ direction and require the presence of a preexisting $3'$-hydroxyl primer (see Figs. 2-9 and 2-10). Pol I has a bound zinc ion and, in addition to the polymerase activity, possesses a $3' \rightarrow 5'$ exonuclease activity used for "proofreading" (see below) and a $5' \rightarrow 3'$ exonuclease activity. Pol II and Pol III also have $3' \rightarrow 5'$ exonuclease activities.

DNA Polymerase Functions as a Dimer

The fact that polymerases synthesize DNA only in the $5' \rightarrow 3'$ direction raises a logistics problem when you consider that replication of both strands occurs simultaneously in what appears to be the same direction. The appearance is that one of the newly synthesized strands violates the $5' \rightarrow 3'$ directive and elongates in the $3' \rightarrow 5'$ direction. DNA Pol III, the polymerase primarily responsible for replication, manages to replicate both strands simultaneously and maintains the $5' \rightarrow 3'$ rule because it works as a **dimer** at the replication fork. Two polymerase holoenzyme complexes are held together by the θ protein (Fig. 2-10). One member of the dimer continuously synthesizes one strand in a $5' \rightarrow 3'$ direction while the other

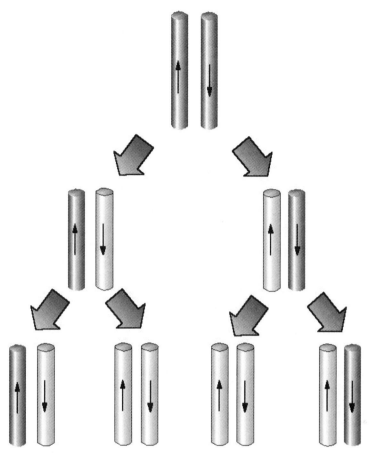

Fig. 2-8. Semiconservative DNA replication. Note that after the first round of replication, each strand of DNA is composed of one new and one old strand. Only after the second round of replication are two completely new strands formed.

polymerase replicates the other strand discontinuously in small fragments of 1000 to 2000 nucleotides (Okazaki fragments). Thus, the template strands are referred to as the leading and lagging strands, respectively. Figure 2-10 illustrates how this is accomplished. The lagging strand loops out and around one of the polymerases. The Pol III dimer can now synthesize the two new DNA strands in the same direction within the moving replication fork while actually moving in opposite directions along the template strands. To avoid difficulties that would result from the generation of very long DNA loops, lagging strand synthesis proceeds discontinuously.

Since DNA polymerases cannot initiate new chain synthesis in the absence of an RNA primer, a primer-generating polymerase (DNA primase) is required for the initiation of each new Okazaki fragment. The primase in *E. coli* is the product of the *dnaG* locus. The short (11 bp) RNA primers are elongated by DNA polymerase III, "erased" by the $5' \rightarrow 3'$ exonuclease activity of RNAse H (an RNA-degrading enzyme that recognizes DNA : RNA hybrids), and resynthesized against the template as DNA by DNA polymerase I. The single-strand nick in the phosphodiester backbone is then resealed by DNA ligase as shown in Figure 2-11.

DNA Polymerase Action

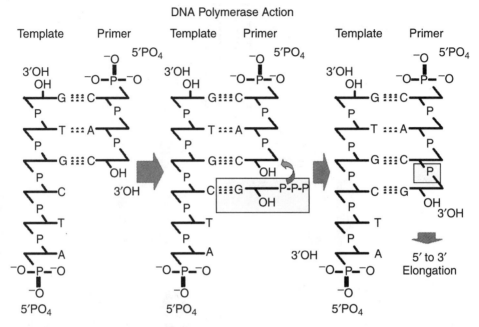

Fig. 2-9. Schematic representation of the action of DNA polymerase in the synthesis of DNA. A DNA template, a complementary RNA primer with a free 3′-hydroxyl end, and Mg^{2+} are required. Replication by DNA polymerase takes place in the 5′ → 3′ direction.

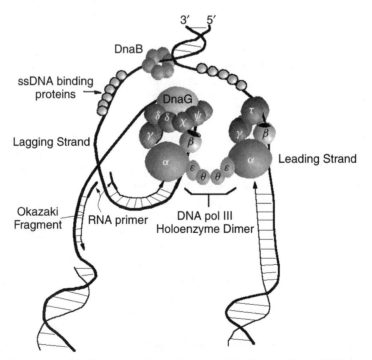

Fig. 2-10. Diagrammatic representation of the replicating fork in DNA synthesis in microbial cells.

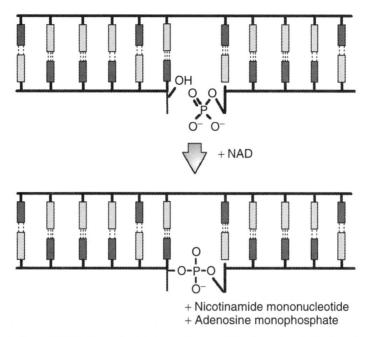

Fig. 2-11. Action of DNA ligase in joining adjacent 3′-hydroxyl and 5′-phosphoryl groups in duplex DNA.

Model of DNA Replication

It is amazing to consider that the bacterial chromosome (4,639,221 bp) replicates at a rate of 800 to 1000 nucleotides per second, yet the frequency of error only amounts to 1 in 10^{10} base pairs replicated. Thus, a high degree of fidelity is maintained under normal circumstances. To bring some perspective to the magnitude of this feat, consider the following. If the DNA duplex was 1 meter in diameter, the replication fork would have to move at about 600 km/hour (375 mph) and the replication machinery would be about the size of a truck. Each replication machine — remember, there are two — would only make a mistake every 170 km (106 miles) over a 400 km (250 mile) trip.

Once initiated, replication is an elaborate process requiring a large number of proteins (Table 2-3). Figure 2-10 presents a detailed model of the replication fork. DNA synthesis is actually a collaboration of two polymerases: primases (DnaG), which start chains, and replicative polymerases, which add to those chains, synthesizing the whole DNA molecule. The following steps comprise the current model of how replication occurs:

1. Prepriming (Primosome). The first step in the replication process is to separate the DNA strands to allow access to the DNA polymerase complex. The DnaB hexameric helicase is required for this process. A DnaC hexamer (loading factor) escorts and loads a DnaB helicase to the origin of replication (or, in combination with preprimosome proteins n, n′, n″, i, to sites where replication has arrested). The DnaB hexamer encircles one strand of DNA, and, using the energy derived from ATP hydrolysis, unwinds the DNA duplex to form a replication fork. Once the primase and DNA polymerase complex are on board, interactions between DnaB helicase and

TABLE 2-3. Genes and Proteins Involved in DNA Replication

Protein	Gene	Map Location	Molecular Weight	Function
	oriC	83.5	245 bp	Origin of replication
Protein i (X)	*dnaT*	99	19,000	Prepriming
Protein n (Z)	*priB*	95	11,300	Prepriming
Protein n' (Y)	*priA*	89	81,500	Prepriming (DNA-dependent ATPase)
Protein n''	*priC*	10	20,200	Prepriming
DnaA	*dnaA*	83.6	52,300	Initiation, binds *oriC*, DnaB loading
DnaB	*dnaB*	91.9	52,200	helicase (hexamer), prepriming priming, DNA-dependent rNTPase
DnaC	*dnaC*	99.1	27,800	Loading factor for DnaB
Pol III (α)	*dnaE*	4.4	129,700	DNA Pol III holoenzyme, elongation
Primase	*dnaG*	69.2	65,400	Priming, RNA primer synthesis, rifampin resistant
γ and τ subunits	*dnaX*	10.6	47,500; 71,000	Synthesis, part of the γ complex, promotes dimerization of Pol III
β (EFI)	*dnaN*	82.5	40,400	β-clamp, processivity
Helix destabilizing	*ssb*	92.1	19,000	Single-stranded binding protein
Helix unwinding	*rep*	85	76,800	Strand separation; not essential for chromosome replication
δ (EFIII)	*holA*	14.4	38,500	Part of the γ complex, clamp loading
δ'	*holB*	24.9	36,700	Part of the γ complex, clamp loading
ψ	*holD*	99.3	15,000	Part of the γ complex, clamp loading
χ	*holC*	99.6	16,500	Part of the γ complex, clamp loading
θ	*holE*	41.5	8,700	Pol III dimerization
ε subunit	*dnaQ* (*mutD*)	5.1	27,000	Proofreading, 3' to 5' exonuclease
DNA Pol I	*polA*	87.2	103,000	Gap filling
Ligase	*ligA*	54.5	73,400	Ligation of single-strand nicks in the phosphodiester backbone
DNA gyrase (subunit α)	*gyrA* (*nalA*)	50.3	97,000	Supertwisting
DNA gyrase (subunit β)	*gyrB*	83.5	89,800	Relaxation of supercoils
DNA Pol II	*polB*	1.4	120,000	Repair
RNA Pol, β subunit	*rpoB*	90	150,000	Initiation

components of the polymerases stimulate the unwinding rate 10-fold. The helicase moves into the fork $5'$ to $3'$ on the lagging strand (opposite direction to elongation on the lagging strand). Strand separation enables replication to proceed and requires helix destabilizing proteins (single-strand DNA-binding proteins) that prevent unwound DNA from reannealing and protect ssDNA from intracellular nucleases

2. Unwinding. This results in the introduction of positive supercoils in the dsDNA ahead of the fork. Since excessive supercoiling can slow fork movement, DNA gyrase removes these positive supercoils by introducing negative ones.

3. Priming. In addition to unwinding DNA, the DnaB protein is thought to "engineer" ssDNA into a hairpin structure that could serve as a signal for the primase (*dnaG* gene product) to synthesize a RNA primer (beginning with ATP) of about 10 nucleotides. This RNA primer is needed because, unlike RNA polymerases, no DNA polymerase can synthesize DNA de novo (i.e., without a preexisting $3'$-OH priming end). The RNA polymerase (primase) activity of DnaG must be activated by interacting with DnaB helicase. Unlike DnaB and Pol III holoenzyme, DnaG does not travel as a stable component of the protein complex at the fork. It is recruited from solution for each priming event.

4. β-clamp loading. Once the RNA primer is formed, DNA polymerase III can bind to the $3'$-hydroxyl terminus of the primer and begin to synthesize new DNA. However, this is not a straightforward process. Before the Pol III-α subunit (polymerase activity proper) can bind, a β-clamp protein is loaded onto the primed, single-stranded DNA. The β-clamp is a ring-shaped homodimer that encircles the DNA (Fig. 2-10). It is loaded onto DNA by the γ-complex clamp loader composed of five different polypeptides with the stoichiometry $\gamma_2\delta\delta'\chi\psi$. The β-clamp then binds the α subunit (polymerase), forming a "sliding clamp" that allows DNA polymerase to move along the template without easily being dislodged. This increases the processivity of DNA synthesis (i.e. increases the number of bases added before DNA polymerase leaves the template). On the leading strand, where DNA is synthesized in one continuous strand, loading of the clamp only needs to be done once on an undamaged DNA template. On the lagging strand, where DNA is synthesized in short 1000 bp fragments, β dimers must be loaded and unloaded many times. This is facilitated by the clamp loader being part of the replication machine, moving with the fork.

5. Completion of lagging strand. Once started, synthesis on the lagging strand will continue until the polymerase meets with the $5'$ terminus of a previously formed RNA portion of an Okazaki fragment. At this point, the polymerase will dissociate from the DNA.

6. Proofreading. Recalling the fidelity with which DNA is replicated in spite of the speed, leads to the question of how fidelity is maintained. Polymerases in general incorporate a relatively large number of incorrect bases while replicating. This, however, sets up a base pair mismatch situation that can be recognized as incorrect by the $3' \rightarrow 5'$ exonuclease proofreading activity of the DNA polymerase holoenzyme. The component of DNA polymerase responsible for proofreading is DnaQ. This subunit allows polymerase to hesitate, excise the incorrect base, and then insert the correct base. However, even with this elegant proofreading system, some mistakes still occur. Chapter 3 covers how the cell deals with mistakes.

7. Replacing the primer. Once Pol III has replicated DNA up to the point where a preexisting RNA primer resides (on the lagging strand), the cell must remove that

RNA and replace it with DNA. RNAse H cleaves the RNA primer and DNA Pol I uses its $5' \rightarrow 3'$ polymerase activity to replace the cleaved RNA with DNA.

8. Repairing single-strand nicks on the lagging strand. When Pol I is finished, there remains a phosphodiester break(or nick) between the 3'-hydroxyl end of the last nucleotide synthesized by Pol I and the 5'-phosphoryl end of the adjacent DNA segment. DNA ligase will "seal" this nick by using NAD as a source of energy (Fig. 2-11).

Train Versus Factory Model of Replication. Two general models have been proposed for DNA replication. In one model, DNA polymerase moves along the DNA like a train on a track. In the second model, DNA polymerase is stationary (like a factory) in the cell and DNA is pulled through it. Support for the factory model came from visualizing that a fluorescently tagged polymerase in *Bacillus* remained localized at discrete locations at or near the midcell rather than being distributed randomly throughout the cell.

Initiation of DNA Replication

Replication of the bacterial chromosome proceeds bidirectionally from a fixed origin (*oriC*) located at 83.7 minutes on the standard *E. coli* linkage map (3881.5 kb) (Fig. 2-12). It takes about 40 minutes to completely replicate the *E. coli* genome. Yet, the generation time of *E. coli* can be as short as 20 minutes. To accomplish this feat, a single cell of *E. coli* will initiate a second and sometimes a third round of replication before the first round is complete. That way, after round one is finished and the cell divides, each daughter cell is already halfway through replicating its chromosomes.

In rapidly growing *E. coli* containing multiple origins of replication, all origins fire at precisely the same time in the cell cycle, and each origin only fires once per cell cycle. It is crucial to cell viability that initiation of chromosome replication is tightly controlled. This control is focused at the origin of replication (*oriC*), a 245 base pair sequence located at 83.68 minutes (3881.8 kb) on the circular *E. coli* map (Figs. 2-12 and 2-13). The region contains several 13-mer $A + T$−rich sequences that enable easy strand separation during initiation (the amount of energy required to separate $A + T$−rich DNA strands is considerably less than that required to separate $G + C$−rich DNA strands because A/T pairs are connected by two hydrogen bonds whereas G/C pairs are connected by three bonds). A number of potential RNA polymerase-binding sites have also been found in this region (see "RNA Synthesis" in this chapter). Transcription from some of these sites (e.g., *moiC*) helps separate the strands during assembly of the initiation complex.

Initiation of DNA synthesis is linked to the ratio of cell mass to the number of origins present in the cell. When a certain ratio is reached, the origins fire. As a result, slow-growing cells (one origin per cell) are smaller (less mass) than fast-growing cells (two or more origins per cell). The rule seems simple, but the molecular details of initiation are still being worked out today. Efforts to understand initiation started with the **replicon** theory proposed in 1963 which predicted that initiation could be controlled negatively and/or positively and that negative control might involve transient association of the DNA origin with the cell membrane. Amazingly, these predictions have proven true. A diagram of the origin is presented in Figure 2-13. The positive regulator of initiation is now known to be DnaA, 20 molecules of which must bind

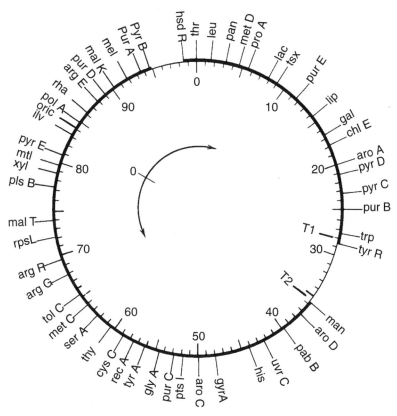

Fig. 2-12. Partial linkage and replication maps of _Escherichia coli_. Circular reference map of _E. coli_ K-12. The large numbers refer to map positions in minutes, relative to the _thr_ locus. From the complete linkage map, 52 loci were chosen on the basis of familiarity as longstanding landmarks of the _E. coli_ K-12 genetic map.

to five repeated, 9 bp sequences called DnaA boxes located within the origin. DnaA binding will promote strand opening of the AT-rich 13-mers and facilitate loading of the DnaB helicase. DnaA protein levels remain relatively constant during the cell cycle, but the activity of DnaA and its accessibility to the origin are thought to be regulated (see below).

Preventing Premature Reinitiation. The negative regulator of initiation is a protein called SeqA that will sequester the _E. coli_ chromosome origin to the membrane (actually the outer membrane). SeqA selectively binds to origins and inhibits reinitiation after replication has begun. This negative control mechanism hinges on when newly made origins become methylated at specific GATC sequences. The chromosome of _E. coli_ contains many GATC palindromic sequences that are targets for an enzyme called deoxyadenosine methylase (Dam methylase). Discussed further in Chapter 3, Dam methylase adds a methyl group to the adenine in GATC sequences. However, immediately after a segment of chromosome is replicated, only the parental strand will contain GA^{Me}TC sequences. The newly synthesized strand remains unmethylated until it encounters Dam methylase. As a result, newly replicated DNA segments are

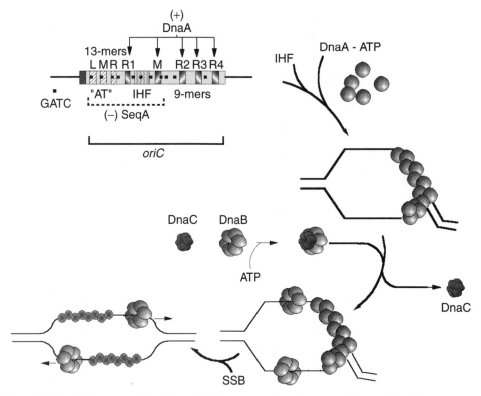

Fig. 2-13. Initiation of replication. The diagram depicts the area of the origin of replication. L, M, R, indicate left, middle, and right 13-mers; R1–4, 9-mer DnaA boxes. DnaA proteins bind to the DNA boxes to initiate replication. In collaboration with the histone-like protein HU and IHF, DnaA helps form an open complex. DnaC guides DnaB to the open complex, where it forms a sliding clamp. Its unwinding activity helps elongate the open complex and serves as a loading site for DNA polymerase and primosome components.

temporarily hemimethylated. It so happens that the SeqA protein has a high affinity for hemimethylated DNA. The *oriC* region contains 11 such GATC sites, so after the origin is replicated, SeqA will bind and sequester the hemimethylated daughter origins to the membrane, preventing more DnaA binding. As a result of its membrane association, methylation of *oriC* is delayed about 13 minutes after initiation.

Premature reinitiation is prevented not only by SeqA-dependent sequestration of the origin but through a timed reduction of DnaA activity. DnaA can bind either ATP or ADP but only the ATP form is active. As the replication machinery begins synthesizing DNA, the β subunit of DNA polymerase III stimulates ATP hydrolysis of DnaA–ATP, forming an inactive DnaA–ADP complex, thus decreasing the possibility of reinitiation. Regulation of initiation also depends on the titration of DnaA protein by a competing binding site called *datA* located close to *oriC*. Once replicated, the two copies of *datA* are predicted to bind DnaA molecules, thereby drastically reducing the concentration of free DnaA. Since *dat* does not contain any GATC sites, SeqA will not bind this area, leaving *dat* open to DnaA binding. In contrast, the *dnaA* promoter contains GATC sites that become hemimethylated after replication. As a consequence,

SeqA will bind and transiently repress *dnaA* transcription, further lowering the free DnaA concentration.

Initiation Hyperstructure. So what actually triggers initiation? The answer is complex but appears to involve increasing the level of active DnaA protein (DnaA–ATP) and the assembly of a so-called DnaA-dependent **initiation hyperstructure** at the cell membrane. In slow-growing cells, once replication stops, the β-clamp protein will no longer hydrolyze ATP bound to DnaA, and DnaA–ATP can build up. In rapidly growing cells, the increased copy number of *dnaA* genes relative to the number of actively replicating forks may lead to higher levels of DnaA and thus more DnaA–ATP, resulting in more frequent initiations. DnaA has been localized to the membrane, displaying an affinity for anionic phospholipids (e.g., cardiolipin) that are localized to cell poles and the septum. Interaction with these acidic phospholipids is also a requirement for DnaA activation. However, activation of DnaA alone is not sufficient for the initiation of DNA synthesis.

Immediately before initiation, the DNA-binding protein called IHF (integration host factor; see "λ Phage" in Chapter 6) binds to a region in the origin and creates a bend in the DNA. The bend imposed by IHF brings DnaA box R1 in close proximity to the weaker DnaA-binding sites in the central part of *oriC*. This facilitates binding of multiple DnaA proteins to form the initiation complex. What might determine the timing of IHF binding is not understood.

Replication Hyperstructure. Distinctly different DnaA-dependent and SeqA-dependent giant hyperstructures at the origin have been proposed. The envisioned SeqA **replication hyperstructure** would bring both replication forks together and include DNA polymerases as well as enzymes responsible for making and delivering DNA precursors. This complex would confer efficiency to the replication process and assure bidirectional replication. SeqA may do this by recruiting specific genes (by binding to strategically located hemimethylated regions) to a localized membrane site. The origin is localized to the center of the cell just before replication, but after replication is complete the daughter origins move to the cell poles. SeqA protein, however, only moves halfway to the poles, meaning there must be a separation at some point. How the origin returns to midcell just prior to replication is not clear. The terminus follows the exact opposite pattern, starting out at the pole before initiation and migrating to midcell after initiation (see "Termination of DNA Replication and Chromosome Partitioning").

The following list summarizes a proposed sequence of events leading to initiation:

1. Increased levels of DnaA–ATP. Origin moves to midcell.
2. Integration host factor binds to *oriC* and introduces bend.
3. DnaA–ATP binds to 9-mer DnaA boxes.
4. Open complex formation at 13-mer region, aided by transcription from *moiC*.
5. Formation of prepriming complex with DnaB helicase and DnaC loading protein.
6. ATP hydrolysis releases DnaC.
7. Helicase unwinds DNA and allows DnaG primase to enter and synthesize RNA primers.
8. DNA replication by DNA Pol III leads to hemimethylation of origin. β-clamp protein hydrolyses DnaA–ATP.

9. Sequestering of hemimethylated origin to membrane via SeqA.
10. Migration of origin toward poles.
11. Detachment of origin from SeqA membrane hyperstructure; methylation.
12. Slow increase in DnaA–ATP.

Termination of DNA Replication and Chromosome Partitioning

Once initiated, replication proceeds to the terminus (*terA, B, C*) between 30 and 35 minutes on the linkage map. The time required to complete one round of replication is about 40 minutes. There are two groups of *ter* sites arranged with opposite polarity that serve to inhibit the two converging replication forks. The T1 terminator (*ter A, D, E*; at 28 minutes) permits clockwise-traveling replication forks to enter the terminus region but inhibits counterclockwise-traveling forks that might exit the region. T2 (*terF, B, C*; at 35 minutes) does the opposite, allowing counterclockwise forks to enter the region but inhibiting clockwise forks from exiting (Fig. 2-12). Another gene essential for termination encodes the **terminus utilization substance**, Tus. A monomer of Tus protein will bind to each *ter* site. The Tus-DNA complex can then act as a "contrahelicase," inhibiting the DnaB helicase activity of an approaching replication fork. Since DnaB helicase drives replication fork movement, inhibiting the helicase will block further movement.

After termination, the two daughter chromosomes will form a linked concatemer due to the topological constraints inherent when separating the strands of a double-stranded helical circle. These linked molecules must be resolved through recombination if the cell is to divide (see Fig. 2-14). In *E. coli*, two tyrosine family site-specific recombinases, XerC and XerD, bind cooperatively at the 33 bp chromosomal site, *dif*, located at the replication terminus. XerC first recognizes *dif* and initiates a recombination (crossover) event between the two strands. The result is a structure, called a Holliday junction, which must be resolved (see Fig. 2-14 and "Recombination" in Chapter 3). A membrane protein, FtsK, located at the septum of a dividing cell, is hypothesized to modify the Holiday structure into a suitable substrate for the second recombinase, XerD. XerD acts on the Holiday structure, leading to completed recombination products and resolution of the concatemer.

Partitioning of the two chromosomes into separate daughter cells is essential for successful cell division. The process is not fully understood, but it appears that the SeqA protein pool, localized to the midcell, binds to newly hemimethylated DNA, then separates into two foci that migrate in opposite directions toward the poles. Type IV topoisomerase (ParC and ParE) is believed to decondense the nucleoid at the center of the cell by decreasing the supercoiling of the chromosome. The chromosome passes

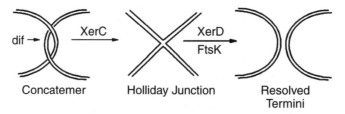

Fig. 2-14. Resolution of chromosome concatemers.

through a stationary replication machine and then is drawn to the cell poles by the action of the MukBKF complex, which is thought to supercoil DNA into a more condensed form. The FtsK protein may contribute to the segregation process as a DNA pump (see further discussion in Chapter 17).

RNA SYNTHESIS: TRANSCRIPTION

RNA Synthesis

Transcription is the process by which the information contained within genes is converted into RNA. The process is the same regardless of whether the gene encodes messenger RNA (mRNA), transfer RNA (tRNA), or ribosomal RNA (rRNA). The transcription of DNA to RNA requires a DNA-dependent RNA polymerase and proceeds in a manner similar to DNA synthesis but uses ribonucleic acid triphosphates (rNTP) rather than deoxyribonucleic acid triphosphates (dNTP).

RNA polymerase (RNAP) is an extremely complex machine that senses signals coming from numerous regulatory proteins as well as signals encoded in the DNA sequence. RNAP consists of four polypeptides: α, β, β', and σ. Core polymerase consists of two α subunits plus one β and one β' subunit. A fifth subunit called omega is now believed to be involved in assembling the RNA polymerase complex. The core enzyme can bind to DNA at random sites and will synthesize random lengths of RNA. Holoenzyme, consisting of core enzyme plus a σ subunit, binds to DNA at specific sites called promoters and transcribes specific lengths of RNA. σ^{70} (70 kDA molecular weight) is considered the "housekeeping" σ, but there are several specialty σ factors that direct RNAP to specific promoters. Thus, the σ subunit plays an important role in promoter recognition by RNA polymerase.

The β subunit carries the catalytic site of RNA synthesis as well as the binding sites for substrates and products. The β' subunit plays a role in DNA template binding while the two α subunits assemble the two larger subunits into core enzymes ($\alpha_2\beta\beta'$). Rifampicin, a commonly used antibiotic, inhibits transcription by interfering with the β subunit of prokaryotic RNA polymerase. Figure 2-15 presents a view of RNA polymerase structure and important binding regions for each subunit. Electron microscopy and X-ray crystallography reveal the RNAP complex is shaped like a crab with a groove or trough that fits double-stranded helical DNA. Figure 2-16 represents an approximate density map of RNAP transcribing DNA.

Within a transcribing region, the coding strand refers to the DNA strand that is not transcribed but consists of the same sequence as the mRNA product (with a T substituting for U). The antisense or template strand is the strand that is transcribed (read) by RNA polymerase. Insofar as RNA polymerase binds to a promoter region, then progressively moves along the template strand from the region encoding the N terminus to the carboxy terminus, the terms *upstream* and *downstream* are used to describe regions relative to the direction of RNA polymerase movement. Thus, the promoter is upstream from the structural gene. It should be noted that RNA polymerase moves along the DNA template strand in the $3' \rightarrow 5'$ direction while synthesizing RNA $5' \rightarrow 3'$ (Fig. 2-17).

As with DNA replication, transcription can be described in three main steps: initiation, elongation, and termination. Initiation involves the binding of polymerase to the promoter with the formation of a stable RNA polymerase–DNA initiation

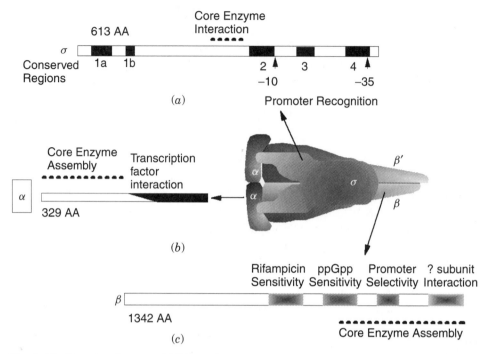

Fig. 2-15. Functional map of RNA polymerase subunits. The molecular model of RNA polymerase holoenzyme is shown in the middle. Areas of each subunit involved with various interactions between subunits, other protein factors, effector molecules, and promoter regions are indicated.

complex and the catalysis of the first $3' \rightarrow 5'$ internucleotide bond. Elongation is the translocation of RNA polymerase along the DNA template with the concomitant elongation of the nascent RNA chain. Termination, obviously, involves the dissociation of the complex. The overall process of transcription is illustrated in Figure 2-17.

Initiation. The site referred to as the promoter includes two DNA sequences that are conserved in most promoters. In general, highly conserved sequences are referred to as consensus sequences. The promoter consensus sequences are centered at -10 and -35 base pairs from the transcriptional start point (designated $+1$) and have been implicated in normal promoter function. The consensus -10 and -35 sequences for the σ^{70} housekeeping σ factor are

$$5' \text{ TTGACA} \qquad 5' \text{ TATAATPu}$$
$$\text{AACTGT } 5' \qquad \text{ATATTAPy } 5'$$
$$-35 \qquad\qquad -10$$

The -10 sequence is referred to as Pribnow's box while the -35 region is called the recognition site. It should be noted that different σ factors recognize different -10 and -35 sequences. One question that must be addressed, however, is how RNA polymerase can recognize these sequences when it would appear that direct interaction between RNA polymerase and DNA is precluded by the double-stranded nature of

Taq RNA Polymerase

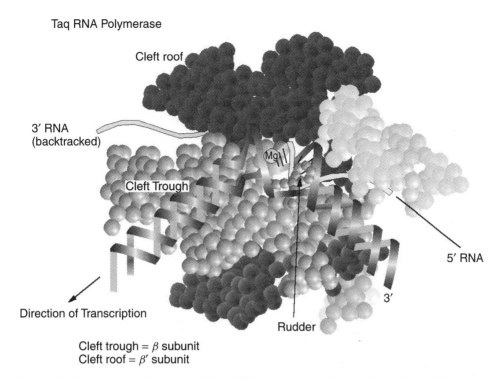

Cleft roof

3′ RNA
(backtracked)

Cleft Trough

Mg

5′ RNA

3′

Direction of Transcription

Rudder

Cleft trough = β subunit
Cleft roof = β' subunit

Fig. 2-16. Electron density map of Taq RNA polymerase. The cleft in which DNA fits is formed between the β subunit (cleft roof) and the β' subunit (cleft trough). Note the Mg^{2+} at the active site and the rudder that helps separate the strands. The position of a backtracked RNA is also shown.

DNA. Association is possible for any protein−nucleic acid interaction through base-specific groups that can be recognized in the major or minor grooves. RNA polymerase can interact with groups in the major groove and recognize the proper sequence upstream (−35 region) from the Pribnow box, then form a stable complex (closed complex) by moving laterally to the −10 region. σ factors are involved in recognizing both regions.

DNA Supercoiling and Transcription. The superhelical nature of the chromosome also plays a role in promoter function. In general, a negatively supercoiled chromosome is a better transcriptional template than a relaxed chromosome. Presumably, the torsional stress imposed by supercoiling makes certain areas of DNA easier to separate by RNA polymerase (i.e., lowers the melting temperature). However, it appears that supercoiling affects the expression of some genes more than others. There is some thought that the cell, through the use of topoisomerases Top I and DNA gyrase, can affect supercoiling in localized areas of the chromosome, facilitating the transcription of some genes while impeding the transcription of others.

Alternate Sigma Factors. Obviously, σ-70 plays an important role in normal transcription initiation. Alternate σ factors (e.g., σ-32, 32 kDa molecular weight) can change the promoter recognition specificity of core enzyme. Thus, σ-32 factor

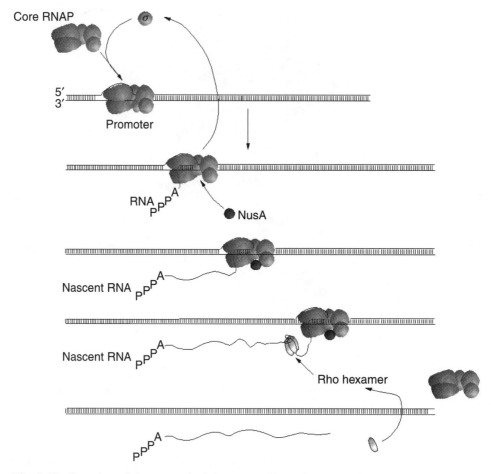

Fig. 2-17. Overview of the transcription process. Sigma factor (σ) directs core polymerase to the promoter region of a gene. After transcription of several nucleotides, σ disassociates and is replaced by NusA. When transcription is complete, RNA polymerase encounters transcriptional stop signals in the mRNA. One type of signal recruits ρ-factor protein to the RNA. The mRNA wraps around a ρ hexamer, which enables ρ to interact with RNA polymerase and trigger disassociation.

is utilized when cells undergo heat shock and is believed to be responsible for the production of proteins required to survive that stress (see Chapter 18). Other alternate σ factors are listed in Table 2-4 and in the discussion on sporulation in Chapter 19. Among the various sigmas there are highly conserved regions designated 1 to 4 as shown in Figure 2-15. The C-terminal part of region 2 (2.3, 2.4, and 2.5) is primarily involved in the use of the -10 region. Important strategies to control gene expression include (1) anti-σ factor proteins to inhibit the activity of a specific σ until needed and (2) regulated proteolysis of the σ factor. These mechanisms are discussed later.

DNA UP Element and Protein–Protein Interactions with RNA Polymerase. Promoter recognition by bacterial RNAP also involves interactions between the C-terminal domain of the RNAP α subunit and DNA sequences upstream of the core promoter.

TABLE 2-4. RNA Polymerase Subunits and Other Transcription Factors of
Escherichia coli

Subunit	Structural Gene	Map Position	Molecular Weight[a]	Function
α	*rpoA*	74.1	36,000	Initiation
β	*rpoB*	90.1	150,000[b]	Initiation, elongation, termination
β'	*rpoC*	90.1	160,000	Initiation
σ-^{70}(F^D)	*rpoD*	69.2	83,000	Initiation
σ-^{32}(F^H)	*rpoH*	77.5	32,000	Initiation, heat-shock response
σ-^{28}(F^F)	*fliA*	43.1	28,000	Flagellar genes, chemotaxis
σ-^{24}(F^E)	*rpoE*	58.4	24,000	Extreme heat shock
σ-^{42}(F^S)	*rpoS*	67.1	42,000	Stationary phase
σ-^{54}(F^N)	*rpoN*	72.0	54,000	Nitrogen
Omega	*rpoZ*	82.3	10,000	RNAP assembly
CRP	*crp*	75.1	22,500	Initiation enhancement
Rho (ρ)	*rho*	85.4	50,000	Termination
NusA	*nusA*	71.4	69,000	Pausing, transcription termination
NusB	*nusB*	9.4	14,500	Pausing, transcription termination
NusG	*nusG*	90.0	20,000	Pausing, transcription termination
GreA	*greA*	71.7	17,000	$3' \rightarrow 5'$ RNA hydrolysis proofreading
GreB	*greB*	76.2	8,384	$3' \rightarrow 5'$ RNA hydrolysis, proofreading

[a]Daltons.
[b]Zn^{2+} metalloenzyme.

Thus, bacterial promoters consist not only of σ-binding sites but α-binding sites as well. The α-binding sites have been named **UP elements** (for upstream). As noted in Figure 2-15, the N-terminal portion of α associates with the other subunits of RNAP. The C-terminal domain (CTD) is linked to the N-terminal domain by a flexible linker. As such, the CTD can interact with other transcriptional activator proteins and with DNA UP sequences. Part of the role of some regulatory proteins is to increase curvature of DNA to allow CTD access to the UP elements.

Many regulatory factors bound to DNA also directly contact RNA polymerase at the promoter region. These proteins can be divided into two classes. Class I transcriptional activators bind upstream from the promoter (e.g., CRP, AraC, Fnr, OmpR). Class II transcription factors overlap the promoter region. Many of these regulators contact the C-terminal end of the α subunit (Fig. 2-15) and activate transcription by engineering interactions with UP elements. Many of these factors are discussed in Chapter 5.

Conversion from a Closed to an Open Complex. The closed promoter complex, in which RNA polymerase-bound DNA remains unmelted, must be converted to an open promoter complex. This involves the unwinding of about one helix turn from the middle of Pribnow's box to just beyond the initiation site (the transcription bubble). Two aromatic amino acid residues within region 2 of σ^{70} are critical to unwinding. Note in Figure 2-16 that there is also a "rudder" structure in β' that separates the strands once unwound. The open complex then allows tight binding of RNA polymerase with the subsequent initiation of RNA synthesis. The first triphosphate in an mRNA chain is usually a purine. Initiation ends after the first internucleotide bond is formed (5'pppPupN).

Elongation. After the formation of transcripts eight to nine nucleotides in length, σ factor is abruptly released from the complex, suggesting that RNA polymerase undergoes a conformational change that causes a decrease in the affinity of σ for the RNA polymerase–DNA–nascent RNA complex (ternary complex). The released σ factor can be reused by a free core polymerase for a new initiation (Fig. 2-17).

Once elongation begins, transcription proceeds at a rate of between 30 and 60 nucleotides per second (at $37\,°C$). The elongation reaction as a whole includes the following steps: (1) nucleotide triphosphate binding; (2) bond formation between the bound nucleotide and the 3′-OH of the nascent RNA chain; (3) pyrophosphate release; and (4) translocation of polymerase along the DNA template. An essential feature of the β' catalytic site is a bound Mg^{++} ion (Fig. 2-16). Movement of RNA polymerase, by necessity, involves the melting of the DNA template ahead of the transcription bubble (ca. 17 nucleotides) as well as reformation of the template behind the bubble. Part of the β subunit forms a "flap" thought to hold the DNA.

The rate of elongation along a template is not uniform. Regions in a template where elongation rates are very slow are called **pausing sites**. Pausing sequences in general are GC-rich regions possessing dyad symmetry that form hairpin loops in the RNA (an RNA–RNA stem-loop structure). The base of a typical pause site stem is located approximately 11 base pairs from the active center of RNAP (an Mg^{++}-binding site) and is thought to cause the β subunit flap to open. The RNA–DNA hybrid would then move away slightly from the active site, slowing elongation.

The **NusA** and **NusG** proteins are believed to associate with core polymerase after σ disassociates and modulate RNAP elongation rates. The Nus interaction with the ternary complex is nonprogressive, involving rapid association and disassociation. These proteins enhance pausing at some sites and may be involved with Rho-dependent transcription termination (see below). Recent evidence suggests that NusA stabilizes the pause hairpin–flap interaction, thereby lengthening the pause period.

The transcription process is not without its flaws. Sometimes RNAP inappropriately backtracks along a nascent message, leaving the 3′ end of the message protruding. This backtracking obviously disengages the 3′ end of RNA from the catalytic site. In this event, transcription is arrested. Two proteins, **GreA** and **GreB**, associated with RNAP, trigger an endonucleolytic activity at the catalytic center that cleaves and removes the 3′ end of the nascent message. This releases RNA polymerase, allowing transcription to proceed.

It should be noted at this point in the discussion that the RNA transcript is normally longer than the structural gene. Most messages contain a promoter-proximal region of mRNA called the **leader sequence** (or untranslated region) that exists between the promoter and the translation start codon. The leader sequence carries information for ribosome binding (see "Shine-Dalgarno Sequence" in Initiation of Polypeptide Synthesis) and in some cases is important in determining whether RNA polymerase will proceed into the structural gene(s) proper (see "Attenuation Controls" in Chapter 5).

Termination. Transcriptional termination includes the following events: (1) cessation of elongation, (2) release of the transcript from the ternary complex, and (3) dissociation of polymerase from the template. There are two classes of termination: ρ-independent and ρ-dependent transcription termination. The ρ-independent termination signal includes a GC-rich region with dyad symmetry, allowing formation of an RNA stem-loop structure about 20 bases upstream of the 3′-OH terminus and a stretch of four to eight consecutive uridine residues. The transcript usually ends within or just distal to the uridine string. The RNA stem-loop structure causes RNA polymerase to pause and disrupts the $5'$ portion of the RNA–DNA hybrid helix. The remaining $3'$ portion of the RNA–DNA hybrid molecule includes the oligo(rU) sequence. The relative instability of rU-dA base pairs causes the $3'$ end of the hybrid helix to melt, releasing the transcript.

As the term implies, termination at ρ-dependent terminators requires the ρ-gene product. ρ-factor termination only occurs at strong pause sites apparently located at specific distances from a promoter region — that is, a strong pause site located closer to the promoter will not serve as a termination site. ρ-utilization (*rut*) sequences on RNA are GC rich with no secondary structure. ρ will assemble as a hexamer at *rut* sites (Fig. 2-17) after which the single-stranded mRNA is thought to wind around the outside of ρ-factor, facilitated by ρ-dependent ATP hydrolysis (Fig. 2-17). This process serves to bring ρ in contact with RNA polymerase. The extended dwell time of RNA polymerase elongation complexes at pause sites allows the ρ protein translocating along the nascent RNA to catch up with the transcription complex. The association of ρ to core enzyme may occur via NusG protein. Subsequently, activation of a ρ RNA–DNA helicase activity causes the mRNA chain and core RNA polymerase to disassociate from the DNA. Free core RNA polymerase can then interact with a σ factor and be used to initiate a new round of transcription.

RNA Turnover

Cellular RNA, thus transcribed, can be classed into two major groups relative to their decay rates in vivo: stable and unstable RNA. Stable RNA consists primarily of ribosomal RNA (rRNA) and transfer RNA (tRNA) while unstable RNA is mRNA. In *E. coli*, 70 to 80% of all RNA is rRNA, 15 to 25% is tRNA, and 3 to 5% is mRNA. Factors that contribute to stability include the association of rRNA in ribonucleoprotein complexes (ribosomes) and the extensive secondary structure exhibited by both rRNA and tRNA. Both of these structural considerations serve to protect stable RNA at the $5'$ terminus from ribonucleases.

The average mRNA (about 1200 nucleotides) has a **half-life** (time required to reduce the mRNA population by one-half) of approximately 40 seconds at $37\,^\circ$C. This number represents an average as each mRNA species has a unique degradation rate. Degradation occurs overall in the $5' \rightarrow 3'$ direction; however, all exoribonucleases present in the cell only degrade in the $3' \rightarrow 5'$ direction. One theory that attempts to resolve this paradox is that an initial, random endonucleolytic event occurs in the mRNA near the $5'$ end that would allow for exonucleolytic degradation of the $5'$ end. Since no new ribosomes can bind to mRNA that has lost its $5'$ end, the remaining message is progressively exposed as the already bound ribosomes move toward the $3'$ end. Thus, additional endonucleolytic events followed by $3' \rightarrow 5'$ exonucleolytic degradation would be possible. Most RNA degradation appears to occur via a multicomponent **degradosome complex** involving polynucleotide phosphorylase (*pnp*), RNAse E (*rnaE*), an ATP-dependent RNA helicase (RhlB), a DnaK chaperone (see Chapter 18, Heat Shock response), and the enzyme enolase.

Poly A Tails. A feature originally thought to be unique to eukaryotic cells is the addition of poly(A) tails to the $3'$ ends of many of their mRNAs. It is now evident that poly(A) tails also occur in bacterial mRNA. At least two poly(A) polymerases have been identified in *E. coli*: Pap I (*pcnB*) accounts for 80% of poly(A) and polynucleotide phosphorylase (Pnp) accounts for the remainder. It has been proposed that polyadenylation provides the basis for the binding of the degradosome to mRNAs. Thus, Pnp serves not only as a 3' exonuclease but will add poly(A) tails to messages facilitating their degradation.

Message stability can be enhanced by the presence of stem-loop structures present at the end or beginning of the mRNA. Stem loops at the 3' end of the message will inhibit 3'–5' exonuclease processivity while the 5' loops may interfere with endonuclease association. However, Pnp-dependent polyadenylation of mRNAs containing 3' stem-loop structures is suggested to increase the susceptibility of those messages to degradation.

RNA Processing

All stable RNA species and a few mRNAs of *E. coli* must be processed prior to their use. For example, each of the seven rRNA transcription units is transcribed into a single message as follows:

5' leader–16S rRNA–spacer–23S rRNA–5S rRNA–trailer-3'

The spacer always contains some tRNA gene. Obviously, this long multigene (polycistronic) precursor RNA molecule must be processed to form mature 16S rRNA, 23S rRNA, and 5S rRNA species. Another example of processing involves the tRNA genes, which are frequently clustered and cotranscribed in multimers of up to seven identical or different tRNAs. These multimers must be processed to individual tRNAs. Along with the required cutting and trimming, the initial transcripts of rRNA and tRNA lack the modified nucleosides of the mature species. Thus, processing of stable RNAs (ie., tRNA and rRNA) includes the modification of bases in the initial transcript.

There are four basic types of processing. The first involves the precise separation of polycistronic transcripts into monocistronic precursor tRNAs. Second, there must be a mechanism by which the mature 5' and 3' termini are recognized followed by the removal of extraneous nucleotides. The third type of processing involves the addition of terminal residues to RNAs lacking them (e.g., the 3' CCA end of some bacteriophage tRNAs such as T4 phage). Finally, the appropriate modification of base or ribose moieties of nucleosides in the RNA chain must be accomplished. Obviously, not every RNA molecule is subject to all four of these processes. The following constitutes a brief description of several enzymes involved with RNA processing in *E. coli*:

> **RNase P, a Ribozyme.** This endonucleolytic enzyme is required for the maturation of tRNA (Fig. 2-18). RNase P removes a 41-base-long fragment from the 5-side of a wide variety of tRNA precursor molecules. The enzyme appears to recognize tRNA secondary and tertiary configurations rather than a specific nucleotide sequence. There is an absolute requirement for the presence of the CCA-terminal sequence within the precursor (see below). The subunit structure of this enzyme is unusual in that there is a polypeptide component (C5) encoded by the *rnpA* locus and an RNA component (M1/M2) encoded by *rnpB*. The RNA component

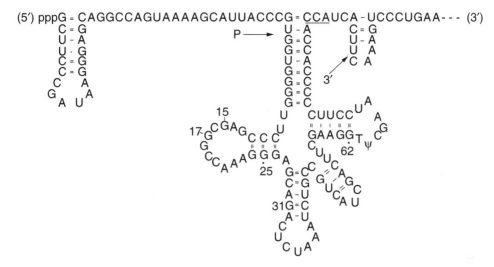

Fig. 2-18. tRNA precursor structure. The sequence shown is that of the tRNA$^{tyr}_1$ transcript. Transcriptions in vivo proceed through a second tRNA$^{tyr}_1$ gene. Sites of in vivo endonuclease cleavage are shown on the 5′ side of the mature tRNA (RNaseP) and near the 3′ end. The mature 3′ terminus is underlined. Numbered nucleotide positions are those at which base changes result in reduced RNase P cleavage in vivo.

is integral to the mechanism by which RNase P cleaves tRNA precursor. It is, in fact, an example of a catalytic RNA. Hence, RnaseP is described as a ribozyme.

RNase II. Encoded by *rnb*, it is one of the major 3′ → 5′ exonucleases in *E. coli*.

RNase III. Encoded by the *rnc* locus, this enzyme contains two identical polypeptide units of 25,000 molecular weight with no RNA component. RNase III cleaves dsRNA by making closely spaced ss breaks (ca. every 15 bases). Other than recognizing perfect double-stranded RNA stems, there does not appear to be a unique recognition sequence for this enzyme.

RNase D. Maturation of tRNA precursors requires the removal of extra nucleotides distal to the CCA sequence located within the precursor tRNA destined to become the mature 3′-OH end of the tRNA molecule. The CCA–OH sequence is required for amino acid acceptor activity on all tRNAs. RNase D (*rnd*) is a monomeric protein with a molecular weight of 38,000. It is a nonprocessive 3′ exonuclease that releases mononucleotide 5′ phosphates from the RNA substrate.

RNaseE. The *rne* locus is the structural gene for RNase E (molecular weight 70,000). The enzyme cleaves p5s precursor rRNA from larger transcripts. The enzyme first cleaves between the ss region of 5s precursor and the double-stranded region. The second cleavage occurs within the double-stranded area (Fig. 2-19). In addition to ribosomal RNA processing, RNase E is part of the RNA degradosome noted above.

RNase H. Encoded by *rnh*, this enzyme degrades the RNA strand of a DNA–RNA hybrid molecule.

RNase R. A major 3′ → 5′ exoribonuclease, (*rnr* or *vacB*), mutants defective in both *rnr* and *pnp* are inviable.

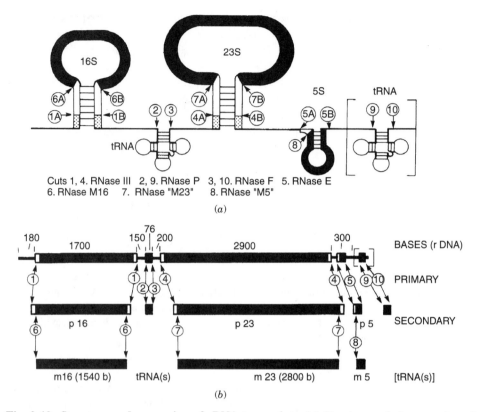

Fig. 2-19. Structure and processing of rRNA transcripts. (*a*) Structure and cleavage sites of the rRNA primary transcript (not to scale). Distal (trailer) tRNAs are bracketed because not all rDNAs contain them. Transcripts may contain one or two spacer tRNAs, and, one or two trailer tRNAs or none. Arrows indicate endonucleolytic cleavage sites. Each cutting event is given a separate number, referring to the enzyme involved; A and B indicate that two (or more) separate cuts may be required. Thick, solid segments represent mature rRNA sequences; thick, open segments represent precursor-specific sequences removed during secondary processing steps; stippled segments are sequences found only in p16b and p23b of RNase III⁻ cells; and thin lines (except for tRNAs) represent nonconserved sequences discarded during primary processing. Enzymes are discussed in the text. (*b*) Processing in wild-type strains. The first line shows the transcriptional map of a representative rDNA unit, drawn approximately to scale. Distances in bases are between verticle bars above the map. The primary and secondary cuts, numbered as in (*a*), are shown above the products they generate. Open and solid segments are as in (*a*).

tRNA nucleotidyltransferase. This enzyme can repair prokaryotic tRNAs in which the CCA sequence is missing. The product of the *cca* locus (67 minutes), nucleotidyltransferase, will sequentially add a 3′-CCA-terminal sequence to these tRNAs. Although it is probably not essential, mutants with a reduced level of this enzyme grow at a rate slower than normal.

Modifying enzymes. These enzymes chemically modify nucleotides present within tRNA and rRNA precursor transcripts and include methylases, pseudouridyllating enzymes, thiolases, and others. There are several examples that illustrate the importance of these modified bases. For example, mutants resistant to the antibiotic kasugamycin have lost an enzyme that specifically methylates two adenine residues to form dimethyl adenine in the sequence A-A-C-C-U-G near the 3′ end of 16S rRNA. Lack of this modification diminishes the affinity of kasugamycin for 30S ribosomal subunits.

In addition, there are several modifications in tRNA that are important. Modifications in the third base of the anticodon loop can contribute to the "wobble" of anticodon–codon recognition (see "Transfer RNA").

Intervening sequences.   Some organisms, such as *Salmonella enterica*, do not contain intact 23S rRNA in their ribosomes. The 23S rRNA locus of *S. enterica* contains an extra 90 base pair sequence that is not present in the *E. coli* homolog. This sequence, called an intervening sequence (IVS), is excised from the original transcript much as an intron is excised from within the mRNA of most eukaryotic genes. However, unlike intron processing, the remaining pieces of the *S. enterica* 23S rRNA are not spliced together. Since the rRNA does not encode a protein, there is no need to splice the fragments. The pieces function normally in all aspects of ribosomal assembly and protein synthesis. The rRNA superstructure is unaffected because the fragments remain associated through secondary and tertiary contacts. RNase III is the enzyme responsible for processing the IVS as proven by observing normal processing of *S. enterica* 23S rRNA in *rnaE*⁺ but not *rnaE*⁻ mutants of *E. coli*. The origins of IVSs are unknown. They may be the "footprints" of transposable elements inserted and then excised from the 23S rRNA genes (see Chapter 3). A summary of RNA processing in *E. coli* is shown in Figure 2-20.

PROTEIN SYNTHESIS: TRANSLATION

During translation, the information coded in mRNA is specifically read and used to form a polypeptide molecule with a specific function. A given nucleotide sequence in mRNA codes for a given amino acid. Since there are 20 naturally occurring amino acids, there must be at least three nucleotides in the code for each amino acid (triplet code). Since there are only four bases, a doublet code would account for only 4^2 or 16 amino acids. With a triplet code, 4^3 or 64 combinations are possible. Since there are only 20 amino acids. There is usually more than one triplet codon that can code for a given amino acid (see Table 2-5). For this reason, the code is termed **degenerate**. On the other hand, a given triplet codon cannot code for any other amino acid. For example, although the triplets UUU or UUC can both code for the amino acid phenylalanine, they cannot code for any other amino acid, indicating that the code is **nonoverlapping**. Of considerable significance is the fact that three codons, UAG, UAA, and UGA do not usually code for any amino acid, and tRNA molecules with the corresponding anticodons are uncommon. [There are exceptions in mitochondria and selenocystine-containing proteins (see below).] These codons (**nonsense** codons) serve as termination signals, stopping the translation process. The code is commaless; that is, there is no punctuation. If a single base is deleted or if another base is added in the sequence, then the entire sequence of triplets will be altered from that point on (**frameshift**).

Although the code is considered to be universal, there are some important differences in detail that are considered later. At this point it is sufficient to note just a few specific

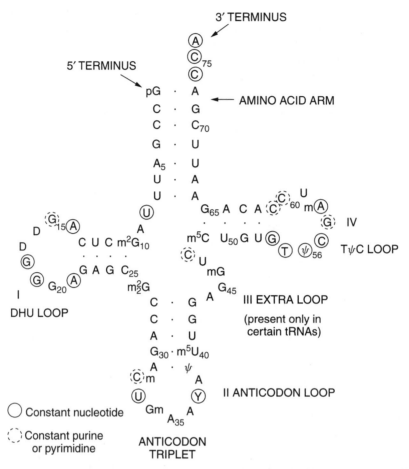

Fig. 2-20. Nucleotide sequence of yeast phenylalanyl tRNA. The structure is shown in the extended cloverleaf configuration. Loop I is termed the DHU loop since it contains dihydroxyuridine. The anticodon loop (II) contains the triplet bases which recognize the triplet code words in mRNA. Loop III varies in size and may even be absent from certain tRNAs. Loop IV is termed the TψC loop, since it contains (ribo)thymidylate, pseudouridine (ψ), and cytosine. Circled bases are constant in all tRNAs, and dashed circles indicate positions that are occupied constantly by either purines or pyrimidines. However, eukaryotic initiator tRNAs do not have the same constant nucleotides. (Modified from Kim et al., 1974.)

aspects. Methionine, rather than n-formylmethionine, is the initiating amino acid in eukaryotic cells as well as the archebacteria. The termination process is similar in both types of organisms. The process of protein synthesis in mitochondria is more closely analogous to that in bacteria than it is to what occurs in the cytoplasm of eukaryotes. Those differences that do exist between eukaryotic and prokaryotic protein synthesis may form the basis for selective inhibition by antibiotics or toxins on one cell type in the presence of the other. For example, some toxins have been shown to preferentially inhibit protein synthesis in eukaryotes. Diphtheria toxin specifically inactivates the elongation factor 2 of eukaryotic cells while having no effect on the analogous factor in prokaryotic cells.

TABLE 2-5. Amino acid code[a]

Second →	U	C	A	G	Third ↓
First ↓	Phe	Ser	Tyr	Cys	U
	Phe	Ser	Tyr	Cys	C
U	Leu	Ser	Ochre	Opal	A
	Leu	Ser	Amber	Trp	G
	Leu	Pro	His	Arg	U
C	Leu	Pro	His	Arg	C
	Leu	Pro	Gln	Arg	A
	Leu	Pro	Gln	Arg	G
	Ile	Thr	Asn	Ser	U
A	Ile	Thr	Asn	Ser	C
	Ile	Thr	Lys	Arg	A
	Met	Thr	Lys	Arg	G
	Val	Ala	Asp	Gly	U
G	Val	Ala	Asp	Gly	C
	Val	Ala	Glu	Gly	A
	Val	Ala	Glu	Gly	G

[a]First, second, and third refer to the order of the bases in the triplet sequence. Ochre, amber, and opal are termination codons.

Transfer RNA

The incorporation of amino acids into polynucleotide chains in a specified order involves the formation of an activated complex between an amino acid and a specific RNA molecule. This activation (charging) of specific transfer RNA (tRNA) molecules is accomplished by aminoacyl tRNA synthetases (aminoacyl tRNA ligases). Each amino acid has its specific synthetase and specific tRNA that can read the code in the mRNA and thereby direct the incorporation of the amino acid into the growing peptide chain. Each tRNA has the capacity to recognize a specific triplet code by virtue of a **codon recognition site** and a site that can recognize the specific tRNA synthetase (ligase) that adds the amino acid to the tRNA (**ligase recognition site**). Additional specificities are required: a specific **amino acid attachment site** and a **ribosome recognition site**. The amino acid attachment site is the 3′ hydroxyl of a terminal adenine ribose. All tRNA molecules have a 3′ terminus consisting of a cytosine–cytosine–adenine (CCA) nucleotide sequence and a 5′-terminal guanosine nucleotide as shown in Figure 2-20.

Transfer RNAs consist of approximately 80 nucleotides. A characteristic feature is the high content of unusual nucleotides (inosine, pseudouridine, and various methylated bases) that occupy specific positions. Figure 2-21 illustrates a number of these unusual bases. They are formed by modification of the structures by specific **modification enzymes** after the synthesis of the tRNA molecule has been completed (posttranscriptional modification). The secondary structure of tRNA appears as a

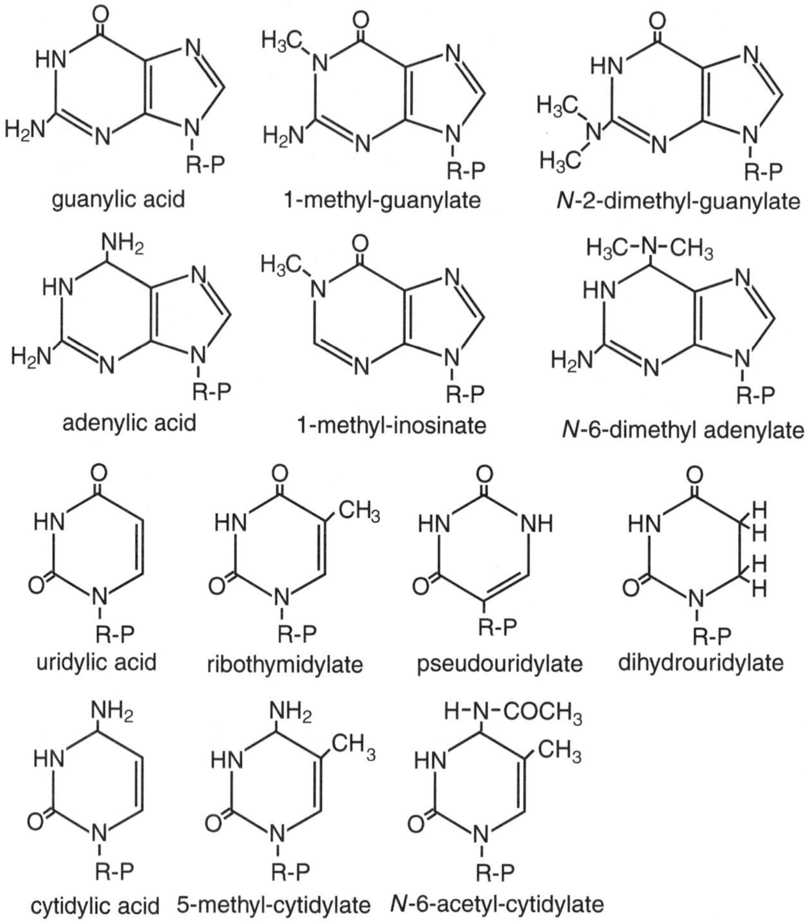

Fig. 2-21. Examples of some of the modified bases occurring in tRNAs.

cloverleaf with three loops containing the unusual bases. Loops result from the fact that some bases do not form double-stranded base-paired regions.

Loop I contains dihydroxyuridine (DHU) and is termed the DHU loop. Loop II contains the **anticodon**, a triplet of bases that recognizes the code in mRNA. Loop III varies in size, and, in those tRNAs in which it is large, it is double-stranded. Loop IV contains (ribo)thymidylate, pseudouridine, and cytosine (Fig. 2-21), a constant feature of all tRNAs that have been examined.

The activity of tRNA is dependent on the secondary and tertiary structures. Yeast phenylalanyl–tRNA was the first tRNA in which the complex tertiary structure was elucidated (Fig. 2-22). The 3′-terminal CCA required in all tRNAs is present upon transcription but requires processing (see "RNA Processing"). While it was generally considered that the DHU loop serves as the recognition site for tRNA ligase and loop III as the ribosome recognition site, it is now clear that the unusual tertiary structure of many sites in tRNA is involved in recognition. In the anticodon loop (loop II), each tRNA has a specific anticodon that directs the attachment of the proper amino acid as specified by the codon in the mRNA. Recognition is dependent on base pairing as in

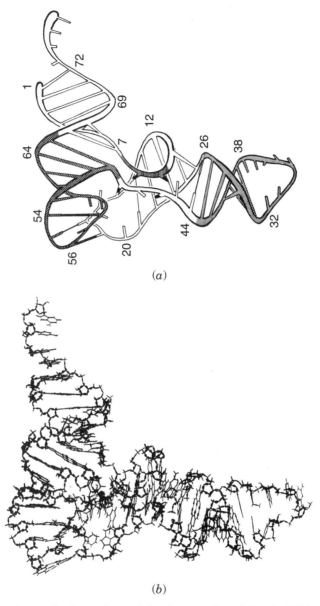

(a)

(b)

Fig. 2-22. Molecular and schematic models of yeast phenylalanyl tRNA. (*a*) Photograph of a molecular model built with Kendrew wire models at a scale of 2 cm per angstrom. The molecule is oriented with the anticodon loop at the bottom, the acceptor stem at the upper right, and the TψC loop at the upper left. The view is approximately perpendicular to the molecular plane. (*b*) Schematic model drawn from the coordinates of the molecular model shown in (*a*). The ribose phosphate backbone is drawn as a continuous cylinder with bars to indicate the hydrogen-bonded base pairs. The positions of single bases are indicated by rods, which are intentionally shortened. The TψC arm is heavily stippled, and the anticodon arm is marked by vertical lines. The black segments of the backbone include residues 8 and 9 as well as 26. Tertiary structure interactions are illustrated by black rods. The numbers indicate various nucleotides in the sequence as shown in Figure 2-20. (Modified from Kim et al., 1974.)

DNA. For example:

tRNA anticodon:	3′AAA	AUG	UUU	GGG	UGA
mRNA codon:	5′UUU	UAC	AAA	CCC	ACU
Amino acid added:	phe	tyr	lys	pro	thr

Since a number of amino acids have more than one codon, one question that must be addressed is how many tRNAs are required. There are, in a number of instances, isoaccepting species of tRNA — that is, different tRNAs that are charged by the same amino acid. This accommodates, in part, the degeneracy of the code. However, when there is more than one codon for a single amino acid, it is usually only the last nucleotide of the triplet that differs. The tRNA anticodon often contains one of the unusual nucleotides noted in Figure 2-21. These odd bases have "sloppy" base-pairing properties that permit them to bind with several different bases. This capability of recognizing more than one base is termed the "wobble" in codon recognition.

Charging of tRNA

The transfer of an amino acid to tRNA occurs via a two-step process. The amino acid must first be activated by ATP to form an aminoacyl–AMP complex with tRNA synthetase. The aminoacyl–AMP complex is then transferred to tRNA:

$$R-\underset{\underset{NH_2}{|}}{CH}-COOH + ATP + tRNA\ synthase \longrightarrow R-\underset{\underset{NH_2}{|}}{CH}-\overset{\overset{O}{\|}}{C}-OAMP-synthase + PP_i \qquad (1)$$

Amino Acid   Aminoacyl–AMP–synthetase complex

Aminoacyl–AMP–syntetase complex + tRNA

$$\longrightarrow aminoacyl-tRNA + AMP + tRNA\ synthetase \qquad (2)$$

Each amino acid has a specific aminoacyl–tRNA synthetase (charging enzyme) and a specific tRNA. Although there may be more than one species of tRNA for a specific amino acid, there is only one charging enzyme for each amino acid. The synthetase must have at least two recognition properties. It must be able to differentiate one amino acid from another and it must also be able to recognize a specific tRNA. These recognition properties are necessary to assure that the specific amino acid is charged on the proper tRNA molecule. Recognition of the codon on mRNA by the specific anticodon on the tRNA is required for the incorporation of an amino acid into the proper position in the polypeptide sequence.

Ribosome Structure and Synthesis

Ribosomes are the protein-synthesizing machines of the cell. They conduct the remarkable task of choosing which amino acids must be added to a growing peptide chain by reading successive mRNA codons. The complex problem of converting information (mRNA) into product (protein) is reflected in the elaborate structure of the ribosome. The prokaryotic ribosome is composed of two subunits that together form the 70S ribosome. The smaller subunit is 30S (900,000 mwt), containing 21 r-proteins (designated S1 through S21 in *E. coli* for small subunit proteins) and one 16S rRNA. The larger subunit is 50S (1.6 million mwt), containing 31 r-proteins (designated L1 through L34 for large subunit proteins) and one copy each of two rRNA species, 23S rRNA and 5S rRNA. Thus, there are 52 different proteins (r-proteins) and three distinct RNAs (rRNA) that make up an intact ribosome.

The 50S subunit contains **peptidyl transferase** activity that forms peptide bonds in a growing protein chain. The 30S subunit contains the mRNA **decoding site** where complimentary interactions between the mRNA codon and tRNA anticodon ensure translational fidelity. The ribosome also has three tRNA-binding sites formed at the interface of the two subunits: the aminoacyl–tRNA–binding or decoding site (A site), the peptidyl–binding site (P site), and the exit (E) site. The A site accepts all incoming charged tRNAs while the P site contains the previous tRNA attached to the nascent polypeptide (peptidyl-tRNA). The E site is discussed later.

X-ray crystallographic and cryomicroscopic analyses of ribosomal subunits from several organisms have given us an amazingly detailed view of the ribosome and provided unique insights into the catalysis of peptide bond formation. One of the more stunning discoveries is that the ribosome is, in fact, a **ribozyme**. The previous assumption was that rRNA simply existed as a framework upon which the catalytic proteins imbued with the real work of peptide bond formation were assembled. The fact is that the peptidyl transferase active center in the 50S subunit is devoid of any protein within an 18 Å radius. That, coupled with the finding that the part of 23S rRNA located within the active center is highly conserved across species, led to the conclusion that RNA (and not protein) is responsible for catalysis. Consequently, it is protein that helps shape rRNA into its proper form to catalyze peptide bond formation, not the other way around.

The r-protein genes of *E. coli* are scattered among 19 different operons (Table 2-6; see also "Translational Repression" in Chapter 5), but each ribosomal protein is only represented by one gene. In striking contrast, the rRNA genes exhibit a considerable amount of redundancy. There are seven rRNA operons each arranged as

$$5' \text{ leader}-16s \text{ rRNA}-\text{spacer}-23s \text{ rRNA}-5s \text{ rRNA}-\text{trailer}-3'$$

(see "RNA Processing"). The redundancy may be a protective strategy designed to reduce the potential danger associated with mutations in a single rRNA gene. Note that the various rRNA operons cluster near the origin of replication and that transcription of a given operon occurs in the same direction as replication. This is thought to minimize collisions between the two polymerases.

Figure 2-23 illustrates the association of the two subunits and their relationships with some translational factors and mRNA. The ribosome is constructed by sequentially adding subunit r-proteins to the processed rRNA (Fig. 2-24). It is a highly organized process involving a series of protein–RNA and protein–protein interactions. For example, protein S7 is a primary binding protein that organizes the folding of the 3' major domain of 16S rRNA and enables subsequent binding of other ribosomal proteins to form the head of the 30S. Protein S8, encoded by *rpsH*, also plays a critical role in both the assembly of the 30S ribosomal subunit and in the translational regulation of ribosomal proteins encoded by the *spc* operon (see below and Chapter 5). During 30S assembly, S8 binds to the central domain of the 16S rRNA and interacts cooperatively

TABLE 2-6. Genetic Nomenclature for Components of the Translation Apparatus

Gene Symbol	Gene Product
Ribosome Components	
rplA to *rplY*	50S ribosomal subunit proteins L1 to L25
rpmA to *rpmG*	50S ribosomal subunit proteins L27 to L33
rpsA to *rpsU*	30S ribosomal subunit proteins S1 to S21
rrnA to *rrnH*	rRNA polycistronic operons
rrf	5S rRNA (encoded within *rrn* operons)
rrl	23S rRNA (encoded within *rrn* operons)
rrs	16S rRNA (encoded within *rrn* operons)
Accessory Factors	
cca	tRNA nucleotidyl transferase
fusA	Translation factor EF-G
prmA, *prmB*	Methylation of ribosomal proteins
rimB to *rimM*	Maturation of ribosome
rna to *rnp*	Ribonuclease (I to P)
trmA to *trmD*	tRNA methyltransferase
infA to *infC*	Initiation factors IF-1 to IF-3
tufA, *tufB*	Translation factor EF-T$_U$
rrf	Ribosome recycling factor
Transfer RNAs and Their Charging[a]	
alaT, *alaU*	Alanine tRNA$_{1B}$
valT	Valine tRNA$_1$
alaS	Alanyl–tRNA synthetase
valS	Valyl–tRNA synthetase
etc.	

Source: From Berlyn. 1998. *Microbiol. Mol. Biol. Rev.* **62**:814.
[a] Generally, capitals of T and further in the alphabet indicate an RNA gene, whereas S and earlier letters are used for synthetases.

with several other small subunit proteins to form a well-defined ribonucleoprotein neighborhood. A similar strategy is employed for 50S subunit assembly using large subunit proteins.

Because rRNA molecules are innately capable of folding into a variety of different nonfunctional secondary structures, the sequential addition of ribosomal proteins direct the RNA into the proper folding pathway. As noted earlier, 16S and 23S rRNAs contain about 30 posttranscriptionally modified nucleotides. All of the modified nucleotides in 16S rRNA form a group at the mRNA decoding site surrounding the anticodon stem loops of tRNA-binding sites (A and P sites; see below). The modified bases in 23S rRNA cluster around the peptidyl transferase center of the 50S subunit. It has been proposed that these modifications fine-tune interactions between tRNA and rRNA. To see more detail of rRNA secondary structure visit http://www-rrna.uia.ac.be/secmodel/ or http://get a.life.uiuc.edu/~nikos/Ribosome/rproteins.html. [Internet addresses should not normally be hyphenated. How can this be handled?]

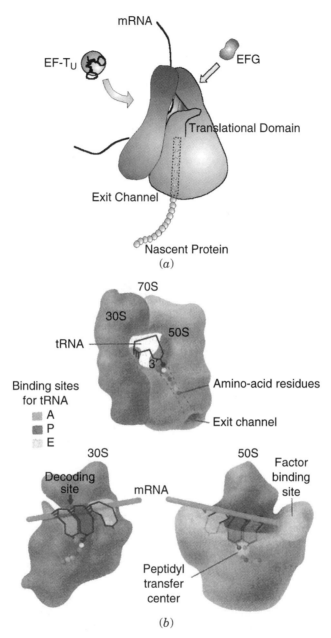

Fig. 2-23. (*a*) **Diagrammatic representation of the ribosome exit and translational domains.**
(*b*) **Structure of the 70S ribosome and its functional center.** *Top*: The tRNA molecules span
the space between the two subunits. The channel in the 50S subunit through which the growing
peptide protrudes is shown in dashed lines. *Bottom*: The 30S (left) and 50S (right) subunits
have been opened up to give a better view of the three binding sites for tRNA; the A, P, and
E sites. The 30S subunit shows the approximate location of the site where the mRNA codons
are read by the tRNA anticodons. The 50S subunit has the tRNA sites shown from the opposite
direction. The acceptor ends of the A and P sites are close to each other in the peptidyl transfer
site, which is close to the exit channel located behind a ridge in the 50S subunit. The binding
site for EF-G and EF-T_U is located on the right-hand protuberance of the 50S subunit. From
L. Ijas, A. 1999. Function is structure. Science **285**:2077–2078.

Growth Rate–Dependent Control of Ribosome Synthesis. Bacteria such as *E. coli* can grow exponentially at a wide range of growth rates depending on the carbohydrate source and how hard it has to work to make things such as amino acids. The ability to grow at different rates, of course, has a consequence in terms of protein synthesis. Basically, the faster the organism grows (the shorter the generation time), the faster it must make proteins. Since the peptide elongation rate per ribosome is constant, faster protein synthesis requires more ribosomes per cell. As a result, de novo synthesis of ribosomes must be linked to the bacterial growth rate. The cell accomplishes this by linking rRNA synthesis to ribosomal protein synthesis.

The first step in the process is growth-rate control of rRNA synthesis. The rRNA operons have two promoters, P1 and P2, with growth rate control focused on P1.

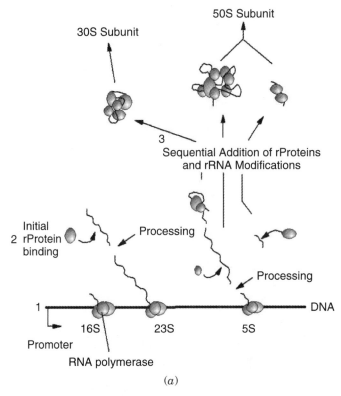

(*a*)

Fig. 2-24. Model for maturation of ribosomal RNAs in *Escherichia coli*. (*a*) The events occur in the following general sequence: (1) 16S, 23S, 5S genes are transcribed sequentially. (2) Key ribosomal proteins associate with 16S and 23S precursors before release from DNA. Transcript is cleaved into separate RNAs before transcription is complete. (3) 16S precursor is 10 to 20% methylated; 23S precursor is 60% methylated. First precursor particles sediment at 22S and 30S and are probably slightly folded. Sequential addition of other proteins gives intermediate precursor particles, sedimenting at 26S and 43S; these are folded into more compact structures. Final set of proteins added to give mature subunits. (*b*) *E. coli* 16S rRNA secondary structure. The binding site for ribosomal protein S8 is shown. From Zimmermann et al. 2000. How ribosomal proteins and rRNA recognize one another. In: The Ribosome: Structure, function antibiotics and cellular interactions. Ed. R. A. Gurrett et al. ASM Press, Washington, DC. pp 93–104.

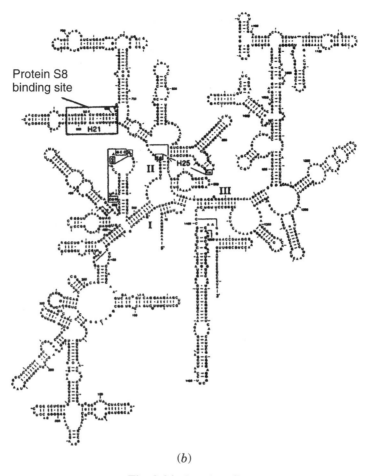

Protein S8
binding site

(b)

Fig. 2-24. *(continued)*

Transcription from P1 requires a high nucleotide triphosphate concentration (ATP or GTP). If growth slows, the high concentration of ribosomes will overtranslate relative to need and will consume more GTP (see below) than can be replenished at the slower growth rate. GTP levels then drop, causing decreased transcription of rRNA operons. With less rRNA scaffolding to bind, the concentration of free ribosomal proteins increases. How does this lower r-protein synthesis? Most ribosomal protein operons are subject to translational repression (see Chapter 5) in which a ribosomal protein binds to its own mRNA, decreases translation, and lowers its own production. This control is distinct from ppGpp-dependent stringent control discussed in Chapter 18. The end result is less rRNA, less ribosomal protein, and less ribosomes per cell.

Evolutionary Consideration. Comparative analysis of the nucleotide sequences of ribosomal RNA has provided a means of direct measurement of genealogical relationships. Ribosomal RNAs, particularly 16S rRNA, serve as phylogenetic markers for prokaryotic organisms. Detailed technical and statistical considerations have revealed that an oligonucleotide of six bases is unlikely to recur more than once in

16S rRNA. When the 16S rRNAs from different organisms include the same six bases, this reflects true homology.

By comparing the occurrence of these six-base sequences in various organisms, it is possible to compile a "dictionary" of them for a given organism. The data can then be analyzed to compare the degree of relatedness among organisms. This type of analysis from a large number of organisms has shown that certain classical criteria (e.g., gram stain) used in bacterial classification distinguish valid phylogenetic relationships, whereas other criteria (e.g., morphology) do not. Interestingly, this type of analysis has added weight to the theory that mitochondria in eukaryotic organisms evolved from prokaryotes. The nucleotide sequence of 16S rRNA from mitochondria relates more closely to bacteria than to the cytoplasmic ribosomes of eukaryotes.

Initiation of Polypeptide Synthesis

The processes of initiation and elongation of the polypeptide chain involve several specific and complex mechanisms (Figs. 2-25 and 2-26). The first stage in initiation finds the 30S and 50S subunits separated from each other with the 30S subunit complexed with two initiation factors (IF-1, mwt 9000, and IF-3, mwt 22,000). In the absence of mRNA, IF-1 and IF-3 function to prevent the association of the 30S and 50S ribosomal subunits. The next events involve the association of the mRNA and initiator tRNA to the 30S subunit (Fig. 2-25.2).

IF-3 binds to both the 30S subunit and to mRNA. As such, IF-3 helps bring mRNA to the ribosome. The 30S-[IF-3][IF-1] complex binds to mRNA around the initiation codon (in order of preference, AUG, GUG, UUG, CUG, AUA, or AUU). IF-3 inhibits initiation on any codon other than the three canonical initiation codons, AUG, GUG, or UUG. How does IF-3 recognize an initiation codon? Every mRNA molecule also includes within the untranslated region a ribosome-binding site for each polypeptide in a polycistronic message. The consensus sequence (5′)AGGAGGU(3′), called the **Shine-Dalgarno (S-D) sequence**, is important in the binding of mRNA to the 30S complex. The S-D sequence has been shown to base pair to a region located at the 3′ end of 16S rRNA. This pairing will position the initiating AUG codon to bind the anticodon of the appropriate initiator tRNA and presumably allows unambiguous distinction between initiation and internal met codons. IF-3 also binds to the ribosome in this area.

Next, cytosolic initiation factor 2 (IF-2, mwt 115,000) complexes with GTP and the initiator tRNA, N-formylmethionyl–tRNA (fMet–tRNA). In this form, IF-2 directs the binding of fMet–tRNA directly to the P site within the 30S subunit. Thus, initiator tRNA bypasses the A site where all other aminoacyl–tRNAs first bind the ribosome (Fig. 2-25.3). Binding of N-formylMet–tRNA then permits the association of the 30S and 50S ribosomal subunits (Fig. 2-25.4). Removal of IF-3, which occurs at the same time, is essential, since its presence prevents the association of the two ribosome subunits. The initiation complex, at this point, consists of an association of mRNA with the 30S ribosomal subunit, IF-1, IF-2–GTP, and fMet–tRNA. Union of the 30S initiation complex with the 50S ribosomal subunit causes the immediate hydrolysis of the bound GTP to GDP. The union itself does not require this hydrolysis, since it can be accomplished in the presence of an analog of GTP, 5′-guanylyl (methylene diphosphonate), GDPCP. The hydrolysis of GTP is required to eject IF-1 and IF-2 from the ribosome, which is required before proceeding to the elongation stage of translation (Fig. 2-25.5).

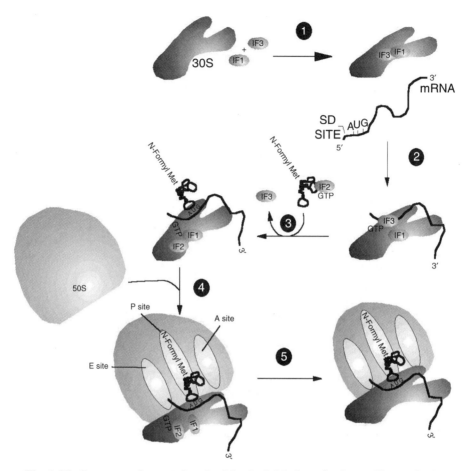

Fig. 2-25. Sequence of events involved in the initiation of polypeptide synthesis.

Elongation

The addition of amino acids to the growing polypeptide chain occurs at approximately 16 residues per second (at $37\,^{\circ}$C) in the order specified by the mRNA code. At the completion of initiation, the 70S ribosome contains fMet–tRNA in the P site while the A site is unoccupied and free to accept the next aminoacyl–tRNA directed by the triplet base code in the mRNA. Elongation factor T is involved in shuttling the proper aa-tRNA into the A site (Fig. 2-26). EF-T has been shown to consist of two subunits. One subunit (EF-T$_U$, mwt 44,000) is unstable while the other (EF-T$_S$, mwt 30,000) is stable. EF-T$_U$ is the most abundant protein in *E. coli*, comprising 5 to 10% of the total cellular protein. GTP binds to EF-T and causes its dissociation into EF-T$_U$–GTP and EF-T$_S$ (Fig. 2-26.1). EF-T$_U$–GTP complexes with all aminoacyl–tRNAs except the initiator tRNA. The resulting ternary complex (GTP–EF-T$_U$–aminoacyl–tRNA) is a required intermediate in the binding of aminoacyl–tRNA to the ribosome. EF-T$_S$ plays no role in this process. How ternary aminoacyl–tRNA complexes initially find the A site is due in part to EF-T$_U$ and 5S rRNA molecules recognizing sequences in the TψC loop (loop IV)of tRNA (Fig. 2-20).

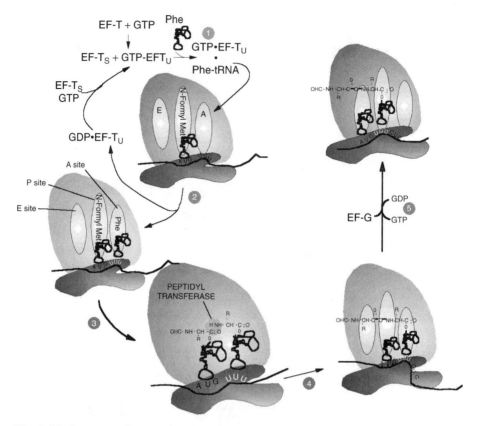

Fig. 2-26. Sequence of events involved in the continuation of peptide chain formation.

Although the correct selection of aminoacyl–tRNA–EF-T_U–GTP at the A site depends on the codon:anticodon match, the binding energy of this interaction is not sufficient to account for the exquisite selectivity of this process. The ribosome must sense and amplify the tRNA fit. Information regarding fit is probably conveyed between the 50S and 30S subunits by alterations in tRNA structure and by a relay of conformational changes through 16S rRNA to 23S rRNA. Residues 1492 and 1493 of 16S rRNA are in a position within the ribosome to monitor codon:anticodon fit via hydrogen bonding. If the fit is not perfect, the tRNA complex is rejected. One supportive piece of evidence for this model is that the antibiotic paromomycin alters the structure of 16S rRNA to mimic conditions of perfect codon:anticodon fit so that incorrect aminoacyl–tRNAs may be selected.

Once aminoacyl–tRNA has bound to the A site, GTP is hydrolyzed and a complex of EF-T_U–GDP is released from the ribosome. One GTP is hydrolyzed per aminoacyl–tRNA bound as shown in Figure 2-26.2. This GTP hydrolysis is not essential for aminoacyl–tRNA binding, since nonhydrolyzable GDPCP can substitute. Although the aminoacyl–tRNA can bind to the A site in the absence of GTP hydrolysis, EF-T_U cannot be released. Therefore, one result of GTP hydrolysis is the release of EF-T_U from the ribosome. After release of EF-T_U–GDP, EF-T_S will catalyze a nucleotide exchange that yields active EF-T_U–GTP. GTP hydrolysis, however, may serve another critical function at this stage. Following GTP hydrolysis, EF-T$_U$ undergoes an unusually large conformation change predicted to move the amino acid–carrying CCA end of the tRNA from the codon:anticodon testing site to the peptidyl transferase site (Fig. 2-26.3).

Peptide Bond Formation

Once the incoming aminoacyl–tRNA has bound to the A site, peptide bond formation occurs (see Fig. 2-26.3 and 2-26.4). The peptide bond is formed between the amino group of the incoming amino acid and the C terminal of the elongating polypeptide bound to the tRNA at the P site. As already noted, the activity responsible for peptide bond formation, the peptidyl transferase center, is not a protein but is a ribozyme comprised of a segment of the 23S rRNA in the 50s ribosomal subunit. No soluble factors appear to be necessary.

Translocation

After the formation of the peptide bond, the growing peptide chain becomes bound to the tRNA that was carrying the incoming amino acid (see Fig. 2-26.4) and occupies the A site. This peptidyl tRNA must now be moved from the A site to the P site to allow the next EF-T$_U$–aminoacyl–tRNA to enter the A site. In addition, the mRNA must be moved by one codon so that appropriate codon-anticodon recognition takes place (decoding). Elongation factor G (EF-G, mwt 80,000), in which G = GTPase, and GTP hydrolysis are required for these processes (Fig. 2-26.5). EF-G binds to the same site as EF-T$_U$; hence the requirement noted above for T$_U$ recycling. Once bound, EF-G catalyzes the hydrolysis of another GTP. Therefore, during elongation two molecules of GTP are hydrolyzed per peptide bond. One hydrolytic step is EF-T dependent and the other is EF-G dependent. Binding of EFG combined with GTP hydrolysis triggers a rotation of the 30S subunit relative to the 50S subunit (Fig. 2-27) and a widening of the

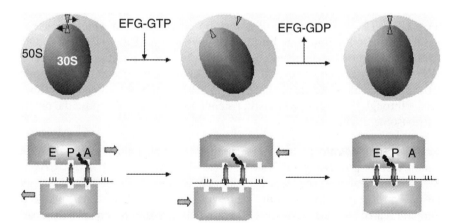

Fig. 2-27. Ratchet movement of ribosome along mRNA. *Top*: Top view illustrating rotation of 30S and 50S subunits relative to each other. *Bottom*: Side view illustrating how the rotation moves uncharged and nascent peptide-containing tRNA molecules from the P and A sites (respectively) to the E and P sites.

mRNA channel. This rotation moves the tRNAs in the A and P sites to the P and E sites, respectively. Thus, EF-G has been suggested as the "motor" for translocation, with the energy for this process coming from GTP hydrolysis. To complete one translocation event, EF-G must be released from the ribosome. Unless EF-G is released, the 30S subunit will not fully ratchet back to its original position relative to the 50S subunit and elongation cannot continue.

At the end of a translocation cycle, the deacylated tRNA resides in the E site, the peptidyl–tRNA sits in the P site, and the A site is open to accept the next incoming aa-tRNA ternary complex. How, then, is the deacylated t-RNA expelled from the E site? The next incoming charged tRNA binds to the unoccupied A site, which causes a conformational change in the ribosome that reduces affinity of the E site for the deacylated tRNA, allowing it to leave.

Termination

When translocation brings one of the termination codons (UAA, UGA, or UAG) into the A site, the ribosome does not bind an aminoacyl–tRNA–EF-T_U–GTP complex. Instead it binds one of two peptide release factor proteins — RF-1, mwt 44,000, or RF-2, mwt 47,000 — which then activate peptidyl transferase, hydrolyzing the bond joining the polypeptide to the tRNA at the P site. Specific recognition of the stop codon by release factors involves protein side chains mimicking nucleic acid base–base interaction. The result is release of the completed polypeptide.

Finally, ribosome disassembly and recycling must occur. RF-3 is known to catalyze the disassociation of RF-1 or RF-2 from ribosomes after polypeptide release. The next step is the disassociation of 50S subunits from the 70S posttermination complex catalyzed by the binding of ribosome release factor (RRF) and EF-G to the A site. This process also requires GTP hydrolysis. Subsequent removal of deacylated tRNA from the resulting 30 S : mRNA : tRNA posttermination complex is then needed to permit rapid 30S subunit recycling. This step requires IF-3 binding. However, some 30S subunits are not ejected from mRNA but are allowed to slide onto the next cistron in a polycistronic message. This has been shown to occur when the beginning of a downstream ORF lies within 3 or 4 nucleotides of the translational stop of the upstream ORF. This is an example of **translational coupling** in which the successful translation of one ORF is required to achieve translation of the neighboring ORF. A very different form of translational coupling is when the translation of one gene in a message relieves secondary structure that sequesters a downstream ribosome-binding site. If the secondary structure includes the 3′ end of the preceding gene, ribosome movement through that region will disrupt stem base pairing and expose the previously hidden ribosome-binding site.

Macromolecular Mimicry. It is interesting to note that the three extraribosomal translation factors possessing GTPase activity all interact with L7/L12. Cryomicroscopic structure analyses have revealed that the overall shape of the EF-T_U ternary complex is surprisingly similar to the shape of EF-G : GDP, with part of the tRNA structure mimicked by portions of EF-G (Fig. 2-28). In addition, sequence comparisons of EF-T_U, EF-G, IF-2, and RF-3, based on the known structures of EF-T_U and EF-G, indicate they all contain similar domains, called 1 and 2, with strikingly similar folds. It is implied that they all bind to the ribosome in similar modes, and that the GTPase-activating center on the ribosome is the same for all of them. This macromolecular

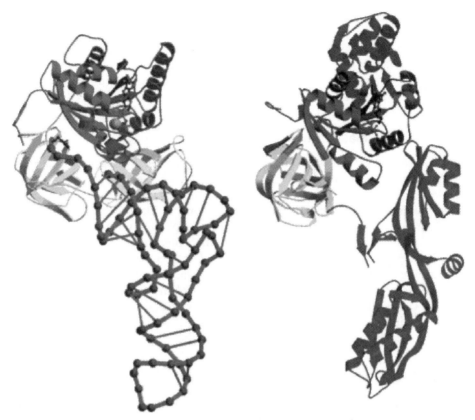

Fig. 2-28. Molecular mimicry, structural comparison of elongation factors. On the left is the ternary complex of Phe-tRNA : EF-T$_U$: GDPNP and on the right is EF-G : GDP. Nucleotides bound to domain 1 are shown in ball-and-stick models, and an Mg^{2+} ion in EF-T$_U$ is shown as a gray ball. The anticodon of tRNA is at the bottom of the ternary complex, and Phe attached to the terminal ribose is seen in black between domains 1 and 2 of EF-T$_U$. From: Nissen et al. *EMBO J.* **19:**489 (2001).

mimicry appears important in maintaining the proper order of events by forcing one factor to leave the ribosome before another can bind. However, there are other potential roles for structural mimicry in translation beyond simple site sharing. For example, the part of IF-2 that resembles tRNA may mimic a tRNA bound to the A site. This would ensure the exclusive binding of initiator tRNA to the P site. Another possibility is that by binding to the A site of the 30S subunit via its macromolecular mimicry of tRNA, EF-G could prevent the peptidyl–tRNA from rebinding to the A site, thereby controlling the correct progress of the elongation cycle.

Posttranslational Processing

The free polypeptide, in vivo, undergoes one of two posttranslational processing steps: (1) the formyl group of the N-terminal fMet may be removed by the enzyme methionine deformylase, or (2) the entire formylmethionine residue may be hydrolyzed by a formylmethionine-specific peptidase (methionyl amino peptidase or MAP). There is apparently some discrimination involved in channeling different polypeptides through these two alternative steps, since not all N-terminal methionines are removed. The discriminating factor appears to be the side-chain length of the penultimate amino acid. The longer the side chain, the less likely MAP will remove the methionine. Other processing can occur such as acetylation (e.g., acetylation of L12 to give L7) or adenylylation (see "Regulation of Nitrogen Assimilation and Nitrogen Fixation" in Chapter 5).

When Nonsense Makes Sense

There are several examples of proteins such as formate dehydrogenase (FdhF) that contain a modified form of cysteine called selenocysteine. The gene for formate dehydrogenase contains an in-frame TGA (UGA in message) codon that encodes the incorporation of selenocysteine instead of terminating translation as a nonsense codon. The requirements for this process include a specific tRNA^{SEC} (selC) containing a UCA anticodon, a unique translation factor (selB) with a function analogous to elongation factor T$_U$ (see "Translocation"), and a specific mRNA sequence (SECIS) downstream of the internal UGA codon in the fdh gene that signals selenocysteine incorporation. This region forms a loop that could, in part, shield the UGA from release factor. SelB is a GTPase that binds specifically to selenocysteyl–tRNA^{SEL} and to the SECIS sequence. These interactions direct selenocysteyl–tRNA to the ribosome only when the SECIS sequence is present in the ribosome A site.

Coupled Transcription and Translation

Prokaryotes differ significantly from eukaryotes in that prokaryotic ribosomes can engage mRNA before it has been completely transcribed (see Figs. 1-1 and 1-8 in Chapter 1). This coupled transcription-translation affords bacteria novel mechanisms for regulation (see "Attenuation Controls" in Chapter 5). Because the transcription process is faster than translation, the cell must protect itself from situations in which a long region of unprotected mRNA might stretch between the ribosome and RNA polymerase. To avoid that situation, most mRNAs contain RNA polymerase pause sites (discussed earlier) that allow the ribosome to keep pace with polymerase, thereby protecting the message from endonucleolytic attack.

PROTEIN FOLDING AND CHAPERONES

A continuing challenge for biochemists is to unravel the processes required to convert the one-dimensional information encoded within genes (i.e., an amino acid sequence) into three-dimensional protein structures that contain biochemical activity. Historically, protein biogenesis was thought to involve only spontaneous folding of polypeptide domains. We now realize that the process is more complex than previously envisioned. Most, if not all, proteins in the living cell require assistance to fold properly. This assistance comes from proteins that are not final components of the assembled product. These "foldases" are called chaperones (or chaperonins, depending on their structure).

The proposed function of chaperone proteins is to assist polypeptides to self-assemble by inhibiting alternative assembling pathways that produce nonfunctional structures. During protein synthesis, for example, the amino-terminal region of each polypeptide is made before the carboxy-terminal region. The chance of incorrect folding of a nascent polypeptide is reduced through interaction with chaperones. Another process in which chaperones can be invaluable is protein secretion or translocation. Proteins that cross membranes do so in an unfolded or partially folded state. Often they are synthesized by cytosolic ribosomes and must be prevented from folding into a translocation-incompetent state.

Cells also take advantage of chaperonins when faced with an environmental stress that will denature proteins into inactive forms that aggregate. To protect against such stresses, cells accumulate proteins that prevent the production of these aggregates or unscramble the aggregates so that they can correctly reassemble (see "Thermal Stress and the Heat Shock Response" in Chapter 18).

Key members of the major chaperone pathway are GroEL(60KDa), GroES(10KDa), DnaK(70KDa), and trigger factor (*tig*) of *E. coli*. They are all abundant proteins that increase in amount after stresses such as heat shock and therefore are called heat shock proteins (HSPs). Their importance to life in general is evident in that they have been preserved throughout evolution. Representatives of the major chaperone families have been found in all species examined. DnaK-like chaperone proteins (called HSP70s) are found in prokaryotes and most compartments of eukaryotes. HSP60s, also called chaperonins, are found in bacteria, mitochondria, and chloroplasts.

How do chaperones identify proteins in need of folding? The specifics are not clear, but it is apparent that most chaperones recognize exposed hydrophobic surfaces on target proteins. These regions of a protein should be buried within the protein structure and held away from water. Surface exposure indicates improper folding.

Folding Stages

There are several stages in the folding pathway of newly synthesized proteins (Fig. 2-29). The ribosome-associated **trigger factor** is the initial chaperone followed by the DnaK and the GroEL/GroES systems. Trigger factor is a 48 kDa protein with an amino terminus that binds ribosomes, a central domain with peptidyl proline isomerase activity, and a carboxy-terminal domain whose function is not defined. As a peptide is synthesized, it first encounters trigger factor on the ribosome. This chaperone is supposed to help the nascent peptide orient proline residues. The peptidyl proline isomerase activity mediates cis-trans conversions of peptidyl proline bonds. Because of their structure, proline residues introduce critical turns or bends in the secondary structure of a protein. Thus, rotating the bond between the COOH end of proline and the NH_2 end of the adjacent amino acid dramatically alters protein structure.

If the protein comes off the ribosome still incompletely folded, DnaK intervenes. DnaK is monomeric with two domains. Its N-terminal portion contains ATPase activity and binds the cochaperone **GrpE**. The C-terminal domain binds polypeptide substrates and the cochaperone **DnaJ** (HSP40). DnaK binds to surface-exposed hydrophobic areas on the target peptide and DnaJ binds DnaK to modulate DnaK ATPase activity. After hydrolysis of ATP, DnaK binds and releases small hydrophobic regions of misfolded proteins, allowing the protein another chance to fold. Release of the peptide often requires an ADP/ATP exchange mediated by GrpE.

The third, and basically last, chance for a peptide to fold properly involves the HSP60 GroEL/HSP10 GroES chaperonin system. Hsp60s assemble into large double-ring structures whose central cavity allows proteins up to 60–65 kDA to fold in a

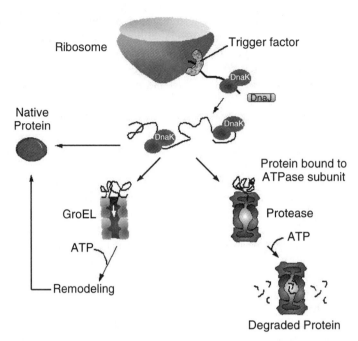

Fig. 2-29. Protein folding versus proteolysis pathways.

protected environment. Reversible association of the small capping protein GroES controls entrance to the chamber. Cycles of ATP binding and hydrolysis produce conformational changes within the chamber that enable binding and release of partially folded substrates.

There is some redundancy between the function of DnaK and trigger factor, since cells remain viable if one or the other remains active. GroEL, however, is essential for viability.

Protein Folding and Chaperone Mechanisms Outside the Cytoplasm

Proteins that move into the periplasm have special folding considerations. Generally, periplasmic proteins are secreted in their unfolded forms because attempts to export folded proteins clog most transport systems. Consequently, soon after entering the periplasmic space, the exported protein must be folded. Since the periplasm is an oxidizing environment whereas the cytoplasm is a very reducing environment, sulfhydryl groups in periplasmic proteins are subject to oxidation and formation of disulfide bonds. To prevent this, there are a number of periplasmic disulfide bond reductases (Dsb) that convert disulfide bonds into sulfhydryl groups and keep them that way. In addition, there are several peptidyl prolyl isomerases (Skp, SurA, FkpA, PpiA, PpiD) that, as noted earlier, rotate prolines within a peptide to achieve a proper bend. As opposed to the broad substrate chaperones noted above, there are other chaperones dedicated to a single purpose. For example, the FimC chaperone of *E. coli* protects the FimH pilin component until the FimH subunits are assembled into a mature pilus.

Quality Control

After initial folding and assembly, proteins may suffer damage as a consequence of environmental stress (e.g., acid pH). These damaged proteins go through a quality-control process that either targets the protein for refolding or degradation (Fig. 2-29). Chaperones and proteases both recognize exposed hydrophobic regions on unfolded peptides. Damaged proteins go through a kind of triage process where they first interact with chaperones that attempt to refold the protein. If the protein is released from the chaperone without being fully renatured, it remains in the pool of nonnative proteins and can either be rebound by a chaperone for another attempt at remodeling or will bind a protease that will target it for degradation. One or the other outcome for each damaged protein is critical if the cell is to survive the stress.

Future research on molecular chaperones will focus on how chaperones work and the structural basis by which a given chaperone recognizes and binds to some features present in a wide variety of unrelated proteins but which is accessible only in the early stages of assembly.

PROTEIN TRAFFICKING

The bacterial inner membrane, periplasmic space, and outer membrane all contain proteins not found in the cytoplasm. There are other proteins exported completely out of the cell into the surrounding medium or into eukaryotic host cells. All of these proteins are first synthesized in the cytoplasm but somehow find their way out. How does this happen?

Proteins destined to be integral membrane proteins are tagged with a very hydrophobic N-terminal **signal sequence** that anchors the protein to the membrane. N-terminal signal sequences range from 15 to 30 amino acids and include 11 hydrophobic amino acids preceded by a short stretch of hydrophilic residues. In contrast to integral membrane proteins, proteins predestined for the periplasm or outer membrane have their signal sequences cleaved following export. The roles of the signal sequence are to mediate binding of nascent polypeptides to the membrane and confer a conformation on the precursor that renders it soluble in the membrane. Signal sequences alone, however, are not enough to export proteins. There are actually seven general mechanisms of protein export that manage the remarkable feat of inserting proteins into membranes and passing hydrophilic proteins through hydrophobic membrane barriers. It is important to note that each system traffics different sets of exported proteins.

Insertion of Integral Membrane Proteins and Export of Periplasmic Proteins

Trigger Factor Versus Signal Recognition Particle. The process begins on the ribosome where either trigger factor (see "Protein Folding and Chaperones") or SRP (signal recognition particle) can bind nascent proteins (Fig. 2-30). These two factors help direct proteins toward different transport systems. Recall that trigger factor is associated with the ribosome. When trigger factor binds to a nascent protein emerging from the ribosome, it will prevent SRP from binding and allow the protein to be completely synthesized into the cytoplasm. Presecreted proteins that bind trigger factor generally have signal sequences targeting them for secretion rather than for

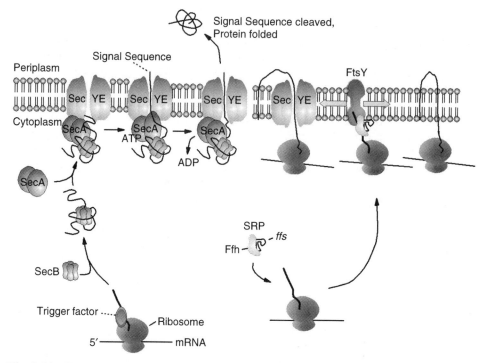

Fig. 2-30. General secretion pathways (Sec dependent, FtsY dependent). SRP, signal recognition particle.

membrane insertion (see "Sec-Dependent Protein Translocation"). On the other hand, SRP, composed of a 54 kDa protein called Ffh (with GTPase activity) and a 4.5S RNA molecule encoded by *ffs*, interacts with integral inner membrane precursor proteins (not presecretion proteins) by binding to their very hydrophobic, noncleaved signal anchor sequences (see SRP database http://carrier.gnf.org/dbs/html/gkd023_gml.html). SRP delivers these precursor proteins, still being translated, to a membrane-bound receptor called FtsY. Secretion from this point occurs while the protein is still being translated (**cotranslational export**). Binding of Ffh to FtsY, a protein–protein interaction actually mediated by *ffs* RNA, greatly enhances the FtsY GTPase activity. Once bound to FtsY, translation of the nascent protein resumes, and, depending on the signal sequence, the protein, as it is translated, has two possible fates. Some FtsY-bound nascent proteins are delivered to the SecYE translocon for insertion into the membrane (see "Sec-Dependent Protein Translocation"). Other FtsY-bound nascent proteins can be cotranslationally inserted directly into the cytoplasmic membrane in a Sec-independent manner.

Sec-Dependent Protein Translocation. The central feature of this system is the SecYE translocon located in the cytoplasmic membrane (Fig. 2-30). Two protein delivery systems converge at this translocon, one of which delivers proteins that will be secreted to the periplasm while the other escorts proteins destined for inner membrane insertion. Signal recognition particle FtsY interacts with SecY, as already noted, in a process that will insert proteins into the membrane. In contrast, proteins with cleavable signal sequences (thus secreted into the periplasm rather than inserted into the membrane) interact with trigger factor on the ribosome and are completely translated into the cytoplasm, where they bind to a chaperone called SecB. SecB is a piloting protein that delivers the fully translated presecretory protein in an unfolded state to the SecA ATPase associated with the SecYE translocon. Upon binding of ATP and the preprotein, SecA deeply inserts into the membrane at the SecYE translocon. Part of SecA actually inserts into the SecYE channel.

Subsequent ATP hydrolysis leads to disinsertion of SecA after which SecA can rebind ATP, precursor protein, and SecY, then reinsert into the SecYE channel, promoting the translocation of the next 20 to 30 amino acids of the preprotein. Following ATP-dependent insertion of the precursor protein into the translocon, proton motive force alone can energize translocation of the remaining polypeptide chain (see Chapter 9). Other components of the system have been identified (SecDF YajC), but their functions are unclear and their roles are dispensable. It should be noted that SecYE channel serves a dual purpose in mediating translocation of secretory proteins to the periplasm (SecA dependent) and the insertion of membrane proteins.

During or immediately following translocation, a **signal peptidase** (LepB) will cleave the signal sequence of the exported protein if it is destined for the periplasm, outer membrane, or external environment. Anchor signal sequences of integral cytoplasmic membrane proteins are usually not cleaved by LepB. Table 2-7 summarizes many of the genes involved with the basic protein translocation system.

Twin Arginine Translocation (TAT). Proteins such as TorA (molybdoprotein trimethylamine N-oxide reductase, a component of an anaerobic respiratory chain) contain a twin arginine motif in their N-terminal signal sequence (RRXFXK). These proteins are secreted in a Sec-independent fashion using the twin arginine translocase comprised of integral membrane proteins TatA, B, C, and E. In contrast to the Sec translocase, which is ATP driven (see above), transport via TAT is dependent on proton motive force (see Chapter 9). This is an example of a Sec-independent protein export system that also does not rely heavily on SRP. The Tat system is unique because it can transport prefolded, active proteins across the inner membrane; the Sec-dependent system only transports unfolded proteins. The function of the Tat proteins, especially how they contribute to the formation of a pore and whether they are involved in recognizing Tat signal sequences, is currently not known.

Traveling to the Outer Membrane. Outer membrane proteins such as OmpA or the porins OmpC and F are first deposited in the periplasm and then make their way to the outer membrane. How they do this remains a matter of debate. There are basically two schools of thought. The first is that these proteins weave their way, perhaps associated with some form of pilot protein, through the periplasm and peptidoglycan to the outer membrane. The second model is that they do not really travel through the periplasm but through zones of adhesion between the inner and outer membranes called Bayer junctions. However, the existence of Bayer junctions has been hotly contested. Regardless of the path, OMPs must contain information that direct them to the outer membrane. The nature of that information has remained elusive. The studies seem to suggest that the overall structure of OMPs is responsible for their targeting to the outer membrane. Whether other factors exist remains unknown.

TABLE 2-7. Components of the *E. coli* Cytoplasmic Membrane Secretory Protein Translocation System and Their Presence in Other Bacteria

Group and Characteristics	Name	Size (kDa)	Location	Presence in Other Bacteria
Secretory chaperonins (pilot protein[a])	SecB	18	Cytoplasm	Widespread in members of Enterobacteriaceae
General chaperones	DnaK	69	Cytoplasm	Universal
	GroEL	62	Cytoplasm	Universal
	GroES	11	Cytoplasm	Universal
Secretory ATPase (pilot protein)	SecA	102	Cytoplasm, ribosome, peripheral cytoplasmic membrane	*Bacillus subtilis*, widespread in members of Enterobacteriaceae
Translocase	SecD	67	Cytoplasmic membrane	*Brucella abortis*, archea
	SecE	14	Cytoplasmic membrane	*B. subtilis*
	SecF	39	Cytoplasmic membrane	*B. subtilis*, archea
	SecY	48	Cytoplasmic membrane	*B. subtilis*[b]
Signal peptidases	LepB[c]	36	Cytoplasmic membrane	*Salmonella enterica*, *Pseudomonas fluorescens*, *B. subtilis* (probably widespread)
	LspA[d]	18	Lipoprotein signal peptidase; cytoplasmic membrane	*Pseudomonas fluorescens*, *Enterobacter aerogenes* (probably widespread)
	Ppp[e]	ca. 25	Cytoplasmic membrane	*Pseudomonas aeruginosa* (PilD/XcpA), *Vibrio cholerae* (TcpJ), *K. oxytoca* (PulO), *B. subtilis* (ComC)
Others[h]	4.5S RNA (*ffs*)		Cytoplasm	*B. subtilis* (6S), *Thermus thermophilus*, *P. aeruginosa*, *Halobacterium halobium* (7S), *Mycoplasma pneumoniae*
	Ffh	48	Cytoplasm	*Streptococcus mutans*, *Neisseria gonorrhea*, others
	FtsY	54	Cytoplasmic membrane	*Sulfolobus solfataricus*, *B. subtilis*, Archaebacteria, others

[a] See text.

[b] Genes coding for proteins with >20% overall sequence identity to SecY have been cloned and sequenced from a variety of different sources, ranging from a plastid genome to *Saccharomyces cerevisiae*. Only in *S. cerevisiae* and *B. subtilis* has any direct participation in protein traffic been demonstrated.

[c] Also called leader peptidase or signal peptidase I.

[d] Also called prolipoprotein signal peptidase, lipoprotein signal peptidase, or signal peptidase II.

[e] Prepilin; the gene-based designation varies according to the organism from which it is derived and its function.

Secretion of Proteins Across the Outer Membrane

Type I Secretion: ABC Transporters. This is the simplest of the protein secretion systems. It consists of just three protein components: an outer membrane channel, an inner membrane ABC (**ATP-binding cassette**) transporter, and a periplasmic protein attached to the inner membrane. This system generates no periplasmic intermediates because there is a continuous channel from the cytoplasmic side of the inner membrane to the extracellular side of the outer membrane. The inner membrane and periplasmic components of type I secretion systems are substrate specific, but the outer membrane component, TolC, couples to many different systems to transport a variety of substrates ranging from the very large *E. coli* HlyA hemolysin (110 kDa) to small organic molecules (multidrug efflux). Figure 2-31a illustrates the story with the protein HlyA (*E. coli* hemolysin). The inner membrane transporter (HlyB$_2$) contains an ATP-binding cassette (the ABC transporter). The periplasmic component is HlyD. Energy is required at two stages in the secretion process. Proton motive force is required early, probably for subunit assembly, and ATP hydrolysis is necessary to secrete HlyA.

TolC contains a β-barrel channel that passes through the outer membrane and a long α-helical tunnel protruding into the periplasm that forms a periplasmic bridge with the inner membrane HlyB transporter. The α-helical channel opens as much as 30 Å once it contacts the inner membrane and periplasmic components of the system. After HlyA has been transported, the three-component system disassembles.

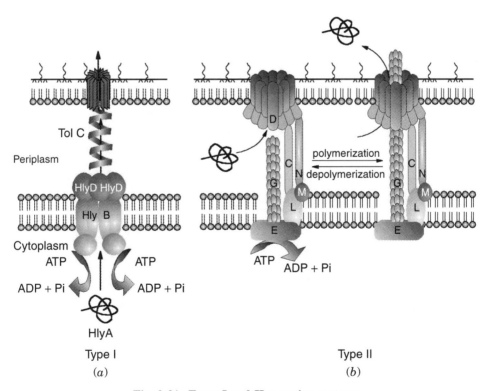

Fig. 2-31. Types I and II secretion systems.

Type II Secretion (Similar to Pilus Assembly). This system is also referred to as the main terminal branch of the general secretory pathway (the Sec-dependent system). Type II secretion differs from other systems in that the substrates consist of folded rather than unfolded proteins (Fig. 2-31). Each system is comprised of at least 12 different components that span the periplasmic space. Type II systems assemble a pilus-like structure, thus sharing many features with type IV pilus biogenesis systems, and are thought to extrude proteins via a piston-like motion of the pilus. Targets for type II secretion have N-terminal signal sequences that allow Sec-dependent translocation across the inner membrane (see above). This is followed by removal of the signal peptide, then folding and release of the mature peptide in the periplasmic space. Type II systems are very specific, and are able to distinguish proteins to be secreted from resident periplasmic proteins. It is not known, however, how type II systems recognize their substrates.

Type III Secretion (Similar to Flagella Assembly). Type III systems are syringe-like structures designed to inject proteins from the cytoplasm of a bacterial cell directly into a eukaryotic cell (Fig. 2-32). They are commonly used by pathogens to deliver toxins and pharmacologically active proteins into the host cell (see Chapter 20). As such, these systems are activated upon contact with the eukaryotic cell. The key distinguishing feature of this class is that the components bear remarkable homology to proteins involved in flagellar assembly. Pathogens that harbor these systems include *Yersinia* species, *Salmonella enterica* serovar Typhimurium, *Shigella flexneri*, Enterohemorrhagic *E. coli*, *Bordetella pertussis*, and *Chlamydia trachomatis*. Proteins secreted by this system are typically escorted to the system by specific chaperone proteins.

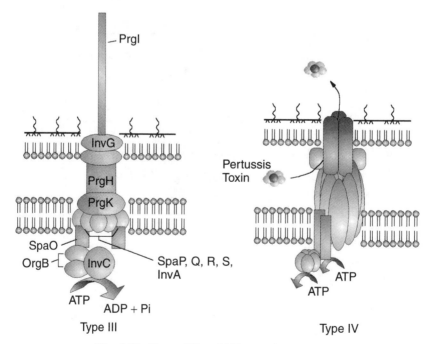

Fig. 2-32. Types III and IV secretion systems.

Type IV Secretion (Related to Conjugation Systems). Conjugation is a highly promiscuous DNA exchange mechanism in which DNA from a donor cell is moved into a recipient cell. The exchange mechanism requires cell–cell contact and an elaborate membrane protein superstructure (see "Conjugation" in Chapter 3). Several pathogens use secretion systems whose subunits are evolutionarily related to those conjugation systems for delivering effector proteins to eukaryotic cells. For example, *Agrobacterium tumefaciens*, a plant pathogen that induces tumorous growth of infected tissues, uses such a mechanism to transfer oncogenic T-DNA and several effector proteins to cell plant nuclei (Fig. 2-32). *Helicobacter pylori*, a causative agent of gastric ulcers, uses this type of system to deliver the 145 kDa CagA protein to mammalian cells.

Autotransporters. Certain extracellular proteins such as gonococcal IgA protease mediate their own transport across the outer membrane (see Chapter 20). Like the type II system, these proteins use the Sec-dependent system to move across the inner membrane followed by cleavage of an N-terminal signal sequence. Once in the periplasm, these proteins can bind to the outer membrane and form their own pores. Autoproteolytic cleavage at the carboxy-terminal end of the surface-bound protein releases IgA into the medium. Other type IV secreted proteins include the vacuolating toxin from *Helicobacter pylori* and the EspC protein produced by enteropathogenic *E. coli*.

Two-Partner Secretion. A collection of large virulence exoproteins produced in *Bordetella pertussis*, *Serratia marcescens*, and *Proteus mirabilis* are secreted by two-partner secretion (TPS) pathways. For each exoprotein there is a specific channel-forming β-barrel transporter protein (TspB). Thus, a noteworthy feature of this secretion pathway is its lack of versatility. Computer-assisted scans of sequenced genomes have identified over 25 potential TspB proteins. Initial transport of the exoprotein across the inner membrane is Sec dependent. The secretion domain of the exoprotein then enters the outer membrane TspB channel and is exported.

PROTEIN DEGRADATION

Degradation of Abnormal Proteins

Bacteria contain a variety of cytoplasmic, periplasmic, and membrane-bound proteases that are critical to the cell (Table 2-8). Under what circumstances would bacteria have to degrade proteins they worked so hard to make in the first place? One class of targets are nonfunctional proteins produced as the result of mutations or misfolding. These abnormal proteins are degraded much more rapidly than their normal counterparts. Stress in the form of high temperature, low pH, or other environmental extremes can also cause partial or complete unfolding of proteins. In order to clear the cell of debris, the levels of several proteases increase in response to conditions such as high temperature (see "Thermal Stress and the Heat Shock Response" in Chapter 18).

Another class of protease targets are timing proteins that regulate complex regulatory pathways. Their functions are required for limited periods of time within a growth cycle. Bacteria with developmental pathways provide striking examples of this class. In addition, proteases often target "sentry sigmas" that direct the expression of stress response genes only in times of trouble (see "Regulated Proteolysis"). These proteins

TABLE 2-8. Proteases and Peptidases of *Escherichia coli* and *Salmonella typhimurium*

	Location	Gene (Map Position)	Nature	Mwt × Subunit	Inhibitors
Proteases					
Do (HtrA or DegP)	Cytosol	*degP* (3.9)	Serine protease	52,000 × 10	D
Re	Cytosol + periplasm		Serine protease	82,000	D, E, O, TPCK
Mi			Serine protease	110,000	D, E, O
Fa	Cytosol		Serine endoprotease	110,000	D, E, O, TPCK
So	Cytosol		Serine protease	77,000 × 2	D, TPCK
La (Lon)	Cytosol	*lon* (10)	ATP-dependent serine protease	87,000 × 4	D, E, NEM
Ti(ClpAP)	Cytosol	*clpA, clpP*	ATP-dependent serine protease	(84,000 × 6) + (21,500 × 14)	
ClpXP	Cytosol	*clpX, clpP*	ATP-dependent serine protease	(46,000 × 6) + (21,500 × 14)	
ClpYQ		*hslU, V* (88.8 m)	ATP dependent, proteasome-like	49,000 + 19,000	
HflA	Membrane	*hflK, C, X* (94.8 m)		45,000, 37,000, 48,000	
HflB (FtsH)		*hflB* (71.6 m)	ATP-dependent protease	70,000	
II	Cytoplasm	*ptrB* (41.5 m)	Tryp-like protease	79,000	D, TLCK
III (Pi)	Periplasm	*ptrA* (63.7 m)	Metalloprotease	107,000	E, O
IV	Inner membrane	*sppA* (39.8 m)	Signal peptide peptidase	67,000	D
V	Membrane	—	—	—	D
VI	Membrane	—	Serine protease	43,000	
OmpT(VII)	Outer membrane	*ompT* (12.5 m)	Tryp-like protease	36,000	D
Ci	Cytoplasm		Metalloprotease	54,000	
DegQ	Periplasm	*hhoA* (72.8 m)	Serine protease	47,000	

Peptidases

Dipeptidases					
PepD	Cytoplasm	*pepD* (5.5 m)	Broad specific	53,000	
PepQ	Cytoplasm	*pepQ* (86.8 m)	X-Pro	50,000	
PepE	Cytoplasm	*pepE* (91.0 m)	Asp-X	24,000	
Aminotripeptidase					
PepT	Cytoplasm	*pepT* (25.5 m)	Tripeptides	45,000	
Signal Peptidases					
I	Inner membrane	*lep* (55 m)	Precursors of secreted proteins	36,000	
II	Inner membrane	*lsp* (0.5 m)	Prolipoproteins	18,000	—
Amino Peptidases					
PepN	Cytoplasm	*pepN* (21.3 m)	Aminoexopeptidase, broad	99,000	Amino acids
PepA (PepI)	Cytoplasm	*pepA* (96.6 m)	Broad	55,000 × 6	E, Zn^{2+}
PepP	Cytoplasm	*pepP* (65.8 m)	Proline aminopeptidase II, X-Pro-Y	50,000 × 4	E
PepM	Cytoplasm	*map* (4.1 m)	N-terminal methionine aminopeptidase, Met-X-Y	29,000	E
C-terminal Peptidases					
Dipeptidylcarboxypeptidase	Periplasm	*dcp* (35 m)	Ac-Ala3	77,000	C

Abbreviations: C, Captopril; D, Diisopropyl fluorophosphate; E, EDTA; O, o-phenanthroline; TPCK, N-tosyl-phenylalanine chloromethyl ketone; NEM, N-ethylmaleimide; Tryp, trypsin.

85

are constantly synthesized and degraded in the absence of an environmental threat. However, a sudden encounter with stress immediately stops degradation of the sentry regulator, allowing it to immediately direct synthesis of protective proteins. This seemingly wasteful strategy actually allows a very rapid response to potentially lethal threats and helps ensure cell survival, a consideration of paramount importance to the population.

A final set of targets has been called "proteins without partners." These are individual components of multiprotein complexes. As a complex, these components are usually quite stable, but when the stoichiometry is disrupted by mutation or overproduction of one subunit, rapid degradation of the uncomplexed protein may occur.

Abnormal proteins are progressively degraded into smaller and smaller fragments by a series of proteases starting with ATP-dependent endoproteases such as Lon (see below). The polypeptide products (mwt > 1500) are then degraded to smaller and smaller peptides by endoproteases, tripeptidases, and dipeptidases until they have been reduced to individual amino acids. Table 2-8 lists many of the proteases identified in *E. coli* and *Salmonella*.

Energy-Dependent Proteases

Many proteases in the cell require ATP to function. The purpose of ATP hydrolysis is to help make cytoplasmic proteases selective for specific substrates and to contribute to the unfolding of substrates in a manner that will facilitate their degradation. For example, the **Lon protease** (also called protease LA) is a cytoplasmic ATP-dependent protease that plays a primary role in the degradation of many abnormal proteins as well as unstable regulatory proteins such as the cell division inhibitor SulA (see SOS inducible repair in Chapter 3) and the positive regulator of capsule synthesis RcsA. It is a homotetrameric protein with each subunit carrying an active site serine (**serine protease**). In the presence of ATP, Lon protease will degrade intracellular proteins to acid-soluble peptides usually greater than 1500 daltons. The fact that the Lon protease only recognizes and acts upon unfolded proteins partially explains how intracellular proteases can exist free in the cytosol without destroying essential cell proteins. In addition, ATP hydrolysis results in repeated autoinactivation after each proteolytic event. The ADP remains bound to inactive Lon protease until a new protein substrate induces its release. This phenomenon will help ensure against indiscriminate proteolysis in vivo. The Lon protease also plays a role in the SOS response to DNA damage (see Chapter 3) and other regulatory systems (see Chapter 5).

The ClpA, ClpP, and ClpX proteins comprise two mix-and-match proteases in which the ATPase and protease active sites reside in different subunits. ClpP, the protease subunit, can combine either with ClpA or ClpX ATPase subunits to form active cytoplasmic protease **ClpAP** (protease Ti) or **ClpXP**. These proteases are composed of two seven-member rings of ClpP arranged back to back, forming a proteolytic barrel (see inset Fig. 2-33). This structure is capped, top and bottom, with six-member rings of ClpX or ClpA. The ATPase subunits appear to be the gatekeepers for substrate translocation into the proteolytic chamber. The role of the ClpXP protease in cell physiology is discussed below.

Degradation of Nascent Proteins Permanently Stalled on Ribosomes. A ribosome that translates a damaged or prematurely truncated mRNA molecule will

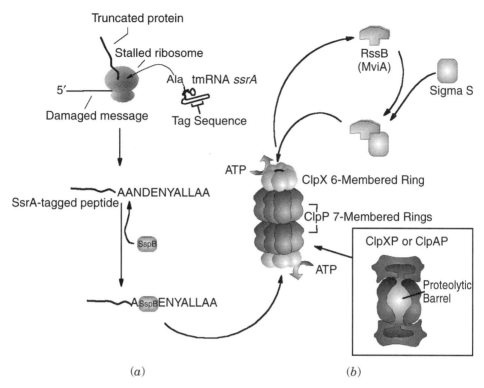

Fig. 2-33. Targeted protein degradation to the ClpXP proteosome. (*a*) SsrA tagging system. (*b*) RssB(MviA) recognition of σS. Inset shows a cutaway of the ClpXP protease revealing the internal proteolytic barrel.

often find itself at the end of the damaged message with no stop codon in sight. Without a stop codon the ribosome cannot release the message. *E. coli* and other prokaryotes have designed a clever strategy that circumvents this problem. A small RNA molecule called *smpA* directs a new type of RNA molecule to the stalled ribosome. This new RNA is called tmRNA because it has the properties of tRNA and mRNA. The tmRNA encoded by the *ssrA* gene helps the ribosome disengage from the message and targets the faulty protein for degradation (Fig. 2-33). The tRNA part of *ssrA* is charged with alanine, finds the ribosome A site, and triggers peptidyl transferase activity, thus adding alanine to the stuck nascent peptide. The mRNA part of *ssrA* can then be translated by the ribosome. This places a specific amino acid tag (AANDEYALLAA) on the end of the truncated protein. This amino acid tag is recognized by another protein called SspB, which presents the doomed protein to the ClpXP protease for degradation.

Degradation of Normal Proteins. The level of a particular protein in the cell is determined not only by its rate of synthesis but also by its rate of degradation. Degradation of a normal protein, however, is not a stochastic or random process. The cell goes to considerable lengths to regulate which proteins are degraded and when. During starvation for a carbon or nitrogen source, for example, the overall rate of protein degradation increases several-fold. This degradation of preexisting cell proteins can provide a source of amino acids for continued protein synthesis or for a source of energy. The enhanced proteolysis occurs coordinately with the cessation of stable RNA synthesis and the accumulation of the signal molecule guanosine tetraphosphate (see "Stringent Control" in Chapter 18).

Another role of proteolytic pathways is to confer short half-lives on proteins whose concentrations must vary with time or alterations in cellular state. This contributes to the overall metabolic economy of the cell, since unbridled persistence of a regulatory protein will not allow target genes to react quickly to changing conditions. How is it that some proteins resist degradation whereas others are degraded rapidly? Certain amino acid sequences, conformational determinants, or chemically modified protein structures confer degradation signals, called **degrons**, on proteins. One example is the N-degron, which is the N-terminal residue of a protein. Manifested as the N-end rule, the in vivo half-life of a protein was found to be related to the identity of its N-terminal residue. Several gene products of *E. coli* have been associated with enforcing the N-end rule. For example, the *aat* product is required to degrade proteins with N-terminal Arg or Lys. Mutations in *clpA* will stabilize proteins with N-terminal Phe, Leu, Trp, Thr, Arg, or Lys (see Table 2-8).

Regulated Proteolysis. A recurring theme emerging from proteolysis research is that key regulatory proteins are often presented to a given protease by a specific adaptor or chaperone protein. For example, the sigma factor σS (important to survival under many environmental stress conditions) is rapidly degraded in logarithmically growing cells by the ATPase protease ClpXP. However, the ClpX chaperone does not directly recognize σS. An adaptor protein called RssB (MviA in *Salmonella*) binds to σS and ClpX, presenting σS in a form that ClpX can then feed to the actual protease ClpP (Fig. 2-33). Upon encountering stress, the degradation of σS slows by virtue of some unknown effect on RssB or other component of the RssB/ClpX/ClpP proteosome. Decreased degradation allows σS to accumulate and trigger expression of genes required to survive the stress. This strategy allows the cell to respond quickly to potentially lethal environments by cutting out the need to synthesize the regulator in the first place.

There are many other proteases listed in Table 2-8 that we have not discussed. However, the physiological roles of several of them are explained within other sections of this book.

ANTIBIOTICS THAT AFFECT NUCLEIC ACID AND PROTEIN SYNTHESIS

Much of our knowledge concerning the synthesis of macromolecules was gleaned from studies designed to elucidate the modes of action of several antimicrobial agents. The following section describes some of these agents and their mechanisms of action.

Agents Affecting DNA Metabolism

Intercalating Agents. Many rigidly planar, polycyclic molecules interact with double-stranded DNA by inserting, or intercalating, between adjacent stacked base pairs of the double helix (see Fig. 2-34). To permit this intercalation, there must be a preliminary unwinding of the double helix, providing space between the bases into

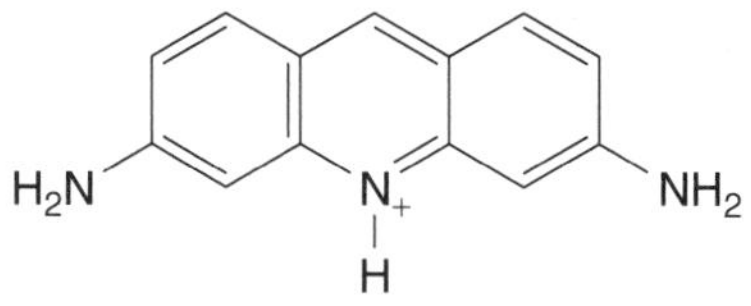

proflavine

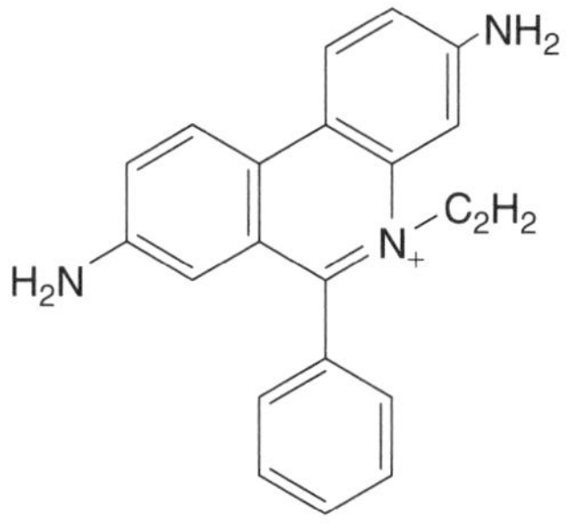

ethidium

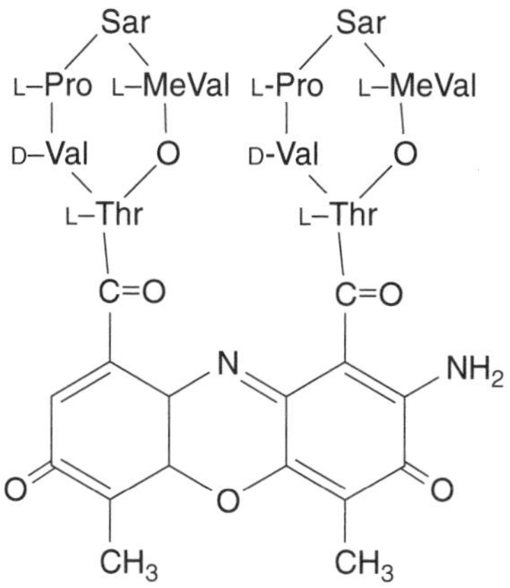

actinomycin D

Fig. 2-34. Three agents that intercalate with DNA. Actinomycin D is sometimes called dactinomycin. Thr, threonine; Val, valine; Pro, proline; Sar, sarcosine; MeVal, N-methylvaline.

which the drug can move. Hydrogen bonding remains undisturbed; however, some distortion of the phosphodiester backbone occurs.

Examples of intercalating agents are proflavine, ethidium, and actinomycin D (Fig. 2-34). Treatment of cells harboring extrachromosomal elements called plasmids (Chapter 8) by proflavine or ethidium bromide will often lead to the disappearance of the plasmids from the cells, which is apparently due to the preferential inhibition of plasmid replication.

Actinomycin D inhibits DNA replication but inhibits transcription at lower drug concentrations. DNA must have certain features to interact with actinomycin D: (1) the DNA must contain guanine; (2) the DNA must be a double-stranded helical structure; and (3) the sugar moiety must be deoxyribose. Double-stranded RNA does not interact with this drug.

Mitomycins. The mitomycins are a group of antitumor and antibacterial agents. Mitomycin C (Fig. 2-35) is one of the more active and more widely studied members of this group. It causes cross-linking between complementary strands of DNA but produces a number of toxic side effects in mammalian host tissues, making it impractical as a chemotherapeutic agent against bacterial infections. Several lines of evidence suggest that mitomycin C and related derivatives (e.g., porfiromycin, the ziridine N-methyl analog of mitomycin) form cross-links by bifunctional alkylation of guanine residues in opposing DNA strands. By binding to sites on each of the two complementary DNA strands, replication of a segment of the DNA in the immediate vicinity of the cross-link fails to occur and a loop of DNA is formed. This inhibition of replication appears to be an antimicrobial mode of action of mitomycin C.

DNA Gyrase Inhibitors. Nalidixic acid has been widely used in the treatment of urinary tract infections caused by *E. coli*, *Klebsiella pneumoniae*, *Enterobacter aerogenes*, *Proteus*, and other gram-negative bacilli. Oxolinic acid is a chemically related derivative of nalidixic acid (Fig. 2-35) that has been found to be more potent than nalidixic acid in its antimicrobial action. Both compounds have an identical mode of action. The primary effect of nalidixic acid in intact, living bacteria is an immediate cessation of DNA synthesis. DNA gyrase is the target protein for the action of nalidixic and oxolinic acids. The *gyrA* or "swivelase" component of DNA gyrase is sensitive

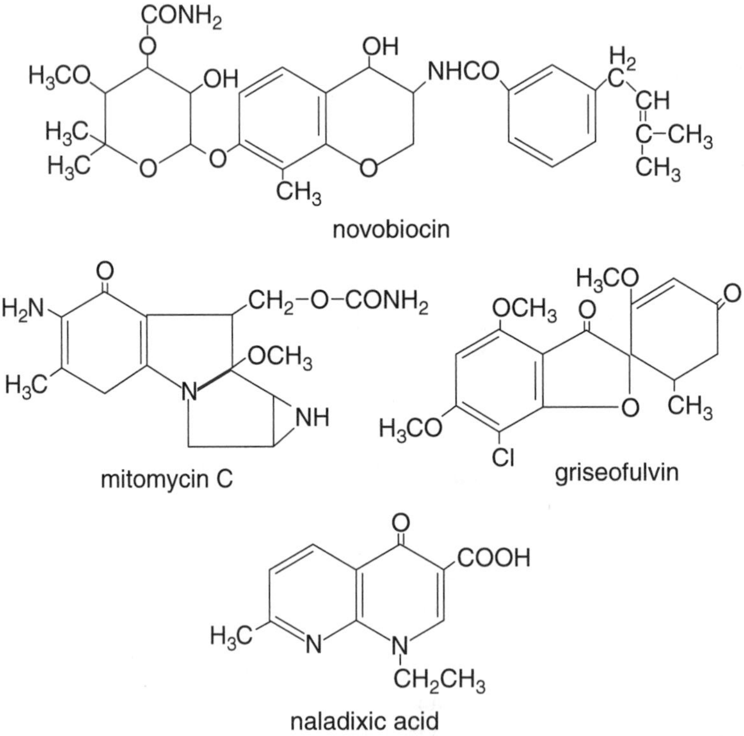

Fig. 2-35. Structures of antimicrobial agents affecting the metabolism, structure, or function of DNA.

to the action of nalidixic and oxolinic acids. DNA gyrase isolated from nalidixic acid–resistant mutants is resistant to the action of nalidixic and oxolinic acids but not to novobiocin or the related compound, coumermycin. Conversely, novobiocin-resistant mutants yield a DNA gyrase that is sensitive to nalidixic acid. Novobiocin binds to the B subunit of DNA gyrase (*gyrB*) and inhibits its ATPase activity.

The **quinolone** antibiotics (e.g., ciprofloxacin) are structurally related to naladixic acid based on the presence of a quinolone ring. The structural modifications in the quinolones increase their potency and utility against a broad spectrum of microorganisms, both gram positive and gram negative. They are also useful clinically because mutants resistant to naladixic acid remain quite sensitive to the quinolones.

Agents Affecting Transcription

Rifamycin/Rifampin. The rifampicin or rifamycin group of agents and the streptovaricins are chemically similar (Fig. 2-36). Rifampin and streptovaricin are semisynthetic compounds, while the rifamycins are naturally occurring antibiotics. All of these agents affect the initiation of transcription by specific inhibition of bacterial RNA polymerase (see Fig. 2-15). The RNA polymerase of eukaryotic cells is not affected. Rifamycin forms a tight, one-to-one complex with the β subunit of RNA polymerase (*rpoB*). Rifamycin-resistant variants of *E. coli* contain mutations in *rpoB* that reduce affinity of the protein for the chemical. The drug acts to inhibit initiation of transcription by inhibiting formation of the second phosphodiester bond. Once transcription is allowed to proceed past the third nucleotide (i.e., to the elongation stage), rifamycin is no longer inhibitory.

Streptolydigin. Streptolydigin is similar to rifamycin in that it inhibits transcription by binding to the β subunit of RNA polymerase. However, this antibiotic will inhibit

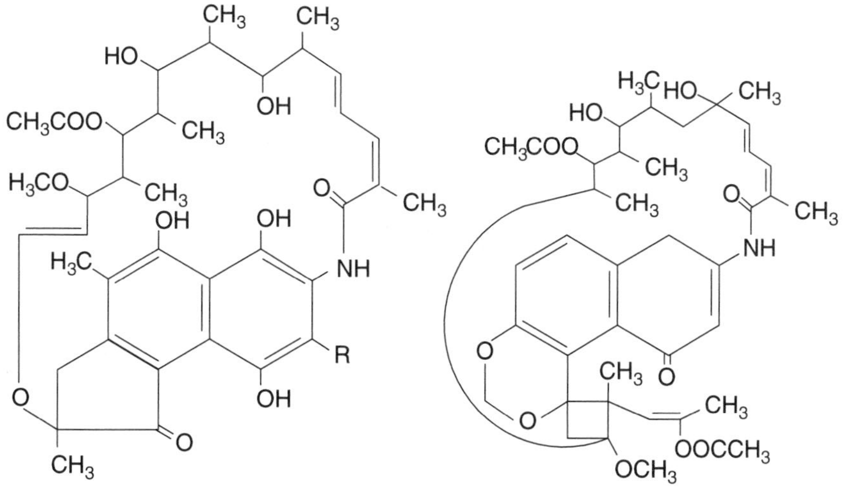

Rifamycin: R = H
Rifampin: R = −CH=N−N⟋‾⟍N−CH

Streptovaricin A

Fig. 2-36. Comparison of the structures of rifamycins and streptovaricins.

chain elongation as well as the initiation process in vitro. In vivo, streptolydigin appears to accelerate the termination of RNA chains. The binding of the drug to RNA polymerase may destabilize the transcription complex, permitting premature termination.

Agents Affecting Translation

Chloramphenicol. One of the simplest antibiotic structures (Fig. 2-37), chloramphenicol was originally isolated as a product of *Streptomyces venezuelae* but is now produced commercially by organic synthesis. Inhibition of peptidyl transferase activity by chloramphenicol is mediated by binding to the 50S subunit of bacterial ribosomes. Its bacteriostatic effect is readily reversed with removal of the drug. The drug binds to several ribosomal proteins including S6, L3, L6, L14, L16, L25, L26, and L27. Protein L16, which is located near the peptidyl transferase active center and forms part of the tRNA acceptor site, is preferentially labeled. This drug does not affect mammalian cells because they contain 60S ribosomal subunits. However, chloroplasts

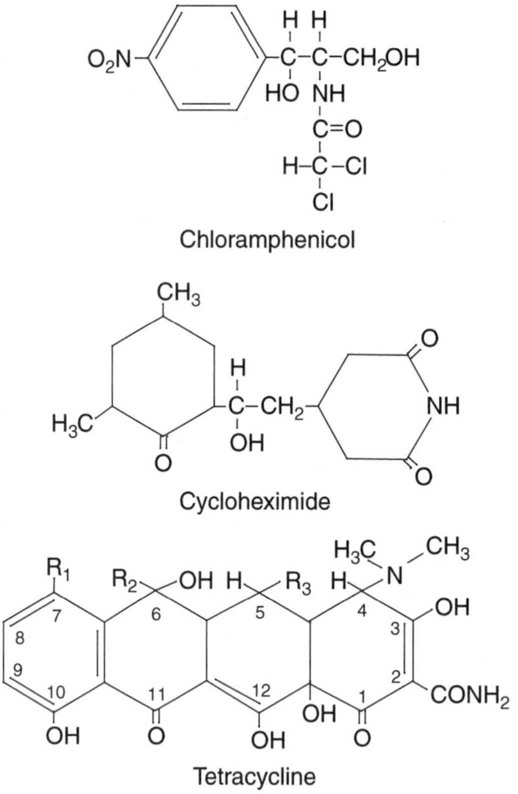

Chloramphenicol

Cycloheximide

Tetracycline

Fig. 2-37. Structures of chloramphenicol, cycloheximide, and tetracycline derivatives. Chloramphenicol bears the chemical name, D-threo-1-p-nitrophenyl-2-dichloroacetamido-1,3-propanediol. Cycloheximide is B-2-(3,5-dimethyl-2-oxocyclohexyl)-2-hydroxyethyl-glutarimide. In tetracycline (R_1, H; R_2, CH_3; R_3, H). In chlortetracycline (R_1, Cl; R_2, CH_3; R_3, H). In oxytetracycline (R_1, H; R_2, CH_3); R_3, OH). In demethylchlor-tetracycline (R_1, Cl; R_2, H; R_3, H.)

and the mitochondria of mammalian cells and lower eucaryotes contain 70S ribosomes comparable to those of prokaryotes. Thus, chloramphenicol also affects protein synthesis in these eukaryotic organelles. This finding explains the toxicity of chloramphenicol toward mammalian cells. Bone marrow toxicity, resulting in aplastic anemia, provides a strong contraindication for the routine use of chloramphenicol, particularly in situations where alternatives in the choice of a therapeutic agent are available.

Cycloheximide (Actidione). Cycloheximide (Fig. 2-37) inhibits protein synthesis in eukaryotic cells (yeasts, fungi, higher plants, and virtually all mammalian cells), but not in prokaryotic organisms (bacteria), by interfering with the activity of cytoplasmic ribosomes but not the ribosomes in mitochondria. The mode of action of cycloheximide resembles that of chloramphenicol. Cycloheximide binds to the 80S ribosomes of eukaryotic cells and prevents ribosome movement along the mRNA. The general effectiveness of cycloheximide on eukaryotic cells of all types accounts for the lack of applicability to the treatment of fungal infections, despite its high in vitro activity against a wide variety of yeasts and molds.

Tetracyclines. This group of antibiotics exhibits broad-spectrum activity (bacteriostatic) against both gram-positive and gram-negative bacteria as well as rickettsiae, chlamydiae, and mycoplasmas (Fig. 2-37). The mechanism of action of all the tetracyclines is based on their ability to prevent the binding of aminoacyl–tRNA to the A site on the 30S ribosome. Although the antibiotic can bind to many sites on both ribosomal subunits, the strongest binding occurs to S7, which helps define an area involved in contact between the two subunits.

Resistance to tetracycline is plasmid (R factor) mediated in many bacteria (e.g., *E. coli*, *S. aureus*). Plasmid-mediated tetracycline resistance is an inducible property that prevents the active accumulation of the drug in the bacterial cell by actively pumping the compound out of the cell. In the enteric bacteria, expression of tetracycline resistance following induction appears to result from the insertion of one or more proteins into the cell envelope. Similar proteins have been reported to be present in *S. aureus*.

Macrolides. This large group of antimicrobial agents includes angolamycin, carbomycin (magnamycin), chalcomycin, erythromycin, kujimycin, leucomycin, macrocin, megalomycin, and several others including oleandomycin and spiramycin. All contain a large lactone ring (an aglycone) containing from 12 to 22 carbon atoms with few or no double bonds and no nitrogen atoms. Erythromycin A, more widely recognized because of its extensive clinical application, is shown in Figure 2-38. The macrolide antibiotics are active mainly against gram-positive bacteria and have relatively limited activity against gram-negative bacteria. All the macrolides are inhibitors of protein synthesis in bacterial but not in eukaryotic systems, as a consequence of their interaction with the 50S ribosomal subunit. Erythromycin binds at or near the peptidyl transferase center by blocking peptide bond formation when the donor site is occupied by a peptidyl moiety of a certain length. Erythromycin has been shown to bind to protein L15 and to 23s rRNA with contacts at the central loop of domain V and residues in domain II, which form part of the peptidyltransferase center. It is proposed that macrolide antibiotics stimulate the disassociation of peptidyl–tRNA from the ribosome by triggering

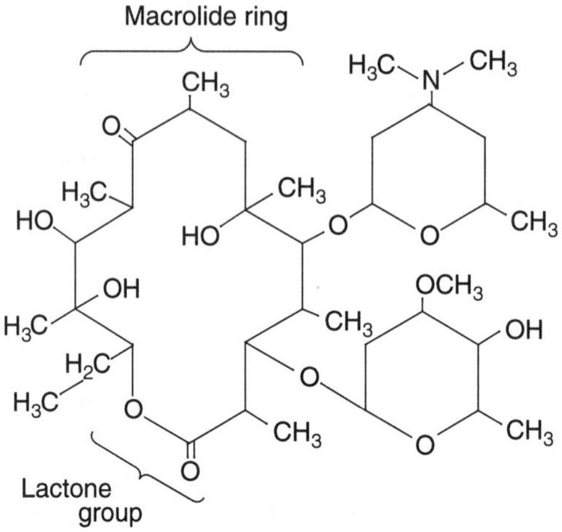

Fig. 2-38. Basic structure of macrolides. Structure of erythromycin A is shown.

an abortive translocation step. It is known that erythromycin is bacteriostatic in its action, and microbial resistance develops rather readily.

Resistance to erythromycin in *E. coli* depends on an alteration in L4 or L12 of the 50S ribosomal subunit, causing reduced affinity for the drug. In *S. aureus*, however, the 23S rRNA of the 50S subunit is modified by an inducible, plasmid-mediated ribosomal RNA methylase.

Aminoglycosides. This large group of antibiotics includes the streptomycins, gentamicins, sisomycin, amikacin, kanamycins A and B, neomycins, paromomycins, tobramycins, and spectinomycins. These agents are all chemically similar (Fig. 2-39). The ultimate mechanism of action is considered to be through binding to ribosomes, causing misreading of the amino acid code. In fact, at low concentrations, streptomycin can suppress some missense mutations by allowing misreading of the mutant codon. Streptomycin partially inhibits the binding of aminoacyl–tRNA and peptide synthesis. The effect of streptomycin on ribosome activity appears due to interactions with 16S rRNA.

Resistance (*rpsL* or *strA*) appears to develop through alteration of the S12 protein of the 30S ribosomal subunit in such a manner that the drug can no longer bind to the 16S rRNA. The drug clearly does not bind to S12. Streptomycin dependence has been explained on the basis of a mutationally altered site in S12 that functions normally only with the addition of streptomycin. Other aminoglycosides apparently are sufficiently dissimilar in their action to have little effect on these dependent mutants. Mutations of another type may render microorganisms resistant on the basis of altered permeability to streptomycin or other aminoglycosides.

Resistance to the aminoglycosides may also develop as a result of the ability to produce enzymes that inactivate various derivatives by the addition of phosphate groups (aminoglycoside phosphotransferases), the addition of nucleotide residues (aminoglycoside nucleotidyltransferases), or the acetylation of various positions (aminoglycoside acetyltransferases). These inactivating enzymes do not react equally with all aminoglycoside derivatives. For this reason, it may be possible to substitute one compound for

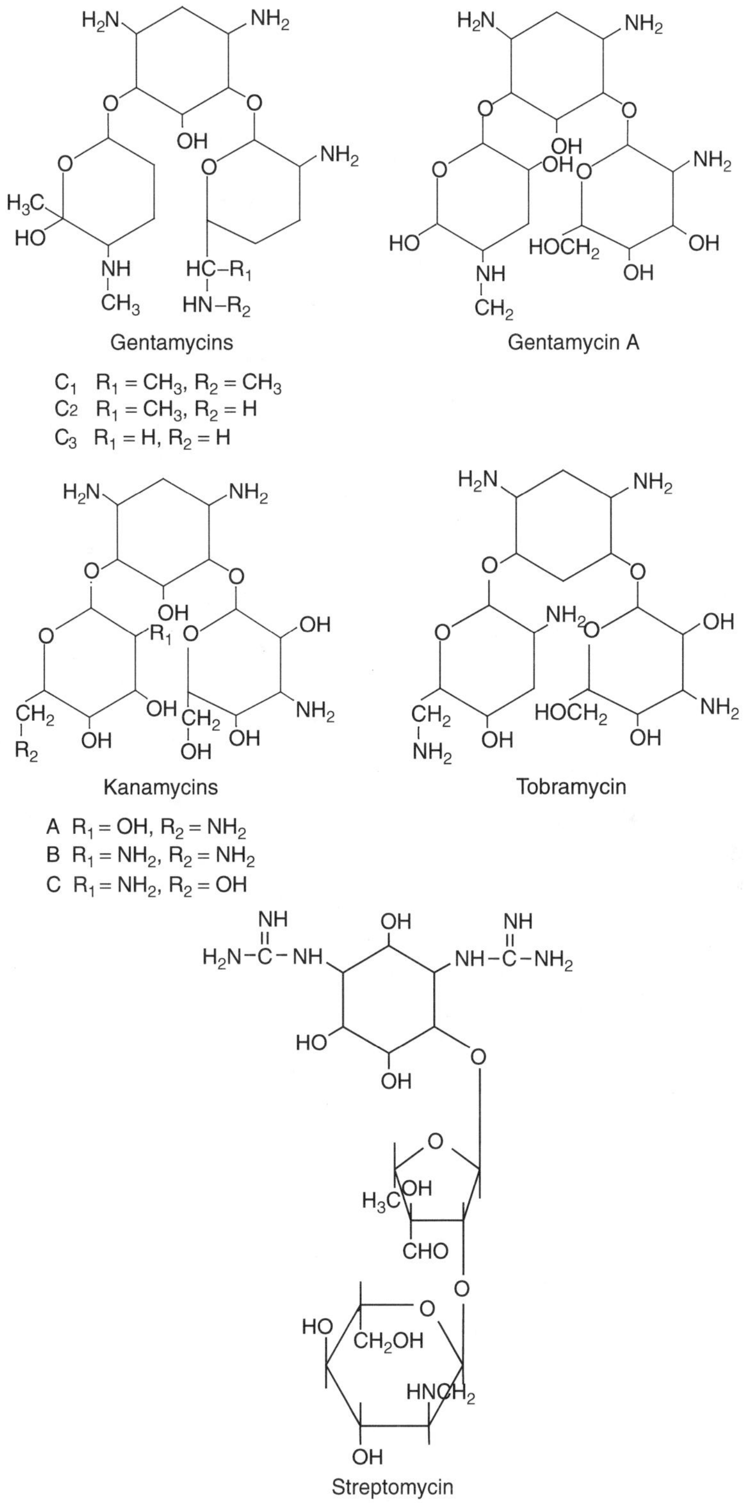

Gentamycins

C_1 $R_1 = CH_3$, $R_2 = CH_3$
C_2 $R_1 = CH_3$, $R_2 = H$
C_3 $R_1 = H$, $R_2 = H$

Gentamycin A

Kanamycins

A $R_1 = OH$, $R_2 = NH_2$
B $R_1 = NH_2$, $R_2 = NH_2$
C $R_1 = NH_2$, $R_2 = OH$

Tobramycin

Streptomycin

Fig. 2-39. Comparison of structures of aminoglycosides.

another. The aminoglycosides exhibit a broad spectrum of activity, being only slightly more effective against gram-negative bacteria than against gram-positive bacteria.

Kasugamycin is an aminoglycoside that acts on the 30S subunit of 70S ribosomes. It inhibits protein synthesis, but it does not induce misreading, nor can it cause phenotypic suppression, as does streptomycin. Kasugamycin-resistant mutants (*ksgA*) are unusual in that they have an altered rRNA instead of protein. The mutation actually affects the activity of an enzyme (S-adenosylmethionine-6-N', N' adenosyl (rRNA dimethyltransferase) that specifically methylates two adenine residues to dimethyl adenine in the sequence AACCUG near the 3' end of 16S rRNA, the region that interacts with the Shine-Dalgarno region of mRNAs.

Lincomycin. Lincomycin and a chemically related derivative, clindamycin (Fig. 2-40), inhibit peptidyl transfer by binding to 23S rRNA in the 50S ribosomal subunit. These agents are effective against both gram-positive and gram-negative organisms. Clindamycin is especially effective against Bacteroides and other anaerobes because of its ability to penetrate into deeper tissues.

Fusidic Acid. Fusidic acid is a member of the steroidal antibiotics (Fig. 2-41). It inhibits the growth of gram-positive but not gram-negative bacteria. The failure to act upon gram-negative bacteria may be due to the inability of fusidic acid to penetrate these cells, since it will inhibit protein synthesis from gram-negative ribosomes in vitro. Addition of fusidic acid to 70S ribosomes in vitro prevents the translocation of peptidyl–tRNA from the acceptor to the donor site. This drug also inhibits the EFG-dependent cleavage of GTP to GDP. The data point to EFG as the target for fusidic acid. The drug forms a stable complex with EFG, GDP, and the ribosome that is unable to release EFG for subsequent rounds of translocation and GTP hydrolysis.

Puromycin. Puromycin inhibits protein biosynthesis by forming a peptide bond with the C-terminus of the growing polypeptide chain on the ribosome, thereby prematurely terminating the chain. A structural similarity between puromycin and the terminal 3' aminoacyl adenosine moiety of tRNA (Fig. 2-42) allows puromycin to bind to the ribosomal A site where it triggers peptidyl transferase activity, linking the amino group of puromycin to the carbonyl group of peptidyl–tRNA.

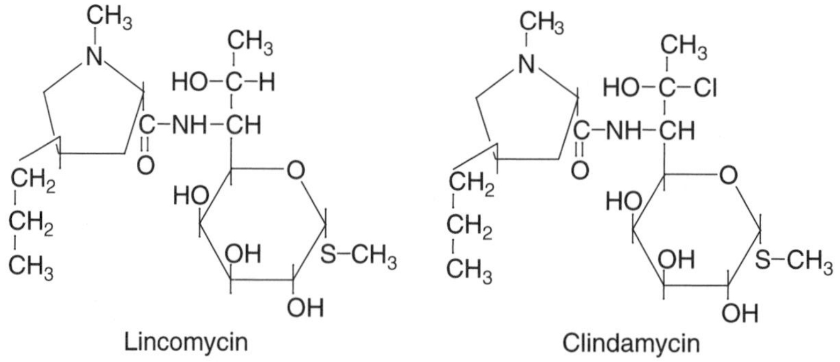

Fig. 2-40. Lincomycin and clindamycin structures.

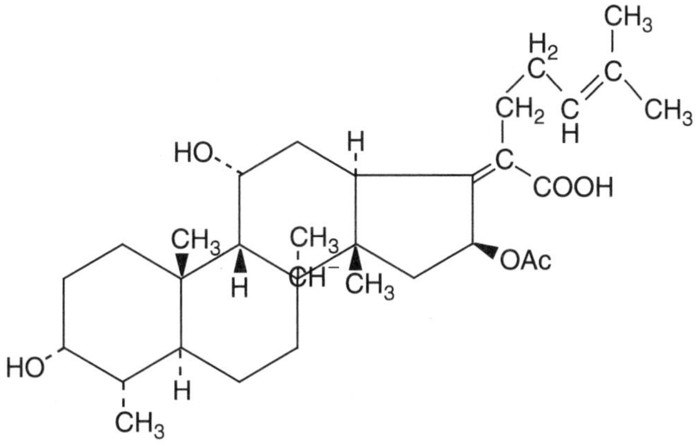

Fig. 2-41. Fusidic acid, an antibiotic with a steroid-like structure.

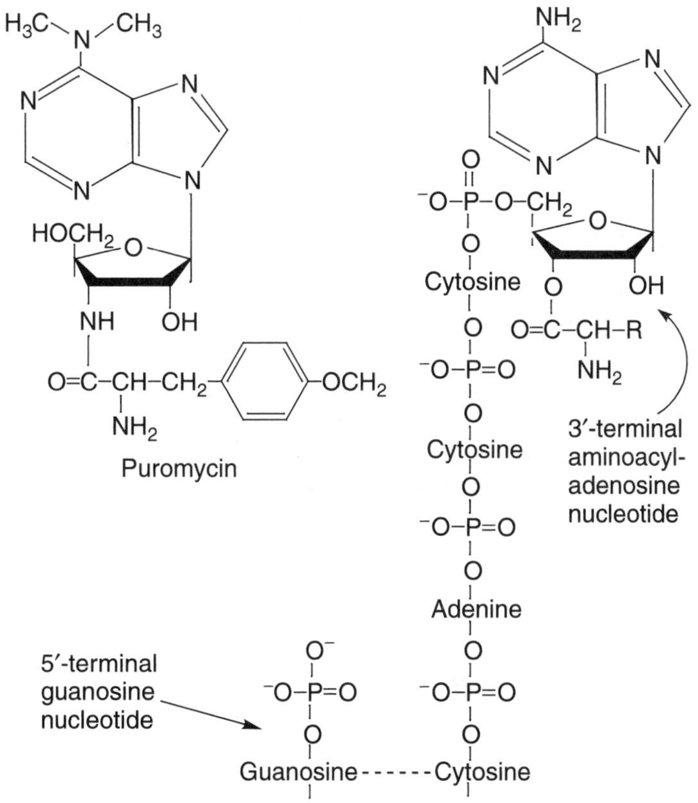

Fig. 2-42. Structure of the terminal end of the amino acid arm of transfer RNA. The -cytosine-cytosine-adenosine (-CCA) at the 3' terminus and the 5'-terminal guanosine nucleotide are features that are common to all tRNAs. The structure of puromycin, which binds to the 50S ribosomal subunit and replaces aminoacyl tRNA as an acceptor of the growing polypeptide, is shown for comparison.

BIBLIOGRAPHY

Nucleoids

Bachellier, S., J. M. Clement, M. Hofnung, and E. Gilson. 1997. Bacterial interspersed mosaic elements (BIMEs) are a major source of sequence polymorphism in *Escherichia coli* intergenic regions including specific associations with a new insertion sequence. *Genetics* **145:**551–62.

Berlyn, M. K. B. 1998. Linkage map of *Escherichia coli* K-12, Edition 10: The traditional map. *Microbiol. Mol. Bio Reviews.* **62:**985–1019.

Dame, R. T., C. Wyman, and N. Goosen. 2000. H-NS mediated compaction of DNA visualised by atomic force microscopy. *Nucleic Acids Res.* **28:**3504–10.

Keck, J. L., and J. M. Berger. 1999. Enzymes that push DNA around. *Nat. Struct. Biol.* **6:**900–2.

Robinow, C., and E. Kellenberger. 1994. The bacterial nucleoid revisited. *Microbiol. Rev.* **58:**211–32.

DNA Replication

Baker, T. A., and S. P. Bell. 1998. Polymerases and the replisome: machines within machines. *Cell* **92:**295–305.

Blakely, G. W., A. O. Davidson, and D. J. Sherratt. 2000. Sequential strand exchange by XerC and XerD during site-specific recombination at dif. *J. Biol. Chem.* **275:**9930–6.

Boye, E., A. Lobner-Olesen, and K. Skarstad. 2000. Limiting DNA replication to once and only once. *EMBO Rep.* **1:**479–83.

Bussiere, D. E., and D. Bastia. 1999. Termination of DNA replication of bacterial and plasmid chromosomes. *Mol. Microbiol.* **31:**1611–8.

Cairns, J. 1963. The chromosome of *Escherichia coli. Cold Spring Harbor Symp. Quant. Biol.* **28**: 43.

Gordon, G. S., and A. Wright. 2000. DNA segregation in bacteria. *Annu. Rev. Microbiol.* **54:**681–708.

Jacob, F., S. Brenner, and F. Cuzin. 1963. On regulation of DNA replication in bacteria. *Cold Spring Harbor Symp. Quant. Biol.* **28:**329–48.

Jensen, R. B., and L. Shapiro. 1999. Chromosome segregation during the prokaryotic cell division cycle. *Curr. Opin. Cell Biol.* **11:**726–31.

Kornberg, A. 1980. *DNA Replication.* W. H. Freeman and Co., San Francisco.

Lemon, K. P., and A. D. Grossman. 1998. Localization of bacterial DNA polymerase: evidence for a factory model of replication. *Science* **282:**1516–9.

Meselson, M., and F. W. Stahl. 1958. The replication of DNA in *Escherichia coli. Proc. Natl. Acad. Sci. USA* **44:**671.

Newman, G., and E. Crooke. 2000. DnaA, the initiator of *Escherichia coli* chromosomal replication, is located at the cell membrane. *J. Bacteriol.* **182:**2604–10.

Norris, V., J. Fralick, and A. Danchin. 2000. A SeqA hyperstructure and its interactions direct the replication and sequestration of DNA. *Mol. Microbiol.* **37:**696–702.

Seitz, H., C. Weigel, and W. Messer. 2000. The interaction domains of the DnaA and DnaB replication proteins of *Escherichia coli. Mol. Microbiol.* **37:**1270–9.

von Freiesleben, U., M. A. Krekling, F. G. Hansen, and A. Lobner-Olesen. 2000. The eclipse period of *Escherichia coli. EMBO J.* **19:**6240–48.

Transcription and Translation

Apirion, D., and A. Miczak. 1993. RNA processing in procaryotic cells. *Bioessays* **15:**113–20.

Brimacombe, R. 1999. The structure of ribosomal RNA. *ASM News.* **65**:144–51.

Brimacombe, R. 2000. The bacterial ribosome at atomic resolution. *Structure Fold Des.* **8**:R195–200.

Burgess, B. R., and J. P. Richardson. 2000. RNA passes through the hole of the protein hexamer in the complex with the *Escherichia coli* rho factor. *J. Biol. Chem.* **276**:4182–9.

Cech, T. R. 2000. Structural biology. The ribosome is a ribozyme. *Science* **289**:878–9.

Crick, F. H. 1966. Codon–anticodon pairing: the wobble hypothesis. *J. Mol. Biol.* **19**:548–55.

Darst, S. A. 2001. Bacterial RNA polymerase. *Curr. Opin. Struct. Biol.* **11**:155–62.

Garen, A. 1968. Sense and nonsense in the genetic code. Three exceptional triplets can serve as both chain-terminating signals and amino acid codons. *Science* **160**:149–59.

Gourse, R. L., W. Ross, and T. Gaal. 2000. UPs and downs in bacterial transcription initiation: the role of the alpha subunit of RNA polymerase in promoter recognition. *Mol. Microbiol.* **37**:687–95.

Helmann, J. D., and M. J. Chamberlin. 1988. Structure and function of bacterial sigma factors. *Annu. Rev. Biochem.* **57**:839–72.

Kim, S. H., F. L. Suddath, G. J. Quigley, A. McPherson, J. L. Sussman, A. H. Wang, N. C. Seeman, and A. Rich. 1974. Three-dimensional tertiary structure of yeast phenylalanine transfer RNA. *Science* **185**:435–40.

Korzheva, N., and A. Mustaev. 2001. Transcription elongation complex: structure and function. *Curr. Opin. Microbiol.* **4**:119–25.

Korzheva, N., A. Mustaev, M. Kozlov, A. Malhotra, V. Nikiforov, A. Goldfarb, and S. A. Darst. 2000. A structural model of transcription elongation. *Science* **289**:619–25.

Maguire, B. A., and R. A. Zimmermann. 2001. The ribosome in focus. *Cell* **104**:813–6.

Nissen, P., M. Kjeldgaard, and J. Nyborg. 2000. Macromolecular mimicry. *EMBO J.* **19**:489–95.

Surratt, C. K., S. C. Milan, and M. J. Chamberlin. 1991. Spontaneous cleavage of RNA in ternary complexes of *Escherichia coli* RNA polymerase and its significance for the mechanism of transcription. *Proc. Natl. Acad. Sci. USA* **88**:7983–7.

Protein Folding, Trafficking, and Degradation

Buchanan, S. K. 2001. Type I secretion and multidrug efflux: transport through the TolC channel-tunnel. *Trends Biochem. Sci.* **26**:3–6.

Driessen, A. J., P. Fekkes, and J. P. van der Wolk. 1998. The Sec system. *Curr. Opin. Microbiol.* **1**:216–22.

Fekkes, P., and A. J. Driessen. 1999. Protein targeting to the bacterial cytoplasmic membrane. *Microbiol. Mol. Biol. Rev.* **63**:161–73.

Jacob-Dubuisson, F., C. Locht, and R. Antoine. 2001. Two-partner secretion in gram-negative bacteria: a thrifty, specific pathway for large virulence proteins. *Mol. Microbiol.* **40**:306–13.

Karzai, A. W., M. M. Susskind, and R. T. Sauer. 1999. SmpB, a unique RNA-binding protein essential for the peptide-tagging activity of SsrA (tmRNA). *EMBO J.* **18**:3793–9.

Levchenko, I., M. Seidel, R. T. Sauer, and T. A. Baker. 2000. A specificity-enhancing factor for the ClpXP degradation machine. *Science* **289**:2354–6.

Muller, M., H. G. Koch, K. Beck, and U. Schafer. 2000. Protein traffic in bacteria: multiple routes from the ribosome to and across the membrane. *Prog. Nucleic Acid Res. Mol. Biol.* **66**:107–57.

Nunn, D. 1999. Bacterial type II protein export and pilus biogenesis: more than just homologies? *Trends Cell Biol.* **9**:402–8.

Plano, G. V., J. B. Day, and F. Ferracci. 2001. Type III export: new uses for an old pathway. *Mol. Microbiol.* **40**:284–93.

Schmidt, M., A. N. Lupas, and D. Finley. 1999. Structure and mechanism of ATP-dependent proteases. *Curr. Opin. Chem. Biol.* **3:**584–91.

Antibiotics

Bottger, E. C., B. Springer, T. Prammananan, Y. Kidan, and P. Sander. 2001. Structural basis for selectivity and toxicity of ribosomal antibiotics. *EMBO Rep.* **2:**318–23.

Franklin, T. J., and G. A. Snow. 1991. *Biochemistry of Antimicrobial Action*. Chapman & Hall, New York.

CHAPTER 3

BACTERIAL GENETICS: DNA EXCHANGE, RECOMBINATION, MUTAGENESIS, AND REPAIR

The plasticity of bacterial genomes is nothing short of amazing! The processes of DNA exchange, replication, and repair are major contributors to gene shuffling and critical players in the evolutionary history of present-day bacteria. The molecular footprints of ancient DNA exchanges between diverse genetic species are evident in most, if not all, of the bacteria for which complete DNA sequences are known. In addition to their evolutionary significance, these processes are crucial for bacterial survival in natural environments where chemical insults constantly threaten to damage DNA and destroy chromosome integrity. The goal of this chapter is to describe these processes, ultimately placing them in the context of evolution and survival.

TRANSFER OF GENETIC INFORMATION IN PROKARYOTES

Genetic recombination is the production of new combinations of genes derived from two different parental cells. Several processes have been described by which prokaryotic cells can exchange genetic information to yield recombinants. They are **conjugation**, **transformation**, and **transduction**. Conjugation requires physical contact between two cells of opposite mating type, resulting in the transfer of DNA from the donor cell to the recipient cell. The transfer of cell-free DNA into a recipient cell is called transformation, whereas transduction involves bacteriophage-mediated transfer of genetic information from a donor cell to a recipient cell.

In the following section, genetic exchange and DNA recombination processes are considered in detail. Often, only a small portion of the donor cell genome is transferred to the recipient cell. The recipient cell temporarily becomes a partial diploid (**merodiploid**) for the short region of homology between the transferred DNA and the DNA of the recipient chromosome. The ultimate fate of the donated DNA, whether it becomes incorporated into the recipient genome by recombination processes, degraded by nucleases (host restriction), or maintained as a stable extrachromosomal fragment, occupies a considerable portion of the ensuing discussion.

PLASMIDS

Bacteria are sexually active organisms! This ability to transfer DNA through cell–cell contact was first discovered in the 1950s using *Escherichia coli*. Transfer of genetic information between different *E. coli* strains was found to depend on the presence in some cells (called donor cells) of a small "extra" chromosome called F (fertility) factor. F factor encodes the proteins necessary for the sexual process and is discussed in detail in this chapter. Subsequent to the discovery of F, many other extrachromosomal DNA elements, called plasmids, were discovered. All plasmids share some common features. They are generally double-stranded, closed circular DNA molecules capable of autonomous replication (i.e., independent of chromosomal replication). One exception to circularity is the linear plasmids found in the Lyme disease bacterium *Borrelia burgdorferi*.

Some plasmids, called episomes, commonly integrate into the bacterial chromosome. F factor is one example of an episome. The size of plasmids can range from 1 to 2000 kb. Compare this to the size of the *E. coli* bacterial chromosome, which is approximately 4639 kb. A plasmid that can mediate its own transfer to a new strain is called a **conjugative** plasmid, whereas one that cannot is referred to as **nonconjugative**. Plasmids that have no known identifiable function other than self-replication are often referred to as cryptic plasmids. However, they are cryptic only because we have not been clever enough to elucidate their true function.

It is important to examine and understand plasmids because of the enormous role they play in biotechnology and medicine. Plasmids are the workhorses of recombinant DNA technology, serving as the vehicles by which individual genes from diverse organisms can be maintained separate from their genomic origins. In addition, many plasmids harbor antibiotic resistance genes. Promiscuous transfer of these plasmids among diverse bacterial species has led to the rapid proliferation of antibiotic-resistant disease-causing microorganisms.

Partitioning

Why have plasmids been maintained throughout evolution? One reason is that they often provide a selective advantage to organisms that harbor them, such as toxins to kill off competing organisms (e.g. Colicins) or resistance genes to fend off medical antibiotics. On a more basic level, the reason plasmids persist in a population is due to their ability to partition. **Partitioning** assures that after replication each daughter cell gets a copy of the plasmid. For plasmids present in high copy numbers (50 to 100 copies per cell; e.g., pBR322), random diffusion may be enough to get at least one copy of the plasmid to each daughter cell. However, random segregation of low-copy-number plasmids (only 2 to 4 copies per cell) would likely mean that, following cell division, one of the daughter cells would not receive a plasmid. The plasmid would eventually be diluted from the population. Consequently, regulated partitioning mechanisms are essential for these plasmids. The mechanism used for partitioning differs depending on the plasmid.

Partitioning, especially of low-copy-number plasmids such as F factor and P1 (see below), usually involves genes called *par* organized on plasmids as gene cassettes. Almost all plasmid-encoded *par* loci consist of three components: a cis-acting centromere-like site and two trans-acting proteins that are usually called Par A and B, which form a nucleoprotein complex at the centromere-like site. **Cis-acting** means that, for a plasmid to replicate, the DNA sequence in question must be part of that plasmid. It does not encode a diffusible product. In contrast, a **trans-acting** replication gene encodes a protein that does diffuse through the cytoplasm. Thus, the gene does not have to reside on the plasmid to help that plasmid replicate.

The upstream gene in a typical *par* operon encodes an ATPase essential to the DNA segregation process. The downstream gene encodes the protein that actually binds to the centromere. The organization of the partitioning loci in F factor and P1 are the same, but the genes are called *par* and *sop*, respectively. ParB/SopB proteins bind as dimers to the downstream *parS/sopS* centromere-like regions. ParA /SopA (the ATPases) then bind through protein–protein interactions with the B proteins.

Genetic fusions between the Par proteins and green fluorescent protein (GFP) from the jellyfish *Aequorea victoria* have been used to visualize the location and movement of plasmids in the cell during segregation. The data suggest that the plasmids are intrinsically located at midcell where they replicate. After replication, the daughter plasmids pair with ParA and B proteins at the *parC/sopC* site, then move from midcell to quarter-cell positions, ensuring each daughter cell receives a plasmid copy. This movement presumably requires interaction with a host-encoded mitotic-like apparatus. What comprises that apparatus is not known.

Incompatibility

Incompatibility is a property of plasmids that explains why two very similar yet distinct plasmids might not be maintained in the same cell. Two plasmids that share some regulatory aspect of their replication are said to be incompatible. For example, if two different plasmids produce similar repressors for replication initiation, then the repressor of one could regulate the replication of the other and vice versa. The choice as to which plasmid will actually replicate is random. So, in any given cell, copies of one plasmid type could outnumber copies of a second plasmid. Because of this incongruity, cell division could result in a daughter cell that only contains one of the two plasmid types. This is called **segregational incompatibility**.

Nonconjugative, Mobilizable Plasmids

Many plasmids that are nonconjugative by themselves nevertheless possess a system that will allow conjugal transfer when present in the same cell as a conjugative plasmid. The colicin E1 (ColE1) plasmid is an example of a nonconjugative, mobilizable plasmid. It has a *dnaA*-independent origin of replication but contains a site called *bom* (basis of mobility) that functions much like *oriT* of the F factor (see below). While some ColE1 genes are required for mobilization (*mob* genes), transfer will not occur unless some functions are provided by F factor. Not all conjugative plasmids will fill this role, however. Where members of the incompatibility groups IncF, IncI, or IncP conjugative plasmids will efficiently mobilize ColE1, members of IncW will do so inefficiently.

Resistance Plasmids

Many extrachromosomal elements have been recognized because of their ability to impart new genetic traits to their host cells. One important factor, resistance (R) factor, was first recognized by the fact that organisms in which it was present were resistant to a number of chemotherapeutic agents. A single R factor may carry traits for resistance to as many as seven or more chemotherapeutic or chemical agents. R factors harbored by organisms in the normal flora of human beings or animals may be transferred to pathogenic organisms, giving rise to the sudden appearance of multiple resistant strains.

There is evidence that R factors did not arise as a direct result of the widespread use of antibiotics. Individuals living in highly sequestered geographic areas who have not been subjected to antimicrobial agents possess normal bacterial flora harboring R factors. Also, organisms maintained in the lyophilized state from time periods prior to the widespread use of antimicrobial agents have been shown to carry R factors. Thus, the origin of these plasmids probably occurred in response to encounters with antimicrobials produced in the natural environment. Regardless of their origins, the fact that R factors are transferred from normal flora to pathogenic microorganisms under natural conditions (e.g., the gastrointestinal tract) generates considerable concern with respect to the widespread and indiscriminant use of antimicrobial agents that could foster the further spread of these plasmids.

Plasmids in Other Bacterial Genera

The association of resistance to chemotherapeutic agents and other genetic traits with extrachromosomal elements has been well established in the Enterobacteriaceae. It is also possible to demonstrate intergeneric transmission of plasmids among members of this closely related group of organisms. Plasmid-associated resistance factors have been well characterized in most bacteria (one notable exception is *Rickettsia*). For example, resistance of *Enterococcus hirae* (previously called *E. faecalis*) to erythromycin and lincomycin is plasmid associated. Self-transferable plasmids encoding hemolysins and bacteriocins as well as multiple antibiotic resistance have also been found in *E. hirae*. As is described in this chapter, transfer of plasmids among the streptococci appears to take place through conjugal mechanisms.

Plasmid Replication

There are three general replication mechanisms for circular plasmids: theta (θ) type, rolling circle, and strand displacement. **Theta (θ)-type replication** has been examined most extensively among circular plasmids of gram-negative bacteria. The mechanism involves melting of the parental strands at a plasmid origin, synthesis of a primer RNA, and then DNA synthesis by covalent extension of that primer. θ-type DNA synthesis can start from one or multiple origins, and replication can be either uni- or bidirectional, depending on the plasmid. Electron microscopy revealed these replication intermediates look like θ-shaped molecules. Initiation generally requires a plasmid-encoded Rep initiator protein that binds to a series of directly repeated DNA sequences within the origin called **iterons** (see below).

Rolling circle replication (RC) is unidirectional. The current model posits that replication initiates by the plasmid-encoded Rep protein, introducing a site-specific nick on one strand (the plus strand) of the DNA. The resulting 3'-OH end serves as a primer for leading strand synthesis involving host replication proteins (DNA pol III, ssb, and a helicase). Elongation from the 3'-OH end accompanied by displacement of the parental strand continues until the replicome reaches the reconstituted origin at which point a DNA strand transfer reaction takes place to terminate leading strand replication. The result is a double-stranded DNA molecule comprised of the parental minus strand and the newly synthesized plus strand, and a single-stranded DNA intermediate corresponding to the parental plus strand. The generation of this single strand is the hallmark of plasmids replicating by the RC mechanism. Finally, the parental strand is converted into dsDNA by the host replication machinery initiating at a second origin physically distant from the RC origin.

The best-known examples of plasmids replicating by the **strand displacement** mechanism are the promiscuous plasmids of the IncQ family. Replication is promoted by the joint activity of three plasmid-encoded proteins, RepA, B, C, with, respectively, 5'-3' helicase, primase, and initiator activities. RepC binds to iteron sequences within the origin and probably with the RepA helicase, promoting exposure of two small palindromic sequences (*ssi*) located on opposite strands at the origin. DNA synthesis occurs with opposite polarities from these two sites. Stem-loop structures generated by the *ssi* sequences are probably required for assembly of the RepB primase, an event required to initiate replication.

Although we tend to think of plasmids as circular, there are notable examples of linear plasmids. These plasmids have two basic structures. The spirochete *Borrelia* contains **linear plasmids** with a covalently closed hairpin loop at each end. This group replicates via concatameric intermediates. Linear plasmids in the gram-positive filamentous *Streptomyces* have a covalently bound protein attached at each end. Representatives of the second group have a protein covalently bound at their 5' ends and replicate by a protein priming mechanism.

Plasmid Copy Number Control. Plasmids are selfish genetic elements. They must ensure that they are stably inherited despite constituting an energy burden on the host cell. In addition to partitioning systems, plasmids contain mechanisms that maintain a certain number of plasmid copies per cell. Too few copies can result in plasmid loss after division, whereas too many copies resulting from runaway replication will overburden the cell, leading to cell death. As noted earlier, plasmids can be characterized into separate groups based on the number of copies normally present in a cell. F factor is a low-copy-number plasmid whose replication is stringently controlled. ColE1, in contrast, is a high-copy-number plasmid that can be present at 50 to 100 copies per cell. Even though there is a wide difference in their copy number, both types of plasmids have mechanisms designed to limit and maintain copy number.

The ColE1 plasmid undergoes unidirectional replication (one replication fork) and can replicate in the absence of de novo protein synthesis. It does this by utilizing a variety of stable host replication proteins (e.g., Pol I, DNA-dependent RNA polymerase, Pol III, DnaB, DnaC, and DnaG) rather than making its own. Because replication initiation of the bacterial chromosome requires de novo protein synthesis, the addition of chloramphenicol to ColE1-containing cells results in the amplification of the plasmid copies relative to the chromosome. But even though the ColE1 replication system allows for a fairly high copy number per cell, the system is regulated so as not to exceed a given number (Fig. 3-1).

Replication of ColE1 actually begins with a transcriptional event that occurs about 500 bp upstream of the replication origin (*ori*). As the transcript progresses through *ori*,

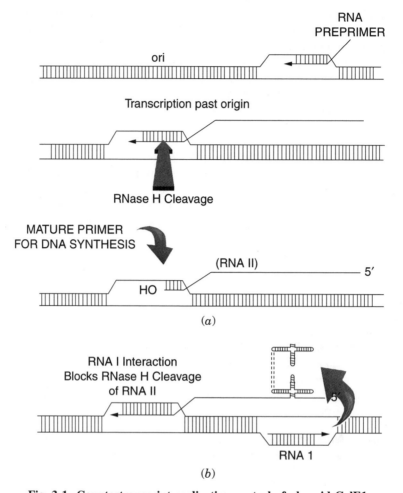

Fig. 3-1. Countertranscript replication control of plasmid ColE1.

an RNA–DNA hybrid molecule forms that is a suitable substrate for the bacterial host enzyme RNaseH. RNaseH cleaves the RNA primer, leaving a 3′-OH end suitable for DNA Pol III and replication. The processed RNA primer is called RNA II. Replication and, thus, copy number, is actually negatively controlled by an antisense RNA, a partial countertranscript of RNA II (100 nucleotide).

The countertranscript (RNA I) is transcribed near the start of the prepriming transcript (the RNA II precursor) but in the opposite direction (Fig. 3-1B). The RNA I and RNA II precursor molecules each contain a significant amount of secondary stem-loop structure, and because the two RNA molecules are complementary, the loop areas can base pair with the help of the plasmid Rom protein. In order for RNA II to prevent replication, it must achieve a concentration that will favor its interaction with the RNA II precursor. It is this aspect of the interaction that allows the plasmid to achieve 20 to 50 copies per cell before RNA I concentrations are high enough to prevent further replication. Following the initial molecular "kiss" of the loop areas, the molecules melt into each other, which extends the base pairing. Formation of the RNA–RNA complex disrupts the RNA–DNA hybrid at *ori* and interferes with processing by RNaseH, thereby aborting replication initiation.

The RNA I–preprimer complex is now also susceptible to RNaseE (*rnaE*) digestion. RNaseE is part of a complex host RNA degradosome that includes polynucleotide phosphorylase, RNA helicase, and poly(A) polymerase (*pcnB,* as noted in Chapter 2, the addition of poly(A) tails to mRNAs may stimulate degradation). Thus, a lack of RNA I degradation decreases plasmid copy number because the resulting higher RNA I level reduces the number of possible replication events. So, under normal circumstances, when the plasmid copy number in a cell decreases, less RNA I is made and preexisting RNA I is degraded. This allows a greater number of initiation events, and the plasmid copy number will rise. Other plasmid systems (F factor, R1 plasmid) utilize a different type of countertranscript (antisense) control of copy number where the target is the mRNA for a rate-limiting initiator protein (see "Fertility Inhibition").

Another type of plasmid copy number control is exemplified by **P1**, which is actually a bacteriophage DNA that exists either as a lytic phage or as a low-copy-number plasmid in its prophage state (Fig. 3-2). How does P1 maintain a low copy number as a plasmid? To understand this process you must first know that the P1-encoded RepA protein serves to initiate P1 plasmid replication. RepA protein must bind to sites within a region of the P1 genome called *oriR* in order to trigger P1 replication. Within *oriR* there are five copies of a 19 bp sequence, called an **iteron**, to which RepA can bind. Another nine iterons are found in a separate region called *incA*. Replication will only initiate if RepA binds to the iterons in *oriR*. The nine iterons present at *incA* were initially thought to titrate the level of available RepA to a point insufficient for binding to *oriR*, especially when multiple copies of P1 are present in the cell. The theory was that with multiple copies of P1 in the cell, the number of competing iterons would increase and thus decrease the probability that RepA would fully occupy the *oriR* iterons and initiate replication. However, artificially increasing RepA did not increase copy number.

It now appears that the monomeric forms of RepA that bind to iterons can subsequently bind through protein–protein interactions with other RepA monomer–iteron complexes, even those on sister plasmids. The result, called handcuffing, effectively

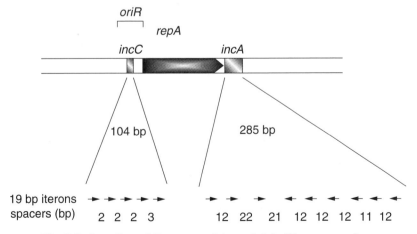

Fig. 3-2. Location of iterons used to maintain P1 copy number.

ties up the plasmids, preventing movement of the replication fork. As the cell volume increases, the RepA dimers disassociate and handcuffing decreases. Since RepA autorepresses its own expression, the increase in cell volume also derepresses *repA*. The resulting increase in RepA monomers leads to full occupancy of iterons in *oriR* and replication ensues, so after cell division, copy number in each daughter cell is maintained. F factor also utilizes an iteron system for copy number control.

Addiction Modules: Plasmid Maintenance by Host Killing: The *ccd* Genes

In addition to the partitioning systems already described, several plasmid systems utilize genes called addiction modules to prevent the development of a plasmid-less population. Low-copy-number plasmids such as F factor, as well as R1, R100, and P1, all have the remarkable ability to kill cells that have lost the plasmid. In F, two plasmid-associated genes are responsible: *ccdA* and *ccdB* (coupled cell division). The *ccdB* product (11.7 kDa) is an inhibitor of cell growth and division. Its activity is countered by the unstable *ccdA* product (8.7 kDa). If plasmid copy number falls below a certain threshold, the loss of CcdA through degradation will unleash the killing activity of CcdB. Affected cells form filaments, stop replicating DNA, and die. This phenomenon is referred to as **programmed cell death**. The population has the appearance of being addicted to the plasmid. The target of CcdB is DNA gyrase. CcdB toxin, in the absence of the CcdA antidote, binds to GyrA, thereby inhibiting DNA gyrase activity. Cell death may be due to DNA cleavage that results from CcdB trapping the DNA gyrase–DNA complex in a poststrand passage intermediate (see Chapter 2). Alternatively, the DNA gyrase–CcdA–DNA complex will inhibit movement of RNA polymerase.

In addition to proteic postsegregational killing systems, there are antisense RNA-regulated systems. The role of antidote is played by a small, unstable RNA transcript that binds to toxin-encoding RNA and inhibits its translation. The best-characterized system of this type is the *hok/sok* locus of plasmid R1. Translation of mRNA encoding the Hok (host-killing) protein is regulated by *sok* (suppressor of killing) antisense RNA. The unchecked production of Hok (52 amino acids) causes a collapse of the transmembrane potential (see Chapter 9).

In the orthodox view of bacteria as single-celled organisms, programmed cell death does not make adaptive sense. However, it is becoming increasingly clear that in nature bacteria seldom behave as isolated cells. They often take on characteristics normally associated with multicellular organisms. Biofilms are one obvious example. In this context, adaptive sense can be made for sacrificing an individual cell that fails to meet the community standards required for survival of the species.

CONJUGATION

Conjugation involves direct cell-to-cell contact to achieve DNA transfer (Fig. 3-3). For this process, certain types of extrachromosomal elements called plasmids are usually required. The prototype conjugative plasmid is the F, or fertility, factor of *E. coli*.

F Factor

F factor is an *E. coli* plasmid (100 kb) with genes coding for autonomous replication, sex pili formation, and conjugal transfer functions. In addition, there are several insertion sequences situated at various sites. F is considered an episome, since it replicates

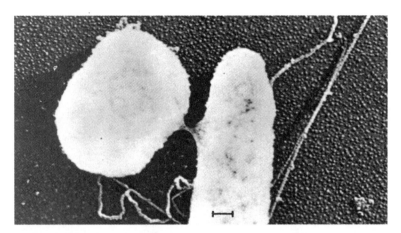

Fig. 3-3. Conjugation: cell-to-cell contact between bacterial cells. The long cell on the right is an Hfr donor cell of *E. coli*. It is attached by the specific F-pilus to a recipient (F⁻) on the left. Bar equals 100 nm. (*Source*: From Anderson et al., 1957.)

either independently of the host chromosome or as part of the host genome. Cells containing an autonomous F are referred to as F^+ cells. Replication of the F factor in this situation requires host proteins but is independent of the *dnaA* gene product (see Chapter 2, initiation of DNA replication). There are only one to three copies of this plasmid per cell; thus, F is an example of a plasmid whose replication is stringently controlled.

Conjugal Transfer Process. A large portion of F-factor DNA is dedicated to the transfer process. Figure 3-4 presents a portion of the genetic map for F encompassing the *tra* region. It is useful to refer to this diagram during the following discussion of F conjugation.

In the Enterobacteriaceae, the presence of specific structural appendages (sex pili) on the cell surface is correlated with the ability of the cell to serve as a donor of genetic material. Of the 25 known transfer genes, 14 are involved with F-pilus formation (*traA, -L,-E,-K,-B,-V,-W,-C,-U,-F,-H,-G, Q, X*). TraA is the F-pilin protein subunit while all the other genes are involved in the complex process of assembling the sex pilus (see Chapter 7). Sex pilus assembly occurs in the membranous layers of the cell envelope. F-donor cells possess only one to three sex pili. The tip of the pilus is involved in stable mating pair formation (TraN and TraG involvement), interacting with the OmpA protein on the outer membrane of the recipient. Once initial contact between the donor pilus and the recipient is established, the pilus is thought to contract, bringing the cell surfaces of the donor and recipient cells into close proximity. This wall-to-wall contact causes a fusion of cell envelopes and a conjugation bridge that includes a pore formed by TraD. Mating mixtures of *E. coli* actually involve aggregates of from 2 to 20 cells each rather than only mating pairs.

Following mating-aggregate formation, a signal that a mating pair has formed is relayed from the TraD pore protein to TraM, a protein that binds the TraD pore, and ultimately to TraY, a site-specific DNA-binding protein that binds *oriT* (origin of conjugal transfer). The transfer of F^+ DNA initiates from *oriT* as opposed to *oriV*, the vegetative replication origin used for plasmid maintenance. TraY and integration

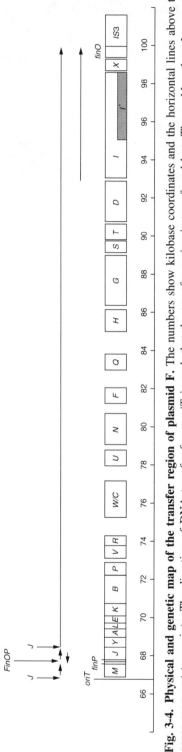

Fig. 3-4. Physical and genetic map of the transfer region of plasmid F. The numbers show kilobase coordinates and the horizontal lines above the genes represent transcripts. The direction of DNA transfer from *oriT* is such that the transfer region is transferred last. The *traM* and *traJ* promoter regions have been sequenced and the *traY-I* operon has been shown to have its own separate promoter. *finP* may be transcribed from the DNA strand opposite the long leader sequence of the *traJ* mRNA. Transcription from the promoters for *traM* and for the *traY* operon is dependent on the product of *traJ*, which is in turn negatively regulated by the FinOP repressor. The *traJ* and *traI** genes are transcribed constitutively from a second promoter at about 18% of the level from the *traJ*-induced *traY-I* operon promoter. Roles attributed to the genes are regulation, finP and traJ; pilus formation, traA, -L, -E, -K, -B, -V, -W/C, -U, -F, -Q, and -G; stabilization of mating pairs, traN and traG; conjugative DNA metabolism, traM, -Y, -D, -I, and I*; surface exclusion, traS and traT.

host factor (IHF, an *E. coli* DNA-binding protein) change the architecture of *oriT* to a structure suitable for binding the bifunctional TraI endonuclease/helicase. The DNA transfer machinery, known as a **transferosome** or **relaxosome**, is then assembled at this origin. The plasmid-encoded TraI endonuclease/helicase nicks the F plasmid at *oriT*, and the 5′ end of the DNA is attached to the protein (Figs. 3-5 and 3-6). The

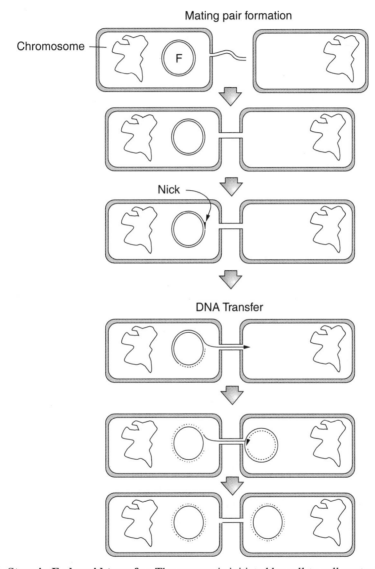

Fig. 3-5. Steps in F plasmid transfer. The process is initiated by cell-to-cell contact, mediated by the F-plasmid-coded pilus on the donor cell. A nick at the origin of transfer site, oriT, supplies the 5′ terminus that invades the recipient cell (F DNA synthesis can, but need not, occur simultaneously on the intact covalently closed circular molecule of single-stranded DNA). Transfer of a genetic element capable of autonomous replication requires production of the complementary molecule on the transferred strand (continuous DNA synthesis is shown by dashed lines) and circularization.

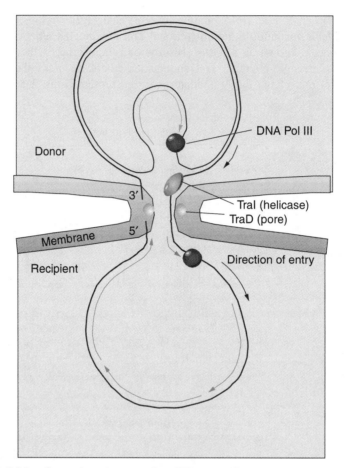

Fig. 3-6. Model for the conjugative transfer of F. A specific strand of the plasmid (thick line) is nicked at *oriT* by the traYI endonuclease and transferred in the 5′ to 3′ direction through a pore, involving the TraD protein, formed between the juxtaposed donor and recipient cell envelopes. The plasmid strand retained in the donor cell is shown by a thin line. The termini of the transferred strand are attached to the cell membrane by a complex that includes the endonuclease. DNA helicase I (traI gene product) migrates on the strand, undergoing transfer to unwind the plasmid duplex DNA; if the helicase is, in turn, bound to the membrane complex during conjugation, the concomitant ATP hydrolysis might provide the motive force to displace the transferred strand into the recipient cell. DNA transfer is associated with synthesis of a replacement strand in the donor cell and a complementary strand in the recipient cell (broken lines); both processes require de novo primer synthesis and the activity of DNA polymerase III holoenzyme. The model assumes that a single-strand-binding protein coats DNA, to aid conjugal DNA synthesis; depending on the nature of the pore, this protein might even be transferred from donor to recipient cell, bound to the DNA.

nicked DNA, containing TraI as a pilot protein, travels through the TraD pore at the membrane bridge, not through the pilus itself, as originally believed.

The 5′-ended strand is transferred to the recipient via a rolling circle type of replication with the intact strand serving as a template. The *traI* gene product (DNA helicase I) unwinds the plasmid duplex (1000 bp/s), pumping one strand into the recipient cell. DNA pol III synthesizes a replacement strand in the donor. The $5'$ end of the strand, upon entering the recipient, becomes anchored to the membrane. As the donor strand is transferred into the recipient cell, it, too, will undergo replication and become circularized. The circularization process does not rely on the host RecA synaptase (see "Recombination") but may require a plasmid-encoded recombination system. At this point, the recipient becomes a donor cell capable of transferring F to another cell. Figure 3-5 illustrates the overall process of conjugation while Figure 3-6 presents a more detailed analysis of the proteins involved.

Barriers to Conjugation. It has been known for some time that cells carrying an F factor are poor recipients in conjugational crosses (termed **surface exclusion** or entry exclusion). A related phenomenon, incompatibility (see above), operates after an F′ element enters into a recipient cell already carrying an F factor and is expressed as the inability of the superinfecting F′ element and the resident F factor to coexist stably in the same cell (see "F′ Formation" for an explanation of F′).

Two genes, *traS* and *traT*, are required for surface exclusion, with TraT being an outer membrane protein. It is believed that the TraT protein might block mating pair stabilization sites or, alternatively, might affect the synthesis of structural proteins (other than OmpA) necessary for stabilization. If mating pair formation occurs in spite of the efforts of TraT, TraS is thought to block DNA-strand transfer, although how it does this is not known. An odd phenomenon never fully explained is that growth of F^+ cells into late stationary phase imparts a recipient ability almost equivalent to an F^- cell. These cells are called F^- phenocopies and are nonpiliated.

Fertility Inhibition. Most conjugative plasmids transfer their DNA at a markedly reduced rate as compared with F because these plasmids (e.g., R100) possess a regulatory mechanism (the FinOP *f*ertility *in*hibition system) that normally represses their *tra* genes. The *finO* and *finP* products interact to form a FinOP inhibitor of *tra* gene expression. In contrast to these other plasmids, the F factor is *finO finP*$^+$ (due to an IS3 insertion in *finO*; see "Insertion Sequences and Transposable Elements") and so is naturally derepressed for conjugation. However, if another plasmid that is *finO*$^+$ resides in the same cell as F, then the F *finP* gene product and the coresident plasmid's *finO* gene product can interact, producing the FinOP$_F$ inhibitor. This will act as a negative regulator of the F factor's *traJ*, which itself is the positive regulator for the other *tra* genes. Consequently, the fertility of F will be inhibited.

FinP is a 79 nucleotide untranslated RNA molecule that is complementary (antisense) to the untranslated leader region of *traJ* mRNA. FinP antisense RNA is thought to form an RNA/RNA duplex with *traJ* mRNA, occluding the *traJ* ribosome-binding site and thereby preventing translation of the TraJ protein. The FinO protein (21 kDa) in F-like plasmids other than F stabilizes FinP antisense RNA by preventing degradation by RNAse E. This increases the in vivo concentration of the antisense RNA. In addition, FinO binds to the FinP and TraJ RNA molecules, promoting duplex formation and thereby enhancing inhibition of TraJ translation.

Hfr Formation. The F factor is an example of a plasmid that can exist autonomously in a cell or can integrate into the bacterial chromosome. A cell that contains an integrated F factor is referred to as an **Hfr cell** (high frequency of recombination). The frequency of insertion occurs at about 10^{-5} to 10^{-7} per generation—that is,

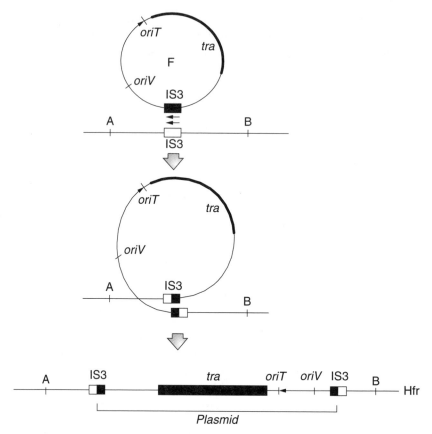

Fig. 3-7. Mechanism of integration of the F factor into the bacterial chromosome. IS3 is an insertion element present on the *E. coli* chromosome and F factor.

among a population of 10^7 F$^+$ cells, 1 to 100 cells will have an integrated F. The mechanism of integration is illustrated in Figure 3-7. Integration involves homologous recombination between two covalently closed circular DNA molecules forming one circular molecule that contains both of the original DNA structures. It is thought that the insertion sequences (IS) present in the F genome (F carries two IS3 and one IS2 sequence) and those in the host chromosome (*E. coli* contains five each of IS2 and IS3) serve as regions of homology for the insertional event. F integration is predominantly RecA dependent (see "Recombination"), but rare RecA-independent Hfr's can be formed based on the transposition functions of the insertion elements. Consequently, Hfr formation is mainly a nonrandom event, primarily occurring in regions of the chromosome containing an insertion sequence element.

Once integrated, the F DNA is replicated along with the host chromosome. However, in situations where the host *dnaA* gene is inactive, replication of the entire chromosome can initiate from an integrated F in a process called **integrative suppression**.

An integrated F factor still has active transfer functions such that an Hfr to F$^-$ cross will transfer host DNA to the recipient where the donor DNA can recombine with the recipient DNA. In Figure 3-7, note that the direction of transfer from *oriT* is such that the *tra* genes are always transferred last. Thus, there is directionality to conjugal

DNA transfer. In this illustration, the proximally transferred host gene is B while the distally transferred gene is A (the tip of the arrowhead represents the 5′ leading end of the transferred strand). If the orientation of the chromosomal IS element was in the opposite direction, then the resulting Hfr would transfer gene A as a proximal marker and gene B would be one of the last genes transferred. Since conjugal transfer of the host chromosome in an Hfr cell is time dependent (it takes approximately 100 minutes to transfer the entire *E. coli* chromosome), a gene can be mapped relative to the position of the integrated F factor simply by determining how long it takes for the gene to be transferred to a recipient. The map position assignments for genes are often given in minutes; an assignment is based on the 100 minutes required for conjugal transfer.

F′ Formation. As indicated in Figure 3-7, integration of F factor is a reversible process. Normally, excision of the F factor restores the host chromosome to its original state. However, improper or aberrant excision can occur at a low frequency, forming a plasmid containing both F and bacterial DNA. This type of plasmid is called an F-prime (F′). There are two types of faulty genetic exchanges that can result in F′ formation. The first involves recombination between a region on the bacterial chromosome and one within the integrated F factor. The resulting F′ has lost some F sequences but now carries some host DNA originally located at one or the other side of the integrated F. This is a type I F′.

As an alternative, host sequences located on both sides of the integrated F can undergo genetic exchange. The F′ formed in this situation (type II) contains all of F plus some host DNA from both sides of the point where F was integrated. In both situations the host DNA contained in the F′ is deleted from the host chromosome. However, when F′ (type I or type II) is transferred to a new host, a partially diploid (merodiploid) situation occurs for the host genes contained on the F′. By constructing merodiploids, information can be derived regarding dominance of certain mutations over wild-type alleles of specific genes (see "cis/trans Complementation Test").

cis/trans complementation Test

When conducting a genetic analysis of mutations that affect a given phenotype, the investigator wants to know whether closely linked mutations reside in a single gene or in different genes of an operon and whether a given mutation resides within a control region (promoter/operator) or within a structural gene. Often these questions can be answered by performing complementation tests using merodiploids. For example, two different mutations are considered to lie in the same gene if their presence on duplicate genes in a merodiploid (one mutation on the plasmid, the other on the chromosome) fails to restore the wild phenotype — that is, the two mutations fail to complement each other in trans.

If the mutations lie in two different genes, then they should complement each other in trans — that is, an F′ with the genotype *trpA*⁺ *trpB*⁻ (two genes involved in tryptophan biosynthesis) will produce a functional TrpA protein that can complement a host with the genotype *trpA*⁻ *trpB*⁺. This premise depends, of course, on whether the gene produces a diffusible gene product — that is, protein or RNA. Regulatory genes such as operators or promoters do not produce diffusible products and so mutations in these genes cannot be complemented in trans. An exception occurs with intragenic complementation. Two different mutations in a gene whose product functions as a multimeric enzyme (homodimer or tetramer) can sometimes complement each other in trans. The resulting active enzyme is composed of two different mutant polypeptides where the defect in one polypeptide sequence is compensated by the presence of the correct sequence on the other polypeptide. A wonderful example of this is called **α complementation** with β-galactosidase. Here a small N-terminal piece of β-galactosidase (called the α peptide) can form an active multimeric enzyme complex with the C-peptide part of β-galactosidase even when the genes encoding each peptide reside on separate plasmids.

Conjugation and Pheromones in Enterococci

For a long time it was believed that conjugation only occurred in the Enterobacteriaceae. Evidence accumulating since 1964, however, has shown that the streptococci also possess conjugative plasmids and conjugative transposons. A number of conjugative plasmids have been characterized, some of which can mobilize nonconjugative plasmids and even some chromosomal markers.

In *Enterococcus hirae* (*E. faecalis*) conjugative plasmids can be placed into two general categories. Members of the first group (e.g., pAD1, pOB1, pPD1, pJH2, pAM1, pAM2, pAM3) transfer at a relatively high frequency in broth (10^{-3} to 10^{-1} per donor cell). Members of the second group transfer poorly in broth but are fairly efficient when matings are carried out on filter membranes (e.g., pAC1, pIIP501, and p5M15346). The reason for this difference involves the production of sex pheromones by streptococci. Plasmids that transfer efficiently in broth use the pheromones to generate cell-to-cell contact. Recipient cells excrete small, soluble peptides (seven to eight amino acids long) that induce certain donor cells to become adherent. This adherence property facilitates formation of donor–recipient mating aggregates that arise from random collisions. Once the aggregate forms, transfer of the plasmid begins. The genes encoding the pheromone peptides have not been unambiguously identified, but computer searches of genomic databases have found several pheromone sequences located within the signal sequence segments of putative lipoproteins (all of unknown function).

The model for induction of pAD1 conjugation starts with a product of pAD1, TraA, repressing many other pAD1 genes (Fig. 3-8). The cAD1 pheromone produced from a potential recipient binds to the PrgZ surface component of the donor cell's oligopeptide permease. The peptide is then transported into the donor cell through a transmembrane channel formed by two other members of the permease: OppB and OppF. Once inside, cAD1 is thought to bind TraA, causing the repressor to release from its DNA-binding site. Transcription then proceeds through TraE, a positive regulator of many pCD1 transfer genes including the *asa1* gene (a.k.a. *prgB* in pCF10) encoding aggregation substance. Aggregation substance (AS) is a cell-surface microfibrillar protein that will recognize a compound known as binding substance (BS) produced on all potential recipient cells. The interaction between AS and BS causes aggregation between donor and recipient and stimulates conjugal transfer.

Once the plasmid is received, how does the new cell keep from stimulating itself? After all, it is still making the cCD1 pheromone. Autostimulation is prevented because the plasmid also contains a cAD1-inducible gene (*iad*; *prgY* in pCF10) that encodes an inhibitor of AD1 peptide. The iAD1 peptide binds to extracellular cAD, effectively neutralizing its effect. However, this only works at low concentrations of cAD1. In

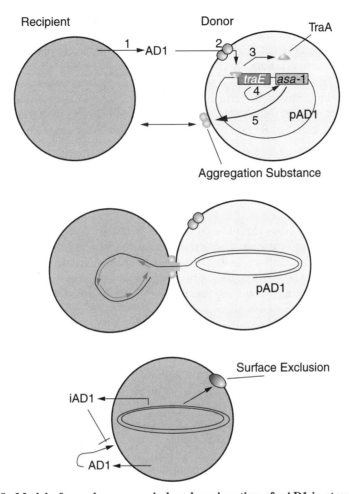

Fig. 3-8. Model of sex pheromone–induced conjugation of pAD1 in streptococci.

addition, cAD1 autoinduces genes required for cell surface exclusion, so a cell already containing a pAD1 plasmid will not receive another copy from a second cell.

It should be apparent that the role of pheromones goes beyond simple aggregate formation. This can be illustrated through donor–donor matings with two donors each having a different plasmid. When only one of the donors is induced, transfer occurs only in the direction from the induced strain to the uninduced strain. Thus, the pheromone also helps trigger transfer of the plasmid.

Conjugation, Cell–Cell Signaling, and Bacterial-Induced Tumors

The Ti plasmid in *Agrobacterium tumefaciens* provides an excellent example whereby environmental conditions influence conjugative functions. The Ti plasmid is very large (1.2×10^8 daltons) but can be transferred from this bacterium to dicotyledonous plants, resulting in production of a plant tumor (e.g., crown gall disease). When wounded, plant cells release specific phenolic signal molecules called opines (e.g., acetosyringone) that are sensed by a two-component signal transduction system, VirA/VirG (see Chapter 5).

In addition, monosaccharide and acid pH in the environment potentiate expression by the VirA/VirG system. Activation of the VirA/VirG system signals transcription of other *vir* genes whose products transfer a defined segment of the Ti plasmid to the plant nuclear genome, transforming the plant cell and leading to tumorous growth. Conjugal transfer of Ti between bacteria or between bacteria and plants is also regulated by cell–cell signaling. Donor bacteria induced by the conjugal opines release a soluble homoserine lactone signal molecule (autoinducer) that interacts with TraR (a homolog of the *Vibrio fischeri* LuxR) and induces conjugal functions (see "Quorum Sensing" in Chapter 5).

TRANSFORMATION

Not all bacteria are capable of conjugation, but that does not stop them from exchanging genetic information. The first demonstrated system of gene transfer actually did not require cell-to-cell contact. Griffith first discovered the phenomenon, called transformation, in 1928 in the course of his investigations of *Streptococcus pneumoniae* (pneumococcus). Capsule-producing pneumococci were shown to be virulent for mice, while nonencapsulated strains were avirulent. Griffith discovered that if mice were injected with mixtures of heat-killed encapsulated (smooth = S) and live uncapsulated (rough = R) cells, a curious phenomenon occurred:

$$\begin{array}{c} \text{Living} \\ \text{R cells} \end{array} + \begin{array}{c} \text{Heat-killed} \\ \text{S cells} \end{array} \xrightarrow[\text{into mice}]{\text{Injected}} \begin{array}{c} \text{Dead mice} \\ \text{(recovered living, virulent S cells)} \end{array}$$

The R cells recovered the ability to produce capsules and regained the capacity of virulence! There are different antigenic types of capsular material produced by different strains of pneumococci. Consequently, Griffith showed that if the avirulent R cells injected into the mice were derived from capsular type II cells and the heat-killed cells were of capsular type III, the viable, encapsulated cells recovered from the dead mice were of capsular type III. This indicated a transformation took place, so type II R cells now produced capsules of antigenic type III.

Later, in 1944, Avery, MacLeod, and McCarty demonstrated that it was DNA from the heat-killed encapsulated strain that was responsible for the transformation. They found that if living, rough type II cells were exposed to DNA isolated from type III encapsulated pneumococci, viable type III encapsulated organisms that were virulent for mice could be recovered. These findings were of exceptional importance because they showed that DNA had the ability to carry hereditary information. Conventional wisdom prior to this time held that hereditary traits were more likely to be borne by protein molecules.

Transformation occurs in many bacterial genera including *Haemophilus*, *Neisseria*, *Xanthomonas*, *Rhizobium*, *Bacillus*, and *Staphylococcus*. Considerable effort has been exerted to elucidate the nature of competence for transformation on the part of recipient cells. Competence is defined as a physiological state that permits a cell to take up transforming DNA and be genetically changed by it. Organisms that undergo natural transformation can be divided into two groups based on development of the competent state. Some organisms become transiently competent in late exponential phase — for example, *Streptococcus pneumoniae*. Others, such as *Neisseria*, are always competent. These different patterns of competence development belie a complex series of regulatory processes required to control this process. The specific physiological and genetic factors involved are discussed in relation to specific groups of organisms.

Transformation processes in different microorganisms can be categorized into two main DNA uptake routes known as the *Streptococcus–Bacillus* model (gram-positive bacteria) and the *Haemophilus–Neisseria* model (gram-negative bacteria). Keep in mind that separation into these two basic models is artificial, since some of the bacteria in one group share features of the transformation process typically associated with the other group.

Gram-Positive Transformation

Streptococcus. Transformation in streptococci has been studied primarily in *Streptococcus pneumoniae* and in members of the streptococci belonging to serological group H. The competent state is transient and persists for only a short period during the growth cycle of a culture of recipient bacteria. The competent state in pneumococci is induced by a specific protein, the **competence-stimulating peptide** (**CSP**, 17 amino acids). Binding of this activator protein to receptors on the plasma membrane triggers the synthesis of at least a dozen new proteins within minutes. After induction by CSF, cells develop the capacity to bind DNA molecules. After binding to recipient cell membranes (there are approximately 30 to 80 binding sites per cell), donor DNA molecules are acted upon by a **translocasome** (also known as **transformasome**) complex located at the cell surface of competent recipient cells. The translocasome includes the endonuclease EndA. This nuclease attacks and degrades one strand of DNA while facilitating the entry of the complementary strand into the cell (Fig. 3-9). After entry, the single-stranded DNA can be recombined into the chromosome of the transformed cell.

The development of competence in *S. pneumoniae* is a wonderful example of a phenomenon known as **quorum sensing** (see Chapter 5). Under appropriate conditions, the 17 amino acid CSP (encoded by *comC*) is processed, exported from the cell, and accumulates in the growth medium until a certain cell density is reached. Once CSP accumulates to a certain level, it binds to the membrane-localized ComD histidine kinase, which autophosphorylates and subsequently transphosphorylates its cognate response regulator protein ComE. ComE-P then activates transcription of competence-specific genes, including the *comCDE* operon in an autocatalytic feedback loop. ComE-P also appears to trigger a regulatory cascade that induces expression of an alternative σ factor ComX (σH), which in turn drives the expression of numerous other operons. Some of these new expressed gene products form the DNA uptake transformasome while other products are thought to remodel the cell wall around the transformasome. Other genes regulated by σH, such as those encoding single-strand binding protein and the recombination enzyme RecA, act on DNA after its entry.

Hex Mismatch Repair. Following recombination between single-stranded donor DNA and the recipient chromosome, nucleotide mismatches occur. A mismatch repair system called Hex (analogous to the MutLS system of *E. coli*; see below) repairs different base–base mismatches with different efficiencies. Thus, the efficiency of integration of genetic markers into the genome of the recipient cell varies drastically with different types of mismatches. Transition mutations (G/T or

120

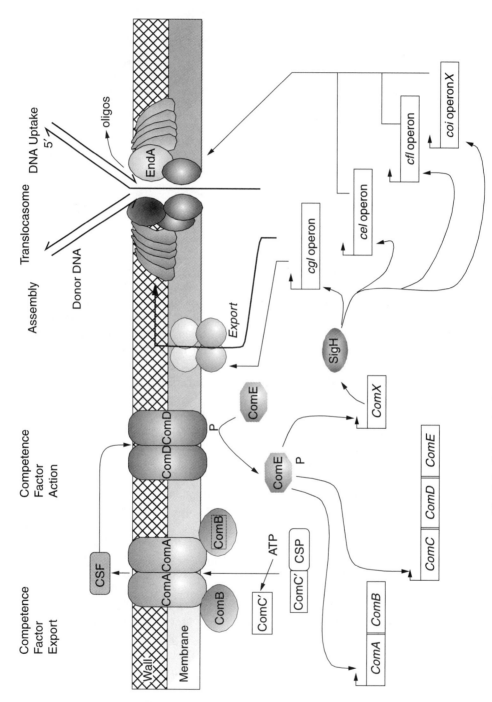

Fig. 3-9. Schematic view of transformation in streptococci.

A/C mismatches) exhibit a low efficiency of transformation (LE markers) because the Hex system is very efficient at repairing the mutations. The HexA protein recognizes the mismatch in heteroduplex DNA resulting from transformation and selectively removes the entire donor strand, which can be as large as up to several kilobases. It is also proposed that HexB protein is activated by binding HexA–DNA mismatch complexes and in some manner increases the efficiency of the system.

In contrast to LE transition mutations, transversion mutations (G/C or C/C mismatches) are not recognized by the Hex system. DNA fragments containing these mutations exhibit 10- to 20-fold higher transformation efficiency and are referred to as high-efficiency (HE) markers. How does the Hex system recognize the donor and recipient strands? Strand discrimination may be determined by identifying the presence of DNA ends in the donor DNA fragments.

B. subtilis. Like *Streptococcus*, *Bacillus* spp can indiscriminantly bind any DNA to its surface. Donor DNA becomes associated with specific sites on competent cell membranes (ComEA protein involvement) and remains associated with these membranes until it is integrated. Competence development in *B. subtilis* is subjected to three types of regulation: nutritional, growth stage specific, and cell-type specific. The regulatory pathway of competence intersects with those of several other stress responses including sporulation, motility, and secretion of macromolecule-degrading enzymes. It is useful to think of competence development as consisting of two modules. Module 1 is a quorum-sensing machine whose major input is population density signaled by two pheromones (CSF and ComX). Once the pheromones are sensed, a small peptide, ComS, is produced. The second module senses ComS and produces the transcriptional regulator ComK in response. ComK activates transcription of the late competence genes required to import DNA and the recombination genes required to incorporate DNA into the chromosome.

As illustrated in Figure 3-10, the ComQ protein is required to process pre-ComX into the smaller peptide pheromone ComX, which is secreted into the medium. As cell density increases, so does the extracellular concentration of ComX. Once it reaches a critical concentration, ComX will interact with a membrane histidine kinase sensor protein called ComP. This triggers a regulatory cascade in which ComP autophosphorylates itself on a conserved histidine residue and then transphosphorylates a cytoplasmic regulatory protein called ComA. Phosphorylated ComA (ComA-P) is a transcriptional activator that activates the *comS* gene. Production of ComS completes the pheromone module.

The ultimate goal of this regulatory cascade is to increase the level of ComK, the transcriptional activator of the late competence genes. However, ComK is subject to degradation by the ClpCP protease. An increase in ComS will interfere with this degradation, allowing ComK to accumulate. Regulation of ComK proteolysis is actually very similar to what we described previously for the *E. coli* σS protein. In *Bacillus*, an adaptor protein called MecA binds to ComS and to the ClpC ATPase chaperone of the ClpCP protease, thus expanding the substrate specificity of ClpC (Fig. 3-10). ComK is then fed into the ClpP barrel and degraded. ComS, however, competes with ComK for binding MecA. Thus, at high enough concentrations, ComS will preferentially bind ClpC and effectively prevent the degradation of ComK. ComK levels increase as a result and induce expression of the genes required for uptake and processing of DNA. The uptake system appears to be very similar to that of *Streptococcus*.

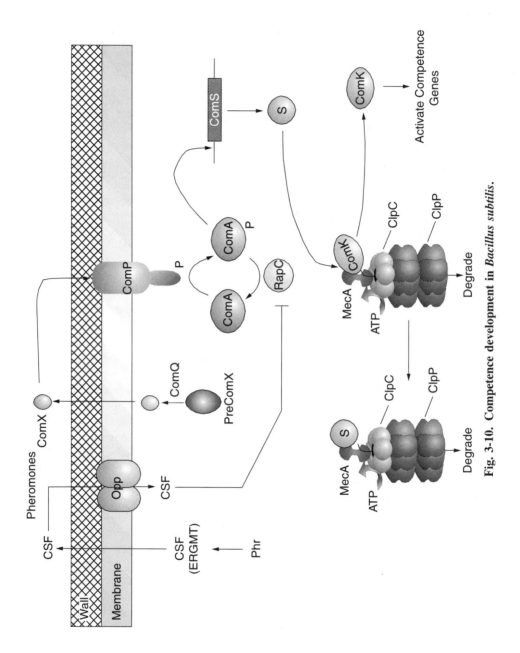

Fig. 3-10. Competence development in *Bacillus subtilis*.

122

Gram-Negative Transformation

In addition to needing a mechanism to transport DNA intact across the cytoplasmic membrane, gram-negative organisms have to move DNA through the outer membrane. The best-studied gram-negative systems are *Neisseria gonorrheae* and *Haemophilus influenzae*. Unlike gram positives, these organisms are naturally competent (no CSF), but competence is induced by entry into the stationary phase. In another departure from the gram-positive systems, transformation in gram-negative systems is sequence specific.

Neisseria. DNA first nonspecifically binds to the neisserial cell surface by electrostatic forces (probably via the Opa outer membrane proteins). Then an unknown protein or protein complex recognizes a specific sequence (GCCGTCTGAA) within the bound DNA. This sequence, resembling a bar code, indicates what DNA is suitable for uptake. Transport of the tagged DNA across the membrane involves an amazing type IV pilus structure similar to that described previously for type II protein export (Chapter 2). The pilus biogenesis factors involved are PilD, T, F, and G, located at the inner membrane. The pilus, comprised of PilE protein, is assembled and protrudes through an outer membrane pore formed by PilQ (a.k.a. Omc). Another protein, PilC, helps gate the polymerized pili through the pore. This same pore serves as the import channel for DNA during transformation.

Recall that type IV pili polymerize and depolymerize, moving like a piston in and out of the outer membrane (see Fig. 2-31B in Chapter 2). One model for DNA uptake is that depolymerization of the pilus would somehow drag DNA into the periplasm. Once in the periplasm, the net-like peptidoglycan sacculus in the periplasm is likely to represent a significant barrier for incoming DNA. Two competence proteins (ComL and Tpc) bind to and cleave peptidoglycan, respectively. It is presumed that these proteins either transiently cleave murein or guide DNA through the sacculus net. The final stage, transport across the inner membrane, requires the inner membrane protein ComA.

Haemophilus influenzae. If exponentially growing *H. influenzae* cells are shifted to nongrowth starvation conditions, 100% of the cells become competent. No secreted competence factors have been reported, so competence development is internally regulated. Changes in intracellular cyclic AMP (cAMP) levels play an important role in this process. As is discussed more fully in Chapter 5, cAMP is an intracellular signal molecule that accumulates as energy sources become scarce. A complex between this molecule and cAMP receptor protein serves as a potent gene regulator in many organisms including *Haemophilus*. As the cells begin to starve upon entering stationary phase, cAMP levels increase and trigger expression of a variety of genes including those for competence. Another competence regulatory gene in *Haemophilus* is called *sxy*. Mutations in this gene either eliminate competence or render it constitutive. An operon of six genes, *comA–comF*, is induced during competence development, but the activator of expression is unknown. Other genes associated with competence in this organism are Por, a periplasmic foldase that is probably involved in the maturation of periplasmic proteins required for transformation, and DprA, a protein required for DNA transport across the inner membrane.

In *H. influenzae*, changes in the cell envelope accompany the development of the competent state. Envelopes from competent cells exhibit elevated levels of lipopolysaccharide with a composition different from that of log-phase cell envelopes.

Six apparently new polypeptides are found in envelopes from competent cells. Most of the polypeptide changes are confined to the outer membrane, although one new polypeptide is associated with the inner cytoplasmic membrane. Structural changes in the envelope also occur in competent cells. Numerous vesicles called *transformasomes* bud from the surface and contain proteins that react specifically with conserved sequences (5′AAGTGCGGTCA3′) present at 4 kb intervals on *Haemophilus* DNA. These vesicles appear to mediate the uptake of transforming DNA. The DNA uptake site is made up of two proteins: 28 kDa and 52 kDa. Several DNA receptor proteins are also produced that recognize the conserved DNA sequence. After binding, the receptor proteins present the donor DNA to the membrane-associated uptake sites. Consequently, DNA binding and uptake are very specific for *Haemophilus* DNA.

Why Competence? There are three proposals that attempt to explain the intrinsic cellular value of developing transformation competence. One model is that cells can use the DNA as an energy source. This may be possible for organisms such as *Streptococcus* and *Bacillus*, which indiscriminantly take up DNA from many different sources. However, the *H. influenzae* and *N. gonorrheae* systems exhibit substrate specificity with respect to DNA uptake. This does not suggest a food-gathering mechanism.

A second proposed function is that transformation plays a role in DNA repair where lysed cells provide a source of DNA that can be taken up and used for the repair of otherwise lethal mutations. The finding that DNA repair machinery is induced as part of competence regulon supports this proposal.

The third popular hypothesis proposes that transformation is a mechanism for exploring the "fitness landscape." For example, it is likely that transformation has played a role in horizontal gene transfer in which a gene from one species of bacteria becomes part of the genome of a second species. If these horizontally acquired genes improve the fitness of the organism, the genes will be retained and the more fit strain will ultimately dominate the population.

Transfection and Forced Competence

Transfection involves the transformation of bacterial cells with purified bacteriophage DNA. The transformed viral DNA will replicate and ultimately produce complete virus particles. Transfection has been demonstrated in a number of bacteria, including *B. subtilis*, *H. influenzae*, *Streptococcus*, *Staphylococcus aureus*, *E. coli*, and *Salmonella typhimurium*.

It should also be noted that organisms not considered naturally transformable (e.g., *E. coli* and *S. typhimurium*) can be transformed under special laboratory conditions. Alterations made in the outer membrane with $CaCl_2$ or through an electrical shock (electroporation) can be used to transfer DNA such as plasmids into cells. This has been an important factor in the success of recombinant DNA research.

TRANSDUCTION

Transduction is the transfer of bacterial genetic markers from one cell to another mediated by a bacteriophage. There are two types of transduction: generalized and specialized.

Generalized transduction is the phage-mediated transfer of any portion of a donor cell's genome into a second cell. The transducing viral particle contains only bacterial DNA, without phage DNA. During normal loading of nucleic acid into virus protein heads (see Chapter 6), the packaging apparatus occasionally makes a mistake, packaging chromosomal DNA into the phage rather than phage DNA. When a transducing bacteriophage binds to a bacterial cell, the donor DNA is injected into the bacterium by the phage and becomes integrated into the genome of the new cell through generalized recombination. Thus, all genes from a donor cell can potentially be transduced to a recipient cell population. Phages that can mediate generalized transduction include P1 (*E. coli*) and P22 (*Salmonella*).

A well-studied example of generalized transduction is that of *Salmonella* phage P22. To understand how generalized transduction occurs, you must have some understanding of how this phage replicates and packages DNA into its head. Phage DNA in P22 is linear but circularly permuted and terminally redundant. Thus, the termini of a P22 DNA molecule contain duplicate DNA sequences, but different P22 molecules within a population contain different terminal sequences. This seemingly odd chromosome structure is due to the packaging method P22 uses to insert DNA into its head. After infecting a cell, the linear P22 DNA circularizes through recombination between the terminal redundant ends. The circular molecule replicates via a rolling circle type of DNA synthesis, forming a long concatemer. A concatemer is a long DNA molecule containing multiple copies of the genome.

Packaging of P22 DNA into empty heads initiates at a specific region in P22 DNA called the *pac* site, where the DNA is first cut, and then proceeds along the concatemer. Once the P22 head is full (headful packaging), a second, nonsequence-specific cut is made. This second cut also defines the start of packing for the next phage head. However, to assure that each head receives a complete genome, the P22 system packs a little more than one genome length of P22 DNA. This is the source of the terminal redundancy of the P22 genome. One consequence of packing extra DNA is that the packing start site changes for each subsequent phage DNA molecule packaged from the concatemer. Each duplicated sequence at the ends of the P22 genome is different for each packed P22 molecule.

When the packaging system encounters a sequence in the bacterial chromosome that is similar to a *pac* site, it does not distinguish this site from P22 *pac* sites. It will use this homologous site to package chromosomal DNA such that progressive packaging will generate a series of phage particles that carry different parts of the chromosome. The size of packaged DNA is approximately 44 kb, which is about one-one hundredth the size of the *Salmonella* chromosome. Therefore, a given P22 will package the equivalent of one conjugation minute worth of chromosomal DNA.

Cotransduction is the simultaneous transfer of two or more traits during the same transduction event, which enables the mapping of genes relative to each other. Cotransduction of two or more genes requires that the genes be close enough to each other on the host chromosome such that both genes can be packaged into the same phage head. The closer the two genes are to each other, the higher the probability they will be cotransduced. The only thing that could separate the two genes would be a recombinational event occurring between them. Thus, the closer the two genes are to each other, the smaller the recombinational target that could separate them (Fig. 3-11).

In **abortive transduction**, DNA that is transferred to the recipient cell does not become integrated into the genome of the recipient cell. Since this DNA is not

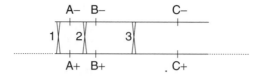

Fig. 3-11. Cotransduction of linked markers. Cross-over events 1,3 will occur more frequently than 1,2.

replicated, it is transmitted unilaterally from the original cell to only one of the daughter cells and only transiently expresses the function of the genetic information in these cells. It is not stably inherited, since passage of this DNA eventually is diluted out through subsequent cell divisions.

Specialized transduction is mediated by bacteriophage that integrate (lysogenize) into a specific site (*att*) on the bacterial chromosome. This specificity limits transfer of genetic material to host markers that lie within the immediate vicinity of this site. Lambda (λ) phage is the most actively studied phage of this type. λ phage almost invariably integrates at an *att* site near the galactose (*gal*) region of the chromosome and therefore can mediate specialized transduction of genes in that area to recipient cells (see Lambda Phage in Chapter 6).

Both low-frequency and high-frequency transduction have been described for λ (Fig. 3-12). An infecting λ DNA integrates into the bacterial chromosome (A). In low-frequency transduction (LFT), the integrated phage (prophage) may be induced through some forms of stress (e.g., DNA damage; see "SOS-Inducible Repair") to enter a cycle of lytic infection. As a relatively rare event (10^{-5}–10^{-6} per cell), a portion of the genome of the phage is replaced by a specific segment of the host chromosome (B). This occurs due to improper excision of integrated prophage DNA in a manner similar to the formation of type I F′ factors (see Fig. 3-7). These are defective phages because

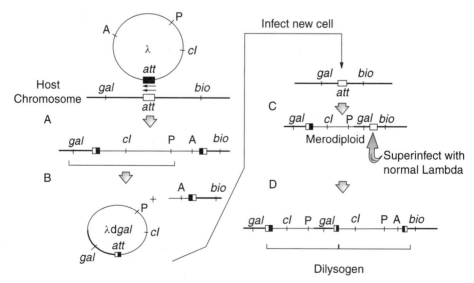

Fig. 3-12. Specialized transduction with λ phage showing the production of low-frequency transducing (LFT) and high-frequency transducing (HFT) lysates.

they lack some portion of the phage genome, but they are capable of transducing and integrating into a new host, carrying the original host genes with them (C). This establishes a merodiploid situation. At this point, the defective λ cannot produce more phage particles on its own. Nevertheless, a high-frequency transducing lysate can be made following an LFT event. This happens because a cell doubly lysogenized with a defective λ (carrying donor DNA) and a normal λ can be induced to yield new phage progeny (D). The normal copy of λ provides replication proteins missing from the defective λ. Thus, an HFT lysate is produced in which approximately half of the particles will be specialized transducing particles.

As with conjugation, transduction usually occurs most readily between closely related species of the same bacterial genus (intrageneric). This is due to the need for specific cell surface receptors for the phage. However, intergeneric transduction has been demonstrated between closely related members of the enteric group of organisms — for example, between *E. coli* and *Salmonella* or *Shigella* species. Various genetic traits such as fermentation capabilities, antigenic structure, and resistance to chemotherapeutic agents are transducible. Keep in mind that transduction is not limited to the bacterial chromosome. Genetic information residing on plasmids may also be transferred by transduction.

Lysogenic conversion is dependent on the establishment of lysogeny between a bacteriophage and the host bacterial cell. Lysogeny occurs when a bacteriophage coexists with its bacterial host without lysing it. In lysogenic conversion, a new phenotypic trait acquired by the host cell is due to a phage gene. Because the gene is part of the normal phage genome, every cell in a population that has been lysogenized acquires the genetic property. This mass conversion of the cell population distinguishes lysogenic conversion from transduction and other genetic transfer mechanisms.

One of the most interesting and thoroughly investigated examples of lysogenic conversion is the relationship of lysogeny to the production of toxin by the diphtheria bacillus, *Corynebacterium diphtheriae*. Cells that are lysogenized by *β* phage are designated *tox*$^+$. The ability to produce toxin is inherent in the genome of the bacteriophage rather than in any trait that may have been transduced from the original host cell. Thus, cells that are "cured" of the lysogenic state no longer produce toxin. The *tox*$^+$ gene is considered to be a part of the prophage genome; however, it is not essential for any known phage function and may be modified or eliminated without any effect on *β* phage replication. The production of *erythrogenic toxin* by members of the group A streptococci and toxins produced by *Clostridium botulinum* types C and D have also been shown to be the result of lysogenization by bacteriophage.

RECOMBINATION

The success of bacterial genetics has been due, in great measure, to the fortuitous ability of bacteria to integrate donor DNA into their genomes. Without this ability, we would be considerably more ignorant of life's secrets. Of course, recombination is not really a philanthropic activity of bacteria toward science but a rather important component of their survival and evolution (see "DNA Repair Systems"). There are basically two types of recombination that can occur in *E. coli* and many other bacteria. These are **RecA-dependent** general recombination (or homologous recombination) and **RecA-independent** or nonhomologous recombination. General recombination requires a large region of homology between the donor and recipient DNAs but does not produce a net gain of DNA by the recipient genome. Homologous sequences are merely exchanged. In contrast, the nonhomologous recombinational mechanisms require very little sequence homology (as little as 5 or 6 bp) but actually produce a net gain of DNA by the recipient. Nonhomologous recombination may be divided into two types: **site specific**, where exchange occurs only at specific sites located on one or both participating DNAs (see "λ Phage" in Chapter 6), and **illegitimate**, which includes other RecA-independent events such as transposition (see "Insertion Sequences and Transposable Elements"). Because these types of recombination increase the length of the recipient's DNA, they have been referred to as **additive**.

General Recombination

Most recombination events that occur in *E. coli* following DNA exchanges are mediated by RecA-dependent pathways that require large regions of homology between donor and recipient DNA. The *recA* product is a synaptase that facilitates the alignment of homologous DNA sequences. Loss of *recA* through mutation reduces recombination frequencies by 99.9% The process of recombination can be viewed in six steps: (1) strand breakage, (2) strand pairing, (3) strand invasion/assimilation, (4) chiasma or crossover formation, (5) breakage and reunion, and (6) mismatch repair.

Recombination events in *E. coli*, or in any other organism for that matter, are not really a random process. A specific DNA sequence called chi (χ) is the primary target for initial recombination events. χ sequences, which are naturally present in *E. coli*, clearly enhance recombination and define recombinational hotspots. The χ-consensus sequence is an octomer, 5'-GCTGGTGG-3', and occurs every 5 to 10 kb (there are 1009 sites in the *E. coli* genome).

χ sequences are important for accomplishing the first goal in the process of recombination, which is to take a double-stranded donor molecule and produce a single-strand from it that can invade and displace one strand of the recipient DNA (see Fig. 3-13). The process begins with the *recBCD* or *recJ* products entering at the ends of a double-stranded DNA molecule and unwinding the DNA using their helicase activities. RecJ is part of an alternative recombination pathway called the RecF pathway. When the sliding RecBCD complex nears a χ site, it introduces a single-strand nick a few nucleotides from the 3' side of χ. RecBCD then degrades the 3'-terminated strand during further DNA unwinding. When RecBCD reaches χ, the enzyme is altered and degradation is reduced.

Once the enzyme is altered by interaction with χ, it does not recognize other χ sites on the same molecule even though it continues to unwind the DNA. RecBCD then loads the new single-stranded 3' end with the RecA synaptase. RecD actually inhibits recombination at sites other than χ by inhibiting the loading of RecA on single-stranded DNA. It has been proposed that upon encountering χ, RecD undergoes an as-yet undocumented change that either ejects RecD from the complex or exposes a domain for RecA loading. Once the resultant single-strand tail is bound to RecA, the complex is poised to invade the donor DNA and begin the recombination process. (Figs. 3-13C and 3-14c). The RecA protein (active species is a tetramer of 38,000 dalton monomers) is known to promote rapid renaturation of complementary single strands, hydrolyzing ATP in the process. It is believed that the RecA synaptase binds to ssDNA, which spontaneously increases RecA ability to bind dsDNA. RecA protein bound to ssDNA then aids in a search for homology between the donor strand and the recipient molecule.

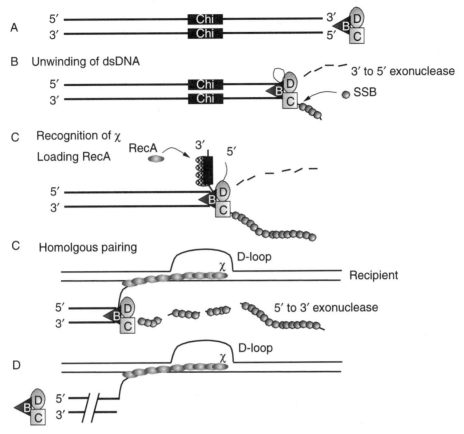

Fig. 3-13. Recombination: D-loop formation.

Experiments suggest a mechanism in which the RecA-mediated interaction between the searching single strand and the recipient double helix occurs randomly at regions of nonhomology, forming nonspecific triple-stranded complexes. The search for homology continues as the RecA protein either reiteratively forms such complexes or translates the two DNA molecules relative to each other until areas of homology are found. Once homologous regions are encountered and the single- and double-stranded DNAs are complexed, a stable displacement loop (D loop) is formed (see Figs. 3-13C and D, 3-14e).

Next, strand assimilation occurs where the donor strand progressively displaces the recipient strand (also called branch migration). Some data suggest that RecA protein will partially denature the recipient DNA duplex and allow the assimilating strand to track in with a 5' to 3' polarity. Extensive assimilation seems to involve RecA binding cooperatively to additional areas in the crossover region to form a "protein–DNA filament." Subsequent to RecA binding, the RuvAB complex is responsible for the branch migration phase of recombination that occurs after the crossover site, known as the **Holliday junction**, has formed (Fig. 3-14g and h). RuvA targets the complex to the Holliday junction, enabling RuvB to assemble as a hexamer around two opposite arms of the Holliday junction. The DNA is then pumped out of the RuvAB complex in opposite directions to drive branch migration (animation at

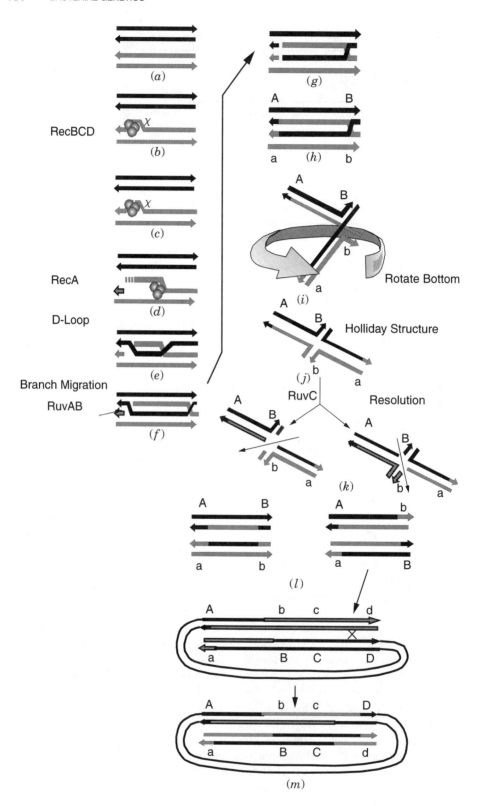

http://www.sdsc.edu/journals/mbb/ruva.html). The RuvAB complex has the ability to promote branch migration for at least 1 kb.

D-loop formation can be coupled with replication of the unpaired donor strand. The fact that certain *dna* mutations cause a decrease in recombination frequencies tends to support the involvement of replication in genetic exchange. Alternatively, a simultaneous assimilation of the sister strand may occur as depicted in Figure 3-14e. While strand assimilation occurs, it is proposed that an endonuclease such as exoV (*recBCD* gene product) attacks the D loop. ExoV possesses both exo- and endonuclease activity. D loops are known to be susceptible to ExoV. A general model that accounts for most of the experimental evidence is shown in Figure 3-14.

The resolution of recombination intermediates is catalyzed by RuvC protein, which binds the Holliday junction as a dimer. This process effects the crossover. Cleavage requires homology and occurs at a preferred sequence [5′(A/T)TT ↓(G/C)3′]. Nicks are introduced symmetrically across the junction to give ligatable products. Mentally rotating the bottom half of the molecule in Figure 3-14i to make the molecule depicted in Figure 3-14j helps to visualize the final crossover event. Breakage and reunion are required to resolve this structure and can occur in either the vertical or horizontal plane. Note that the exchange of markers A/a and B/b occur following vertical but not horizontal cleavage. Figure 3-14l illustrates the end products of the right-hand resolution pathway as it would appear in a circular DNA molecule (e.g., chromosome). Note that to retain continuity of the circular DNA molecule following resolution event 2, a second crossover event is required (Fig. 3-14m). A short animation of recombination may be found at http://www.wisc.edu/genetics/Holliday/holliday3.html.

Genetics of Recombination

Many genes that participate in recombination have been identified in *E. coli*. Table 3-1 provides a list of the known loci associated with recombination in this organism.

Fig. 3-14. Model for the recombination of genetic material. (*a*) The RecBCD enzyme moves along a donor DNA duplex until it encounters a χ sequence (*b*) at which point a single-stranded nick (*c*) is introduced, creating a single-stranded DNA tail (*d*). (*e*) RecA protein, along with single-stranded-binding proteins, can bind to the ssDNA region and subsequently bind to recipient duplex DNA searching for regions of homology. (*f*) Upon encountering a region of homology, D-loop formation occurs. Concomitant to D-loop formation, replication may occur on the unpaired donor strand. Some branch migration of the D loop occurs by cooperative binding of additional RecA tetramers to the donor ssDNA. At some point, a nuclease cleaves and partially degrades the D loop (*f*). The single-stranded tail remaining from the D loop is trimmed (*g*) and the exchanged strands ligated (by DNA ligase) to their recipient molecules (*h*). This structure as shown in (*i*) is referred to as the Holliday intermediate or chiasma formation and has been observed in electron micrographs. Further branch migration at this point occurs by the action of RuvAB. Mentally rotating the bottom half of the molecule while keeping the top half stationary (*i*) helps to visualize the final crossover event (*j*). Subsequent breakage and reunion, catalyzed by RuvC or RecG, is required to resolve this structure and can occur either in the vertical or horizontal plane (*k*). Note the exchange of markers A/a and B/b occur following vertical but not horizontal cleavage. The products of the *ruvC* and *recG* loci are thought to be alternative endonucleases specific for Holliday structures. In (*m*), the second crossover event is shown that maintains the circularity of the bacterial chromosome following the vertical resolution event in (*l*).

TABLE 3-1. Recombination Genes

Gene	Map Location[a]	Function of Gene, Distinguishing Characteristics of Mutants, or Other Pertinent Information
recA	60.8	Complete recombination deficiency and many other phenotype defects including suppression of tifl, DNA-dependent ATPase, 38 kDa
recB	63.6	Structural gene of exonuclease V, 135 kDa, couples ATP hydrolysis to DNA unwinding
recC	63.7	Structural gene of exonuclease V, 125 kDa
recD	63.6	α subunit of exo V, 67 kDa
recE (sbcA)	30.5	Exonuclease VIII (140 kDa), $5' \rightarrow 3'$ dsDNA
recF	83.6	ATPase stimulated by RecR
recJ	64.6	$5' \rightarrow 3'$ exonuclease
recG	82.4	ATPase, disrupts Holliday structures, 76 kDa
recN	59.3	Unknown function, 60 kDa
recO	58.2	Promotes renaturation of complimentary ssDNA, 31 kDa (interacts with RecR)
recR	10.6	Helps RecA utilize SSB–SSDNA complexes as substrates, 22 kDa
recT (sbcA)	30.4	DNA annealing, in rac phage, activated by sbcA mutation
ruvA	41.9	Complexes with Holliday junctions, branch migration
ruvB	41.9	ATPase, branch migration at Holliday junctions
ruvC	41.9	Endonuclease, Holliday junction, resolvase
recQ	86.3	DNA helicase, 74 kDa
recL (uvrD)	86.1	Recombination deficiency of recB⁻ recC⁻ sbcB⁻ strain
rus	12.3	Holliday junction resolvase
sbcB (xon)	44.8	The structural gene for exonuclease I; recB⁻ recC⁻ sbcB⁻
sbcC	8.9	ATP-dependent dsDNA exonuclease

[a]Map positions are derived from the *E. coli* K-12 linkage map at http://ecocyc.pangeasystems.com or http://genolist.pasteur.fr/Colibri/.

The identification of these genes has led to the discovery of multiple recombinational pathways. Mutations in *recA* reduce recombination frequencies to 0.1% of normal levels commensurate with its central role in many homologous recombination systems. One of these pathways was defined by mutations in *recBC*, which reduce the frequency to 1%. Other pathways were discovered following a search for revertants of these *recBC* mutants. Among the revertants found were two additional genes that can suppress the Rec⁻ phenotype. The suppressors of *recB* and *recC* (*sbc*) include *sbcA* and *sbcB*. The *sbcA*⁺ locus negatively controls the expression of exoVIII, the *recE*⁺ gene product. The *sbcA* mutations increase exoVIII activity, which presumably can substitute for exoV (*recBCD*) in the RecBC pathway. It turns out that both *sbcA*(*recT*) and *recE* are genes present on a lambdoid prophage called *rac* that maps near the replication terminus.

The fact that *recBC* mutants retain some recombination ability suggested there was at least one RecBCD-independent recombinational pathway. Support for this theory came from another suppressor of *recBC* mutations, *sbcB*, which encodes exonuclease I. Mutations in *sbcB* decrease exonuclease I activity. Exo I is proposed to shuttle a DNA recombinational intermediate from an alternative pathway to the RecBCD pathway. However, since *recBC* mutants cannot use the RecBC intermediate, very little recombination would occur because the important intermediate in the alternate RecBCD-independent pathway would be siphoned off by Exo I. Exo I mutants (*sbcB*) would accumulate the alternative intermediate rather than siphon it off, allowing the intermediate to serve as a substrate for the RecBCD-independent system, thereby increasing recombination through this pathway. Actually, the *sbcB* mutants contain a second mutation in the unlinked *sbcC* gene. Apparently both mutations are required to suppress the *recBC* mutant phenotype, although the reason for this is unclear. The RecBCD-independent pathway is now called the RecF pathway, since mutations in *recF* eliminate this alternative recombinational route. The recF pathway also requires the participation of the *recJ*, *N*, *R*, *O*, *Q* and *ruv* gene products.

Mismatch Repair. Following recombination, the initial duplex product may contain unpaired regions resulting from genetic differences. These regions are unstable and susceptible to mismatch repair. The mechanism involves the excision of one or the other mismatched base along with up to 3000 nucleotides (see below). RecFJO is also involved in short patch mismatch repair and thus is involved not only in the initial stages of the recombinational process but also in its ultimate resolution.

Restriction and Modification

Although bacteria have elaborate methods to exchange DNA, they have also developed a means to protect themselves from potentially harmful foreign DNA (e.g., bacteriophage DNA). Restriction and modification of prokaryotic DNA were discovered when bacteriophage λ was grown on one strain of *E. coli* (*E. coli* K-12) and then used to infect a different strain (*E. coli* B). It was noticed that the plating efficiency of the virus was less on the strain B relative to the K strain. Plating efficiency is determined by diluting the phage lysate (λ, K) and adding aliquots of diluted phage to tubes containing soft agar seeded with either *E. coli* K-12 or B. The soft agar is then poured onto a normal nutrient agar plate and incubated. After incubation, the plate will be confluent with growth of *E. coli* except where a phage particle has infected a cell. This area will appear clear and is called a plaque. The plaque is visible to the naked eye because progeny phage from the initial infected cell will progressively infect and lyse adjacent cells.

The unusual observation was that equivalent amounts of λ, K yielded more plaques on *E. coli* K-12 than on *E. coli* B. However, if λ was isolated from the plaques on *E. coli* B and propagated on *E. coli* B, this lysate (λ, B) produced more plaques on *E. coli* B than on *E. coli* K, the exact opposite of the original observation. The underlying mechanism for this phenomenon is the presence of specific endodeoxyribonucleases known as restriction endonucleases in each strain of *E. coli*. Invading bacteriophage DNA originating from the K-12 strain of *E. coli* will undergo cleavage (restriction) by the B endonuclease and then will be degraded rapidly to nucleotides by subsequent exonuclease action. Bacterial host DNA is not degraded because the nucleotide sequence recognized by the restriction endonuclease has been modified by methylation. The few molecules of bacteriophage DNA that survive the initial infection do so because they were modified by the host methylase before the restriction enzyme had a chance to cleave the DNA. The lucky methylated phage is thus protected by methylation.

TABLE 3-2. Characteristics of Types I, II, and III Restriction-Modification Systems

Characteristics	Type I	Type II	Type III	Type IV
Protein structure	Three different subunits	One or two identical subunits	Two different subunits	Two peptides
Endonuclease and methylase activities performed by:	One multifunctional enzyme	Separate methylase and endonuclease enzymes	One multifunctional enzyme	Bifunctional endonuclease/separate methylase
Cofactor requirements for endonuclease	ATP, SAM, Mg^{2+}	Mg^{2+}	ATP	Mg^{2+}
Stimulatory cofactors (not required) for endonuclease	—	—	SAM, Mg^{2+}	SAM
Cofactor requirements for methylase	SAM	SAM	SAM	SAM
Stimulatory cofactors (not required) for methylase	ATP, Mg^{2+}	—	ATP, Mg^{2+}	Ca^{2+}
Cleavage and modification sites	Random, from 1000 bp from recognition site	At or near recognition site	24–26 bp from recognition site	—
Recognition site	EcoK: AACN^{6}GTGC EcoB: TGAN^{8}TGCT	Mostly at sites with dyad symmetry	EcoP1: AGACC EcoP15: CAGCAG HinfIII CGAAT	Eco571 CTGAAG
DNA translocation	Yes	No	No	—

134

λ DNA that has been modified in *E. coli* K-12 carries the K modification and is referred to as λ, K while lambda grown on *E. coli* B carries the B modification and is referred to as λ, B. The genetics of this system is dealt with in more detail below. There are four classes of restriction endonucleases (Table 3-2). The enzymes in the *E. coli* K-12 and B systems belong to class I while most of the other known restriction enzymes belong to class II.

Class I enzymes (e.g., *Eco*B, *Eco*K, and *Eco*PI) are complex multisubunit enzymes that cleave unmodified DNA in the presence of S-adenosylmethionine (SAM), ATP, and Mg^{2+}. These enzymes are also methylases. Each contains two α subunits (mwt 135,000), two β subunits (mwt 60,000), and one γ subunit (mwt 55,000). Prior to binding to DNA, EcoK enzyme binds rapidly to SAM. This initiates an allosteric conformation of the enzyme to an active form that will interact with DNA at random, nonspecific sites. The enzyme then moves to the recognition site with subsequent events depending on the state of the site (Fig. 3-15).

The recognition sequence for both *Eco*K and *Eco*B includes a group of three bases and a group of four bases separated by six (*Eco*K) or eight (*Eco*B) nonspecific bases. Four of the specific bases are conserved (boxes in Fig. 3-15). The adenines with an asterisk are methylated by the methylase activity. If both strands of the recognition site are methylated when encountered, the enzyme does not recognize it. If one strand is methylated, as would be found immediately upon replication, the enzyme binds and methylates the second strand. However, if both strands are unmethylated, the restriction capability is activated. Restriction does not occur directly at the binding site. Rather, the DNA loops past the enzyme, which does not leave the recognition site (DNA translocation), forming supercoils that are cleaved at nonrandom sites approximately 100 bp from the recognition site.

Elegant genetic experiments have demonstrated the existence of three genes whose products are required for these systems. These include *hsdM* (β subunit, methylase activity), *hsdR* (α subunit, restriction endonuclease activity), and *hsdS* (γ subunit, site recognition). Whether *Eco*K recognizes the K recognition site appears to be dependent on the *hsdS* gene product. Hybrid proteins consisting of HsdM and HsdR from *Eco*K and HsdS from *Eco*B have B-site specificity. When characterizing the phenotype of strains carrying mutations in one or more of these genes, r$^+$m$^+$ indicates both restriction and modification activities are functioning, r$^-$m$^+$ indicates a restriction-less strain that can still modify its DNA, and r$^-$m$^-$ indicates the loss of both activities. An r$^-$m$^-$ phenotype can be the result of *hsdR hsdM hsdS*$^+$

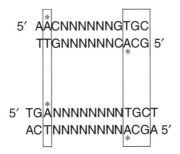

Fig. 3-15. Recognition sequences for the EcoK and EcoB restriction modification enzymes. Asterisks mark sites of methylation.

or *hsdR*⁺ *hsdM*⁺ *hsdS* genotypes, since the *hsdS* product is required for site recognition.

Class II restriction-modification enzymes are less complex and only require Mg²⁺ for activity. These enzymes typically recognize palindromic sequences four to eight base pairs in length, depending on the enzyme. Where the type I enzymes function as a complex including endonuclease and methylase activities, the type II endonucleases and methylases are distinct enzymes. Type II restriction enzymes, some of which are listed in Table 3-3, will only cleave DNA that is unmodified by the cognate type II methylase. If even one of the strands of the sequence is modified, the endonuclease will not recognize it. As opposed to type I enzymes, these endonucleases cleave at the recognition site, forming either blunt ends (e.g., *Hae*III) by cutting both strands in the center of the site or cohesive ends by introducing staggered, single-strand nicks. An example of cohesive ends is as follows:

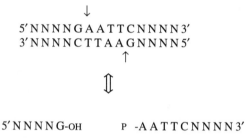

TABLE 3-3. Examples of Class II DNA Restriction and Modification Enzymes

Enzyme	Restriction and Modification Site[a]	Bacterial Strain
*Eco*RI	G↓A*ATTC	*E. coli* RY13
*Eco*RII	↓C*CTGG	*E. coli* R245
*Ava*I	C↓Py°CGRG	*Anabaena variabilis*
*Bam*HI	G↓A*ATTC	*Bacillus amyloliquefaciens* II
*Bgl*II	A↓GATCT	*Bacillus globiggi*
*Hah*I	G°CG↓C	*Haemophilus haemolyticus*
*Hae*III	GG↓CC	*Haemophilus aegyptius*
*Hind*III	*A↓AGCTT	*Haemophilus influenzae* Rd
*Hinf*I	G↓ANTC	*Haemophilus influenzae* Rf
*Hpa*II	C↓*CGG	*Haemophilus parainfluenzae*
*Hga*I	GACGCNNNNN↓	*Haemophilus gallinarum*
*Sma*I	CCC↓GGG	*Serratia marcescens* Sb
Dpn	*AC↓TC	*Diplococcus pneumoniae*

[a]↓Shows the site of cleavage and *shows the site of methylation of the corresponding methylase where known. °shows the site of action of a presumed methylase — that is, methylation at this site blocks restriction, but the methylase has not yet been isolated.

Note that the two cohesive ends can reanneal to each other and, in the presence of DNA ligase, will form an intact molecule. These enzymes are heavily responsible for the current explosion in molecular biology in that they are essential in forming recombinant DNA molecules (cloning). The reason is that DNA from two unrelated organisms will both contain sequences that can be recognized by these enzymes in vitro. Thus, DNA from evolutionarily distinct organisms (e.g., humans and *E. coli*) can be cut with EcoRI, for example, their DNAs mixed and ligated to each other because of the identical cohesive ends. This forms a recombinant molecule. Figure 3-16 illustrates how a recombinant molecule can be formed in vitro. This molecule, because it is a plasmid, can be used to transform *E. coli*.

Class III restriction-modification systems include those from prophage P1, the *E. coli* plasmid P15, and *Haemophilus influenzae* RF. Unlike the type I systems, the type III enzymes do not require SAM for restriction and cleave target DNA to fragments of distinct size, typically 25 to 27 bp away from the recognition sequence.

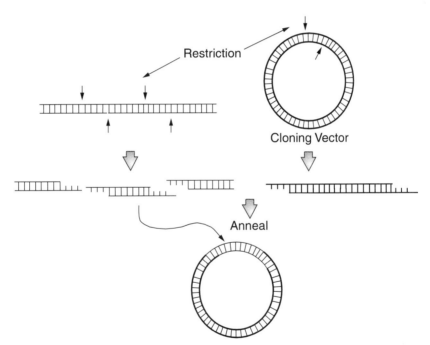

Fig. 3-16. Cloning a fragment of DNA in a plasmid vector. DNA to be cloned (thin lines) and cloning vector (thick lines) are treated with the same type II restriction enzyme to generate molecules with complementary "sticky" end sequences. The cloning vector should only contain a single restriction site for the enzyme being used as is the case in this example. A recombinant molecule is formed by in vitro annealing and ligation with a DNA ligase (usually phage T4 DNA ligase). The recombinant molecules are then transformed into suitable *E. coli* hosts (usually r$^-$ m$^+$ to avoid degrading the foreign DNA insert). An antibiotic resistance gene on the vector will allow selection of cells that received the vector. Clones carrying recombinants may be selected when the inserted DNA complements a mutant defect (e.g., cloning of arg genes), insertionally inactivates a readily selectable marker (e.g., tetracycline resistance), or can be detected using labeled DNA probes.

Class IV restriction-modification systems (e.g., *Eco*571) include a large bifunctional polypeptide with endonuclease and methylase activities and a separate methyltransfersase polypeptide. The large protein methylates only one strand of the asymmetric recognition site, whereas the smaller methyltransferase methylates both strands. Cleavage has an absolute requirement for Mg^{2+} and is stimulated by SAM. A key distinction between type III and type IV endonucleases is that the latter does not depend on ATP.

INSERTION SEQUENCES AND TRANSPOSABLE ELEMENTS

Transposable elements are discrete sequences of DNA that encode functions to catalyze the movement (**translocation**) of the transposable element from one DNA site to a second, *target* site (Fig. 3-17A). The target site is duplicated during the transposition event with a copy found to either side of the transposed element. There are essentially two types of transposition. Replicative transposition involves both replication and recombination with a copy of the element remaining at the original site. Conservative

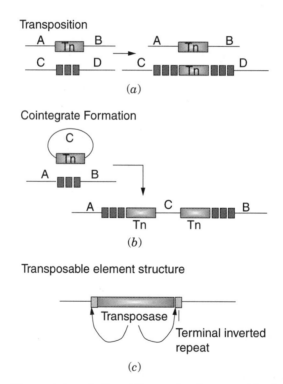

Fig. 3-17. Transposition, cointegrate formation, and a transposable element. (*a*) Transposition. A schematic presentation of replicative transposition in which a transposable element (Tn) translocates to a new site (■ ■ ■) with concomitant duplication of the site. (*b*) Cointegrate formation. Replicon fusion mediated by a transposable element. (*c*) A transposable element. The minimal components of a transposable element include a gene encoding a transposase and terminal target sites for the transposase. These sites are typically the same sequence in inverted orientation.

transposition does not involve replication. The element is simply moved to a new location. When the target site occurs within a gene, either type of transposition will generate insertion mutations. In addition to causing insertion mutations, transposition of these elements can cause deletions, inversions, and cointegrate formation in which two distinct replicons (e.g., plasmids) are joined (replicon fusion). Cointegrate formation is illustrated in Figure 3-17B.

There are essentially three classes of transposable elements in *E. coli*:

1. **Insertion sequences (IS)**, which encode no function other than transposition, and can be simply diagrammed as possessing inverted repeats at either of its ends and a gene encoding the "transposase" responsible for recognizing the terminal repeats and catalyzing the transposition process.

2. **Transposons**, which possess additional genetic information encoding properties such as drug resistance unrelated to the transposition process. There are two subclasses of transposons. **Class I transposons** are composite elements in which two insertion sequence elements flank an intervening sequence encoding antibiotic resistance or some other function (e.g., Tn*10*). **Class II transposons** are complex elements that do not appear to be related to IS elements (e.g., Tn*3*).

3. **Bacteriophage Mu**, a lysogenic bacteriophage that employs transposition as a way of life.

Table 3-4 lists the characteristics of several transposable elements.

Before discussing specific examples of transposable elements, it is useful to discuss a model for transposition. The diagram in Figure 3-18 shows the origin of the short, duplicated regions of DNA that flank the insertion element. The target DNA is cleaved with staggered cuts by the transposase, the transposon is attached to the protruding

TABLE 3-4. Examples of Transposable Elements

Element	Size	Terminal Repeat Element	Target (bp)	Drug Resistance
Insertion Sequences				
IS*1*	768	30 bp inverted	9	None
IS2	1,327	32 bp inverted	5	None
IS*3*	1,400	32 bp inverted	3–4	None
Transposons				
Tn*1*	4,957	38 bp inverted	5	Ap^R
Tn5	5,400	1,450 bp (IS50) inverted	9	Km^R
Tn9	2,638	768 bp (IS1 direct)	9	Cm^R
Tn*10*	9,300	1,400 (IS10) inverted	9	Tet^R
Tn*3*	4,957	38 bp inverted	5	Ap^R
Bacteriophages				
Mu	38,000	11 bp inverted	5	None

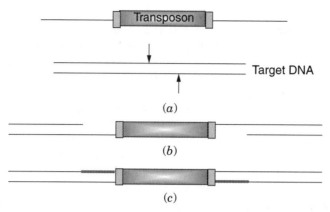

(a)

(b)

(c)

Fig. 3-18. Origin of the short, flanking duplications of target DNA. The target is cleaved with staggered cuts and the extended single strands of the target are then joined to the transposon termini. Repair of the single-stranded gaps completes the duplication.

single-stranded ends, and the short ends are filled in by repair synthesis, leading to the flanking duplications.

A model that explains the transposition process is presented in Figure 3-19. A **transpososome** complex forms between the transposase, the paired ends of the transposon, and the DNA possessing the intended target site (A). The transposase hydrolytically nicks one strand at each end of the transposon (B). In replicative transposition, the 3'-OH ends of the nicked element attack the target DNA, resulting in a transesterification reaction covalently joining the transposon ends to the target DNA (C). Attachment of both ends of the transposon to the target immediately forms two replication forks (D). The transposon is completely replicated, with final sealing of the replicated DNA to flanking sequences generating a cointegrate (E). Resolution of the cointegrate by genetic exchange between the two transposon copies results in a simple insertion and regeneration of the donor replicon (F). While this model appears adequate for Tn*3*-like transposons, it does not completely hold for IS elements or Mu.

Transposons such as Tn*10* or Tn*5* generate simple insertions by a **nonreplicative** cut-and-paste mechanism that does not involve cointegrate formation. The model can be seen in Figure 3-19G, H, and I. Nonreplicative transposition involves, at each end of the transposon, the 3'-OH end of the nicked strand attacking the opposite unnicked strand (G). The resulting transesterification reaction yields a hairpin at the transposon termini and a double-strand cleaved flanking DNA (H). The hairpins are then nicked to regenerate the 3'-OH ends, which then carry out a second transesterification reaction with DNA strands at the target site (I). The result is movement ("jumping") of the transposon from one site to another without making a copy of the transposon.

Transposon Tn*10*

While there are several examples of transposons that could be discussed in greater detail, the one most clearly understood is Tn*10*. Tn*10* is actually a composite element in which two IS-like sequences cooperate to mediate the transposition of the entire element. The basic structure of Tn*10* can be seen in Figure 3-20. There are two IS*10* elements that flank a central region containing a tetracycline-resistant locus. The two

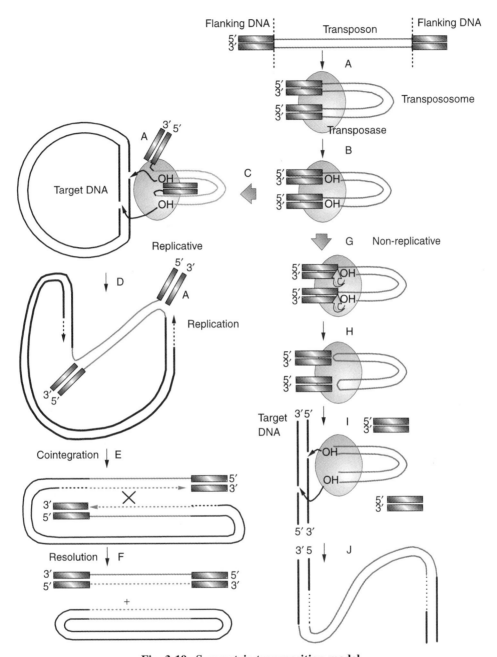

Fig. 3-19. Symmetric transposition model.

IS*10* elements are not identical, however. It is believed that both started out the same, but IS*10*-left has evolved to the point where its ability to mediate transposition is very low relative to IS*10*-right.

IS*10*-right produces a trans-acting function required for transposition (the transposase). There is an inward promoter called p-IN that is required for transcription of the

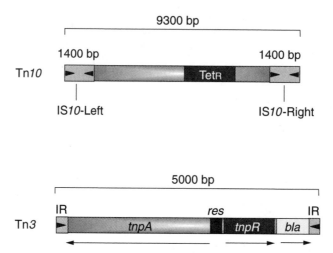

Fig. 3-20. Basic structure of transposons Tn*10* and Tn*3*. IR represents the inverted repeats located at either end; *tnpR*, resolvase locus; *bla*, ß-lactamase locus; *res*, site-specific resolution site; transcripts are represented by arrows.

transposase, but there is also an outward promoter (p-OUT) that produces an antisense RNA that regulates transposition. This 69 nucleotide RNA binds to the ribosome-binding site of the transposase message, preventing the overproduction of transposase. The p-OUT promoter can also be used to activate genes adjacent to where a Tn*10* has inserted. The IS*10*-right transposase acts on sites located within the outermost 70 bp of both IS*10* elements. This was determined by showing that deletion of this region prevented transposition even when the transposase function was supplied in trans.

The activity of IS*10* is regulated by *dam* methylation (see "Initiation of DNA Replication" in Chapter 2). IS*10* has two GATC methylation sites: one overlapping the pIN −10 region and the other located at the inner end of IS*10* where transposase might bind. Methylation of the *dam* site within pIN reduces promoter efficiency of the transposase gene. The reduction in transposase will lower Tn*10* transposition frequency. Methylation of the second *dam* site will lower independent transposition of the constituent IS*10* elements, thereby increasing the cohesiveness of the composite element. Consequently, the element can only transpose after replication by the host machinery and before the newly synthesized strand is methylated. The transposition process for Tn*10* is nonreplicative.

Although Tn*10* can insert at many different sites in both *E. coli* and *S. enterica* chromosomes, the sites of insertion are not totally random. There is approximately one Tn*10* insertion "hot spot" per thousand base pairs of DNA. The specificity sequence is a 6 bp consensus sequence (GCTNAGC). This consensus specificity is located within the 9 bp target site, which is cleaved by staggered nicks during Tn*10* insertion. As originally illustrated in Figure 3-18, the 9 bp target site is duplicated during the insertion process.

As noted earlier, in addition to promoting the movement of a transposon from one site to another, transpositional recombination can result in a variety of genetic rearrangements such as inversions or deletions. Figure 3-21 illustrates how Tn*10* can generate either deletions (A) or inversions (B). Note in both cases that the central

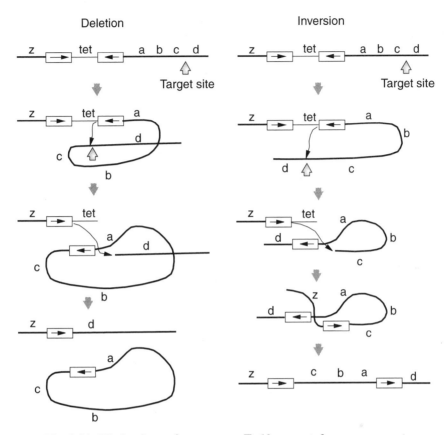

Fig. 3-21. Mechanisms of transposon Tn*10*-promoted rearrangements.

portion of Tn*10* containing the Tet[R] determinant is lost after the rearrangement. In both situations, the inner aspects of the IS*10* elements are utilized rather than the outermost sequences, as is the case for transposition. The only difference for whether an inversion or a deletion occurs is the orientation of the target site relative to the transposon.

Transposon Tn*3*

Tn*3* is an example of a complex transposon. It does not have a modular structure like Tn*10*, does not appear to be based on IS elements, and has no obvious evolutionary link to them. But even though Tn*10* and Tn*3* move by different mechanisms, they both go through the same intermediates as illustrated in Figure 3-19. Figure 3-20 illustrates the basic structure of Tn*3*. The *tnpA* locus codes for the transposase, which works at the inverted repeats located at either end of the element. TnpR is the repressor for *tnpA* but also acts as a site-specific resolvase. The replicative model for transposition presented in Figure 3-19 involves cointegrate formation in which the transposon has duplicated, joining the donor and recipient molecules. Resolution of this structure requires a recombinational event between the two transposons (Fig. 3-19E). For Tn*3*, the res site is the point where recombination, carried out by the TnpR resolvase, occurs.

Conjugative Transposition

Conjugative transposons are characterized by their ability to move between bacterial cells by a process that requires cell-to-cell contact. This phenomenon was first observed in plasmid-free strains of the gram-positive organism *Enterococcus hirae* (formerly *E. faecalis*). This organism contains an 18 kb Tet^R element designated Tn916 that mediates mating and self-transfer to other Enterococci. Now known to be more widespread in nature, conjugative transposons can excise from the chromosome, form a circular intermediate, and be transferred to a new cell via conjugation. Thus, conjugative transposons are plasmid-like in that they have a covalently closed circular transfer intermediate and they are transferred by conjugation, but unlike plasmids, the circular intermediate of a conjugative transposon does not appear to replicate. However, the possibility that circular forms might undergo autonomous replication in some hosts cannot be ruled out.

The conjugative transposons are also phage-like in that their excision and integration resembles excision/integration of temperate bacteriophages such as λ (see Chapter 6). An interesting feature of Tn916 is that the excision process involves dissimilar sequences at either end of the transposon. During excision, staggered cuts are made at both ends of the element, leaving six nucleotide single-strand overhangs called coupling sequences. The Tn916 circle is formed by covalently coupling the ends, but since the coupling sequences are not homologous, they do not base pair but form a bubble. The mismatched joint region undergoes some form of repair, since all the joint regions of circular intermediates end up as homoduplexes, with 80% of the joints reflecting the left coupling sequence and 20% reflecting the right coupling sequence.

Little is known about the actual process of cell-to-cell transfer of conjugative transposons and their establishment in the recipient. The precise excision of Tn916 suggests a conservative mechanism of transposition in which replication does not leave a copy of the transposon at the donor site. However, following mating, the recipient chromosome will end up with multiple copies of the transposon, suggesting the possibility of replicative transposition.

Although initially discovered in gram-positive organisms, examples of conjugative transposons are also found in the gram-negative genera *Bacteroides* and *Neisseria*. The transfer of many conjugative plasmids is regulated, but only in a few cases is the signal known. The signal for a *Bacteroides* conjugative plasmid named Tc^R Em^R Dot is tetracycline. Like the bacterial equivalent of an aphrodisiac, exposure to a low level of tetracycline stimulates mating 1000-fold. Curiously, *Bacteroides* conjugative transposons transfer only when the donor and recipient reside on solid surfaces. Mating in liquid culture does not occur. How mating became regulated by tetracycline is an evolutionary mystery. Tet has only been in use as an antibiotic for a short time — not long enough for such a complex system to develop in response. One theory is that *Bacteroides* ancestors encountered tetracycline produced in nature by the actinomycetes. Alternatively, tetracycline may bear a resemblance to the true original inducer, perhaps a plant phenolic compound.

Evolutionary Consideration

Insertion sequences and transposons are thought to have been, and probably continue to be, an important component of evolution, since they can catalyze major rearrangements of chromosomes as well as serve to introduce new genes into a given organism. For example, there is evidence suggesting that the chromosome of *E. coli*, as we know it, is the result of perhaps two major duplications (which IS elements can also initiate). It only needs to be noted that genes with related biochemical functions often reside in positions 90° to 180° apart from each other for this hypothesis to be appreciated. Also, even though *S. enterica* and *E. coli* are closely related organisms, a comparison of their genetic maps reveals several inversions and deletions. Consequently, it would appear that insertion elements may be an important key to understanding evolutionary processes.

Integrons

One of the consequences of the microbial genome sequencing era is the realization that bacteria are promiscuous gene swappers. This is not restricted to passing antibiotic resistance genes around. There are numerous unexplained resemblances between DNAs of evolutionarily distinct species. Genetic elements called integrons are the reason for this high level of gene trafficking. Integrons are elements that encode a site-specific recombination system that recognizes and captures mobile gene cassettes. Integrons include an integrase (Int) and an adjacent recombination site (*attI*).

A glimpse into how this type of gene trafficking occurs comes from studies with *Vibrio cholerae*, the cholera agent. This organism has a versatile integron acquisition system embedded in its own genome that appears capable of capturing genes serendipitously encountered from other organisms. The proposed integron contains an integrase gene and adjacent gene(s) flanked by DNA repeats that are targets for the integrase. The integrase, however, can act on similar DNA repeats that may be found flanking genes in DNA from heterologous organisms. *Vibrio* may encounter this heterologous DNA as a result of conjugation or transformation. The integrase can then "capture" this new DNA by recognizing these DNA repeats and can integrate the heterologous gene at one of *Vibrio*'s own DNA repeats. Although not proven, there is plenty of evidence that many pathogens have acquired new virulence genes using this strategy.

MUTAGENESIS

During growth of an organism, DNA can become damaged by a variety of conditions. Any heritable change in the nucleotide sequence of a gene is called a mutation regardless of whether there is any observable change in the characteristics (**phenotype**) of the organism. We will now discuss the various mechanisms by which mutations are introduced and repaired. You may be surprised to learn that it is not usually a chemical agent that causes heritable mutations but rather the bacterial attempt to repair chemically damaged DNA.

But before delving into the molecular details of this process, the terminology of mutations must be defined. For example, a bacterial strain that contains all of the genetic information required to grow on a minimal salts medium is called wild-type or **prototrophic**, whereas a mutant strain requiring one or more additional nutrients is **auxotrophic**. Mutations themselves come in a variety of different forms. A change in a single base is a **point mutation**. The most common type of point mutations are **transition mutations** that involve changing a purine to a different purine (A ↔ G) or a pyrimidine to a different pyrimidine. A **transversion mutation** occurs when a purine is replaced by a pyrimidine or vice versa.

A point mutation can change a specific codon, resulting in an incorrect amino acid being incorporated into a protein. The result is a **missense mutation**. The codon change could also result in a translational stop codon being inserted into the middle of a gene. Since a mutation of this sort does not code for an amino acid, it is a **nonsense mutation**. A missense mutation allows the formation of a complete polypeptide, whereas a nonsense mutation results in an incomplete (truncated) protein. Nonsense mutations (also referred to as amber, UAG; ochre, UGA; or opal, UAA mutations), if they occur in an operon involving several genes transcribed from a single promoter, may have polar (distal) effects on the expression of genes downstream from the mutation. A mutation of this type is a **polar mutation**. Figure 3-22 compares a polar nonsense mutation with a nonpolar nonsense mutation.

Ordinarily, you would not expect a translational stop codon to interfere with the transcription or translation of downstream members of an operon, since each gene member of the operon has its own ribosome-binding site. The key to understanding polar mutations is that some nonsense mutations will cause premature transcription termination, lowering the amount of downstream message produced and available for translation. Premature transcription termination occurs when the nontranslated mRNA downstream from the nonsense codon possesses a secondary structure (i.e., stem-loop) that mimics transcription termination signals.

If a process causes the removal of a series of bases in a sequence, the result is a **deletion mutation**. Likewise, the addition of extra bases into a sequence is an addition or **insertion mutation**. Both additions and deletions can result in changing the translational reading frame, causing all of the amino acids situated downstream of the mutation to be incorrect. The offending mutation in this event is a **frameshift mutation**.

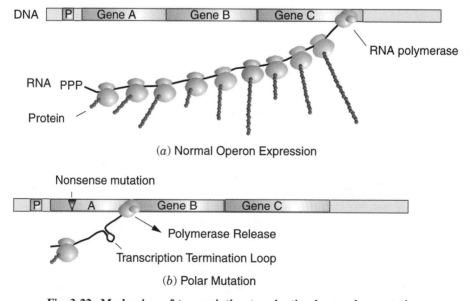

Fig. 3-22. **Mechanism of transcription termination by a polar mutation.**

One type of mutation that has proven extremely useful is the **conditional mutation**, an example of which is the temperature-sensitive (ts) mutant. These mutants grow normally at a low, permissive temperature (30 °C) but exhibit a mutant phenotype at a higher, nonpermissive temperature (42 °C). An amino acid replacement altering the conformational stability of the gene product is usually thought to be the cause. The mutant protein, being less stable, unfolds at the nonpermissive temperature and becomes inactive.

Spontaneous Mutations

In a population of cells, mutations can arise spontaneously without overt treatment with a mutagen. Spontaneous mutations are rare, ranging from 10^{-6} to 10^{-8} per generation depending on the gene and organism. **Mutation rate** is calculated as the number of mutations formed per cell doubling according to the formula $a = m$/cell generations $= m \ln 2/n - n_0$, where $a =$ mutation rate and m is the number of mutations that occur as the number of cells increases from n_0 to n.

There are many causes of spontaneous mutations. Even though the replication apparatus is very accurate with various "proofreading" and repair functions, such as the $3' \rightarrow 5'$ exonuclease activity of DNA polymerase III, mistakes do occur at a very low rate. One reason that misincorporation by DNA polymerase can go undetected involves tautomeric considerations discussed below. Besides misincorporation mistakes, endogenously occurring DNA damage is an important source of spontaneous mutations. For example, cytosine deaminates spontaneously to yield uracil (G : C→A : T). In addition, purines are particularly susceptible to spontaneous loss from DNA by breakage of the glycosidic bond between the base and sugar. The result is the formation of an **apurinic site** in the DNA.

Metabolic activities can yield DNA-damaging intermediates (e.g., H_2O_2, O_2^-, $^\bullet OH$), and endogenous methylation agents (e.g., S-adenosylmethionine) can spontaneously methylate DNA bases. The spontaneous methylation of the N7 position of guanine, for example, weakens the glycosidic bond, resulting in spontaneous loss of the base (apurinic site) or opening of the imidazole ring forming a methyl formamide pyrimidine. In addition to mispairing, some of these spontaneous events can lead to major chromosomal rearrangements such as duplications, inversions, and deletions. Mutagens tend to increase the mutation rate by increasing the number of mistakes in a DNA molecule, as well as by inducing repair pathways that themselves introduce mutations.

THE NATURE OF MUTATIONAL EVENTS

Mutations can arise by a number of molecular events depending in part on the nature of the mutagen. Table 3-5 lists various mutagens, the type of mutation(s) that can arise from each, as well as an indication of the molecular event involved. This section discusses some of the commonly employed mutagens and their mechanisms for causing mutations. Ultraviolet (UV) irradiation at 254 nm causes the production of pyrimidine dimers between thymine-thymine (Fig. 3-23) or thymine-cytosine or cytosine-cytosine pairs. Ultraviolet irradiation may also result in distortion of the backbone of the

TABLE 3-5. Mutagen Action

Mutagen	Specificity	Mechanism[a]
Spontaneous	Substitution	Mispairing
	Frameshift	Slipping
	Multisite	Recombination
	All types	Misrepair
	All types	Misrepair
UV radiation		
Base analogs		
5-Bromouracil	A-T ↔ G-C[b]	Mispairing
2-Aminopurine	A-T ↔ G-C[c]	Mispairing
Base modifiers		
Nitrous acid	A-T ↔ G-C	Mispairing
Hydroxylamine	G-C → A-T	Mispairing
Alkylating agents	Mainly transitions[d]	Mispairing
	All types	Misrepair
Intercalators	Frameshift	Slipping
	All types	Misrepair

[a]Mispairing — that is, nonstandard base pairing — may arise spontaneously or through the presence of nucleotide derivatives. Slipping refers to imperfect pairing between complementary strands due to base sequence redundancy. All types of mutations may be induced indirectly by faulty repair mechanisms — that is, misrepair.

[b]G-C → A-T is favored.

[c]A-T → G-C transitions occur 10–20-fold more frequently.

[d]EMS (ethylmethane sulfonate) and MNNG (*N*-methyl-*N*'-nitro-*N*-nitrosoguanidine).

[e]The wide range of lesions induced by ICR compounds may stem from their alkylating side chain.

DNA helix, causing replication errors. Ionizing radiations cause instability in the DNA molecule, resulting in single-strand breaks.

Chemical mutagens may directly modify the purine or pyrimidine bases, causing errors in base pairing. Local distortions in the helix may result in replication or recombination errors. Compounds such as nitrous acid cause the deamination of adenine and cytosine (Fig. 3-23B), and alkylating agents such as ethylethane sulfonate produce base analogs that result in transitions in base pairing in the nucleic acid structure. Acridine dyes may intercalate between the stacked bases (Fig. 3-24), distorting the structure of the DNA and causing frameshift errors during replication. Abnormal base pairing may also result from tautomeric shifts in the chemical structure of the bases as shown in Figure 3-25.

Normally, the amino and keto forms predominate (ca. 85%). Tautomeric shifts occurring during the replication of DNA will increase the number of mutational events. Certain analogs of the purine or pyrimidine bases may be incorporated into DNA in place of the normal base. For example, 5-bromouracil, upon incorporation into DNA, may pair with either adenine or guanine, as shown in Figure 3-26, resulting in base transitions in the DNA. Figure 3-27 illustrates the target sites on bases that yield common modified bases. It should also be recognized that some modified bases are mutagenic (altered base pairing) while others are lethal (stop replication).

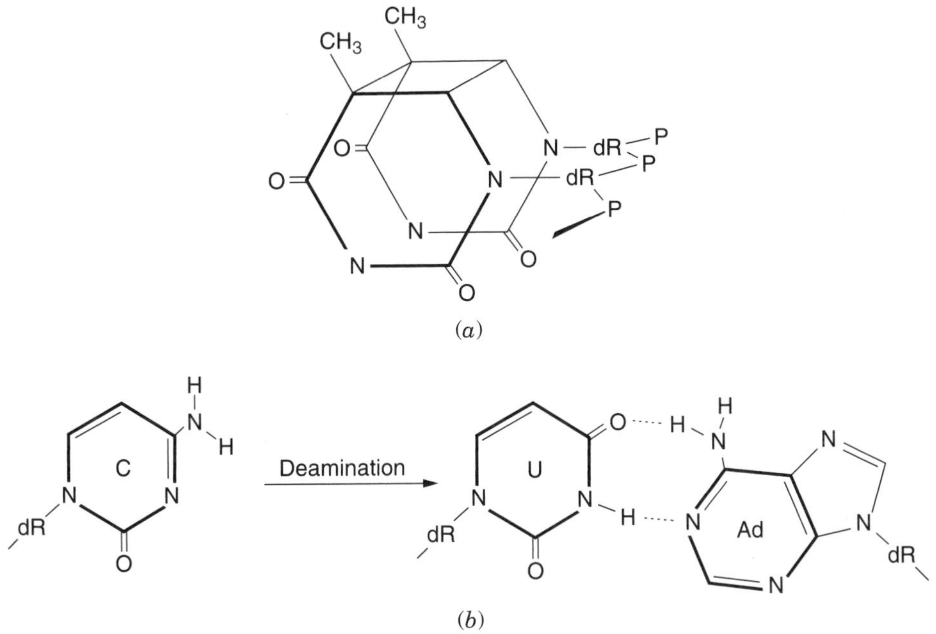

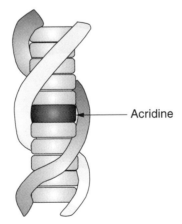

(a)

(b)

Fig. 3-23. (*a*) **Photodimer of thymine found in ultraviolet-irradiated DNA.** Of four possible stereoisomeric photodimers of thymine, only the cis 5,5 : 6,6 isomer shown is obtained by ultraviolet irradiation of DNA. Thymine-cytosine and cytosine-cytosine dimers are also produced. (*b*) **Altered base-pairing as a result of deamination by nitrous acid.**

— Acridine

Fig. 3-24. Diagrammatic view of the manner in which acridine orange and ethidium bromide intercalate between the stacked base pairs in the double-stranded DNA molecule.

Suppressor Mutations

Suppression is the reversal of a mutant phenotype as a result of another, secondary mutation. The second mutation may occur in the same gene as the original mutation (**intragenic suppressor**) or in a different gene (**extragenic suppressor**). Intragenic

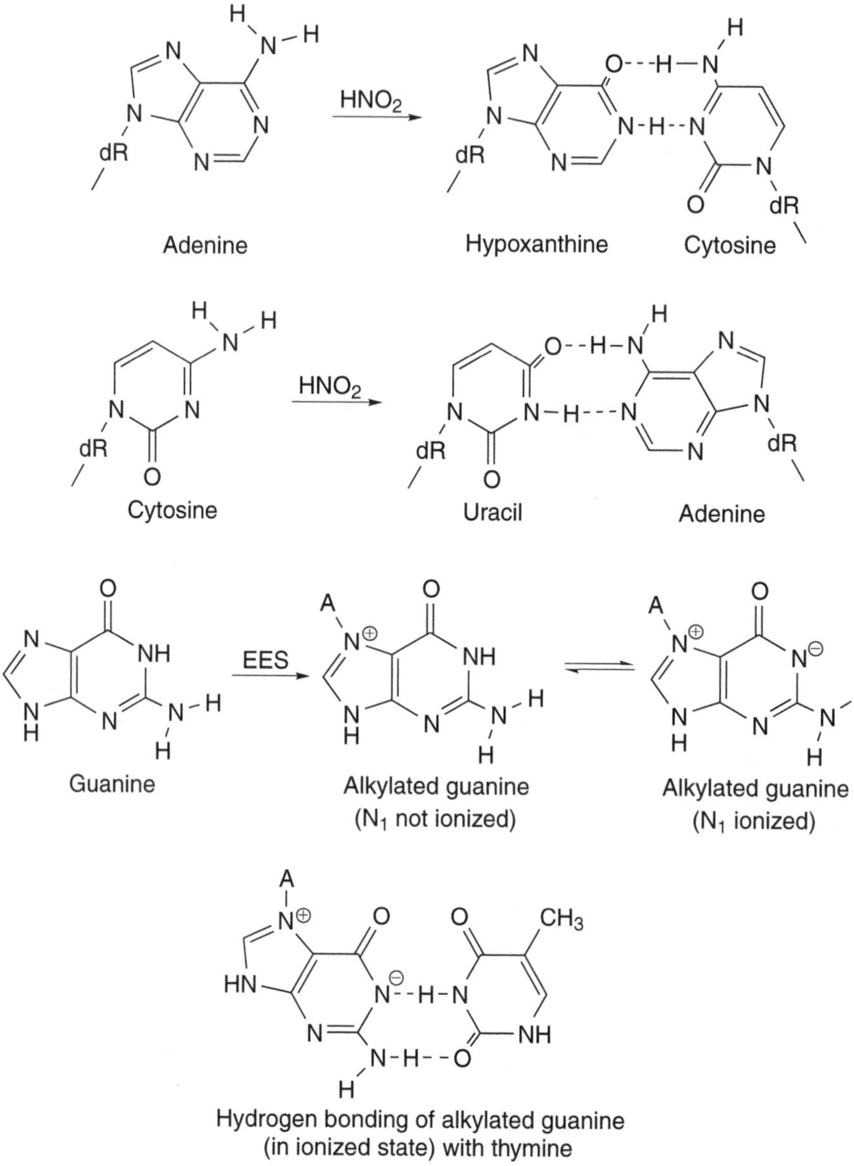

Fig. 3-25. Normal (A=T; G≡C) and abnormal (A=C; G≡T) base pairing as a result of tautomeric shifting.

suppressors may cause an amino acid substitution that compensates for the primary missense mutation, thereby partially restoring the lost function, or they may be the result of an insertion or deletion that compensates for the original frameshift mutation. Extragenic nonsense suppressors are usually mutations that alter the genes encoding tRNAs. The revertant phenotypes may be found to contain tRNAs that recognize nonsense codons (e.g., UAA and UAG) and result in the insertion of a particular amino acid at the site of the nonsense mutation.

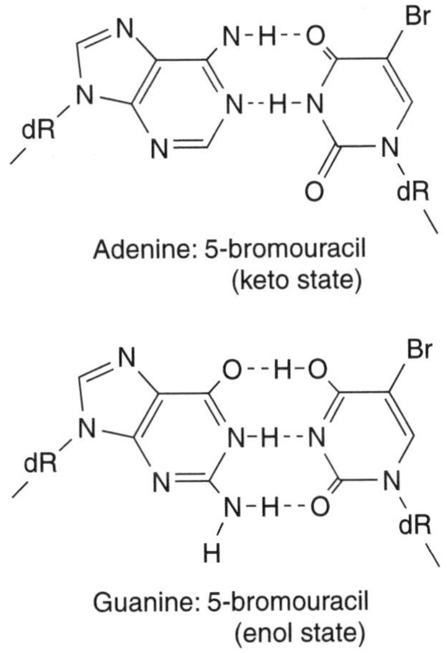

Adenine: 5-bromouracil
(keto state)

Guanine: 5-bromouracil
(enol state)

Fig. 3-26. Base pairing of 5-bromouracil (keto state) with adenine and 5-bromouracil (enol state) with guanine.

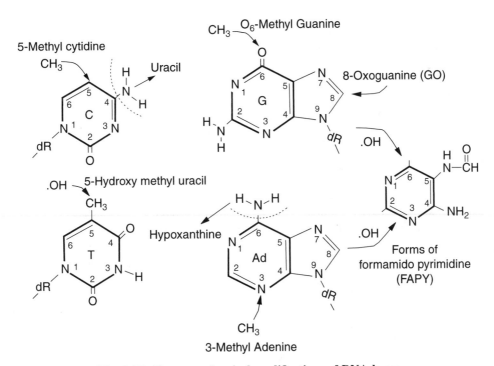

Fig. 3-27. Common chemical modifications of DNA bases.

Revertants that are due to suppressor mutations are usually less efficient than the wild type. For example, in the reversion of an auxotroph to wild type as a result of a tRNA suppressor mutation (*sup*), the revertants usually grow more slowly and produce smaller colonies on minimal agar than the wild type. The direct demonstration of a suppressor mutation can be accomplished by genetic crosses, since a suppressor mutation does not result in any alteration of the original mutation. The original mutant type can be recovered among the progeny of a cross between wild type and the revertant. Also, by means of appropriate crosses, suppressor mutations can be introduced into other mutants in order to assess their effect on other alleles.

DNA REPAIR SYSTEMS

Preventing an unacceptably high mutation rate is of extreme importance to the cell. Consequently, a variety of mechanisms have evolved to repair misincorporated residues or bases altered by exposure to radiation or chemical mutagens. Repair pathways can be viewed as either **prereplicative** or **postreplicative** and as **error-proof** or **error-prone**. The error-prone pathways are often responsible for producing heritable mutations. Table 3-6 lists the genes associated with various aspects of DNA repair in *E. coli*.

Photoreactivation

The cyclobutane ring structure in pyrimidine dimers produced during UV irradiation can be removed enzymatically by the product of the *phr* locus. The *E. coli phr*$^+$ gene product (photolyase) will bind to pyrimidine dimers in the dark. However, the reaction that monomerizes the dimer requires activation by visible light (340–400 nm); hence the term *photoreactivation*. The absorption responsible for photoreactivation is developed only while the enzyme is bound to UV-damaged DNA. The reaction does not require the removal of any bases, just the monomerization of dimers (Fig. 3-23). This activity is, therefore, considered an error-proof repair pathway.

Nucleotide Excision Repair

Bulky lesions such as UV-induced pyrimidine dimers can be excised by a complex exinuclease encoded by the *uvrA*, *uvrB*, and *uvrC* genes of *E. coli*. The UvrABC nuclease recognizes distortions created by cyclobutane rings (Fig. 3-28). Two single-strand incisions are made: one at the eighth phosphodiester bond 5′ to the dimer and the other at the fourth or fifth phosphodiester bond 3′ to the dimer. The net result is the release of a 12–13 nucleotide DNA fragment that contains the site of damage. Subsequently, DNA polymerase I uses the 3′-OH end of the gapped DNA to synthesize a new stretch of DNA containing the correct nucleotide sequence. Finally, DNA ligase seals the remaining single-strand nick.

The *uvrA* gene product is a damage-recognition DNA-binding protein that forms a UvrA$_2$B complex. UvrA initially recognizes DNA lesions, binds, and delivers UvrB to the damaged sites. UvrC protein then binds to the UvrB-bound damaged DNA to induce dual incisions as indicated above. The *uvrD* product (DNA helicase II) is required for release of the UvrABC nuclease. This *uvr*$^+$-dependent excision produces a relatively short patch of repair and is very accurate, since UV treatment is far more

TABLE 3-6. Selected Genetic Loci Associated with Repair

Gene	Map Location	Type of Mutation	Size	Name and/or Function
ada	49.7	$G:C \rightleftharpoons A:T$	39,166	Regulates adaptive response, 0^6–alkylguanine–DNA alkyl transferase
alkA	46.2		31,241	3-methyladenine glycosylase II; hypoxanthine DNA glycosylase; sensitive to methylmethanesulfonate
dcm	43.7	$GC \rightleftharpoons AT$	53,300	DNA cytosine methylase, 5'-CC(A/T)GG target site
dam	74		31,947	DNA adenine methylase (mismatch correction), 5'-GATC target site
endA	66.6		26,560	DNA-specific endonuclease I
fpg (*mutM*)	82		30,139	Formamidopyrimidine DNA glycosylase, 8-hydroxyguanine, 8-oxodeoxyguanine glycosylase
lexA	91.7		21,212	Repressor of SOS mutations stimulate transversions;
mutC (*glyY*)	42.9			tRNA^{gly} inserts at ASP codons; mutant induces unknown error-prone polymerase
mutD (*dnaQ*)	5.1	Base substitutions; frameshifts	26,949	Pol III subunit ε
mutH	63.9	$G:C \rightleftharpoons A:T$ frameshifts	25,380	Increased rate of frameshifts (methyl-directed mismatch repair)
mutL	95	$G:C \rightleftharpoons A:T$ frameshifts	67,750	Increased rate of AT $\leftrightarrow$ GC transitions, methyl-directed mismatch repair
mutS	61.5	$G:C \rightleftharpoons A:T$ frameshifts	95,050	Increased rate of AT $\leftrightarrow$ GC transversions, methyl-directed mismatch repair
mutT	2.4	$A:T \quad G:C$	14,786	Hydrolyzes 8-oxo dGTP; prevents incorporation into DNA
mutY	66.8	$G:C \rightarrow T:A$	38,992	Adenine glycosylase excises A from G-A mispair; mutants stimulate G-C $\rightarrow$ T-A transversions
mutR (*topB*)	39.7	Deletions	73,038	Topoisomerase III, increase in spontaneous deletions

(continued overleaf)

153

TABLE 3-6. (*continued*)

Gene	Map Location	Type of Mutation	Size	Name and/or Function
nei	16.0	G:C ⇌ A:T	29,695	Endo VIII, 8-oxoguanine DNA glycosylase; oxidatively damaged pyr.
nfo	48.4	Base substitutions	31,326	EndoIV; AP-specific (BER)
nth	36.9	G:C ⇌ A:T	23,415	Endo III (BER), oxidatively damaged pyr.
ogt	30.1	G:C ⇌ A:T	19,035	O^6–alkylguanine–DNA transferase
phrB	15.9	CC, TT dimers	53,501	Photolyase
recA	60.8		22,212	Recombination, effector of SOS regulon
recBCD	63.7		Table 3-1	Postreplication repair, Exo V
recF	83.6		40,355	Post replication repair, replication restart
ruvA	41.9		21,939	See Table 3-1
ruvB	41.9		37,017	See Table 3-1
ruvC	41.9		18,603	See Table 3-1
tag	79.9		20,955	3-methyladenine glycosylase I
ung	58.5	G:C ⇌ A:T	25,540	Uracil DNA glycosylase
uvrA	92.0	CC or TT dimers	103,668	
uvrB	17.5	CC or TT dimers	76,044	ATP-dependent endonuclease (NER)
uvrC	42.9	CC or TT dimers	68,013	
uvrD	86.1	G:C ⇌ A:T, frameshifts	81,803	Helicase II (mutU, uvrE, recL)
umuDC	26		14,922/47,516	Error-prone repair; DNA pol V
vsr	43.7		17,872	Endonuclease, cleaves 5′ of the T in T/G mismatches [5′C(T/G)(A/T)GG]
xseA	56.7	Base substitutions	51,669	ExoVII; 5′ → 3′ exo ss specific
xthA	39.5	Base substitutions	30,818	ExoIII, EndoII (AP endonuclease)

BER, involved in base excision repair; NER, involved in nucleotide excision repair.

Source: Data derived from: http://genolist.pasteur.fr/Colibri/.

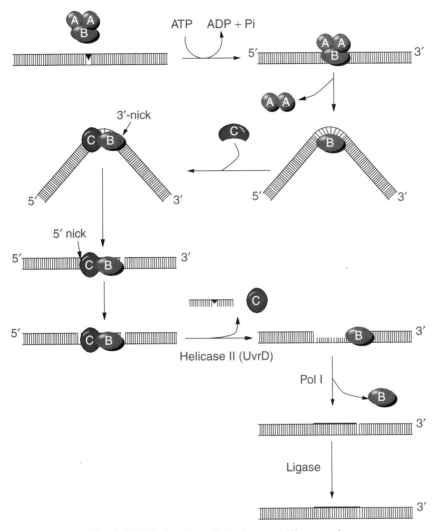

Fig. 3-28. Mechanism of the base excision repair.

mutagenic in *uvr* mutants. Thus, nucleotide excision repair (NER) is often referred to as **short-patch repair** and is error-proof, since a proofreading DNA polymerase (Pol I) fills in the gap. Short patch-type repair systems can also recognize other structural alterations such as missing bases (AP sites, see "DNA Glycosylases and Base Excision Repair").

Transcription-Coupled Repair

An ability by bacteria and mammalian cells to repair actively transcribing genes more rapidly than nontranscribed (silent) genes can be described as a form of targeted NER. A current model to explain this phenomenon is based on the observation that the movement of RNA polymerase is blocked by lesions in the template but not in the coding strand of a gene. The product of the *mfd* (mutation frequency decline) gene, called transcription repair coupling factor (TRCF), binds to the RNA pol–mRNA–DNA complex, causing release of RNA pol and the truncated transcript. TRCF is thought to replace RNA polymerase at the lesion site and attract the UvrABC complex by its affinity for UvrA. TRCF and UvrA then simultaneously disassociate, leaving the preincision UvrB–DNA complex to bind with UvrC, which makes the dual incisions.

Another model for transcription-coupled repair (TCR) is based on the observation that UvrA can associate with the β subunit of RNA polymerase. The resulting complex may act as a lesion sensor in which recognition of a lesion by UvrA could stall RNA polymerase. The stalled complex is proposed to localize to the inner membrane, where TRCF coaxes RNA polymerase off of the lesion site, allowing the Uvr repair proteins to function.

Transcription-coupled repair in *E. coli* is very important for gene expression in cells that have been exposed to DNA-damaging agents. For example, expression of *lacZ* (encoding β-galactosidase; see Chapter 4) during UV irradiation is dramatically enhanced in cells with a functional TCR. Consequently, targeting DNA repair enzymes to transcriptionally active genes is clearly advantageous for cell survival, since it permits the selective repair of genes deemed essential by virtue of their need to be transcribed. The TCR process may be the only means of repairing transcription-blocking damage at active genes.

Methyl-Directed Mismatch Repair

Methyl-directed mismatch repair (long-patch repair) is a postreplicative DNA repair system that recognizes mismatched bases produced in newly synthesized DNA. These misincorporated bases have eluded the $3' \rightarrow 5'$ proofreading function of DNA polymerase (DnaQ subunit). Correction of this sort of mismatch must occur prior to a subsequent round of replication in order to prevent a mutation. The mismatch repair system requires the products of the *mutH*, *mutL*, *mutS*, and *uvrD* (*mutU*) genes. Mutations in these genes cause a tremendous increase in the spontaneous mutation rate (10^4 times higher than wild type) referred to as a **mutator phenotype**. To assure that only the newly misincorporated base is removed, this system must discriminate newly synthesized strands from parental strands. Discrimination is accomplished by recognizing methylation at the N6 position of adenine at GATC sequences, the target sequence for DNA adenine methylase (*dam*). Newly synthesized DNA is hemimethylated (methylated on only one strand) while parental DNA is methylated on both strands. The methyl-directed mismatch repair system recognizes hemimethylated DNA and repairs the mismatch on the unmethylated strand. The system works as represented in Figure 3-29.

The *mutS* gene (so named because mutations in these genes cause a mutator phenotype) encodes a 97 kDa protein that specifically binds to single base-pair mismatches and directs subsequent repair. The resulting conformational change in DNA-bound MutS allows a homodimer of MutL to bind. DNA is then translocated through the MutS–MutL recognition complex to form a loop. This translocation may allow the MutSL complex to interact with the nearest hemimethylated GATC site. The result is association and activation of the latent MutH endonuclease, which incises at unmethylated GATC sites and initiates the excision process. DNA helicase II (UvrD) then unwinds the DNA, exposing ssDNA to attack by exonuclease I (*xon*),

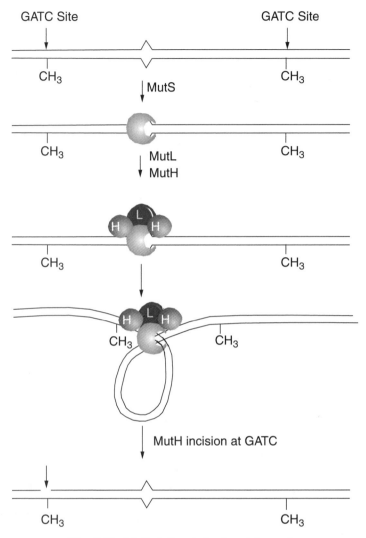

Fig. 3-29. Methyl-directed mismatch repair.

exonuclease VII (*xseA*), or the RecJ exonuclease. A strand break 3′ to the mismatch requires Exo I for repair while either RecJ or Exo VII will suffice for processing strand breaks 5′ to a mismatch. Once excised, DNA polymerase III will fill in the gap. The methyl-directed mismatch repair system is very effective in preventing transition and frameshift mutations but is less efficient in preventing transversions. Thus, methyl-directed mismatch repair is well suited to correct replication errors because the replicative polymerase Pol III produces primarily frameshifts and transitions (75% of all point mutations).

As you might predict, mutants lacking the Dam methylase also have a mutator phenotype — that is, since there is no discrimination between unmethylated parent and daughter strands, either the incorrect or correct base may be excised. If the correct base is repaired, then a mutation occurs.

Very Short-Patch Mismatch Repair

This system in *E. coli* efficiently corrects the T in G : T mismatches that occur within the Dcm methylase recognition sequence CC(A/T)GG. Dcm methylates the internal C's to generate 5-methyl cytosine. Dcm recognition sites are hot spots for mutations because of elevated spontaneous deamination of 5-methyl cytosine over that of cytosine. Deamination of 5-methyl cytosine in a G : 5MeC base pair will lead to a G : T mispair. Very short-patch (VSP) mismatch repair involves MutS and MutL (but not MutH or UvrD) and requires the *vsr* product as well as DNA Pol I. Vsr (18 kDa) is a strand-specific mismatch endonuclease. It recognizes G : T mismatches in 5'CT(A/T)GG and NT(A/T)GG, making incisions 5' of the underlined T. DNA polymerase I removes the T with its $5' \rightarrow 3'$ exonuclease activity and then carries out repair synthesis. MutS and MutL, though not required, stimulate Vsr function, enhancing very short-patch repair.

DNA Glycosylases and Base Excision Repair

DNA glycosylases catalyze the cleavage of sugar-base bonds in DNA and act only on altered or damaged nucleotide residues. The products of these glycosylase reactions are apurinic or apyrimidinic sites (AP sites) whose phosphodiester backbone can be cleaved by AP-specific endonucleases. The AP endonucleases cleave either at the 3' (class I) or the 5' (class II) side of the AP site (Fig. 3-30). Some glycosylases are monofunctional and require a separate endonuclease, whereas others are bifunctional, possessing both glycosylase and AP lyase activities. The bifunctional glycosylases remove the base and cleave the phosphodiester backbone. AP-specific endonucleases in *E. coli* include the following: (1) exonuclease III (*xth*), although initially identified as an exonuclease, it is also a class II endonuclease, in which mutations result in an increased sensitivity to hydrogen peroxide; (2) endonuclease IV (*nfo*), a class II endonuclease with no associated exonuclease activity; and (3) endonuclease V, a class I endonuclease with an unclear role.

There are a variety of DNA glycosylases that recognize a broad range of damaged bases. **Uracil–DNA glycosylase**, the product of the *ung* locus, removes deaminated cytosine residues and is responsible for repairing misincorporated uridine residues during replication. **Hypoxanthine–DNA glycosylase** removes spontaneously deaminated adenine residues. No mutant has been found lacking this enzyme, which suggests a critical role in maintaining cell viability. A major alkylation product in DNA treated with methylating agents such as methyl methane sulfonate or MNNG is 3-methyladenine. This modified base is very lethal because it interferes with DNA replication. The enzyme **3-methyladenine–DNA glycosylase I** (*tag*) removes 3-methyladenine from DNA. There is a second, inducible enzyme, **3-methyladenine–DNA glycosylase II** (*alkA*), produced during the adaptive response to methylating agents (see below). 3-Methyladenine glycosylases can release 3-methylguanine, 7-methylguanine, 7-methyladenine, hypoxanthine, 8-oxoguanine, and 5 hydroxylmethyurea as well as 3-methyladenine from methylated DNA.

A major advance in understanding how 3-methyladenine DNA glycosylases operate was achieved by solving the crystal structure of AlkA. AlkA protein has three domains bordering on an adjustable cleft that contacts DNA substrates. The enzyme is thought to flip the nucleotide out of the DNA helix and into an active site lined with amino acids that interact with the target base. Although the AlkA substrates differ in their

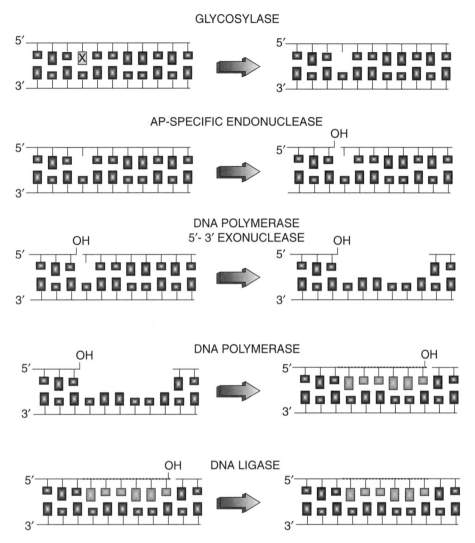

Fig. 3-30. Enzyme activities implicated in DNA repair. These functions are presented in the sequence in which they are thought to act. Resynthesis (DNA polymerase $5' \to 3'$ activity) may or may not take place in conjunction with nucleotide excision (associated $5' \to 3'$ exonuclease activity of DNA polymerase I). DNA ligase would, of course, be required to complete repair.

chemical structures, they share the trait of being electron-poor. The AlkA active site pocket is lined with hydrophobic tryptophan and tyrosine residues with electron-rich conjugated rings that stack against the target bases. These electron-rich amino acids may be the sensors for electron-deficient AlkA substrates.

A common DNA lesion, 7, 8 dihydro-8-oxoguanine (8-oxoG, GO), results from attack by various oxidizing chemical species (see Fig. 3-27). **MutY** is a bifunctional adenine DNA glycosylase that removes adenines misincorporated across from 8-oxoG as well as from AG and AC mismatches. 8-OxoG itself, along with other derivatives of A and G, are removed by another bifunctional enzyme, **FAPY glycosylase** (*fpg*).

Adaptive Response to Methylating and Ethylating Agents

The major mutagenic lesion in cells exposed to alkylating agents such as MNNG is O^6-methylguanine. Such lesions give rise to mutations by mispairing during replication. However, two O^6**–alkylguanine–DNA alkyltransferases**, Ada (O^6-MGT I) and Ogt (O^6-MGT II), specifically remove the methyl group from the O^6 position of guanine. The adaptive response occurs when *E. coli* is exposed to low concentrations of methylating and ethylating agents (e.g., nitrosoguanidine). Following this exposure, the cells become resistant to the mutagenic and lethal effects of higher doses of these agents. The *ada$^+$* gene product, one of the methyl transferases already discussed, is also an AraC-like regulatory protein that transcriptionally controls this adaptive response.

Ada is able to transfer methyl groups from DNA to two of its Cys residues: Cys-69 and Cys-321 (Fig. 3-31). Methyl and/or ethyl groups from either the O^6 position of guanine or the O^4 position of thymine are transferred to Ada cysteine residue 321, a reaction that irreversibly inactivates the methyltransferase activity, making Ada an example of a suicide enzyme. However, Ada has a second methyltransferase activity that removes methyl groups from DNA methylphosphotriesters and transfers them to residue Cys-69. Methylation at Cys-69 results in a conformational change that converts Ada into a strong transcriptional activator. The residue Cys-69 is important in binding Zn^{2+}. Once Cys-69 is methylated, the Zn^{2+} is released. This form of Ada stimulates transcription of the other adaptive response genes including *alkA* (3-methyladenine–DNA glycosylase II), *alkB* (unknown function), *aid* (similarity to acyl coenzyme A dehydrogenases, function unknown), as well as *ada* itself.

An interesting regulatory aspect of this system is that Ada regulates *ada* and *alkA* differently! Ada-Me binds to a sequence called an Ada box located upstream of the *ada* -35 recognition site as well as to a sequence that overlaps the *alkA* -35 region. Ada is classified as a class I transcription factor because it requires interaction with the carboxyl terminal domain of the RNA polymerase α subunit to transcriptionally activate *ada*. Truncated forms of Ada missing 10 to 20% of the C terminus constitutively activate *ada* but require methylation to induce *alkA*. In addition, the amino terminal half of Ada will not activate *ada* expression but is still capable of inducing *alkA* in the presence of methylating agents, indicating that elements in the carboxyl terminal half of Ada are required for activation of *ada* but not *alkA* transcription.

The mechanism by which the adaptive response shuts off appears to reflect a repressor effect of unmethylated Ada at the *ada*, but not the *alkA*, promoters. Once all the repairable methylphosphotriesters have been repaired, the unmethylated Ada that accumulates shuts down *ada* expression (Fig. 3-31).

Postreplication Daughter Strand Gap Repair

Although replication after the introduction of DNA damage often results in a heritable mutation, some DNA damage can be repaired after replication. Figure 3-32 illustrates that repair of DNA damage can be accomplished following replication by two basic routes: an error-proof recombinational route or an error-prone DNA synthesis route. If a pyrimidine dimer, for example, is not repaired prior to replication, a gap occurs opposite the dimer where normal DNA Pol III cannot synthesize DNA because it does not recognize the dimer as a template. The gap can be repaired by recombining the appropriate region from the intact sister chain (Fig. 3-32). This pathway is error-proof but requires the *recA$^+$*, *recBC$^+$*, *polA$^+$*, and *uvr* systems. Following recombination,

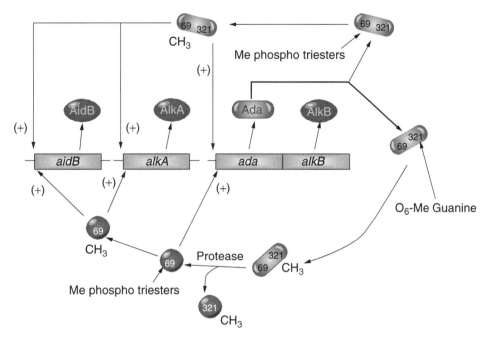

Fig. 3-31. Outline of the adaptive response.

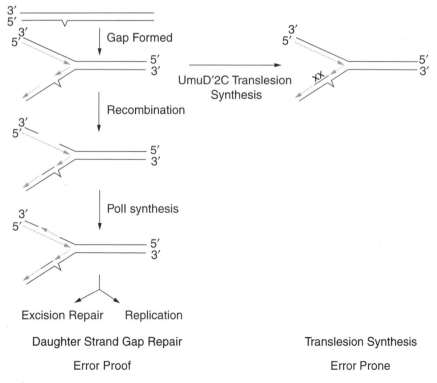

Fig. 3-32. Postreplication repair mechanisms.

the dimer can either be removed by excision repair or can undergo another round of replication. If the mutation still manages to elude error-proof repair, an alternative polymerase produced as a result of SOS induction (see below) can be used to synthesize DNA across the dimer. The bases inserted opposite the dimer are randomly chosen — that is, there is a good chance one or another of the inserted bases will represent a mutation.

SOS-Inducible Repair

The "Save-Our-Ship" Damage Response. Conditions that either damage DNA or interfere with DNA replication result in the increased expression of genes that are members of the SOS regulatory network. This repair system profoundly affects the mutagenic consequences (i.e., whether a heritable mutation is produced) of a number of DNA-damaging treatments. The SOS system is under the coordinate control of *recA* and *lexA*. Aspects of this regulation are outlined in Figure 3-33. In an uninduced cell, the *lexA*$^+$ gene product acts as a repressor for a number of unlinked genes including *recA* and *lexA*. Several of these genes, such as *recA* and *lexA*, are significantly expressed even in the repressed state relative to the other genes under LexA control.

The RecA protein, along with its synaptase activity, possesses a coprotease activity. Cells sense they have experienced DNA damage when RecA forms nucleoprotein filaments with ssDNA produced as a consequence of damage. The coprotease activity of RecA is activated, and, following an interaction with LexA, triggers the cleavage of the LexA repressor (as well as a few other proteins such as λ phage cI repressor; see Chapter 6) at an ala-gly peptide bond, yielding two LexA fragments. RecA actually activates an autocatalytic cleavage of LexA and cI. In any case, as the quantitative levels of LexA decrease in the cell, the expression of the various SOS genes increases. Genes with operators that bind LexA weakly are the first to become induced. During recovery of the cell from the inducing treatment, RecA molecules return to their inactive state, helped by interaction with DNA damage–inducible DinI. A quiescent RecA allows LexA repressor to accumulate and restore repression of the SOS system.

Halting Cell Division. One of the physiological responses that occurs following UV irradiation is filamentation due to a cessation of cell division. The cell continues to grow but fails to divide. A halt in cell division is desirable for DNA-damaged cells in order to allow sufficient time for DNA repair systems to fix damage before replication makes the changes permanent. In addition, replication and cell division of heavily UV-damaged cells is often lethal. Consequently, the SOS response has been engineered to stop cell division by regulating expression of the cell division inhibitor gene *sulA*. The LexA protein normally represses the *sulA* gene, but, following mutagenesis, when LexA is cleaved by RecA, SulA protein is synthesized and binds to the *ftsZ* gene product, inhibiting its activity. The *ftsZ* gene product plays a critical role in defining the cell division plane. However, when SulA protein is produced, cell division does not occur. The result is long, filamentous cells. As the cells recover from irradiation, RecA stops degrading LexA, LexA represses *sulA* expression, and the *lon* gene product, a protease whose general function is to degrade abnormal cellular proteins (see Chapter 2), degrades SulA protein, allowing cell division to resume.

Translesion Synthesis. The major genetic response following SOS induction is the production of mutations. Two systems have been identified that are responsible

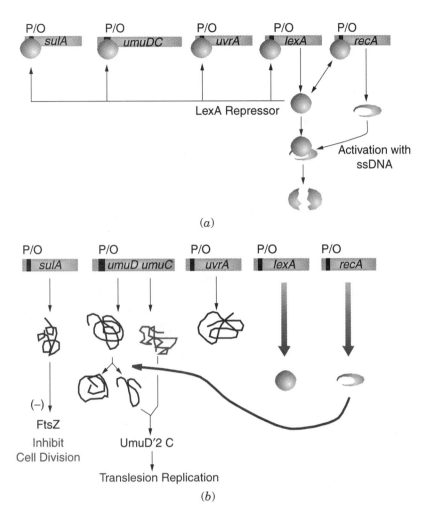

P/O   P/O   P/O   P/O   P/O

sulA   umuDC   uvrA   lexA   recA

LexA Repressor

Activation with
ssDNA

(a)

P/O   P/O   P/O   P/O   P/O

sulA   umuD umuC   uvrA   lexA   recA

(−)
FtsZ

Inhibit
Cell Division

UmuD′2 C

Translesion Replication

(b)

Fig. 3-33. Model of the SOS regulatory system. (*a*) The LexA repressor binds to the operator regions of 20 SOS genes (4 examples are given including *recA*), inhibiting their expression. The coprotease activity of RecA is activated by DNA damage. Activated RecA will trigger autocleavage of LexA. (*b*) The removal of LexA allows the induction of the 20 SOS genes. When DNA damage is repaired, the RecA coprotease is deactivated and LexA can reestablish control. As the cell is recovering from irradiation, the *lon* gene product, a protease whose general function is to degrade abnormal cellular proteins (see Chapter 2), will degrade SulA protein, allowing cell division to resume.

for mutability. The major system requires two genes, *umuD* and *umuC*, which map at 25 minutes and produce 16,000 and 45,000 dalton proteins, respectively. The genes are organized in an operon controlled by the LexA protein with *umuD* located upstream of *umuC*. The UmuD and UmuC proteins are uniquely required for SOS processing and mutagenesis, since mutations in either locus result in a nonmutable phenotype. While other SOS functions are induced in a *umuC* mutant, the mutation rate does not increase. However, elevated levels of UmuCD alone are not sufficient to promote mutagenesis. UmuD must be posttranslationally modified to an active form (UmuD′).

This activation occurs by an autocatalytic cleavage triggered once again by activated RecA and therefore occurs only in the face of DNA damage. The active form is actually a dimer of UmuD′.

The Umu mutagenesis model places a UmuD′$_2$C complex at the site of translesion synthesis (Fig. 3-32). UmuC is a member of a new family of DNA polymerases specialized for lesion bypasss replication by virtue of a relaxed proofreading capability. UmuC is now referred to as Pol V. When a replication fork is blocked at a DNA lesion, Pol V is recruited to the site, becomes activated by UmuD′, and catalyzes translesion DNA synthesis. Translesion synthesis is the nonspecific insertion of a base opposite a lesion in the template that blocks replication by Pol III. A mutation may result immediately across from the lesion, but additional mutations may occur downstream of the lesion as Pol V continues synthesizing DNA. The *dinB* gene encodes another lesion bypass DNA polymerase (Pol IV) regulated by SOS and LexA. The role of Pol IV in *E. coli* mutagenesis is not currently clear. Other genes induced by SOS repair are involved in long-patch excision repair, BER repair, site-specific recombination, and RecF-dependent recombination.

An additional role for the UmuDC proteins in DNA damage tolerance besides catalyzing translesion synthesis has been proposed. Uncleaved UmuD protein and UmuC appear to provide resistance to UV killing by regulating growth after cells have experienced DNA damage in the stationary phase. In starved cells treated with UV, the uncleaved UmuD protein and UmuC delay recovery of DNA replication and cell growth, thereby allowing additional time for accurate repair synthesis (error-proof) to remove or process damage before replication is attempted. RecA-mediated cleavage of UmuD to UmuD′ then acts as a molecular switch, permitting UmuD′$_2$C to carry out translesion synthesis at unrepaired lesions that continue to block replication. The strategy affords the cell ample opportunity to repair damage without resorting to mutagenesis.

To Live or Let Die. It should be stressed that many of the mutations that occur following mutagenic treatments are the result of the error-prone *recA*$^+$-dependent SOS repair systems. Thus, it may seem contradictory that a *recA*$^+$ *uvrA*$^-$ strain with its high mutation rate survives better than a *recA*$^-$ mutant that does not produce mutations. This apparent paradox can be explained by the adage "it is better to live with mutations than to die with dignity." In *recA*$^-$ *uvrA*$^-$ cells, the lack of error-prone repair systems leads to a decreased viability following DNA damage (mutagenic treatments) because of the inability to replicate past gaps in the chromosome — a failure that destroys chromosome integrity. The mutation rate is lower in *recA*$^-$ *uvrA*$^-$ mutants that survive DNA damage because these cells lack the error-prone *rec*A$^+$-dependent SOS system normally required to synthesize past gaps in DNA. Without the SOS system, incorrect bases are not introduced.

Weigle Reactivation and Weigle Mutagenesis. The experiments that first and most clearly indicated the presence of an inducible repair system involved the infection of UV-irradiated cells with UV-irradiated bacteriophage. Preirradiation of *E. coli* with a low dose of UV was found to greatly increase the survival and mutation rate of UV-irradiated λ phage. This indicated a repair system that also produced mutations was induced in *E. coli* following UV irradiation. These findings by Weigle, which are now known to be part of the SOS phenomenon, have been termed Weigle reactivation and Weigle mutagenesis.

Replication Restart

In bacterial cells, replication forks often encounter DNA damage that can inactivate, not just stall, the fork. Replication forks are routinely inactivated under normal growth conditions as well as following environmentally induced DNA damage. In fact, most, if not all, forks originating from *oriC* encounter DNA damage under normal growth conditions. Many of these encounters cause an enzymatic train wreck, leading to replication fork demise. Replication restart involves recombination functions and a primosome distinct from that used at *oriC*.

DNA polymerase II appears to play a pivotal role in rapid replication restart, a process whereby reinitiation of DNA synthesis on UV-damaged DNA allows lesion bypass to occur in an error-free repair pathway. The gene encoding DNA polymerase II (*polB*) is induced as part of the SOS system. One model for how error-free lesion bypass can occur holds that a RecA filament formed downstream of a blocked Pol III holoenzyme triggers a recombination process that allows the blocked strand to associate with the undamaged complementary strand (Fig. 3-34). Pol II then bypasses the lesion by copying the undamaged daughter strand and switches back to the original strand. The PriABC proteins then work to assemble a new primosome that recruits Pol III

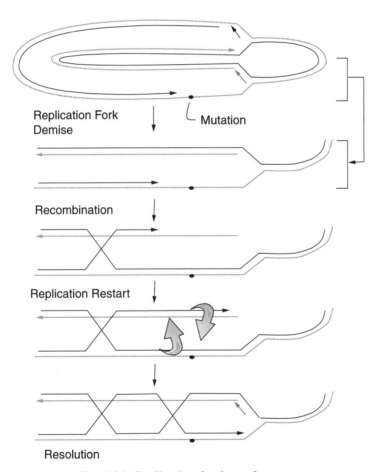

Fig. 3-34. Replication demise and restart.

to take over replication, completing replication restart. Often the result of this process is a concatemer of linked chromosomes requiring resolution by Xer enzymes (see Chapter 2). The repair of stalled replication forks is extremely important for survival and illustrates the intimate relationship between recombination and DNA synthesis. In fact, the most frequent use of bacterial recombination systems is in the rescue of stalled replication forks, not gene exchange.

Adaptive Mutations

In the early 1900s, scientists began to wonder whether mutations occurred spontaneously or were induced or directed when bacteria were exposed to a specific hostile condition. For example, rare phage-resistant variants (mutants) of *E. coli* arise when cells are plated with excess virulent bacteriophage. The question was whether the mutations were caused by the phage or whether the phage simply revealed spontaneous preexisting rare mutants. A classic series of papers published during the 1940s by Luria and Delbruck, as well as Joshua and Esther Lederberg, demonstrated that the phage-resistant mutants were present before exposure to virus. The most elegant proof involved replica-plating *E. coli* to identify which colonies were resistant to the phage. The sibling colony, present on the original master plate, had never been exposed to phage yet also proved to be resistant. These experiments seemed to settle the issue: mutations occurred spontaneously.

However, in 1988, the issue was reopened in a provocative article by John Cairns. Cairns and colleagues argued the flaw in the earlier work was that a lethal selection was used. Consequently, only preexisting mutants could have been found. The early work clearly demonstrated the occurrence of mutations in the absence of selection but did not eliminate the possibility that some mutations might be caused by selection. To support his argument, Cairns examined the ability of a lactose-negative (Lac$^-$) *E. coli* to revert through mutation to Lac$^+$. Lac$^+$-revertant colonies will form when Lac$^-$ cells are plated on minimal lactose medium. Each day after plating, the number of Lac$^+$ colonies that appeared increased. The onset of these mutants could be delayed if they were first plated on medium without lactose and provided with lactose several days later! It seemed many of these late-appearing Lac$^+$ revertants were being caused (directed) by the presence of lactose!

Additional work from other researchers confirmed this phenomenon with genes other than *lac*. Several models have emerged that seek to explain these results. A popular model proposes that at any given instant during prolonged starvation some fraction of cells in a population might enter into a "hypermutable" state, while the remaining cells remain more-or-less immutable. While in a hypermutable state, mutations can occur randomly in any gene. A cell would exit the hypermutable state and form a colony if any one of those mutations solved the problem—for example, allowed it to grow on lactose. However, if none of the mutations solved the problem, then the cell would die from the accumulation of lethal mutations.

In the Cairns experiments, cells hypermutable before the addition of lactose might still generate *lac$^+$* mutations, but because there is no lactose present they will remain stressed and hypermutable, generating secondary lethal mutations. Thus, they will not form colonies once lactose is added. Only cells that have not yet entered the hypermutable state will still be viable when lactose is introduced and thus have potential for rescuing themselves. This model offers a mechanism with an underlying random basis that does not invoke true directed mutations. In part, this is why these types of mutations are now called adaptive rather than directed.

In some circumstances, we should realize that the antimutagenic (error-proof) repair mechanisms are not advantageous. Environmental stress responses are often characterized by the potential for genetic diversification in cell populations by mutation. The hypermutable state would be an example. An alternative model is that growth of cells on carbohydrates like glucose limits expression of error-prone repair pathways via catabolite repression. Thus, growth on carbohydrates that do not repress the error-prone systems will increase mutation rate and the chance that the defect will be corrected. Whatever the true mechanism, we may no longer be able to view mutations and selections as entirely separate processes.

BIBLIOGRAPHY

Plasmids

Abeles, A. L., L. D. Reaves, B. Youngren-Grimes, and S. J. Austin. 1995. Control of P1 plasmid replication by iterons. *Mol. Microbiol.* **18**:903–12.

Casjens, S., N. Palmer, R. van Vugt, W. M. Huang, B. Stevenson, P. Rosa, R. Lathigra, G. Sutton, J. Peterson, R. J. Dodson, D. Haft, E. Hickey, M. Gwinn, O. White, and C. M. Fraser. 2000. A bacterial genome in flux: the twelve linear and nine circular extrachromosomal DNAs in an infectious isolate of the Lyme disease spirochete *Borrelia burgdorferi*. *Mol. Microbiol.* **35**:490–516.

Chattoraj, D. K. 2000. Control of plasmid DNA replication by iterons: no longer paradoxical. *Mol. Microbiol.* **37**:467–76.

del Solar, G., R. Giraldo, M. J. Ruiz-Echevarria, M. Espinosa, and R. Diaz-Orejas. 1998. Replication and control of circular bacterial plasmids. *Microbiol. Mol. Biol. Rev.* **62**:434–64.

Engelberg-Kulka, H., and G. Glaser. 1999. Addiction modules and programmed cell death and antideath in bacterial cultures. *Annu. Rev. Microbiol.* **53**:43–70.

Gerdes, K., J. Moller-Jensen, and R. B. Jensen. 2000. Plasmid and chromosome partitioning: surprises from phylogeny. *Mol. Microbiol.* **37**:455–66.

Hinnebusch, J., and K. Tilly. 1993. Linear plasmids and chromosomes in bacteria. *Mol. Microbiol.* **10**:917–22.

Volff, J. N., and J. Altenbuchner. 2000. A new beginning with new ends: linearisation of circular chromosomes during bacterial evolution. *FEMS Microbiol. Lett.* **186**:143–50.

Wagner, E. G., and R. W. Simons. 1994. Antisense RNA control in bacteria, phages, and plasmids. *Annu. Rev. Microbiol.* **48**:713–42.

Transformation

Avery, O. T., C. M. MacLeod, and M. McCarty. 1944. Studies on the chemical nature of the substance inducing transformation of pneumococcal types. Induction of transformation by a deoxyribonucleic acid fraction isolated from penumococcus type III. *J. Exp. Med.* **79**:137.

Buttaro, B. A., M. H. Antiporta, and G. M. Dunny. 2000. Cell-associated pheromone peptide (cCF10) production and pheromone inhibition in *Enterococcus faecalis*. *J. Bacteriol.* **182**:4926–33.

Claverys, J. P., and S. A. Lacks. 1986. Heteroduplex deoxyribonucleic acid base mismatch repair in bacteria. *Microbiol. Rev.* **50**:133–65.

Dubnau, D. 1991. The regulation of genetic competence in *Bacillus subtilis*. *Mol. Microbiol.* **5:**11–18.

Dubnau, D. 1999. DNA uptake in bacteria. *Annu. Rev. Microbiol.* **53:**217–44.

Dubnau, D., and R. Provvedi. 2000. Internalizing DNA. *Res. Microbiol.* **151:**475–80.

Dubnau, D., and K. Turgay. 2000. Regulation of competence in *Bacillus subtilis* and its relation to stress response. In G. Storz and R. Hengge-Aronis (eds.), *Bacterial Stress Response*. ASM Press, Washington, D.C., pp. 249–60.

Fussenegger, M., T. Rudel, R. Barten, R. Ryll, and T. F. Meyer. 1997. Transformation competence and type-4 pilus biogenesis in *Neisseria gonorrhoeae* — a review. *Gene* **192:** 125–34.

Griffith, F. 1928. The significance of pneumococcal types. *J. Hygiene* **27:**113.

Humbert, O., M. Prudhomme, R. Hakenbeck, C. G. Dowson, and J. P. Claverys. 1995. Homologous recombination and mismatch repair during transformation in *Streptococcus pneumoniae*: saturation of the Hex mismatch repair system. *Proc. Natl. Acad. Sci. USA* **92:**9052–6.

Kues, U., and U. Stahl. 1989. Replication of plasmids in gram-negative bacteria. *Microbiol. Rev.* **53:**491–516.

Mejean, V., and J. P. Claverys. 1993. DNA processing during entry in transformation of *Streptococcus pneumoniae*. *J. Biol. Chem.* **268:**5594–9.

Conjugation

Anderson, T. F., E. L. Wollman, and F. Jacob. 1957. Sur les processus de conjugaison det de recombinaison chez *Escherichia coli*. *Ann. Inst. Pasteur* **93:**450.

Clewell, D. B. (ed.). 1993. *Bacterial Conjugation*. Plenum Press, New York.

Clewell, D. B. 1993. Bacterial sex pheromone-induced plasmid transfer. *Cell* **73:**9–12.

Dunny, G. M., and B. A. Leonard. 1997. Cell-cell communication in gram-positive bacteria. *Annu. Rev. Microbiol.* **51:**527–64.

Frost, L. S., K. Ippen-Ihler, and R. A. Skurray. 1994. Analysis of the sequence and gene products of the transfer region of the F sex factor. *Microbiol. Rev.* **58:**162–210.

Matson, S. W., J. K. Sampson, and D. R. Byrd. 2000. F plasmid conjugative DNA transfer: the *traI* helicase activity is essential for DNA strand transfer. *J. Biol. Chem.* **27:**27.

Wilkens, B., and E. Lanka. 1993. DNA processing and replication during plasmid transfer between gram-negative bacteria. In D. B. Clewell (ed.), *Bacterial Conjugation*. Plenum, New York, pp. 105–36.

Willetts, N., and B. Wilkins. 1984. Processing of plasmid DNA during bacterial conjugation. *Microbiol. Rev.* **48:**24–41.

Recombination

Amundsen, S. K., A. F. Taylor, and G. R. Smith. 2000. The RecD subunit of the *Escherichia coli* RecBCD enzyme inhibits RecA loading, homologous recombination, and DNA repair. *Proc. Natl. Acad. Sci. USA* **97:**7399–404.

Umezu, K., N. W. Chi, and R. D. Kolodner. 1993. Biochemical interaction of the *Escherichia coli* RecF, RecO, and RecR proteins with RecA protein and single-stranded DNA binding protein. *Proc. Natl. Acad. Sci. USA* **90:**3875–9.

West, S. C. 1992. Enzymes and molecular mechanisms of genetic recombination. *Annu. Rev. Biochem.* **61:**603–40.

West, S. C., and B. Connolly. 1992. Biological roles of the *Escherichia coli* RuvA, RuvB and RuvC proteins revealed. *Mol. Microbiol.* **6:**2755–9.

Restriction Modification

Doronina, V. A., and N. E. Murray. 2001. The proteolytic control of restriction activity in *Escherichia coli* K- 12. *Mol. Microbiol.* **39**:416–29.

Janulaitis, A., M. Petrusyte, Z. Maneliene, S. Klimasauskas, and V. Butkus. 1992. Purification and properties of the *Eco*57I restriction endonuclease and methylase — prototypes of a new class (type IV). *Nucleic Acids Res.* **20**:6043–9.

Murray, N. E. 2000. Type I restriction systems: sophisticated molecular machines (a legacy of Bertani and Weigle). *Microbiol. Mol. Biol. Rev.* **64**:412–34.

Transposition

Clewell, D. B., and C. Gawron-Burke. 1986. Conjugative transposons and the dissemination of antibiotic resistance in streptococci. *Annu. Rev. Microbiol.* **40**:635–59.

Fluit, A. C., and F. J. Schmitz. 1999. Class 1 integrons, gene cassettes, mobility, and epidemiology. *Eur. J. Clin. Microbiol. Infect. Dis.* **18**:761–70.

Grindley, N. D., and R. R. Reed. 1985. Transpositional recombination in prokaryotes. *Annu. Rev. Biochem.* **54**:863–96.

Hall, R. M. 1997. Mobile gene cassettes and integrons: moving antibiotic resistance genes in gram-negative bacteria. *Ciba Found. Symp.* **207**:192–202.

Hall, R. M., and C. M. Collis. 1995. Mobile gene cassettes and integrons: capture and spread of genes by site-specific recombination. *Mol. Microbiol.* **15**:593–600.

Kennedy, A. K., A. Guhathakurta, N. Kleckner, and D. B. Haniford. 1998. Tn*10* transposition via a DNA hairpin intermediate. *Cell* **95**:125–34.

Kleckner, N., R. M. Chalmers, D. Kwon, J. Sakai, and S. Bolland. 1996. Tn10 and IS10 transposition and chromosome rearrangements: mechanism and regulation in vivo and in vitro. *Curr. Top. Microbiol. Immunol.* **204**:49–82.

Lawley, T. D., V. Burland, and D. E. Taylor. 2000. Analysis of the complete nucleotide sequence of the tetracycline-resistance transposon Tn*10*. *Plasmid* **43**:235–9.

Mahillon, J., and M. Chandler. 1998. Insertion sequences. *Microbiol. Mol. Biol. Rev.* **62**:725–74.

Mazel, D., B. Dychinco, V. A. Webb, and J. Davies. 1998. A distinctive class of integron in the *Vibrio cholerae* genome. *Science* **280**:605–8.

Rice, L. B. 1998. Tn*916* family conjugative transposons and dissemination of antimicrobial resistance determinants. *Antimicrob. Agents Chemother.* **42**:1871–7.

Salyers, A. A., N. B. Shoemaker, A. M. Stevens, and L. Y. Li. 1995. Conjugative transposons: an unusual and diverse set of integrated gene transfer elements. *Microbiol. Rev.* **59**:579–90.

Scott, J. R., and G. G. Churchward. 1995. Conjugative transposition. *Annu. Rev. Microbiol.* **49**:367–97.

Mutagenesis

Drake, J. W. 1991. Spontaneous mutation. *Annu. Rev. Genet.* **25**:125–46.

Foster, P. L. 1999. Mechanisms of stationary phase mutation: a decade of adaptive mutation. *Annu. Rev. Genet.* **33**:57–88.

Rosche, W. A., and P. L. Foster. 2000. Mutation under stress: adaptive mutation in *Escherichia coli*. In G. Storz and R. Hengge-Aronis (eds.), *Bacterial Stress Responses*. ASM Press, Washington, D.C., pp. 239–48.

Repair Mechanisms

Cox, M. M. 1991. The RecA protein as a recombinational repair system. *Mol. Microbiol.* **5:**1295–9.

Cox, M. M., M. F. Goodman, K. N. Kreuzer, D. J. Sherratt, S. J. Sandler, and K. J. Marians. 2000. The importance of repairing stalled replication forks. *Nature* **404:**37–41.

Fijalkowska, I. J., R. L. Dunn, and R. M. Schaaper. 1997. Genetic requirements and mutational specificity of the *Escherichia coli* SOS mutator activity. *J. Bacteriol.* **179:**7435–45.

Kowalczykowski, S. C. 2000. Initiation of genetic recombination and recombination-dependent replication. *Trends Biochem. Sci.* **25:**156–65.

Little, J. W. 1993. LexA cleavage and other self-processing reactions. *J. Bacteriol.* **175:**4943–50.

Maor-Shoshani, A., N. B. Reuven, G. Tomer, and Z. Livneh. 2000. Highly mutagenic replication by DNA polymerase V (UmuC) provides a mechanistic basis for SOS untargeted mutagenesis. *Proc. Natl. Acad. Sci. USA* **97:**565–70.

Marra, G., and P. Schar. 1999. Recognition of DNA alterations by the mismatch repair system. *Biochem. J.* **338:**1–13.

Murli, S., T. Opperman, B. T. Smith, and G. C. Walker. 2000. A role for the *umuDC* gene products of *Escherichia coli* in increasing resistance to DNA damage in stationary phase by inhibiting the transition to exponential growth. *J. Bacteriol.* **182:**1127–35.

Pham, P. T., M. W. Olson, C. S. McHenry, and R. M. Schaaper. 1998. The base substitution and frameshift fidelity of *Escherichia coli* DNA polymerase III holoenzyme *in vitro*. *J. Biol. Chem.* **273:**23,575–84.

Rangarajan, S., R. Woodgate, and M. F. Goodman. 1999. A phenotype for enigmatic DNA polymerase II: a pivotal role for pol II in replication restart in UV-irradiated *Escherichia coli*. *Proc. Natl. Acad. Sci. USA* **96:**9224–9.

Reuven, N. B., G. Arad, A. Maor-Shoshani, and Z. Livneh. 1999. The mutagenesis protein UmuC is a DNA polymerase activated by UmuD', RecA, and SSB and is specialized for translesion replication. *J. Biol. Chem.* **274:**31,763–6.

Selby, C. P., and A. Sancar. 1994. Mechanisms of transcription-repair coupling and mutation frequency decline. *Microbiol. Rev.* **58:**317–29.

Tang, M., P. Pham, X. Shen, J. S. Taylor, M. O'Donnell, R. Woodgate, and M. F. Goodman. 2000. Roles of *E. coli* DNA polymerases IV and V in lesion-targeted and untargeted SOS mutagenesis. *Nature* **404:**1014–8.

Witkin, E. M. 1994. Mutation frequency decline revisited. *Bioessays* **16:**437–44.

Woodgate, R., and S. G. Sedgwick. 1992. Mutagenesis induced by bacterial UmuDC proteins and their plasmid homologues. *Mol. Microbiol.* **6:**2213–8.

Wyatt, M. D., J. M. Allan, A. Y. Lau, T. E. Ellenberger, and L. D. Samson. 1999. 3-methyladenine DNA glycosylases: structure, function, and biological importance. *Bioessays* **21:**668–76.

CHAPTER 4

MICROBIAL PHYSIOLOGY IN THE GENOMIC ERA: A REVOLUTIONARY TALE

The field of microbial physiology is undergoing a dramatic revolution. This reformation is the direct result of three technological achievements: the personal computer, the Internet, and rapid DNA sequencing techniques. In the past, examining the physiology of a microorganism was a long, painstaking process. Even now, 60 years after initiating the analysis of the common intestinal microorganism *Escherichia coli*, we are still far from having a completely, integrated picture of its biochemistry and genetics.

Today, however, many organisms that have proven almost intractable to scientific inquiry have had much of their genetics and physiology laid bare. Examples include *Rickettsia prowazekii* and *Mycobacterium leprae*, neither of which can grow outside of a living host. We now know with some certainty what biochemical pathways they have, how they make energy, what possible virulence proteins are in their pathogenic arsenal, and how they relate evolutionarily to other bacteria! All this has come about by modern sequencing techniques that have allowed entire genomes to be decoded in relatively short periods of time and because powerful sequence analysis software has been created that identifies open reading frames [an open reading frame (ORF) is a DNA sequence predicted to encode a protein] and promoter sequences, and compares new sequences with those already known. Comparing the amino acid sequence of an ORF of unknown function with all other known protein sequences (called a homology search) identifies those proteins with sequences and motifs similar to what is present in the ORF. This analysis provides considerable insight as to the possible function of the predicted ORF in the cell without ever having conducted a single, true biochemical experiment! The personal computer and Internet have fueled this renaissance of microbial physiology by allowing unfettered sharing of this information among scientists around the world.

This chapter briefly describes the basic tools molecular biologists use to examine DNA sequences, gene expression, DNA–protein interactions, and protein–protein interactions. Initially, however, we discuss how the sequence of a complete genome is obtained and the means by which that knowledge can be used to study microbial physiology and evolution.

GENOMIC AND PROTEOMIC TOOLS

Cloning a Genome

Every quest to sequence a new genome begins by deconstructing the bacterial chromosome into small (<2 kb) overlapping fragments. This is accomplished through techniques first described in Chapter 3. First, a plasmid or phage vector is chosen and cut with a restriction enzyme. Chromosomal DNA from the subject organism is extracted, digested with the same restriction enzyme, and cloned into the cut vector. Following ligation with DNA ligase, the clones are transformed into *E. coli* (a restrictionless strain). This is known as "shotgun" cloning as opposed to a directed cloning strategy that requires the careful identification of individual DNA fragments encoding a specific gene of interest. The result of shotgun cloning is a random clone library that can be quickly sequenced (some microbial genomes only take days to completely sequence using automated methods).

DNA Sequencing

Once a random clone library is constructed, DNA sequencing efforts can begin. As illustrated in Figure 4-1, a single strand from each clone is used as a template to synthesize a series of fragments, each of which is only one base longer than the previous fragment. The trick is to label the last base synthesized on each fragment (the 3′ end) with a fluorescent tag and to do it in such a way that a different colored tag is used for each of the four nucleotides.

The first step is to use a short oligonucleotide primer (about 20–25 bases) that binds to a specific sequence in the vector molecule. Remember that DNA-dependent DNA

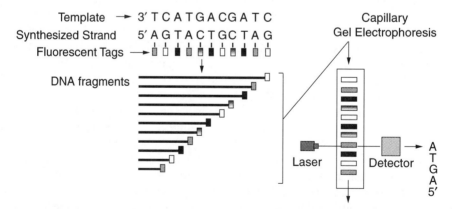

Fig. 4-1. DNA sequencing. A DNA template is sequenced using a DNA oligonucleotide as a primer, DNA polymerase, deoxynucleotides, and a low percentage of dideoxynucleotides. Each dideoxynucleotide is tagged with a different fluorescent tag (differently shaded boxes in figure). Each newly synthesized DNA fragment has only one dideoxynucleotide, and that molecule is incorporated at the 3′-terminal base. As a result, each fragment is labeled with a fluorescent dye that signals the identity of the last base. To read the sequence, the mixture of fragments undergoes electrophoresis through a capillary tube with polyacrylamide that separates the fragments by size. A laser detector positioned near the bottom of the gel reads the fluorescent dye tag of each fragment as it passes.

polymerases only synthesize DNA by adding a new nucleotide to a preexisiting 3′ hydroxyl group of a growing strand. The primer molecule allows DNA sequencing to begin at a predetermined, fixed point in each clone. Generating different-sized fragments while specifically labeling the last nucleotide incorporated into each fragment is accomplished by including small amounts of fluorescently tagged, dideoxy versions of each nucleotide to the DNA synthesis reaction mix, which also contains normal deoxynucleotides. The key to the method is that dideoxy nucleotides lack both the 2′ and 3′ hydroxyl groups. So, if DNA polymerase chooses to add a dideoxy nucleotide to an elongating chain, the chain cannot extend any further because there is no 3′ hydroxyl primer. Because each dideoxy base is tagged differently, the last base in that fragment, and, thus, the fragment itself, is specifically labeled.

To read the sequence, the mixture of different-sized and labeled DNA fragments are separated by size via polyacrylamide gel electrophoresis, usually conducted in a capillary tube. The overall negative charge on DNA causes all fragments to travel toward the positive pole, and the sieving action of the polyacrylamide matrix allows smaller DNA fragments to travel faster than larger fragments. The result is a parade of fragments progressively moving down the gel, with each fragment being exactly one base longer than the one before it. An optical laser positioned toward the end of the gel will excite the fluorescent tag at the end of each fragment, and the color emitted is read as a specific base.

Because shotgun cloning makes fragments with overlapping ends, computer analysis of the many DNA sequences will identify and assemble the overlapping fragments into **contigs** (a continuous, uninterrupted DNA sequence). With some finishing gap filling, the contigs are assembled into a complete genome ready for annotation. **Annotation** is the process whereby potential genes, known as open reading frames (ORFs), are identified by looking for telltale sequence signatures such as translational start codons in proximity to potential ribosome-binding sites. Then the ORF sequence and the predicted protein product are compared to the existing database to look for similarities with known genes and proteins. Based on what is found, the ORF may be annotated as having a potential function.

The Institute for Genomic Research (TIGR) published the first complete microbial genomic sequence, that of *Haemophilus influenzae*, in 1995. Since then, over 50 genomes have been sequenced and many more are in progress (http://www.tigr.org/; click on the TIGR comprehensive microbial resource).

Web Science: Internet Tools for DNA Sequence Analysis

Once you get the DNA sequence for a gene you are studying, what do you do with it? The first stop these days is the World Wide Web, where you can gain amazing insight into the possible function of your gene, even if it has never been sequenced before. How is this level of prognostication possible? It all boils down to hard work and evolution. Before complete genome sequencing efforts became formalized, scientists had already identified the sequences of a large number of genes as well as the functions associated with them. These sequence-function databases have become indispensable in biology because they can be used to assign signature sequences that predict functional motifs. Processes such as gene duplications within an organism, horizontal gene transfers between species, and the mutational divergence of genes are considered to have played major roles in shaping evolution, contributing to the rapid diversification of enzymatically catalyzed reactions and providing material for the invention of new enzymatic activities. These evolutionary relationships are why genomics is possible and serve as the basis for genomic terminology.

Two genes that appear **homologous** in sequence have likely evolved from a single ancestral sequence. Two terms have been used to classify the types of homology. **Orthologous** genes (or proteins) are homologs in two different organisms (species) that evolved from a common ancestral gene. They may or may not retain ancestral function. So, the gene encoding σ-70 in *E. coli* is orthologous to the σ-70 gene in other species, even species not closely related to *E. coli* (e.g., *Staphylococcus*). **Paralogous** genes (or proteins) are those whose evolution reflects gene duplications. For example, the gene encoding σ-38 is a paralog of the gene encoding σ-70. One gene arose as a duplication of the other, followed by mutational divergence to change — in this case, promoter selectivity.

As noted, the first stop once a sequence is in hand is the Web. Major Internet repositories for gene sequences include EMBL (European) and GenBank (USA). These databases can be accessed via a number of Web sites such as through the National Institutes of Health home page (http://www.ncbi.nlm.nih.gov/) or the European Bioinformatics Institute home page (http://www.ebi.ac.uk/Tools/index.html). Upon sequencing part or all of a gene, the researcher can enter these sites, and, through programs such as BLAST (Basic Local Alignment Search Tool) or FASTA, can compare their query DNA sequence, or deduced protein sequence, to all the known DNA or protein sequences. These programs use statistical calculations to identify significant sequence matches between the query sequence and the sequences in the database. If homology is found over a major portion of a known gene/gene product in another organism and if the function of that homolog is known, the putative function of the query gene can be cautiously predicted. Many times only a portion of the query sequence is homologous to a known gene. This may be due to the presence of a common cofactor-binding site such as ATP. This can be analyzed with another program found on the Internet that screens sequences for known motifs (Table 4-1).

Virtual expeditions to the Internet sites listed in Table 4-1 will reveal a wealth of computer analysis tools including those that predict isoelectric points, molecular weights, three-dimensional structures, and proteolytic peptide patterns of deduced proteins, as well as those that reveal potential DNA regulatory sites, promoters, and so on. A particularly intriguing reference is the KEGG (Kyoto Encyclopedia of Genes and Genomes) site (http://www.genome.ad.jp/kegg/), which graphically illustrates the biochemical systems predicted to be present in any microorganism whose genome has been sequenced. To see how this works, go to this site, open KEGG, and under "Pathway Information" click on "Metabolic Pathways." Click on the metabolic system of interest. The reference pathway for this system will appear. In the "Go to" window, select the organism you want to query (e.g., *Staphylococcus aureus*). The reference pathway will reappear but with colored boxes highlighting the enzymes predicted to be present in that organism based on the genome sequence. Clicking on a box reveals the enzyme name with amino acid and DNA sequences from that organism. A short tour around this site will underscore how extensively genomics and the Internet have transformed the field of microbial physiology.

Genomics has provided some interesting insights into niche selection and species survival. For example, two pathogens, *H. influenzae* and *Mycoplasma pneumoniae*, both infect the respiratory tract, yet their strategies for acquiring solutes are distinct.

TABLE 4-1. Web Site Resources for Genomics and Proteomics

Web Address	Description
http://web.bham.ac.uk/bcm4ght6/genome.html	A database collating research on *E. coli* genes whose products have been characterized subsequent to in silico predictions from the completed genome sequence
http://cgsc.biology.yale.edu/	*E. coli* genetic stock center
http://www.ucalgary.ca/%7Ekesander/	Salmonella genetic stock center
http://genome.wustl.edu/gsc/bacterial/ salmonella.shtml	Salmonella genome
http://ecocyc.PangeaSystems.com:1555/ server.html	Ask specifics about different biochemical pathways in *E. coli*; must acquire a password
http://www.genome.ad.jp/kegg/	Ask specifics about biochemical pathways in many sequenced organisms
http://bomi.ou.edu/faculty/tconway/global.html	*E. coli* functional genomics
http://susi.bio.uni-giessen.de/ecdc/ecdc.html	*E. coli* database collection (very useful)
http://www2.ebi.ac.uk/fasta33/	Homology searches
http://www.motif.genome.ad.jp/	Motif searches within proteins
http://www.expasy.ch/	Many tools for sequence analysis
http://www.expasy.ch/cgi-bin/map1	Two-dimensional gel profiles
http://www2.ebi.ac.uk/clustalw/	Nucleotide and protein sequence alignment tools
http://www.ncbi.nlm.nih.gov/PubMed/	Pub Med literature searches
http://www.tigr.org/tdb/mdb/mdbcomplete.html	TIGR microbial genome database
http://bmbsgi11.leeds.ac.uk/bmbknd/DNA/ genomic.html	Genome databases
http://www.biology.ucsd.edu/~msaier/transport/	Information on transport systems

Analysis of the *H. influenzae* genome indicates that the organism has transporters for 13 different amino acids and small peptides. In contrast, the *M. pneumoniae* genome only encodes three potential transporters with specificity for amino acids. This species, which may have evolved through reductive evolution from a gram-positive ancestor, has discarded most of the transporter genes, and, instead, utilizes a smaller set of porters possessing broad substrate specificity. The two species appear to occupy similar niches but use different approaches to solve the problem of solute acquisition. This example illustrates that an ability to compare, annotate, and design experiments across multiple genomes via computational methods will have a profound influence on microbiological experimentation.

However, be aware that computer analysis of a DNA sequence can be seductive. Many students think that just because a sequence in a database has been annotated as having a potential function, that *is* the function. This is truly fuzzy logic. Until a predicted gene function has been confirmed by genetic or biochemical means, it should only be referred to as a *potential* function. This pitfall reveals that continued sequence annotation by experimental work in any organism is important to increase the accuracy of computer-based theoretical predictions. This is true regardless of whether the genome for the organism under study has been completed.

Gene Replacement

One step toward confirming a predicted function is to mutate the gene in the organism by what used to be called reverse genetics—that is, identify the gene, then make the mutation rather than the other way around. A variety of techniques can be used, but most methods rely on delivering a mutated gene into the organism in such a way that easily selects recombination and replacement of the resident gene. One method, **suicide mutagenesis**, involves deleting an internal portion of the gene you wish to mutate and placing the deleted gene into a **suicide vector** (Fig. 4-2). In this example, a drug cassette encoding chloramphenicol resistance has replaced the deleted section of gene. A suicide vector is an antibiotic-resistant plasmid that only replicates in cells containing a specific replication protein. The only way the plasmid can convey drug resistance on a cell missing the plasmid-specific replication protein is if the plasmid recombines

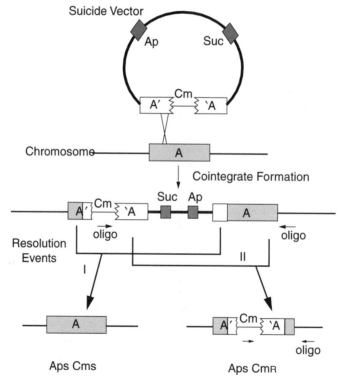

Fig. 4-2. Suicide mutagenesis. A suicide vector plasmid contains the target gene with an internal fragment replaced by a drug resistance marker (chloramphenicol in this case). The plasmid is introduced into a nonpermissive host and Cm resistance is selected. This will only occur if homologous recombination between the mutated gene on the plasmid and the intact copy on the host chromosome cause integration of the plasmid. This plasmid also contains a gene encoding sucrase, an enzyme that confers sensitivity to sucrose on the host cell (*E. coli*). Subsequent selection for sucrose resistance will select for two secondary recombination events (labeled I and II). Both events lead to loss of plasmid-encoded Ap resistance. Resolution event I leaves an intact copy of the gene in the chromosome (loss of Cm resistance) while resolution event II removes the intact gene, leaving the gene A::Cm insertion (The two colons signify the Cm gene is inserted with gene A.)

into the host chromosome. As shown in the figure, recombination occurs at regions of homology determined by gene sequences flanking the deletion. The insertion actually forms a duplication that can be resolved through a second recombinational event. The resolution event can be scored as loss of the plasmid drug marker and retention of the drug marker inserted into the gene (event II in Fig. 4-2). The mutant can now be analyzed for phenotypic characters based on the computer-predicted function.

Gene Arrays

Knowing how the cell coordinates *all* of its biochemical activities will be a major step toward understanding what constitutes life. Unfortunately, a method that would simultaneously examine all the biochemical and structural facets of cell physiology does not yet exist. However, there are powerful techniques available such as gene arrays and two-dimensional separation of cell proteins that provide global macromolecular "snapshots" of the cell. These procedures allow us to view how tweaking one biochemical system impacts the synthesis of all the other systems in the cell.

The first technique, gene array technology, allows us to view gene expression at the whole genome level. The expression of every gene can be monitored simultaneously using a **DNA microchip**. The basic idea is to attach DNA sequences derived from each gene in an organism to a solid support surface in a way that will allow for rapid hybridization to a fluorescently labeled pool of DNA (or cDNA; see below). DNA sequences from individual genes occupy separate spots arranged as a grid on the chip (Fig. 4-3). The entire chip is only about 1 or 2 square inches. Next, the RNA from a cell culture is rapidly extracted and converted into **complimentary DNA (cDNA)** using reverse transcriptase (RNA-dependent DNA polymerase). By incorporating fluorescently tagged nucleotides into the reaction mix, each cDNA molecule is labeled. Individual cDNA molecules will bind to specific locations on the chip that contain their original encoding gene. Laser-scanning and fluorescent-detection devices read the chip surface, and computer analysis reveals which genes were expressed in the original sample and which were not.

Direct comparison of gene expression patterns produced by one strain grown under two *different* conditions can be made using gene array analysis. This is achieved by labeling the cDNA derived from cells grown under different conditions with different fluorescent markers (Fig. 4-3), which is done during the synthesis of cDNA by incorporating different fluorescently tagged dTTP solutions into the reaction mixes. A single array is then used to probe a mixture of the two differentially labeled cDNA preparations. If a specific gene is expressed under only one condition, the corresponding spot on the array will fluoresce one color (e.g., red) because only the RNA extract from cells grown in that condition will have that species of RNA. If a second gene is expressed only when cells are grown under the other condition, its spot on the array will exhibit the other color (e.g., green). When a composite image of the two scans is made via computer, spots corresponding to genes expressed equally under both conditions will exhibit a third color (yellow or orange), which is a blend of the first two colors. This type of analysis can be used to examine the global effects of growth conditions, stresses, or mutations on gene expression.

Proteomics

The proteome encompasses all the expressed proteins of an organism. In contrast to the genome, which is static in composition, the proteome can change moment

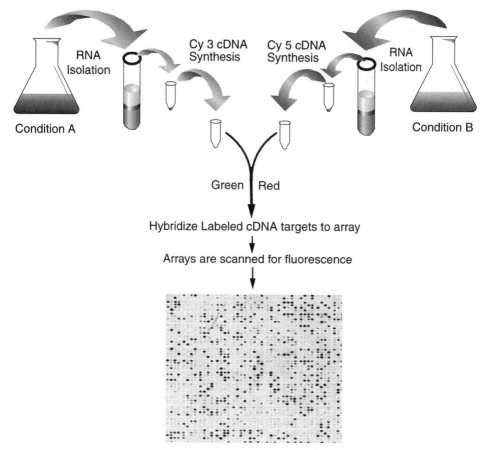

Fig. 4-3. Gene array technology. Dark spots on the array indicate cDNA molecules binding to their corresponding oligos.

by moment in response to changes in the cytoplasmic or extracytoplasmic environments. This is most evident when a cell encounters a stress such as acid shock or heat shock. In response to impending danger, there is a major reorganization of the proteome. Many new proteins are expressed in response to the stress while the levels of other proteins diminish. While gene array technology can predict proteome changes due to transcriptional fluctuations, it fails to predict alterations in the proteome resulting from translational or posttranslational forces — that is, a change in growth condition might not change transcription of a given gene, but the translation of its mRNA might be profoundly affected.

The proteome is typically viewed using two-dimensional polyacrylamide gel electrophoresis (Fig. 4-4). A cell extract containing a mixture of 1000 to 2000 different proteins is subjected first to isoelectric focusing (IEF), a process that separates proteins based on their net charges. In IEF, the proteins are applied to an immobilized pH gradient gel and subjected to a current. Each protein contains many amino acid side groups (e.g., carboxyl or amino groups) that can exhibit a charge depending on whether the group is protonated. A side group changes its protonation state (and, thus, charge)

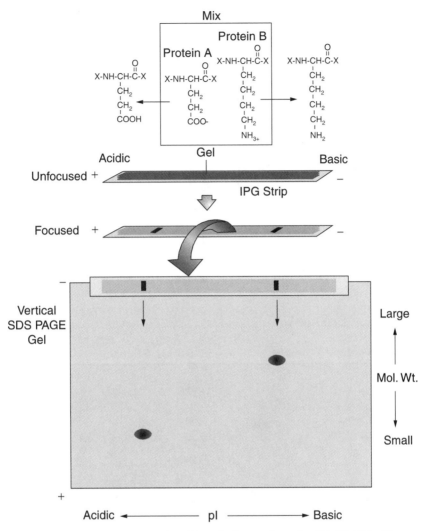

Fig. 4-4. Two-dimensional polyacrylamide gel electrophoresis (2D PAGE). Two proteins with different isoelectric points are depicted in the top box. Horizontal arrows show the direction that each protein travels in the IPG strip.

based on the pKa of the side group(s) and the pH of the surrounding environment. As a rule, side groups will be protonated at pH values below their pKa value.

Side chains of the glutamate and lysine residues in the example have pKa values of 4.3 and 9.8, respectively. So, at pH 7, the carboxyl group of glutamate will be negatively charged (unprotonated) while the amino group of lysine will be positively charged (protonated). The **net charge** of a protein depends on the sum total of the side-group charges. The two proteins in the illustration will begin to move toward opposite poles of the IEF gel because of their initial charges, but, as they move, each protein is exposed to a pH gradient. At some point (called the isoelectric point), the pH encountered at a specific location in the gel results in an equal number of positive and negative charges on the protein, so the net charge is zero. Because there is no charge, the protein stops moving and focuses at that position. Proteins with different isoelectric points focus to different positions along the IEF gel. IEF separation is the first dimension of two-dimensional gel electrophoresis.

As good as this initial separation process is, proteins of very different molecular weights can and do have identical isoelectric points. As a result, one band on an IEF gel could contain 10 or 15 different proteins. To get a clearer picture of the proteome, a second technique must be used that separates proteins based on their molecular weight. The second dimension begins with the IPG strip from the first dimension being soaked in a sodium dodecyl sulfate (SDS) solution. SDS is a negatively charged molecule that will coat all proteins and impart a negative charge on them. The strip is then placed at the top of a vertical slab gel containing a polyacrylamide matrix and again subjected to an electric current (Fig. 4-4). The proteins begin to move through spaces in the matrix toward the positive pole. The smaller proteins (i.e., lower molecular weight) travel easiest through the spaces while the larger proteins travel more slowly. This sieving effect separates the proteins by molecular weight. Combining the two techniques arranges the proteins in a pattern illustrated in Figure 4-5.

The proteins can be visualized either by radioactively labeling the proteins with 35S-methionine during growth or by staining with conventional coomassie blue dyes

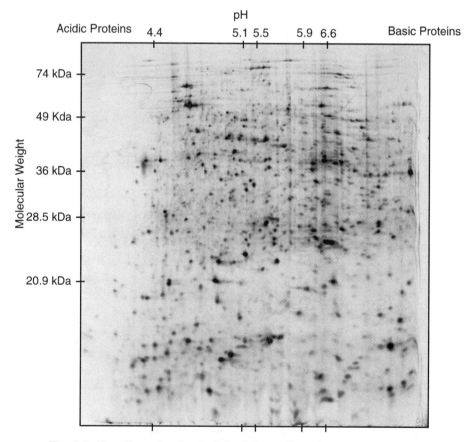

Fig. 4-5. Two-dimensional gel of the *Salmonella typhimurium* proteome.

or the newer fluorescent tagged dyes. Computer analysis can be used to compare the proteomes produced in response to different growth conditions. Once a protein is earmarked for further study, it must be identified. Here, the combined power of genomics and proteomics is most apparent. When the genome of the organism or a closely related organism has already been sequenced, the protein spot of interest can be excised from the gel and the precise mass of the protein can be determined by mass spectrometry. In this technique, a protein is digested with a protease such as trypsin and the fragments are introduced into the mass spectrometer, which precisely measures the mass of each fragment. This can be used to unambiguously assign the identity of the protein when the entire genome of the organism has already been sequenced and the exact molecular weights of all potential peptides have been calculated.

If the genome of the organism has not been sequenced, the spot can be excised and subjected to N-terminal amino acid sequence analysis or cleaved with a protease, such as trypsin, and the fragments separated by HPLC and sequenced. The peptide sequence(s) obtained is then used to query various databases to obtain matches. Alternatively, tandem mass spectrometry, or MS/MS, can be used to derive partial sequence information for a peptide. One of the many tryptic peptides separated by the initial round of MS is selected for further mass analysis. Following further fragmentation, the mass spectrum derived from the overlapping fragments is used to identify positions of specific amino acids in the peptide. The resulting amino acid sequence can be used to query known databases for matches.

TRADITIONAL TOOLS

Mutant Hunts

The classical approach used to expose the details of a biochemical pathway requires the presence of a selectable phenotype, such as an ability to grow on a carbohydrate. Mutants that have lost this phenotype are then sought. Once a mutant is found, the biochemical step in the pathway can be explored and the encoding gene can be mapped, identified, cloned, and sequenced.

It is not unusual for a mutant hunt to screen 10,000 individual mutagenized colonies and only find a handful of mutants in the particular system under study. This is why the selection phenotype is so important. If there is a **positive selection** phenotype [such as resistance to an amino acid analog or a fermentable (red colony) versus nonfermentable (white colony) phenotype on a MacConkey media], thousands of mutants can be screened on a single plate. In the analog example, only mutants that fail to transport or use the amino acid analog will grow and form colonies on a plate containing that analog. In looking for mutants defective in carbohydrate fermentation, mutant colonies that fail to ferment the test sugar will appear white among many nonmutated red colonies.

In the absence of positive selection, the investigator must resort to **negative selection** screens that are more labor intensive. For example, the phenotype of an *E. coli* mutant that cannot make the amino acid alanine will only grow on a defined medium if that medium contains alanine. To screen for *ala* mutants, 10,000 random mutants are inoculated onto 200 rich medium plates in grids of 50 mutants per plate. Each master plate is then replicated to two agar plates containing defined media, one with and one without alanine. Replicating colonies from master to test plates involves making an imprint of the colonies from the master plate onto sterile velveteen pads and using that pad to "stamp" the new plates. The new plates are inspected after allowing the patches time to grow. An *ala* mutant will be seen as a colony patch that grows on defined media containing alanine but will not grow on the same media lacking alanine.

More sophisticated mutant hunts in which individual cells rather than whole colonies can be screened are now possible with the discovery of fluorescent-activated cell sorting (FACS). To appreciate this technique, imagine making a gene fusion between a gene encoding green fluorescent protein (*gfp*) and a gene that is only expressed under conditions where the cell will *not* grow (an acid pH–inducible gene at pH 4.5) and cannot form colonies. You would like to find mutants that do not express this gene at pH 4.5. To find these mutants, the *gfp* fusion strain can be mutated, a pool of mutants grown at normal pH, adjusted to pH 4.5 to allow induction, and a sample taken and passed through a FACs machine. The cells are passed single file through an orifice where instrumentation reads whether a cell is fluorescing. Simultaneously, a charge is placed on the fluorescing cells. Then charged and uncharged cells can be separated, and the mutant cells that do not express the test gene at pH 4.5 are collected and used for further study. This type of single-cell mutant hunt strategy has dramatically expanded the capability of microbial geneticists.

Transcriptional and Translational Gene Fusions (Reporter Genes)

A very powerful tool for analyzing various aspects of gene expression involves fusing easily assayed reporter genes such as *lacZ* (*β*-galactosidase) or *gfp* (green fluorescent protein) to host target gene promoters. Two types of fusions are typically used (Fig. 4-6). Operon or transcriptional fusions are used when a promoterless *lacZ* reporter gene is inserted within a target gene in an orientation that places *lacZ* under the control of the target gene's promoter. Thus, whatever factors control expression from the target gene promoter will also control the production of *β*-galactosidase. Since the *lacZ* message resulting from this fusion still contains its ribosome-binding site, the regulation observed is typically, although not always, due to transcriptional control. The second type of fusion (called a gene or protein fusion) involves inserting a reporter (e.g., *lacZ*) into a target gene that is missing both its promoter and ribosome-binding site. Not only will the messages from the target and reporter gene be fused but the truncated target gene peptide and the reporter gene peptide will be fused when inserted in the proper reading frame. In this case, anything controlling the transcription or translation of the target gene will also control *β*-galactosidase levels.

Construction of these fusions sometimes involves using a genetically engineered *μ* phage (Chapter 6) to randomly insert a reporter *lacZ* gene into the *E. coli* or *Salmonella* genome. Alternatively, the fusion can be constructed in vitro and transferred into recipient cells in a manner that will promote allelic exchange of the reporter fusion for the resident gene. Once constructed, these fusions allow for monitoring of how different growth conditions or known regulators influence the transcription or translation of the gene. In addition, these fusions provide convenient phenotypes (e.g., lactose fermentation or fluorescence) that can be exploited in mutant hunts to screen for new regulators.

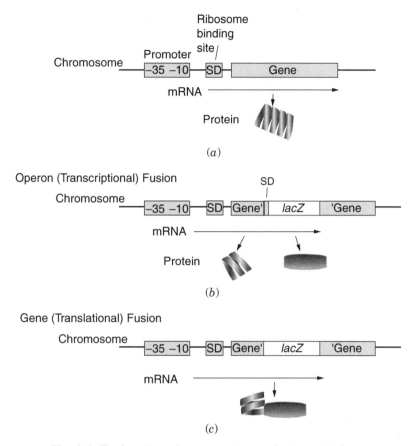

Fig. 4-6. **Engineering of operon and gene fusions with** *lacZ.*

Polymerase Chain Reaction

In the past, obtaining large concentrations of purified DNA used to be a frustrating rate-limiting step in molecular biology. However, in the late 1980s, the discovery of a relatively simple process called the polymerase chain reaction solved the problem and revolutionized molecular biology. Kary Mullis, a scientist and avid surfer from California, conceived the idea that won him the 1993 Nobel prize in chemistry not through painstaking experimentation, as you might expect, but while cruising California Highway 128 in a Honda Civic.

The basic polymerase chain reaction (PCR) process, outlined in Figure 4-7, can produce a $>10^6$ amplification of target DNA. The figure depicts the use of specific oligonucleotide primers (20–30 bp) that anneal to sequences flanking the target DNA. These primers are used by a thermostable DNA polymerase to replicate the target DNA. For the technique to amplify a specific-size DNA, thermostable DNA polymerases must be used because the reaction mixture must be subjected to repeated cycles of heating to 95 °C (to separate DNA strands), cooling to 55 °C (to allow primer reannealing), and heating to 72 °C (the optimum reaction temperature for the polymerase). The polymerase *Taq*, from the thermophile *Thermus aquaticus*, is commonly used for this purpose.

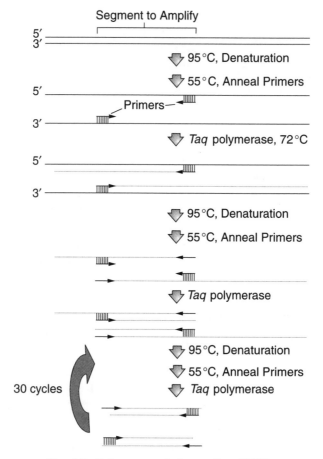

Fig. 4-7. Polymerase chain reaction (PCR).

Since amplification requires 25 to 30 heating and cooling cycles, a machine called a thermocycler is used to reproducibly and quickly deliver these cycles. This basic PCR technique has been modified to serve many different purposes. Primers can be engineered to contain specific restriction sites that simplify subsequent cloning, or if the primers are highly specific for a gene present in only one species of microorganism, PCR can be used to detect the presence of that organism in a complex environment (e.g., the presence of the pathogen *E. coli* O157:H7 in hamburger).

PCR can also be used to engineer specific mutations in a given gene (**site-directed mutagenesis**). For example, computer analysis of a gene reveals a potential functional motif (e.g., Zn^{2+} finger). One step toward determining whether the motif is important to the function of the gene product would be to change one of the key amino acid codons by site-directed mutagenesis and examine the mutant phenotype. The example in Figure 4-8 uses four primers. The two outside primers (#2 and #4) amplify the entire gene. The two inside primers (#1 and #3) overlap the target region and are engineered to contain the desired mutation. There are three PCR reactions required. The first two reactions amplify each end of the gene and include the mutation. To reconstruct the complete gene, the products from the initial reactions are mixed, denatured at 95 °C,

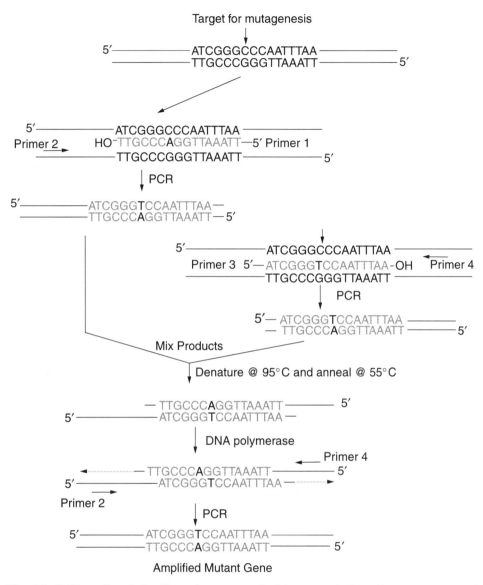

Fig. 4-8. PCR-mediated site-directed mutagenesis. Primers are indicated by arrows or actual sequence for the mutagenic primers (primers 1 and 3). Short arrow points to the target base.

and reannealed to form the hybrid structure shown. Subsequent PCR with the outside primers regenerates the complete gene but with the site-directed mutation.

DNA Mobility Shifts (Gel Shifts and Supershifts)

Once a gene encoding a putative transcriptional regulatory protein is identified, it is imperative to test whether that protein directly binds to the target DNA sequence. One approach is to take purified protein, add it to the putative target DNA fragment, and see if the protein causes a shift in the electrophoretic mobility of that fragment.

An example is illustrated in Figure 4-9. Linear DNA undergoing electrophoresis in an agarose or polyacrylamide gel matrix will travel at a rate based on the size of the DNA molecule. Larger molecules travel slower than smaller molecules. If a protein binds to that DNA molecule, the complex is larger than the DNA alone and will travel more slowly in the gel. This is called a mobility shift or gel shift.

You can determine if a suspected DNA-binding protein is actually part of a mixture of proteins (e.g., a cell extract) that cause a gel shift by performing a **supershift** experiment. A supershift of the target fragment will occur when an antibody directed against the putative DNA-binding protein is incorporated into the binding reaction. The complex of anti-DNA-binding protein, DNA-binding protein, and DNA is even larger than the binding protein–DNA complex, causing an even greater gel shift (supershift). If a supershift does not occur, the suspected protein is not the one causing the original shift.

Finding Transcriptional Starts by Primer Extension

As discussed in Chapter 5, many genes are driven by multiple promoters, each of which produces a different transcript. Documenting whether a gene produces multiple

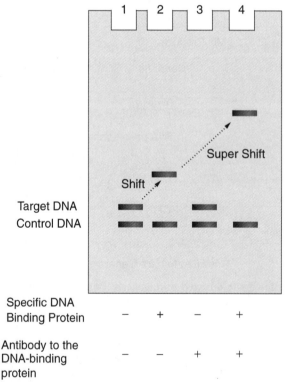

Fig. 4-9. Electrophoretic mobility shift and supershift. Polyacrylamide gel. All lanes contain a target DNA fragment and a control fragment. DNA-binding protein and antibody against the DNA-binding protein are added where indicated *before* running the sample on the gel. The binding of protein to the target fragment (lane 2) makes a larger complex that travels more slowly than the unbound DNA fragment. The addition of DNA-binding protein and antibody makes mobility slower still.

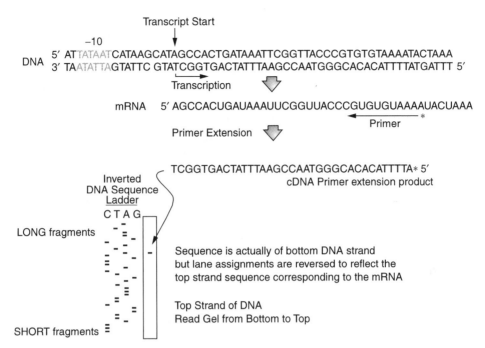

Fig. 4-10. **Primer extension.** The bottom DNA strand is actually read by RNA polymerase. The leftward arrow below mRNA represents the primer used for primer extension and DNA sequencing. The asterisk indicates the fragment is radiolabeled. Each lane of the DNA sequence gel represented in the lower panel detects a different base as indicated above each lane. Sequence starts after the end of the primer.

transcripts includes defining where each transcript begins. Primer extension is one method that achieves this result (Fig. 4-10). A DNA primer that anneals to the mRNA transcript under investigation is used in a reverse transcription reaction that extends the primer to the 5′ end of the message. This produces a cDNA of a precise length that is radiolabeled due to the use of radioactive nucleotides in the reaction mix. The key to this procedure is that the same primer is used in a DNA sequencing reaction with the template DNA. The primer extension fragment is then run alongside a DNA sequencing ladder on a polyacrylamide gel. The size of the primer extension fragment will be identical to one of the "rungs" on the sequencing ladder. That rung represents a base in the sequence, and that base correlates to the transcriptional start site. The RNA from genes that have multiple promoters will produce different-sized primer extension fragments that comigrate with different sequencing fragments.

Detecting DNA, RNA, Protein, and DNA-Binding Proteins by Southern, Northern, Western, and Southwestern Blots

Methods to detect specific DNA, RNA, and protein molecules are critically important tools used to probe gene expression strategies. A **Southern blot**, so named for Ed Southern who developed the technique in 1975, begins with cutting a DNA molecule with one or more restriction endonucleases. The molecules are separated by agarose electrophoresis and then transferred from the agarose gel to a nitrocellulose membrane overlayed onto the gel. The transfer process involves capillary action in which buffer is drawn first through the agarose gel and then through the facing membrane. DNA traveling with the buffer is deposited on the membrane in exactly the same pattern as it is displayed in the gel. Once DNA fragments are affixed to the membrane, the strands are separated chemically and the membrane is probed with a denatured, tagged DNA fragment under hybridizing conditions that allow annealing between the probe and matching DNA fragment. The probe can be labeled with radioactivity, a fluorescent dye, or biotin. The biotin is detected later by a chemiluminescent enzyme assay. DNA fragment bands that have annealed to the probe are visualized by **autoradiography** — that is, by exposing the membrane to X-ray film and photographically developing the film.

Northern blots are used to analyze the presence, size, and processing of a specific RNA molecule in a cell extract. RNA is extracted from the cell using special precautions to avoid contaminating the preparation with RNases that are ubiquitous in the laboratory environment. The RNA is fractionated by electrophoresis on an agarose-formaldehyde gel and the fragments are transferred onto a nylon or nitrocellulose membrane by capillary action as with Southern gels. Once transferred, the RNA is fixed to the membrane either by baking (nitrocellulose) or UV cross-linking (nylon). The membrane can then be probed with a small radiolabled DNA probe specific for a given mRNA. After hybridization, the membrane is autoradiographed or subjected to phosphorimaging (a **phosphorimager** records the energy emanated as light or radioactivity from gels or membranes). Visualized bands can be sized by comparing their location with standard RNA-molecular-weight markers loaded in a parallel lane. This technique is useful to determine the presence and turnover of a given message. By estimating size of the transcript, the method can also be used to evaluate whether two adjacent genes are transcribed as an operon.

Western blots are used to detect specific proteins within a complex mixture of heterologous proteins. In Figure 4-11, the protein extract is subjected to SDS PAGE and the protein bands are electrophoretically transferred from the gel to a PVDF membrane. The membrane is probed with a primary antibody directed against the specific protein. Depending on the source of the antibody (mouse, rabbit, other), the membrane is probed with a secondary antibody that specifically binds to the primary antibody (e.g., anti-mouse IgG antibody). This secondary antibody is tagged with horseradish peroxidase, for example. The result is an antibody "sandwich" where the primary antibody binds the target protein and the secondary antibody binds the primary antibody. Adding hydrogen peroxide and luminol sets off a luminescent reaction wherever an antibody sandwich has assembled. The light emitted can then be detected by autoradiography or phosphorimaging. Only protein bands that bind to the primary antibody will be detected. The technique is used to determine if, and at what levels, specific proteins are present in a cell extract. It can also be used to detect the processing of a protein into smaller fragments (Fig. 4-11).

The **Southwestern blot** is one method that can be used to identify proteins that bind specific DNA molecules. Proteins from the cell are separated using standard SDS polyacrylamide gel electrophoresis. The fractionated protein bands are transferred to a nitrocellulose membrane and the membrane is probed with radiolabeled oligonucleotides. Autoradiography will reveal if a specific protein band binds the labeled DNA. The protein can be excised and identified by a variety of techniques.

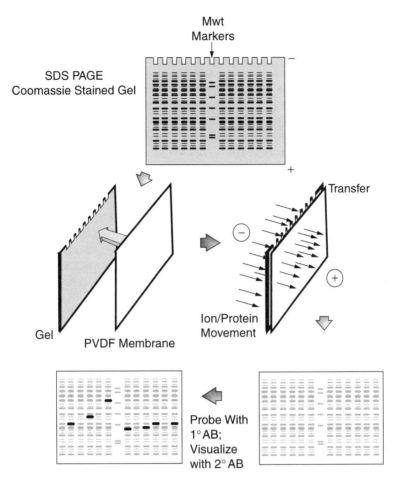

Fig. 4-11. Western blot. Mixtures of proteins from different cultures are loaded onto separate lanes of the SDS polyacrylamide gel (top gel). Standard molecular-weight markers are loaded in the center lane. After electrophoresis, this gel can be stained with Coomassie blue or fluorescent dyes to give the pattern shown. To identify which band in each lane is the protein of interest, the unstained protein bands are electrophoretically transferred to a PVDF membrane as shown. The proteins are deposited on the membrane (the bands are not usually visible but are shown here in gray highlight for convenience). The membrane is then probed with primary antibody directed against the protein of interest. Bands to which these antibodies have bound are then visualized by adding a secondary antibody directed against the first antibody. The secondary antibody is tagged with the enzyme horseradish peroxidase (HRP). After binding of the secondary antibody, hydrogen peroxide and luminol are added to the membrane. HRP, in the presence of H_2O_2, oxidizes luminol, which enters an excited state that decays and emits light. Thus, areas where the primary antibody bind to the gel (detecting the target protein) will emit light that can be detected with autoradiography (bottom left-hand membrane). Bands corresponding to the target protein are in bold. Note that different extracts exhibit different-sized bands. This could be the result of processing of the protein to smaller sizes under different conditions or crossreacting bands that share antigenic epitopes with the protein of interest.

Two-Hybrid Analysis

Protein–protein interactions are an important part of many regulatory circuits, from anti-σ factors to multisubunit enzyme complexes. It is valuable to probe these interactions in vitro but also in vivo where the proteins actually work. Two-hybrid systems are cleverly designed tools that not only measure in vivo protein–protein interaction but can be used to mine for unknown "prey" proteins that interact with a known "bait" protein.

While there are many types of two-hybrid techniques, the principle can be illustrated with two examples (Fig. 4-12). The now classic method is the **yeast two-hybrid system**. This strategy involves fusing genes encoding two potentially interacting proteins to the separate components of the yeast GAL4 transcriptional activator (Fig. 4-12A). These are the DNA-binding domain fragment and the transcriptional activation domain fragment. Both components of the GAL4 activator must interact to induce expression of a yeast chromosomal *GAL1-lacZ* fusion, although other reporter genes can be used. The *lacZ* gene encodes β-galactosidase, whose enzymatic activity is easily measured. Normally, both domains are part of one GAL4 protein. However, when separated, both domains can still interact if each domain is fused to one of a pair of other interacting proteins.

When the two resulting hybrid proteins interact via the two target protein domains, the GAL4 DNA-binding and activation domains are brought together. Thus assembled, the two-hybrid GAL4 complex will bind and activate transcription of *GAL1-lacZ*. Expression of *GAL1-lacZ* is then visualized on agar plates containing a chromophoric substrate of β-galactosidase called X-gal, or it can be directly assayed in vitro. The power of this technique is that it can also be used to find unknown proteins that interact with a known protein. The known protein is fused to one of the GAL4 domains and used as bait in which cells containing the bait fusion are transformed with a plasmid pool containing randomly cloned prey genes fused to the other GAL4 domain. After plating, colonies that express β-galactosidase are the result of protein–protein interactions occurring between the bait and prey fusion proteins. The gene encoding the prey protein can then be identified by sequencing the DNA insertion.

A novel **bacterial two-hybrid system** utilizes N-terminal (T25) and C-terminal (T18) halves of *Bordetella pertussis* adenylate cyclase (CyaA) produced on separate plasmids. Genes encoding suspected interacting proteins are cloned into the two plasmids, creating gene fusions between each tester gene and the *cyaA* moieties (Fig. 4-12B). Both plasmids are introduced into a *cya* mutant of *E. coli*, which is phenotypically Lac⁻ due to the absence of cAMP (see Chapter 5). If the CyaA hybrid proteins interact in vivo, the two CyaA subunits will be brought into close proximity, forming an active adenylate cyclase. As a result, the cells will make cAMP and the lactose operon will be expressed.

Although bacterial protein–protein interactions have been examined using both systems, there is an advantage to using bacterial two-hybrid systems over yeast systems. For example, if a bacterial-motivated protein modification is required for an interaction to occur, the modification may not happen in the yeast system, so the interaction might be missed.

Using the Two-Hybrid System to Identify a New Regulator. *Helicobacter pylori*, a remarkable gram-negative microbe that can inhabit the human stomach and cause gastric ulcers, provides an interesting example in which a two-hybrid

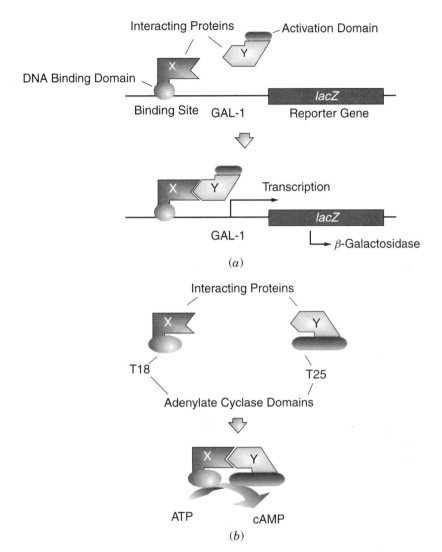

Fig. 4-12. Two-hybrid systems to detect protein–protein interactions. (*a*) Yeast two-hybrid system. Protein–protein interaction between X and Y fusion proteins bring together the DNA-binding and activation domains of GAL4, which activates transcription of a *GAL1-lacZ* fusion. (*b*) Bacterial two-hybrid system involving *Bordetella pertussis* adenylate cyclase. Interaction between the X and Y fusion proteins brings together the T18 and T25 moieties of *B. pertussis* adenylate cyclase. The resulting production of cAMP can be measured through expression of a cAMP-dependent gene (e.g., *lacZ*.)

approach was used in conjunction with proteomic and genomic strategies to reveal a new family of regulators (Colland et al., 2001). Computer analysis of the completed helicobacter genome indicated that the organism possesses a σ-28 ortholog responsible for controlling a cascade of flagellar genes. In *E. coli*, the activity of σ-28 is kept in check by a specific anti-σ factor, FlgM (see "Anti-σ" in Chapter 5). However, scrutiny of the helicobacter genome failed to reveal a similar gene. Convinced that this organism must contain a FlgM-like molecule, Colland and colleagues used a yeast two-hybrid system to fish out a gene whose product binds to *H. pylori* σ-28 in vivo. DNA sequence analysis of this gene exposed its precise location in the *H. pylori* genome. Once confirmed as an anti-σ factor, BLAST comparison of the deduced amino acid sequence of the *H. pylori* FlgM to known FlgM homologs revealed a carboxy terminal amino acid motif common to other known σ-28 molecules. With this knowledge in hand, the authors used computer analysis to identify potential FlgM orthologs in three other microorganisms. This is a wonderful example of how genomics has transformed the field of microbial physiology.

SUMMARY

There are numerous commonly used molecular tools for studying microbial physiology. This chapter presents only a few key methods chosen to underscore the impact that genomic and proteomic strategies have had on this field. Other techniques such as DNA footprinting (where a DNA-binding protein is shown to protect its target DNA from in vitro nuclease digestion) or reverse transcriptase-mediated PCR (where low levels of an RNA molecule can be visualized using reverse transcriptase to make cDNA and PCR primers to amplify the cDNA to visible levels) have not been discussed. Details of these and many other tools of the trade are readily available in other publications such as *Applied Molecular Genetics* by Roger L. Miesfeld (Wiley-Liss).

As discussions progress in this textbook, you should begin to understand how many of the techniques covered in this chapter are used to answer important biochemical and genetic questions. Perhaps the most striking observation that can be made when comparing the present state of science to that of only 10 or 15 years ago is that many investigations that took decades to complete can now be accomplished in a mere fraction of that time using the current tools of genomics, molecular biology, and biochemistry.

BIBLIOGRAPHY

Colland, F. et al. 2001. Identification of the *Helicobacter Pylori* anti σ-28 factor. *Mol. Microbiol.* **41**:477–87.

Field, D., D. Hood, and R. Moxon. 1999. Contribution of genomics to bacterial pathogenesis. *Curr. Opin. Genet. Dev.* **9**:700–3.

Gogarten, J. P., and L. Olendzenski. 1999. Orthologs, paralogs and genome comparisons. *Curr. Opin. Genet. Dev.* **9**:630–6.

Hu, J. C., M. G. Kornacker, and A. Hochschild. 2000. *Escherichia coli* one- and two-hybrid systems for the analysis and identification of protein–protein interactions. *Methods* **20**:80–94.

Karimova, G., J. Pidoux, A. Ullmann, and D. Ladant. 1998. A bacterial two-hybrid system based on a reconstituted signal transduction pathway. *Proc. Natl. Acad. Sci. USA* **95**:5752–6.

Karp, P. D., M. Riley, M. Saier, I. T. Paulsen, S. M. Paley, and A. Pellegrini-Toole. 2000. The EcoCyc and MetaCyc databases. *Nucleic Acids Res.* **28**:56–9.

Lin, J., R. Qi, C. Aston, J. Jing, T. S. Anantharaman, B. Mishra, O. White, M. J. Daly, K. W. Minton, J. C. Venter, and D. C. Schwartz. 1999. Whole-genome shotgun optical mapping of *Deinococcus radiodurans*. *Science* **285**:1558–62.

Miesfeld, R. L. 1999. *Applied Molecular Genetics*. Wiley-Liss, Inc., New York, NY.

Mullikin, J. C., and A. A. McMurragy. 1999. Techview: DNA sequencing. Sequencing the genome, fast. *Science* **283:**1867–9.

Ouzounis, C. A., and P. D. Karp. 2000. Global properties of the metabolic map of *Escherichia coli*. *Genome Res.* **10:**568–76.

Richmond, C. S., J. D. Glasner, R. Mau, H. Jin, and F. R. Blattner. 1999. Genome-wide expression profiling in *Escherichia coli* K-12. *Nucleic Acids Res.* **27:**3821–35.

Rogers, J. 1999. Sequencing. Gels and genomes. *Science* **286:**429.

Rudd, K. E. 2000. EcoGene: a genome sequence database for *Escherichia coli* K-12. *Nucleic Acids Res.* **28:**60–64.

Salgado, H., G. Moreno-Hagelsieb, T. F. Smith, and J. Collado-Vides. 2000a. Operons in *Escherichia coli:* genomic analyses and predictions. *Proc. Natl. Acad. Sci. USA* **97:**6652–7.

Salgado, H., A. Santos-Zavaleta, S. Gama-Castro, D. Millan-Zarate, F. R. Blattner, and J. Collado-Vides. 2000b. RegulonDB (version 3.0): transcriptional regulation and operon organization in *Escherichia coli* K-12. *Nucleic Acids Res.* **28:**65–7.

Tao, H., C. Bausch, C. Richmond, F. R. Blattner, and T. Conway. 1999. Functional genomics: expression analysis of *Escherichia coli* growing on minimal and rich media. *J. Bacteriol.* **181:**6425–40.

Dougherty, T. J., J. F. Barrett and M. J. Pucci. 2002. Microbial genomics and novel antibiotic discovery: New technology to search for new drugs. *Current Pharmaceutical Design* **8:**99–110.

CHAPTER 5

REGULATION OF PROKARYOTIC GENE EXPRESSION

The cardinal rule of existence for any organism is economy. A cell need not waste energy by simultaneously synthesizing 20 different carbohydrate utilization systems if only one carbohydrate is available. Likewise, it is wasteful to produce all the enzymes required to synthesize an amino acid if that amino acid is already available in the growth medium. Extravagant practices such as these will jeopardize survival of a bacterial species by making it less competitive with the more efficient members of its microbial microcosm. However, there are many instances when economy must be ignored. Microorganisms in nature more often than not find themselves in suboptimal environments (stationary phase) or under chemical (e.g., hydrogen peroxide) or physical (e.g., ultraviolet irradiation) attack. Consequently, in any given ecological niche, the successful microbe not only needs regulatory systems designed to maximize the efficiency of gene expression during times of affluence but must also sense danger and suspend those safeguards in favor of emergency systems that will remove the threat or minimize the ensuing damage. Economy is important, but backup plans ensure survival of the species.

Prokaryotic gene expression is classically viewed as being controlled at two basic levels: DNA transcription and RNA translation. However, it will become apparent that mRNA degradation, modification of protein activity, and protein degradation also play important regulatory roles. This chapter deals with the basics of gene expression, describes some of the well-characterized regulatory systems, and illustrates how these control systems integrate and impact cell physiology. The strategies used by bacteria to avoid or survive environmental stresses are considered in Chapter 18.

TRANSCRIPTIONAL CONTROL

The most obvious place to regulate transcription is at or around the promoter region of a gene. By controlling the ability of RNA polymerase to bind to the promoter, or, once bound, to transcribe through to the structural gene, the cell can modulate the amount of message being produced and hence the amount of gene product eventually synthesized. The sequences adjacent to the actual coding region (**structural gene**) involved in this control are called regulatory regions. These regions are composed of the promoter, where transcription initiates, and an operator region, where a diffusible regulatory protein binds. Regulatory proteins may either prevent transcription (**negative control**) or increase transcription (**positive control**). The regulatory proteins may also require bound effector molecules such as sugars or amino acids for activity (see "The *lac* Operon" in this chapter). Repressor proteins produce negative control while activator proteins are associated with positive control. Transcription initiation requires three steps: RNA polymerase binding, isomerization of a few nucleotides, and escape of RNA polymerase from the promoter region, allowing elongation of the message. Negative regulators usually block binding while activators interact with RNA polymerase, making one or more steps, often transition from closed to open complexes, more likely to occur.

An **operon** is several distinct genes situated in tandem, all controlled by a common regulatory region (e.g., *lac* operon). The message produced from an operon is **polycistronic** in that the information for all of the structural genes will reside on one mRNA molecule. Regulation of these genes is coordinate, since their transcription depends on a common regulatory region—that is, transcriptions of all components of the operon either increase or decrease together. Often genes that are components in a specific biochemical pathway do not reside in an operon but are scattered around the bacterial chromosome. Nevertheless, they may be controlled coordinately by virtue of the fact that they all respond to a common regulatory protein (e.g., the arginine regulon). Systems involving coordinately regulated, yet scattered, genes are referred to as **regulons**.

Figure 5-1 presents a schematic representation of negative-control versus positive-control regulatory circuits. Whether an operon is under negative or positive control, it can be referred to as **inducible** if the presence of some secondary effector molecule is required to achieve an increased expression of the structural genes. Likewise, an operon is **repressible** if an effector molecule must bind to the regulatory protein before it will inhibit transcription of the structural genes.

DNA-Binding Proteins

It is evident from Figure 5-1 that regulation at the transcriptional level relies heavily on DNA-binding proteins. Studies conducted on many different regulatory proteins have revealed groups based on common structural features. The following are brief descriptions of several common families.

Helix-turn-helix (HTH) DNA recognition motifs were the first discovered (see "λ Phage" in Chapter 6). They include the repressors of the *lac*, *trp*, *gal* systems; λCro and CI repressors; and CRP. The proteins consist of an α helix, a turn, and a second α helix. A glycine in the turn is highly conserved as is the presence of several hydrophobic amino acids. A diagram of how an HTH motif interacts with DNA can be found in Figure 6-14. It should be noted that, unlike some DNA-binding motifs, the HTH motif is not a separate, stable domain but rather is embedded in the rest of the protein. Commonalities among HTH proteins are (1) repressors usually bind as dimers, each monomer recognizing a half-site at a DNA-binding site; (2) the operator sites are B-form DNA; (3) side chains of the HTH units make site-specific contacts with groups in the major groove.

Zinc fingers were first discovered in the Xenopus transcription factor IIIA. Proteins in this family usually contain tandem repeats of the 30 residue zinc-finger motif (Tyr/Phe-x-Cys-X_2 or $_4$-Cys-X_{12}-His-X_{3-5}-His-X_3-Lys). They contain an antiparallel β sheet and an α helix. Two cysteines, near the turn in the β sheet, and two histidines in the α helix, coordinate a central zinc iron and hold these secondary structures together,

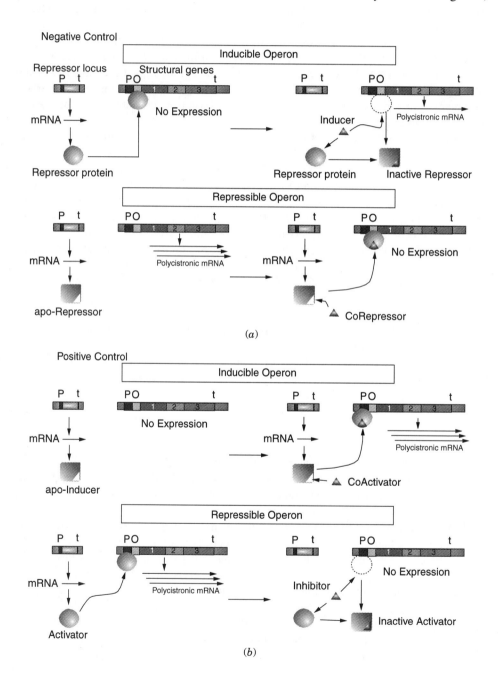

forming a compact globular domain. The zinc finger binds in the major groove of β-DNA, and because several fingers are usually connected in tandem, they are long enough to wrap partway around the double helix, like a finger.

Leucine zipper DNA-binding domains generally contain 60 to 80 residues with two distinct subdomains: the leucine zipper region mediates dimerization of the protein while a basic region contacts the DNA. The leucine zipper sequences are characterized by a heptad repeat of leucines over 30 to 40 amino acids:

$$L-X_3-L-X_2-L-X_6-L-X_3-L-X_2-L-X_6-L$$

β**-sheet DNA-binding proteins** bind as a dimer with antiparallel ß helices filling the major groove (e.g., MetJ). The arrangement is a 7-residue β sheet, a 14–16-residue α helix, and a 15-residue α helix. The β sheet enters the major groove and side chains on the exposed face of the protein contact the base pairs. IHF and HU proteins may use β-sheet regions for DNA binding.

The *lac* Operon: A Paradigm of Gene Expression

The operon responsible for the utilization of lactose as a carbon source, the *lac* operon, has been studied extensively and is of classical importance. Its examination led Jacob and Monod to develop the basic operon model of gene expression nearly half a century ago. Even now, understanding the many facets of its control underscores the economics of gene expression and reveals that no single operon acts in a metaphorical vacuum. It will become apparent that aspects of cell physiology beyond the simple notion of lactose availability govern the expression of this operon.

Lactose is a disaccharide composed of glucose and galactose:

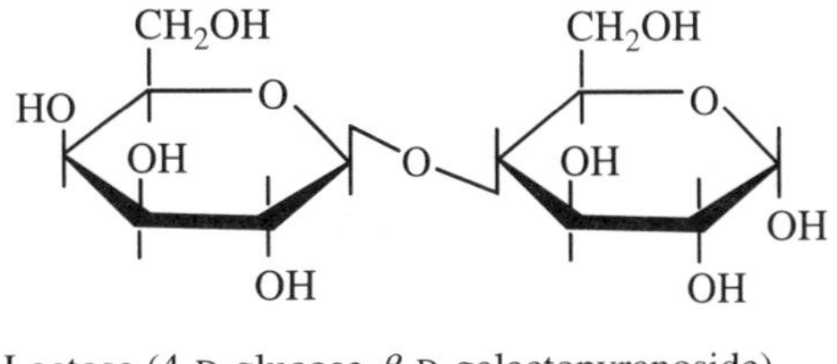

Lactose (4-D-glucose-β-D-galactopyranoside)

Fig. 5-1. Types of genetic regulatory mechanisms. (*a*) Negative control regulators (repressors) can govern inducible or repressible operons. Negatively controlled inducible systems utilize a repressor molecule that prevents transcription of structural genes unless an inducer molecule binds to and inactivates that repressor. Negatively controlled repressible systems also involve a repressor protein, but in this instance the repressor is inactive unless a corepressor molecule binds to and activates the repressor protein. Thus, negatively controlled inducible operons are normally "turned off" while repressible operons are normally "turned on" unless a secondary molecule interacts with the respective regulatory proteins. (*b*) Positive control mechanisms can also regulate inducible or repressible operons. Inducible systems utilize an activator protein that requires the presence of a coactivator molecule in order to enhance transcription. Positively controlled repressible systems produce an activator protein that will enhance the transcription of target structural genes unless an inhibitor molecule is present. t, terminator; P, promoter; O, operator.

The product of the *lacZ* gene, ß-galactosidase, cleaves the ß-1,4 linkage of lactose, releasing the free monosaccharides. The enzyme is a tetramer of four identical subunits, each with a molecular weight of 116,400. Entrance of lactose into the cell requires the *lac* permease (46,500), the product of the *lacY* gene. The permease is hydrophobic and probably functions as a dimer. Mutations in either the *lacZ* or *lacY* genes are phenotypically Lac⁻ — that is, the mutants cannot grow on lactose as a sole carbon source. The *lacA* locus is the structural gene for thiogalactoside transacetylase (mwt 30,000) for which no definitive role has been assigned. The promoter and operator for the *lac* operon are *lacP* and *lacO*, respectively. The *lacI* gene codes for the repressor protein, and in this system is situated next to the *lac* operon. Often, regulatory loci encoding for diffusible regulator proteins map some distance from the operons they regulate. The *lacI* gene product (mwt 38,000) functions as a tetramer.

The *lac* operator is 28 bp in length and is adjacent to the ß-galactosidase structural gene (*lacZ*). Figure 5-2 presents the complete sequence of the *lacP-O* region, including the C terminus of *lacI* (the repressor gene) and the N terminus of *lacZ*. The operator overlaps the promotor in that the *lac* repressor, when bound to the *lac* operator in vitro, will protect part of the promoter region from nuclease digestion. The mechanism of repression remains somewhat controversial. There is evidence that under some in vitro conditions (e.g., low salt), RNA polymerase can bind to the promotor in the presence of *lac* repressor. Thus, one proposal is that the binding of repressor to the operator region situated between the promoter and *lacZ* physically blocks transcription by preventing the release of RNA polymerase from the promoter and its movement into the structural gene. Other evidence indicates that RNA polymerase and *lac* repressor cannot bind simultaneously, leading to a model whereby *lac* repressor simply competes with RNA polymerase for binding in the promoter/operator region.

Whatever the mechanism, induction of the *lac* operon, outlined in Figure 5-3, occurs when cells are placed in the presence of lactose (however, see "Catabolite control"). The low level of ß-galactosidase that is constitutively present in the cell will convert lactose to **allolactose** (the galactosyl residue is present on the 6 rather than the 4 position of glucose). Allolactose is the actual inducer molecule. The *lac* repressor is an allosteric molecule with distinct binding sites for DNA and the inducer. The binding of the inducer to the tetrameric repressor occurs whether the repressor is free in the cytoplasm or bound to DNA. Binding of the inducer to the repressor, however, allosterically alters the repressor, lowering its affinity for *lacO* DNA. Once the repressor is removed from *lacO*, transcription of *lacZYA* can proceed. Thus, the *lac* operon is a negatively controlled inducible system. It should be noted that experiments designed to examine induction of the *lac* operon usually take advantage of a **gratuitous inducer** such as isopropyl-ß-thiogalactoside (IPTG) — a molecule that will bind the repressor and inactivate it but is not itself a substrate for ß-galactosidase. This eliminates any secondary effects that the catabolism of lactose may have on *lac* expression (see "Catabolite control").

There are several regulatory mutations that have been identified in the *lac* operon that serve to illustrate general concepts of gene expression. Mutations in the *lacI* locus can give rise to three observable phenotypes. The first and most obvious class of mutations results in an absence of, or a nonfunctional, repressor. This will lead to a constant synthesis of *lacZYA* message regardless of whether the inducer is present. This is referred to as **constitutive expression** of the *lac* operon. The second class of *lacI* mutations, *lacI*ˢ, produces a **superrepressor** with increased operator binding

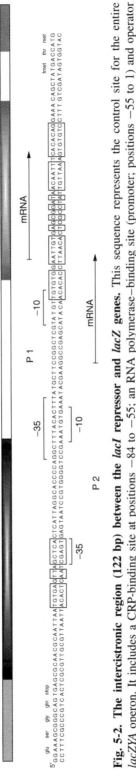

Fig. 5-2. The intercistronic region (122 bp) between the *lacI* repressor and *lacZ* genes. This sequence represents the control site for the entire *lacZYA* operon. It includes a CRP-binding site at positions −84 to −55; an RNA polymerase–binding site (promoter; positions −55 to 1) and operator region (1–28). Symmetrical regions in the CRP and operator are boxed. The −10 (Pribnow box) and −35 regions that constitute the sequences recognized by RNA polymerase are shown as overlapping the promoter.

199

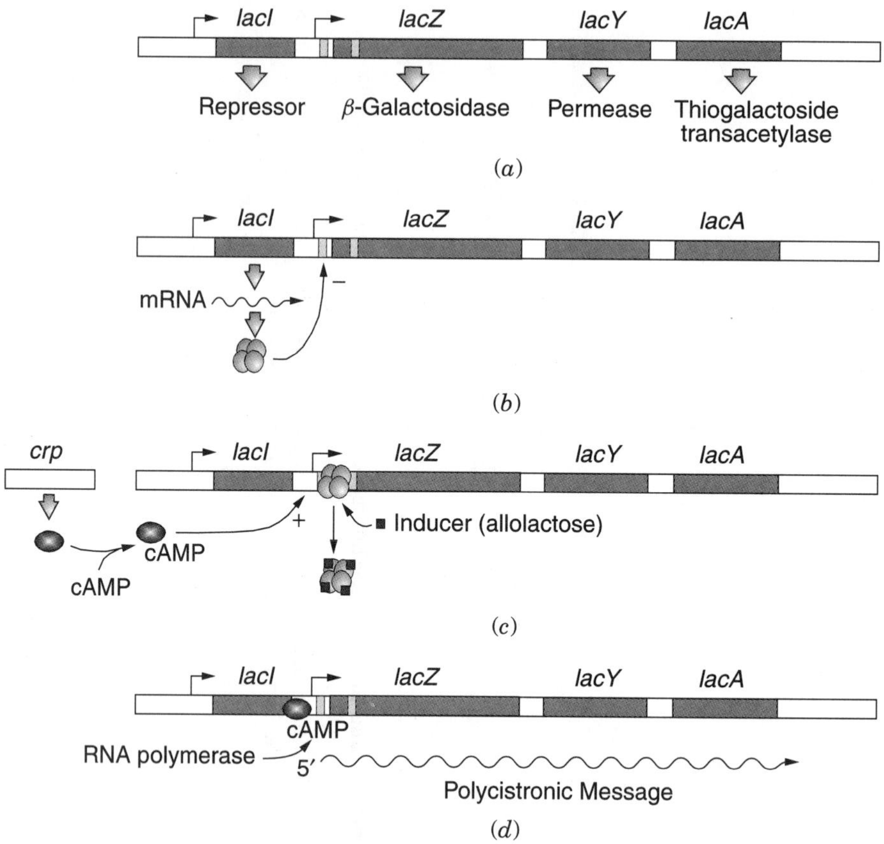

Fig. 5-3. Regulation of the lactose operon. (*a*) Genetic organization and products of the *lac* operon. (*b*) Transcription and translation. The *lacI* gene produces the repressor protein, which, as a tetramer, binds to the *lac* operator region and prevents transcription of the *lac* structural genes. (*c, d*) In the presence of inducer (allolactose), the inducer binds to and allosterically changes the conformation of the LacI repressor such that it will no longer bind to the operator. This allows transcription to proceed. Under conditions of high cAMP concentrations, cAMP will bind to cAMP receptor protein (CRP). The CRP–cAMP complex binds to a specific CRP site on *lacP*. This facilitates polymerase binding and, through interactions with the a subunit, stimulates open complex formation and transcription of the *lac* structural genes.

and/or diminished inducer (IPTG)-binding properties. These mutations result in an uninducible *lac* operon. The *lacI* locus contains its own promoter, *lacI*^{p}, distinct from *lacP*. The third class of *lacI* mutations occurs in its promoter. Mutations in the *lacI* promoter, called *lacI*^{q}, have been identified as promoter-up mutations that increase the level of *lacI* transcription. The region of the promoter affected by *lacI*^{q} mutations is in the −35 region. Increased transcription due to the *lacI*^{q} mutation occurs by virtue of facilitated RNA polymerase binding. The net result is a 10–50-fold increase in the *lacI* gene product (*lac* repressor). This increased production of LacI repression leads to an almost total absence of basal expression of *lacZYA* yet still allows induction in the presence of the inducer. Certain mutations in the *lac* operator (*lacO*^{c}) can also result in constitutive expression of the *lac* operon by causing decreased affinity for the repressor.

DNA Looping. Although Figure 5-3 suggests the presence of a single DNA-binding site for the tetrameric LacI repressor, there are actually three DNA-binding sites, all of which are involved in repression. This introduces the concept of **DNA looping** in the control of transcription. Besides the *lacO* site already introduced, the two other binding sites are O_I at -100 bp (placing it within *lacI*) and O_Z at $+400$ bp (placing it within *lacZ*). Binding of LacI to these sites tethers them together, forming competitive loops, meaning either an O_I loop or an O_Z loop can form. The O_I loop represses initiation while the O_Z loop inhibits mRNA synthesis, both directly by preventing transcription elongation and indirectly by stabilizing repressor binding at the primary operator, O. These auxiliary operators may also simply increase the relative concentration of LacI around the principal operator. This is important because there are only 10 tetrameric LacI molecules in the cell. The binding of one dimer of a tetrameric LacI to O_I leaves the other dimer free to search for O. Since O_Z and O are 400 bp apart, this tethering effectively increases the concentration of LacI in the vicinity of O about 200-fold.

"Priming the Pump". If the *lac* repressor prevents expression of the *lac* operon, then how does the cell allow for production of the small amount of β-galactosidase needed to make allolactose inducer and enough LacP permease to enable full induction when the opportunity arises? A reexamination of Figure 5-2 reveals the presence of a second *lac* promoter called P2. This promoter binds RNA polymerase tightly but initiates transcription poorly. Once initiation occurs from P2, RNA polymerase can transcribe past the LacI-bound operator and into the structural *lacZYA* genes. Note also that the message produced from P2 will contain a large palindromic region corresponding to the operator-binding site. The stem-loop structure that forms will sequester the *lacZ* SD sequence, allowing only minimal production of β-galactosidase. This palindromic sequence is not produced from P1.

Catabolite Control: Sensing Energy Status

The *lac* operon has an additional, positive regulatory control system. The purpose of this control circuit is to avoid wasting energy by synthesizing lactose-utilizing proteins when there is an ample supply of glucose available. Glucose is the most efficient carbon source for *E. coli*. Since the enzymes for glucose utilization in *E. coli* are constitutively synthesized, it would be pointless for the cell to also make the enzymes for lactose catabolism when glucose and lactose are both present in the culture medium. The phenomenon can be visualized in Figure 5-4. This classic experiment shows *E. coli* initially growing on succinate with IPTG as an inducer of β-galactosidase activity (*lacZ* gene product). The early part of the graph shows an increase in β-galactosidase activity due to the induction. Then, at the point indicated, glucose is added to one culture. A dramatic cessation of further β-galactosidase synthesis is observed (**transient repression**) followed by a partial resumption of synthesis (**permanent repression**). It was presumed that a catabolite of glucose was causing this phenomenon, hence the term catabolite repression. However, adding cyclic AMP simultaneously with glucose reduces catabolite repression.

The phenomenon of catabolite repression is partly based on the fact that when *E. coli* is grown on glucose, intracellular cAMP levels decrease, but when grown on an alternate carbon source such as succinate, cAMP levels increase. The enzyme responsible for converting ATP to cAMP is adenylate cyclase, the product of the *cya*

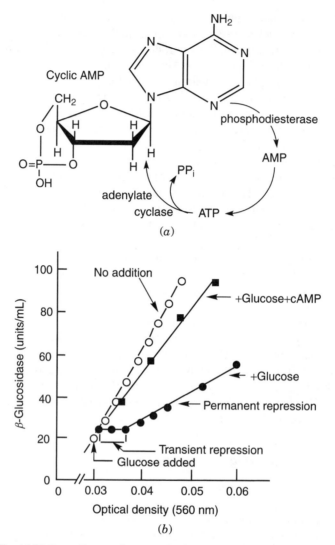

Fig. 5-4. Cyclic AMP formation and reversal of catabolite repression of β-galactosidase synthesis by cyclic AMP. (*a*) Formation of cyclic AMP by adenylate cyclase and its degradation by cAMP phosphodiesterase. (*b*) Catabolite repression of β-galactosidase synthesis and its reversal by cyclic AMP. Glucose, with or without cAMP, was added to a culture of *E. coli* growing on succinate in the presence of an inducer of the enzyme, isopropyl-β-o-thiogalactoside (IPTG). The addition of cAMP overcomes both the transient (complete) and the permanent (partial) repression by glucose. The brief lag before repression reflects the completion and the translation of the already initiated messenger.

locus. How does cAMP increase transcription of the *lac* operon? The evidence indicates that cAMP binds to the product of the *crp* locus termed the cAMP receptor protein (CRP, sometimes referred to as catabolite activator protein, CAP). The CRP–cAMP complex then binds to the CRP-binding site on the *lac* promoter. CRP represses transcription from *P*2 but activates *P*1. (Fig. 5-2). Several studies have shown that

CRP causes DNA bending of around 90° or more. The bend enables CRP to directly interact with RNA polymerase (see below). CRP interaction with RNP increases RNP promoter binding as well as allows RNA polymerase to escape from the promoter and proceed through elongation.

Many operons are affected by cAMP and CRP in a similar manner. These include the *gal*, *ara*, and *pts* operons involved with carbohydrate utilization. The genes controlled by CRP are referred to as members of the **carbon/energy regulon**. Other operons not directly related to carbohydrate catabolism are positively regulated by cAMP as well. The CRP–cAMP complex can also act as a negative effector for several genes including the *cya* and *ompA* loci.

What, then, is the mechanism by which glucose reduces cAMP levels? What is the catabolite responsible? Actually, it is not a catabolite per se but rather the inducible component of the phosphotransferase system, IIA^{glu} (the product of the PTS *crr* gene), which is at the center of this control system. IIA^{glu} controls the activity of preexisting adenylate cyclase rather than its synthesis. The model is presented in Figure 5-5. The PTS system employs several proteins, some of which are specific for a given sugar, to

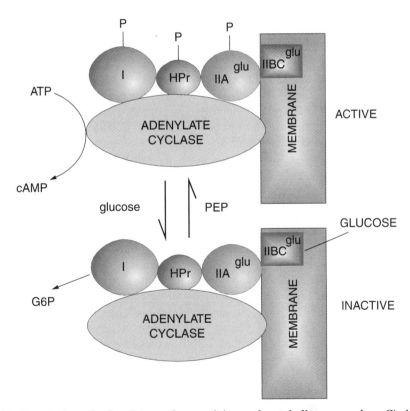

Fig. 5-5. Regulation of adenylate cyclase activity and catabolite repression. Circles and squares represent components of the phosphotransferase system; IIA, inducible component of phosphotransferase system. **Top**: Disuse of the PTS will lead to the accumulation of phosphorylated-IIA^{glu} and thus active adenylate cyclase. **Bottom**: Transport of glucose through the PTS will deplete phosphorylated III^{glu}. The dephosphorylated form of III^{glu} will inhibit adenylate cyclase activity.

transfer a phosphate from PEP to a carbohydrate during transport of that carbohydrate across the membrane. The phosphate group is transferred from protein to protein until, in the case of glucose, it reaches enzyme IIA^{glu}. In the absence of glucose, IIA^{glu} remains phosphorylated. It then interacts with and activates adenylate cyclase in the presence of inorganic phosphate. The result is a dramatic increase in cAMP levels.

However, when glucose is available and transported across the membrane, the phosphate on IIA^{glu} is transferred to the sugar forming glucose-6-phosphate. The dephosphorylated IIA^{glu} binds and inhibits adenylate cyclase activity, causing intracellular cAMP levels to diminish. So, when cells are grown on non-PTS sugars (e.g., glycerol), cAMP levels are high and the CRP–cAMP complex forms. This complex then acts at the transcriptional level to alter the expression of several operons either positively or negatively, depending on the operon. There is also evidence that when CRP–cAMP levels get too high, phosphatases dephosphorylate IIA^{glu}-P, Which will downregulate adenylate cyclase by decreasing IIA^{glu}-P activator.

CRP–cAMP actually plays two different roles in controlling the *lac* operon. Aside from activating *lacZYA* expression when lactose is present, it actually appears to increase the affinity of LacI to *lacO* by cooperative binding! So, when lactose is present, CRP–cAMP activates *lacZYA* transcription, but when lactose is absent, CRP–cAMP increases repression by LacI.

Another property of catabolite repression is **diauxie growth**. When *E. coli* is given a choice of glucose or lactose in the medium, the organism preferentially uses the glucose. The use of lactose is prevented until the glucose is used up. While part of the reason for this is due to the cAMP effect, the major cause of diauxie is inducer exclusion in which nonphosphorylated IIA^{glu} plays a role. In inducer exclusion, glucose prevents the uptake of lactose, the inducer for the *lac* operon. It appears that nonphosphorylated IIA^{glu} inhibits the accumulation of carbohydrates via non-PTS uptake systems (e.g., lactose). It has been shown that a direct interaction occurs between IIA^{glu} and the lactose carrier (LacP permease), the result being the inhibition of lactose transport.

Class I and Class II CRP-Dependent Genes

CRP, acting as a dimer, regulates more than 100 genes. The cAMP–CRP complex binds to target 22 bp sequences located near or within CRP-dependent promoters. At these promoters, the cAMP–CRP complex activates transcription by making direct protein–protein interactions with RNA polymerase. Simple CRP-dependent promoters contain a single CRP-binding site and are grouped into two classes. Class I CRP-dependent promoters have the CRP-binding site located upstream of the RNAP-binding site (e.g., *lacP1*, −61 bp from transcriptional start). Class II CRP-dependent promoters (e.g., *galP1*) overlap the RNAP-binding site. At class I promoters, CRP activates transcription by contacting RNAP via a surface-exposed β turn on the downstream CRP subunit. The contact patch on CRP, called activating region 1 (AR1), makes contact with the C-terminal domain to the RNAP α subunit (α-CTD). This serves to recruit RNAP to the promoter (Fig. 5-6).

The process of transcription activation at class II promoters is more complex. RNAP binds to promoter regions both upstream and downstream of the DNA-bound CRP, making multiple interactions with CRP. In this situation, AR1 in the upstream CRP subunit interacts with α-CTD situated upstream of CRP. The second activating region (AR2) is a positively charged surface on the downstream CRP subunit. AR2 interacts with the N-terminal domain of the α subunit (α-NTD) but only at class II promoters.

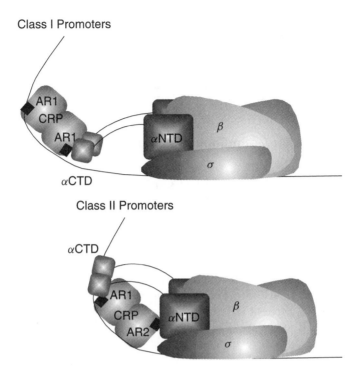

Fig. 5-6. Classes of CRP-dependent promoters. The position of the CRP-binding site on DNA relative to the RNAP-binding site influences which regions of CRP are available to interact with the α-subunit C-terminal domain. The patches on CRP that can bind the α-CTD are called activating regions (AR1 and AR2).

The Catabolite Repressor/Activator Protein Cra

The cAMP–CRP story only partly explains the extent by which catabolite repression affects cell physiology. The rest of the story involves another global regulator called Cra (catabolite repressor/activator) protein. The physiological significance of this regulator is the control of so-called peripheral pathways that allow growth on substrates that are not intermediates in a central pathway such as glycolysis. These peripheral pathways can take intermediates in the TCA cycle, such as malate, and convert them to pyruvate and back up to F-6-P (a reverse glycolysis). This process provides the carbon compounds needed for the synthesis of critical cell components (e.g., nicotinamide coenzymes). Details can be found in Chapter 8.

The Cra regulator controls numerous genes, some positively, others negatively, in response to fructose-6-phosphate or fructose-1,6-bisphosphate. Following transport of glucose, the resulting G-6-P is converted first to F-6-P and then to F-1,6-bP through glycolysis. The presence of a Cra-binding site [consensus TGAA(A/T)C·(C/G)NT(A/C)(A/C)(A/T)] upstream of the RNA polymerase-binding site is found in genes activated by Cra [PEP synthase (*ppsA*), PEP carboxykinase (*pckA*), isocitrate dehydrogenase (*icd*), fructose-1,6-bisphosphatase (*fdp*)]. Genes repressed by Cra have Cra-binding sites that overlap or follow the RNAP-binding site [Entner-Douderoff enzymes (*edd-eda*), phosphofructokinase (*pfk*), pyruvate kinase (*pykF*)]. Cra is displaced from binding sites by binding F-1-P or F-1,6-bP. Thus, as glucose is used, F-6-P accumulates, which then causes derepression of Cra-repressed genes such as *pfk* and *pykF* and deactivation of Cra-activated genes, PEP synthetase, and PEP carboxykinase. As a consequence, when cells are growing on glucose, gluconeogenesis genes are switched off, but genes involved in glucose utilization (e.g., Entner-Douderoff or Embden-Meyerhof; see Chapter 8) are turned on. Figure 5-7 illustrates some of the relationships between the PTS system, adenylate cyclase, and Cra.

Catabolite Control: The Gram-Positive Paradigm

Bacillus subtilis and other gram-positive microorganisms utilize a different system for catabolite repression. A protein called CcpA (catabolite-control protein) is able to differentially bind the phosphorylated forms of HPr. As part of the PTS transport system (see Chapter 9), HPr is phosphorylated at residue His-15 in the absence of carbohydrate transport. When a suitable carbohydrate is transported through the PTS system, HPr donates the phosphate at His-15 in a cascade leading to phosphorylation of the carbohydrate (Fig 5-5). Thus, the presence of phosphorylated HPr indicates the absence of a PTS sugar. In contrast, when glucose is present, the level of phosphorylated HPr is very low. Glucose, however, activates an HPr kinase (PtsK) that phosphorylates a different HPr residue, Ser-46. Phosphorylation of HPr at Ser-46 indicates the presence of glucose. The CcpA protein binds to HPr (ser-P). The complex then acts as a transcriptional regulator of the catabolite-sensitive genes.

The *gal* Operon: DNA Looping with a Little Help from Hu

The *gal* operon of *E. coli* consists of three structural genes, *galE*, *galT*, and *galK*, transcribed from two overlapping promoters upstream from *galE*, P_{G1} and P_{G2} (Fig. 5-8). Regulation of this operon is complex because aside from being involved with the utilization of galactose as a carbon source, in the absence of galactose the *galE* gene product (UDP–galactose epimerase) is required to convert UDP-glucose to UDP-galactose, a direct precursor for cell wall biosynthesis. While transcription from both promoters is inducible by galactose, it is imperative that a constant basal level of *galE* gene product be maintained even in the absence of galactose. The *gal* operon is also a catabolite-repressible operon. When cAMP levels are high the CRP–cAMP complex binds to the −35 region (*cat* site), promoting P_{G1} transcription but inhibiting transcription from P_{G2}. When grown on glucose, however, where cAMP levels are low, transcription can occur from P_{G2} assuring a basal level of *gal* enzymes (Fig. 5-8).

Both of the *gal* promoters are negatively controlled by the *galR* gene product, a member of the LacI family of repressors. The *galR* locus is unlinked to the *gal* operon. There are two operator regions to which the repressor binds: an extragenic operator (O_E), located at −60 from the P_{G1} transcription start site (S1), and an intragenic operator (O_I), located at position +55 from S1 start. A *galR* repressor dimer binds independently to O_E and O_I. For repression to occur, the two DNA-bound GalR dimers must interact with each other, forming a tetramer (Fig. 5-8c) that engineers the DNA into a repression loop. However, GalR dimers only weakly interact. In order to stabilize that interaction, the histone-like protein Hu binds to a specific site (*hbs*) between the two operators. The DNA loop sequesters P_{G1} and P_{G2} from RNA polymerase access. Some access must occur, however, since, as mentioned above, basal levels of the *gal* enzymes are synthesized even under repressed conditions.

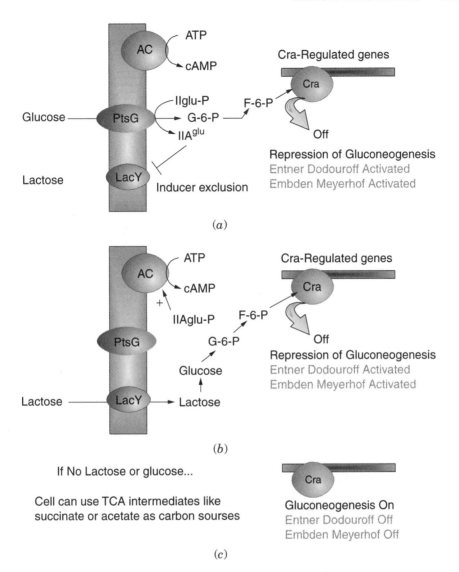

Fig. 5-7. Regulation of carbohydrate metabolism in *E. coli* by Cra. (*a*) Growth on glucose. The cell will want to inhibit gluconeogenesis, stimulate glycolysis, and prevent use of other carbon sources such as lactose. Consequently, the synthesis of cAMP is low, limiting expression of other carbohydrate operons; the PTS protein II^{glu} accumulates, inhibiting lactose transport; and F-6-P is produced. F-6-P binds Cra, releasing it from target genes. The results are repression of gluconeogenesis, which is not required because glucose is plentiful, and an increase in glycolytic enzymes required to use glucose. (*b*) When lactose is present (no glucose), II^{glu}-P accumulates, stimulating cAMP synthesis, which increases *lac* operon expression. The catabolism of lactose also produces significant levels of F-6-P, so, as with glucose catabolism, gluconeogenesis in low while glycolysis is high. (*c*) If no carbohydrates are available, the cell can use TCA cycle intermediates as carbon sources. This is made possible because F-6-P is not made at high levels, allowing Cra to bind to its target genes. This stimulates the gluconeogenesis pathways needed to use the TCA intermediates and limits glycolysis.

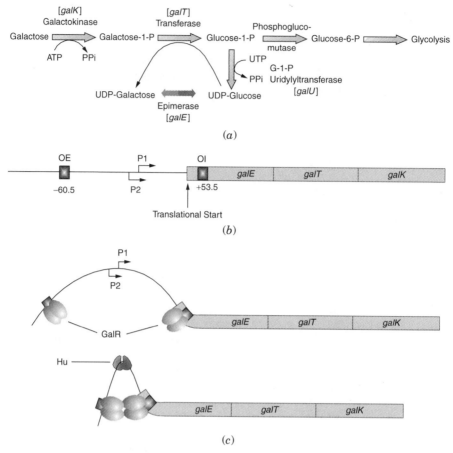

Fig. 5-8. Regulation of the galactose operon. (*a*) Biochemical pathway. (*b*) Genetic organization of the *galETK* operon including transcription and translation starts. (*c*) Architecture of the GalR–Hu *gal* DNA complex. Repression of the *gal* operon by the *galR* product binding to the internal and external operator regions and Hu binding between the two operators.

The Arabinose Operon: One Regulator, Two Functions

Catabolism of the carbohydrate L-arabinose by *E. coli* involves three enzymes encoded by the contiguous *araA*, *araB*, and *araD* genes (Fig. 5-9). The product, xylulose-5-phosphate (xl-5-p), can be catabolized further by entering the pentose phosphate cycle as outlined in Figure 1-10 (see also Chapter 8). The transcriptional activity of these genes is coordinately regulated by a fourth locus, *araC*. The *araC* locus and the *araBAD* operon are divergently transcribed from a central promoter region (Fig. 5-10b). Both the *araC* promoter (P_C) and the *araBAD* promoter (P_{BAD}) are stimulated by cAMP and CRP. Regulation by *araC* is unusual, however, in that the *araC* product activates transcription of *araP*$_{BAD}$ in the presence of arabinose (inducing condition) yet represses transcription of both promoters in the absence of arabinose (repressive condition). Thus, *araC* protein can assume two conformations: repressor and activator (Fig. 5-10). The *araC* product also regulates its own expression in the presence or absence of arabinose. The arabinose operon was one of the first examples where looped DNA structures were shown to regulate transcription initiation.

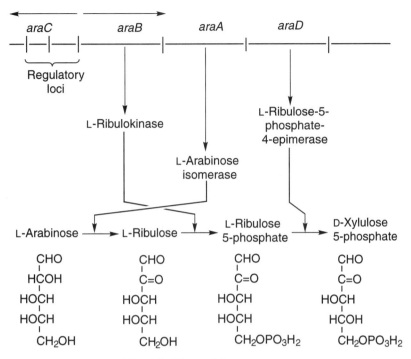

Fig. 5-9. The arabinose operon.

AraC functions as a dimer in vivo. Figure 5-10a presents the basic domain structure of an AraC monomer and two alternative dimer arrangements. The protein consists of a C-terminal DNA-binding domain connected by a flexible linker to an N-terminal dimerization domain. The AraC molecule ends with an N-terminal arm that can bind either the C-terminal domain of the same molecule or the N-terminal domain of a partner AraC molecule depending on whether arabinose is present. Without arabinose, the N-terminal arm binds to the C terminus, which, combined with the linker, forms a very rigid structure. Dimerization of this form results in the rather long complex shown in Figure 5-10a. In the presence of arabinose, the N-terminal arm preferentially binds the N-terminal domain of the partner molecule in an AraC dimer. As a result, only the flexible linker connects the main domains, which allows the dimer to attain the third, more compact, structure in Figure 5-10a. Understanding these very different structures is the key to understanding regulation of this system.

Figure 5-10b presents the *ara* operator and promoter regions. The AraC protein can bind to four operators: O_1, O_2, I_1, and I_2 (I for induction). In the absence of arabinose, the rigid AraC dimer can only bind to O_2 and I_1, forming a negative repression loop that prevents *araBAD* expression (Fig. 5-10c). Upon binding to arabinose, it becomes energetically more favorable for the N-terminal arm to bind to the dimerization domain of the AraC partner than to the DNA-binding domain. The dimer is now flexible enough to bind to I_1 and I_2. Occupancy of the I_2 site places AraC near the RNP-binding site, allowing interactions to occur between these two proteins. It is important to note that AraC can sample arabinose while remaining bound to DNA. Having AraC already

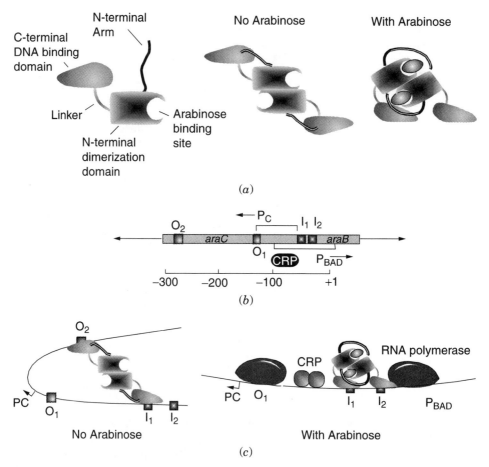

Fig. 5-10. Model for control of the *araBAD* and *araC* promoters. (*a*) Alternative structures of an AraC dimer formed in the presence or absence of arainose. (*b*) The divergent promoter region of the *ara* operon. O_1, O_2, I_1, and I_2 are binding sites for AraC. P_c and P_{BAD} represent the promoter regions for *araC* and *araBAD*, respectively. The CRP box marks the binding region for CRP. (*c*) Alternative AraC binding architectures that allow differential regulation of this divergently promoted operon.

bound to the regulatory region in the absence of arabinose allows it to respond very quickly to the addition of arabinose, since the potentially slow DNA-binding step has already taken place.

Transcription from P_{BAD} is also stimulated by cAMP–CRP through protein–protein interactions with RNA polymerase. The α-CTD of RNA polymerase α subunit contacts a third activating region of CRP called activation region 3 (AR3). The combination of arabinose binding AraC and binding of cAMP–CRP to its recognition site causes a dramatic increase in *araBAD* expression.

There are now over 100 transcriptional regulators that contain regions homologous to the helix-turn-helix DNA-binding domain of AraC. These proteins are classified as the **AraC family** of regulators.

ATTENUATION CONTROLS

Transcriptional Attenuation Mechanisms

The trp Operon. The genes that code for the enzymes of the tryptophan biosynthetic pathway in *E. coli* are assembled in an operon (Fig. 15-17). There are two regulatory mechanisms that control the expression of this operon at the transcriptional level: transcriptional end-product repression and transcriptional attenuation. Repression involves the product of the *trpR* gene (repressor) binding tryptophan. This increases the affinity of the repressor for the *trp* operator region, thus blocking transcription. However, there is a second level of regulation placed on this and many other amino acid biosynthetic operons.

This control mechanism, called **attenuation**, depends on the coupling between transcription and translation as illustrated in Figure 1-8. Between the *trpO* and *trpE* genes is the attenuator region. This region codes for a small (14 amino acid) nonfunctional polypeptide called the leader polypeptide. The key to attenuation is understanding the competitive stem-loop structures that can form in the leader mRNA. Figure 5-11a presents the sequence of the *trp* leader mRNA along with the amino acid sequence of the leader polypeptide. The brackets above the mRNA sequence show regions of mRNA capable of base pairing with each other, leading to transcription termination. Region 1 can form a stem structure (Fig. 5-11) with region 2 (1:2 stem), and region 3 can form a stem structure with region 4 (3:4 stem). The 3:4 stem (called the **attenuator stem-loop**) principally acts as a transcription termination signal. If it is allowed to form, transcription terminates prior to the *trpE* gene. The brackets below the message show regions that can base pair in an **antiterminator** configuration (2:3 stem-loop). The formation of a 2:3 loop prevents the formation of the 3:4 terminator stem; thus, transcription is allowed to proceed into the *trp* structural genes.

How, then, does the cell regulate the formation of the 2:3 or 3:4 secondary structures? This is accomplished by varying the rate of translation of the leader polypeptide relative to the rate of transcription (Fig. 5-12). Notice the two adjacent Trp codons in the leader sequence. If tryptophan levels within the cell are low, then the levels of charged tRNA^{trp} will also diminish. As the ribosome translates the leader polypeptide, it will stall over the region containing the Trp codons (Fig. 5-12c), overlapping and thereby preventing region 1 from binding to region 2 and allowing RNA polymerase to synthesize more of the leader transcript. With the 1:2 stem unable to form and once RNA polymerase has transcribed past region 3, the 2:3 stem-loop (called the **antiterminator loop**) can form, which will prevent formation of the 3:4 terminator. Consequently, RNA polymerase can proceed into the *trp* structural genes, ultimately leading to an increase of tryptophan levels. (There is another Shine-Dalgarno sequence in *trpE*, allowing for ribosome binding).

Once intracellular tryptophan (and, therefore, charged tRNA^{trp}) levels are high, the ribosome will proceed to the translational termination codon present in the leader (Fig. 5-12d). In this situation, the ribosome overlaps region 2 and prevents the formation of a stable 2:3 loop. Consequently, when RNA polymerase transcribes past region 4, the 3:4 transcription terminator stem forms and halts transcription prior to the *trp* structural genes. Attenuation is a common regulatory mechanism for amino acid operons. Figure 5-11b illustrates the attenuator region and alternative stem-loop structures for the *his* operon.

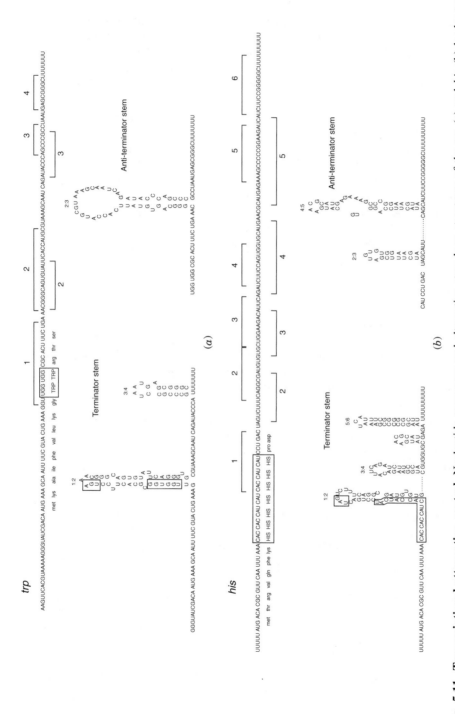

Fig. 5-11. Transcriptional attenuation control. Nucleotide sequences and alternative stem-loop structures of the *trp* (a) and *his* (b) leader transcripts amino acid biosynthetic operons regulated by attenuation. Brackets above and below the sequences show the regions that can base pair in the termination and antitermination configurations, respectively. The regulatory codons (capital letters) are boxed.

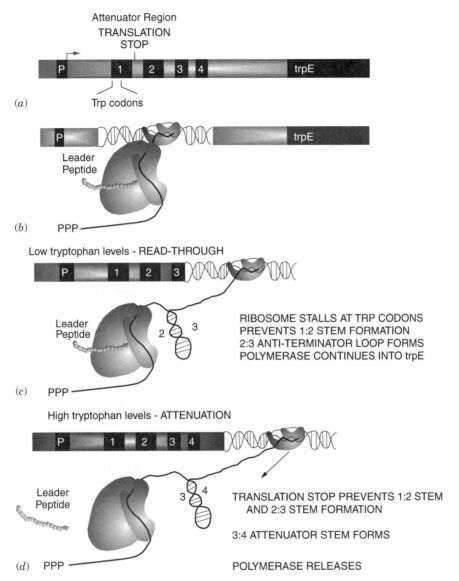

Fig. 5-12. Model for attenuation of the tryptophan operon. (*a*) The attenuator region illustrating positions of promoter (P), tryptophan codons (vertical lines), and regions capable of forming stem-loop structures (1, 2, 3, 4). (*b*) Coupled transcription and translation of leader polypeptide. (*c*) Under conditions where internal tryptophan levels are low, the ribosome stalls over the *trp* codons in the message due to a depletion of charged tryptophanyl tRNA. This prevents 1 : 2 stem formation but allows the 2 : 3 stem to form. Consequently, the 3 : 4 attenuator loop cannot form and will permit RNA polymerase to read through into the *trp* structural genes. (*d*) At high levels of internal tryptophan, the ribosome encounters the translational stop codon, preventing the formation of the 1 : 2 and 2 : 3 stem structures. RNA polymerase continues, allowing 3 : 4 stem formation, which is a transcriptional stop signal. Transcription does not proceed into the *trp* structural genes.

Attenuation by TRAP in Bacillus subtilis. In the discussion of the *E. coli* tryptophan operon, the concept of attenuation involving ribosome stalling was introduced. However, there are other ways a cell can utilize attenuation that do not involve ribosome stalling. The first is an RNA-binding protein-dependent attenuation system that regulates the formation of competing mRNA stem structures in a 5′ untranslated region. In Bacillus, a protein called TRAP (tryptophan RNA-binding attenuation protein) forms an 11-member ring structure that can bind tryptophan. Tryptophan-activated TRAP binds to 11 closely spaced (G/U)AG repeats present in the nascent *trpEDCFBA* leader transcript. TRAP binding prevents formation of an antiterminator structure in the leader region, which leaves an overlapping Rho-independent terminator free to form. Thus, TRAP binding promotes transcription termination before RNA polymerase reaches the structural genes. In the absence of TRAP, the antiterminator structure prevents formation of the terminator, resulting in transcriptional readthrough into the *trp* structural genes.

The pyrBI Strategy. In *E. coli* or *Salmonella enterica*, the de novo synthesis of UMP, the precursor of all pyrimidine nucleotides, is catalyzed by six enzymes encoded by six unlinked genes and operons (see Chapter 16). One operon within this system is *pyrBI*, which codes for the catalytic (PyrB) and regulatory (PyrI) subunits of aspartate transcarbamylase. The operon is negatively regulated over 100-fold by intracellular UTP concentration. Attenuation in this system involves RNA polymerase pausing rather than ribosome stalling. The basic model is presented in Figure 5-13. As with the tryptophan operon, regulation depends on the degree of coupling between transcription and translation. Within the *pyrBI* regulatory region there is an RNA polymerase pause site flanked by U-rich (T-rich in the DNA) areas (Fig. 5-13a). When cellular UTP levels are low (meaning the cell will want to synthesize more UMP), RNA polymerase undergoes a lengthy pause as it tries to find UTPs to extend the message (Fig. 5-13c). This strong pause allows the translating ribosome to translate up to the stalled RNA polymerase. When the polymerase finally escapes the pause region and reaches the transcriptional attenuator region, the ρ-independent terminator hairpin cannot form due to the close proximity of the ribosome. This permits RNA polymerase to transcribe into the *pyrBI* structural genes. However, when UTP levels are high, RNA polymerase undergoes only a weak pause, then quickly releases to transcribe the attenuator before the ribosome has caught up (Fig. 5-13b). Therefore, the transcription termination loop forms, terminating transcription before RNA polymerase has entered structural genes.

It may seem that attenuation is a very costly and inefficient regulatory system for the cell. Why would nature provide a regulatory mechanism that involves the synthesis of nonfunctional polypeptides or transcripts? Remember that all regulatory systems require information and so of necessity require the expenditure of energy. The energetic cost of attenuation may be less than that required to synthesize a larger trans-acting regulatory protein. A clear advantage of attenuation is that it uses information present in the nascent RNA transcript. This is a strategy that quickly and directly translates what is sensed (i.e., low tryptophan levels) into action (i.e., increased transcription).

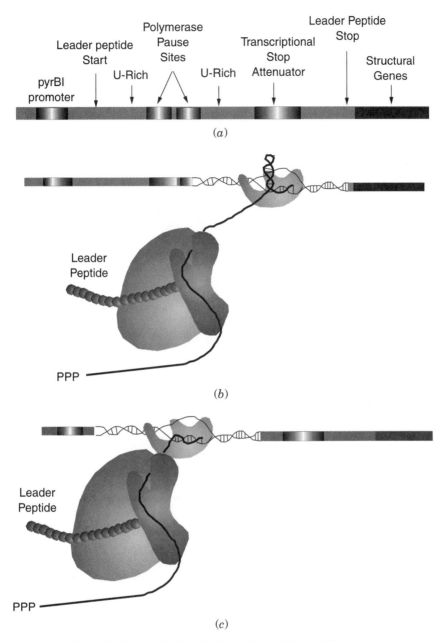

Fig. 5-13. Transcriptional attenuation of the *pyrBI* operon.

Translational Attenuation Control: The *pyrC* Strategy

The *pyrC* locus encodes the pyrimidine biosynthetic enzyme dihydroorotase (see Chapter 16). Expression of this enzyme varies over 10-fold depending on pyrimidine availability. Regulation is not due to the modulation of mRNA levels but is the result of an elegant translational control mechanism. The promoter region for *pyrC* is shown in Figure 5-14. Transcription starts at one of the bases indicated by an asterisk and depends on pyrimidine availability. When grown under pyrimidine excess, transcription starts predominantly at position C_2. Transcripts started at this site can form a hairpin loop due to base pairing between regions indicated by arrows. This hairpin blocks ribosome binding by sequestering a Shine-Dalgarno sequence and lowers production of the *pyrC* product.

Alternatively, when pyrimidine availability is low, transcripts preferentially start at G_4. This shorter transcript cannot form the hairpin and thus will have an SD sequence available for ribosome binding. This translational control of *pyrC* is a striking example of a regulatory mechanism employing only the basic components of the transcriptional and translational machinery. RNA polymerase directly senses nucleotide availability and by differential transcription initiation will control translational initiation of the transcript.

Membrane-Mediated Regulation: The *put* System

Salmonella enterica can degrade proline for use as a carbon and/or nitrogen source. The process requires two enzymes: a specific transport system encoded by *putP* and the *putA*-coded bifunctional membrane-bound proline oxidase that converts proline

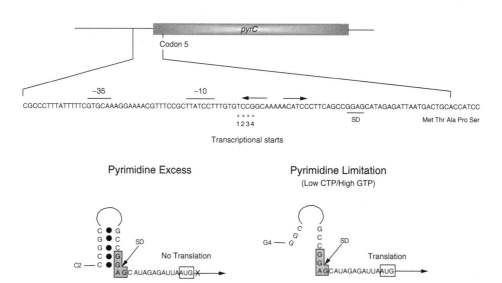

Fig. 5-14. Translational control of *pyrC* expression. The top figure shows the *pyrC* locus with the promoter region expanded. Arrows above the sequence indicate the areas capable of base pairing in the mRNA. The numbered asterisks indicate potential transcriptional start sites. The choice of start site varies with the ratio of CTP to GTP in the cell. The bottom pair of illustrations indicate the mRNA stem-loop structure that forms during pyrimidine excess but fails to form during pyrimidine limitation due to the shorter transcript. This stem-loop structure will sequester a ribosome-binding site and prevent synthesis of the *pyrC* product.

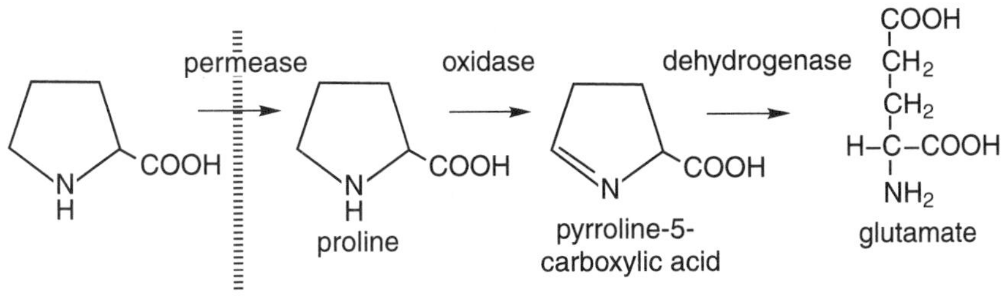

Fig. 5-15. Proline degradation pathway. P-5-C, pyrroline-5-carboxylic acid.

to glutamate (Fig. 5-15). The two genes are adjacent but are transcribed in opposite directions from a common controlling region. The *putA* gene product also possesses regulatory properties in that it represses the expression of both *putA* and *putP*. The regulatory and enzymatic properties of *putA* are separable by mutation — that is, some *putA* point mutations result in the constitutive expression of *putAP* yet do not affect proline oxidase activity.

The PutA protein is a flavoprotein that, when bound to proline, exists as a membrane-bound enzyme associated with the electron transport chain. When associated with the membrane, PutA will not repress *putAP* expression. The regulatory switch centers on the redox state of the FAD prosthetic group, which in turn depends on the presence of proline. Proline interaction with PutA reduces the FAD. PutA containing reduced FAD has a greater affinity for membrane lipids. Thus, when proline is present, PutA is sequestered away from *putAP* operator sequences and the system is induced. As the levels of free proline drop, the PutA FAD moiety oxidizes. This form of PutA is released from the membrane and accumulates in the cytoplasm where it will bind *putAP* and repress expression. Other examples of redox-sensitive genetic switches include Fnr (anaerobic control) and OxyR (oxidative stress response, See Chap 18).

Recombinational Regulation of Gene Expression (Flagellar Phase Variation)

Salmonella periodically alters the antigenic type of the polypeptide flagellin, the major structural component of bacterial flagella. It does this as a form of misdirection designed to hinder mammalian immune systems from getting a "fix" on the organism. As the initial antibody response begins to attack, the invader has changed its appearance. Two genes, *fliC* (formerly H1) and *fljB* (formerly H2), encode different forms of the flagellin protein, but only one gene is expressed in the cell at any one time. The organism can switch from the expression of one gene to the other at frequencies from 10^{-3} to 10^{-5} per cell per generation. The two flagellin genes do not map near each other, with *fliC* and *fljB* mapping at 40 and 56 min, respectively. Figure 5-16 presents the model for the flagellar phase variation switch.

The principle of the control involves the protein-mediated inversion of a 993 bp region of DNA called the H region. The H region is flanked by two 26 bp sequences designated *hixL* (repeat left) and *hixR* (repeat right). Each 26 bp sequence is composed of two imperfect 13 bp inverted repeats. Note in Figure 5-16 that the promoter for the *fljB* and *fljA* (the *fliC* repressor) operons resides within the H region. Thus, when this region occurs in one orientation (Fig. 5-16a), the promoter is positioned to transcribe

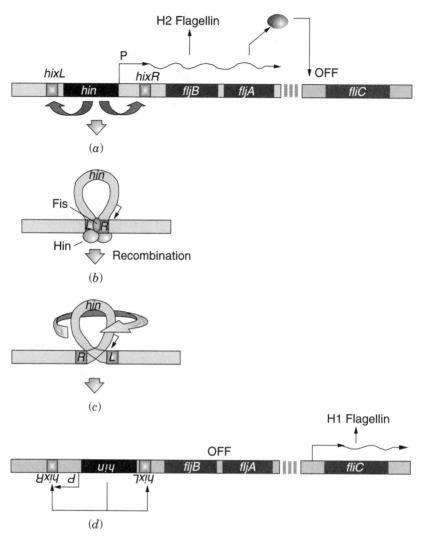

Fig. 5-16. Molecular mechanism of flagellar phase variation in *Salmonella typhimurium*. P, promoter regions.

fljB and *fljA*. Consequently, the H1 locus is repressed by the *fljA* product (also known as rH1) and H2 flagellin is synthesized. The *hin* gene product (22 kDa) is the recombinase that mediates the inversion of this region. When the region between *hixL* and *hixR* is inverted (Fig. 5-16d), the promoter is reoriented and now reads away from the H2 operon, preventing the synthesis of H2 flagellin and FljA repressor. Without FljA, the *fliC* gene can be expressed, producing H1 flagellin.

Within the H region are **enhancer sequences** that stimulate the inversion process 100-fold. Enhancers are cis-acting DNA sequences that magnify the expression of a gene. They do not require any specific orientation relative to the gene and can function at great distances from it. The stages of inversion include Hin binding to the *hix* sites and Fis (a 12 kDa dimeric protein) binding to the enhancer. The *hix* sites are brought together by protein–protein interactions (Hin–Hin and Hin–Fis). The intermediate looped structure is called an **invertasome**. Hin then makes a two-base staggered cut in the center of each *hix* site. Strand rotation probably occurs through subunit exchange, since the strands are covalently attached to Hin. After inversion, Hin relegates the four strands. Similar mechanisms are used by other organisms to regulate the expressions of alternative genes. For other aspects of flagellar synthesis and function, see Chapter 7.

Translational Repression

The synthesis of ribosomes in *E. coli* is regulated relative to the growth rate of the cell. The synthetic rates of the 52 ribosomal proteins and the three rRNA components of the ribosome are balanced to respond coordinately to environmental changes. The set of ribosomal protein operons comprise approximately 16 transcriptional units having between 1 and 11 genes (Fig. 5-17). The model for the autogenous control of ribosomal

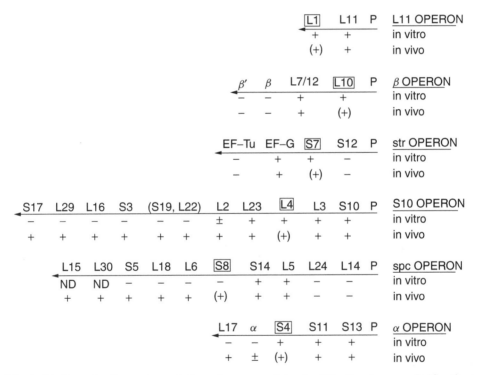

Fig. 5-17. Organization and regulation of genes contained within the *str-spc* and *rif* regions of the bacterial chromosome. Genes are represented by the protein product. (→) For each operon, the direction of transcription from the promoter (P). It has been shown that the L11 and β operons are probably a single operon. The L11 promoter functions as the major promoter for all genes contained within the L11 and β operons in exponentially growing cells. Regulatory ribosome proteins are indicated by boxes. Effects of the boxed proteins on the in vitro and in vivo synthesis of proteins from the same operon are shown. L10 can function as a repressor in a complex with L12. + indicates specific inhibition of synthesis; −, no significant effect on synthesis; ±, weak inhibition of synthesis; (+), inhibition presumed to occur in vivo; ND, not determined. It has not been established how the regulation of the synthesis of ribosomal proteins S12, L14, and L24 is achieved.

protein synthesis states that the translation of a group of ribosomal protein genes encoded on a polycistronic message is inhibited by one of the proteins encoded within that operon. It appears, for example, that L1 of the L11 operon can bind to its own message, preventing its translation.

Many of the proteins that serve as translational repressors also bind specifically to either 16S or 23S rRNA during ribosome assembly. Figure 5-18 illustrates the similarities in the secondary structure of the S4 and S8 protein-binding sites on rRNA versus mRNA. Thus, if growth slows, leading to a decrease in rRNA, the translational repressors will bind instead to their own message. Homologies have been found between the binding sites on the rRNA and the target site on the mRNA for several operons. From a single target site on a polycistronic message, a repressor can affect the translation of all the sensitive downstream citrons through sequential translation — that is, translation of downstream citrons can be dependent on the translation or termination of the upstream target citron. Notice that in each operon presented in Figure 5-17 not all of the citrons in a polycistronic message are affected.

Anti-σ Regulation by Molecular Hijacking

Since σ factors contain the information needed to recognize promoters, cells will often use alternative σ factors to coordinately transcribe related families of genes in response to environmental or cellular signals. Because different σ factors bind competitively to core RNA polymerase, gene expression can be modulated by regulating the availability of the various σ factors. The cellular levels of many σ factors are controlled by transcription, translation, or proteolysis, but other alternative σ factors are synthesized and maintained in an inactive state until they are actually needed. Proteins that inhibit σ-factor activity are known as **anti-σ factors**. These negative regulatory proteins interact directly with their cognate σ factors, preventing them from stably associating with core RNAP. Three σ/anti-σ examples are considered: σ^F/SpoIIAB involved in the Bacillus sporulation cascade; σ^{70} and the *E. coli* phage T4 protein AsiA (see Chapter 6, T4 phage, middle gene expression); and σ^{28}/FlgM involved with flagellar assembly in *Salmonella enterica*.

As noted in Chapter 2, σ factors contain four major conserved regions designated 1 through 4. The SpoIIAB protein of **Bacillus** binds to conserved σ^F regions 2.1, 3.1, and 4.1, sterically hindering the binding of σF to core RNAP. The T4 product AsiA, however, binds to region 4.2 of σ^{70}, which recognizes the −35 promoter region. AsiA does not hinder σS association with core RNAP but will prevent the σ from recognizing a target promoter. Details of how these regulatory mechanisms integrate into sporulation and phage development are discussed in Chapters 6 and 19, respectively.

FliA. Flagellar biogenesis is an extraordinarily complex process involving many different gene products. Separate systems are involved in assembling the basal body, motor, and the flagellum. Each system must be expressed at the appropriate time and in the correct order. Attempting to synthesize flagella before the basal body is ready won't work. Consequently, the genes for flagellar biogenesis in *Salmonella* are organized into a transcriptional hierarchy composed of three classes. Class 1 consists of the master regulators FlhD that, together with σ^{70}, transcribe the Class 2 genes. The Class 2 genes encode proteins required for the assembly and structure of the hook−basal

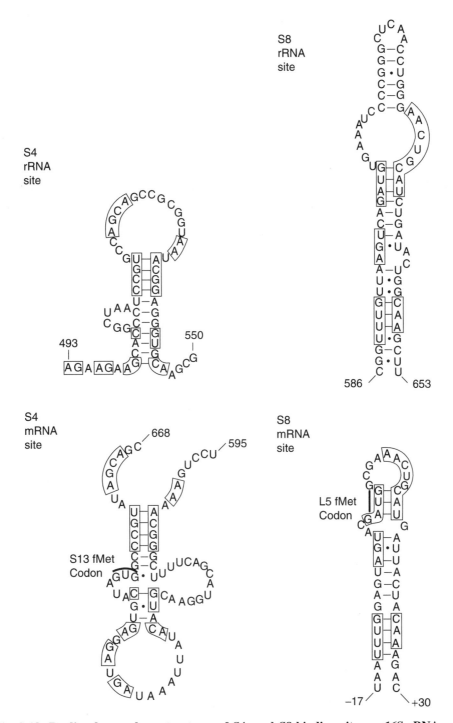

Fig. 5-18. Predicted secondary structures of S4- and S8-binding sites on 16S rRNA and their respective mRNAs. The binding site of S4 on the mRNA overlaps the S13 initiation region; the mRNA site for S8 overlaps the L5 initiation region. Boxed sequences indicate homologies.

body (see Chapter 7 for more details) as well as for the FliA σ (σ^{28}) and its anti-σ FlgM. The Class 3 genes encoding the flagellar filament, motor, and chemotaxis proteins are transcribed exclusively by σ^{28}. Although σ^{28} is produced simultaneously with the hook–basal body proteins, expression of Class 3 genes does not occur until the basal body structure is complete. The reason is that FlgM keeps σ^{28} in check during the assembly process. Once assembly is complete, however, the basal body structure secretes FlgM into the surrounding medium through its core channel. This lowers the intracellular concentration of the anti-σ to a level that permits σ^{28}-dependent transcription of the Class 3 genes.

Titrating a Posttranscriptional Regulator: The CsrA/CsrB Carbon Storage Regulatory Team

During the transition from exponential phase growth into the stationary phase, nonsporulating bacteria readjust their physiology from allowing robust metabolism to

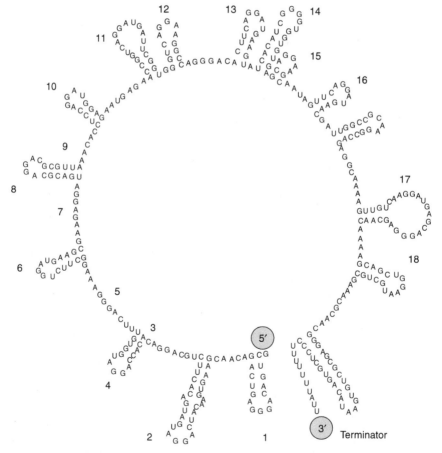

Fig. 5-19. CsrB untranslated RNA. There are 18 copies of the consensus sequence 5'-CAGGA(U,C,A)G-3' embedded in the 350 bp RNA (copies are numbered). These sequences bind 10 molecules of CsrA protein.

providing greater stress resistance and an enhanced ability to scavenge substrates from the medium. One induced pathway is the glycogen (*glg*) biosynthesis pathway, which may provide carbon and energy needed to survive the rigors of the stationary phase (see Chapter 8). The novel regulatory mechanism that in large part controls the expression of glycogen synthesis genes involves a 61 amino acid protein called CsrA and a 350 bp RNA called CsrB. As shown in Figure 5-19, CsrB contains 18 copies of the sequence 5′-CAGGA(U,C,A)G-3′ that appear to bind 18 molecules of CsrA protein. CsrB RNA, in effect, serves as a CsrA "sink," lowering the concentration of free CsrA. CsrA is itself a negative regulator of glycogen synthesis genes. Free CsrA binds to regions around the ribosome-binding sites of target RNAs and seems to facilitate message degradation. Since the stationary phase increases *glg* expression, entry into it might increase CsrB RNA levels to "soak up" more CsrA. Alternatively, CsrA protein levels could themselves decrease. Either way, the prediction is that the ratio of CsrA to CsrB is what controls CsrA activity, although this has not been tested to date.

CsrA also activates some glycolysis genes (e.g., *pfkA*, phosphofructokinase) and represses expression of some genes associated with gluconeogenesis (e.g., *fbp*, fructose-1,6-bis-phosphate phosphatase). Therefore, CsrA has an effect on these carbon utilization pathways opposite that of Cra (see above). How CsrA activates the expression of a gene is not clear but may involve CsrA negatively regulating a negative regulator of the actual target. In this case, the loss of CsrA through mutation would increase production of the negative regulator, which would decrease expression of the ultimate target. This is in contrast to the direct negative control of *glg* where loss of CsrA increases *glg* mRNA stability.

GLOBAL CONTROL NETWORKS

Escherichia coli has the ability to grow in or survive numerous environmental stress conditions including (1) shifts from rich to minimal media, (2) changes in carbon source, (3) amino acid limitations, (4) shifts between aerobic and anaerobic environments, (5) heat shock, (6) acid shock, and (7) numerous starvations such as phosphate, nitrogen, and carbon. Through all of these stresses, the cell modulates DNA replication, cell growth, and division in order to remain viable. The cell must coordinate many different regulatory circuits that control seemingly disparate aspects of cellular physiology. This integrated response to environmental stress is referred to as **global control**. Global regulatory networks include sets of operons and regulons with seemingly unrelated functions and positions scattered around the chromosome, yet they all are coordinately controlled in response to a particular stress.

Some attention should be given to the terminology used to describe regulatory networks. A **regulon** is a network of operons controlled by a single regulatory protein (e.g., the arginine regulon). A **modulon** refers to all the operons and regulons under the control of a common regulatory protein (e.g., the CRP modulon). The simplest way for all of the operons in a modulon to respond to a stress is through the production of a signal molecule (**alarmone**) that accumulates during the stress. Alarmones are often small nucleotides such as cyclic AMP or guanosine tetraphosphate (ppGpp). The term **stimulon** is used to describe all of the genes, operons, regulons, and modulons that respond to a common environmental stimulus. This word is useful when describing all of the coinduced or corepressed proteins that form a cellular response without knowing the mechanism of the response.

Communication with the Environment: Two-Component Regulatory Systems

Bacteria constantly sense and adapt to their environment in order to optimize their ability to survive and multiply. We have covered a variety of regulatory systems, each of which utilizes a regulatory protein (repressor/activator) as a sensor of the internal environment. We now discuss how bacteria sense and interpret changes in the external environment. The general theme here is that bacteria sense and respond to changes in the outside world through a network of two-component signal transduction mechanisms.

Stimuli sensed by two-component systems include pH, osmolarity, temperature, the presence of repellents and attractants, nutrient deprivation, nitrogen and phosphate availability, plant wound exudates, and others. In two-component systems, a protein spanning the cytoplasmic membrane (**sensor/kinase component**) will sense an environmental signal and then transmit a signal through phosphorylation/dephosphorylation reactions to regulatory protein components (**response regulators**) located in the cytoplasm. The phosphorylated cytoplasmic component then regulates target gene expression (Fig. 5-20). Although each system only responds to specific stimuli, they are functionally similar. This conclusion is based on the conservation of amino acid sequences identified within key domains of these proteins (Fig. 5-20).

As illustrated in Figure 5-20, the sensor kinase (or transmitter) protein is usually, but not always, a transmembrane protein. Typically, the extracytoplasmic domain (sensor) senses a change in the environment, which triggers an intrinsic histidine

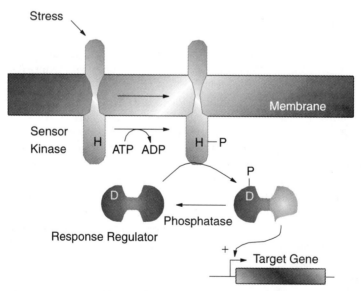

Fig. 5-20. General model for two-component signal transduction. An external stimulus interacts with the periplasmic (usually N-terminus) portion of the sensor/transmitter protein. This activates an autokinase activity in the C-terminal cytoplasmic region, which phosphorylates a conserved histidine residue. The phosphoryl group is then transferred from the sensor kinase histidine to a conserved aspartate residue in the receiver module of the response regulator protein. This phosphorylation induces a conformational change in the output domain (usually DNA binding), activating its DNA-binding function.

kinase activity that phosphorylates a conserved histidine residue in the cytoplasmic domain (**transmitter module**). The sensor/kinase transmitter domain then transfers the phosphate to a conserved aspartate residue in the amino-terminal end (**receiver domain**) of the regulator protein. Phosphorylation of the response regulator receiver domain induces a conformational change in the regulator or **output domain** located at the carboxyl end of the protein. The output domain of most of the known response regulators typically binds target DNA sequences, although some output domains are involved with protein–protein interactions (see "Chemotaxis" in Chapter 7).

Most, if not all, two-component systems are downregulated by specific phosphatases that dephosphorylate the aspartyl phosphate in the response regulator. Phosphatase activity can reside in a protein separate from the signal transduction proteins (e.g., SixA) or may be part of the sensor kinase itself (e.g., EnvZ). Thus, the regulatory switch for two-component signal transduction systems equates to changes in the ratio of kinase activity to phosphatase activity brought about by sensing an environmental signal. The result is a change in phosphorylated versus dephosphorylated levels of the cognate response regulator.

It is also important to recognize that phosphorylation of the response regulator does not necessarily lead to a downstream response, as might be assumed. The role of a given histidine kinase may be to maintain the response regulator in its phosphorylated state, which prevents it from triggering a response. The dephosphorylated regulator might, in fact, be the species that promotes subsequent events.

Sensor Kinases. Several examples of two-component signal transduction systems are listed in Table 5-1. The sensor kinase proteins can be grossly lumped into two classes with respect to membrane location. The integral membrane proteins show typical transmembrane topology. These include, among many others, EnvZ, which is discussed in detail in "Osmotic Control of Gene Expression" in Chapter 18. The second group of sensor kinase proteins are cytoplasmically located with no obvious transmembrane domains. NtrB and CheA are examples that are discussed in this chapter under "Regulation of Nitrogen Assimilation and Nitrogen Fixation" and in Chapter 7 under "Chemotaxis."

In addition to membrane location, sensor kinases can be grouped by their domain structure. There are classic sensor kinases such as EnvZ that possess a sensor or **input domain** at the amino terminus and a **transmitter domain** at the carboxyl terminus containing the conserved histidine phosphorylation residue. Another group is referred to as the "hybrid" sensor kinase proteins. These resemble the classic sensor kinases except they also contain a response regulator–type receiver domain (Fig. 5-21). The hybrid sensor kinases undergo a sequential series of reactions starting with phosphorylation of the conserved histidine in the transmitter domain, which can either directly transfer the phosphate extrinsically to a cognate response regulator aspartate residue or undergo intrinsic phosphotransfer to a receiver module aspartate residue within the sensor kinase itself. A subsequent phosphotransfer must be made to a histidine in a Hpt domain (histidine phosphotransfer) that is either located at the carboxyl terminus of the sensor kinase (e.g., ArcA; see Chapter 18, oxygen regulation) or is a separate protein. In either case, a final phosphotransfer takes place from the histidine in the Hpt domain to the aspartate residue in the response regulator protein. Examples of classic and hybrid systems are shown in Figure 5-21. A useful Web site for signal transduction histidine kinases is http://wit.mcs.anl.gov/WIT2/Sentra/HTML/sentra.html.

TABLE 5.1. Some Properties of Various Two-Component Systems

Organism	Sensor/ Transmitter	Regulator/ Receiver	Response or Regulated Gene	Environmental Stimulus
E. coli/S. typhimurium	EnvZ	OmpR	ompC, ompF	Changes in osmolarity
E. coli	PhoR	PhoB	phoA, phoE, etc.	Phosphate limitation
E. coli	PhoM	PhoM-ORF2	phoA, etc.	Extracellular glucose
E. coli	CpxA	CpxR	traJ, etc.	Dyes and toxic chemicals
E. coli	UhpB	UphA	uhpT, etc.	Glucose-6-phosphate
E. coli	?	UrvC-ORF2	Unknown	Unknown
E. coli	NarX	NarL	narGHJI	Nitrate concentration
E. coli/S. typhimurium	NtrB(NRII)	NtrC(NRI)	glnA, etc.	Ammonia limitation
E. coli/S. typhimurium	CheA	CheB/CheY	Chemotaxis	Repellents and attractants
S. typhimurium	PhoQ	PhoP	phoN, etc., virulence	Magnesium, pH
Klebsiella pneumoniae/Bradyrhizo hium parasponiae	NtrB	NtrC	nifLA	Nitrogen limitation
Rizobium meliloti	FixL	FixJ	nifLA, fixN	Nitrogen limitation
R. leguminosarum	DctB	DctD	dctA	C4-dicarboxylic acids
Pseudomonas putida	?	XylR	xylCAB, xylS	m-xylene or m-methylbenzyl alcohol
Pseudomonas aeruginosa	?	AlgR	algD	Not known
Bordetella pertussis	BvgC	BvgA/BvgC	ptx, fha, cya, etc., virulence	Temperature, nicotinic acid, MgSO$_4$
Agrobacterium tumefaciens	VirA	VirG	virB, virC, etc., virulence	Plant wound exudate
Bacillus subtilis	?	SpoOA/SpoOF	Control of sporulation	Nutrient deprivation
B. subtilis	DegS	DegU	Regulation of degradative enzymes	Not known

Response Regulators. The second components (receiver/regulator proteins) of the sensory transduction systems can be grouped into at least four families. First, there are typical DNA-binding transcriptional activator proteins such as OmpR from *E. coli* (see "Osmotic Control of Gene Expression" in Chapter 18). A second family has, in addition to an N-terminal receiver module with conserved regions to the OmpR family, a conserved region in the C terminus (presumably involved with DNA binding) that is different from OmpR. Examples include BvgA of *Bordetella pertussis* and NarL and UhpA of *E. coli*. The third family of the receiver group, exemplified by DNA-binding proteins NtrC and NifA, has C-terminal regulator regions similar to each other but distinct from the other families. A fourth family, including CheY from *S. typhimurium* and SpoOF from *B. subtilis* only, has a receiver module. CheY is not a DNA-binding protein but interacts with other proteins at the flagellar motor (see "Chemotaxis" in Chapter 7).

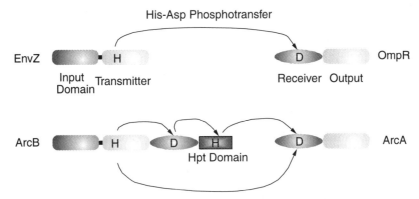

Fig. 5-21. Phosphotransfer reactions in classic (top) and hybrid (bottom) two-component systems. EnvZ and ArcB are sensor kinases. Their cognate response regulators are OmpR and ArcA, respectively. H and D indicate conserved histidine and aspartate phosphorylation target residues.

A one-component sensory transduction system has also been characterized that combines the features of the two-component system into one protein. The ToxR system from *Vibrio cholerae* mediates regulation of several virulence components for this intestinal pathogen in response to environmental stimuli. ToxR is a transmembrane sensor that also binds specifically to the promoter regions of virulence-factor genes through its cytoplasmic domain. Because it is a single protein, it does not require transmitter or receiver modules. There is homology at the DNA-binding cytoplasmic N-terminus region with the C terminus of the OmpR-like regulators. Therefore, it is classified as a member of the OmpR family.

Cross-Talk. Even though signal transduction proteins respond to different stimuli, they still exhibit a high degree of amino acid sequence conservation within key domains. Thus, you can gaze at the completed genome sequence and predict with some confidence the number of potential two-component systems contained in an organism's sensory array. *E. coli*, for example, has 32 potential two-component systems based on sequence analysis. The relatively high degree of sequence conservation between these different systems leads one to wonder about their fidelity. Can different systems end up "cross-talking" to each other? Under normal in vivo circumstances each system is able to specifically transduce its own signal. However, if a system is disturbed by mutations, cross-talk between systems can be observed whereby a component of one system can, at a low level, substitute for the missing component of another. It seems likely, then, that the various two-component systems present in the same bacterial cell may communicate with each other and form a complex and highly balanced regulatory network.

Regulation of Nitrogen Assimilation and Nitrogen Fixation: Examples of Integrated Biochemical and Genetic Controls

Discussion of several specific two-component regulatory systems are discussed later in the book where they are viewed in the context of their physiological role (see "Chemotaxis" in Chapter 7; Chapters 18 and 19). However, two basic examples are presented here. The acquisition of carbon, nitrogen, and phosphate are major nutritional challenges for any organism. We have already discussed some regulatory aspects of carbon source utilization (e.g., CRP above) and now examine regulatory features governing the utilization of nitrogen and phosphate sources, both of which rely heavily on two-component signal transduction mechanisms.

The preferred nitrogen source for enteric bacteria is ammonia, with the principal product of ammonia assimilation being glutamate. Glutamic acid, in turn, is the precursor of several amino acids and ultimately furnishes the amino group of numerous amino acids via transamination (see Chapters 1 and 14). Glutamate is also the precursor of glutamine, which is involved in the synthesis of other amino acids as well as purines and pyrimidines. A total of 85% of the cellular nitrogen is derived from the amino nitrogen of glutamate and 15% from the amide nitrogen of glutamine.

In the absence of ammonia, other nitrogen-containing compounds such as histidine or proline can serve as alternate sources of nitrogen. Rapid synthesis (induction) of these alternate nitrogen assimilatory pathways occurs during ammonia deprivation. This chapter deals with the regulatory mechanisms controlling various pathways of nitrogen assimilation and nitrogen fixation. Because of the ingenious way that *E. coli* entwines the biochemical and genetic controls of nitrogen metabolism, both forms of regulation are discussed here.

Ammonia Assimilation Pathways. Glutamine synthetase (GlnS) is centrally important to cells growing in medium containing less than 1 mM ammonia (NH_4^+). The pathway is illustrated below:

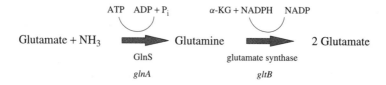

Under conditions where the NH_4^+ concentration is above 1 mM, the GlnS/GS system is inactive, but glutamate dehydrogenase (*gdh*) can be used for ammonia assimilation via the following reaction:

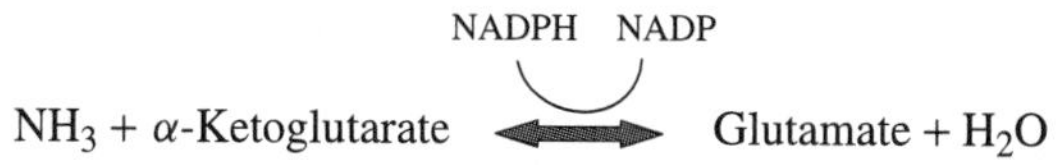

$$\text{NH}_3 + \alpha\text{-Ketoglutarate} \overset{\text{NADPH} \quad \text{NADP}}{\rightleftharpoons} \text{Glutamate} + \text{H}_2\text{O}$$

Left unabated, GlnS converts all of the cellular glutamate to glutamine, which is detrimental to the cell. As a result, GlnS activity and synthesis must be highly regulated. The precise regulation of GlnS is exerted by regulating expression of its structural gene, *glnA*, and regulating GlnS activity by reversible covalent adenylylation. The link between genetic and biochemical controls is a protein called PII encoded by *glnB*. PII communicates with the glutamine synthetase adenylation enzyme that modulates the biochemical activity of GlnS. PII also communicates with the sensor kinase of a two-component signal transduction system that regulates *glnA* expression.

Biochemical Regulation. PII activity is regulated by small molecules that signal carbon and nitrogen status. The nitrogen signal molecule is glutamine while the carbon signal is α-ketoglutarate. The sensing of both molecules keeps the assimilation of nitrogen in balance with the rate of carbon assimilation. The nitrogen signal (glutamine) is sensed by the GlnE adenylyltransferase (ATase) and the bifunctional GlnD uridylyl transferase/uridylyl–removing enzyme (Utase/UR). As illustrated in the center of Figure 5-22, Low levels of glutamine stimulate GlnD UTase activity, which uridylylates PII. PII–UMP binds to GlnE ATase and stimulates the deadenylylation of GlnS–AMP. This activates glutamine synthetase activity and increases ammonia assimilation.

High levels of glutamine do the opposite by stimulating GlnD uridylyl removal from PII. The unmodified form of PII activates the GlnE ATase, which adenylylates and inactivates GlnS, decreasing ammonia assimilation. However, when α-KG levels are high (meaning a large sink is still available to assimilate more ammonia), the PII activation of ATase activity is overridden even if glutamine levels are high. As a result, GlnS remains active and ammonia assimilation continues. This maintains a balance between available nitrogen and the available carbon sink for nitrogen (α-KG).

Genetic Control. As stated above, PII protein links the biochemical control of nitrogen assimilation to the genetic regulatory system, governing synthesis of the assimilating enzymes. The *glnA* (GlnS), *ntrB*, and *ntrC* loci form an operon as shown in Figure 5-23. Note that expression of the *glnA ntrB ntrC* operon is driven by three promoters: *glnAp1* (transcriptional start 187 bp upstream of *glnA* translational start), *glnAp2* (73 bp upstream of *glnA* translational start), and *ntrBp* (also known in the literature as *glnLP*; transcriptional start 256 bp downstream of *glnA* translational termination and 33 bp upstream of *ntrB* translational start). The *glnAp1* and *ntrBp* promoters both utilize σ-70 for initiation and have the appropriate −10 and −35 regions. They are not strong promoters but allow the cell to maintain a low level of glutamine synthetase as well as low levels of the nitrogen regulators NtrC (NRI) and NtrB (NRII) during growth in high nitrogen. In contrast, the *glnAp2* promoter is a

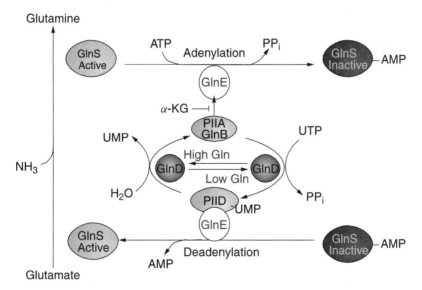

Fig. 5-22. Biochemical regulatory circuit controlling nitrogen assimilation.

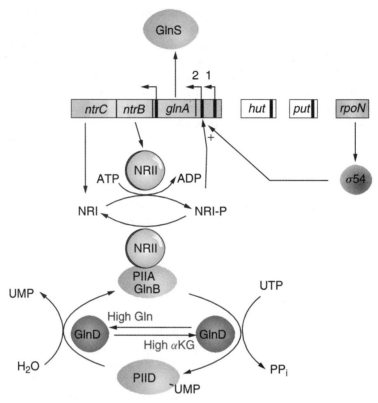

Fig. 5-23. Genetic regulation of nitrogen assimilation under high or low intracellular glutamine concentrations.

strong promoter that utilizes an alternate σ factor, σ-54, the product of *rpoN*. It contains the -10 consensus sequence TTGGCACAN$_4$TCGCT common among other nitrogen-regulated promoters. Nitrogen and carbon regulation of *glnA* expression centers on the activation of this promoter.

The NtrB and NtrC proteins, called NRII and NRI, respectively, form a two-component signal transduction system that regulates the expression of the *glnA* operon as well as other operons encoding nitrogen assimilation proteins (e.g., histidine or proline utilization). The NtrB (NRII) sensor kinase phosphorylates and activates the NtrC (NRI) response regulator when glutamine levels are low. The resulting NtrC-P is a transcriptional activator of the *glnA* operon. Increasing expression of this operon, of course, increases glutamine synthetase production and, thus, glutamine. When glutamine levels rise too high, the intrinsic phosphatase activity of NRII removes the phosphate from NRI and activation of *glnA* transcription stops. Regulation of this system hinges on changing the relative kinase and phosphatase activities of NRII in response to glutamine and carbon flux.

The regulatory cascade leading to NtrC phosphorylation once again begins with GlnB (PII). Along with interacting with the ATase as described, PII will modulate the kinase/phosphatase activities of the NRII sensor kinase, providing an elegant link between biochemical control of GlnS activity and the transcriptional activation of the *glnAntrBC* operon. Figures 5-23 and 5-24 illustrate this regulatory network.

When glutamine levels are low, PII is uridylylated and does not bind to the NRII sensor kinase, leaving NRII free to autophosphorylate and then transphosphorylate the response regulator NRI. NRI-P then activates transcription at *glnAp2*, producing a tremendous increase in glutamine synthetase, which enables the cell to better assimilate NH$_3$ (Fig. 5-23).

How is the system shut off? As the level of glutamine rises in the cell, the UTase activity of GlnD is activated, which converts PII-UMP to PII (GlnB). In addition to *inactivating* GlnS ATase as described earlier, PIIA will activate the phosphatase activity of NRII that removes phosphate from NRI-P (Fig. 5-23). Since dephosphorylated NRI cannot activate the *glnAp2* promoter but will repress *glnAp1* and *glnLp*, the operon is deactivated and returns to background levels of expression.

This is not the whole story, however. As with biochemical control, the level of α-KG (the nitrogen sink) binds to and influences PII effects on the phosphatase/kinase

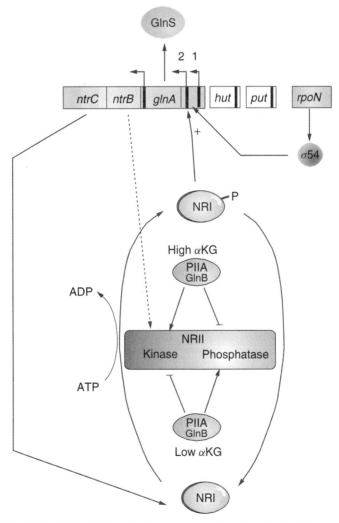

Fig. 5-24. Modulation of nitrogen regulation by α-ketoglutarate.

activities of NRII (Fig. 5-24). Thus, when glutamine levels are high and α-KG levels are low (not much of a nitrogen sink), a PII trimer binds a single α-KG molecule. This complex stimulates NRII phosphatase, inhibits NRII kinase, and, because the resulting NRI-P levels are low, shuts down *glnA* transcription. If the α-KG level remains high even in the presence of high glutamine, PII will bind three α-KG molecules. The PII:3 α-KG complex inhibits NRII phosphatase but stimulates kinase activity. This leaves NRI-P levels high and *glnA* transcription active. Thus, the cell can continue to assimilate ammonia even though intracellular glutamine levels are high.

Enhancer Sequence. It is of considerable importance that the binding site for NRI at *glnAp2* can be moved over 1000 bp away from the original site without diminishing the ability of NRI to activate transcription from *glnAp2*. Thus, the NRI-binding sites do not function as typical operator regions but resemble **enhancer sequences** found in eukaryotic cells. Enhancers work by increasing the local concentration of an important regulator near a regulated gene by tethering it to the region. NtrC(NRI)-P initially binds its DNA target as a dimer but eventually ends up as a large oligomer in which some of the NRI molecules are bound only through protein-protein interactions. Activation of *glnp2* expression does not involve recruitment of $E\sigma^{54}$ by NtrC-P but actually involves DNA-bound NtrC-P interacting with an $E\sigma^{54}$ complex already bound to *glnp2*. This interaction, via a DNA loop, converts the $E\sigma^{54}$-promoter complex from a closed to an open form, thereby initiating transcription.

The cell has other systems designed to capture nitrogen from amino acids that may be available in the environment. These include the *hut* (histidine utilization) and *put* (proline utilization) operons. Both systems require NRI-P and σ-54 RNA polymerase for their expression and thus are regulated in a manner similar to the *glnA* operon.

Control of Nitrogen Fixation. Along with the general nitrogen assimilation regulatory system (*ntr*) just described, *Klebsiella pneumoniae* can fix atmospheric nitrogen (N_2). Nitrogen fixation in *Klebsiella* involves 17 genes all participating in some way with the synthesis of nitrogenase, the enzyme complex that converts N_2 to NH_3 (see Fig. 14-2). The *nifHDK* transcriptional unit codes for nitrogenase proper while the remaining *nif* genes are responsible for (1) synthesizing the Mo-Fe cofactor, (2) protein maturation, (3) regulation, or (4) unknown functions. The *ntrC* and *ntrB* products will also regulate *nifLA* transcription in a manner analogous to what was described above. The *nifAL* operon is activated by NRI-P, which results in increased intracellular concentration of NifA. NifA can then activate expression of the other *nif* operons. NifA activity is regulated by the NifL product in response to oxygen and probably ammonia, although the mechanism is not known. The *nif* system is active only under anaerobic and low NH_4^+ conditions. For additional discussion of nitrogen fixation and its regulation, see Chapter 14.

Phosphate Uptake: Communication Between Transport and Two-Component Regulatory Systems

Another global regulatory system studied extensively is the phosphate-controlled stimulon of *E. coli*. Expression of the majority of phosphate-regulated loci is specifically **phosphate-starvation inducible** (*psi*). However, other phosphate-controlled loci are **phosphate-starvation repressible**. Many of the *psi* genes encode proteins associated with the outer membrane or the periplasmic space and function either in the transport of inorganic phosphate across the cell membrane or in scavenging phosphate from organic phosphate esters. Two of these inducible genes, *phoA* (alkaline phosphatase, PhoA) and *phoE* (outer membrane porin protein, PhoE), have been studied and shown to be under the control of a complex regulatory circuit. This circuit involves the two-component regulators *phoR* (sensor kinase) and *phoB* (response regulator), *phoU*, and a high-affinity phosphate transport system.

There are actually two transport systems for inorganic phosphate: the low-affinity PIT (phosphate inorganic transport) and an inducible high-affinity PST (phosphate-specific transport). In addition to *phoU*, *phoR*, and *phoB*, the *pst* operon also plays a role in regulation. This was first evident upon observing that *pst* mutants lacking the PST transport system constitutively synthesize alkaline phosphatase (derepressed) even though they grow well and maintain a high internal phosphate concentration using the PIT system. Clearly, phosphate itself is not the corepressor for this system. The system really works through communication between the PST system and PhoR.

A model for phosphate-controlled gene regulation is outlined in Figure 5-25. PhoR and PhoB are members of a two-component regulatory system. PhoR is an integral membrane-sensing protein that can trigger induction of the Pho regulon under limited phosphate conditions (less than 4 μM). When the cell senses low P_i levels, the PhoR protein is autophosphorylated. PhoR-P then transphosphorylates PhoB. PhoB-P activates transcription of target genes by binding to the consensus "Pho-box" sequences (CTTTTCATAAAACTGTCA) located upstream of *pho* regulon promoters. As extracellular phosphate levels increase, transport through PST increases, and PstB,

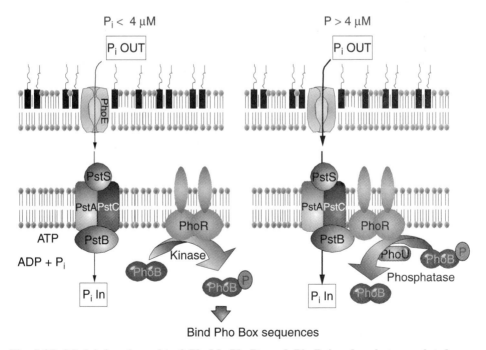

Fig. 5-25. Model for the role of PhoM, PhoR, and PhoB in phosphate-regulated gene expression.

a member of the high-affinity transport system, will interact with PhoR and inhibit its kinase activity. Next, PhoR, together with another protein, PhoU, will dephosphorylate PhoB, thereby turning off transcription of the *pho* regulon. So, this system successfully links the level of inorganic phosphate transport to the expression of phosphate-scavenging systems through protein–protein interactions between a member of the transport system (PstB) and the governing sensor kinase (PhoR).

Quorum Sensing: How Bacteria Talk to Each Other

Thirty years ago, A bacterium was thought to be a self-contained bag of protoplasm that had no sense of community. This concept was shattered with the discovery of a microbial cell-to-cell signaling phenomenon known now as **quorum sensing**. *Vibrio fischeri* is a free-living, bioluminescent bacterium that can become a symbiont of squid, colonizing their light organs. The bacteria only emit light, however, when they are present at a high cell density as can be found in the contained space of the light organ. The gene cluster (*lux*) responsible for bioluminescence is shown in Figure 5-26. LuxI is an enzyme that synthesizes the autoinducer molecule N-(3-oxohexanoyl)-L-homoserine lactone. At low cell densities, there is very little expression of the *luxICDABEG* operon and, thus, very slow production of the autoinducer. As the cell density increases, the autoinducer will accumulate in the medium. As the concentration of autoinducer increases, it will diffuse across the membrane back into the cell and bind the LuxR protein. The LuxR–AI complex activates transcription of the divergent *lux* operon, increasing the production of enzymes responsible for bioluminescence. The organism then emits light.

In contrast to the relatively simple system in *V. fischeri*, *Vibrio harveyi* possesses two autoinducer-response systems that function in parallel to control the density-dependent expression of the luciferase structural operon *luxCDABE* (Fig. 5-27). Autoinducer-1 (AI-1) is a homoserine lactone molecule, but the structure of AI-2 is not fully elucidated. Another difference with the *V. fischeri* system is that the *V. harveyi lux* inducer systems utilize complex two-component response regulators to sense levels of extracellular autoinducer rather than simple diffusion of AI across the cell membrane. As is typical of two-component regulatory systems, signals are transmitted via a series of phosphorylation/dephosphorylation reactions to a DNA-binding response regulator protein. However, the two transmembrane histidine kinases that detect each autoinducer (LuxN for AI-1 and LuxQ for AI-2) channel their signal (via phosphorylation) to a

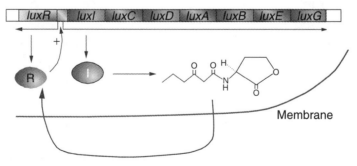

Fig. 5-26. Cell-to-Cell communication. The *lux* operon operon of *Vibrio fischeri*.

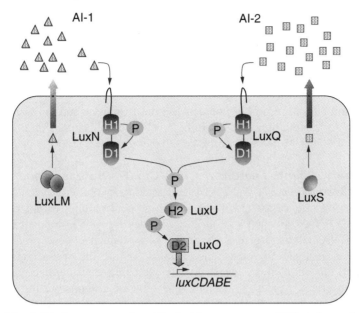

Fig. 5-27. Quorum sensing. The luciferase systems of *Vibrio harveyi*.

shared integrator protein (LuxU). LuxU then transphosphorylates the LuxO response regulator, which in turn activates *luxCDABE* expression. Complex two-component systems generally have a hybrid transmembrane protein with two modules: a sensor kinase that becomes phosphorylated at a conserved histidine residue and a C-terminal response regulator module that becomes phosphorylated at a conserved aspartate residue. The phosphate is then transferred to a conserved histidine on a phosphorelay protein (LuxU in this case) and then finally to a conserved aspartate on a second response regulator (LuxO).

The reason for dual systems appears to involve whether the organism is communicating with its own species or with other species. Homologs of LuxS have been identified in many gram-negative and gram-positive species. Amazingly, the AI-2 autoinducers produced from many of these species can "talk" to *V. harveyi* via the LuxQ AI-2 sensing system. Mutants of *V. harveyi* that do not make AI-2 (*luxS* mutants) will bioluminesce when in the presence of other species making AI-2 molecules! The capacity to respond to both intra- and interspecies signals could allow *V. harveyi* to know not only its own cell density but also its proportion in a mixed population. Although evidence indicates that *E. coli* and *S. enterica* produce AI-2-like molecules, their role in the physiology of these organisms is unclear at present. Recall that other autoinducer systems were discussed in the section on transformation in Chapter 3.

Proteolytic Control

Lon Protease. The *lon* locus of *E. coli* codes for a protease that degrades "abnormal" proteins (see Chapter 2). Mutations in the *lon* locus have pleiotropic effects including an increased UV sensitivity, filamentous growth, and mucoid colonies due to capsule overproduction. UV sensitivity and filamentous growth due to *lon*, explained in

Chapter 3, results from an inability to degrade the SOS-inducible cell division inhibitor SulA. The mucoid phenotype is the result of decreased degradation of RcsA, a positive regulator of capsular polysaccharide synthesis. In both cases, the proposed role of Lon protease is to assure that certain regulatory proteins persist for only a short time. However, if we define a global regulator in terms of initiating a coordinated response of a given set of operons to a specific stimulus, then *lon* does not qualify as such. Rather than initiating a response, Lon acts to limit responses and may be important as a mechanism for maintaining and returning a cell to equilibrium. For a description of the interaction of *lon* protease with the cell division process, see Chapter 17.

The Alternate Sigma Factor σ^S, the ClpXP Protease, and Regulated Proteolysis. As *Salmonella* and *E. coli* enter stationary phase or when exponentially growing cells experience a sudden stress such as acid or high osmolarity, the level of an alternate σ factor, called σ^S (38 kDa, encoded by *rpoS*), dramatically increases and directs RNA polymerase to a variety of stress-inducible gene promoters. While the *rpoS* gene is subject to some transcriptional control, the primary reasons for the stress-induced elevation in σ^S levels involve increased translation of *rpoS* message and decreased proteolysis of σ^S protein. The protease responsible for degrading σ^S is ClpP and its associated ClpX recognition subunit (see Chapter 2). Under rapidly growing conditions, ClpXP rapidly degrades σ^S, but following encounters with stress, degradation slows, which allows σ^S to accumulate.

However, ClpXP alone cannot recognize σ^S. An adaptor recognition protein called RssB (regulation of sigma S, also known in *Salmonella* as *mviA*, mouse virulence gene) is required to present σ^S to ClpXP. Recent evidence indicates that the amino terminus of RssB (*mviA*) binds σ^S while the carboxy terminus binds ClpX. Regulation of σ^S degradation is mediated by RssB, apparently through phosphorylation of a conserved aspartate residue within an amino-terminal domain that bears homology to the receiver domains of two-component response regulators. Once phosphorylated, RssB efficiently presents σ^S to ClpXP. The cognate sensor kinase has not been identified, but acetyl phosphate has been implicated as a phospho donor.

SUMMARY

In this chapter, the basic concepts of how cells regulate gene expression are introduced. It should be more and more apparent that no single system operates independently in the cell. There are several layers of control that monitor and modulate the expression of any given gene. The reason for this overlapping control is clearly the rule of economy. Ensuing chapters discussing the biochemical aspects of cellular metabolism can now be viewed in the overall context of regulatory networks where one metabolic system impacts other seemingly far-removed systems.

BIBLIOGRAPHY

Babitzke, P., J. T. Stults, S. J. Shire, and C. Yanofsky. 1994. TRAP, the trp RNA-binding attenuation protein of *Bacillus subtilis*, is a multisubunit complex that appears to recognize G/UAG repeats in the *trpEDCFBA* and *trpG* transcripts. *J. Biol. Chem.* **269**:16,597–604.

Bassler, B. L. 1999. How bacteria talk to each other: regulation of gene expression by quorum sensing. *Curr. Opin. Microbiol.* **2:**582–7.

Becker, D. F., and E. A. Thomas. 2001. Redox properties of the PutA protein from *Escherichia coli* and the influence of the flavin redox state on PutA–DNA interactions. *Biochemistry* **40:**4714–21.

Beynon, J., M. Cannon, V. Buchanan-Wollaston, and F. Cannon. 1983. The nif promoters of *Klebsiella pneumoniae* have a characteristic primary structure. *Cell* **34:**665–71.

Busby, S., and R. H. Ebright. 1999. Transcription activation by catabolite activator protein (CAP). *J. Mol. Biol.* **293:**199–213.

Campbell, K. M., G. D. Storms, and L. Gold. 1983. Protein-mediated translational repression. In J. Beckwith, J. Davies, and J. A. Gallant (eds.), *Gene Function in Procaryotes.* Cold Spring Harbor Press, Cold Spring Harbor, NY.

Freeman, J. A., B. N. Lilley, and B. L. Bassler. 2000. A genetic analysis of the functions of LuxN: a two-component hybrid sensor kinase that regulates quorum sensing in *Vibrio. harveyi. Mol. Microbiol.* **35:**139–49.

Geanacopoulos, M., G. Vasmatzis, D. E. Lewis, S. Roy, B. Lee, and S. Adhya. 1999. GalR mutants defective in repressosome formation. *Genes Dev.* **13:**1251–62.

Geanacopoulos, M., G. Vasmatzis, V. B. Zhurkin, and S. Adhya. 2001. Gal repressosome contains an antiparallel DNA loop. *Nat. Struct. Biol.* **8:**432–6.

Geiduschek, E. P. 1997. Paths to activation of transcription. *Science* **275:**1614–6.

Grebe, T. W., and J. B. Stock. 1999. The histidine protein kinase superfamily. *Adv. Microbiol. Physiol.* **41:**139–227.

Gross, R., B. Arico, and R. Rappuoli. 1989. Families of bacterial signal-transducing proteins. *Mol. Microbiol.* **3:**1661–7.

Harmer, T., M. Wu, and R. Schleif. 2001. The role of rigidity in DNA looping-unlooping by AraC. *Proc. Natl. Acad. Sci. USA* **98:**427–31.

Helmann, J. D. 1999. Anti-sigma factors. *Curr. Opin. Microbiol.* **2:**135–41.

Hughes, K. T., and K. Mathee. 1998. The anti-sigma factors. *Annu. Rev. Microbiol.* **52:**231–86.

Kimata, K., H. Takahashi, T. Inada, P. Postma, and H. Aiba. 1997. cAMP receptor protein-cAMP plays a crucial role in glucose-lactose diauxie by activating the major glucose transporter gene in *Escherichia coli. Proc. Natl. Acad. Sci. USA* **94:**12914–9.

Landick, R., and C. L. J. Turnbough. 1992. *Transcriptional Attenuation: Transcriptional Regulation.* Cold Spring Harbor Press, Cold Spring Harbor, NY.

Liberek, K., T. P. Galitski, M. Zylicz, and C. Georgopolos. 1992. The DnaK chaperone modulates the heat shock response of *Escherichia coli* by binding to the σ-32 transcription factor. *Proc. Natl. Acad. Sci. USA* **89:**3516–20.

Lobell, R. B., and R. Schleif. 1990. DNA looping and unlooping by AraC protein. *Science* **250:**528–32.

Magasanik, B. 1993. The regulation of nitrogen utilization in enteric bacteria. *J. Cell. Biochem.* **51:**34–40.

Mizuno, T. 1998. His-Asp phosphotransfer signal transduction. *J. Biochem.* **123:**555–63.

Muller-Hill, B. 1998. The function of auxiliary operators. *Mol. Microbiol.* **29:**13–8.

Ninfa, A. J., and M. R. Atkinson. 2000. PII signal transduction proteins. *Trends Microbiol.* **8:**172–9.

Ninfa, A. J., P. Jiang, M. R. Atkinson, and J. A. Peliska. 2000. Integration of antagonistic signals in the regulation of nitrogen assimilation in *Escherichia coli. Curr. Top. Cell. Reg.* **36:**31–75.

Perego, M., and J. A. Hoch. 1996. Protein aspartate phosphatases control the output of two-component signal transduction systems. *Trends Genet.* **12:**97–101.

Peterkofsky, A., I. Svenson, and N. Amin. 1989. Regulation of *Escherichia coli* adengulate cyclase activity by the phosphoenol pyruvate: sugar phosphotransferase system. *FEMS Microbiol. Rev.* **63:**103–8.

Pirrung, M. C. 1999. Histidine kinases and two-component signal transduction systems. *Chem. Biol.* **6:**R167–75.

Reizer, J., C. Hoischen, F. Titgemeyer, C. Rivolta, R. Rabus, J. Stulke, D. Karamata, M. H. Saier, Jr., and W. Hillen. 1998. A novel protein kinase that controls carbon catabolite repression in bacteria. *Mol. Microbiol.* **27:**1157–69.

Reznikoff, W. S. 1992. Catabolite gene activator protein activation of *lac* transcription. *J. Bacteriol.* **174:**655–8.

Romeo, T. 1998. Global regulation by the small RNA-binding protein CsrA and the non-coding RNA molecule CsrB. *Mol. Microbiol.* **29:**1321–30.

Saier, M. H., Jr., and T. M. Ramseier. 1996. The catabolite repressor/activator (Cra) protein of enteric bacteria. *J. Bacteriol.* **178:**3411–7.

Salmond, G. P., B. W. Bycroft, G. S. Stewart, and P. Williams. 1995. The bacterial "enigma": cracking the code of cell-cell communication. *Mol. Microbiol.* **16:**615–24.

Schlax, P. J., M. W. Capp, and M. T. Record, Jr. 1995. Inhibition of transcription initiation by lac repressor. *J. Mol. Biol.* **245:**331–50.

Schleif, R. 2000. Regulation of the L-arabinose operon of *Escherichia coli*. *Trends Genet.* **16:**559–65.

Storz, G., L. A. Tartaglia, and B. N. Ames. 1990. Transcriptional regulator of oxidative stress-inducible genes: direct activation by oxidation. *Science* **248:**189–94.

Surber, M. W., and S. Maloy. 1999. Regulation of flavin dehydrogenase compartmentalization: requirements for PutA-membrane association in *Salmonella typhimurium*. *Biochim. Biophys. Acta* **1421:**5–18.

Surette, M. G., M. B. Miller, and B. L. Bassler. 1999. Quorum sensing in *Escherichia coli*, *Salmonella typhimurium*, and *Vibrio harveyi:* a new family of genes responsible for autoinducer production. *Proc. Natl. Acad. Sci. USA* **96:**1639–44.

Wanner, B. L. 1993. Gene regulation by phosphate in enteric bacteria. *J. Cell Biochem.* **51:**47–54.

Wei, B., S. Shin, D. LaPorte, A. J. Wolfe, and T. Romeo. 2000. Global regulatory mutations in *csrA* and *rpoS* cause severe central carbon stress in *Escherichia coli* in the presence of acetate. *J. Bacteriol.* **182:**1632–40.

Wilcox, G., S. Al-Zarban, L. G. Cass, D. Clarke, L. Heffern, A. H. Horwitz, and C. G. Miyada. 1982. DNA sequence analysis of mutants in the *araBAD* and *abaC* promoters. In R. Rodrigues and M. Chamberlin (eds.), *Promoters: Structure and Function*. Praeger, New York, pp. 183–94.

Wilson, H. R., C. D. Archer, J. K. Liu, and C. L. Turnbough, Jr. 1992. Translational control of *pyrC* expression mediated by nucleotide-sensitive selection of transcriptional start sites in *Escherichia coli*. *J. Bacteriol.* **174:**514–24.

Yanofsky, C. 2000. Transcription attenuation: once viewed as a novel regulatory strategy. *J. Bacteriol.* **182:**1–8.

Zinkel, S. S., and D. M. Crothers. 1991. Catabolite activator protein-induced DNA binding in transcription initiation. *J. Mol. Biol.* **219:**201–15.

CHAPTER 6

BACTERIOPHAGE GENETICS

Even after nearly 100 years of study, bacterial viruses continue to play an important role in the development of bacterial genetics. Their study has been instrumental in revealing many important biological processes applicable to prokaryotic and eukaryotic organisms alike. For example, T4 and ϕX174 have been important tools used to reveal cardinal features of the DNA replication fork. Likewise, λ phage was integral to studies dealing with the initiation of DNA replication. Interchangeable σ factors were discovered using T4 while λ proved central to our understanding of transcriptional repressors and activators.

Another phage we discuss, μ, has taught us much regarding DNA rearrangements. As you will see, bacteriophages remain critically important research tools. For example, the λ family of phages has enjoyed a resurgent interest in the genomic era. The availability of complete genome sequences has revealed a pivotal role for members of this phage family in the development of *E. coli* virulence. The genomes of many pathogens contain entire sections, called **pathogenicity islands**, that encode virulence genes with an AT content different from the rest of the chromosome. The DNA flanking these islands often contain remnants of viral genes, like scars marking how the organism may have acquired these genes. (see "Type III secretion" in Chapter 20). In addition to their evolutionary significance, the unique recombination systems of λ have been exploited to simplify prokaryotic and eukaryotic genetic engineering strategies. Even more remarkable is the renewed debate over an old idea: the potential therapeutic use of bacterial viruses as antimicrobial agents. Clearly, phage genetics has shaped modern biology and continues to be a critical factor in its development.

GENERAL CHARACTERISTICS OF BACTERIOPHAGES

The first step toward understanding bacteriophages is to learn something about their classification. However, characterization and classification of the bacteriophage viruses poses many problems. From a practical point of view, they can be distinguished on the basis of their natural host, host range, and other similar characteristics. In addition, they can be characterized on the basis of their RNA or DNA content. Each phage generally contains only one kind of nucleic acid (the filamentous viruses have been shown to be an exception to this general rule). Structural symmetry and susceptibility or resistance to ether and other solvents and to other chemical agents also provide additional criteria for classification (Table 6-1). There are over 5000 phage types that comprise 13 phage families (Table 6-2).

Bacteriophages range in size from 20 to 300 nm. Large phages are barely within the realm of resolution of the light microscope. For visualization of any details of viral structure, the electron microscope is essential. The nucleic acid may be either double stranded, single stranded, circular, or linear. Morphologically, phages generally display a considerable degree of geometric symmetry. They generally have a head or **capsid** composed of identical protein subunits called **capsomeres**. The nucleic acid **core** is housed within the capsid. Some bacteriophages have a prominent tail structure. The capsid structure may assume an icosahedral form (20 sides with 12 vertices). The entire infectious unit is generally referred to as a **virion**. The smallest phages contain nucleic acid (either DNA or RNA) with a molecular weight on the order of 1×10^6 daltons. This relatively short-chain nucleic acid can obviously only code for a small number of genes. These genes, of necessity, must encode information that governs the formation of the basic viral subunits. The presence of such a small number of genes reflects the degree to which viruses are dependent on the host cell.

Each species of microorganism is susceptible to a limited subset of viruses. Put another way, each phage has a specific (and limited) host range of susceptible microorganisms that it can infect. For example, a hypothetical phage A might only infect one strain of *E. coli* while phage B could infect many different strains of *E. coli*. Phage A is viewed as having a narrow host range relative to that of phage B. Host range depends on the presence of specific viral receptors on the host cell surface. These receptors are usually composed of specific carbohydrate groups on lipopolysaccharides (LPS) of cell surface structures. Some viruses, such as M13, attach to sex pili and are referred to as male-specific phages. The salient point here is that receptors are actually normal cellular proteins with a specific function that are co-opted by the phage.

Two types of infection cycles may occur following the initial infection. One is a **virulent** or **lytic** infection that ends in lysis and death of the host cell. The other type of infection is called **lysogenic**, which may be quite unapparent since the host cell does not die. In fact, the general appearance and activity of the lysogenized host cell may not be altered in any overt manner. A lytic type of phage only undergoes virulent infection. Temperate phages, on the other hand, can undergo either lytic or lysogenic infection.

Bacteriophage infections of bacteria follow a characteristic pattern. The growth curve of bacteriophages, sometimes referred to as the one-step growth curve, is depicted in Figure 6-1. Once a population of bacterial cells has been inoculated with a given number of bacteriophages, the number of detectable infectious particles rapidly decreases. This stage is termed the **eclipse phase** and represents that portion of the time following infection during which bacteriophages cannot be detected either in the culture medium or within the cell. The overall period encompassing adsorption and eclipse is referred to as the **latent period**. After the latent period has been completed (approximately 20 minutes for *E. coli* cells infected with coliphage), infectious viruses

TABLE 6-1. Characteristics of Some Phages of *Escherichia coli*

Phages	Virion Morphology Head (nm)	Tail (nm)	Nucleic Acid, Type and Amount (daltons)	Latent Period (min)	Average Yield per Cell	Growth Cycle at 37°C Lysogeny	Peculiarities
T1	Icosahedral 50	10 × 150	DNA, 2.5×10^7	13	150	−	Resistant to drying
T2, T4, T6	Prolated icosahedral 65 × 95	25 × 110	DNA, 1.2×10^8	21–25	150–400	−	Contain glucosylated HMC
T3, T7	Icosahedral 47	10 × 15	DNA, 2.4×10^7	13	300	−	Give semitemperate mutants
T5	Icosahedral 65	10 × 17	DNA, 7.5×10^7	40	200	−	DNA injection in two steps
λ, φ80	Icosahedral 54	10 × 140	DNA, 3.3×10^7	35	100	+	DNA circularizes, *in vitro* or *in vivo*
P1	Icosahedral 65	12 × 150	DNA, 6×10^7	45	80	+	General transduction
P2	Icosahedral 50	10 × 150	DNA, 2.2×10^7	30	120	+	Multiple chromosomal locations
φX174 S13	Icosahedral 30	None	DNA, 1 strand 1.7×10^6	13	180	−	Circular DNA
f2, MS2	Icosahedral 24	None	RNA, 9×10^5	22	20,000	−	Male specific attachment to F pili
f1, fd	None	6 × 800	DNA, 1 strand 1.3×10^6	30	100–200 (continuous release)	−	Male specific, circular DNA
χ[a]	Icosahedral 67.5	12.5 × 230	DNA	60	200	−	Attaches to motile flagella

[a] Not a coliphage; grows on many strains of *Salmonella*.

TABLE 6-2. Families of Bacterial Viruses

Corticoviridae
  icosahedral capsid with lipid layer,
  circular supercoiled dsDNA

Cystoviridae
  enveloped, icosahedral capsid, lipids,
  three molecules of linear dsRNA

Fuselloviridae
  pleomorphic, envelope, lipids, no
  capsid, circular supercoiled dsDNA

Inoviridae, genus Inovirus
  long filaments with helical symmetry,
  circular ssDNA

Leviviridae
  quasi-icosahedral capsid, one molecule
  of linear ssRNA

Lipothrixviridae
  enveloped filaments, lipids, linear
  dsDNA

Microviridae
  icosahedral capsid, circular ssDNA

Myoviridae
  icosohedral capsid, dsDNA

See T4 an μ Phages

Plasmaviridae
  pleomorphic, envelope, lipids, no
  capsid, circular supercoiled dsDNA

(Image not available)

Podoviridae
  tail short and noncontractile, head
  isometric

Rudiviridae
  helical rods, linear dsDNA

(Image not available)

TABLE 6-2. (*Continued*)

Siphoviridae, morphotype B1
  tail long and noncontractile, head
  isometric

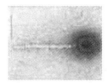

Tectiviridae
  icosahedral capsid with inner
  lipoprotein vesicle, linear dsDNA,
  "tail" produced for DNA injection

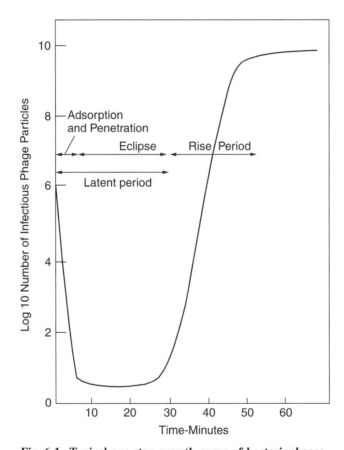

Fig. 6-1. Typical one-step growth curve of bacteriophages.

begin to be released from the cell. The average number of infectious bacteriophage particles released per cell is referred to as the **burst size**. The replication cycle may be depicted as follows:

1. Attachment (see Fig. 6-2).
2. Injection of nucleic acid or penetration of the entire virus into the cell.

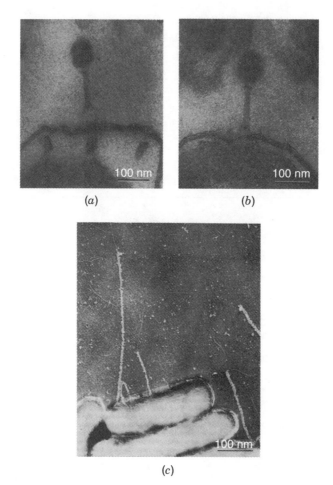

(a)   (b)

(c)

Fig. 6-2. Bacteriophage attachment to cells and pili. (a, b). Attachments of T-even virions to points of adhesion between cytoplasmic membrane and outer membrane (cell wall) of *E. coli.* Bars equal 100 nm. (*C*) *E. coli* cells with F pili showing attached particles of MS2 (icosahedral) and M13 (filamentous) phages. (*Source*: From Luria et al., (1978). General Virology, 3rd Edition, Wiley, NY.

3. Uncoating (i.e., removal of the capsid surrounding the nucleic acid "core").

4. Intracellular synthesis of viral subunits (capsid and nucleic acid).

5. Assembly of the subunits into complete virus particles (maturation). Complete bacteriophages within an *E. coli* cell are shown in Figure 6-3.

6. Release of viruses (burst or lysis).

Steps 1 through 4 comprise the latent period of which steps 2, 3, and 4 represent eclipse.

Lysogenic infection occurs when the nucleic acid of the virus enters the host cell, becomes aligned with the genome of the cell, and eventually is integrated into the cell genome. In this way, the virus is replicated in unison with replication of the host cell genome. At some later time, information coded within the viral genome may be **induced** to function. This triggers replication of the virus and entry into the lytic stage.

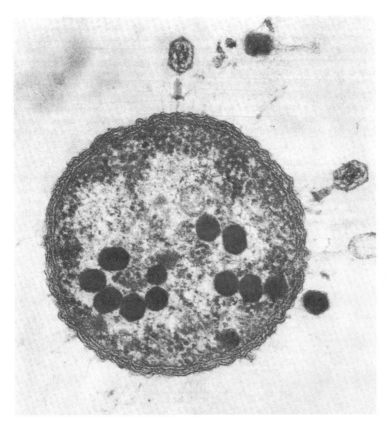

Fig. 6-3. A section of a cell of *E. coli* strain B from a culture infected with bacteriophage T2. Clearly visible are the cell envelopes (wall and cytoplasmic membrane); the clear area of the phage DNA pool containing many condensed phage DNA cores; and attached to the cell surface, three phages, one empty and two still partially filled. The phage at the top shows the long tail fibers and the spikes of the tail plate in contact with the cell wall. The tail sheath is contracted and the tail core has apparently reached the cell surface but has not penetrated it. (Courtesy of Dr. Lee D. Simon.)

Tumor viruses may initiate lysogenic infections of animal cells in a manner comparable to lysogenic infections of bacteria by bacteriophages. Additional discussion regarding the genetics of four specific bacterial viruses, each of which represents a different lifestyle, may be found next.

T4 PHAGE

Structure

T4 phage is one member of the Myoviridae family of T-even (T2, T4, T6) lytic phages of *E. coli*. Figure 6-4 presents the structure of a typical T-even phage of *E. coli* (T4). T4 possesses an oblong head (80 × 120 nm) and a contractile tail (95 × 20 nm). Within the capsid is a linear, double-stranded DNA molecule (1.3×10^8 daltons or 1.69×10^5 base pairs). Associated with the DNA are a number of polyamines (putrescine, spermidine, and cadaverine). The base composition of T4 DNA is unusual in that all of

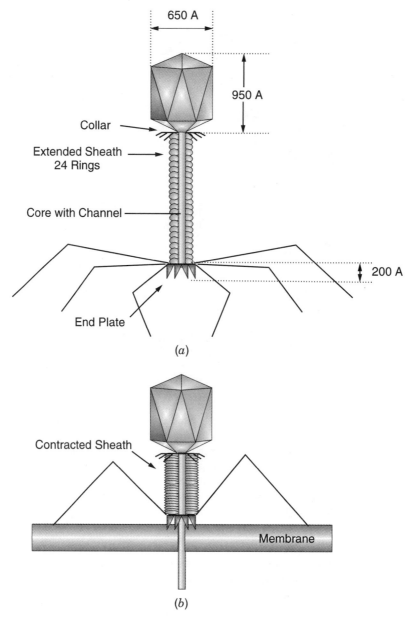

650 A

950 A

Collar

Extended Sheath
24 Rings

Core with Channel

200 A

End Plate

(*a*)

Contracted Sheath

Membrane

(*b*)

Fig. 6-4. Schematic representation of phage T4 virion, its component parts and penetration of the phage core through the bacterial envelope. (*a*) Virion with extended tail fibers. (*b*) Virion with the tail sheath contracted and the spikes of the tail plate pointing against the bacterial cell wall. (*Source*: Adapted from Luria et al., 1978.)

the cytosine residues have been replaced by a modified base, hydroxymethylcytosine (HMC). The reason for this replacement will become apparent from subsequent discussion. While virions in general are thought not to possess enzymes, a number of enzymatic activities have been found to be associated with the T4 virion. Several of these are listed in Table 6-3 along with their possible roles in T4 infection.

TABLE 6-3. **Virus-Specific Enzymatic Activities Associated with T4 Virions**

Enzyme	Location and/or Function
Dihydrofolate reductase	Located in base plate, may have role in unfolding tail fibers
Thymidylate synthetase	Found in base plate; necessary for infectivity
Lysozyme	Possible role in penetration through cell wall
Phospholipase	Possible role in host cell lysis
ATPase	Associated with the tail sheath; presumed to be involved with the contractile process
Endonuclease V	Excision repair (see 13.1A) of phage or host DNA
alt function	Alteration of the host RNA polymerase

The one-step growth experiment described earlier was originally developed with T4. The curve reflects a series of events beginning with the tail fibers of T4 binding to specific *E. coli* cell surface receptors (adsorption). Once the pins of the base plate contact the outer surface of the cell, the tail sheath contracts and the tail core penetrates the cell wall (Fig. 6-4). The T4 DNA molecule and associated proteins are then injected into the host. Animations of T4 phage attachment and injection can be found within the phage images section at www.phage.org.

General Pattern of T4 Gene Expression

Early Gene Expression. It is a general strategy of viruses to defer some gene expression to later stages of infection and progeny virus maturation. T4 phage is no exception. Transcription of certain T4 genes begins almost immediately. The **immediate–early genes** are transcribed at 30 seconds postinfection and the **delayed–early** or **middle genes** are transcribed at 2 minutes postinfection (model for T4 infection in Fig. 6-5). The products of these genes establish the infection. Since transcription of the immediate early genes is directed by σ^{70} promoters, normal host RNA polymerase is used. Among the early gene products are nucleases and other T4 proteins that stop host transcription, unfold the host chromosome, and degrade host DNA to nucleotides. Several enzymes are involved with the synthesis of T4 DNA including a T4 DNA polymerase (gene product 43, gp43).

Stealth Technology. A unique feature of T4 DNA is that its DNA contains the unusual base 5-hydroxymethylcytosine (hm-dCTP):

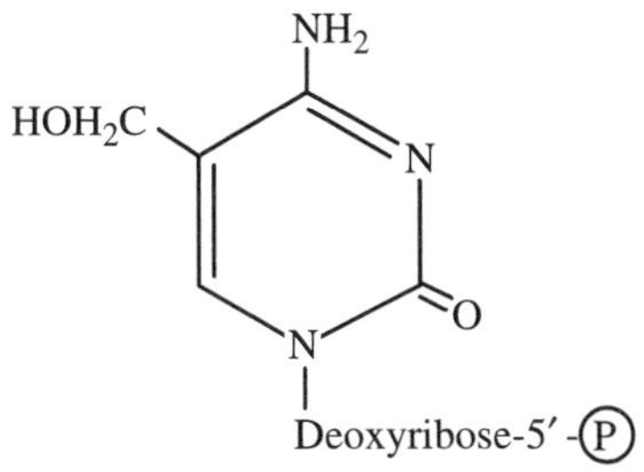

$$NH_2$$

HOH$_2$C

N

N

O

N

Deoxyribose-5'-(P)

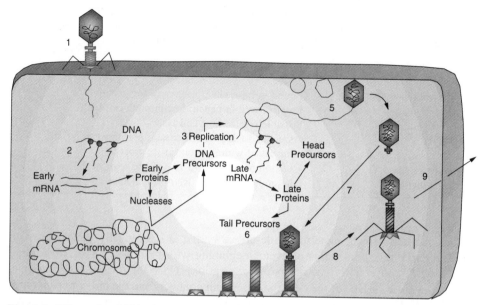

Fig. 6-5. Life cycle of T4 phage. (1) Attachment and injection of DNA; (2) transcription of early genes; (3) replication and concatemer formation; (4) transcription of late genes; (5) head assembly; (6) tail assembly; (7) attachment of head to tail; (8) attachment of tail fibers; (9) cell lysis and release of mature phage.

This base is synthesized as hm-dCMP from dCTP by a T4 enzyme prior to incorporation of the base into T4 DNA. Following incorporation, this nucleotide is glucosylated by a T4-encoded HMC glucosyltransferase. These modifications serve two basic functions: (1) phages with HMC-modified DNA can use cytosine-specific endonucleases to scavenge deoxyribonucleotides from host DNA without damaging their own DNA; and (2) the HMC-modified phages can shuttle back and forth undetected between strains carrying cytosine-specific restriction-modification systems, since the phages do not contain cytosine proper (see "Restriction and Modification" in Chapter 3).

Middle Gene Expression. The transcriptional switch from immediate–early to middle genes is associated with a series of T4-directed modifications as shown in Table 6-4. Alt and Mod proteins catalyze the ADP ribosylation of Arg265 in the RNA polymerase α subunits. This modification lowers the affinity of core polymerase for the housekeeping σ^{70} protein and is part of a strategy to reduce host RNA synthesis. The Alt protein is actually packaged in the T4 capsid and injected with the DNA during infection. Mod, on the other hand, is synthesized after infection. Transcription of middle genes requires the MotA protein, a positive regulator that binds to a TAT/AGCTT DNA sequence 13 bp upstream of a −10 box, converting the region into an active promoter even though the region lacks a −35 box. T4 also produces a 10 kDa anti-σ-70 protein (AsiA) that prevents transcription of the T4 σ^{70} early gene promoters. AsiA has been shown to bind region 4.2 of σ^{70} (Chapter 2), blocking interaction between σ^{70} and the −35 upstream promoter element of early T4 genes and most *E. coli* genes.

TABLE 6-4. T4-Evoked Changes of *E. coli* RNA Polymerase

Change	Time First Detectable in RNAP (min after infection at 30°)	Effect or Function
ADP ribosylation of α subunits (Arg-265); encoded by *alt* (70K)	<0.5, synthesized late, packaged, and injected	Lowers affinity for σ; participation in host shutoff.
ADP ribosylation of α subunits (Arg-265); encoded by *mod* (27K)	1.5–2.0	Lowers affinity for σ; selective shutoff of some early genes around 4 min; participation in host shutoff?
DNA-binding activator of middle transcription (MotA)	<5	
Binding of 10K protein (AsiA)	<5	σ Antagonist, anti-σ-70.
Binding of 15K protein	5	T4 gene 60 codes for a subunit of T4 DNA topoisomerase and may also code for the 15K protein.
Binding of gp33 (12K)	5–10	Positive control of late transcription, helper of σ^{gp55}.
Binding of σ^{gp55} (22K)	5–10	Positive control of late transcription, alternate σ.
Interaction with gp45 (25K)	(Does not copurify with RNA polymerase)	Gp45 is a component of the core of the T4 replisome and is also directly involved in *late* transcription.

Late Gene Expression. Expression of the early and middle genes ceases about 12 minutes postinfection. However, at 5 minutes postinfection, DNA replication has begun and transcription of the **late genes** begins. Further modifications of RNA polymerase accompany this transition. For example, gp55 is a viral-encoded σ factor that is required for late gene transcription. It essentially hijacks the host RNA polymerase by altering its promoter sequence recognition properties. This works because the molecular structure of T4 late promoters is quite different from other known prokaryotic promoters. It uses the consensus sequence TATAAATACTATT at the -10 region but exhibits no consensus sequence at the -35 region. The alternate σ gp55 and the unusual structure of late gene promoters directs T4 transcription toward the late genes.

Another curiosity of T4 transcription is that expression of late genes is closely tied to DNA replication. One of the proteins involved with late gene transcription (gp33) serves as a link between RNA polymerase-σ^{gp55} and components of the T4 replication machinery. Apparently the T4 replication complex serves as a "mobile enhancer," although how this works is not clear.

Capsid Assembly and Phage Release. The late gene products are primarily components of viral capsids. Approximately 40 have been identified. In spite of the complexity of the T4 capsid, the assembly of the various components (e.g., heads, tail fibers, base plate) is remarkably efficient. It is fascinating that final assembly of the various phage component parts occurs spontaneously, without guidance. This feature is referred to as self-assembly and is due to the remarkable affinity each component has for its mate. The inner portion of Figure 6-6 illustrates the pathways involved with capsid formation and points out where the genes responsible are positioned on the T4 genome. The entire process results in the production of at least 200 particles per cell by 25 to 30 minutes postinfection (at 37 °C, in rich medium). At the completion of the T4 multiplication cycle the cell will lyse. Lysis is accomplished by a sudden cessation of respiration and requires the phage t gene product plus the phage e product (T4 lysozyme). The t protein is an example of a **holin protein** that assembles in the cell membrane and allows passage of the e lysozyme through the inner membrane, allowing e lysozyme access to the outer membrane to degrade lipopolysaccharide carbohydrates.

Holin proteins in general have to be regulated; otherwise, the cell will prematurely lyse. One piece of the timing puzzle has finally been exposed. It involves **lysis inhibition**, an intriguing and historically important feature of T4 phage. Over 50 years ago, it was observed that the normal lysis of T4-infected cells could be inhibited by secondary T4 infection occurring at least 3 minutes after the primary infection. Lysis inhibition by superinfecting phages is effective even if the superinfecting DNA does not replicate. T4 mutants in the rI, rII, and rIII genes are not subject to lysis inhibition and exhibit unusual plaque morphology. It now appears that the r genes are part of the holin timing mechanism. The key role appears to be rI protein, located in the periplasm. Injection of superinfecting DNA appears to activate rI, which in turn interacts with the t holin, inhibiting its pore forming ability. The rest of the timing mechanism, however, remains a mystery.

T4 Genome

Terminally Redundant and Circularly Permuted. A useful website dedicate to T4 phage can be found at (http://www.evergreen.edu/user/T4/home.html). It should be noted at this point that while the DNA of T4 phages is linear both in the virus and within the host cell, the genetic map is circular (i.e., genes situated at opposite ends of the linear structure, which should not map close to each other, nevertheless, upon genetic mapping, are found to be linked). The resolution of this paradox can be found in the realization that the ends of each T4 DNA molecule are **terminally redundant**—that is, a duplication of genetic information occurs at both ends of a single molecule. This phenomenon is the result of the **headful mechanism** utilized for the packaging of T4 DNA into the T4 capsid. After DNA replication (discussed more fully below), replica molecules can recombine with each other at their terminal redundancies, creating very long molecules (concatemers) many genomes in length. Each mature virion is then made by packaging a "headful" of DNA cut from such a T4 concatemer. One headful, however, is 105% of a single genome. Therefore, the packaging process will generate a population of phages carrying all possible **circular permutations** of the T4 genome (Fig. 6-7). Genetic crosses between members of this population will result in a circular map.

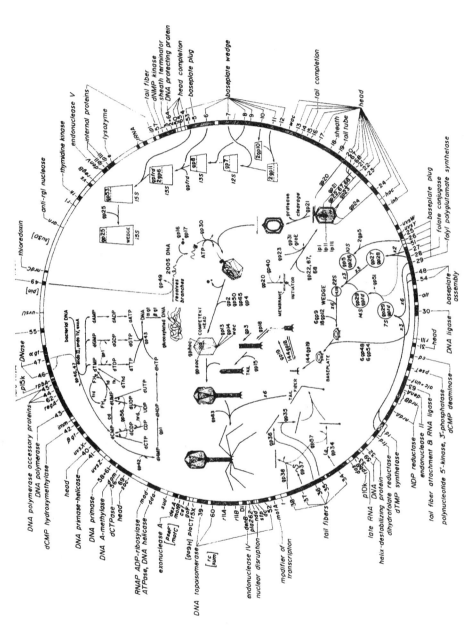

Fig. 6-6. Genomic map of bacteriophage T4. (Courtesy of Drs. B. S. Guttman and E. M. Kutter, Evergreen State College, Olympia, WA.)

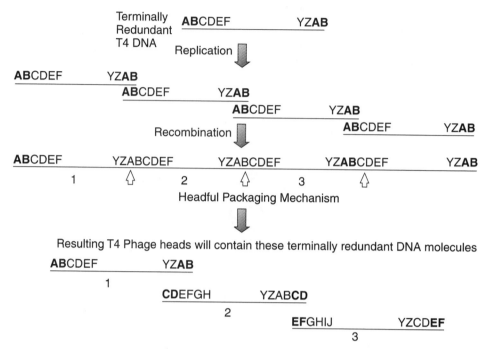

Fig. 6-7. Formation of terminally redundant, circularly permuted T4 DNA.

T4 DNA Replication. There are approximately 30 T4 genes known to be involved with some aspect of T4 DNA synthesis. Many of these are not directly involved with the replication fork mechanism but encode enzymes required for nucleotide precursor synthesis. In this section we deal only with those genes directly involved with the replication process as well as analyze the replication machinery itself.

Because of its rapid infection cycle, T4 uses multiple replication origins and initiation mechanisms. There is an origin-dependent "early" mode of initiation involving multiple origins spaced around the T4 genome. Transcription in these areas is essential to initiating replication. As illustrated in Figure 6-8, transcription at an origin creates a substrate for RNAse H. Following cleavage of the RNA in the RNA:DNA hybrid, the remaining RNA serves as a replication primer. One major initiation site occurs between genes rII and 42, and there are up to four additional sites at other locations. As with host DNA replication, there is a leading or continuously synthesized strand and a lagging strand. Okazaki fragments in T4 are about 2000 nucleotides long. The major protein responsible for nucleotide synthesis is gp43, the T4 DNA polymerase (mwt 110,000). Accessory proteins include g44, g45, and g62, which collectively form a sliding clamp on the DNA much as DnaB protein does on chromosomal DNA (see Chapter 2).

The gene 41 protein is a DNA helicase that will also serve to unwind dsDNA at the replication fork. A complex of gp41 and gp61 is envisioned to move along the lagging strand in the 5' to 3' direction dependent on nucleotide hydrolysis. The gene 32 product, a helix-destabilizing protein (ssDNA-binding protein), will bind to the single-stranded regions. At intervals of about 2000 nucleotides, the gp41/61 complex stops at specific sites to make pentaribonucleotide primers (pppApCpNpNpNp). The primers probably

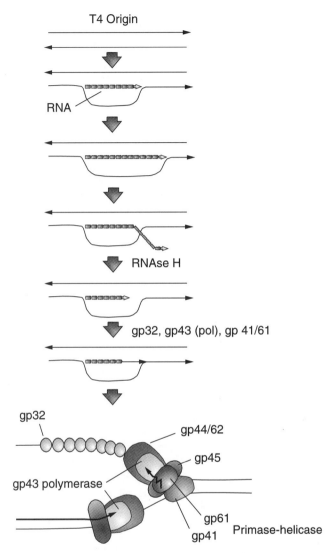

Fig. 6-8. Origin-dependent T4.

remain attached to the priming complex until elongated by T4 DNA polymerase (gene 43 product). Joining the Okazaki fragments to make a continuous DNA chain requires a special DNA repair system that includes T4 RNase H (44 kDa) to degrade the RNA primer, T4 DNA polymerase to replace the RNA with DNA, and T4 DNA ligase to form a phosphodiester bond between the 3′ end of each fragment and the 5′ end of the preceding one (Fig. 6-8). The gene 44/62 complex and gene 45 proteins appear to keep the polymerase continuously on the growing 3′-OH chain.

After a few rounds of DNA synthesis, the T4 UvsW protein inactivates origin-dependent replication. UvsW is a helicase that unwinds RNA primers from these origins, effectively halting this form of initiation. Subsequently, T4 switches to a recombination-dependent initiation mechanism. This is an essential feature of T4

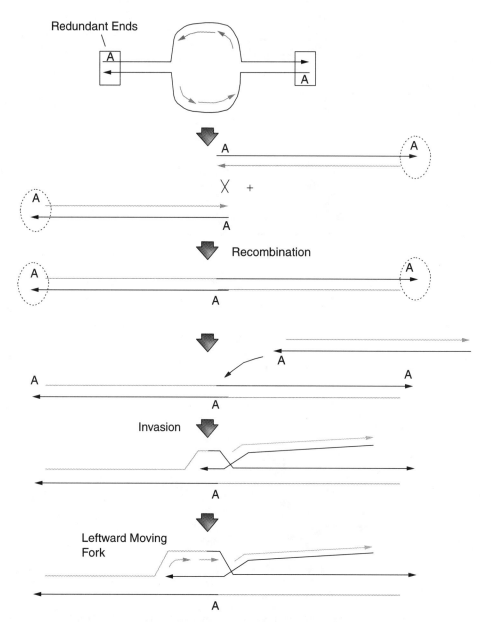

Fig. 6-9. Recombination-dependent initiation of T4 DNA replication. Redundant ends are marked by circles and boxes. Recombination between redundant areas of two molecules creates a concatemer. A third molecule can invade the internal redundancy of the concatemer and generate a leftward-moving replication fork.

development because its genome is linear. No known DNA polymerase can initiate chains de novo. As a result, the 3' ends of linear DNA molecules replicated from specific origins will remain unreplicated. The first clue as to how T4 solves this problem was the discovery that its genome is terminally redundant (see above). The redundancy extends 3 to 4 kb. As diagrammed in Figure 6-9, daughter T4 molecules can recombine at the homologous blunt ends to form a two-copy concatemer. The 3′ single-stranded end of another T4 molecule can then invade the homologous region positioned in the middle of the concatemer, forming a new replication fork.

The expression of four of the DNA replication genes (g43, g44, g45, and g62) is regulated by controlling the initiation and termination of transcription as well as translation (Fig. 6-10). Although these four genes are transcribed by the host RNP, the T4 *mot* product influences promoter selection by RNAP, stimulating transcription of g43 and g45 while inhibiting transcription of g44. The gp43 product (T4 DNA polymerase) also competes with Mot for binding, thereby autoregulating gene 43 transcription. Translational control by *regA*, also indicated in Figure 6-10, is explained below.

Regulation of T4 Gene Expression. Regulation of T4 gene expression has been shown to occur both at the transcriptional and translational levels. At the translational level two systems are recognized; autogenous translational repression by the gene 32 product (helix-destabilizing protein) and translational repression of many T4-induced early mRNAs by the phage *regA* product. In the systems of translational repression already discussed (see "Translational repression" in Chapter 5) each protein repressor was specific to only one species of mRNA. This mechanism appears the same for the T4 gene 32 product. However, translational regulation by RegA (12,000 mw) is unique in that it appears to repress the translation of different mRNA species. The mechanism appears to be the simple recognition of and binding to a consensus mRNA

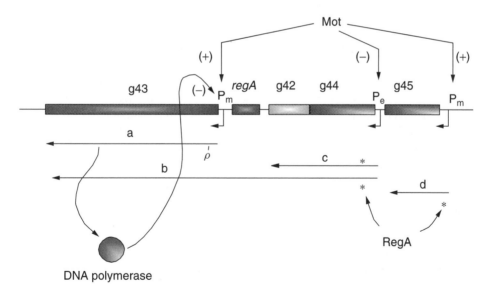

Fig. 6-10. Schematic summary of proposed mechanisms controlling expression of replication genes. P_e indicates an immediate–early promoter recognized by *E. coli* RNA polymerase that is independent of *mot*. P_m indicates a middle–early promoter that is dependent on *mot* for transcription. The wavy lines indicate length and direction of mRNA synthesis. Transcript *a* accounts for 75% of the cellular level of gene 43 mRNA and the *b* transcript contributes 25%. Translation of transcripts *c* and *d* may be controlled by the product of the *regA* gene. ρ indicates a potential site for ρ-mediated termination that the *mot* protein may modulate. (*Source*: Modified from Spicer and Konigsberg, 1983, *Bacteriophage T4*, in Matthews, et al., 1983. pp. 291–301.

sequence [(AUG)UACAAU-3′)] thus blocking ribosome access. Secondary mRNA structures appear to be unnecessary. Although the regulatory significance of regA remains a mystery, it would seem to play a role in coordinating the early events in T4 development.

Hyphenated Genes: Introns in T4. Many eukaryotic genes include one or more nontranslated intervening sequences (IVS), or **introns**. These intervening sequences must be excised from the primary transcripts prior to translation. Few examples of introns have been documented in prokaryotes. The first was a 1 kb intron within the thymidylate synthase (*td*) gene of bacteriophage T4. Since then, two more T4 genes, *nrdB* and *sunY*, have been shown to contain introns. The T4 introns are all self-splicing, in which a series of transesterifications, or phosphodiester bond transfers, occurs with the RNA functioning as an enzyme. The introns within *td* and *sunY* are actually mobile introns that can catalyze their transfer to another DNA molecule.

A novel form of gene splicing is found in gene 60, which codes for an 18 kDa subunit of the T4 topoisomerase. Within the message is a 50 nucleotide untranslated region. However, this region is not removed as an intron from the message. It seems a pair of 5 nucleotide direct repeats flanking the IVS can base pair, forming a hairpin that brings codons on either side close together. A ribosome can presumably move through or jump across this structure, ignoring the nucleotides in the loop.

λ PHAGE

The λ (lambda) phage (http://www.ncbi.nlm.nih.gov/cgi-bin/Entrez/framik?db= Genome&gi =10119) (Fig. 6-11) of *E. coli* is a *Siphoviridae* and a classic example of a temperate phage — that is, it can undergo either a lytic or a lysogenic cycle upon infecting a sensitive *E. coli*. This property manifests itself in the form of a cloudy plaque when λ is grown on a lawn of *E. coli*. The cloudy center is due to growing bacterial cells that have been lysogenized by the phage DNA and to the fact (discussed below) that these lysogens are resistant to superinfection. Mutations in certain λ genes (*cI, cII, CIII*) will prevent lysogeny, so the mutant phage will produce clear plaques in which all infected cells are lysed. The λ genome (48.5 kDa) expresses approximately 50 proteins. The genes of this phage can be grouped into four categories based on their roles in either lysis or lysogeny: (1) genes involved in lytic development, (2) genes involved with the development of lysogeny, (3) genes participating in both, and (4) genes with unknown functions. The genes in the second and fourth groups are nonessential for phage growth. Figure 6-12A presents a general map of the λ genome.

The DNA within a λ capsid is linear (48,502 bp) but contains cohesive ends designated *cos* sites. Upon phage attachment and injection, the *cos* ends of λ DNA anneal and are ligated to form a circular molecule. There are two phases of λ replication as the phage undergoes lytic development: an early phase in which the circular molecules replicate bidirectionally and a late phase, approximately 10 minutes postinfection, consisting of a rolling circle mechanism that generates long concatemeric molecules containing multiple copies of the λ genome. This is discussed in detail below. The concatemers are then acted upon by the λ-packaging system and are used to fill preformed capsids with λ DNA. A precise amount of λ DNA is packaged, corresponding to one complete genome, following cleavage by λ terminase at the unique *cos* sites mentioned above (Fig. 6-13).

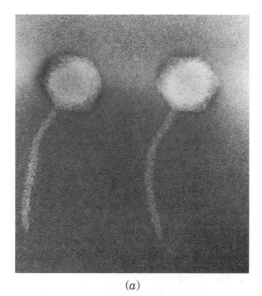

(a)

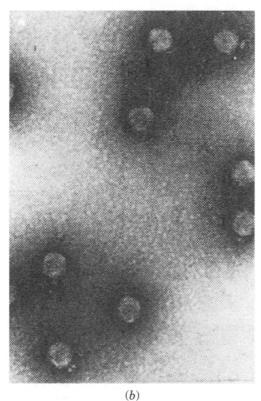

(b)

Fig. 6-11. Electron micrographs of some temperate bacteriophages. (a) λ phage negatively stained with potassium phosphotungstate. (Micrograph by E. A. Birge.) (b) Phage P22, also negatively stained. Bar equals 100 nm. (*Source:* From King and Casjeus, Nature (London) **251:**112–119, 1974.)

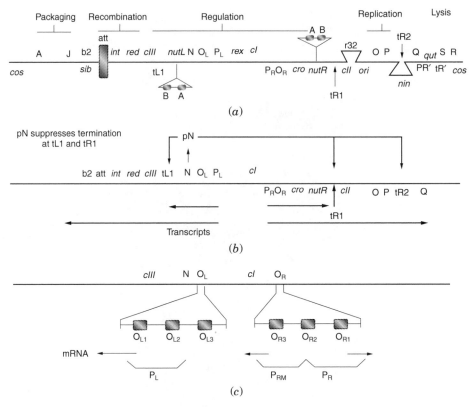

Fig. 6-12. λ phage. (a) Genetic organization. (b) Transcriptional organization. (c) Structure of the leftward and rightward promoter/operator regions.

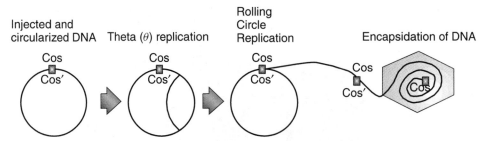

Fig. 6-13. Infective pathway of bacteriophage λ. The phage DNA forms a covalently closed circle through pairing of the cohesive ends, (cos), followed by ligation. Early DNA replication by the mode generates circular monomeric molecules. The switch to the rolling circle mode of replication is dependent on the product of the *gam* gene, which inhibits the host *recBC* nuclease. The multimeric DNA is the substrate for encapsidation. (Modified From Murray, in Hendrix et al., 1983.) Lambda II. Cold Spring Harbor Laboratory, Cold Spring Harbor, N.Y.

The Lysis-Lysogeny Decision

Upon infection, λ phage must make a molecular decision to multiply in a lytic manner or "hibernate" as an integrated prophage in the lysogenic state. As will become evident, it is impossible for λ to go down one path (i.e., toward lysis) without partially going down the other (i.e., toward lysogeny). The multiplicity of infection greatly influences the final decision. In general, a high multiplicity of infection leads to lysogeny, whereas a low ratio of phage to cells results in lytic multiplication. This correlation should be kept in mind while reading the following sections.

Transcription

Transcription of λ initiates at P_L (leftward promoter) and P_R (rightward promoter), with the early transcripts terminating at the ρ-dependent terminators at t_{L1} and t_{R1} (Fig. 6-12B). The leftward transcript results in the production of the antiterminator N while the rightward transcript codes for Cro protein. Transcription into genes beyond the terminators requires N protein acting as an antiterminator at t_{L1} and t_{R1}. Six factors produced by *E. coli* are required for effective N utilization: *nusA* (71 min), *nusB* (9 min), *nusC* (*rpoB*, 90 min), *nusD* (*rho*, 84 min), *nusE* (*rpsJ,* 74 min), and *nusG* (90 min). The *nus* acronym stands for N-utilization substance. *E. coli nus* mutants fail to support λ growth by blocking the action of pN rather than its synthesis. The *nusA* and *nusG* gene products and their normal roles in *E. coli* are discussed in Chapter 2 with regard to transcription. As noted above, *nusC*, *nusD*, and *nusE* mutations occurred in the genes for RNA polymerase ß subunit, Rho, and ribosomal protein S10 respectively.

The model for N action involves a modification of RNA polymerase at the *nut* sites (see Fig. 6-12A and B). The presence of NusA, NusB, and/or NusG on core polymerase enables pN to modify RNA polymerase when a *nut* site is encountered. The modified RNA polymerase can then transcribe past t_{R1} and t_{L1}, resulting in the synthesis of *CII* and *CIII* message, respectively, and past t_{R2}, yielding another antiterminator pQ. The *nut* site in λ mRNA is divided into three parts: BoxA, a site that binds NusB and S10; a spacer region; and a stem-loop structure called BoxB that binds N and NusA. NusB binds to the BoxA sequence of nascent message and communicates with RNA polymerase through the S10 collaborator. N-antiterminator protein binds to the *nut* BoxB hairpin loop and interacts with RNA polymerase through NusA, an interaction stabilized by NusB–S10 and NusG. NMR studies have revealed that the N–BoxB interaction involves the aromatic ring of a tryptophan residue in N actually stacking with an unpaired adenine in Box B. This multifactorial complex ultimately suppresses the pausing of RNA polymerase at the downstream transcriptional termination signal.

Several minutes after λ infection, the synthesis of early (N and Cro) and delayed early (*CII, P, O, Q* and *CIII, red, int*) transcripts diminishes due to the action of Cro repressor on P_L and P_R following its binding to O_L and O_R. However, the Q antiterminator will allow a P'_R transcript to extend through the late genes involved with capsid production and lysis. In contrast to N, which binds mRNA, Q binds to a DNA site called *qut*. RNA polymerase initiating at P'_R pauses at $+16/+17$, which allows Q to bind and modify RNAP, shortening the pause. The P'_R extended transcript eventually terminates at *b* (remember the λ genome is circular at this point). As a general rule, when the multiplicity of infection is low, there is an insufficient quantity of CII or CIII to establish lysogeny. However, enough Q is synthesized to transcribe through the late genes, producing the lytic cycle of λ phage.

Function of Cro Versus CI Repressor and the Structure of O_L and O_R

As opposed to CI, which promotes lysogeny, Cro protein promotes lytic multiplication of λ by preventing the synthesis of CI. Although both Cro and CI bind to the same operators (O_L and O_R), these repressor proteins serve distinct physiological functions due to several unique features of the operators as illustrated in Figure 6-12C. First, within each operator there exist three CI (and Cro)-binding sites. Each binding site (designated O_{L1}, O_{L2}, etc.) is similar in sequence and possesses axes of 2-fold hyphenated symmetry. The two repressors have different relative affinities for each of these binding sites. At low concentrations, dimers of CI bind to O_{R1} and O_{R2}, blocking transcription from P_R. Actually, CI repressor binds cooperatively to O_{R1} and O_{R2}, binding first at O_{R1}, which increases the affinity of O_{R2} for CI, then at O_{R2}. Binding of CI to O_{R2} will activate P_{rm} (promoter for repressor maintenance), ensuring continued production of CI. In contrast, Cro protein binds more tightly to O_{R3}. This prevents transcription from P_{rm}, halting synthesis of CI. Thus, Cro binding to O_R will promote the lytic cycle of development by preventing lysogeny.

The study of both Cro and CI repressors revealed a motif that has proven to be common to many repressor proteins — that is, the helix-turn-helix motif. The DNA-binding domain of CI resides at the amino-terminal end of this protein. In this region are several alpha helices, two of which participate in DNA binding. As shown in Figure 6-14, the two helices interact, forming a helix-turn-helix structure that can fill the major groove of the operator DNA. Both Cro and CI bind the operator as dimers. Figure 6-14 illustrates how two of the HTH regions of a dimer fill the major groove and also shows the importance of the N-terminal arms (last six amino acids) in reaching around the major groove to contact the backside of the operator. The affinity of an armless CI repressor for DNA is reduced about 1000-fold.

Establishment of Repressor Synthesis

If *cI* transcription from P_{rm} requires CI as a positive regulator, how is synthesis of CI established in the first place? After infection, *cI* transcription originates from a different promoter called P_{re} (promotor for repressor establishment). Transcription from P_{re} is positively regulated by the CII protein (Fig. 6-15). CII also activates the integrase gene whose product is required for λ DNA to integrate into the host chromosome (prophage state). Thus, CII coordinately regulates the two transcriptional units required for lysogeny.

One of the main criteria for whether lysis or lysogeny occurs is the amount of CII activator produced upon infection. This amount is dependent on several factors: (1) the level of *cII* mRNA, (2) the rate of *cII* mRNA translation, and (3) the rate of CII processing. The level of *cII* mRNA is controlled by CI and Cro repressors as well as by pN, which is required for transcription past t_{R1} into *cII*. Translational control of *cII* expression involves the host HimA protein. Efficient translation of CII mRNA requires HimA, which appears to allow ribosome access to the sequestered Shine-Dalgarno sequence and translation initiation codon in CII mRNA. There is a potential stem-loop structure found at the 5′ end of the CII message that protects these sequences.

In addition to the amount of CII message produced and the efficiency with which it is translated, the amount of active CII produced in the cell is dependent on its processing. The first two amino acids (N-formyl methionine and valine) must be removed in order to form an active CII protein. Also, turnover (or degradation) of CII is very

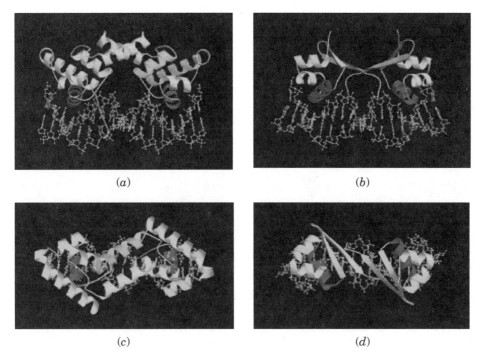

(a)   (b)

(c)   (d)

Fig. 6-14. Helix-turn-helix region of the λ CI and Cro repressors. Comparison of the structures of complexes of λ repressor (5, 6) and Cro (unpublished results) with operator DNA. In both cases, the proteins were crystallized with a 19 bp duplex of which the central 17 bp are shown. (*a*) Headpiece of λ repressor bound to DNA. The consensus half is to the left. As can be seen, there is substantial asymmetry, especially in the location of the amino-terminal arm. The bulk of repressor contacts occur on one side of the DNA, but the N-terminal arms reach around to contact the other side. (*b*) Cro bound to operator DNA. (*c*) View down the pseudo-2-fold axis of λ repressor bound to operator DNA. (*d*) Related view of Cro bound to operator DNA. (*Source:* From Albright and Mathews, 1998. Copyright, National Academy of Sciences, USA.)

important. The products of the *hflA* and *hflB* loci (94 and 72 centisomes, respectively) are responsible for CII degradation. HflB is actually a membrane-bound protease called FtsH. FtsH is responsible for degrading CII as well as σ 32 (see "Thermal Stress and the Heat Shock Response" in Chapter 18) and the SecY component of the *E. coli* protein secretory system. The *hflA* locus consists of three genes: *hflK*, *hflC*, and *hflX*. HflK and C are membrane proteins (periplasmic side) that interact with the FtsH protease and regulate its activity. It is the apparent role of the CIII protein to inhibit this protease, thus allowing sufficient accumulation of CII to promote transcription from P_{re} (i.e., CI production). Once again, a high multiplicity of infection will result in a large amount of CIII. This inhibits FtsH (HflB) protease from degrading CII. The accumulation of CII promotes the transcription of CI and integrase, which allows λ to integrate and represses all but CI transcription—that is, establishes the lysogenic state.

It is interesting, too, that the overall energy state of the cell affects the process of lysogenization. Under conditions where cAMP levels are low, lysogeny is the preferred developmental route, while the reverse is true where cAMP levels are high. It was discovered that cAMP and CRP appear to inhibit the transcription or activity of HflA. Thus, high cAMP levels lead to lowered HflA protease activity. This allows for a greater

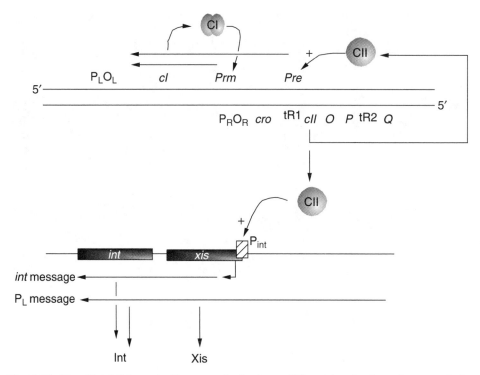

Fig. 6-15. Top: Establishment of lysogeny by λ phage. CII protein stimulates the transcription of *cI* from *pre*, the promoter for repressor establishment. Once CI protein is produced, it will stimulate its own transcription from *prm*, the promoter for repressor maintenance. Note that *prm* overlaps O_R (Fig. 6-12C) such that when CI binds to O_{R1} and O_{R2} transcription is inhibited from P_R but stimulated from *prm*. **Bottom: Regulating integration: the P_{int} promoter region and the mechanism of regulation by CII.** The CII protein recognizes and activates transcription from P_{int}. Positive regulation of *int* (integrase) but not *xis* occurs because the CII-stimulated P_{int} RNA transcript lacks the translation start for Xis.

level of CII, thereby stimulating CI production. However, there is some evidence that contradicts this model, so it is not precisely clear how cAMP affects this system.

Control of Integration and Excision

Once molecular events lead toward the lysogenic state, the circularized genome must integrate into the host chromosome as depicted in Figure 6-16. Integration involves site-specific recombination between the phage *att* and bacterial *att* attachment sites. The integration requires the *int* gene product plus integration host factor (IHF; see below). Prophage attachment normally occurs at the *att* site located between the *gal* and *bio* operons at 17 minutes. Note the permutation of gene order that occurs during the integration — that is, A and R are distal loci in the linear λ DNA present at infection but end up as proximal loci in the prophage.

The *int* and *xis* loci are tightly linked, and their coding regions partially overlap (Fig. 6-15). The *int* locus is controlled by the P_{Int} promoter while transcription of *xis* usually results from P_L. However, as illustrated in Figure 6-15, activation of P_{Int} by CII results in transcription starting beyond the beginning of *xis*. Thus, an incomplete

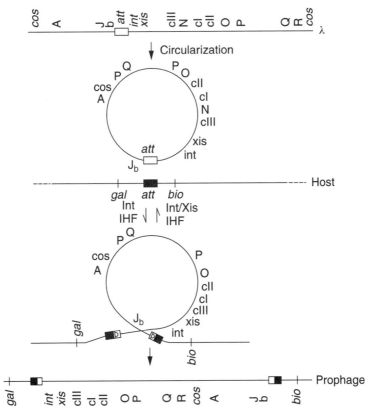

Fig. 6-16. Integration and excision of the λ genome. Note the permutation of gene order as a result of integration (i.e., *b* and *int* are proximal in the vegetative form but distal in the prophage.)

Xis message results from P$_{Int}$. The end result of CII activation, then, is the preferential synthesis of Int over Xis, which is obviously preferred for integration.

Negative Retroregulation of *int* by *sib*

Upon induction of prophage by UV mutagenesis, it is preferred that more Xis be produced than Int in order to achieve lytic replication. The *sib* locus within the *b* region is responsible for this. Upon exposure to UV light, activation of the host *recA* coprotease results in the autocleavage of CI protein (see "Inducible repair" in Chapter 3). Transcription of λ is allowed to occur from P$_L$ and will proceed through *xis* and *int*. Both Int and Xis are needed for prophage excision (host integration factor is also needed). Examination of Figure 6-16 reveals that in the prophage state *sib* (in region *b*) is removed from its normal proximal location to *int*. However, upon excision, the *sib* region is situated near *int* once again (separated by about 200 bp). At this point, Int is no longer required and could prove detrimental if it causes reintegration of λ in a cell ravaged by UV irradiation.

Regulation by *sib* is posttranscriptional and occurs only in *cis*. The basis for *sib* retroregulation (retro because *sib* is located downstream from *int*) is the formation of a stem-loop structure near the 3′ end of *int* message (Fig. 6-17). This stem-loop structure

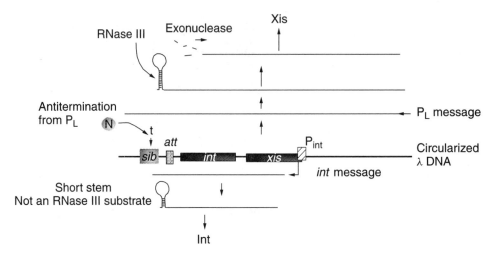

Fig. 6-17. Retroregulation by the *sib* region. Top: Abandoning ship. After excision of λ prophage from the chromosome, the *sib* region is repositioned downstream from *int*. This allows λ to make more Xis than Int, a situation that will prevent reintegration and favor lytic growth. Transcription of the *xis int* region from P_L, which only occurs when CI is inactive, allows antitermination of the downstream terminator by N protein. The longer mRNA includes the *sib* region. The *sib* RNA sequence forms a stem-loop structure that can be degraded by RNase III; which triggers additional degradation of *int* message. Thus, little integrase is produced. Xis mRNA sequence integrity is maintained and XIS is produced. **Bottom:** Setting up camp. This is the situation with a newly infected, circularized λ DNA that made a large amount of CII and enough CI to stop transcription from P_L. When CII initiates transcription of *xis int* at P_{int}, N is not available to antiterminate the *int* terminator. The shorter stem loop is not an RNase III substrate, so integrase is made and integration occurs. Recall from Figure 6-15 that no Xis is made from the P_{int} transcript, making integration the only option.

is a substrate for RNase III; thus, RNase III cleaves at the stem-loop region and the 3′ end is degraded by a 3′ exonuclease. Retroregulation, however, only regulates the P_L transcript but not the transcript resulting from P_I. The reason is that transcription from P_L is regulated by N protein, which, upon binding RNA polymerase, renders it resistant to most termination signals. Thus, the P_L transcript will proceed through the *sib* region, bypassing a termination signal. The CII-activated P_I transcript terminates within *sib*. The longer stem-loop structure formed from the P_L transcript is probably a more suitable substrate for RNase III allowing retroregulation.

λ-Phage Replication

Upon infection and circularization, early replication is bidirectional, with the origin of replication (*ori*) situated within the O gene. Evidence suggests that the amino-terminal end of the O protein (as four dimers) interacts with *ori* at four iteron sequences as the first step in a chain of interactions that control replication. The result is a nucleosome-like structure called the O-some. The λ P protein's job is to compete with DnaC in the host replication machinery for the DnaB helicase. The λ P–DnaB complex then interacts with the O-some, forming a preprimosome complex. The recruitment of DnaB into the λ-replication complex sequesters DnaB away from host replication machinery, with the consequence of decreasing *E. coli* DNA replication. Activation of DnaB within the λ *ori* complex is different from its activation at the chromosomal origin. Three *E. coli* heat shock proteins, DnaK, DnaJ, and GrpE, release λ P from the complex, triggering DnaB activation (see "Protein folding and chaperones" in Chapter 2). Following unwinding of λ *ori* by DnaB, DNA replication can proceed upon the addition of primase and DNA polymerase III.

Transcriptional activation of *ori* is also required for replication with transcription through the origin from P_R as a specific requirement. It is thought that DnaB can only unwind DNA in one direction from a preprimosome complex that includes λ O protein. Transcription directed toward the complex on the opposite side dislodges λ O, which is then degraded by the ClpXP protease. Because of this transcriptional requirement, replication is directly inhibited by CI repressor binding to the right operator. This is desirable, of course, when choosing lysogeny as a lifestyle.

As noted previously, early bidirectional replication of λ only occurs for a brief period after which a rolling circle mode of replication, referred to as σ replication, begins to make the long concatemeric DNA required for packaging (Fig. 6-13). Searches for a gene that would mediate the switch from θ- to σ-replication modes have led to the theory that DnaA protein is involved. Transcription from P_R is required for initiation of bidirectional replication, and DnaA protein (see "Initiation of DNA Replication" in Chapter 2) activates transcription from P_R. Since λ DNA contains many DnaA-binding sites, as the λ copy number increases, so, too, does the number of DnaA-binding sites, leading to the hypothesis that the titration of DnaA by these sites is part of the switch to σ replication.

The *gam* and *red* genes are clearly important for the switch to σ replication. The host *recBC* product, exonuclease V, will degrade the tails produced during σ-mode replication. The λ *gam* product inhibits exonuclease V activity, thus allowing concatemers to form. A *gam⁻* mutant, therefore, will only undergo θ-mode replication. However, concatemers are required for packaging to occur. The *red* product of λ mediates recombination between θ molecules, resulting in a large, circular concatemer also suitable for packaging. The packaging process itself is illustrated in Figure 6-18.

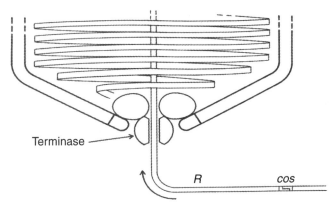

Fig. 6-18. Scanning for the terminal *cos* site, showing part of a nearly filled head with terminase positioned at the connector. The DNA strand being packaged is "scanned" by terminase until the cohesive end symmetry segment of the terminal *cos* site is recognized and cut. Other aspects of packaging, such as the terminase–connector interaction, are purely hypothetical. (*Source:* From Feiss and Becker, in Hendrix et al., 1983.) Lambda II. Cold Spring Harbor Laboratory, Cold Spring Harbor, N.Y.

The linear DNA molecule is inserted into an empty procapsid until the terminase enzyme (gene A product) identifies a *cos* site, at which point the enzyme introduces a staggered double-strand cut. The filled capsid then has a tail assembly attached and is ready for release from the cell. Release of mature phage particles from an infected cell involves λ lysozyme (the R gene product), which is a 17 K protein with murein transglycosylase activity, and the 11 K S gene product. The S product is believed to interact with the cell membrane, causing pore formation.

μ PHAGE: TRANSPOSITION AS A LIFESTYLE

The genome of μ (mu) phage (http://www.ncbi.nlm.nih.gov/cgi-bin/Entrez/framik?-db=Genome&gi=15122) is a linear molecule of about 38 kb consisting of 36 kb of μ DNA and 1.5 kb of host DNA at one end and 50 to 150 of host DNA at the other. The attached host sequences reflect the μ life cycle. When μ replicates, it undergoes repeated cycles of transposition in the cell (approximately 100 events per cell during the lytic cycle) and is ultimately packaged along with adjacent host sequences into virus particles. While μ transposition results in the duplication of a 5 bp target sequence, μ is different from other transposons in that the ends of μ are not inverted repeats.

The basic map of μ can be found in Figure 6-19 and Table 6-5. Two gene products are required for the transposition process. The A gene product (mwt 70,000) is the transposase, and the B gene product (mwt 33,000) is an accessory protein necessary for replication and full transpositional activity. During the lytic cycle where this activity is greatest, μ transposes by a cointegrate process. This fact indicates that during replication and cointegrate formation, inversion insertions or reciprocal adjacent deletions will occur (i.e., the host chromosome will be completely rearranged).

Fig. 6-19. Map of μ DNA. The DNA is 37 kb long. Lengths are shown in kb (not drawn to scale). Mature μ DNA in phage particles is flanked by heterogeneous host sequences that result from packaging of μ inserted into the host chromosome at different sites. α, G, ß: Different segments of the μ genome. The α segment contains the early genes and the genes for head and tail morphogenesis. The early region contains the repressor gene *c* and a negative regulator (*ner*). In addition to transposase gene *A*, the main replicator gene *B*, and a gene for amplified replication of μ (*arm*), other genes in this region are involved in DNA metabolism. Gene *C* appears to control the late functions. The G segment contains genes for the μ host range and is flanked by 34 bp inverted repeats. The *gin* protein, whose gene is located to the right of G, acts on the 34 bp inverted repeats to invert the G segment; hence, G is known as the flip-flop segment. The rightmost gene in μ encodes the DNA modification function (*mom*). (Modified From Toussaint and Resibois, 1983.) In, *Mobile genetic elements*, J. A. Shapiro, Ed., Academic Press, New York, pp 105–158.

TABLE 6-5. Characteristics of the μ Genes and Functions

Gene	Localization with Respect to the c End (Kb)	Function	Molecular Weight of Protein
c	0–0.9	Repressor	26,000
ner (negative regulation)		Shutoff of early transcription	6,000
A	1.3–3.3	Transposase	75,000
B	3.3–4.3	Replication, stimulates MuA	33,000
cim (control of immunity)		When inactivated in prophage lacking the S end, no repressor synthesis	7,000
kil (killing)		Killing of the host even in the absence of replication	8,000
gam	5–6	Protects DNA from exoV digestion by binding to DNA	14,000
sot (stimulation of transfection)	6–7	When present and expressed in a bacterium, stimulates efficiency of transfection with μ DNA	
arm (amplified replication of μ)		When deleted, less replication of μ DNA	
lig (ligase)		Expression of an activity that can substitute for $E.\ coli$ and T4 ligase	
C	9–10	Positive regulator of late gene expression	15,500
lys (lysis)		Lysis of host cell	18,900
D		Head protein	
E		Head protein	
H		Head protein	64,000
F (gp30)			49,400
G			17,200
I		Possible protease	38,900
T		Major head protein	
J		Head protein	
K		Tail protein	
L		Tail protein	55,000
M		Tail protein	12,800
Y		Tail protein	12,500
N		Tail protein	60,000
P		Tail protein	41,800
O		Tail protein	
V		Tail protein	
W		Tail protein	
R		Tail protein	
S	α and G	Tail protein	55,300
S'	α and G	Tail protein	48,000

(*continued*)

TABLE 6-5. (*Continued*)

Gene	Localization with Respect to the *c* End (Kb)	Function	Molecular Weight of Protein
U	In G	Tail protein	22,000
U'	In G	Tail protein	22,000
com		Regulates expression of DNA modification function	7,408
gin (G inversion)	Left of β	Inversion of the G region	21,500
mom (modification of μ)	Middle of β	Modifies μ DNA, protecting against restriction; is not a methylase	28,000
ner		Negative regulator of early transcription	8,500

The μ-phage particle, another myoviridae, is comprised of an icosahedral head 600 Å long bound to a contractile tail that ends in a base plate containing supporting spikes and fibers (Fig. 6-20). Infection with μ involves adsorption of the phage tail to a lipopolysaccharide component of the bacterial outer membrane followed by injection of viral DNA into the host cytoplasm. The G segment of μ is an invertible region that affects host range by changing the type of tail fiber produced. Inversion of the G region occurs via a reciprocal recombination between the short inverted repeats that flank this region. This switching occurs during the prophage state and is controlled by the *gin* product (a recombinase). The switch determines which of two alternative tail fibers are produced in a manner similar to what was described for *S. enterica* flagellar phase variation (see Chapter 3). μG(+) phages, with Sv and U at the left end of G, will infect *E. coli* K-12 or *S. enterica*. μG(−) phages, with S'v and U' at the left end of G, will infect such strains as *E. coli* C, *Shigella sonnei*, *Erwinia cloacae*, or other *Erwinia* species.

Following infection, the linear μ DNA circularizes, but the circular DNA is proteinase sensitive, suggesting that circularization may be the result of proteins covalently attached to the DNA. Ten minutes after infection, much of the infecting μ DNA is associated with the chromosome. Transposition to the chromosome from infecting phage involves a nonreplicative simple insertion as opposed to replicative cointegrate formation (see Fig. 6-21). There is a striking difference between μ and other temperate phages in that μ integrates into the host chromosome whether or not it enters the lytic or lysogenic cycle. Integration of μ into the bacterial chromosome occurs at a more-or-less random location, although there are preferred sites for insertion.

μ DNA synthesis via replicative transposition commences 6 to 8 minutes after infection or after induction of prophage. The original prophage copy is not excised prior to the onset of replication. The newly replicated μ copies also remain associated with the chromosome. Each new round of replication involves additional convalent attachment of the ends of μ to a distant target site. In effect, μ replicates by transposition to numerous sites on the bacterial chromosome. Both the *A* and *B* gene products are required for replication and transposition. In addition, transposition

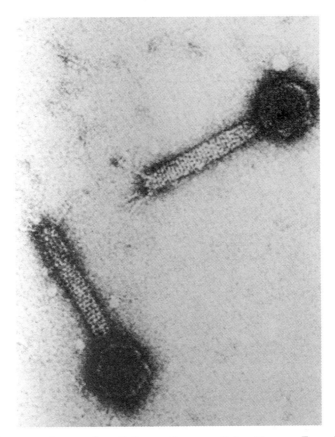

Fig. 6-20. Electron micrographs of the μ-phage particle. (*Source:* From Toussaint and Resibois, 1983.) In, *Mobile genetic elements*, J. A. Shapiro, Ed., Academic Press, New York, pp 105–158.

requires both the c and S ends of μ. Two *c* ends or two *S* ends are not suitable substrates for the transposition enzymes. The *A* gene product (MuA) recognizes the ends of the phage genome, assembles as a tetramer, and catalyzes the two basic transposition reactions. The protein introduces single-strand nicks at each end of μ and promotes strand transfer where the 3′ ends of the nicked DNA attack a new DNA site. MuA alone, however, only promotes a low level of transposition.

The *B* product (MuB) stimulates transposition by binding to target DNA and to MuA, essentially recruiting target DNA to the transpososome complex. In addition, MuB binding to MuA stimulates the strand transfer activity of MuA and prevents premature disassociation of the MuA tetramer by the ClpX chaperone (see Chapter 2). Recent evidence indicates that MuB is an ATPase and that ATP hydrolysis acts as a switch, regulating the affinity of MuB for target DNA. Multiple rounds of transposition into the same target molecule are actively discouraged, a phenomenon known as *target immunity*. If MuB–ATP binds to target DNA near a preexisting Mu, MuA stimulates MuB to hydrolyze the bound ATP. The resulting MuB–ADP disassociates from the inappropriate target. The presumed purpose of target immunity is to prevent one copy of μ from inserting

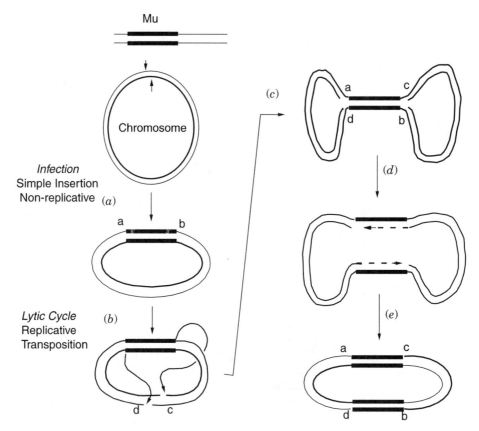

Fig. 6-21. Integration and replication of infecting μ DNA. (*a*) Integration. The paired filled rectangles represent μ DNA sequences; the lines extending from either side represent variable host DNA that was packaged into the infecting μ capsid. The circle represents the host chromosome; the arrowheads mark the staggered cut at the target site for integration. Integration occurs as a nonreplicative transposition event. (*b*) Lytic cycle. *a* and *b* represent the 5 bp repeat formed during integration as a result of the staggered cut in the chromosomal DNA. *c* and *d* represent host DNA sequences that flank a target for replicative transposition. Curved arrows indicate the attachment points of specific μ strands at the new target site. (*c, d*) represent a transposition intermediate and its replication to form two μ insertions. (*e*) illustrates the final transposition product. Note the chromosomal rearrangement that has occurred with respect to markers *a, b, c,* and *d*.

into another during the extensive rounds of transposition that occur during lytic growth.

μ transposition is regulated by the Mu *c* repressor protein. Mu repressor binds to an operator region 1 kb from the left end of the phage genome. When bound, μ repressor negatively controls the synthesis of the transposase and other functions involved in transposition, thus maintaining μ in a lysogenic state. Accumulation of a second repressor called Ner can initiate the lytic cycle by binding to another operator that controls production of μ repressor. The μ repressor is also selectively degraded by the ClpXP protease. This will trigger lytic growth by allowing synthesis of the μA and μB proteins. ClpX, therefore, has a dual role in controlling transposition. It limits transposition by disassembling μA (see above) and can stimulate transposition by targeting μ repressor for degradation.

At the end of the lytic cycle, the various μ–DNA copies scattered around the chromosome are cut and packaged into phage particles. Packaging of μ chromosomes into capsids occurs by a headful mechanism starting from a *pac* site located at the *c* end (see "Transduction" in Chapter 3 for *pac* sites). One cut is made 100 nucleotides outside this end, with the second cut made at a fixed length past the *S* end. Thus, cuts are made in the flanking bacterial DNA, so the DNA from each phage particle contains different terminal sequences. These particles are subsequently released into the environment to seek new hosts.

ΦX174

The ΦX174 phage (http://www.ncbi.nlm.nih.gov/cgi-bin/Entrez/framik?db=Genome-&gi=10126) is an example of single-stranded DNA phages (Microviridae) that also include M13 and G4. ΦX174 has been used extensively in studies designed to unravel the mysteries of how the *E. coli* chromosome, 1000 times its size, is replicated. The viral strand of ΦX174 contains 5386 bases and is packaged within an icosahedral capsid with a knob at each vertex (Fig. 6-22). The viral strand is designated (+) because it

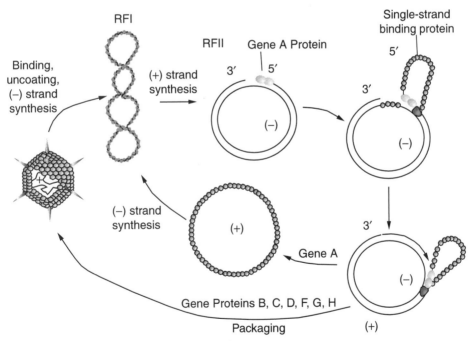

Fig. 6-22. Life cycle of ϕ174. Continuous replication initiated by gene *A* protein cleavage generates viral (+) circles, and discontinuous replication of the viral circles by the SS → RF system (host replication machinery) produces RF. In the presence of phage-encoded maturation and capsid proteins, viral circles are encapsulated rather than replicated, DNA making it unavailable for (–) strand synthesis. This strategy only produces new + strands, which are used in the new progeny phage. Following morphogenesis of the phage particle (40 min postinfection), cell lysis ensues.

is the template strand and can be transcribed directly by RNA polymerase. The strand contains genes for 11 gene products. However, the combined molecular weights of these proteins exceed the limit imposed by the available DNA. How is this possible? The apparent incongruity is explained by **overlapping genes** where one gene sequence encodes more than one protein.

The genetic map of ΦX174 is shown in Figure 6-23 and Table 6-6. The region originally assigned to gene *A* is now known to also contain all of gene *B*. This phenomenon of overlapping genes is possible because gene *B* is translated in a different reading frame from gene *A*. In a similar fashion, gene *E* is encoded in its entirety within gene *D*. Another translational control mechanism further expands the use of gene *A*. The 37 kDa gene *A** protein is formed by reinitiating translation at an internal AUG codon within the gene *A* message. The same translational phase is used, but the functions of the two proteins are distinct. The manner in which this virus extends the one gene–one protein hypothesis demonstrates a frugal usage of genetic information in small phage genomes.

During infection (Fig. 6-22), the phage adsorbs irreversibly to the lipopolysaccharide in the outer membrane of rough strains of *E. coli* and *S. enterica*. Rough strains lack the O antigen outer chains but retain intact core and lipid A (see Chapter 7). Attachment is accomplished via the spikes on the capsid. Following attachment, the virus enters the

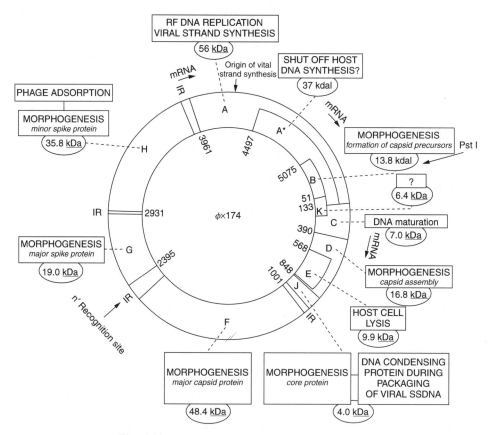

Fig. 6-23. Genetic map of bacteriophage ϕX174.

TABLE 6-6. ΦX174 Genes and Functions

Gene	Function	Protein Mass, kDa	
		SDS Gel	Sequence
A	RF replication; viral strand synthesis	59	56[a]
A*	Shutoff host; DNA synthesis	37	
B	Capsid morphogenesis	20	13.8[b]
C	DNA maturation	7	9.4[a]
D	Capsid morphogenesis	14	16.8[b]
E	Cell lysis	10	9.9[a]
F	Major coat protein	48	48.3[a]
G	Major spike protein	19	19.1[b]
H	Major spike protein; adsorption	37	35.8[a]
J	Core protein; DNA condensation	5	4.1[b]
K	Unknown	6	

[a]Calculated from DNA sequence $\left(\dfrac{\text{number of nucleotides}}{3 \times 0.00915} \right)$.

[b]Calculated from amino acid sequence.

eclipse stage in which the eclipsed virus has lost infectivity for new cells. Uncoating of the virus is coupled to replication of the viral genome, forming a double-stranded replicative form (RFI). The binding of ΦX174 particles and RFI occurs at areas of adhesion between inner and outer membranes (see Chapter 7) and possibly affords direct access to the host chromosomal replicative apparatus fixed at the inner surface. This SS ⟹ RFI replication is dependent solely on host replication proteins and includes prepriming, priming, elongation, gap filling, ligation, and supercoiling. RFI synthesis commences at the n' recognition site in the intervening region between genes G and F (Fig. 6-23).

The first stage of replication (SS ⟹ RF) occurs within 1 minute of infection. The second replication stage involves RF ⟹ RF synthesis and occurs up to 20 minutes postinfection. Two additional proteins are required for RF ⟹ RF replication. These are the phage-encoded gene A protein and E. coli rep product. The gene A product nicks supercoiled RFI in the (+) strand between residues 4305 and 4306, forming the open circular form RFII. The gene A protein remains attached to the 5′ end of the (+) nicked strand by one of two tyrosine residues: Tyr-343 or Tyr-347 (Fig. 6-22). By complexing with RepA (a DNA helicase), the gene A protein remains indirectly attached to the (−) strand and is dragged around the genome with the helicase. Leading strand replication then occurs from the 3′ end of the open (+) strand, regenerating a new duplex, while the old (+) strand is displaced, extended, and coated with single-stranded DNA-binding proteins. After traversing the entire length of the template, the gene A protein cleaves the regenerated origin, again attaching to the 5′ phosphoryl end by one of the two tyrosines noted above. The now free 3′-OH end of the (+) strand can attack the original gene A protein-5′-phosphoryl bond, generating a single-strand circle coated with SSB. This single strand can be reconverted into a new RFI using host polymerase machinery as before to generate another (−) strand.

Once approximately 35 double-stranded molecules of ΦX174 DNA have accumulated, lagging strand (i.e., minus-strand) synthesis stops and new progeny plus (+)

strands are encapsidated to form progeny phage particles. The switch is brought about by the accumulation of capsomeres and phage protein C. Concomitantly, host DNA synthesis is arrested either by the sequestration of a limiting host replication protein to phage DNA replication or by a phage-induced interference protein. This third stage of RF $\Rightarrow$ SS replication and encapsidation (20–30 minutes postinfection) depends on seven virus-encoded proteins. These proteins complex and encapsidate the (+) strand, making it unavailable for minus-strand synthesis.

ΦX174 does not encode an endolysin or holin and has no way to attack preexisting bacterial wall structures. So, how does it escape the cell? This phage interferes with bacterial enzymes that make precursors of peptidoglycan. The lack of coordination between peptidoglycan synthesis and bacterial wall construction leaves the wall weak, leading to its ultimate collapse due to osmotic pressures from within. ΦX174 protein E, associated with the bacterial membrane, blocks the activity of MraY, a bacterial protein that catalyzes the transfer of peptidoglycan precursors to lipid carriers so that the precursors can be transported through the membrane (see Chapter 7). As the cell disintegrates, ΦX174 particles are released to find new, unsuspecting hosts.

SUMMARY

Most, if not all, free-living microorganisms are susceptible to attack by one or more viruses. Learning how viruses usurp their host cells has provided tremendous insight not only into viruses but into the molecular machinery of the bacterial cell. Take, for example, the N-utilization proteins discovered as a result of investigations with λ phage or the developing model of bacterial DNA replication pieced together from studying the replication of a variety of phage systems. And who could have predicted that a study examining how bacteria attack viral DNA (restriction-modification systems) would lead to the era of recombinant DNA techniques and future gene replacement therapies for human disease (see Chapter 4)? Certainly continued explorations of phage–host interactions promise to yield additional knowledge of the intricacies of life.

BIBLIOGRAPHY

General

Birge, E. A. 2000. *Bacterial and Bacteriophage Genetics*, 4th ed. Springer-Verlag, New York.

Miller, J. H. E. 1991. *Bacterial Genetic Systems. Methods in Enzymology*. Academic Press, San Diego.

T4 Bacteriophage

Benzer, S. 1955. Fine structure of a genetic region in bacteriophage. *Proc. Natl. Acad. Sci. USA* **41**:349–54.

Colland, F., G. Orsini, E. N. Brody, H. Buc, and A. Kolb. 1998. The bacteriophage T4 AsiA protein: a molecular switch for sigma 70–dependent promoters. *Mol. Microbiol.* **27**:819–29.

Geiduschek, E. P. 1991. Regulation of expression of the late genes of bacteriophage T4. *Annu. Rev. Genet.* **25**:437–60.

Karam, J. D. 1994. *Bacteriophage T4*. ASM Press, Washington, DC.

Orsini, G., A. Kolb, and H. Buc. 2001. The *Escherichia coli* RNA polymerase.anti-sigma 70 AsiA complex utilizes alpha-carboxyl-terminal domain upstream promoter contacts to transcribe from a −10/−35 promoter. *J. Biol. Chem.* **276:**19, 812–9.

Wang, I. N., D. L. Smith, and R. Young, 2000. Holins: the protein clocks of bacteriophage infections. *Annu. Rev. Microbiol.* **54:**799–825.

Williams, K. P., G. A. Kassavetis, D. R. Herendeen, and E. P. Geiduschek. 1994. Regulation of late gene expression. In J. D. Karam (ed.), *Molecular Biology of Bacteriology T4.* American Society of Microbiology, Washington, DC, pp. 161–75.

λ Phage

Albright, R. A., and B. W. Matthews. 1998. How Cro and lambda-repressor distinguish between operators: the structural basis underlying a genetic switch. *Proc. Natl. Acad. Sci. USA* **95:**3431–6.

Baranska, S., M. Gabig, A. Wegrzyn, G. Konopa, A. Herman-Antosiewicz, P. Hernandez, J. B. Schvartzman, D. R. Helinski, and G. Wegrzyn. 2001. Regulation of the switch from early to late bacteriophage lambda DNA replication. *Microbiology* **147:**535–47.

Friedman, D. I., and D. L. Court. 2001. Bacteriophage lambda: alive and well and still doing its thing. *Curr. Opin. Microbiol.* **4:**201–7.

Gottesman, M. 1999. Bacteriophage lambda: the untold story. *J. Mol. Biol.* **293:**177–80.

Hendrix, R. W., J. W. Roberts, F. W. Stahl, and R. A. Weisberg. 1983. *Lambda II.* Cold Springs Harbor Laboratory, Cold Springs Harbor, NY.

Kihara, A., Y. Akiyama, and K. Ito. 1996. A protease complex in the *Escherichia coli* plasma membrane: HflKC (HflA) forms a complex with FtsH (HflB), regulating its proteolytic activity against SecY. *EMBO J.* **15:**6122–31.

Kihara, A., Y. Akiyama, and K. Ito. 1997. Host regulation of lysogenic decision in bacteriophage lambda: transmembrane modulation of FtsH (HflB), the cII degrading protease, by HflKC (HflA). *Proc. Natl. Acad. Sci. USA* **94:**5544–9.

φX174

Baas, P. D. 1985. DNA replication of single-stranded *Escherichia coli* DNA phages. *Biochim. Biophys. Acta* **825:**111–39.

Kornberg, A., and T. Baker. 1992. *DNA Replication.* W. H. Freeman and Co., San Francisco.

Novick, R. P. 1998. Contrasting lifestyles of rolling-circle phages and plasmids. *Trends Biochem. Sci.* **23:**434–8.

Reinberg, D., S. L. Zipursky, P. Weisbeek, D. Brown, and J. Hurwitz. 1983. Studies on the phi X174 gene A protein-mediated termination of leading strand DNA synthesis. *J. Biol. Chem.* **258:**529–37.

μ Phage

Bukhari, A. I. 1983. Transposable genetic elements: the bacteriophage Mu paradigm. *ASM News* **49:**275–80.

Lavoie, B. D., and G. Chaconas, 1996. Transposition of phage Mu DNA. *Curr. Top. Microbiol. Immunol.* **204:**83–102.

Mizuuchi, K. 1992. Transpositional recombination: mechanistic insights from studies of mu and other elements. *Annu. Rev. Biochem.* **61:**1011–51.

Roldan, L. A., and T. A. Baker. 2001. Differential role of the Mu B protein in phage Mu integration vs. replication: mechanistic insights into two transposition pathways. *Mol. Microbiol.* **40:**141–55.

Williams, T. L., E. L. Jackson, A. Carritte, and T. A. Baker. 1999. Organization and dynamics of the Mu transpososome: recombination by communication between two active sites. *Genes. Dev.* **13:**2725–37.

Yamauchi, M., and T. A. Baker. 1998. An ATP-ADP switch in MuB controls progression of the Mu transposition pathway. *EMBO J.* **17:**5509–18.

CHAPTER 7

CELL STRUCTURE AND FUNCTION

As discussed briefly in Chapter 1, prokaryotic cells differ in a number of major ways from eukaryotic cells (Table 7-1). Lower eukaryotes (protozoa, algae, and fungi) lack some of the distinguishing features of metazoan cells. Nevertheless, all eukaryotic cells appear to be similar versions of the same overall plan. In the past, some cytologists considered that bacterial cells were merely smaller replicas of the cells of higher forms and that failure to distinguish comparable subcellular structures in bacteria could be attributed to the limitations of the microscopic and cytological techniques available. This may prove to be the case, since many recent investigations with highly improved techniques and equipment have shown that there may be a closer similarity between eukaryotes and prokaryotes than previously thought. It has also become apparent that the cells of archaebacteria (*Archaea*) display important structural differences from the cells of either eubacteria or eukaryotes.

THE EUKARYOTIC NUCLEUS

Eukaryotes (*Eukarya*) display a cytologically distinct **nucleus**, the organizational and regulatory center for virtually all of the biochemical and hereditary processes of the cell. With the aid of the electron microscope it has been possible to demonstrate many structural details of the eukaryotic nucleus. Figure 7-1 shows a mammalian cell with a well-defined nuclear membrane composed of at least two distinct layers. The outer surface contains pores with tubules that transcend both membrane layers. Amoebae, protozoa, algae, and fungi (yeasts and molds) also contain a discrete membrane-bound nucleus (Fig. 7-2). The myxomycete *Acyria cinerea* contains a double-layered nuclear membrane that exhibits pores (Fig. 7-3).

TABLE 7-1. Major Components of Cells of Various Classes of Organisms

| Higher Eukaryotes | Lower Eukaryotes | | | Prokaryotes |
Metazoan	Protozoan	Algae	Fungi	Eubacteria and Archaebacteria[a]
Nucleus	Macronucleus	Nucleus	Nucleus	Nucleoid
Nuclear membrane	Nuclear membrane	Nuclear membrane	Nuclear membrane	No nuclear membrane[b]
Nucleolus	Nuclear elements			
Ribosomes	Ribosomes	Ribosomes	Ribosomes	Ribosomes
40S, 60S/80S	40S, 60S/80S	40S, 60S/80S	40S, 60S/80S	30S, 50S/70S
Cell respiration in mitochondria 70S ribosomes	Cell respiration in mitochondria 70S ribosomes	Cell respiration in mitochondria 70S ribosome	Cell respiration in mitochondria 70S ribosomes	Cell respiration in cytoplasmic membrane
Endoplasmic reticulum	Endoplasmic reticulum	Endoplasmic reticulum	Cytoplasm	Cytoplasm
Golgi apparatus	Dictyosomes	Dictyosomes		
Inclusions	Specialized organelles	Chloroplasts	Inclusions	Inclusions
Lysosomes				
Peroxisomes				
Phycobilisomes				
Plasma membrane	Plasma membrane	Plasma membrane	Cytoplasmic membrane	Cytoplasmic membrane
No cell wall	No cell wall	Cell wall chitin, glycans	Cell wall chitin, glycans	Cell wall peptidoglycan[a]
Undulating flagella or cilia	Undulating flagella or cilia	Undulating flagella or cilia	Undulating flagella or cilia	Rotating flagella

[a] Although comparable to eubacteria in many respects, the archaebacteria (*Archaea*) do not produce a cell wall peptidoglycan comparable to that produced by eubacteria. They also differ from eubacteria in a number of other characteristics not included in this table.

[b] Some species of eubacteria and archaebacteria have now been shown to have a nuclear membrane.

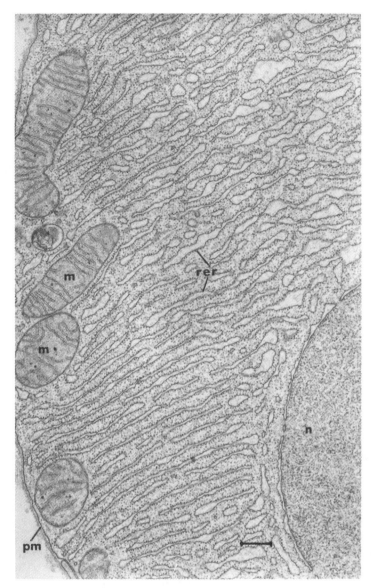

Fig. 7-1. Electron micrograph of a portion of a mammalian cell. The region of this pancreatic exocrine cell between the nucleus (n) on the lower right and the plasmalemma on the lower left is occupied by numerous cisternae of the rough endoplasmic reticulum (rer) and a few mitochondria (m). Prominent pores can be seen in the nuclear membrane. The ribosomes appear as small black dots lining the rer. The nucleolus is not visible in this photograph. Bar equals 1 μm. (*Source*: From Palade, G. E. *Science* **189**:347, 1975.)

BACTERIAL NUCLEOIDS

In most bacteria the DNA-containing region of the cell (the chromosome) is folded into a cytologically distinct region that does not appear to be bound by a nuclear membrane and is generally termed the **nucleoid** to distinguish it from the eukaryotic nucleus.

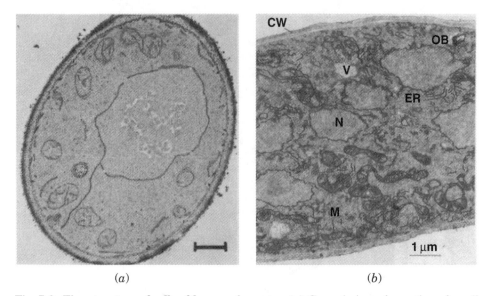

Fig. 7-2. Fine structure of cells of lower eukaryotes. (*a*) General view of a section of a cell of *Saccharomyces cerevisiae*. Prominent features include the cell wall, the nucleus with distinct pores in the nuclear membrane, mitochondria with cristae, and intermembranous structures. Bar equals 100 nm. (*Source*: From Avers, C. J., et al., *J. Bacteriol.* **100:**1044, 1969.) (*b*) Fine structure of a hyphal element of the fungus *Mucor genevensis*. Nuclei (N) with pores, vacuoles (V), mitochondria (M), endoplasmic reticulum (ER), cell wall (CW), and dark bodies (DB) are visible. Bar equals 1 μm. (*Source*: From Clark-Walker, C. D. *J. Bacteriol.* **109:**299, 1972.)

The compaction process is aided by a number of DNA-binding proteins, sometimes called **histone-like proteins**, since they resemble the histones found in the nucleus of eukaryotic cells. Most of the DNA is in the form of supercoils (see Chapter 2) constrained by these binding proteins. Interaction between the DNA supercoils and crowding within the cell is presumed to lead to further compaction and subsequent phase separation between the nucleoid and the cytoplasm.

As shown in Figure 7-4 (a, b, and c), growth in normal or high salt medium and fixation with OsO_4 results in visualization of a condensed nucleoid, whereas glutaraldehyde fixation (Fig. 7-4d) results in a dispersed appearance of the nucleoid. The development of confocal scanning light microscopy made it possible to observe the shape and substructure of the nucleoid and to compare these images with phase-contrast and electron microscope images. In Figure 7-5, a confocal scanning light microscope image of an OsO_4-fixed cell is compared with an electron micrograph and a reconstruction model based on a study of serial sections. Some of the differences observed might reflect the presence of transcription–translation complexes with ribosomes and proteins in the more dense preparations observed with OsO_4 fixation.

The nucleoid is observed as a coralline (coral-like) shape as shown in Figure 7-6a. The branches of the coralline nucleoid spread far into the cytoplasm and throughout the entire interior of the cell. By viewing serial sections, it has been possible to reconstruct the ribosome-free area of the nucleoid. Figure 7-6b presents a three-dimensional model of how the coralline nucleoid may appear.

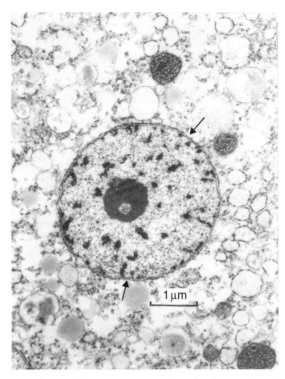

Fig. 7-3. Interphase nucleus of the myxomycete *Arcyria cinerea*. Prominent pores interrupt the typical nuclear envelope. The nucleolus near the center is well defined, while the chromatin is randomly dispersed throughout the nucleus. (*Source*: From Mims, C. W. *J. Gen. Microbiol.* **71**:53, 1972.)

Fast-growing cells harvested and fixed for observation under the electron microscope often display multiple nucleoids. This multinucleate condition occurs because fast-growing cells contain DNA in a state of multifork replication (see Chapter 2). Moreover, it has now become apparent that **multiploidy** (multiple copies of nucleoids or entire chromosomes), a frequent occurrence in eukaryotes, is also common in many bacterial species. The presence of a high copy number of chromosomes (nucleoids) is thought to facilitate rapid cell doubling during fast growth. Separation of the duplicating nucleoid during cell division must occur during the S phase (the period between the initiation of DNA replication and the completion of segregation of the two nucleoids; see Chapter 17).

Fluorescence microscopy studies reveal a mitosis-like movement of immunolabeled nucleoids during replication and cell division in bacteria. Protein components of a chromosomal segregation apparatus have been shown to direct the *oriC* region toward the poles. These proteins (ParB in *Caulobacter crescentus* and SpoOJ in *Bacillus subtilis*) appear to move toward the poles in a mitosis-like manner in concert with *oriC*. In bacteria such as *Escherichia coli* and *Haemophilus influenzae* that lack ParB/SpoOJ homologs, it has been suggested that other systems participate in a mitosis-like mechanism that effects nucleoid segregation. Many of these activities are considered in more detail in relationship to the cell division process discussed in Chapter 17.

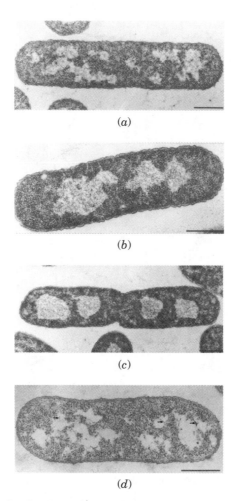

(a)

(b)

(c)

(d)

Fig. 7-4. Standard OsO₄ fixation of *E. coli* by the Ryter-Kellenberger technique of prefixation directly in the culture medium. (*a*) Cell was grown in low-salt medium. Bar equals 0.5 μm. (*b*) Same as (*a*), except that the cell was grown in normal salt medium (6 g KCl/L). The nucleoid is more confined and localized. Bar equals 0.5 μm. (*c*) Same as (*a*), except that the cell was grown in a high-salt medium (21 g KCl/L). The nucleoid is still more confined. (*d*) Fixation of *E. coli* B with glutaraldehyde embedded in Epon. The cell was grown in high-salt medium. The nucleoid is dispersed into small patches that contain DNA precipitates (arrows). Bar equals 0.5 μm. (*Source*: From Hobot, J. A., et al. *J. Bacteriol.* **182:**960–971, 1985.)

Isolation of the intact bacterial nucleoid has provided some insight into its properties. Two types of nuclear bodies can be obtained: an envelope-associated nucleoid and an envelope-free nucleoid. Large amounts of RNA, proteins, lipids, and peptidoglycans are found in the envelope-associated nucleoid. The free nucleoid contains lesser amounts of non-DNA components such as protein and RNA. Examination of the isolated nucleoid under the electron microscope (Fig. 7-7) reveals a number of supercoiled loops of DNA extending from the center. The central core appears to contain RNA that can be diminished by washing the grid with RNAse.

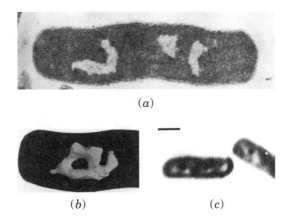

(a)

(b)   *(c)*

Fig. 7-5. Fast-growing *E. coli* B/r H266 (generation time, 21 min). The photographs compare a confocal scanning light micrograph (*c*) with a reconstruction of an OsO_4 fixed nucleoid (*b*), based on serial sections studied with the electron microscope. Only one of these sections is shown here (*a*). Bar equals 1.0 μm. (*Source*: From Valkenberg, J. A. C., et al. *J. Bacteriol.* **161:**478–483, 1985.)

NUCLEOSOMES

Eukaryotic chromosomes are organized into DNA–protein complexes termed **nucleosomes**. The proteins, mainly **histones**, play an important role in the structure of eukaryotic chromosomes by determining the conformation referred to as **chromatin**. The nucleosomes in the repeating unit of DNA organization are frequently seen in a "beads-on-a-string" configuration. Extensive digestion of chromatin with micrococcal nuclease releases the nucleosome core, a small, well-defined particle that has been crystallized. The particle mass is equally distributed between 146 bp of DNA and an octamer formed by two each of four major histone proteins (H2A, H2B, H3, and H4). With the aid of X-ray diffraction it has been possible to develop a three-dimensional model of the structure of the eukaryotic nucleosome core (Fig. 7-8).

The genomic DNA of *E. coli* is compacted into a nucleosome-like structure by interaction with several DNA-binding proteins. As many as 12 species of nucleoid proteins have been identified as shown in Table 7-2. The histone-like protein HU contains two closely related 10 kDa subunits, HU-α and HU-β, encoded by *hupA* and *hupB*. Fluorescein-labeled HU taken up by EDTA-treated cells of *E. coli* concentrates in the nucleoid and is uniformly distributed throughout this structure. The proposed role of HU protein is in stabilizing higher-order nucleoprotein structures to confer specificity during DNA interactions. The fact that HU exhibits little sequence specificity suggests that it may be involved in a variety of DNA protein activities requiring coiling of specific DNA sequences. In vitro studies have implicated HU protein in the initiation of DNA replication, binding of repressors, and transposition of bacteriophage Mu. In *E. coli*, *hupAhupB* double mutants that lack HU protein entirely display severe defects in cell division, DNA folding, and DNA partitioning.

Fis is a small, basic DNA-binding protein that was originally identified by its ability to effect flagellar phase variation in *E. coli* by stimulating DNA inversion. Fis has also been implicated as a participant in rRNA and tRNA transcription. It may play a major role in chromosomal DNA replication as evidenced by its binding to the origin of

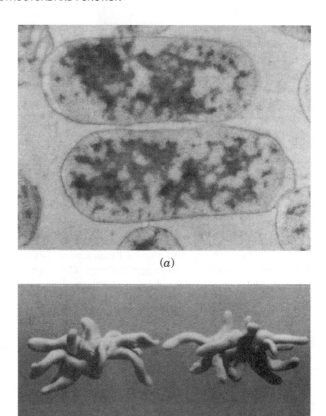

(a)

(b)

Fig. 7-6. Coralline shape of the bacterial nucleoid. (*a*) Immunostaining of DNA in exponentially growing *E. coli* B using antidouble-stranded DNA mouse IgM, followed by goat IgG
antibodies to mouse IgM stained with a mixture of 1 volume of 2% KMno$_4$ and 2 volumes of 2%
uranyl acetate. Label is observed as gray-stained patches over DNA-rich area. In this micrograph
the upper cell shows two distinct nucleoids. (*b*) Schematic models of bacterial nucleoids show
a pair of nucleoids of growing cells deduced from thin sections of cryofixed, freeze-substituted
E. coli B. For practical reasons, the models were flattened on the lower side. (*Source*: From
Bohrmann, B., et al. *J. Bacteriol.* **173**:3149–3158, 1991.)

replication (*oriC*) in *E. coli*. Fis levels vary dramatically during the course of cell
growth and in response to changing environmental conditions. When stationary-phase
cells are subcultured into a rich medium, Fis levels increase from less than 100 to over
50,000 copies per cell prior to the first cell division. As cells commence exponential
growth, Fis synthesis is reduced and the intracellular concentration drops as a function
of the rate of cell division. Concurrent alteration in *fis* mRNA levels suggests that
regulation is at the transcriptional level. Fluctuation in Fis levels may serve as an
early signal of a nutritional upshift and may play an important role in governing
physiological activities. Fis appears to autoregulate its own expression as shown by
the fact that binding to its promoter excludes binding of RNA polymerase.

The bacterial histone-like protein H-NS, encoded by *hns*, inhibits site-specific DNA
recombination as evidenced by the dramatic increase in inversion of the *pilA* promoter

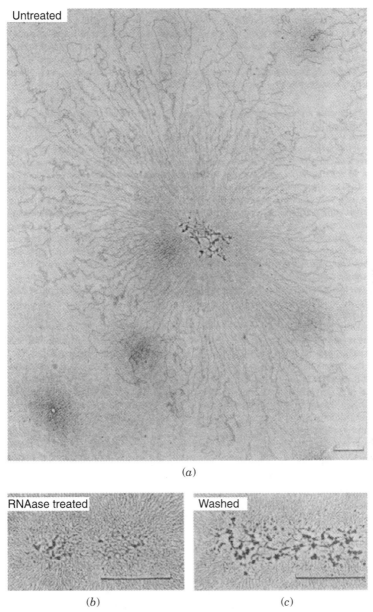

Fig. 7-7. **Cell envelope-free nucleoid from** *E. coli* **15** τ**-bar.** For preparation of the cell-associated nucleoid, cells are first suspended in 20% sucrose, treated briefly with lysozyme and ethylene-diaminetetraacetic acid (EDTA), and then lysed with a mixture of the nonionic detergents Brij-58 and deoxycholate in the presence of I *M* NaCl. Centrifugation at 17,000 g for 15 min on a 10–50% sucrose gradient permits isolation of the cell envelope–associated nucleoid. Preparation of the envelope-free nucleoid requires longer lysozyme treatment and use of an ionic detergent such as Sarcosyl. The sedimentation coefficients are 3200S for the envelope-associated nucleoid and 1600S for the envelope-free nucleoid. The hypophase contained 0.4 *M* salt. The chromosome had 141 ± 3 loops and possibly a fork on the loop at about 1 o'clock. In (*b*), the grid was washed with the control buffer. Bars equals 1.0 μm. (*Source*: From Kavenoff, R. and B. C. Bowen, *Chromosoma* **59**:89–101, 1976.)

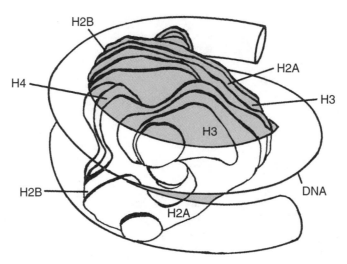

Fig. 7-8. Model of the nucleosome core. The model was made by winding a tube simulating the DNA superhelix on a model of the histone octomer, which was built from a three-dimensional map derived from electron micrographs of the histone octomer. The ridges of the periphery of the octomer form a more-or-less continuous helical ramp on which a 146-nucleotide pair length of DNA can be wound. The locations of individual histone molecules (whose boundaries are not defined at this resolution) are proposed here on the basis of chemical cross-linking data. (*Source*: From Kornberg, R. D., and A. King, *Sci. Am.* **250**:52–64, 1981.)

TABLE 7-2. Nucleoid (Histone-like) Proteins in Bacteria

Protein	Identified Function
CbpA	Curved DNA-binding protein A
CbpB	Curved DNA-binding protein B
	[also known as Rob (right origin-binding protein)]
DnaA	DNA-binding protein A
Dps	DNA-binding protein from starved cells
Fis	Factor for inversion stimulation
Hfq	Host factor for phage Q_β
H-NS	Histone-like nucleoid structuring protein
HU	Heat-unstable nucleoid protein
IciA	Inhibitor of chromosome initiation A
IHF	Integration host factor
Lrp	Leucine-responsive regulatory protein
StpA	Suppressor of *td* mutant phenotype A
H1	?

region in mutants lacking H-NS (see further discussion in "Pili or Fimbriae" in this chapter).

In the *Archaea*, chromosomal DNA also exists in protein-associated form. Histone-like proteins have been isolated from stable nucleoprotein complexes in *Thermoplasma acidophilum* and *Halobacterium salinarum*. Whole chromosomes of *H. salinarum* consist of regions of both protein-associated DNA and protein-free DNA. In electron micrographs the protein-associated DNA region appears as nucleosome-like fibrous structures. However, the suggestion that these observations represent a major difference between eubacteria and *Archaea* seems untenable in view of the widespread occurrence of protein-associated DNA and nucleosome-like structures in a variety of bacteria.

MITOCHONDRIA

Mitochondria are subcellular organelles that carry out oxidative metabolism (respiration) in eukaryotic cells. The highly membranous interior of the mitochondrion contains the cytochrome systems and oxidative phosphorylation processes that yield energy in the form of ATP (see Chapter 9). Lower eukaryotes (yeasts, algae, filamentous fungi, and protozoa) have all been shown to contain mitochondria, although their number and distribution vary widely from one cell type to another. Even within a given species, the number of mitochondria per cell and their intracellular arrangement may vary with the cultural conditions and growth phase.

Mitochondria possess two distinct membranes, **outer** and **inner membranes**, which may be closely associated with each other at certain contact sites. The inner membrane is highly invaginated, forming pleats referred to as **cristae**. Yeasts produce mitochondria actively when grown in an aerobic environment. Under anaerobic conditions, mitochondria are not observed; yet upon shifting back to aerobic conditions, they quickly reappear. Yeast cells grown under strict anaerobiosis contain mitochondria-like particles designated as **promitochondria**. These structures lack most of the characteristic components of the respiratory chain but still contain ATPase, mitochondrial DNA (mtDNA), and various structural proteins. Anaerobic growth apparently induces a dedifferentiation of yeast mitochondria. When yeast cells are grown in the presence of high concentrations of glucose, the respiratory and energy transduction systems are repressed even under highly aerobic conditions. The system responsible for mitochondrial gene replication and expression is completely blocked, and the cell satisfies all its energy requirements by glycolysis.

Mitochondrial DNA (mtDNA) maps as a closed circular molecule, although linear forms may also exist. Despite the presence of its own genetic material, the mitochondrion is not genetically self-sufficient. Most of the mtDNA-specified proteins are components of the complexes that carry out the respiratory functions of the mitochondrion and the RNA components of their protein synthetic apparatus (ribosomal RNAs and tRNAs). In yeast and possibly other lower eukaryotes, the structural gene coding for the ribosome-associated protein VarI is located on mtDNA. Detailed genetic studies indicate that DNA sequences have been transferred from the mitochondrial to the nuclear genome. Nuclear genes code for most other mitochondrial components, including the enzymes of the tricarboxylic acid (TCA) cycle. Transcription of mitochondrial genes appears to be catalyzed by a nucleus-encoded RNA polymerase. One subunit of the yeast RNA polymerase functions as a core enzyme while a second subunit serves as a specificity factor that facilitates recognition of one of the approximately 20 specific mitochondrial transcription initiation signals on mtDNA. Yeast cells lacking functional mtDNA (ρ^- mutants) do not synthesize mitochondrial proteins. They exhibit irregular shapes, fewer cristae, and aberrant multilamellar configurations. On the other hand, most nuclear mutations that result in respiratory deficiency (*pet*) do not show a marked effect on the morphology of yeast mitochondria.

A considerable body of evidence supports the hypothesis that the mitochondrion originated from a primitive bacterial form (α-proteobacterium) through development of an endosymbiotic relationship. As discussed in Chapter 2, the primary nucleotide sequences of rRNA have been highly conserved throughout evolution. A comparison of the nucleotide sequence of various classes of mt-rRNA molecules from human beings and other mammals as well as from lower eukaryotes reveals a very close resemblance to bacterial rRNA. Also, it has been shown that a homolog of the bacterial cell division gene *ftsZ* is present in the mitochondrion of the alga *Mallomonas splendens*. Homologs of FtsZ have also been found in chloroplasts. This information, together with the revelation that the mitochondrial genome closely resembles that of bacteria, suggested a prokaryotic origin for mitochondria and chloroplasts. Recent studies with *Saccharomyces cerevisiae* and *Rickettsia prowazekii* suggest that mitochondrial genomes are descended from α-proteobacterium. However, studies with other organisms provide evidence that challenges the endosymbiosis-based view of the origin of the eukaryotic mitochondrion. These studies have been invoked to support an alternate hypothesis that the mitochondrion may have arisen at the same time that the eukaryotic nucleus originated rather than from some primitive bacterial form.

In eukaryotic cells, programmed cell death (apoptosis) has been shown to involve the mitochondrion. In response to stress (nutrient deprivation or exposure to irradiation), mitochondria rupture or leak and release factors that trigger caspases (protein-splitting enzymes), which disrupt proteins in the cell membranes and the nucleus, resulting in rapid cell death. Two proteins, BAK and BAX, have been shown to be gatekeepers for apoptotic signals that act through the mitochondrion. Activation of a series of reactions leads to release of cytochrome *c* from mitochondria and ultimately to cell death. Only mutants lacking both BAK and BAX proteins fail to undergo apoptosis. The Fas death receptor pathway involves binding of a ligand to the cell surface protein Fas. Activation of Fas results in the production of caspases that ultimately destroy the cell. Recent evidence indicates that the Fas receptor pathway also requires the help of the mitochondrial pathway to trigger cell death.

MICROBIAL CELL SURFACES

Eukaryotic Cell Surfaces

As discussed briefly in Chapter 1, both eukaryotic and prokaryotic microorganisms are often enclosed within a rigid cell wall. This characteristic is not unique to microorganisms nor is it strictly confined to either the plant or animal kingdom. The cell walls of vascular plants are largely composed of **cellulose,** a β-1,4-linked polymer of glucose (Fig. 7-9). Many algae, as well as higher green plants, contain cellulose as a major cell wall constituent.

The cell walls of some protozoa and many fungi contain **chitin,** a linear polymer of *N*-acetyl glucosamine (Fig. 7-9). Most fungi, algae, and higher plants contain microfibrils of either cellulose or chitin as a prominent skeletal component of their cell walls. Production of these microfibrils occurs at or near the surface of the cell. Among prokaryotes, some species of *Acetobacter* and a few others produce cellulose.

The yeast, *S. cerevisiae*, contains a cell wall that appears as a layered structure when viewed under the electron microscope. The chemical composition of the yeast cell wall is rather uniform over most of the cell surface. Chemical analysis of yeast

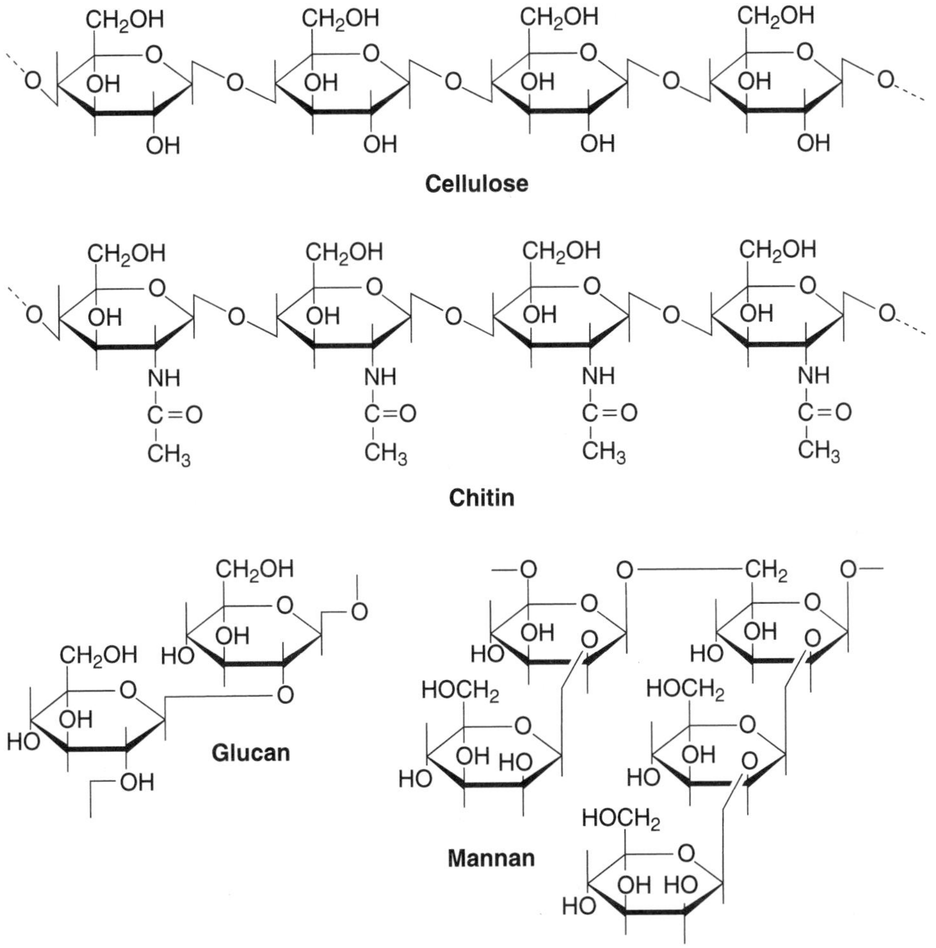

Fig. 7-9. Chemical structures of cell wall polymers. Cellulose, a β-1,4-linked polymer of glucose; chitin, a linear polymer of N-acetyl-D-glucosamine; glucan, a branched polymer of glucose; and mannan, a branched polymer of mannose.

walls reveals the presence of 29% β-glycans (both 1,6-β-glycan and 1,3-β-glycan are found), 31% mannan, and 13% protein (Fig. 7-9). The mannan-rich outer layer of the wall contains a higher proportion of 1,6-β-glycan. Yeast cell walls also contain small percentages of lipids and other materials. Chitin (poly-N-acetyl glucosamine) is present in small amounts in vegetative cells of yeast, being confined almost entirely to the ring encircling the septum of budding yeast (bud scar) in mother cells. However, the septum and the chitin ring encircling it show a different chitin : glycan ratio. In the cell walls of all filamentous fungi so far examined, chitin is the only common constituent found.

Prokaryotic Cell Surfaces

In Chapter 1 it was stated that a high-molecular-weight, sugar-containing, rigid structure called **peptidoglycan** formed the major backbone of the murein sacculus of the cell wall of both gram-positive and gram-negative bacteria. Indeed, the production of peptidoglycan is considered to be the hallmark of the eubacteria. Archaea produce a **pseudomurein** and an associated surface layer (**S-layer**) composed of protein or glycoprotein. In many archaebacteria the S-layer may represent the only surface component outside the plasma membrane as shown in Figure 7-10.4. In *B. subtilis* the S-layer is associated with the peptidoglycan-containing sacculus (Fig. 7-10.1). Gram-negative bacteria such as *E. coli* produce an outer membrane (Fig. 7-10.2). The S-layer, if produced, is attached to this outer membrane (Fig. 7-10.3).

Surface Layers of Bacteria

Bacteria from all of the major phylogenetic groups have been shown to produce a crystalline cell surface layer (S-layer) as the outermost component of the cell (Fig. 7-11). S-layers are composed of identical protein or glycoprotein subunits. They assemble into two-dimensional crystalline arrays showing oblique, square, or hexagonal symmetry in gram-negative or gram-variable bacteria and are associated with the surface of the outer membrane (Fig. 7-10.4). The S-layer may appear quite different in various organisms. In gram-positive bacteria the subunits are linked to the peptidoglycan-containing layer or to secondary cell wall polymers attached to the peptidoglycan layer. In *B. sphaericus* a specific cell wall polymer has been characterized. It contains N-acetylglucosamine, N-acetylmannosamine, and pyruvic acid and serves as the specific binding site for the N-terminal portion of the S-layer protein. In gram-negative bacteria the S-layer is bound to the lipopolysaccharides of the outer membrane.

Chemical analysis of purified S-layers shows that they are composed of a single protein or glycoprotein species. In a few organisms, such as *Clostridium dificile* or *B. anthracis*, two types of subunits are observed. S-layers often contain a high proportion of acidic and hydrophobic amino acids. Lysine is the predominant basic amino acid. Arginine, histidine, and methionine are generally low in amount, and the presence of cysteine is rare, being found in only a few S-layer proteins.

Diverse functions have been attributed to the S-layer, including mediation of bacterial attachment to surfaces and to host tissues. In *Campylobacter* and *Aeromonas*, the S-layer serves as a virulence factor. Since the S-layer represents the outermost layer external to the cell membrane in archaea, it must contribute substantially to the shape of the cell (as in *Thermoproteus tenax*). However, there is still a great deal to be learned about the true function(s) of the S-layer in many of the organisms that have been shown to produce this external protein complex.

Peptidoglycans of Bacterial Cell Walls

Peptidoglycan is a heteropolymer of repeating units of β-1,4-N-acetylglucosamine (NAG) and N-acetyl-muramic acid (NAM). The glycan linkages of peptidoglycan are considered to be uniform in all bacteria with every D-lactyl group of the NAM being peptide substituted. All glycans have short tetrapeptide units terminating with D-alanine or occasionally tripeptide units lacking the terminal D-alanine. The L-alanine at the N terminus can be replaced by glycine.

The interpeptide bridges linking peptidoglycans are of four major types as shown in Figure 7-12 and Table 7-3. Although the proportion of peptide cross-linking varies

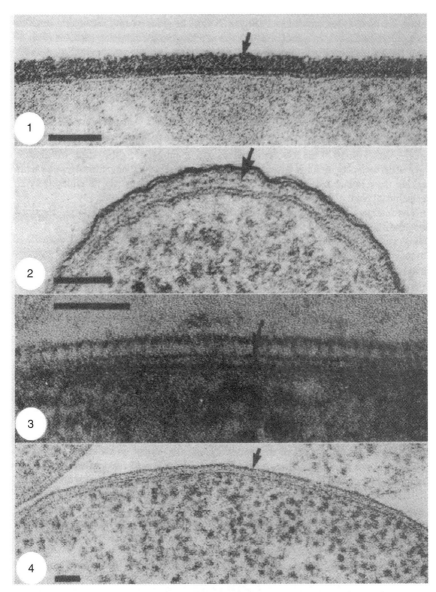

Fig. 7-10. Envelope profiles of conventionally fixed and embedded bacteria. Bar equals 50 nm. (**1**) Amorphous wall fabric of *B. subtilis* 168 (arrow) lies directly above the plasma membrane. (**2**) *Escherichia coli* possesses a thin peptidoglycan murein layer above the plasma membrane (arrow), over which lays a wavy outer membrane. The region between the outer and plasma membranes is called the periplasm or periplasmic space. The waviness of the outer membrane is believed to be an artifact of the conventional fixation-embedding technique, and much of the periplasm has been leached out. (**3**) *Clostridium thermosaccharolyticum* is an example of a gram-variable bacterium that has a wall profile intermediate between the gram-positive and gram-negative formats in 1 and 2. The peptidoglycan layer (arrow) is thinner than that of *B. subtilis* but thicker than that of *E. coli*. Above this layer is a proteinaceous S-layer of periodically arrayed subunits. (**4**) *Methanococcus voltae* is an archaebacterium and possesses only a thin S-layer (arrow) above the plasma membrane as its sole wall layer. (*Source*: From Beveridge, T. J. and L. L. Graham, *Microbiol. Rev.* **55**:684–705, 1991.)

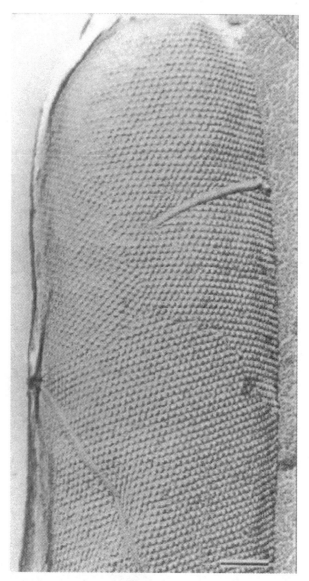

Fig. 7-11. Electron micrograph of a freeze-etched preparation showing a whole bacterial cell with a hexagonally ordered S-layer lattice. Bar equals 100 nm. (*Source*: From Sàra, M. and Sleytr, U. B. *J. Bacteriol.* **182**:859–868, 2000.)

considerably from one type to another, in all cases a continuous bag-like sacculus completely surrounds the cell. There is now ample evidence to show that the glycans form helically twisted chains from which the peptide bridges radiate in all directions from the axis of the backbone. In this conformation the peptide cross-links are arranged so that they protrude at ~90° to one another. In a given horizontal plane approximately 50% of the peptide stems will be cross-linked. This orientation permits interpeptide bridging between many different neighboring chains, forming a supramolecular network (mosaic) of peptidoglycan.

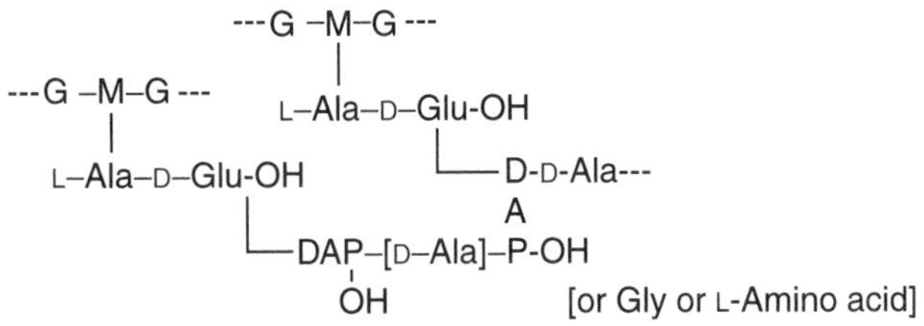

```
                              ---G –M–G ---
                                   |
  ---G –M–G ---                L–Ala–D–Glu-OH
        |                                 └——D-D-Ala---
   L–Ala–D–Glu-OH                               A
           └—DAP–[D–Ala]–P-OH
                   |
                   OH              [or Gly or L-Amino acid]
```

Type I. Direct D-alanyl-R₃ peptide bond. Found in *E. coli* and most other gram-negative bacteria; may be found in many bacilli.

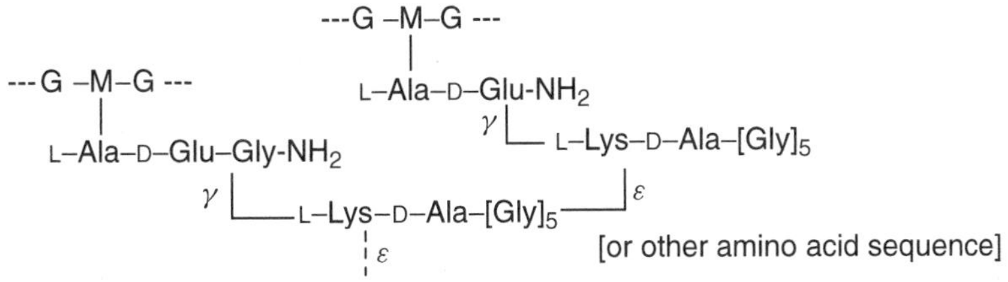

```
                              ---G –M–G ---
                                   |
  ---G –M–G ---                L–Ala–D–Glu-NH₂
        |                        γ |
   L–Ala–D–Glu–Gly-NH₂              └— L–Lys–D–Ala–[Gly]₅
        γ |                                         | ε
          └——L–Lys–D–Ala–[Gly]₅———
                        ¦ ε          [or other amino acid sequence]
```

Type II. Pentaglycine or other L- or D-amino acid sequence.
This type of interpeptide bridge varies from organism to organism as shown in Table 6-3

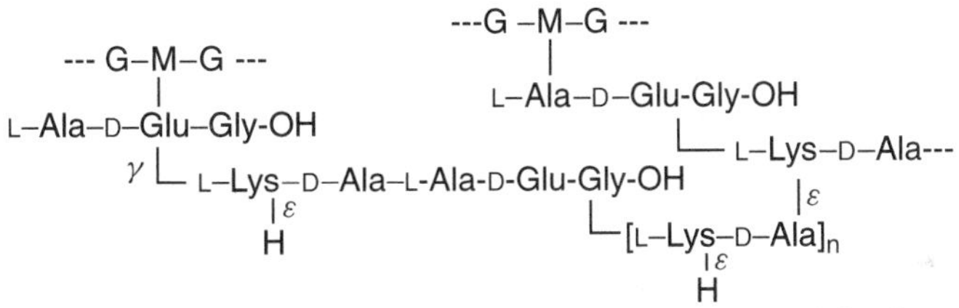

```
                                 ---G –M–G ---
    --- G–M–G ---                      |
         |                        L–Ala–D–Glu-Gly-OH
   L–Ala–D–Glu–Gly-OH                       └— L–Lys–D–Ala---
        γ └ L–Lys–D–Ala–L-Ala-D–Glu-Gly-OH               | ε
               | ε                        └—[L–Lys–D–Ala]ₙ
               H                                  | ε
                                                  H
```

Type III. Bridge composed of one to several peptides, each having the same amino acid sequence as the peptide unit attached to muramic acid; found in *Micrococcus luteus.*

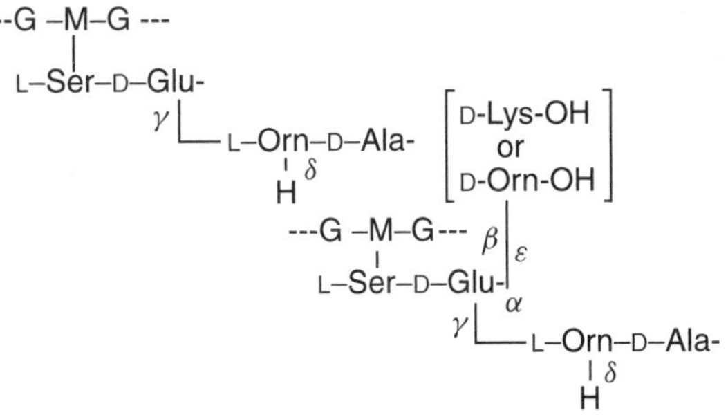

```
   ---G –M–G ---
        |
   L–Ser–D–Glu-
        γ └— L–Orn–D–Ala-    ⎡ D-Lys-OH ⎤
               | δ           ⎢    or    ⎥
               H             ⎣ D-Orn-OH ⎦
                  ---G –M–G---  β | ε
                       |
                  L–Ser–D–Glu-|
                             γ | α
                               └—L–Orn–D–Ala-
                                      | δ
                                      H
```

Type IV. Bridge extending between carboxyl groups belonging to either D-alanine or to D-glutamate and a diamino acid residue or a diamino acid- containing short peptide; found in *B. rettgeri.*

Fig. 7-12. Major types of interpeptide bridges in peptidoglycan. G, *N*-acetylglucosamine; M, *N*-acetylmuramic acid; DAP, diaminopimelic acid. Other amino acid sequences replacing [Gly]5 (pentaglycine) in type II bridges are shown in Table 6-3. (*Source*: Redrawn from Moat, A. G. 1985. Biology of the lactic, acetic, and propionic acid bacteria. In *Biology of Industrial Microorganism*, A. L. Demain and N. A. Solomon (Eds.). Benjamin-Cummings, Menlo Park, CA, pp. 143–188.)

TABLE 7-3. Types of Interpeptide Bridges in Various Bacteria

Amino Acid(s)	Organism
-[Gly]$_5$-	*Staphylococcus aureus*
-[L-Ala]$_3$-L-Thr-	*Micrococcus roseus*
-[Gly]$_5$-[L-Ser]$_2$-	*Staphylococcus epidermidis*
-L-Ser-L-Ala	*Lactobacillus virdescens*
-L-Ala-L-Ala-	*Streptococcus pyogenes*
-L-Ala-	*Arthrobacter crystallopoietes*
	Enterococcus hirae
-Asp-MH2-	*Enterococcus faecium*
	Enterococcus hirae
	Lactobacillus casei

A fundamentally different concept of the organization of the cell wall has been presented that suggests the glycan strands run perpendicular to the plasma membrane, each strand being cross-linked by peptide bridges with four other strands. This arrangement would permit the formation of a structured matrix pierced with ordered ionophoric channels potentially harboring lipoprotein or teichoic acid in gram-negative organisms or lipoteichoic acid in gram-positive bacteria. Arrangement of the wall components in a perpendicular manner would allow murein synthesis to progress in unison with formation of the associated synthesis machinery and other multicomponent systems. Further studies should resolve the question of whether this altogether different perspective of the structure of the peptidoglycan layer represents a realistic and provable concept.

Peptide cross-bridging between peptidoglycan strands is considered to provide structural rigidity to the cell. This is certainly the case in gram-positive cells in which the multilayered network of cross-linked peptidoglycan ranges between 20 and 40 nm in thickness. In gram-negative cells the peptidoglycan layer is much thinner (2–6 nm), and in most organisms it is probably only a few monolayers in depth. This difference in thickness between gram-positive and gram-negative peptidoglycan layers is generally invoked to explain the fundamental difference in staining with crystal violet (gram stain). In gram-negative bacteria, support from other structures such as the outer membrane with its embedded proteins may provide the additional support required to withstand the turgor pressure from within the cell.

Some organisms possess cell envelopes that are intermediate between these two general types. The cyanobacteria possess envelopes with a combination of these features. Although cyanobacteria have a general structure more closely resembling gram-negative bacteria, their peptidoglycan layers range from 10 to 35 nm, depending on the species examined. The degree of cross-linking is far higher (56–63%) than the usual 20 to 33% cross-linking observed in most other gram-negative organisms. Most cyanobacteria contain the typical *meso*diaminopimelic acid found in gram-negative bacteria as opposed to the L-diaminopimelic acid or L-lysine of gram-positive cells. As an exception, L-lysine is found in *Anabaena cylindrica*. With the great diversity of bacteria known to exist, it would not be surprising to find other major differences in cell structure and function as more and more organisms are examined more thoroughly.

Peptidoglycan (Murein) Hydrolases

Almost all bacteria produce peptidoglycan (murein) hydrolases — enzymes that hydrolyze bonds in the peptidoglycan structure. These enzymes are of three basic types:

1. **Glycan-strand hydrolyzing**
   a. *Endo-N*-acetylmuramidases
   b. *Endo-N*-acetylglucosaminidases
2. **Endopeptidase hydrolyzing**
   a. Peptide bonds in the interior of the peptide bridges
   b. Bonds involving the C-terminal D-alanine residue
3. *N*-**acetylmuramoyl-L-alanine amidase** acting at the junction between the glycan strands and the peptide units

Enlargement of the peptidoglycan sacculus must require an intricate interplay of synthesizing and hydrolyzing enzymes to allow new peptidoglycan subunits to be inserted into the preexisting structure. However, direct proof of the participation of these hydrolases in the formation of the murein of the bacterial cell wall has not been fully verified. Nevertheless, these enzymes appear to play an important role in a number of cellular activities including septum and wall extension during cell growth, cell separation, turnover of wall components, sporulation, competency for transformation, and the excretion of toxins and exoenzymes.

Activation of certain of these hydrolases, particularly under conditions of external stress, may result in cellular autolysis. In fact, it is now considered that triggering of murein hydrolases or other autolysins by environmental stress is responsible for programmed cell death (sometimes referred to as apoptosis).

Peptidoglycan (Murein) Synthesis

Murein biosynthesis involves a number of cytoplasmic, membrane, and periplasmic steps (Fig. 7-13). *N*-acetylglucosamine (GlcNAc) is first coupled with UDP. A portion of the UDP-GlcNAc is converted into UDP-MurNAc (*N*-acetylmuramic acid), and the peptide chain is developed by sequential addition of amino acids. The growing chain is then coupled with undecaprenyl-phosphate, enabling its transfer across the cytoplasmic membrane where it is incorporated into the growing peptidoglycan. At the interface between the growing cell wall and the cell membrane, transglycosidation reactions polymerize the growing chain and transpeptidases introduce cross-linking.

The 2 min region on the chromosome map of *E. coli* contains a large cluster of genes that code for proteins involved in various aspects of peptidoglycan synthesis and cell division. Seven genes mapping in this region (*murC*, *murD*, *murE*, *ddl*, *murF*, *mraY*, *murG*) participate in the pathway of peptidoglycan synthesis from UDP-GlcNAc to the formation of the C_{55}-prenol intermediate undecaprenyl-PP-MurNAc-pentapeptide. However, other genes mapping at sites removed from this region also participate in murein synthesis. For example, the *E. coli* enzyme UDP-*N*-acetylglucosamine enolpyruvate transferase (reaction 4 in Fig. 7-13) catalyzes the first committed step in peptidoglycan formation. This enzyme (encoded by *murZ*, *E. coli* map position 69.3 min) is inhibited by the bactericidal antibiotic

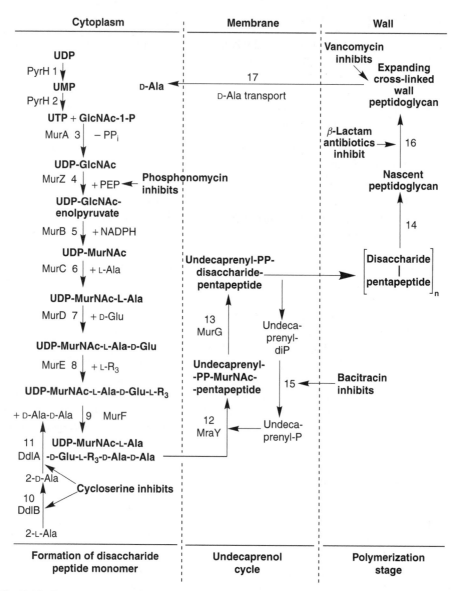

Fig. 7-13. Peptidoglycan biosynthesis. The three stages (cytoplasmic, membrane bound, and wall bound) are separated by the dashed vertical lines. GlcNAc = N-acetylglucosamine; MurNAc = N-acetylmuramic acid; L-R_3, for example, *meso*-diaminopimelic acid. The sites of action of antimicrobial agents affecting peptidoglycan synthesis are shown. The structural genes and names of the enzymes are (1, 2) *pyrH*, UMP kinase; (3) UDP-N-acetylpyrophosphorylase; (4) *murZ*, UDP-N-acetylglucosamine enolpyruvate transferase; (5) UDP-N-acetylglucosamine enol-pyruvate reductase; (6) *murC*, L-alanine adding enzyme; (7) *murD*, D-glutamate adding enzyme; (8) *murE*, *meso*-diaminopimelate adding enzyme; (9) *murF*, D-alanyl:D-alanine adding enzyme; (10) alanine racemase; (11) *ddl*, D-alanine:D-alanine ligase; (12) *mraY*, UDP-N-acetyl-muramoyl-pentapeptide transferase (first step in lipid carrier cycle); (13) *murG*, N-acetyl-glucosaminyltransferase (final step in lipid carrier cycle); (14) transglycosylases and transpeptidase; (15) membrane-bound pyrophosphatase; (16) membrane-bound transpeptidase (target of β-lactam antibiotics); (17) D-ala transport system.

phosphomycin (L-*cis*-1,2-epoxypropylphosphonic acid; phosphonomycin), a structural analog of phosphoenolpyruvate:

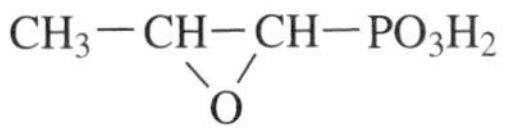

$$CH_3-CH-CH-PO_3H_2 \qquad\qquad CH_2=C-CO_2H$$
$$\diagdown\diagup O \qquad\qquad\qquad\qquad\qquad\quad OPO_3H_2$$

Phosphonomycin 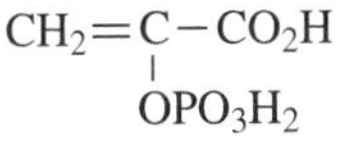   Phosphoenolpyruvate

D-glutamic acid, a specific component of peptidoglycan, is added to the UDP-*N*-acetyl-muramyl-L-alanine by the product of the *murD* gene. Another gene, *murI* (*E. coli* map position 90 min), is required for the synthesis of D-glutamate from α-ketoglutarate.

The *mraY* gene (2 min on the *E. coli* map) encodes the enzyme UDP-*N*-acetylmuramoyl-pentapeptide:undecaprenyl-phosphate. The MraY enzyme is involved in the first step of the cycle of reactions leading to the synthesis of C_{55}-undecaprenyl intermediates that aid in transport of peptide intermediates across the cell membrane (reactions 12 to 15 in Fig. 7-13). The glycopeptide antibiotics vancomycin and ristocetin bind to the D-Ala-D-Ala terminus of peptidoglycan precursors, preventing cross-linking of adjacent strands. This action prevents the addition of *N*-acetylglucosamine-*N*-acetyl-muramoylpentapeptide-pyrophospholipid to the growing point of the peptidoglycan backbone chain. Bacitracin, a member of a group of low-molecular-weight cyclic peptides, inhibits the dephosphorylation of the lipid-P-P carrier involved in the transfer of precursors into the peptidoglycan structure (reaction 15 in Fig. 7-13). This action prevents the lipid carrier from functioning in the reaction cycle.

Muramyl pentapeptide is an integral structural element of the walls of both gram-positive and gram-negative cells. Peptidases catalyze transpeptidase reactions involving the incorporation of the terminal D-alanyl-D-alanine during the final stages of peptidoglycan biosynthesis (reaction 16 in Fig. 7-13).

Bacterial cells contain a variety of **penicillin-binding proteins (PBPs)**, as shown in Table 7-4. These membrane-bound PBPs interact covalently with penicillin and other β-lactam antibiotics. The five high-molecular-weight PBPs (1A, 1B, 1C, 2, 3) are involved in peptidoglycan biosynthesis. Both PBP1A and PBP1B are associated with cell elongation. PBP2 aids in determination of cell shape, and PBP3 mediates septum formation. The structural similarity between penicillin and the pentapeptide precursor of the bacterial cell wall (Fig. 7-14) allows the β-lactam antibiotics to interact with the active site of a transpeptidase, terminating cross-linking (reaction 16 in Fig. 7-13). Mecillinam, a β-lactam with an amidino side chain, as shown in the structure below, appears to have a different mode of action from that of other penicillins:

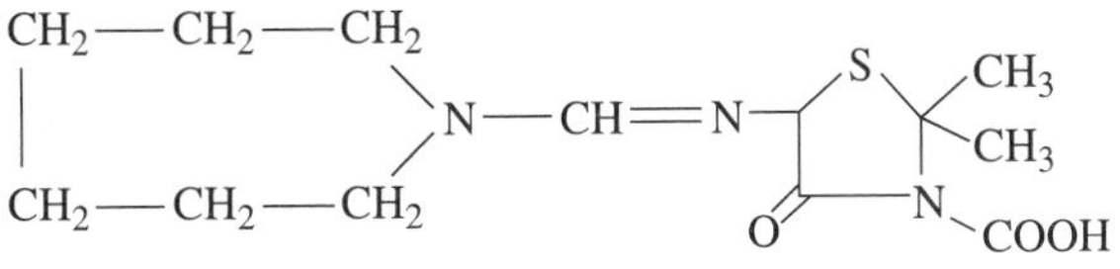

$$CH_2-CH_2-CH_2$$
$$| \qquad\qquad\qquad\qquad\qquad\qquad S \quad CH_3$$
$$\qquad\qquad\qquad N-CH=N \qquad\qquad CH_3$$
$$| \qquad\qquad\qquad\qquad\qquad\qquad N$$
$$CH_2-CH_2-CH_2 \qquad\qquad O \qquad COOH$$

6-β-amidino-penicillanic acid
(Mecillinam or FL-1060)

Gram-negative organisms such as *E. coli* are greater than 100 times more sensitive to this agent than gram-positive organisms such as *S. aureus* or *B. subtilis*. Other

TABLE 7-4. Penicillin-Binding Proteins (Peptidoglycan-Synthesizing Enzymes in *E. coli*)[a]

PBP No.	Gene Map Position	Synonyms, Biological Functions, Other Properties
1A	*mrcA* 75.9 min	(*ponA*) High molecular weight; transpeptidase and transglycosylase in wall elongation; inhibition results in lysis; increase in mecillinam-resistant mutants; deficiency in both PBP1A and PBP1B required for lethality. Inhibition of DNA replication by nalidixic acid prevents lytic blockade of PBP1A and PBP1B by β-lactams such as cefsulodin.
1B	*mrcB* 2 min	(*ponB*) High molecular weight (90,000); transpeptidase and transglycolase activity helps maintain rod shape during wall elongation; lysis triggered by β-lactams (cefsulodin) that bind to PBP1A and PBP1B.
1C		High-molecular-weight bifunctional PBP with transpeptidase and transglycolase activity. Existence revealed by genome analysis.
2	*mrdA* 14.3 min	(*pbpA, rodB*) High molecular weight; with RodA, conducts transpeptidation and transglycosylation; maintains rod shape; binds mecillinam specifically; action essential to prevent lysis in PBP1B$^-$ strains; *rodA* and *pbpA* constitute *rodA* operon.
3	*ftsI* 2 min	(*pbpB, sep*) High molecular weight (63,850). Mediates septum formation; transpeptidation and transglycosylation; interacts with FtsA, FtsQ, FtsW, and RodA; helps maintain rod shape; overproduced in mecillinam-resistant (*mre*) strains; only transpeptidase activity is β-lactam sensitive.
4	*dacB* 2 min	Low molecular weight; DD-carboxypeptidase and endopeptidase; postinsertional modification of new murein.
5	*dacA* 2 min	Low molecular weight (41,337); major PB component; DD-carboxypeptidase removes terminal D-Ala from pentapeptide side chains; regulates peptide cross-linkage.
6	*dacC* 2 min	Low molecular weight (40,804); D-alanine carboxypeptidase; PBP 5 and PBP 6 account for 85% PB; binds to CM by C-terminal end; bulk of enzyme is in periplasm; regulates peptide cross-linkage; stabilizes PG during stationary phase.
6B	*dacD* 44.8 min	D-alanine carboxypeptidase.
7	*pbpG* 47.9 min	PBP8 is a proteolytic artifact of PBP7.
—	*ampC* 94.3 min	(*ampA*); β-lactamase; ampicillin (penicillin) resistance; affects peptido glycan synthesis; cell morphology.
—	*ampH* 8.5 min	β-Lactamase; ampicillin resistance; probable role in peptidoglycan cell wall synthesis; cell morphology.

[a] All proteins listed fulfill the classic definition that they bind covalently to a β-lactam agent.

Muramyl–N, AcGl–(Muramyl–N, AcGl)$_{\overline{n}}$

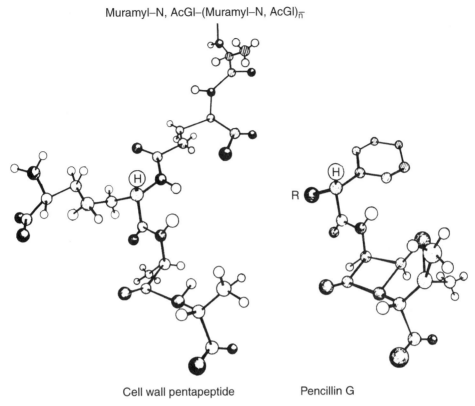

Cell wall pentapeptide   Pencillin G

Fig. 7-14. Structural similarity between penicillin and the pentapeptide precursor of the bacterial cell wall. (*Source*: From Butler, K., et al., *J. Inf. Dis.* **122:**Supplement S1, 1970.)

penicillins may bind with more than one PBP, but mecillinam binds specifically to PBP2 of *E. coli*, causing the formation of protoplasts without effect on the activity of other known penicillin-sensitive sites.

The action of different classes of penicillins at different sites suggests that combinations of agents from the two classes of compounds may act synergistically. Synergy has been demonstrated between mecillinam and a variety of other penicillins and cephalosporins both in vitro and in vivo. Similarly, *N*-formimidoyl thienamycin and cephoxitin exhibit synergistic killing action on *E. hirae*, suggesting at least two sites of action of these β-lactams. One block in the cell division cycle, induced by *N*-formimidoyl thienamycin or methicillin, occurs before the completion of chromosome replication. A second block, induced by cephoxitin or cephalothin, takes place later in the cell division cycle. Studies with *E. coli* suggest that on a quantitative basis *N*-formimidoyl thienamycin and methicillin preferentially inhibit PBP3, whereas cephoxitin and cephalothin preferentially inhibit PBP2. The synergistic action of these agents in *E. hirae* suggests a similar mode of action in this organism.

Staphylococci resistant to the action of methicillin and other β-lactam antibiotics produce β-lactamase, encoded by *blaZ*, which hydrolyzes the β-lactam ring. They also produce the penicillin-binding protein PBP2A, a product of the *mecA* gene. The PBP2A protein substitutes for other PBPs, but because of its low affinity for methicillin and other β-lactams, it does not bind these agents at a clinically effective level. The sensor-transducer proteins BlaR1 and MecR1, and their respective repressors, BlaI and MecI, regulate transcription of blaZ and mecA. Proteolytic cleavage inactivates the BlaI repressor, resulting in derepression of the *blaZ* gene and increased β-lactamase production. Similarly, induction of MecR1 results in the overexpression of PBP2A. The two events render the organism resistant to β-lactam agents. This appears to be the first demonstration in bacteria of a transmembrane signal transmitted by sequential proteolytic events. It has been proposed that these regulatory pathways may be attractive targets for the development of new drugs to combat resistant strains of staphylococci.

Cycloserine, a compound chemically related to D-alanine, prevents the incorporation of D-alanine into muramyl pentapeptide through its inhibitory action on dipeptide synthetase and alanine racemase (reactions 10 and 11, Fig. 7-13). These combined activities result in the accumulation of mucopeptides lacking the terminal D-alanine.

Insertion of new material into the expanding murein layer requires the activity of enzymes that degrade the glycan strand and endopeptidases that degrade the intact murein polymer. *E. coli* produces two soluble lytic transglycosylases (encoded by *slt* and *mlt*) that catalyze the cleavage of the β-1,4-glycosidic bond between *N*-acetylmuramic acid and *N*-acetylglucosamine. At the same time, there is an intramolecular transglycosylation at the *N*-acetylmuramic acid residue, resulting in the formation of a 1,6-anhydro bond. These enzymes apparently play a major role in recycling peptidoglycan components during expansion of the murein layer. For additional discussion regarding the role of genes involved with murein synthesis during the process of growth and cell division, see Chapter 17.

Teichoic Acids and Lipoteichoic Acids

Teichoic acids (TAs) are found in all gram-positive organisms but are absent in gram-negative bacteria. Teichoic acids are polymers of either ribitol phosphate or glycerol phosphate in which the repeating units are joined together through phosphodiester linkages (Fig. 7-15). Sugars, amino sugars, or amino acids may be condensed to the hydroxyl groups of the ribitol or glycerol to provide wide variations in overall structure (Fig. 7-16). By definition, teichoic acids include all polymers containing glycerol phosphate or ribitol phosphate associated with the membrane, cell wall, or capsule. Wall TAs are covalently linked to peptidoglycan through muramic acid and the phosphate group of ribitol or glycerol phosphate. The ribitol TAs of *S. aureus* and some strains of *B. subtilis* are linked to peptidoglycan through a common linkage unit: (glycerol phosphate)$_3$-GlcNAc. The glycerol TA of *B. cereus*, several strains of *B. subtilis*, and *B. licheniformis* are joined to peptidoglycan through a common linkage disaccharide, *N*-acetylmannosaminyl-1,4-*N*-acetylglucosamine, irrespective of the structural diversity in the glycosidic branches and backbone chains. Wall TAs are effective antigens and have been identified as the group- or type-specific substances of many organisms.

The wall ribitol TAs of *S. pneumoniae* that serve as the C antigen are more complex than those described above. Choline is present along with glucose, ribitol, phosphorus, galactoseamine, and 2,3,6-trideoxy-2,4-diaminohexose. The walls of *Lactobacillus plantarum* contain a mixture of one polymer of glucosylglycerol phosphate and two polymers of isomeric diglucosylglycerol phosphates. In *L. acidophilus* the TA is a

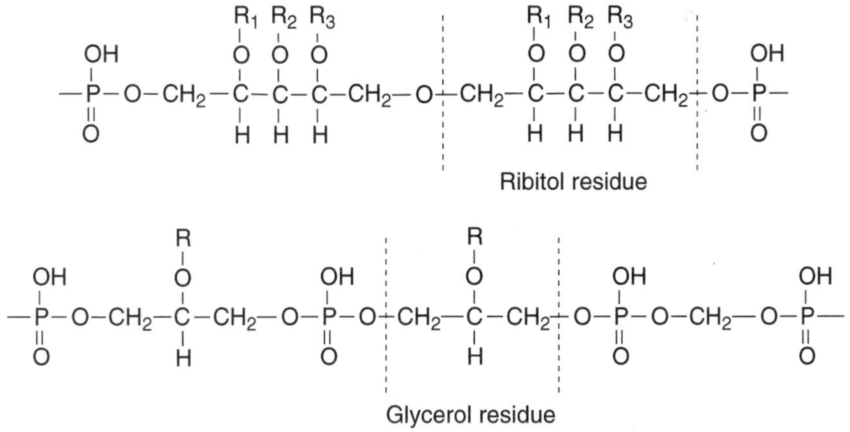

Ribitol residue

Glycerol residue

Fig. 7-15. Teichoic acid structures. In glycerol teichoic acid, R may be alanine, glucose, or glucosamine. In ribitol teichoic acid, R1 may be glucose or N-acetylglucosamine; R2 and/or R3 may be alanine. Some of the variations in the substitutions on teichoic acids are shown in Figure 7-16.

mixture of α- or β-linked polyglucose polymers with monomeric α-glycerol phosphate side chains attached at C-2 or C-4.

Other anionic polymers that lack polyol phosphate and may or may not be present in cell walls are not, by definition, teichoic acids. Nevertheless, such components are important. For example, the group-specific polysaccharide of *S. pyogenes* group A contains phosphate, glycerol, rhamnose, and N-acetylglucosamine in a molar ratio of 1 : 1 : 2 : 1. Rhamnose and N-acetylglucosamine are incorporated from thymidine-5′-diphosphorhamnose and UDP-N-acetylglucosamine into the group A polysaccharide of *S. pyogenes*. Assembly of the group A polysaccharide occurs at the cell membrane with the participation of a lipid anchor or acceptor molecule. Immunological specificity of the carbohydrate group antigen of streptococci is based on the relative amounts of N-acetylglucosamine or N-acetylgalactosamine present.

Lipoteichoic acids (LTAs) are membrane-associated polymers characteristic of gram-positive bacteria. The LTAs are linear polymers of 16 to 40 phosphodiester-linked glycerol phosphate residues covalently linked to a membrane anchor (generally a glycolipid or glycophospholipid). Physiological roles postulated for LTA include regulation of autolysin activity, scavenging of divalent cations such as Mg^{2+}, electromechanical properties of the cell wall, and interaction of bacteria with cells of infected hosts.

The proposed pathway for the synthesis of LTA occurs in three phases: (1) the glycolipid anchor, (2) the poly(glycerophosphate) component, and (3) the D-alanyl esters linked to poly(glycerophosphate). In *S. aureus* the diglucosyldiacylglycerol moiety of LTA functions as the membrane anchor. The gene (*ypfP*) that encodes diglucosyldiacylglycerol synthase (YpfP) has been cloned from *S. aureus* as well as from *B. subtilis*. The YpfP from *S. aureus* appears to be the only diglucosyldiacylglycerol synthase providing glycolipid for LTA assembly. The YpfP-deficient mutants of *S. aureus* display pleomorphic cells and the glycolipid anchor is replaced by diacylglycerol. When staphylococci are grown under the conditions of phosphate limitation,

Glycerol teichoic acids

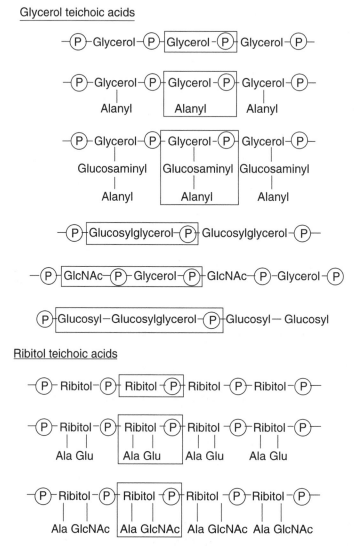

Ribitol teichoic acids

Fig. 7-16. Variations in teichoic acid structures. In each structure the repeating unit is enclosed in a rectangle. Encircled P, interconnecting phosphate group; GlcNAc, *N*-acetylglucosamine. (*Source*: Redrawn from Moat, A. G. 1985. Biology of the lactic, acetic, and propionic acid bacteria. In *Biology of Industrial Microorganisms*. A. L. Demain and N. A. Solomon (Eds.) Benjamin-Cummings, Menlo Park, CA, pp. 143–188.)

no wall teichoic acid is formed. However, membrane LTAs and teichuronic acids are still produced.

Membrane LTAs of the glycerophosphate polymer type occur in many gram-positive bacteria. These compounds have long polar glycerolphosphate chains linked to a small hydrophobic glycolipid. The LTA of *S. pyogenes* is a polymer-containing glycerophosphate linked to a glycerophosphoryl diglycosyl diglyceride. In *E. hirae* the glycolipid is phosphatidylkojibiosyl diglyceride. The Forssman antigen of *S. pneumoniae* is similar to LTA in that it is found in the cytoplasmic membrane and contains lipids and choline.

Lipoteichoic acids are exposed at the cell surface in many organisms. In *L. plantarum*, specific antiserum to the glycerophosphate sequence shows the label extending from the outer surface through the wall and even outside the boundary of the cell. Spontaneous release of TA and LTA has been described in streptococci and lactobacilli. The LTA of *S. pyogenes* binds via its polyanionic backbone to positively charged surface proteins termed **M proteins**. These proteins have been associated with the virulence of group A, β-hemolytic streptococci. This orientation leaves the lipid end free to interact with fatty acid–binding sites on host cell membranes. In the lactic acid–producing lactobacilli, *L. fermenti*, but not *L. casei*, can be agglutinated by the antisera to LTA. The long polar glycerophosphate chains of LTAs probably extend through the network of the wall to evoke an immune response.

Binding of divalent cations may serve in concentrating Mg^{2+} or other ions at the cell surface. In *Lactobacillus buchneri*, one Mg^{2+} ion is bound for every two phosphate groups in the wall TA. Release of TA and LTA may influence the interaction of bacterial pathogens with cells of the invaded host. In addition to their serological reactivity, TAs may activate the alternative pathway of complement and may play a role in the specific adhesion of bacteria to host epithelial surfaces.

The biosynthesis of D-alanyl-LTA has been studied in *L. rhamnosus*, *B. subtilis*, and *S. mutans*. The D-alanyl-carrier protein (Dcp) is required for incorporation of the activated D-alanine into membrane-associated LTA. This activation and ligation of D-alanine to Dcp is catalyzed by D-alanine:Dcp ligase. Genetic analysis reveals that the proteins for D-alanine incorporation are encoded by four genes present in the *dlt* operon: *dltA*, encoding D-alanine:Dcp ligase; *dltC*, encoding Dcp; *dltB*, encoding a putative transmembrane protein involved in the secretion of activated D-alanine; and *dltD*, encoding a membrane-associated thioesterase for mischarged carrier protein. In *S. mutans*, expression of the *dlt* operon is induced in the exponential phase when cells are grown with sugars transported by the phosphoenolpyruvate sugar phosphotransferase system (PTS) but is expressed constitutively when the cells are grown with the non-PTS sugars raffinose and mellibiose. Defects in D-alanyl-LTA synthesis in *S. mutans* result in sensitivity to acid. Cells of a *dltC*-defective mutant do not initiate growth below pH 6.5 and cannot survive prolonged (3 hours) exposure in medium buffered at pH 3.5.

A number of gram-positive bacteria are known to lack classical LTAs. In their place, polymers with chemical properties similar to those of LTAs have been identified. These polymers are variously referred to as macroamphiphiles, lipoglycans, or cell surface glycolipids.

Outer Membranes of Gram-Negative Bacteria

Gram-negative bacteria display a prominent outer membrane that is peripheral to the periplasmic region and the peptidoglycan sacculus (Fig. 7-17). This outer membrane is covalently attached to the peptidoglycan layer through lipoprotein and serves to reinforce the shape of the cell and to provide a protective barrier against the external environment. Under the electron microscope the outer and inner membranes present a similar appearance as double-track layers. Chemical analysis shows that they are similar in lipid content. Some components of the outer membrane are qualitatively similar to those of the cytoplasmic membrane; however, the outer membrane appears as an asymmetric bilayer, with the external layer being composed primarily of LPS

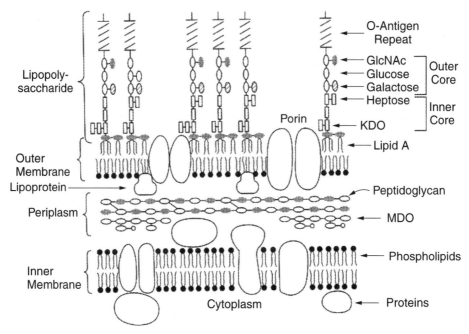

Fig. 7-17. Diagram of gram-negative bacterial cell surface. Ovals and rectangles represent sugar residues; circles depict the polar head groups of glycerophospholipids (phosphatidylethanolamine and phosphatidylglycerol). MDO = membrane-derived oligosaccharides. The core region of LPS is that of *E. coli* K-12, a strain that does not normally contain an O-antigen repeat unless transformed with an appropriate plasmid. (*Source*: From Raetz, C. R. H. *J. Bacteriol.* **175:**5745–5753, 1993.)

and the inner layer containing primarily phospholipids (Fig. 7-17). **Outer membrane proteins (OMPs)**, called porins, form large water-filled pores with diameters of 1 to 2 mm that traverse the membrane and regulate the access of hydrophilic solutes to the cytoplasmic membrane.

Lipopolysaccharides. Lipopolysaccharides consist of three basic components or regions, as shown in Figures 7-17 and 7-18. **Region I**, the outermost portion, contains repeating carbohydrate units that represent the "O" antigen. Alteration in the sugar composition of the O antigen results in a change in the immunological specificity. The sugars found in the O-antigen region can occur in a wide variety of combinations, accounting for tremendous antigenic diversity and many hundreds of chemical types or serotypes of *Salmonella*, *Shigella*, and other *Enterobacteriaceae*.

The **core region (region II)** consists of an outer and an inner core. The outer core shows high-to-moderate structural variability, whereas the inner core shows very low

---

Fig. 7-18. Covalent structure of a model LPS. The O-specific chain (region I) consists of four repeating units and a terminal Gal residue. A hexaacyl lipid A component is shown, and the two main chain heptoses are substituted in position 4 by phosphate and 2-amino-ethylpyrophosphate, respectively. On the right the designations of each of the saccharide residues are given. (*Source*: From Kastowsky, M., et al., *J. Bacteriol.* **174:**4798–4806, 1992.)

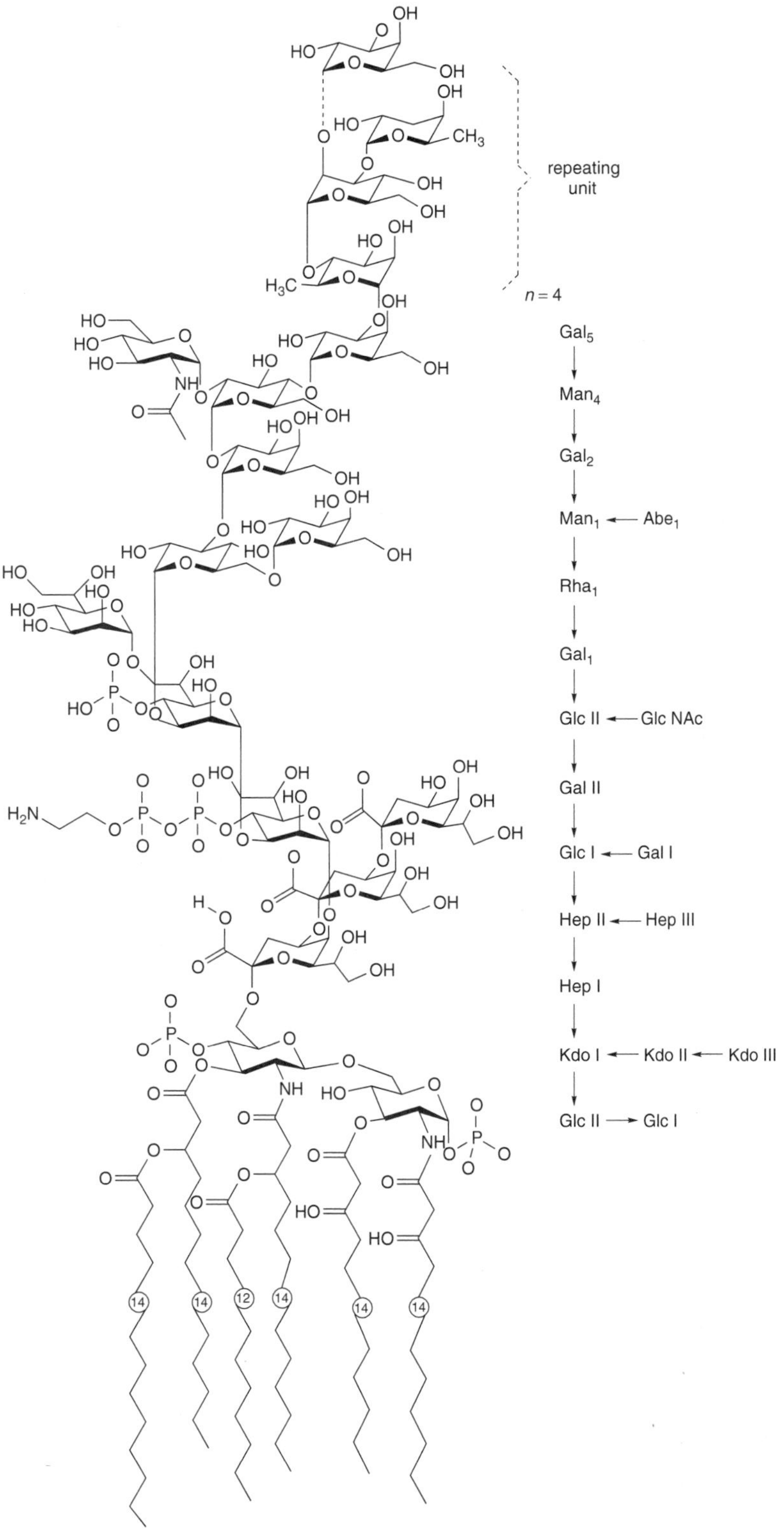

repeating
unit

$n = 4$

Gal$_5$

↓

Man$_4$

↓

Gal$_2$

↓

Man$_1$ ← Abe$_1$

↓

Rha$_1$

↓

Gal$_1$

↓

Glc II ← Glc NAc

↓

Gal II

↓

Glc I ← Gal I

↓

Hep II ← Hep III

↓

Hep I

↓

Kdo I ← Kdo II ← Kdo III

↓

Glc II → Glc I

structural variability, particularly within a very closely related group of organisms, for example, the *Salmonella*. The oligosaccharide subunits of the core region of *E. coli* and *Shigella* differ only slightly from those of *Salmonella*. In other gram-negative bacteria less closely related to the enteric bacteria, a greater diversity in the structure of the core region may be encountered. However, the unique octose sugar, 2-keto-3-deoxyoctulosonic acid (KDO), appears to be a common component of the core region of most gram-negative organisms.

Lipid A (region III), embedded in the outer membrane, has been extensively studied in *Salmonella*. In this organism, the chemical composition of lipid A has been shown to consist of a chain of D-glucosamine disaccharide units with all of the hydroxyl groups substituted. The substituents are the core polysaccharide units on the one hand and long-chain fatty acids on the other. The most commonly observed fatty acid is β-hydroxymyristic acid (3-hydroxy-tetradecanoic acid), a C_{14} saturated fatty acid that is substituted on the amino groups.

The hydroxyl groups are also esterified with other long-chain fatty acids such as lauric, myristic, and palmitic acids. The structure shown in Figure 7-18 is a monomeric unit of LPS. An average of three such subunits are linked together through pyrophosphate bridges between the lipid A molecules. The exact manner in which the LPS is linked to other surface structures is uncertain. However, LPS is very tightly associated with OMPs embedded in the outer membrane, particularly OmpA. The OmpA contributes to the stability of the outer membrane since it spans the membrane and is cross-linked to the underlying peptidoglycan layer. OmpA is exposed at the surface, where it also serves as a receptor for T-even phage and plays a role in conjugation and the action of colicins K and L.

Porins. A variety of other OMPs are found in the *Enterobacteriaceae* and other gram-negative bacteria. As shown in Table 7-5, these OMPs play a variety of overlapping roles in the physiology of the cell. Some of these proteins facilitate the entry of specific metabolites such as vitamin B_{12}, iron, or maltose, while others, such as OmpC and OmpF, constitute components of general porins that allow hydrophilic solutes of a molecular weight less than 700 to traverse the outer membrane. Both OmpF and OmpC have similar functional and structural properties, but expression of their structural genes, *ompF* and *ompC*, is regulated in opposite directions by the osmolarity of the growth medium.

As shown in Figure 7-19, the channel-forming porin trimers of *E. coli* span the outer membrane. Electron microscopy and image reconstruction techniques show that the three channels of OmpF on the outer surface merge into a single channel at the periplasmic face. By comparison, the PhoE porin of *E. coli* forms three separate channels that traverse the width of the membrane. *Pseudomonas aeruginosa*, on the other hand, forms a single, small, highly anion-selective channel (OmpP) in which the permeability is related to the presence of a selectivity filter (S) containing three positively charged lysine molecules. A study of 12 different porins from *E. coli*, *P. aeruginosa*, and *Yersinia pestis* revealed that most porins appear to be cation selective. Only 3 of the 12 showed anion selectivity.

In *E. coli*, a major OMP, LamB, serves as a receptor for bacteriophage λ (Chapter 6). Production of this protein is induced by growth on maltose and is involved in the passage of maltose and maltose-containing oligosaccharides through the outer membrane. Immunoelectron microscopy of newly induced LamB at the surface of *E. coli* reveals that LamB is inserted homogeneously over the entire surface of the cell.

TABLE 7-5. Outer Membrane Proteins of Gram-Negative Bacteria

Protein	Functions
OmpA	Stabilization of outer membrane and mating aggregates in F-dependent conjugation; receptor for phage TuII
Murein lipoprotein (Braun's lipoprotein)	Most abundant surface protein in *E. coli* and *S. enterica*; major structural protein; in conjunction with OmpA, stabilizes cell surface
OmpB (porin)	Diffusion channel for various metabolites including maltose
LamB (maltoporin)	Specific porin for maltose and maltodextrin; receptor for bacteriophage λ
OmpC (porin)	Diffusion channel for small molecules; receptor for phages TuIb and T4
OmpF (porin)	Diffusion channel for small molecules; receptor for phages TuIa and T2
OmpT	Protease
PhoE (protein E)	Anion-selective diffusion channels induced under phosphate limitation
Protein P	Anion-selective diffusion channel in *Pseudomonas aeruginosa*; induced under phosphate limitation
TolA	Maintenance of OM integrity; activity of group A colicins
TonA	Ferrichrome siderophore uptake; receptor for phages T1, T5, 80, and colicin M
TonB	Siderophore-mediated iron transport; B_{12} transport
Tsx	Nucleoside-specific channel; receptor for T-even phages and colicin K

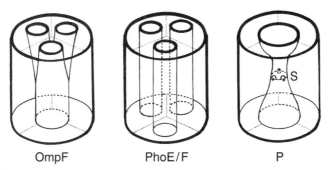

OmpF PhoE/F P

Fig. 7-19. Schematic three-dimensional representations of the structures of three different porin proteins. The proteins are oriented such that the top of the figure is the portion exposed to the external environment, whereas the bottom is the portion extending into the periplasmic space. The channels traverse the width of the outer membrane. The OmpF porin of *E. coli* forms coalescing channels. The PhoE porin of *E. coli* and the protein F porin of *Pseudomonas aeruginosa* form three separate and distinct channels. Protein P of *P. aeruginosa* forms a single anion-specific channel containing a selectivity filter (S) consisting of three charged lysine molecules (+). (*Source*: From Hancock, R. E. W., *J. Bacteriol.* **169**:929–933, 1987.)

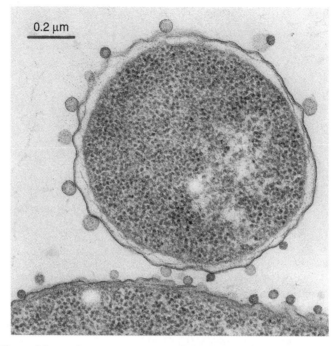

Fig. 7-20. Effect of loss of outer membrane lipoprotein (Lpp) and OmpA on the cell envelope and shape of *E. coli*. (*Source*: From Sonntag, I., et al. *J. Bacteriol.* **136**:280, 1978.)

Murein lipoprotein is a major OMP present in large quantities in *E. coli*. Lipoprotein molecules serve to anchor the outer membrane to the peptidoglycan layer. Mutants lacking *lpp*, the structural gene for lipoprotein, and *ompA*, the structural gene for OmpA, display spherical morphology and abundant blebbing of the outer membrane (Fig. 7-20). In these mutants the murein layer is no longer associated with the outer membrane. These mutants also display an increased growth requirement for Mg^{2+} and Ca^{2+} and are sensitive to hydrophobic antibiotics such as novobiocin, or to detergents, suggesting a protective function of the outer membrane. For additional discussion of the role of porins in nutrient transport, see Chapter 9.

Lipopolysaccharide Biosynthesis

The three components of LPS (**O-antigen, core oligosaccharide**, and **lipid A**) are synthesized independently of each other and later ligated in or on the inner membrane. After assembly, the intact LPS is translocated to the outer membrane. The LPS mutant strains of *S. enterica* and of *E. coli* K-12 have been isolated and their LPS structures defined in terms of polysaccharide composition as shown in Figure 7-21. Most of the genetic determinants responsible for oligosaccharide core biosynthesis in *E. coli* and *S. enterica* have been shown to reside in the *rfa* gene cluster. The genes coding for the five transferases (*rfa*KJIGB) required for assembly of the outer core region of *E. coli* LPS have been identified as shown in Figure 7-21. Similarly, genes coding for enzymes responsible for inner core biosynthesis (*rfaC*, *rfaD*) have been identified. The *rfaD* gene codes for ADP-L-glycero-D-mannoheptose-6-epimerase. Mutation in the

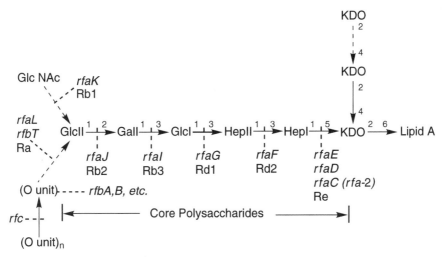

Fig. 7-21. Schematic representation of the genetic determinants of the LPS of *S. enterica* serovar Typhimurium. The genes (*rfaBCDEFGIJK*) and the LPS structures (chemotypes Re, Rd2, Rd1, Rc, Rb3, Rb2, and Ra) of mutants blocked at various stages of LPS biosynthesis are indicated. The dotted lines indicate the defective LPS termination points. KDO, 2-keto-3-deoxyoctulosonic acid; Hep, L-glycero-D-mannoheptose; Glc, glucose; Gal, galactose; GlcNac, *N*-acetylglucosamine; (O unit)$_n$, number of O-antigen side chains. The structural genes presumed to be responsible for LPS core biosynthesis are as follows: *rfaE*, specific function unknown; *rfaC*, ADP-heptose:LPS heptosyltransferase 1; *rfaD*, UDP-glucose:LPS glycosyl-transferase 1; *rfaB*, UDP-galactose:LPS-α-1,3-galactosyltransferase; *rfaJ*, UDP-galactose:LPS glucosyltransferase 1; *rfaK*, UDP-*N*-acetylglucosamine:LPS glucosaminyltransferase. (*Source*: From Chen, L., and W. G. Coleman, Jr. *J. Bacteriol.* **175**: 2534–2540, 1993.)

rfaC gene results in production of a heptoseless LPS structure referred to as chemotype Re. These mutants display increased permeability to both hydrophobic and hydrophilic agents. In *S. enterica*, the *rfaL* gene encodes a component of the O-antigen ligase and *rfaK* encodes the *N*-acetylglucosamine transferase. The order of genes within the *rfa* cluster at 79 units on the *S. enterica* linkage map has been shown to be *cysE-rfaDFCLKZYJIBG-pyrE*.

The major pathway leading to lipid A biosynthesis takes place in three stages: stage 1, UDP-GlcNAc acylation; stage 2, disaccharide formation and 4′ kinase action; and stage 3, KDO transfer and late acylation. The genes and gene products involved in this pathway are shown along with the intermediates in Figure 7-21. UDP-2,3-diacylglucosamine plays a key role in lipid A biosynthesis, as shown in Figures 7-22 and 7-23.

Enterobacterial Common Antigen

Enterobacterial common antigen (ECA) is a cell surface glycolipid synthesized by all members of the *Enterobacteriaceae*. It is a heteropolymer containing *N*-acetyl-D-glucosamine (GlcNAc), *N*-acetyl-D-mannosaminuronic acid (ManNAcA) and 4-acetamido-4,6-dideoxy-D-galactose (Fuc4NAc) linked together to form chains of trisaccharide repeat units. Enzymes encoded by genes in the *rfe* and *rfb* gene clusters,

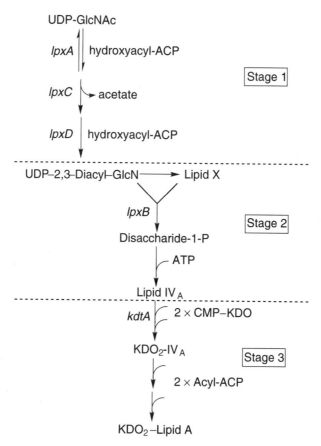

Fig. 7-22. Genes and enzymes involved in lipid A biosynthesis in *E. coli*. The three stages of the pathway are stage 1, UDP-GlcNAc acylation; stage 2, disaccharide formation and 4′ kinase action; and stage 3, KDO transfer and late acylation. (*Source*: Redrawn from Raetz, C. R. H. *J. Bacteriol.* **175**:5745–5753, 1993.)

as well as the *rff* genes, catalyze the biosynthesis of ECA. The *rfe* and *rfb* gene clusters are also involved in LPS biosynthesis. This anomaly is explained by the fact that common intermediates are required for both LPS and ECA synthesis.

CYTOPLASMIC MEMBRANES

The cytoplasmic membrane of bacterial cells is a bimolecular lipid leaflet of phospholipid molecules aligned at their hydrophobic ends. The polar phospholipids are hydrophilic and face the external or outside of the membrane and the internal or cytoplasmic side of the membrane. Various proteins and other components of the cell may be partially or wholly embedded in the membrane layer (Fig. 7-17). The cell membrane serves as a permeability barrier, preventing most solutes from gaining entrance to the cytoplasm except through specific transport proteins present in the membrane layer.

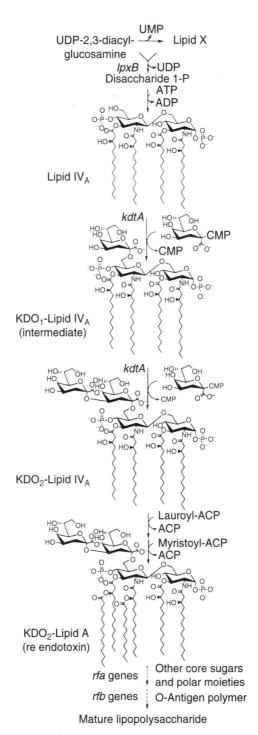

Fig. 7-23. Pathway of lipid A biosynthesis. The *lpxB* and *kdtA* genes code for the lipid A disaccharide synthase and KDO transferase. In wild-type cells, the level of the intermediates is very low (<1000 molecules per cell). (*Source*: From Raetz, A. *J. Bacteriol.* **175:**5745–5753, 1993.)

The phospholipids of bacteria contain ester-linked phospholipids whereas the archaea contain fatty alcohols linked to glycerol via either diether or tetraether bonds:

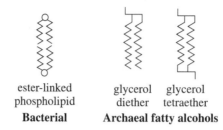

ester-linked phospholipid	glycerol diether	glycerol tetraether
Bacterial	**Archaeal fatty alcohols**	

Further discussion of the structure of ester-linked phospholipids and the diether- or tetraether-linked fatty alcohols can be found in Chapter 13.

Removal of the outer membrane and the peptidoglycan layer by the action of various chemical and enzymatic agents does not alter the integrity of the cell with respect to retention of its internal contents. Thus, the inner membrane of gram-negative cells serves the same function as the cytoplasmic membrane of gram-positive cells. The cytoplasmic membrane of bacteria and extensions of it assume many of the functions attributable to specialized organelles in eukaryotic cells. Systems governing the active transport of metabolites into and out of the cell, oxidative phosphorylation, cell wall biosynthesis, phospholipid biosynthesis, and the secretion of various extracellular enzymes and other proteins have all been associated with the cell membrane. In addition, anchoring of DNA to the membrane to aid its distribution to the daughter cells during cell division has also been described. In gram-positive cells, lipoteichoic acids are anchored hydrophobically in the cytoplasmic membrane. A discussion of the synthesis of the lipid and phospholipid components of the cell membrane can be found in Chapter 13.

Using any of the enzymes that selectively degrade the peptidoglycan and digest away the wall structure allows isolation of the cell membrane of gram-positive bacteria. If cells are treated with lysins while suspended in a solute to which the cell is impermeable (e.g., sucrose) at a concentration that approximately balances the high internal osmotic pressure, then an osmotically fragile body (**protoplast**) is formed. The protoplast can then be lysed osmotically and the membrane isolated. The cytoplasmic membrane contains numerous enzymes that perform a myriad of functions. Very few enzymes are located in the outer membrane of gram-negative bacteria. Terminal electron transport enzymes and metabolite transport functions are found in the cytoplasmic membrane. Only those compounds of a molecular weight less than 700 are aided in their passage through the outer membrane by porins.

Initially, it was not an easy task to identify the cell wall and cell membrane as distinct entities. The ability of cells to remain intact as protoplasts in an osmotically stable environment after the cell wall was removed by treatment with lytic agents provided the most convincing evidence that the wall and membrane were separate structures. Protoplasts are capable of conducting most, if not all, of the various activities attributable to intact cells (e.g., enzyme synthesis, various metabolic activities, bacteriophage synthesis, and spore formation) and are even capable of multiplication in a properly controlled osmotic environment. Some organisms, such as the *Mycoplasma*, exist as stable forms in nature without an outer cell wall. Sterols serve to stabilize the membrane of these organisms. All of the available evidence indicates that the cell membrane is the structure responsible for retention of the intracellular contents and controls the entrance and exit of metabolites.

Electron micrographs of gram-negative cells reveal two unit membranes: an outer membrane composed of phospholipids, protein, and LPS and an inner membrane composed of phospholipids, glycolipids, and phosphatidylglycolipids. Embedded in the lipid bilayer are various proteins and lipoproteins that function in metabolite transport and the synthesis of macromolecular constituents of the cell wall and the outer membrane. Certain proteins serve to maintain the integrity of the wall–membrane complex, while others provide protection against phagocytosis or act as attachment sites for bacteriophage or chemical agents such as antibiotics.

Permeability and Transport

Control mechanisms that regulate the transport of substances into and out of the cell have been thoroughly investigated and we now understand the permeability properties of the membrane and its components reasonably well. At one time, cell physiologists regarded the cytoplasmic membrane as an osmotic barrier equivalent to a collodion membrane. Metabolites were considered to enter or leave the cell primarily on the basis of the concentration gradient of metabolites from one side of the membrane to the other. Membranes were considered to have pores of finite size that permitted penetration of molecules into the cell. Size of the molecule, the relative internal and external concentrations, the charge distribution of the membrane and solute, and the pH and ionic strength of the internal and external environments were considered to be the primary factors governing the entrance of compounds into the cell. While all of these factors have their influences on the passage of materials into and out of the cell, it is now well established that cells have highly specialized transport proteins either inserted in the cytoplasmic membrane or occasionally found in the periplasmic region of gram-negative bacteria. The function of these transport systems in the movement of metabolites into and out of the cell is considered in detail in Chapter 9.

Periplasm

In gram-negative bacteria a region that includes the peptidoglycan layer separates the inner and outer membranes. This region is the periplasmic space. Certain types of cytological evidence suggest that there is no truly empty space between the outer and inner membranes. Peptidoglycan in a hydrated state represents a gel that fills the entire space between the two membranes. The degree of cross-linking of the peptidoglycan probably diminishes toward the cytoplasmic membrane, resulting in the formation of a **periplasmic gel** in which proteins are freely diffusible. This periplasmic region has been estimated to comprise approximately 10% of the total cell volume. Compression of this gel by a sudden reduction of the external osmotic pressure (osmotic shock) results in the abrupt release of enzymes and other periplasmic constituents. Periplasmic proteins include binding proteins for amino acids, carbohydrates, and vitamins; a variety of enzymes; and a number of other proteins with diverse functions. In addition to their function in transport, periplasmic-binding proteins are involved in the chemotactic response to attractants and repellants in the environment (see further discussion in the section on chemotaxis later in this chapter).

Other Membranous Organelles

Under the electron microscope certain organisms have been shown to contain membranous structures other than the cytoplasmic membrane. Membranous structures that appear to be sac-like invaginations of the cytoplasmic membrane of bacteria, termed **mesosomes**, have been observed with considerable regularity. Originally, a variety of physiological functions were attributed to mesosomes. However, these structures have been shown to vary in size and appearance depending on the fixation procedures used in preparing cells for electron microscopy. As a result, some consider these structures to be artifacts that can be manipulated by changes in fixatives, and much of the earlier work on mesosomes has been discredited. Nevertheless, these invaginated structures may represent functionally different membranous entities that will require further study before they are dismissed as artifacts of chemical fixation procedures.

The marine bacterium, *Nitrocystis oceanus*, possesses laminar membranous organelles (Fig. 7-24). These membranous organelles consist of approximately 20 vesicles so flattened that the lumen is only 100 Å thick. The outer surfaces are in contact

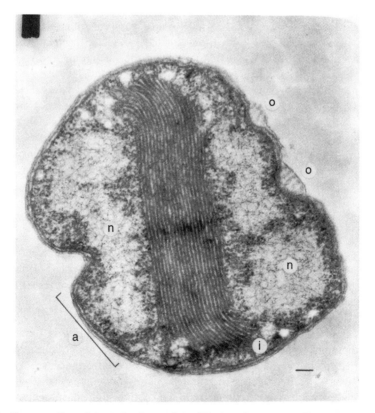

Fig. 7-24. Cross-section of a marine bacterium, *Nitrocystis oceanus*. The electron micrograph shows the cell wall and plasma membrane (area a), wall organelles (o), nuclear material (n), and the surrounding ribosome-packed cytoplasm. The lamellae of the membranous organelle traverse the entire cell. Other inclusions (i) may be seen within the cytoplasm. Bar equals 0.1 m. (*Source*: From Murray, R. G. E. and S. W. Watson, *J. Bacteriol.* **89**:1594, 1974.)

and form a triplet structure with an accentuated center line; these lamellae almost traverse the cell, displace the cytoplasm and the nuclear material, and form the most prominent cytological feature of the cell. *Nitrosomonas europaea* and other nitrifying organisms also produce distinctive membrane systems that differ significantly from one another. The presence of these lamellar organelles in nitrifying organisms (organisms that convert NH_4^+ to NO_2^- and NO_3^-) must somehow be related to the specialized mechanisms for acquiring energy from this process. However, the association of the intricate membrane systems observed in these organisms with the actual oxidation process is largely assumptive. It has been considered that these lamellar membranous organelles may be equivalent to mesosomes, plasmalemmasomes, or chondroids that have been studied in other kinds of bacteria, but their distinctive appearance and high degree of organization suggest that they are unique structures that may ultimately be directly associated with the oxidative processes of the autotrophic nitrifying bacteria.

CAPSULES

Many microorganisms produce an external layer of mucoid polysaccharide or polypeptide material that adheres to the cell with sufficient tenacity to be observed by simple techniques such as negative staining with India ink (Fig. 7-25). When this outer layer of material reaches sufficient size to be readily visible, it is referred to as a **capsule**. Several examples of well-defined capsules are shown in Figure 7-25. Many microorganisms produce a variety of extracellular materials that may be too sparse or too soluble in the medium to be observed by microscopic techniques. These materials are often referred to as a **slime layer**. The production of capsules or slime layers may impart the ability to form biofilms that enhance their ability to attach to surfaces.

The extracellular polysaccharide colanic acid or M antigen produced by *E. coli* strains and other enteric bacteria (Fig. 7-26) is considered to be a slime layer. Most of these surface components can be separated from the underlying cell wall without impairing the function or structural integrity of the cell or its viability. This is not to say that the properties of the cell may not be altered in some manner by removal of the material. For example, strains of *Streptococcus pneumoniae* that fail to produce the polysaccharide capsular material are essentially avirulent. Similarly, the D-polypeptide capsule markedly increases the virulence of the anthrax bacillus (*Bacillus anthracis*).

From time to time, other materials have been shown to be associated with the outer surface of the cell. However, these materials are usually considered to be transient metabolic excretions or chance adherents acquired from the environment and do not represent true capsular material. True capsules are generally composed of high-molecular-weight components, are highly viscous materials, and stain poorly, if at all, with the usual stains. The best staining procedures employ mordants that cause the precipitation of the capsular material by metal ions, alcohol, acetic acid, and so on. Reaction of the capsule with specific antiserum is frequently employed to enlarge the capsular area so that it may be more readily visualized under the microscope (Figs. 7-25.6 and 7-25.7). Techniques have also been developed for visualization of the capsule under the electron microscope (Figs. 7-25.8 and 7-25.9).

The polysaccharide capsular substances produced by *S. pneumoniae* confer immunological specificity and are also associated with virulence. Ninety distinct capsular serotypes, each differing in sugar composition and/or linkages, have been recognized

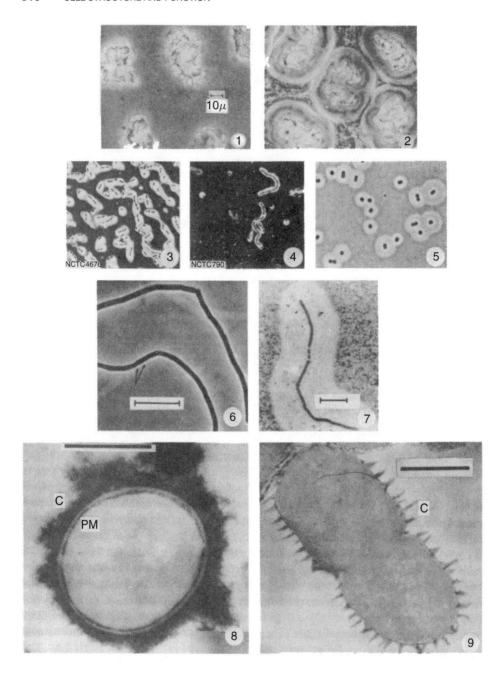

and the structures of a large number of them have been determined. Type 3 pneumococcal polysaccharide is composed of glucopyranose and glucuronic acid in alternating β-1,3- and β-1,4-linkages:

A given *S. pneumoniae* isolate expresses only one capsular polysaccharide, and the genetic basis of this unique attribute lies in the set of genes encoding specific biosynthetic reactions for specific polysaccharides. Genetic exchange of the genes from a donor to a recipient strain results in the replacement of the recipient's serotype-specific genes with those of the donor and subsequent expression of the donor capsular polysaccharide.

Synthesis of the type 3 capsular polysaccharide of *S. pneumoniae* requires UDP-glucose (UDP-Glc) and UDP-glucuronic acid (UDP-GlcUA) for production of the [3-β-D-GlcUA-(1,4)-β-D-Glc-(1$\rightarrow$]$_n$ polymer. The generation of UDP-Glc proceeds by conversion of Glc-6-P to Glc-1-P to UDP-Glc and is mediated by a phosphoglucomutase (PGM) and a Glc-1-P uridylyltransferase, respectively. Genes encoding both a Glc-1-P uridylyltransferase (*cps3U*) and a PGM homolog (*cps3M*) are present in the type 3 capsule locus, but these genes are not essential for capsule production. Spontaneous mutation in a distant gene (*pgm*, encoding a second PGM homolog) resulted in a 4-fold reduction in the amount of capsular material produced as compared with the

---

Fig. 7-25. Representative examples of encapsulated microorganisms. (**1** and **2**) Dark-phase-contrast photomicrographs of a pond alga. In (**2**) skimmed milk run under the coverslip permits visualization of the capsule as a broad hyaline periphery that is virtually unseen in (**1**). Bar equals 1 μm in (1) and (2). (*Source*: From Dondero, N., *J. Bacteriol.* **85**:1171, 1963.) (**3** and **4**) Hyaluronic acid containing capsules of group C streptococci. The dense capsules are visualized on plates of moist 20% horse serum in 1% glucose agar with India ink background stain. The capsules are absent from smears of hyaluronidase cultures. (*Source*: From MacLennan, A. P. *J. Gen. Microbiol.* **15**:485, 1956.) (**5**) Phase-contrast photomicrograph of a gram-negative, capsule-forming coccus. These capsules were formed after the organisms had been stripped of their original capsules. The presence of oxygen and an oxidizable carbon source were the only requisites for resynthesis of the capsular polysaccharide. (*Source*: From Juni, E., and G. A. Heym, *J. Bacteriol.* **87**:461, 1964.) (**6** and **7**) Phase-contrast photomicrographs of *B. anthracis* encapsulated cells. Bars in (6) and (7) equal 10 μm. (**6**) Membrane-like outline of capsule in untreated cells. (**7**) Same cells after treatment with antiserum. Negative staining with India ink. Note the marked swelling of the capsule. (*Source*: From Avakyan, A. A., et al., *J. Bacteriol.* **90**:1082, 1965.) (**8** and **9**) Ultrastructure of the capsules of *Streptococcus pneumoniae* type II and *Klebsiella pneumoniae* type I as seen in the electron microscope using ruthenium red in combination with osmium tetroxide. (**8**) A mat-like capsule (C) surrounds the cell. Ruthenium red penetration is evident at the plasma membrane (PM) and into the cytoplasm. Bar equals 0.5 μm. (**9**) *Klebsiella pneumoniae* capsular fibrils (C) are seen at regular intervals along the wall. Bar equals 0.5 μm. (*Source*: From Springer, E. L. and I. L. Roth, *J. Gen. Microbiol.* **74**:21, 1973.)

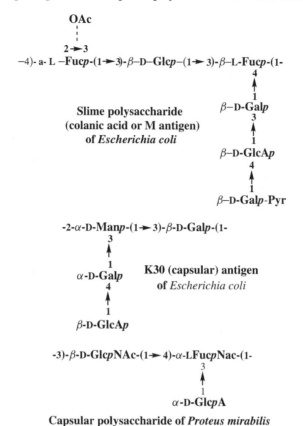

-4-β-GlcUA-1,4-α-GlcNAc-1,-

Repeating unit of K5 capsular polysaccharide of *Escherichia coli*

OAc
⋮
2→3
—4)- a- L —Fuc*p*-(1→3)-β–D–Glc*p*–(1→ 3)-β–L–Fuc*p*-(1-
4
↑
1
Slime polysaccharide   β–D-Gal*p*
(colanic acid or M antigen)   3
of *Escherichia coli*   ↑
1
β–D-GlcA*p*
4
↑
1
β–D-Gal*p*-Pyr

-2-α-D-Man*p*-(1→ 3)-β-D-Gal*p*-(1-
3
↑
1
α-D-Gal*p*   **K30 (capsular) antigen**
4   **of** *Escherichia coli*
↑
1
β-D-GlcA*p*

-3)-β-D-Glc*p*NAc-(1→ 4)-α-LFuc*p*Nac-(1-
3
↑
1
α-D-Glc*p*A

Capsular polysaccharide of *Proteus mirabilis*

Fig. 7-26. Chemical structures of capsules or slime layers of gram-negative bacteria. Glc*p*NAc, *N*-acetylglucosamine (italicized *p* indicates pyranose form of the sugar); Fuc, L-fucose; Gal, D-galactose; Glc, D-glucose; Man, D-mannose; Fuc*p*NAc, *N*-acetylfucosamine.

wild-type parental strain. Thus, it appears that most of the PGM activity required for type 3 capsule biosynthesis is derived from the cellular PGM.

In other *S. pneumoniae* serotypes and in other streptococci, the enzymes expected to be essential for capsule production are often not encoded by genes in the capsule locus. For example, the type 14 locus lacks genes necessary for the synthesis of UDP-glucose, UDP-galactose, and UDP-*N*-acetylglucosamine, precursors of the type 14 polysaccharide. In type 14 and other strains of *S. pneumoniae* or *S. pyogenes* the sugar or other intermediate is an important component of teichoic acid or peptidoglycan, and the cell uses existing cellular pools of these precursors for synthesis of the capsular polysaccharide.

In each individual case, alteration in the immunological specificity results from the incorporation of monosaccharides other than glucose and glucuronic acid. For example, galacturonic acid, rhamnose, *N*-acetyl-amino sugars, and uronic acid may be added in various combinations. Enhancement of virulence by the presence of a capsule results

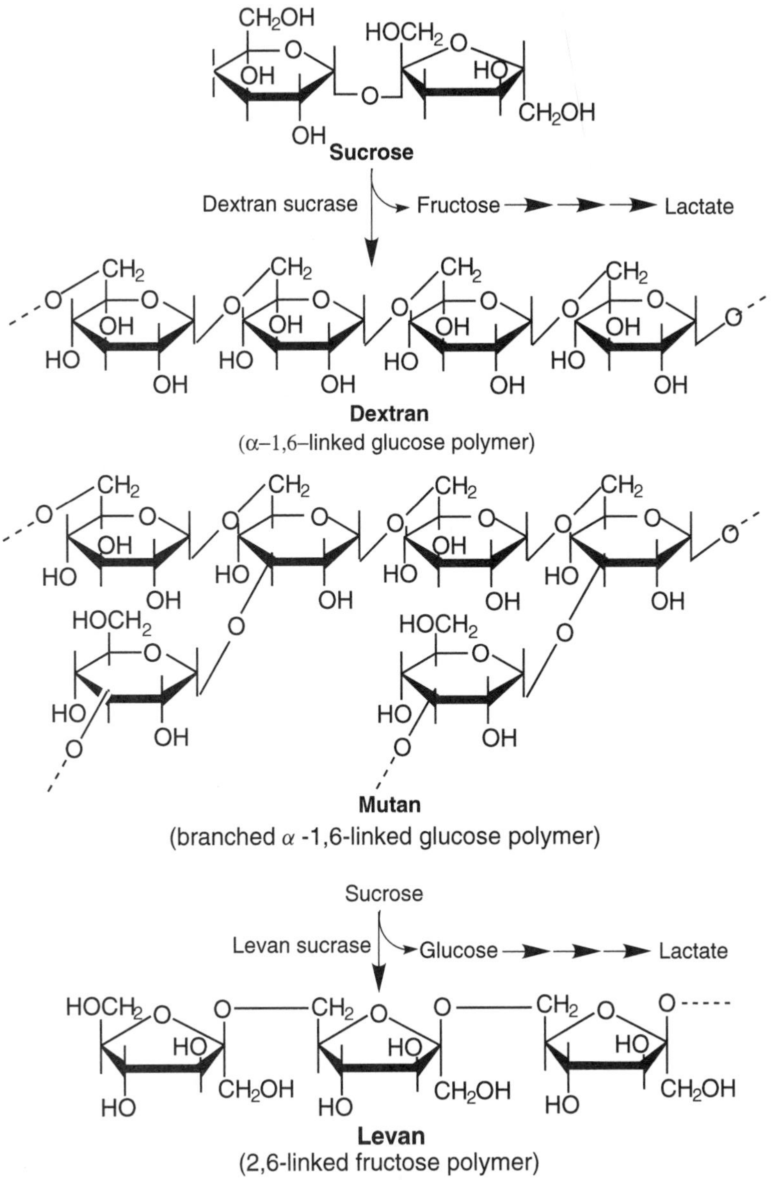

Fig. 7-27. Production of dextran and levan by bacteria. The enzyme dextransucrase splits sucrose to form fructose and dextran. The fructose is fermented to lactic acid. Levansucrase splits sucrose to form glucose and levan. The glucose is fermented to lactic acid. The structure of mutan, a branched dextran polymer, is also shown.

from inhibition of phagocytosis. Although physical size of the capsule correlates with the degree of virulence of the pneumococcus, the precise mechanism whereby capsules interfere with phagocytosis is not known.

Certain members of the streptococci produce both soluble and insoluble forms of dextrans and levans (Fig. 7-27). *S. salivarius* synthesizes levan, a polymer of fructose, from sucrose using levan sucrase (fructosyltransferase). Other streptococci produce both soluble and insoluble forms of dextrans (polymers of glucose) by the action of dextran sucrase. Mutans are branched glucose polymers with linear α-1,3-linked glucose units (Fig. 7-27). *S. mutans* produces FruA, an extracellular fructan hydrolase encoded by *fruA*, which releases fructose from levan, inulin, and raffinose, and cleaves sucrose into glucose and fructose. Expression of *fruA* is regulated by induction and carbon catabolite repression. The ability to synthesize fructan polymers is thought to aid in the formation of oral biofilms (bacterial aggregates referred to as plaque) that aid in the adherence of *S. mutans* and other oral bacteria to the enamel surfaces of teeth.

β-Hemolytic streptococci can be classified serologically on the basis of differences in their Lancefield group antigens. The capsular substances of Lancefield groups A and C streptococci are polysaccharides containing glucosamine and glucuronic acid units, the components of hyaluronic acid. This substance provides protection from the destructive effects of atmospheric oxygen by virtue of its ability to aid in the formation of cell aggregates. Disruption of the aggregates with hyaluronidase results in an increased oxygen uptake and the production of toxic levels of hydrogen peroxide. Unencapsulated variants are sensitive to oxygen.

A number of strains of *E. coli* produce capsular polysaccharides commonly referred to as K antigens. The K antigens are structurally diverse and give rise to serological specificity. There are over 70 recognized K antigens in *E. coli*. The K antigens are placed in either capsular Group I (heat-stable) or capsular group II (temperature regulated, expressed at $37\,^{\circ}$C but not at $\leq 20\,^{\circ}$C). For example, the K30 antigen of *E. coli* is a member of Group I. The K5 antigen is in Group II. The Group II K antigens are also characterized by acidic components, such as 2-keto-3-deoxy-D-mannooctulonic acid, *N*-acetylneurominic acid (sialic acid), and *N*-acetylmannosaminouronic acid (Fig. 7-26). The K5 repeating unit, shown in Figure 7-26, is identical to that of the first polymeric intermediate in the biosynthesis of heparin. As a consequence, the K5 polysaccharide is practically nonimmunogenic and strains of *E. coli* expressing this capsule are quite virulent since host cells do not mount an active immune response to it. A correlation has been drawn between activity and active production of K5 polysaccharide. The capsular polysaccharide 17-kb multi-gene cluster encodes the proteins required for the synthesis, activation, and assembly of the K antigens of *E. coli*. Three distinct regions have been identified in this gene cluster. The central region (Region 2) is composed of *neu* genes concerned with polysaccharide synthesis and polymerization. The *kps* genes in Regions 1 and 3 are involved in polymer assembly and transport and govern the same functions in all *E. coli*–producing group II capsules.

A model for polysialic (poly-*N*-acetylneuraminic) acid synthesis and translocation is presented in Figure 7-28. Two genes important in *N*-acetylneuraminic acid (neuAc) synthesis, *neuA* and *neuC*, are located at the 2.7 kb *Eco* R1-*Hind*III fragment of the *E. coli* K-1 gene cluster. The *neuA* gene encodes CMP-neuAc synthetase, and *neuC* encodes a 45,000-molecular-weight protein (P7) required for the synthesis of neuAc. Region 3 of the *kps* gene cluster contains two genes, *kpsM* and *kpsT*, which encode protein constituents of a system for the transport of polysialic acid.

Under appropriate growth conditions, *E. coli* K-12 and other enteric bacteria produce a slime polysaccharide, colanic acid, termed the M antigen (Fig. 7-26). The *E. coli* $rcsA_{K12}$ and $rcsA_{K30}$ genes are transcriptional activators involved in the expression of colanic acid synthesis. There is evidence that the RcsA protein may be a relatively

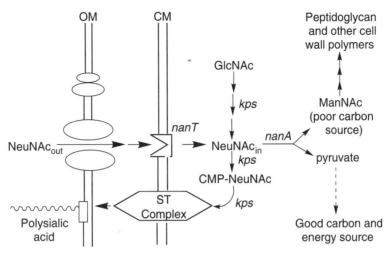

Fig. 7-28. Model for synthesis and translocation of polysialic acid (poly-*N*-acetylneuraminic acid) in *E. coli* K-1. External *N*-acetylneuraminic acid (NeuNAc$_{out}$) diffuses into the periplasm through outer membrane pores and is transported by the sialic acid permease encoded by *nanT*. Internal sialic acid (NeuNAc$_{in}$), transported or produced endogenously by *kps* gene products, is either degraded by the aldolase encoded by *nanA* or activated to the sugar nucleotide precursor CMP-NeuNAc. Sialic acid is polymerized and translocated to the outer membrane as PSA by the sialyltransferase (ST) complex. The aldolase cleavage products either enter intermediary metabolic pathways or are disseminated into surface polysaccharides. (*Source*: From Vimr, E. R. *J. Bacteriol.* **174**:6191–6197, 1991.)

widespread regulatory component for the synthesis of enterobacterial extracellular polysaccharides.

Proteus mirabilis produces an acidic capsular polysaccharide that is a high-molecular-weight polymer of branched trisaccharide units composed of 2-acetamido-2-deoxy-D-glucose (*N*-acetyl-D-glucosamine), 2-acetamido-2,6-dideoxy-L-galactose (*N*-acetyl-L-fucosamine), and D-glucuronic acid (Fig. 7-26). *Proteus mirabilis* 2573 also produces an O:6 serotype LPS in which the O-chain component has the same structure as the homologous capsular polysaccharide. The acidic nature of this polysaccharide may play a role in urinary calculi (stone) formation in a manner comparable to the roles of acid proteins and glycoproteins in mineralization and control of crystal formation in biological tissues. This process is initiated by a minute focus of solid matter that grows continually larger as chemical crystals that precipitate from the urine adhering to it.

High-molecular-weight peptide capsules are produced by *Yersinia pestis* and *Bacillus anthracis*. The *B. anthracis* capsule is a polypeptide of γ-D-glutamyl subunits originally thought to be produced only in vivo during infection but later shown to be produced in culture if an excess of carbon dioxide or bicarbonate is present. Other species of bacilli (e.g., *B. subtilis*) produce a polypeptide capsule containing a mixture of D- and L-glutamic acid subunits. The ratio of the two isomers can be controlled by culture conditions. It is not known specifically what governs the ratio of D : L isomers, but aeration tends to favor an approximately 1 : 1 ratio of D : L subunits. In addition to carbon dioxide, the presence of DL-isoleucine, DL-phenylalanine, and glutamic acid affect *B. anthracis* capsule production. This capsule is of special interest because of its association with virulence of the organism through its interference with phagocytosis.

Acetobacter xylinum synthesizes cellulose, a polymer of β-1,4-linked glucose units:

Cellulose structure showing 180° flip of each glucose unit with respect to its neighbor

Cellulose accumulates as an extracellular aggregate of crystalline microfibrils. Intertwined ribbons from large numbers of cells form a tough pellicle on the surface of the culture medium. Although cellulose synthesis has been studied extensively in higher plants as well as in bacteria, the intermediary steps in the pathway have not been entirely elucidated. Resting cells and particulate membrane-bound preparations of *A. xylinum* incorporate $[1\text{-}^{14}C]$-glucose into glucose-6-phosphate, glucose-1-phosphate, uridine glucose-5′-phosphate (UDPG), and cellulose. Labeling studies and demonstration of enzyme activities in cell-free extracts indicate that the sequence of reactions leading to cellulose synthesis is

$$\text{Glucose} \longrightarrow \text{G-6-P} \longrightarrow \text{G-1-P} \longrightarrow \text{UDPG} \longrightarrow \text{cellulose}$$

Lipid-linked and protein-linked cellodextrins (partially polymerized glucose units) may function as intermediates between UDPG and cellulose. A multienzyme complex orchestrates intracellular cellulose synthesis. Extrusion through pores in the LPS layer results in the aggregation of bundles of cellulose that undergo crystallization by self-assembly at the cell surface.

Microbial Biofilms

Many organisms are known to produce layers of material external to the usual cell surface layers. These external layers may contain carbohydrates, peptides, proteins, lipids, or combinations of these substances, which contribute to the adherence of the organism to solid surfaces to produce **biofilms** — complex communities of microorganisms attached to surfaces. Studies of organisms grown in artificial media in the laboratory may bear very little relationship to what occurs in the natural environment. In many cases, organisms grown in the laboratory lose their ability to produce the protective outer layers observed during growth in nature, apparently because these materials provide no selective advantage in laboratory cultures. It is now widely recognized that most bacteria found in natural, clinical, and industrial settings persist in association with surfaces. These microbial communities may often involve multiple species that interact with each other as well as with the environment. Recent inventive experimental approaches and improved methodologies have permitted exploration of the metabolic interactions, phylogenetic relationships, and competition in these complex communities. Following microbial adhesion to solid surfaces, biofilms can develop on almost any solid surface that comes in contact with aqueous liquids.

A number of factors play a role in adherence of bacteria to solid surfaces. When encountering a solid surface, bacteria experience a totally different set of conditions than in an aqueous environment. Under these circumstances, cells often undergo structural and functional adaptations. For example, in *V. parahaemolyticus* flagellar synthesis is triggered, while in *P. aeruginosa* polysaccharide production is increased. In *E. coli* the adhesion of type 1 fimbriated cells to abiotic surfaces leads to altered composition of the outer membrane proteins. It has been suggested that physical interactions between type 1 fimbriae and the surface are part of a surface-sensing mechanism in which protein turnover may contribute to the observed change in composition of outer membrane proteins. The resultant alteration of the cell envelope may influence adhesion. Further discussion of the physiology of organisms forming biofilms can be found in "Pili or Fimbriae" and also in Chapter 15.

ORGANS OF LOCOMOTION

Many microorganisms are motile — that is, they are able to move about in a concerted manner in an aqueous environment. Motile bacteria and protozoa produce flagella or cilia. Some organisms employ pseudopodal or ameboid movement. Spirochetes swim by virtue of a screw-like movement that is effective even in a viscous (semisolid) medium. Motile organisms are placed at an advantage in seeking food, in avoiding toxic chemicals or predators, and in colonizing favorable ecological niches. As a result, other forms of motility have evolved: swarming, which involves the development of specialized flagella; gliding, which involves cooperation between groups of cells (rafts); and twitching, a pilus-dependent means of translocation over solid surfaces. This last type of movement is discussed in "Pili or Fimbriae." All motile organisms have a common feature: energy is required for locomotion. Eukaryotes use ATP to activate the contractile movement of complex cilia or flagella. In prokaryotes, proton motive force (PMF) activates a motor that turns the flagellum like the propeller of a boat. In some of the other forms of motility, the detailed mechanisms have yet to be elucidated.

Cilia and Flagella of Eukaryotes

Cilia and flagella of virtually all eukaryotic microorganisms have the same basic arrangement. A cross-sectional view under the electron microscope reveals paired tubules arranged in a typical $9 + 2$ pattern that traverse the length of the flagellum of *Chlamydomonas reinhardtii* (Fig. 7-29). The nine peripheral pairs and one central pair of tubules are arranged in a bundle termed the **axoneme**. The individual doublet microtubules are cross-linked by nexin, a protein linker, and connected to the central core by radial spokes (Fig. 7-30). As a result of this rigid structure, when the dynein molecules in one doublet exert a force on the neighboring doublet, the entire structure bends instead of the microtubules sliding against each other. In normally beating flagellar axonemes, the ATPase-containing dynein arms generate longitudinal shear forces between the outer doublets, causing a localized sliding opposed by resistive components that convert it into a transverse bending movement.

Axonemal dynein contains ATPase and functions as a motor enzyme that provides the energy for ciliary and flagellar bending. This motor enzyme complex contains several proteins. Three heavy chains of approximately 500 kDa each contain the motor and ATPase sites. Two intermediate chains of 70 to 80 kDa mediate cargo attachment. The ATP cyclic cross-bridging of the dynein arms appears to involve the interaction of the projections on the globular heads with tubulin subunits in the tubule of the adjacent doublet (Fig. 7-30). The regulatory mechanism responsible for the control of the interdoublet sliding is not thoroughly characterized, but it is considered that

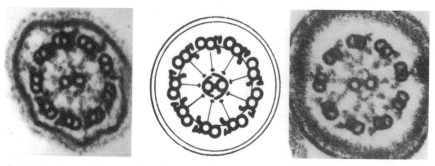

Fig. 7-29. Electron micrographs of cross-sections through eukaryotic flagella and diagram of axonemal components. The electron micrographs show cross sections of a *Chlamydomonas* flagellum (left) and a ram sperm flagellum (right). The ram sperm flagellum has a fibrous sheath between the axoneme and membrane. Otherwise, these flagella from two widely separated genera are remarkably similar, indicating that axoneme structure and composition has been highly conserved throughout evolution. (Photos and diagram courtesy of Dr. G. B. Witman, Department of Cell Biology, University of Massachusetts Medical School, Worcester, MA.)

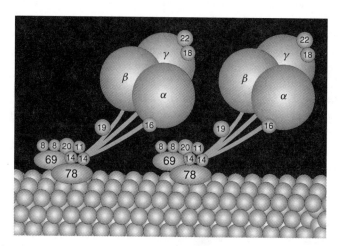

Fig. 7-30. Diagrammatic view of two *Chlamydomonas* dynein arms attached to an A-tubule. Each dynein heavy chain (α, β, and γ) has a head that contains the ATP hydrolytic site and interacts with the B tubule of the adjacent doublet (not shown) to generate interdoublet sliding, and a fibrous stem that extends to the base of the dynein. At the base there is an intermediate chain/light-chain complex containing the two dynein intermediate chains (IC69 and IC78) and several light chains. Additional dynein light chains are associated with the heads and stems. Numbers indicate apparent molecular weights of subunits. (Courtesy of Dr. G. B. Witman, Department of Cell Biology, University of Massachusetts Medical School, Worcester, MA.)

interactions between the radial spokes attached to each doublet tubule and the central complex of the axoneme are involved in some manner. It has been shown that the tubulin in the microtubules of *C. reinhardtii* is acetylated and partially detyrosinated. Detyrosination is largely confined to the outer doublets and has not been detected in the microtubules of the central pair. These posttranslational modifications of the tubulin structure may have some functional role at the molecular level. Also, there are several kinases and phosphorylases in the axoneme that operate to control dynein activity.

Bacterial (Prokaryotic) Flagella

Bacterial flagella consist of three major components: a **basal body**, a **proximal hook**, and a long **helical filament** (Fig. 7-31). The filament and hook are hollow cylindrical structures that extend from the cell surface. The basal body is embedded in the cell surface and contains a motor that turns the helical flagellum to drive the cell through a liquid environment much as a propeller drives a boat. The motor is reversible and is under the control of a chemotactic system that enables the cell to move toward favorable or away from unfavorable environmental conditions.

The genetic system governing the synthesis and function of flagella in *E. coli* is composed of 14 operons arranged in a regular cascade of three classes. The class 1 operon encodes the transcriptional activator of class 2 operons. Class 2 genes include structural components of the rotary motor and hook structure as well as the transcriptional activator for class 3 operons. The genes in class 3 include flagellar filament structural genes and the chemotaxis signal transduction system that directs cellular motion. A checkpoint mechanism ensures that class 3 genes are not transcribed before the basic structural assembly has been completed.

Flagella assembly and function requires over 40 genes (Table 7-6). The biosynthetic processing in *E. coli* and *S. enterica* occurs in a sequence that starts with the assembly of the M and S rings under the influence of regulatory proteins FlhC and FlhD (Fig. 7-31). This step is followed by the addition of switch proteins FliG, FliM, and FliN; assembly of the export apparatus containing FlhA, FliH, and FliI; and formation of the proximal and distal rods containing FlgB, FlgC, FlgF, and FlgG. Mutants defective in any of the genes coding for these proteins result in recovery of the MS ring only. A series of light chains of 8 to 22 kDa are involved in assembly and has been interpreted to mean that intermediary stages of rod assembly are relatively unstable, and, as a result, incomplete rod stages have not been found.

The next simplest isolated structure is the "rivet" portion of the distal rod that contains FliF, FliE, and four rod proteins (FlgB, FlgC, FlgF, and FlgG). The P and L rings (outer cylinder) contain FlgI and FlgH. These proteins appear to be the only flagellar components exported by the primary cellular export pathway, as shown by the presence of signal peptides. The next stages involve formation of the hook under the influence of *flgD* and addition of the hook proteins FlgE and FliK. Addition of the first hook–filament junction protein (FlgK) and the second hook–filament junction protein (FlgL) followed by the filament capping protein (FliD) completes the formation of components necessary for filament assembly. The addition of FliC (filament protein or **flagellin**) completes the biogenic process of flagellar assembly.

The filament portion of common bacterial flagella is composed of 100% protein (flagellin). Electron density maps show that the filament is composed of densely packed subunits (Fig. 7-32). The inner part of the filament forms a dense core of 30 to 50 Å radius. Flagellar growth occurs at the distal end of the filament. The hook portion is composed of at least three types of protein—that is, **hook-associated proteins (HAPS)** HAP1, HAP2, and HAP3. HAP2 forms a cap structure that presumably serves to prevent arriving flagellin subunits from diffusing away before polymerization occurs. Electron density studies show well-defined subunit packing in the core region and a central hole of approximately 60 Å diameter, which is large enough to accommodate the folded flagellin. A flagellum-specific pathway exports most of the flagellar components. Only the P and L proteins of the outer ring seem to be transported by the signal peptide-dependent pathway.

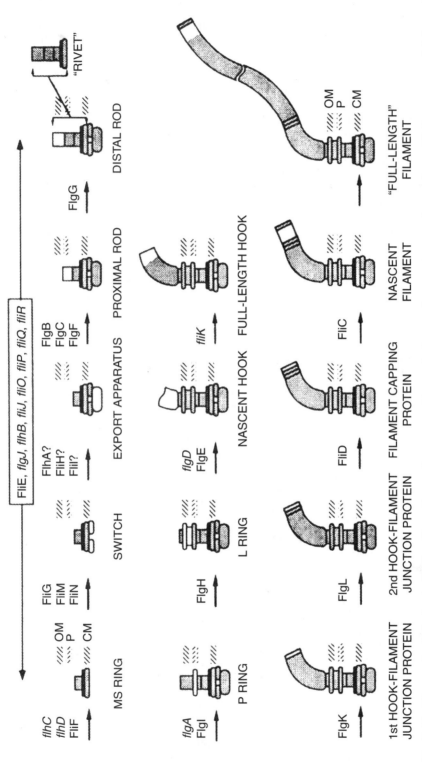

Fig. 7-31. Biogenesis of the bacterial flagellum. Succeeding stages of increasingly complex structure are shown along with the genes needed for each stage. Each incremental feature is shown in white with all preceding structures shown with stippling. The structure known as the rivet is, after the MS ring, the simplest substructure that has been detected by electron microscopy, and has lost the switch and export complex and perhaps other structures during the isolation procedure. The filament does not have a well-defined mature length; "full-length" simply implies that the filament is long enough to function in propulsion. Where the gene product is know to incorporate into structure, its symbol is given in Roman letters; where this is not known, the gene symbol is given in italics. The gene product and genes indicated in the box are needed at approximately the stages shown and certainly prior to the assembly of the distal rod. OM, outer membrane; P, periplasmic space and peptidoglycan layer; CM, cell membrane. (*Source:* From Macnab, R. M. *Annu. Rev. Genet.* **26:**131–158, 1992.)

Certain flagellar components, for example, the Mot proteins (MotA and MotB), are associated with the cytoplasmic membrane and surround the MS-ring component (Fig. 7-33). MotA and MotB are components of torque generators that enable motor rotation. The MotA protein is involved in conducting protons across the cytoplasmic membrane. Although MotB is associated with the cytoplasmic membrane, most of the protein protrudes into the periplasmic space. MotB is considered to be a linker that

TABLE 7-6. Flagellar, Motility, and Chemotaxis Gene Products of *S. enterica* and *E. coli* and Their Known or Suspected Functions

Gene Product	Function/Location
Regulatory Proteins	
FlhC, FlhD	Master regulators of the flagellar region acting in class 2 operons; transcription initiation (σ) factors?
FliA	Transcription initiation (σ) factor for classes 3a and 3b operons.
FlgM flagellar	Anti-FliA (anti-σ) factor. Also known as RflB. Active only when assembly has not proceeded through completion of the hook.
FliS, FliT, FliD?	Repressor of classes 3a and 3b operons (RflA activity).
FljA	Repressor of *fliC* operon.
Hin	Site-specific recombinases, affecting *fljB* promoter.
Proteins Involved in the Assembly Process	
FlhA, FliH, FliI factors	Export of flagellar proteins? FlhA is homolog of various virulence factors.
	FliI is homolog of catalytic subunit of F_0 F_1 ATPase.
FlgA	Assembly of basal body periplasmic P ring.
FlgD	Initiation of hook assembly.
FliK	Control of hook length.
FliB	Methylation of lysine residues on the filament protein, flagellin; functions of this modification unknown.
Flagellar Structural Components	
FliG, FliM, FliN	Components of flagellar switch, enabling rotation and determining its direction (CCW vs. CW).
MotA, MotB	Enable motor rotation. No effect on switching.
FliF	Basal body MS (membrane and supramembrane) ring and collar FliE. Basal body component, possibly at MS ring–rod junction.
FlgB, FlgC, FlgF	Cell-proximal portion of basal body rod.
FlgG	Cell-distal portion of basal body rod.
FlgI	Basal body periplasmic P ring.
FlgH	Basal body outer membrane L (LPS layer) ring.
FlgE	Hook.
FlgK, FlgL	Hook–filament junction.
FliC, FljB	Filament (flagellin protein), FljB (found in *S. enterica* only) is an alternative, serotypically distinct, flagellin.
FliD	Filament cap, enabling filament assembly.
Flagellar Proteins of Unknown Function	
FlgJ, FlhB, FlhE, FliJ, FliL, FliO, FliP, FliQ, FliR	

TABLE 7-6. (*continued*)

Gene Product	Function/Location
Sensory Transduction Components	
CheA	CheY and CheB kinase.
CheZ	CheY phosphatase. Antagonist of CheY as switch regulator.
CheW	Positive regulator of CheA activity.
CheY	Switch regulator, placing it in CW state.
CheR	Methylation of receptors. Sensory adaptation.
CheB	Demethylation of receptors. Sensory adaptation.

Chemoreceptors

Tar (aspartate); Tap (peptide, *E. coli* only); Trg (ribose); Tsr (serine); sugar receptors of the phosphotransferase uptake system.

Source: From Macnab, 1992.

fastens MotA and other components of the torque-generating machinery to the cell wall, keeping the motor components stationary with respect to the rest of the cell. The MotA and MotB proteins and the flagellar switch proteins FliG, FliM, and FliN constitute the flagellar motor. The FliG, FliM, and FliN proteins affect motor rotation and the switch from CW to CCW rotation. Additional aspects of the switch are discussed later in conjunction with chemotaxis. A change in expression from one flagellin gene to another (phase variation) is discussed in some detail in Chapter 5.

Chemotaxis

Many microorganisms, including *E. coli* and *Salmonella*, exhibit a wide range of metabolic and physiological versatility, enabling them to grow and survive in diverse environments. This versatility is employed to greatest advantage when a motile organism is able to direct its movement toward more favorable or away from unfavorable environments. This capability is called **chemotaxis**. But how can bacteria knowingly change direction? The trick is to control the gyration of their flagella. The rotation of the flagellar rotor is reversible, capable of turning in clockwise (CW) or counterclockwise (CCW) directions. When all the motors in a single cell rotate in a CCW direction, the flagella sweep around the cell in a common axis, forming a concerted bundle (Fig. 7-34a). The result is a smooth swimming motion of the cell. However, if the motors reverse and turn in a CW direction, the flagella disperse (Fig. 7-34b) and the cell undergoes a tumbling motion. This run-and-tumble strategy does not hold for all bacteria. For example, in *Halobacterium* spp., CCW rotation causes movement in one direction and CW rotation causes movement in the opposite direction.

Chemotaxis is the result of regulating the switch between CW and CCW rotation. A cell placed in a medium containing a unidirectional gradient of attractant, such as glucose, will suppress the onset of CW rotation if it is moving toward the attractant but will tumble more often to reorientate its direction if it is moving away from the attractant (Fig. 7-35).

The critical question here is to explain how bacteria interface with their environment to sense various attractants and repellants. A first step toward understanding the

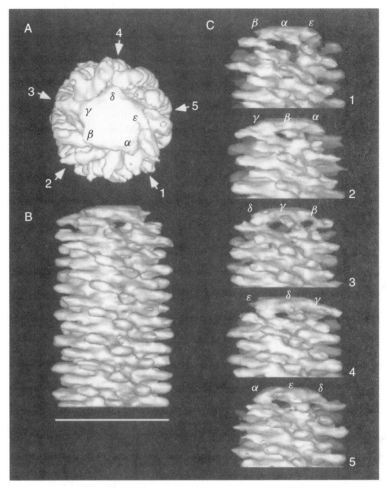

Fig. 7-32. Three-dimensional density map of the cap filament complex and its central section. (*a*) End-on view from the top, showing a pentagonal shape of the plate domain of the cap. The five vertices of the pentagon are labeled with Greek letters, α to ε, which are used to guide the orientation of the pentagon in the side views in (*c*). (*b*) Side view showing a regular helical array of flagellin subunits and the plate domain of the cap. Scale bar, filament diameter of 230 Å. (*c*) Five side views, showing each of the five gaps between the plate and the filament end. The viewing directions are as indicated by arrows in (*a*) labeled with corresponding numbers. Greek letters labeling the vertices as in (*a*) also indicate the orientations of the pentagonal plate. (*Source*: From Yonekura, K., et al. *Science* **290**:2148–2152, 2000.)

chemotactic response is to view the time-course (kinetic) response of a bacterium to the pulse addition of an attractant. As illustrated in Figure 7-36, there is a latency phase (0.2 s) followed by an excitation phase in which CCW flagellar rotation (smooth swimming) is favored. A gradual decrease in the relative probability of CCW versus CW rotation marks the adaptation phase. The adaptation phase may last from seconds to minutes. Basically, the cell recognizes the lack of further change in the concentration of attractant, and it now requires a higher concentration of attractant to trigger another excitation phase.

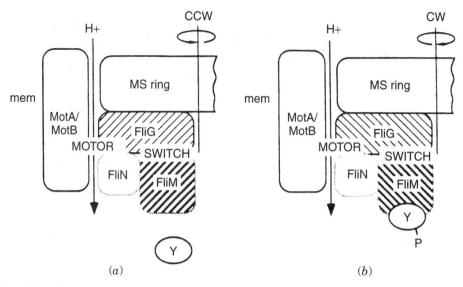

Fig. 7-33. Model for the structure of the flagellar switch and its function. The FliG is located in the cytoplasmic face of the MS ring. Both MotA and MotB are integral membrane (mem) proteins that surround the MS ring. The FliM and FliN are considered to interact with FliG; the locations shown for them are speculative but are consistent with available evidence. The FliM is postulated to be the target for CheY (Y), based on the large number of positions within its sequence that affect the counterclockwise (CCW) versus clockwise (CW) state of the switch. Both FliM and FliG are postulated to constitute the switching function, while FliG and FliN (perhaps with MotA and MotB) constitute the motor function. (*a*) When CheY is not bound to FliM, both fliM and FliG are in their CCW states and the motor rotates CCW. (*b*) When CheY is phosphorylated (Y-P) and binds to FliM, it places FliM in its CW state; this in turn causes FliG to change to the CW state, and so the motor rotates CW. (*Source*: From Irikura, V. M., et al. *J. Bacteriol.* **175**:802–810, 1993.)

In *E. coli*, the mechanism regulating the CW/CCW switch involves a family of membrane receptors and six cytoplasmic proteins called Che proteins that transduce information from the receptors to the motor. The membrane receptors (also called methyl-accepting chemotaxis proteins) are homodimers that span the membrane. The sensory domain lies in the periplasm and the signaling domain resides in the cytoplasm. There are five chemoreceptors (Trg, Tap, Tar, Tsr, Aer; think of them as the five senses of the bacterium), each of which senses a different subset of attractants and repellants. The signal-transducing domains of chemoreceptors interact with CheW and CheA as depicted in Figure 7-37. The CheA protein has an autokinase activity that is stimulated 100-fold when associated with the receptor.

Amazingly, the vast majority of receptor signaling complexes are clustered within a single patch generally localized to one of the poles of *E. coli*. These complexes are huge — 1.4 million molecular weight — with a subunit ratio of approximately 28 receptors to 6 CheW and 4 CheA molecules. This chemotaxis complex is not homogeneous with respect to receptors; each complex contains representatives of all five sensory receptors. In fact, the receptors Trg, Tap, and Aer will not function in the absence of both Tar and Tsr. Thus, the minor receptors function in conjunction with the major receptors within a single structure composed of thousands of receptor

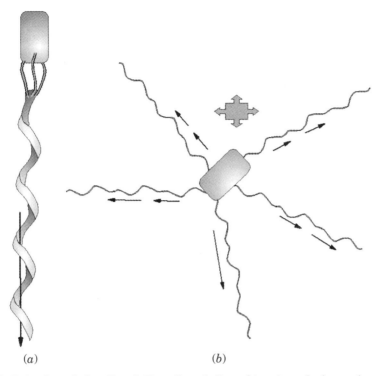

(a)   (b)

Fig. 7-34. Behavior of flagella of *E. coli* and *S. typhimurium* during swimming and tumbling. (*a*) Swimming: with the motors in CCW rotation, the flagellar filaments form a propulsive bundle, with wave propagation proceeding form proximal to distal. (*b*) Tumbling: CW rotation of the motors causes the normal left-handed form of the filament to undergo a polymorphic transition to the curly right-handed form. While the filaments are undergoing such transitions, the bundle is dispersed and the cell body moves chaotically, end over end — that is, the cell tumbles and reorients randomly, ready for the next swimming interval.

monomers organized within an ordered array by the association of CheW and CheA with their common C-terminal extensions. The role of CheW in this complex appears to be architectural.

The natural tendency of the flagellar motor is toward CCW rotation and, thus, smooth swimming. When transphosphorylated by CheA, a protein called CheY can interact with the FliM component of the motor and trigger CW rotation (tumbling). The goal of the chemotactic machinery, then, is to modulate the level of CheY-PO$_4$ and, thus, the frequency of tumbling. When a receptor interacts with an attractant, it introduces an asymmetry in the receptor structure that is thought to alter its ability to pack into an array compatible with formation of a complex between the receptor signaling domain, CheW, and CheA. As a result, CheA autophosphorylation stops, as does transphosphorylation of CheY. The CheZ protein is a CheY phosphatase that lowers the level of CheY-PO$_4$ (Fig. 7-37). Since unphosphorylated CheY does not affect the motor switch, the result is suppression of tumble frequency, an effect that increases the length of smooth swimming.

How does the cell know that it should continue to move toward higher concentrations of attractant? It adapts or desensitizes itself to the most recently encountered

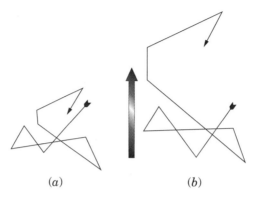

(a)   *(b)*

Fig. 7-35. Idealized trajectory of a swimming cell of *E. coli* or *S. typhimurium* (*a*) in an isotropic medium and (*b*) in a unidirectional gradient of attractant (open arrow). In either case, the cell swims in a straight line, randomizes its direction by tumbling, swims again, tumbles, and so on, yielding a three-dimensional random walk. The effect of an attractant gradient is to extend (by tumbling suppression) the mean duration of swimming segments in an up-gradient direction (heavy lines); segments in the down-gradient direction (fine lines) are not appreciably shortened. To illustrate this, the cells (*a*) and (*b*) start at the same point (arrow tail) and execute the same number of tumbles with the same resulting directional changes, but the cell (*b*) in the gradient has up-gradient segments extended. Therefore, it displays migration in the gradient direction by executing a time-based random walk.

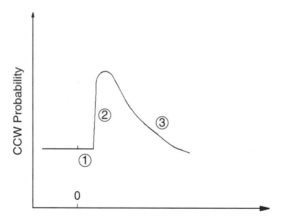

Fig. 7-36. Time course of the response of a bacterium to temporal stimulation by the sudden addition of an attractant at time zero. After an initial latency phase (1) of about 0.2 s, the excitation phase (2) manifests itself as a rapid increase in the probability that the flagellar motors will be in CCW versus CW rotational state. The degree of excitation progressively decreases during the adaptation phase (3), which can last from seconds to minutes, depending on the magnitude of the stimulus.

concentration of attractant. Chemotactic "memory" involves methylation of the membrane receptor signal domain following excitation; hence the designation of these proteins as methyl-accepting chemotaxis proteins (MCPs). Methylation of glutamate residues in the signaling domains of an MCP dimer (after excitation) is thought to neutralize negative charges on those domains. This would reduce the repulsive

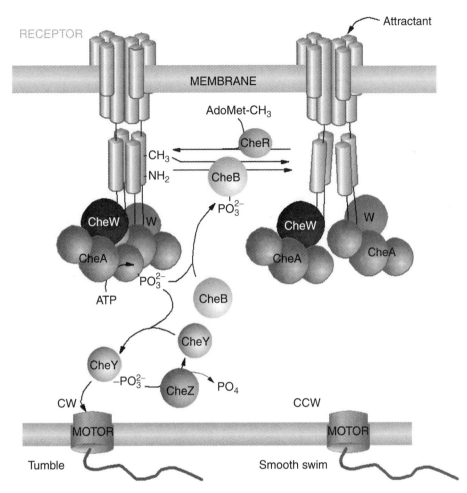

Fig. 7-37. The chemotaxis switch in *E. coli*. *Tumble mode.* Symmetrical assembly of CheA and CheW with the membrane receptors stimulates CheA kinase. Transfer of the phosphate from CheA to CheY and subsequent interaction of CheY-PO$_4$ with the flagellar motor causes clockwise rotation and tumble. *Smooth swimming mode.* Binding of attractant to the periplasmic domain of the membrane receptor causes asymmetry in the receptor structure and decreased kinase activity. Dephosphorylated CheY (catalyzed by CheZ phosphatase) will not interact with the motor, thereby allowing CCW rotation and smooth swimming. *Resetting the switch.* Following excitation, CheR methylates regions on the MCP that allow reassociation of the MCP C-terminal struts and reactivation of CheA kinase, reinstituting tumble. Concomitant phosphorylation of CheB will demethylate and deaminate residues on the MCP struts, once again disrupting the assembly structure downregulating CheA kinase.

electrostatic effects, forcing separation of the two C-terminal struts, allowing them to move closer to each other and reestablish an active CheW–CheA-receptor complex (i.e., restoration of CheA kinase activity). This increases the CheA-PO$_4$ level, and tumbling occurs even though attractant is still present.

Two proteins are involved with adding and removing methyl groups on the MCP receptors. The methyltransferase CheR uses SAM as a methyl donor to methylate the glutamate residues while the methylesterase CheB can remove those methyl groups. CheB can also deamidate glutamine residues in the receptor, which will increase the net negative charge on the receptor as well as provide more substrate for CheR methylase. The CheB methylesterase is really part of a feedback loop that "resets" the receptor for excitation just in case the organism encounters a higher concentration of attractant or moves toward a lower concentration. The amino-terminal end of CheB is homologous to CheY and will compete with CheY for CheA-PO$_4$. The phosphorylation of CheB then activates the esterase activity (Fig. 7-37).

Consequently, the increased kinase activity that occurs following activation of the receptor-CheW–CheA complex not only elevates CheA-PO$_4$, which increases tumble frequency, but increases CheB esterase-dependent removal of MCP methyl groups. The resulting increase in negative charges intensifies the electrostatic force, repelling the MCP struts from each other. Movement of the struts then deactivates the CheA kinase complex, leading to lower CheA-PO$_4$ levels and smooth swimming. In the absence of attractant, this switch occurs at a relatively fixed frequency. However, an encounter with attractant (or with a higher concentration of attractant after an adaptation phase) hastens inactivation of the CheW–CheA-receptor kinase activity and extends the smooth swim toward the attractant.

Swarming Motility

A number of motile gram-negative bacteria that display characteristic swimming and chemotactic behavior when grown in liquid media undergo a differentiation into a swarming type of motility when grown on agar of appropriate consistency (usually 0.5 to 0.8% agar). *Proteus mirabilis*, *Proteus vulgaris*, *Vibrio parahaemolyticus*, *Serratia marcescens*, and some strains of *Bacillus* and *Clostridium* have been shown to display this phenomenon — that is, radially spreading colonies on semisolid agar medium (Fig. 7-38). Swarmer cells undergo group translocation on an agar surface within a slime layer containing a complex mixture of polysaccharides, surfactants, proteins, and peptides. Swarming has been associated with the formation of biofilms and may contribute to bacterial virulence.

When grown in liquid medium, *V. parahaemolyticus* swims by means of a single sheathed polar flagellum that is structurally and functionally similar to the flagella of other gram-negative bacteria. However, on solid surfaces the cell develops into a swarmer cell that is highly elongated and produces extensive lateral flagella (Laf). The polar flagellar system (Fla) is expressed constitutively, whereas the Laf system is expressed only when the organism is grown on a solid surface. Lateral flagella are unsheathed and display a different subunit composition from polar flagella.

The gene systems controlling the two distinct motility systems are large, each composed of 40 or more genes without overlap except for shared chemotaxis components. Genes belonging to the Laf system have been identified and ordered in a hierarchical scheme of gene control. These genes encode structural components of the flagellum and the motor and a σ factor of RNA polymerase important for directing swarmer cell development. One gene (*lafA*) codes for lateral flagella. Other genes required for swarming but not for swimming have been identified by gene replacement mutagenesis. Reversible motors embedded in the cytoplasmic membrane drive both types of flagella. However, as compared with the systems of other gram-negative bacteria, the polar flagellar motor of *V. parahaemolyticus* is powered by sodium-motive

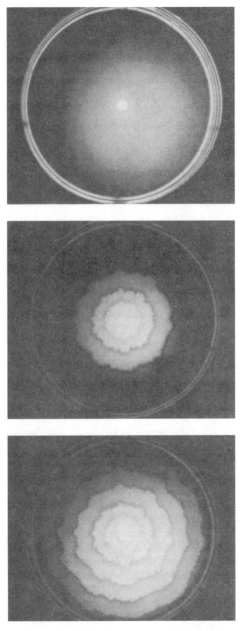

Fig. 7-38. Development of the *P. mirabilis* swarming colony. Multicellular swarming behavior and swarming colony formation were observed by inoculating cells at the center of an agar plate and incubating them at 37 °C. The swarming colony was then photographed at 6 (upper), 24 (middle), and 48 (bottom) hours postinoculation. Consolidation zones are seen as the dark areas separating lighter regions of swarming motility. (*Source*: From Belas, R. *ASM News* **58**:115–22, 1992.)

force. The two independent flagellar organelles appear to be directed by a common chemosensory control system.

Proteus mirabilis is a motile gram-negative bacterium that produces characteristic peritrichous flagella when grown in liquid media. When transferred to solid media, the cells begin to elongate, the normal septation mechanism is inhibited, and giant swarmer cells are produced. The swarmer cells undergo extensive production of new flagella (Fig. 7-39). This alteration enables the swarmer cell to move over solid medium (swarming motility). By definition, individual swarmer cells cannot "swarm." Swarming is the result of the coordinated effort of groups of differentiated swarmer cells that move about in rafts. The swarming process is cyclic or periodic. During swarming, concentric zones of bacterial growth (swarm bands) are formed around a central colony. Swarming occurs for a period of 1 to 2 hours and then stops for several hours before the appearance of a second swarm. At the end of the swarming period, the cells slow their activity and divide into short, sparsely flagellated cells. Transposon mutagenesis techniques have been used to study the behavior of mutant phenotypes of *P. mirabilis* with altered swarmer cell differentiation. These mutants appear to lack the appropriate intercellular signals required to regulate the cyclic swarming process.

The relatively recent demonstration of swarming motility in *Escherichia coli* and *Salmonella enterica* serovar Typhimurium (hereinafter referred to as *S. enterica*) provided the opportunity to study this phenomenon in genetically well-characterized organisms. Mutants obtained by transposon mutagenesis as well as existing mutants have been used to identify genes involved in swarming in *S. enterica*. Some mutants are defective in LPS synthesis while others are defective in chemotaxis or lack some portion of the two-component signaling system. Failure to identify specific swarming signals such as amino acids, pH changes, oxygen, iron starvation, increased viscosity,

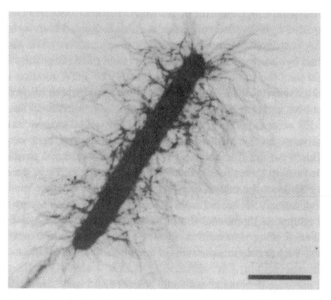

Fig. 7-39. Electron micrograph of a wild-type swarmer cell of *P. mirabilis*. The cell was taken from the periphery of a swarming colony grown on an agar plate at 37 °C for 6 hours and negatively stained with uranyl acetate. Bar equals 5 μm. (*Source*: From Belas, R. *J. Bacteriol.* **173**:6279–6288, 1991.)

flagellar rotation, or autoinducers has led to the conclusion that the production of external slime is a major factor in the development of swarming motility.

Motility in Spirochetes

Spirochetes are helical in shape and are motile by means of flagella enclosed within the periplasmic space. Major species of spirochetes include *Brachyspira*, *Spirochaeta*, *Treponema*, *Borrelia*, and *Leptospira*. The handedness of the helices (right- or left-handed) is species specific, and the organisms vary in shape and size. A right-handed helix rotates clockwise (CW), and a left-handed helix rotates counterclockwise (CCW), going away from an observer. Attached near each end of the protoplasmic cell cylinder are from one to several hundred periplasmic flagella that extend toward the center of the cell where they may or may not overlap. The size of the spirochetes, the number of flagella attached at each end, and the degree of overlap of the flagella at the center vary from species to species.

The flagella of spirochetes are comparable to those of other bacteria in that they contain a basal body, rod, flagellar hook, and filament. Beyond that, however, spirochetes differ from other bacteria in that their organelles for motility reside inside the cell and within the periplasmic space. Surrounding both the protoplasmic cylinder and the periplasmic flagella is an outer membrane sheath. This unique morphological structure allows them to bore through viscous gel-like media that inhibit the motility of most other bacteria. In fact, increasing the viscosity of the medium by the addition of a compound such as methylcellulose increases the speed of spirochete motility. Although they are structurally similar, the swimming motions of *Leptospira* and *Borrelia burgdorferi* appear to be quite different, and models have been developed to explain these differences.

The cell cylinder of a spirochete such as *Leptospira biflexa* forms a right-handed helix. Only one periplasmic flagellum extends from each end of the cell along the center axis of the helix, but flagella do not overlap. In a translating cell, the posterior end is hook shaped and the anterior end is spiral shaped (see Fig. 7-40). When the cells reverse direction, the end of the cell concomitantly changes shape. The periplasmic flagella rotate between the outer membrane sheath and the right-handed protoplasmic cylinder. They are more rigid than the cell cylinder, causing the end of the cell to conform to the shape of its flagellum. As viewed from the center of the cell toward one of the cell ends, CCW rotation of a periplasmic flagellum results in the flagellum at that end forming a left-handed helix whose helical pitch and helical diameter are larger than those of the cell body. This form is referred to as the spiral-shaped end. On the other hand, rotation of that periplasmic flagellum in the CW direction results in the flagellum and cell body at that end being hook shaped. When the periplasmic flagella are rotating in the same direction (both CCW or CW), the cell does not translate.

In the model for rotation of a *Leptospira*, forward thrust results from two rotations. The model can be conveniently viewed from behind a swimming cell (Fig. 7-41). First, both periplasmic flagella rotate CCW. Second, the torque of the rotating flagellum produces a CW counterrotation of the right-handed helical cell body. Two modes of thrust are proposed. CCW rotation of the anterior periplasmic flagellum causes that end of the cell to gyrate (i.e., to bend in a circular manner without necessarily rotating); it generates a backward-moving spiral wave. This wave is predicted to be left-handed and is sufficient for the cell to translate forward in a low-viscosity medium. Concomitantly,

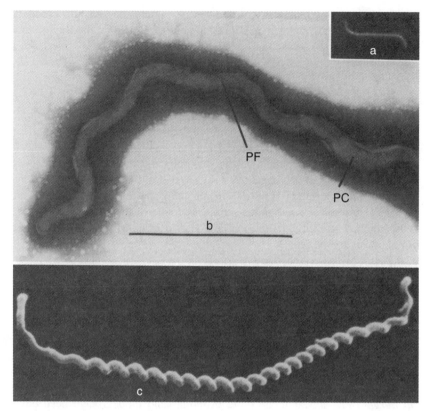

Fig. 7-40. *Leptospira biflexa* **serovar patoc.** (*a*) Dark-field micrograph illustrating the hook-shaped ends and helical cell morphology. Cell length is 10 μm. (*b*) Transmission electron micrograph of one cell end. Note the protoplasmic cell cylinder (PC) and periplasmic flagellum (PF). (*c*) Scanning electron micrograph illustrating the hook-shaped ends and right-handed helical PC. Cell length is 19 μm. (*Source*: From S. F. Goldstein and N. W. Charon, *Cell Motility Cytoskel.* **9**:101–110, 1988.)

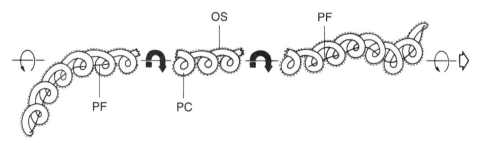

Fig. 7-41. Possible conformations and gyrations of periplasmic flagella and protoplasmic cylinder of *Leptospira.* Motion is toward the right of the figure (open arrow). The anterior spiral end is left-handed. An observer behind the cell (left side of figure) sees the spiral and hook-shaped ends gyrate CCW (thin arrows) because of the rotation of the periplasmic flagella in that direction, while the protoplasmic cylinder rotates CW (wide filled arrows). (*Source*: From Goldstein, S. F. and N. W. Charon, *Cell Motility Cytoskel.* **9**:101–110, 1988.)

CW rolling of the cell cylinder gives the organism the unique capacity to bore through gel-like media or connective tissue. Due to the roll of the cell cylinder, the torque is counterbalanced by the gyrations of the cell ends.

The model for *Leptospira* motility applies to *Treponema phagedenis* as well. This spirochete has a right-handed cell cylinder with four to eight short, nonoverlapping periplasmic flagella attached subterminally at each end. As with *Leptospira*, these structures cause the cell ends to be left-handed or irregular in shape. However, CCW gyration of the bent-shaped ends does not yield sufficient thrust to cause the cell to move in a low-viscosity medium. But because the cell is helical and right-handed, *T. phagedenis* can translate in a highly viscous gel-like medium such as 1% methylcellulose. The CW counterrotation of the cell cylinder is considered to cause cell movement.

Borrelia burgdorferi differs from other spirochetes in that it has a flat-wave morphology (nonhelical; planar wave). Cells swim with the flat waves being propagated from the anterior end of a translating cell to the posterior. These flat waves are reminiscent of eukaryotic cell motility and flagella. In *B. burgdorferi* there are between 7 and 11 periplasmic flagella attached near each end and they overlap in the center. Nonmotile mutants are rod shaped, indicating that the periplasmic flagella dictate the shape of the entire cell.

The flagellar filaments of most species of spirochetes contain three proteins (FlaB1, FlaB2, and FlaB3) and are attached subterminally at each end of the cell cylinder and within the periplasmic sheath that contains the sheath protein FlaA. A separate gene encodes each of these proteins. Periplasmic flagella consist of one to two FlaA proteins and three to four FlaB proteins. *B. burgdorferi* is an exception in that it has one FlaA protein and only one FlaB protein. Studies with mutants of *Brachyspira hyodysenteriae* indicate that the interaction of FlaA with the FlaB core impacts periplasmic flagella helical morphology.

Exclusive of chemotaxis genes, at least 36 motility genes have been identified in *B. burgdorferi* and *T. pallidum*. The motility genes of spirochetes appear to be very tightly clustered. Only 8 motility operons are present in *B. burgdorferi* and nine or 10 in *T. pallidum*. In *B. burgdorferi* there is a large motility gene cluster of 21 kb. The gene order is similar to that of a gene cluster found in *B. subtilis*. This cluster appears to be initiated by a σ^{70}-like promoter mapping upstream of the rod protein *figB*. The direct involvement of this housekeeping protein suggests the importance of motility in all phases of growth and survival of this spirochete in nature.

Gliding Motility

Gliding bacteria display an unusual form of motility in that they glide on solid surfaces without the aid of flagella or other obvious external organs of locomotion. A diverse assortment of bacteria from many different phylogenetic groups displays this form of motility. Gliding motility has been observed in cyanobacteria (*Anabaena*, *Oscillatoria*, and many others), *Flavobacterium*, *Cytophaga*, *Flexibacter*, *Myxococcus*, *Beggiatoa*, *Leucothrix*, *Chloroflexus*, *Helicobacterium*, and *Mycoplasma*. As more studies have been conducted, it has become apparent that there are probably several different mechanisms of gliding motility.

Myxococcus xanthus, one of the most well-studied members of the gliding bacteria, travels by means of coordinated rafts or swarms. No obvious organs of locomotion have been identified. Gliding motility by vegetative cells is controlled by at least two systems of genes. The S or social system is required for the rafts of cells to move out from the edge of a vegetative swarm. The A or adventurous system is required for single cells to move out from the edge of a vegetative swarm.

A number of genes affecting S motility have been identified. Type IV pili (see next section) are components of the S system. Mutants unable to synthesize type IV pili lack S motility. Cell interactions are required for S motility; individual cells are nonmotile. This form of motility may very likely be related to the twitching motility that requires type IV pili as described in the next section. The A system of motility is governed by a large number of genes but does not require pili. These include genes coding for the production of surface LPS. One class of mutants defective in A motility involves several *cgl* (for contact gliding) genes (*cglB* through *cglF*). Mutants in this class become transiently motile after contact with wild-type cells. The *cglB* gene codes for a protein that is essential for movement, but its function has not been elucidated. Another gene, *mglA*, is required for the proper functioning of both the A-motility and S-motility systems, suggesting that MglA (a GTPase) may play a regulatory role in both systems.

Motility defects caused by "frizzy" (*frz*) mutations affect the frequency of reversal in the direction of gliding. These mutants produce frizzy filaments of aggregating cells during development, which would be characteristic of cells unable to migrate to centers of aggregation. The *frz* genes show striking homology to the *che* genes that govern the chemotactic signal transduction systems in flagellated bacteria.

Flavobacterium johnsoniae (formerly known as *Cytophaga johnsonae*) cells move in groups, known as rafts, away from the point of inoculation to form thin, veil-like colonies with feathery edges. On solid media, colonies of motile cells spread much faster than can be accounted for by growth alone. Raft movement can be observed on the edge of the colony. Motility in this organism has been correlated with the ability to move polystyrene-latex beads over the cell surface. Mutants lacking this ability are all nongliding. Secretion of extracellular slime has been shown to mediate surface adhesion during gliding motility. Slime is often deposited on slime tracks on agar surfaces over which the cells move, and these tracks serve as preferred paths of movement for other cells. Extracellular material appears to be secreted ahead and to the sides of the leading organisms in a mass of cells, indicating that the cells are moving through the secreted slime. An alternative hypothesis involves cell surface macromolecules (proteins or glycoproteins) either in or on the outer membrane that promote adherence to the substratum. Some gliding motility mutants of *F. johnsoniae* are defective in the production of sulfonolipids present in the outer membranes. Motility in one of these mutants could be reestablished by the addition of cysteate, a sulfonolipid precursor that restors normal levels of cellular sulfonolipids.

PILI OR FIMBRIAE

Bacteria from many different genera produce a multiplicity of nonflagellar, filamentous, proteinaceous appendages called **pili** (Latin, *pilus*, meaning "hair") or **fimbriae** (Latin, *fimbra*, meaning "fringe"). Conjugative or **F pili** are involved in the mating process described in Chapter 3. A number of other types of pili, sometimes referred to as **common** or **generalized pili**, are also produced by numerous bacterial species. These pili play a role in adherence of microorganisms to surfaces or to specific receptors on eukaryotic cells, aiding in the colonization of these ecological niches through the formation of biofilms. Type IV pili are involved in a form of motility known as **twitching**.

Members of the *Enterobacteriaceae* produce a wide assortment of pili or fimbriae (Table 7-7). In *E. coli*, as well as in most members of the Enterobacteriaceae, the most prevalent class of fimbriae are of type 1. These pili are composed of a short-tip fibrillar structure containing FimG and the Fim H adhesin attached to a rod composed of FimA subunits. The FimH adhesin mediates binding to mannose oligosaccharides (mannose sensitivity). The chaperone-usher pathway, a donor strand complementation mechanism in which the chaperone donates a strand to complete pilin formation, mediates pilus synthesis. Expression of the type 1 fimbriae in *E. coli* exhibits recombination-mediated on-off phase variation. This phase variation correlates with the orientation of a short,

TABLE 7-7. Pili or Fimbriae Produced by *Enterobacteriaceae* and Other Pathogenic Bacteria

Pilus Type	Properties
1	Encoded by genes (*fimA–fimH*) clustered at 98 min on *E. coli* genetic map; 17 kDa pilin subunits encoded by *fimA*; agglutination of fowl or guinea pig erythrocytes inhibited by α-D-mannose (mannose sensitivity); produced by *E. coli*, *Klebsiella*, *Salmonella*, and several other species of *Enterobacteriaceae*.
2	Nonhemagglutinating; observed in *Escherichia* and *Salmonella* strains; may be type 1 antigenic variants.
3	Mediate mannose-resistant agglutination of tannic acid–treated animal erythrocytes; expression in *K. pneumoniae* involves at least six genes (*mrkA–mrkF*). Produced by *Klebsiella*, *Salmonella*, *Yersinia*, *Proteus*, and *Providencia*
IV	Bundle-forming pili of enteropathogenic *E. coli*, other human and animal pathogens including *Neisseria gonorrhoeae*; associated with "twitching" motility (nonflagellar movement). Encoded by *pilE* genes that result from nonreciprocal reactions with numerous silent loci, *pilS*.
P	Produced by uropathogenic or pyelonephritis-associated (Pap) *E. coli*. Mediates binding Pap to digalactoside receptors (galactose sensitive) in urinary tract epithelium. Eleven genes (*papA–pakK*) are involved in biosynthesis and expression of P pili; Pap is an adhesin located at the tips of pili. Expression of Pap pili subject to on-off phase variation involving DNA methylation by deoxyadenosine methylase (Dam).
S	Mediate mannose-resistant hemagglutination; facilitate binding to α-sialic acid-2,3-β-D-Gal residues; encoded by *sfa* gene cluster in *E. coli* K1.
F1C	Nonhemagglutinating; encoded by 6 genes in *foc* gene cluster; *focA* encodes major fimbrial subunit; *focC* gene product required for fimbria formation; *focG*, *focH* encode minor fimbrial subunits; mediate adherence to collecting ducts and distal tubules of human kidney.
K88	Produced by porcine enterotoxigenic *E. coli*. Plasmid-borne structural genes (*faeC–faeH)* are involved.
K99	Produced by bovine enterotoxigenic *E. coli*.
F17	Mediate binding to *N*-acetylglucosamine receptors in calf intestinal mucosal cells; F17 gene cluster contains at least 4 genes.

invertible 314 bp DNA element (switch) located immediately upstream of *fimA*. The promoter for *fimA* is thought to reside within this invertible element and to direct the transcription of *fimA* when the element is in one orientation (on) but not when the element is in the alternate orientation (off).

Two genes, *fimB* and *fimE*, mapping immediately adjacent to the *fim* invertible element, are believed to encode site-specific recombinases. A model for the roles of *fimB* and *fimE* suggests that *fimB* promotes recombination of the *fim* invertible element I in both directions, whereas *fimE* promotes recombination primarily from "on" to "off." However, the on orientation of the *fim* invertible element is necessary but not totally sufficient for fimbrial expression, since strains in which the invertible element is locked in the on orientation continue to exhibit phase-variable expression of type 1 fimbriae. Other genes whose products affect the invertible element include *pilG*, which encodes the histone-like protein H1, and *himA* and *himD/hip*, which together encode IHF.

Immediately 3' to *fimA* are two genes (*fimC* and *fimD*) whose products are required for polymerization of the *fimA* gene product into pili. The *fimD* product is apparently incorporated into the outer membrane and serves as a scaffold or polymerization channel for pilus assembly. Genes encoding minor components of type 1 fimbriae (*fimF*, *fimG*, and *fimH*) are located distally to *fimA* and encode products similar to pilin structure. The *fimH* product appears to be the type 1 pilus adhesion component that interacts directly with the eukaryotic cell receptor and confers mannose sensitivity. It has been suggested that the *fimF* product aids in starting new pili and that the *fimG* product is an inhibitor of pilus polymerization.

An additional gene, *lrp*, is also required for normal activity of the *fim* switch. The *lrp* gene product, Lrp, is a site-specific DNA-binding protein that protects extended regions of DNA in vitro from nuclease digestion. It affects the transcription of many genes, collectively called the leucine-Lrp regulon. Lrp could influence the *fim* inversion either indirectly by altering the expression of other proteins, such as FimB and FimE or IHF, or directly by participating in the *fim* inversion as an auxiliary factor. Lrp is known to influence phase variation of Pap fimbriation by blocking the methylation of two *dam* sites in the promotor region of *papAB*. Lrp may be required either for the expression of an additional factor that promotes the *fim* inversion or as a direct participant in the recombination reaction.

Uropathogenic *E. coli* produce P pili encoded by genes *papA* through *papK* located in the *pap* (pilus associated with pyelonephritis) gene cluster. The P pili are composite fibers containing a rod joined to a thinner-tip fibrillum. Biogenesis of P pili occurs via the highly conserved chaperone-usher pathway. In this system, periplasmic chaperones escort pilus subunits to the usher, a large protein complex that facilitates the translocation and assembly of subunits across the outer membrane.

Neisseria gonorrhoeae and *N. meningitidis* produce type IV pili, which are essential determinants of virulence. These type IV pili, encoded by the *pilE* gene, mediate adhesion to host cells, preventing phagocytosis by polymorphonuclear leukocytes. Expression of *pilE* is controlled transcriptionally by the products of two linked chromosomal genes: *pilA* and *pilB*. In *N. gonorrhoeae*, antigenic variation occurs when DNA sequences from one of several silent pilin gene copies (*pilS*) are transferred unidirectionally to replace variable sequences within the *pilE* gene.

The *pilE* sequence changes alter the amino acid sequence of the pilin protein and the antigenicity of the pilus. Presumably, these antigenic changes aid in the avoidance of the host immune response. DNA recombination is required for natural DNA transformation, DNA repair, and pilus antigenic variation. RecA mediates all these processes, as discussed in Chapter 3. A homologue of the recombination-dependent growth gene, *rdgC*, is involved in gonococcal pilin antigenic variation. There are two routes for generating pilin variation in *N. gonorrhoeae*: intragenomic recombination and recombination following transformation with exogenous DNA from lysed gonococci. The incoming donor sequence can be either a second *pilE* locus or any *pilS* locus. In either case, the antigenicity of the produced pilin differs from that originally present.

Twitching is a form of motility that is expressed as a jerky translocation over solid surfaces. In liquid medium, *P. aeruginosa* swims rapidly by means of flagella. However, once attached to a solid surface, large groups of cells are translocated by means of a flagellum-independent activity dependent on the presence of type IV pili. Other bacteria — for example, *N. gonorrhoeae*, and *E. coli* — also exhibit this unusual form of motility. Twitching is essential for biofilm formation and undoubtedly contributes to the ability of these organisms to form biofilms or to effect attachment to cells, thereby aiding in the invasion of human or animal hosts.

Pseudomonas aeruginosa is freely motile in an aqueous environment by means of polar flagella. However, it can display a swarming type of motility on agar surfaces through the development of alternate forms of flagella. *P. aeruginosa* also produces type IV pili that are involved in twitching motility and are presumed to mediate adherence to surfaces including those of eukaryotic cells. It has been shown that the initiation of biofilm formation in *P. aeruginosa* correlates with the emergence of hyperpiliated and highly adherent phenotypic variants that are deficient in swimming, swarming, and twitching motilities. These variants produced small, rough colonies and autoaggregated in liquid cultures, and rapidly initiated the formation of strongly adherent biofilms. In contrast, the large-colony variant (parent form) was poorly adherent, homogeneously dispersed in liquid cultures, and produced scant polar pili. It has been suggested that phase variation ensures the presence of different phenotypic forms that adapt readily to initiate biofilm formation under the appropriate environmental conditions.

BIBLIOGRAPHY

Nucleus, Nucleosomes, and Nucleoids

Azam, T. A., S. Hiraga, and A. Ishihama. 2000. Two types of localization of the DNA-binding proteins within the *Escherichia coli* nucleoid. *Genes Cells* **5**:613–26.

Bendich, A. J. 2001. The form of chromosomal DNA molecules in bacterial cells. *Biochimie* **83**:1–11.

Bendich, A. J., and K. Drlica. 2000. Prokaryotic and eukaryotic chromosomes: What's the difference? *BioEssays* **22**:481–6.

Bernander, R. 2000. Chromosome replication, nucleoid segregation, and cell division in archaea. *Trends Microbiol.* **8**:278–83.

Graumann, P. J. 2000. *Bacillus subtilis* SMC is required for proper arrangement of the chromosome and for efficient segregation of replication termini but not for bipolar movement of newly duplicated origin regions. *J. Bacteriol.* **182**:6463–71.

Huls, P. G., N. O. E. Vischer, and C. L. Woldringh. 1999. Delayed nucleoid segregation in *Escherichia coli*. *Mol. Microbiol.* **33**:959–70.

Kornberg, R. D., and Y. Lorch. 1999. Twenty-five years of the nucleosome, fundamental particle of the eukaryotic chromosome. *Cell* **98:**285–94.

Lamond, A. I., and W. C. Earnshaw. 1998. Structure and function in the nucleus. *Science* **280:**547–53.

Li, J. Y., B. Arnold-Schultz-Gahmen, and E. Kellenberger. 1999. Histones and histone-like DNA-binding proteins: correlation s between structural differences, properties and functions. *Microbiology* **145:**1–2.

Margolin, W. 1998. A green light for the bacterial cytoskeleton. *Trends Microbiol.* **6:**233–8.

Murphy, L. D., and S. B. Zimmerman. 2000. Multiple restraints to the unfolding of spermidine nucleoids from *Escherichia coli. J. Struct. Biol.* **132:**46–62.

 Pettijohn, D. E. 1996. The nucleoid. In F. C. Neidhardt, et al. (eds.), *Escherichia coli and Salmonella: Cellular and Molecular Biology*, 2nd ed. ASM Press, Washington, DC, pp. 158–66.

Sandman, K., S. L. Pereira, and J. N. Reeve. 1998. Diversity of prokaryotic chromosomal proteins and the origin of the nucleosome. *Cell Mol. Life Sci.* **54:**1350–64.

Sawitzke, J., and S. Austin. 2001. An analysis of the factory model for chromosome replication and segregation in bacteria. *Mol. Microbiol.* **40:**786–94.

Sharpe, M. E., and J. Errington. 1999. Upheaval in the bacterial nucleoid. an active chromosome segregation mechanism. *Trends Genetics* **15:**70–4.

Trun, N. J., and J. F. Marko. 1998. Architecture of a bacterial chromosome. *ASM News* **64:**276–83.

Mitochondria

Beech, P. L., et al. 2000. Mitochondrial FtsZ in a chromophyte alga. *Science* **287:**1276–9.

Brenner, C., and G. Kroemer. 2000. Mitochondria—the death signal integrators. *Science* **289:**1150–1.

Gray, M. W., G. Berger, and B. F. Lang. 1999. Mitochondrial evolution. *Science* **283:**1476–81.

Jacobs, H. T., S. K. Lehtinen, and J. N. Spelbrink. 2000. No sex please, we're mitochondria: a hypothesis on the somatic unit of inheritance of mammalian mtDNA. *BioEssays* **22:**564–72.

Kurland, C. G., and S. G. E. Andersson. 2000. Origin and evolution of the mitochondrial proteome. *Microbiol. Mol. Biol. Rev.* **64:**786–820.

Sarasti, M. Oxidative phosphorylation at the *fin de siècle. Science* **283:**1488–93.

Wei, M. C., et al. 2001. Proapoptotic BAX and BAK: a requisite gateway to mitochondrial dysfunction and death. *Science* **292:**727–30.

Yaffe, M. P. 1999. The machinery of mitochondrial inheritance and behavior. *Science* **283:**1493–7.

Eukaryotic Cell Surface

Cid, V. J., et al. 1995. Molecular basis of cell integrity and morphogenesis in *Saccharomyces cerevisiae. Microbiol. Rev.* **59:**345–86.

Walker, G. M. 1998. *Yeast Physiology and Biotechnology.* Wiley & Sons, Chichester, UK.

Surface (S) layers

Beveridge, T. J., et al. 1997. Functions of S-layers. *FEMS Microbiol. Rev.* **20:**1–49.

Ilk, M., et al. 1999. Structural and functional analyses of the secondary cell wall polymer of *Bacillus sphaericus* CCM 2177 that serves as an S-layer-specific anchor. *J. Bacteriol.* **181:**7643–6.

Navarre, W. W., and O. Schneewind. 2000. Surface proteins of gram-positive bacteria and mechanism of their targeting to the cell wall envelope. *Microbiol. Mol. Biol. Rev.* **63:** 174–229.

Sára, M., and U. Sleytr. 2000. S-Layer proteins. *J. Bacteriol.* **182:**859–68.

Sleytr, U. B., and P. Messner. 2000. Crystalline bacterial cell surface layers. In J. Lederberg (ed.), *Encyclopedia of Microbiology*, 2nd ed. Academic Press, San Diego, pp. 899–906.

Bacterial Cell Wall Peptidoglycan (Murein)

Archer, G. L., and J. M. Bosilevac. 2001. Signaling antibiotic resistance in staphylococci. *Science* **291:**1915–6.

Brennan, P. J., S. Mahapatra, and D. C. Crick. 2000. Comparison of the UDP-*N*-acetyl-muramate: L-alanine ligase enzymes from *Mycobacterium tuberculosis* and *Mycobacterium leprae*. *J. Bacteriol.* **182:**6827–30.

Dimitriev, B. A., S. Ehlers, E. T. Rietschel, and P. J. Brennan. 2000. Molecular mechanics of the mycobacterial cell wall: from horizontal layers to vertical scaffolds. *Int. J. Med. Microbiol.* **258:**251–8.

Dimitriev, B. A., S. Ehlers, and E. T. Rietschel. 1999. Layered murein revisited: a fundamentally new concept of bacterial cell wall structure, biogenesis and function. *Med. Microbiol. Immunol.* **187:**173–81.

Giesbrecht, P., T. Kersten, H. Maidhof, and J. Wecke. 1998. Staphylococcal cell wall: morphogenesis and fatal variations in the presence of penicillin. *Microbiol. Mol. Biol. Rev.* **62:**1371–414.

Goffin, C., and J.-M. Ghuysen. 1998. Multimodular penicillin-binding proteins: an enigmatic family of orthologs and paralogs. *Microbiol. Mol. Biol. Rev.* **62:**1079–93.

Hoiczyk, E., and A. Hansel. 2000. Cyanobacterial cell walls: news from an unusual prokaryotic envelope. *J. Bacteriol.* **182:**1191–9.

Höltje, J.-V. 1998. Growth of the stress-bearing and shape-maintaining murein sacculus of *Escherichia coli*. *Microbiol. Mol. Biol. Rev.* **62:**181–203.

Höltje, J.-V. 2000. Cell walls, bacterial. In J. Lederberg (ed.), *Encyclopedia of Bacteriology*, 2nd ed. Academic Press, San Diego, pp. 759–71.

Koch, A. L. 1995. *Bacterial Growth and Form*. Chapman & Hall, New York.

Koch, A. L. 2000. The bacterium's way for safe enlargement and division. *Applied Environ. Microbiol.* **66:**3657–63.

Labichinski, H., and H. Maidhof. 1994. Bacterial peptidoglycan: overview and evolving concepts. In J.-M. Ghuysen and R. Hakenbeck (eds.), *Bacterial Cell Wall*. Elsevier, Amsterdam, The Netherlands, pp. 23–38.

Matsuhashi, M. 1994. Utilization of lipid-linked precursors and the formation of peptidoglycan in the process of cell growth and division: membrane enzymes involved in the final steps of peptidoglycan synthesis and the mechanism of their regulation. In J.-M. Ghuysen and R. Hakenbeck (eds.), *Bacterial Cell Wall*. Elsevier, Amsterdam, The Netherlands, pp. 55–71.

Mengin-Lecreulx, D., T. Falla, D. Blanot, J. van Heijenoort, D. J. Adams, and I. Chopra. 1999. Expression of the *Staphylococcus aureus* UDP-*N*-acetyl-muramoyl-L-alanyl-D-glutamate:L-lysine ligase in *Escherichia coli* and effects on peptidoglycan biosynthesis and cell growth. *J. Bacteriol.* **181:**5909–14.

Navarre, W. W., and O. Schneewind. 2000. Surface proteins of gram-positive bacteria and mechanism of their targeting to the cell wall envelope. *Microbiol. Mol. Biol. Rev.* **63:**174–229.

Pink, D., et al. 2000. On the architecture of the gram-negative bacterial murein sacculus. *J. Bacteriol.* **182:**6925–30.

Shockman, G. D., and J.-V. Höltje. 1994. Microbial peptidoglycan (murein) hydrolases. In J.-M. Ghuysen and R. Hakenbeck (eds.), *Bacterial Cell Wall*. Elsevier, Amsterdam, The Netherlands, pp. 131–66.

Van Heijenoort, J. 1994. Biosynthesis of the bacterial peptidoglycan unit. In J.-M. Ghuysen and R. Hakenbeck (eds.), *Bacterial Cell Wall*. Elsevier, Amsterdam, The Netherlands, pp. 39–54.

Zhang, H. Z., C. J. Hackbarth, K. M. Chansky, and H. F. Chambers. 2001. A proteolytic transmembrane signaling pathway and resistance to β-lactams in staphylococci. *Science* **291**:1962–5.

Teichoic and Lipoteichoic Acids

Boyd, D. A., D. C. Cvitkovitch, A. S. Bleiweis, et al. 2000. Defects in D-alanyl-lipoteichoic acid synthesis in *Streptococcus mutans* results in acid sensitivity. *J. Bacteriol.* **182**:6055–65.

Fisher, W. 1994. Lipoteichoic acids and lipoglycans. In J.-M. Ghuysen and R. Hackenbeck (eds.), *Bacterial Cell Wall*. Elsevier, Amsterdam, The Netherlands, pp. 199–215.

Kiriukhin, M. Y., D. V. Devarov, D. L. Shinabarger, and F. C. Neuhaus. 2001. Biosynthesis of the glycolipid anchor in lipoteichoic acid of *Staphylococcus aureus* RN4220: role of YpfP, the diglucosyldiacylglycerol synthase. *J. Bacteriol.* **183**:3506–14.

Pooley, H. M., and D. Karamata. 1993. Teichoic acid synthesis in *Bacillus subtilis*: genetic organization and biological roles. In J.-M. Ghuysen and R. Hackenbeck (eds.), *Bacterial Cell Wall*. Elsevier, Amsterdam, The Netherlands, pp. 187–98.

Outer Membrane

Kotra, L. P., et al. 2000. Visualizing bacteria at high resolution. *ASM News* **66**:675–81.

Hancock, R. E. W., D. N. Karunaratne, and C. Bernegger-egli. 1994. Molecular organization and structural role of outer membrane macromolecules. In J.-M. Ghuysen and R. Hackenbeck (eds.), *Bacterial Cell Wall*. Elsevier, Amsterdam, The Netherlands, pp. 263–79.

Reeves, P. 1994. Biosynthesis and assembly of lipopolysaccharide. In J.-M. Ghuysen and R. Hackenbeck (eds.), *Bacterial Cell Wall*. Elsevier, Amsterdam, The Netherlands, pp. 281–317.

Rocchetta, H. L., L. L. Burrows, and J. S. Lam. 1999. Genetics of O-antigen biosynthesis in *Pseudomonas aeruginosa*. *Microbiol. Mol. Biol. Rev.* **63**:523–53.

Whitfield, C. 2000. Lipopolysaccharides. In J. Lederberg (ed.), *Encyclopedia of Microbiology*, 2nd ed. Academic Press, San Diego, pp. 71–85.

Cytoplasmic Membrane

Kadner, R. J. 2000. Cell membrane: structure and function. In J. Lederberg (ed.), *Encyclopedia of Microbiology*, 2nd ed. Academic Press, New York, pp. 710–28.

Periplasm

Bayer, M. E., and M. H. Bayer. 1994. Periplasm. In J.-M. Ghuysen and R. Hackenbeck (eds.), *Bacterial Cell Wall*. Elsevier, Amsterdam, The Netherlands, pp. 447–64.

Capsules

Burne, R. A., Z. T. Wen, Y. Y. M. Chen, and J. E. Penders. 1999. Regulation of expression of the fructan hydrolase gene of *Streptococcus mutans* GS-5 by induction and carbon catabolite repression. *J. Bacteriol.* **181**:2863–71.

Hardy, G. G., M. J. Caimano, and J. Yother. 2000. Capsule biosynthesis and basic metabolism of *Streptococcus pneumoniae* are linked through the cellular phosphoglucomutase. *J. Bacteriol.* **182:**1854–63.

Biofilms

Davey, M. E., and G. A. O'toole. 2000. Microbial biofilms: from ecology to molecular genetics. *Microbiol. Mol. Biol. Rev.* **64:**847–67.

Otto, K., J. Norbeck, T. Larsson, et al. 2001. Adhesion of type 1-fimbriated *Escherichia coli* to abiotic surfaces leads to altered composition of outer membrane proteins. *J. Bacteriol.* **183:**2445–53.

Watnick, P., and R. Kolter. 2000. Biofilm, city of microbes. *J. Bacteriol.* **182:**2675–9.

Cilia and Flagella of Eukaryotes

Benashski, S. E., R. S. Patel-King, and S. M. King. 1999. Light chain 1 from the *Chlamydomonas* outer dynein arm is a leucine-rich repeat protein associated with the motor domain of the γ heavy chain. *Biochemistry* **37:**15,033–41.

Habermacher, G., and W. S. Sale. 1996. Regulation of flagellar dynein by an axonemal type-1 phosphatase in *Chlamydomonas*. *J. Cell Sci.* **109:**1899–907.

Johnson, K. A. 1998. The axonemal microtubules of the *Chlamydomonas* flagellum differ in tubulin isoform content. *J. Cell Sci.* **111:**313–20.

Mitchell, D. R., and W. S. Sale. 1999. Characterization of a *Chlamydomonas* insertional mutant that disrupts flagellar central pair microtubule structures. *J. Cell Biol.* **144:**293–304.

Okada, Y., and N. Hirokawa. 1999. A processive single-headed motor: kinesin superfamily protein KIF1A. *Science* **283:**1152–7.

Perrone, D., P. Yang, E. O'Toole, W. Sale, and M. Porter. 1998. The *Chlamydomonas IDA7* locus encodes a 140-kDa dynein intermediate chain required to assemble the 11 inner arm complex. *Mol. Biol. Cell* **9:**3351–65.

Bacterial Flagella

Aizawa, S.-I. 2000. Flagella. In J. Lederberg (ed.), *Encyclopedia of Microbiology*, 2nd ed. Academic Press, San Diego, pp. 380–9.

Berry, R. M., and J. P. Armitage. 1999. The bacterial flagella motor. *Adv. Microb. Physiol.* **41:**291–337.

Chilcott, G. S., and K. T. Hughes. 2000. Coupling of flagellar gene expression to flagellar assembly in *Salmonella enterica* serovar typhimurium and *Escherichia coli*. *Microbiol. Mol. Biol. Rev.* **64:**694–708.

Kalir, S., J. McClure, K. Pabbaraju, et al. 2001. Ordering genes in a flagella pathway by analysis of expression kinetics from living bacteria. *Science* **292:**2080–3.

Kilhara, M., T. Minamino, S. Yamaguchi, and R. M. Macnab. 2001. Intergenic suppression between the flagellar MS ring protein FliF of *Salmonella* and FlhA, a membrane component of its export apparatus. *J. Bacteriol.* **183:**1655–62.

Lybarger, S. R., and J. R. Maddock. 2001. Polarity in action: asymmetric protein localization in bacteria. *J. Bacteriol.* **183:**3261–7.

Macnab, R. M. 1992. Genetics and biogenesis of bacterial flagella. *Annu. Rev. Genet.* **26:**129–56.

Macnab, R. M. 1999. The bacterial flagellum: reversible rotary propellor and type III export apparatus. *J. Bacteriol.* **181:**7149–53.

Macnab, R. M. 2000. Type III protein pathway exports *Salmonella* flagella. *ASM News* **66**:738–45.

Yonekura, K., et al. 2000. The bacterial flagellar cap as the rotary promoter of flagellin self-assembly. *Science* **290**:2148–52.

Chemotaxis

Aizawa, S. L., C. S. Harwood, and R. J. Kadner. 2000. Signaling components in bacterial locomotion and sensory reception. *J. Bacteriol.* **182**:1459–71.

Grebe, T. W., and J. Stock. 1998. Bacterial chemotaxis: the five sensors of a bacterium. *Curr. Biol.* **8**:R154–7.

Stock, J., and S. Da Re. 1999. A receptor scaffold mediates stimulus-response coupling in bacterial chemotaxis. *Cell Calcium* **26**:157–64.

Swarming Motility

Belas, R. 1998. Characterization of *Proteus mirabilis* precocious swarming mutants: identification of *rsbA*, encoding a regulator of swarming behavior. *J. Bacteriol.* **180**:6126–39.

Harshey, R. M., and T. Matsuyama. 1994. Dimorphic transition in *Escherichia coli* and *Salmonella typhimurium*: surface-induced differentiation into hyperflagellate swarmer cells. *Proc. Natl. Acad. Sci. USA* **91**:8631–5.

McCarter, L. M., and M. E. Wright. 1993. Identification of genes encoding components of the swarmer cell flagellar motor and propeller and a sigma factor controlling differentiation of *Vibrio parahaemolyticus*. *J. Bacteriol.* **175**:3361–71.

Sar, N., L. McCarter, M. Simon, and M. Silverman. 1990. Chemotactic control of the two flagellar systems of *Vibrio parahaemolyticus*. *J. Bacteriol.* **172**:334–41.

Toguchi, A., M. Siano, M. Burkhart, and R. M. Harshey. 2000. Genetics of swarming motility in *Salmonella enterica* serovar Typhimurium: critical role for lipopolysaccharide. *J. Bacteriol.* **182**:6308–21.

Gliding Motility

McBride, M. J. 2000. Bacterial gliding motility: mechanisms and mysteries. *ASM News* **66**:203–10.

Spormann, A. M. 1999. Gliding motility in bacteria: insights from studies of *Myxococcus xanthus*. *J. Bacteriol.* **63**:621–41.

Motility in Spirochetes

Izard, J., W. A. Samsonoff, M. B. Kinoshita, and R. J. Limberger. 1999. Genetic and structural analyses of cytoplasmic filaments of wild-type *Treponema phagadenis* and a flagellar filament-deficient mutant. *J. Bacteriol.* **181**:6739–46.

Li, C., L. Corum, D. Morgan, E. L. Rosey, T. B. Stanton, and N. W. Charon. 2000. The spirochete FlaA periplasmic flagellar sheath protein impacts flagellar helicity. *J. Bacteriol.* **182**:6698–706.

Li, C., A. Motaleb, M. Sal, S. Goldstein, and N. W. Charon. 2001. Gyrations, rotations, periplasmic flagella: the biology of spirochete motility. In M. J. Saier, Jr., and J. Garcia-Lara (eds.), *The Spirochetes: Molecular and Cell Biology*. Horizon Scientific Press, Wymondham, UK.

Limberger, R. J., L. L. Silvienski, J. Izard, and W. A. Samsonoff. 1999. Insertional inactivation of *Treponema denticola tap1* results in a nonmotile mutant with elongated flagellar hooks. *J. Bacteriol.* **181**:3743–50.

Pili or Fimbriae

Bieber, D., et al. 1998. Type IV pili, transient bacterial aggregates, and virulence of enteropathogenic *Escherichia coli*. *Science* **280**:2114–8.

Choudhury, D., et al. 1999. X-ray structure of the FimC-FimH chaperone-adhesin complex from uropathogenic *Escherichia coli*. *Science* **285**:1061–6.

Déziel, E., Y. Comeau, and R. Villemur. 2001. Initiation of biofilm formation by *Pseudomonas aeruginosa* 57RP correlates with emergence of hyperpiliated and highly adherent phenotypic variants deficient in swimming, swarming, and twitching motilities. *J. Bacteriol.* **183**:1195–204.

Howell-Adams, B., and H. S. Seifert. 1999. Insertion mutations in *pilE* differentially alter gonococcal pilin antigenic variation. *J. Bacteriol.* **181**:6133–41.

Kearns, D. B., J. Robinson, and L. J. Shimkets. 2001. *Pseudomonas aeruginosa* exhibits directed twitching motility up phosphatidylethanolamine gradients. *J. Bacteriol.* **183**:763–7.

Long, C. D., S. F. Hayes, J. P. M. van Putten, et al. 2001. Modulation of gonococcal piliation by regulatable transcription of *pilE*. *J. Bacteriol.* **183**:1600–9.

Mehr, I. J., et al. 2000. A homologue of the recombination-dependent growth gene, *rdgC*, is involved in gonococcal pilin antigenic variation. *Genetics* **154**:523–32.

Mulvey, M. A., K. W. Dodson, G. E. Soto, and S. J. Hultgren. 2000. Fimbriae, pili. In J. Lederberg (ed.), *Encyclopedia of Microbiology*, 2nd ed. Academic Press, San Diego, pp. 361–79.

Sauer, F., et al. 1999. Structural basis of chaperone function and pilus biogenesis. *Science* **285**:1058–61.

Virji, M. 1998. Glycosylation of the meningococcus pilus protein. *ASM News* **64**:398–405.

CHAPTER 8

CENTRAL PATHWAYS OF CARBOHYDRATE METABOLISM

The study of carbohydrate metabolism involves consideration of the following factors:

1. Central pathways of carbohydrate metabolism
2. Conversion of compounds to intermediates usable in central pathways
3. Mechanisms of energy (ATP) production
   a. Substrate-level phosphorylation
   b. Oxidative phosphorylation
   c. Other mechanisms of energy transfer
4. Metabolic steps involved in the generation and use of reducing activity
   a. Reduction of pyruvate or other substrates to fermentation end products
   b. Biosynthetic reactions requiring reducing action
5. Oxygen involvement in energy-generating reactions
   a. Aerobic metabolism
   b. Anaerobic metabolism
   c. Facultative metabolism
6. Metabolic intermediates serving as biosynthetic precursors
7. Reactions that replenish biosynthetic intermediates (anapleurotic reactions)
8. Metabolic and genetic regulatory systems

Carbohydrates are not the only compounds utilized as sources of energy by microorganisms. Fatty acids, lipids, amino acids, purines, pyrimidines, and a wide variety of other compounds can also serve as carbon and energy sources. Generally, utilization of an alternate substrate involves its conversion to an intermediate intrinsic to one of the central pathways of carbohydrate metabolism. Some of these alternate pathways are discussed in Chapter 10. The details of energy production and metabolite transport are discussed in Chapter 9.

ALTERNATE PATHWAYS OF CARBOHYDRATE METABOLISM

Fructose Bisphosphate Aldolase Pathway

One major pathway of glucose degradation is accomplished by the series of reactions shown in Figure 8-1. This pathway is often referred to as the Embden-Meyerhof-Parnas (EMP) pathway in recognition of some of the earliest scientists who contributed to its elucidation. Several important reactions occur in the EMP pathway:

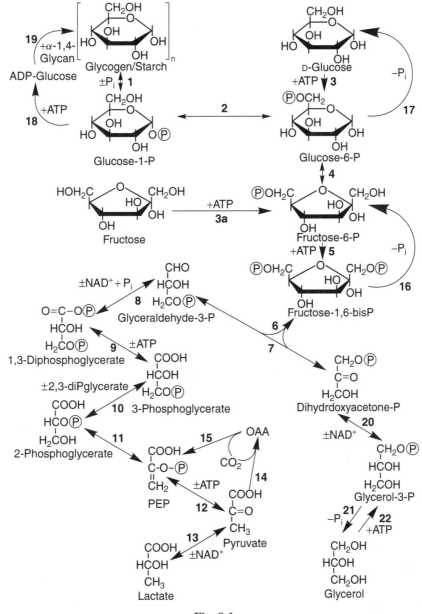

Fig. 8-1

1. Phosphorylation of glucose and fructose-6-phosphate by ATP
2. Cleavage of fructose-1,6-bisphosphate to trioses by a specific aldolase
3. Structural rearrangements
4. Oxidation–reduction and P_i (inorganic phosphate) assimilation

Fig. 8-1 Fructose bisphosphate (FBP) aldolase or Embden-Meyerhof-Parnas (EMP) pathway of glycolysis. Enzymes involved in gluconeogenesis are also shown. The italicized three-letter designation indicates the structural gene for the enzyme in *E. coli*.

Glycolytic enzymes

1. Phosphorylase. Degradation of glycogen or starch to G-1-P.
2. Phosphoglucomutase. Isomerization of G-1-P to G-6-P.
3. Hexokinase. Phosphorylation of glucose to G-6-P, using ATP Hexokinase also phosphorylates fructose to F-6-P using ATP (reaction 3a).
4. Phosphoglucoisomerase (*pgi*). Isomerization of G-6-P to F-6-P.
5. Phosphofructokinase (*pfkA*). Phosphorylation of F-6-P to FBP using ATP.
6. Fructose bisphosphate (FBP) aldolase (*fbaA*). Cleaves FBP to GA-3-P and dihydroxyacetone phosphate.
7. Triose phosphate isomerase (*tpi*). Interconverts GA-3-P and dihydroxyacetone-P.
8. Glyceraldehyde-3-phosphate dehydrogenase (*gap*). Oxidizes GA-3-P to 1,3-diphosphoglycerate using nicotinamide adenine dinucleotide (NAD^+) and inorganic phosphate (P_i) to form $NADP + H^+$.
9. Phosphoglycerokinase (*pgk*). Generates ATP from ADP.
10. Phosphoglyceromutase (*pgm*). Uses 2,3-diphosphoglycerate to convert 3-phosphoglycerate to 2-phosphoglycerate.
11. Enolase (*eno*). Enolization of 2-phosphoglycerate forms high-energy phosphate bond (~P; encircled P) in phosphoenolpyruvate (PEP).
12. Pyruvate kinase (*pykA; pykF*). Generates ATP from ADP.
13. Lactate dehydrogenase. Reduces pyruvate to lactate using $NADP + H^+$.

Gluconeogenesis enzymes

14. Pyruvate carboxylase. Converts pyruvate to oxaloacetic acid (OAA) via carbon dioxide (CO_2) fixation using ATP.
15. PEP carboxykinase. Forms phosphoenolpyruvate from OAA using GTP.
16. Fructose-1,6-bisphosphatase. Removes P_i from F-1,6-bisP to form F-6-P.
17. Glucose-6-phosphatase. Removes P_i from G-6-P to form glucose.
18. ATP-glucose pyrophosphorylase. Forms ADP-glucose from G-1-P and ATP.
19. Glycogen synthase. Adds α-1,4-glycan to ADP-glucose to form glycogen or starch.

Glycerol formation or utilization

20. Glycerophosphate dehydrogenase. Reduces dihydroxyacetone-P to glycerol-3-P.
21. Phosphatase. Removes P_i from glycerol-3-P to form glycerol.
22. Glycerol kinase. Phosphorylates glycerol using ATP.

5. Energy transfer by phosphoglycerokinase and pyruvate kinase
6. Metabolic and genetic regulation of the pathway

The overall equation for the ethanol fermentation in yeast can be shown as

$$C_6H_{12}O_6 + 2P_i + 2ADP \longrightarrow 2CH_2CH_3OH + 2CO_2 + 2ATP + 2H_2O$$

The EMP pathway is common to a great many microorganisms as well as higher forms. The enzyme fructose bisphosphate (FPB) aldolase is one of the most critical steps in the pathway. In the absence of this enzyme, glucose or other hexose sugars must be metabolized via one of several alternative pathways, as discussed later.

In general, glycolysis in muscle tissue, yeast, and many bacterial species appears to be identical in terms of the intermediates involved. Pyruvate is the last common intermediate. In yeast, pyruvate is cleaved to acetaldehyde and carbon dioxide. The acetaldehyde is then reduced to ethanol by alcohol dehydrogenase. In muscle tissue and in lactic acid bacteria (*Streptococcus, Lactococcus, Lactobacillus*), pyruvate is reduced to lactate. Other microorganisms that use the EMP pathway have the capacity to convert pyruvate to a wide variety of other fermentation end products. These fermentation pathways are discussed in more detail in Chapter 10.

The enzymes of the glycolytic pathway and the tricarboxylic acid (TCA) cycle are regulated either positively or negatively by specific metabolic intermediates (**feedback control**). As shown in Table 8-1, AMP or FBP stimulates several enzymes. Some metabolites, such as AMP, phosphoenolpyruvate (PEP), or dihydroxyacetone phosphate (DHAP), or cofactors such as reduced nicotinamide adenine dinucleotide (NADH), may have an inhibitory effect on certain catabolic enzymes. Elevation of the AMP concentration may signal a low-energy state. In the case of DHAP or PEP, AMP may regulate the flow of these metabolites into biosynthetic pathways or transport functions. The FBP-activated lactic acid dehydrogenases are characteristically produced by a number of lactic acid–producing bacteria. In streptococci, 6-phosphogluconate inhibits the activity of phosphohexose isomerase, and the activity of 6-phosphogluconate dehydrogenase is inhibited by FBP (Table 8-1). Details of the regulation of prokaryotic gene expression as they apply to carbohydrate metabolism are discussed in Chapter 5.

The structural genes for the enzymes in the glycolytic pathway of *E. coli* have been identified as shown in Figure 8-1. The expression of *pgi, pfk, fbaA,* and *pykA,* genes that code for enzymes in both the upper and lower portions of the glycolytic pathway, are unaffected by glucose. However, the *gap* gene (coding for glyceraldehyde-3-phosphate dehydrogenase) and the phosphoglucokinase (*pgk*) operon are activated by glucose. This activation is dependent on catabolite-control protein A (CcpA), a global regulator that represses several catabolic operons involved in the degradation of secondary carbon sources. This regulator is also involved in glucose repression of genes encoding enzymes of the TCA cycle (see later discussion). The genes and operons required for the utilization of specific carbon sources are usually expressed only if the carbon source is present and glucose or other sugars that can be used via the glycolytic pathway are absent from the growth medium.

TABLE 8-1. Metabolic Regulation of Enzymes of Glycolysis and Tricarboxylic Acid Cycle

Enzyme	Positive Effector	Negative Effector
6-Phosphogluconate dehydrogenase		FBP
Phosphohexose isomerase		6-PG
Fructose-1,6-bisphosphatase		AMP
Phosphofructokinase	ADP	PEP
Pyruvate kinase 1	FBP	
Pyruvate kinase 2	AMP	
Citrate synthase	AMP	NADH, or α-KG
Malate dehydrogenase		NADH
Malate enzyme		NADH
Phosphoenolpyruvate carboxykinase		NADH
Phosphoenolpyruvate carboxylase	FBP, acetyl-CoA	Aspartate

FBP, fructose-1,6-bisphosphate; 6-PG, 6-phosphogluconate; AMP, adenosine monophosphate; ADP, adenosine diphosphate; PEP, phosphoenolpyruvate; NADH, reduced nicotinamide adenine dinucleotide.

Alternate Pathways of Glucose Utilization

Warburg and Christian provided the first evidence for the existence of an alternative pathway for the utilization of hexose sugars. They described the oxidation of glucose-6-phosphate to 6-phosphogluconate (6-P-G) via G-6-P dehydrogenase (*zwischenferment*). They also described the decarboxylation of 6-P-G to form a pentose sugar. For a long time, relatively little attention was paid to the real significance of this alternative pathway because Meyerhof and others strongly asserted that the FBP aldolase (EMP) pathway was the main route of glucose catabolism. Subsequently, ribulose-6-phosphate was shown to be the first product formed and was converted to ribose-5-phosphate via an isomerase reaction. This series of reactions is common to several alternate pathways of carbohydrate metabolism as shown in Figures 8-2 and 8-4 and discussed below.

Entner-Doudoroff or Ketogluconate Pathway

The Entner-Doudoroff pathway branches from the ketogluconate pathway of hexose oxidation as shown in Figure 8-2. It consists of two enzymes: 6-phosphogluconate dehydratase (encoded by *edd*) and 2-keto-3-deoxy-6-phosphogluconate (KDPG) aldolase (*eda*). Thus, the Entner-Doudoroff pathway differs from the EMP pathway primarily in the form of the 6-carbon intermediate that undergoes C_3-C_3 cleavage (aldol cleavage) to form three-carbon intermediates. Note that the formed glyceraldehyde-3-phosphate is metabolized to pyruvate via the triose phosphate pathway common to the EMP and phosphoketolase pathways. Originally described as a major pathway in *Pseudomonas*, the Entner-Doudoroff pathway is active in *Escherichia coli* and many other gram-negative species. It is also present in many other microorganisms including members of the archaebacteria. The widespread occurrence of the Entner-Doudoroff pathway has led to its consideration as a much more significant pathway than previously recognized and has given rise to the suggestion that it may have predated the EMP pathway.

E. coli utilizes gluconate via the Entner-Doudoroff pathway. Gluconate dehydratase activity is virtually absent in cells grown on glucose and is induced only by gluconate.

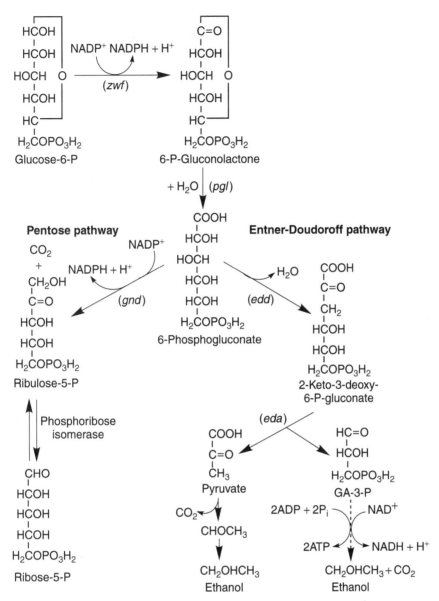

Fig. 8-2. Divergent pathways from 6-phosphogluconate. The structural genes for the enzymes in *E. coli* are indicated by their three-letter designations. In the core pathway, glucose-6-phosphate dehydrogenase (*zwf*, for *zwischenferment*) oxidizes glucose-6-phosphate to 6-phosphogluconolactone. The lactone is dehydrated to 6-phosphogluconate via lactonase (*pgl*). In the pentose pathway, 6-phosphogluconate is oxidized to ribulose-5-phosphate and carbon dioxide by 6-phosphogluconate dehydrogenase (*gnd*). Phosphoribose isomerase maintains ribulose-5-phosphate and ribose-5-phosphate in equilibrium. In the Entner-Doudoroff pathway, 6-phosphogluconate is dehydrated by 6-phosphogluconate dehydratase (*edd*) to yield 2-keto-3-deoxy-6-phosphogluconate (KDPG). The enzyme KDPG aldolase (*eda*) cleaves KDPG to form pyruvate and GA-3-P (glyceraldehyde-3-phosphate). Pyruvate decarboxylase action yields ethanol and carbon dioxide. The GA-3-P is metabolized via the triose phosphate portion of the EMP pathway to yield ethanol and carbon dioxide. The net yield in the Entner-Doudoroff pathway is 2 ethanol + 2 CO_2.

High basal levels of KDPG aldolase activity are present regardless of the carbon source. This enzyme plays a major role in the metabolism of pectin and aldohexuronate by *Erwinia* and other related organisms, as discussed in Chapter 10. In *E. coli* the *edd* and *eda* genes are closely linked to *zwf*, which codes for G-6-P dehydrogenase (Fig. 8-2). However, these genes are regulated under a separate set of regulatory controls. The *zwf* gene is subject to growth rate–dependent regulation at the level of transcription. On the other hand, the *edd-eda* operon is regulated by a gluconate-responsive promoter, P_1, located upstream of *edd*, which is responsible for induction of the Entner-Doudoroff pathway. High basal levels of KDPG aldolase are explained by constitutive transcription of *eda* from additional promoters (P_2, P_3, and P_4) within the *edd-eda* region but not from P_1.

The Entner-Doudoroff pathway is active in several *Pseudomonas* species. It appears to be the sole pathway for the metabolism of glucose in *Zymomonas mobilis* and a major pathway in other members of the pseudomonad group of organisms as well as other gram-negative species. However, *Z. mobilis* is unique in that it is the only genus known to utilize the Entner-Doudoroff pathway anaerobically. This organism lacks an oxidative electron transport system and is, therefore, obligately fermentative. The pathway is inefficient in that it yields only 1 mol of ATP per mol of hexose fermented. It is of special interest for the industrial production of alcohol, since the yield of ethanol approaches the theoretical 2 mol/mol of substrate. Rapid production of ethanol by *Z. mobilis* as the sole product of sugar fermentation results from the presence of pyruvate decarboxylase, an enzyme not frequently observed in bacteria.

A cyclic version of the Entner-Doudoroff pathway is used by *P. aeruginosa* to metabolize carbohydrates. In this pathway, the catabolism of mannitol occurs via glucose-6-phosphate and requires the activity of glucose-6-phosphate dehydrogenase. Mannitol is converted to fructose by mannitol dehydrogenase. Fructose is phosphorylated to F-6-P by fructokinase. Phosphoglucoisomerase converts F-6-P to G-6-P, which is then utilized through the Entner-Doudoroff pathway to form GA-3-P and pyruvate. The GA-3-P is then recycled to F-1,6-BP and F-6-P. The enzymes responsible for conversion of glucose to GA-3-P and pyruvate are coordinately regulated and induced by growth on glycerol, fructose, mannitol, glucose, and gluconate. The genes regulating this pathway, referred to as the *hex* regulon, are clustered in at least three operons near 39 minutes on the chromosome and are under the control of the *hexR* repressor.

Phosphoketolase Pathway

One major fermentation pathway involves the conversion of ribulose-5-phosphate to xylulose-5-phosphate (X-5-P). The X-5-P is then cleaved to form a C_3 compound (glyceraldehyde-3-phosphate) and a C_2 compound (acetyl phosphate) by the action of the phosphoketolase enzyme (Fig. 8-3). Glyceraldehyde-3-phosphate is metabolized via the triose phosphate portion of the EMP pathway to form lactate. Acetyl phosphate is converted to acetyl CoA, which is then reduced to ethanol. The conversion of glucose to pentose sugars serves as a major source of the reduced NADP that drives a myriad of biosynthetic reactions. However, this pathway is not essential for the growth of *E. coli*. Mutants blocked in the pentose phosphate pathway still grow with glucose as the carbohydrate source without other nutritional supplements because other routes are available for the formation of pentoses and reduced NADP (e.g., the malate enzyme and isocitrate dehydrogenase, as discussed later).

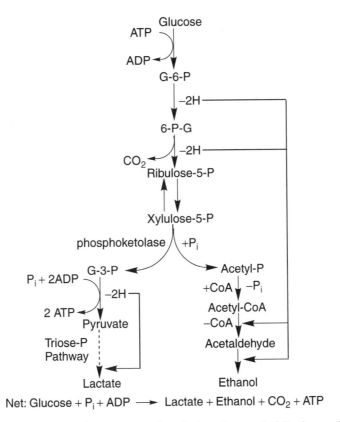

Fig. 8-3. Phosphoketolase or hexose monophosphate pathway. G-6-P, glucose-6-phosphate; 6-P-G, 6-phosphogluconate; G-3-P, glyceraldehyde-3-phosphate.

A number of microorganisms utilize the phosphoketolase pathway as the major route of glucose metabolism. *Leuconostoc mesenteroides*, a typical heterofermentative organism, utilizes this pathway, yielding lactate, ethanol, and carbon dioxide as shown in Figure 8-3. Within the genus *Lactobacillus* it has been possible to clearly differentiate **homofermentative** (*L. casei, L. pentosus*) from **heterofermentative** (*L. lysopersici, L. pentoaceticus, L. brevis*) types. Heterofermentative species are found in the genera *Streptococcus, Lactococcus, Pediococcus, Microbacterium*, some *Bacillus* species, and the mold *Rhizopus*. In *Lactobacillus pentoaceticus* and *Leuconostoc mesenteroides*, the basic pathway of glucose conversion leads to the formation of equimolar amounts of lactate, ethanol, and carbon dioxide. A commonly observed variation involves formation of considerable quantities of glycerol, as discussed in Chapter 11.

The pathway outlined in Figure 8-3 does not reveal some of the details of the reactions involved. In actuality, cleavage of the pentose molecule involves the cofactors thiamine pyrophosphate (TPP) and coenzyme A (CoA) as shown in the following series of reactions:

$$\text{xylulose-5-P} + \text{TPP} \longrightarrow \text{dihydroxyethyl-TPP} + \text{glyceraldehyde-3-P}$$

$$\text{dihydroxyethyl-TPP} + P_i \longrightarrow \text{acetyl-P} + \text{TPP}$$

$$\text{acetyl-P} + \text{ADP} \longrightarrow \text{acetate} + \text{ATP}$$

$$\text{glyceraldehyde-3-P} + P_i + 2\text{ADP} \longrightarrow \text{lactate} + 2\text{ATP}$$

Net : $\text{xylulose-5-P} + 2P_i + 3\text{ADP} \longrightarrow \text{acetate} + \text{lactate} + 3\text{ATP}$

If acetyl phosphate kinase is present, an additional ATP will be generated. In this series of reactions, TPP or diphosphothiamine (DPT) functions as a C_2 carrier in a similar manner to its function in the pyruvate dehydrogenase and α-ketoglutarate dehydrogenase complex of reactions in the TCA cycle.

Oxidative Pentose Phosphate Cycle

In some organisms a cyclic mechanism, the **oxidative pentose phosphate cycle**, accounts for the complete oxidation of carbohydrates (Fig. 8-4). In this cycle, G-6-P

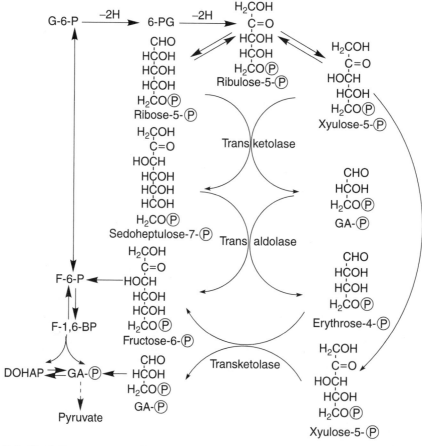

Fig. 8-4. Oxidative pentose phosphate cycle. Encircled P, phosphate group; G-6-P, glucose-6-phosphate; 6-PG, 6-phosphogluconate; F-6-P, fructose-6-phosphate; F-1,6-BP, fructose-1,6-bisphosphate; DOHAP, dihydroxyacetone phosphate; GA-P, glyceraldehyde-3-phosphate.

is converted to ribulose-5-phosphate and CO_2. Ribulose-5-phosphate is maintained in equilibrium with ribose-5-phosphate (R-5-P) and xylulose-5-phosphate (X-5-P) by the action of ribose phosphate isomerase and ribulose phosphate epimerase. Transketolase converts R-5-P and X-5-P to sedoheptulose-7-phosphate (SH-7-P) and glyceraldehyde-3-phosphate (GA-3-P). SH-7-P and GA-3-P are converted to F-6-P and erythrose-4-phosphate (E-4-P) via transaldolase. Transketolase also catalyzes the conversion of E-4-P and X-5-P to F-6-P and GA-3-P. By reversal of the FBP aldolase and G-6-P isomerase reactions, GA-3-P and F-6-P may be converted to G-6-P. The G-6-P can then reenter the oxidative pentose cycle. After one turn of the cycle, the net reaction is

$$\text{G-6-P} + 2\text{NADP}^+ \longrightarrow \text{R-5-P} + CO_2 + 2\text{NADPH} + 2\text{H}^+$$

Three turns of the cycle are required to produce one triose phosphate:

$$3\text{G-6-P} + 6\text{NADP}^+ \longrightarrow 3CO_2 + 2\text{F-6-P} + \text{GA-3-P} + 6\text{NADPH} + \text{H}^+$$

Repetitive action of the cycle could account for the complete oxidation of G-6-P:

$$\text{G-6-P} + 12\text{NADP}^+ \longrightarrow 6CO_2 + 12\text{NADPH} + 12\text{H}^+ + \text{P}_\text{i}$$

However, the cycle does not appear to function in this manner under normal conditions. It is important to note that glycolysis generates NADH, which can be reoxidized by linkage to the electron transport system, or under anaerobic conditions it can be used to reduce an oxidized substrate, such as pyruvate, to lactate.

By contrast, the pentose phosphate pathway generates NADPH, which is used primarily for reducing power in biosynthetic reactions (e.g., the conversion of α-ketoglutarate to glutamate or the incorporation of acetate into fatty acids) and is not linked to the terminal respiratory system. Operation of the oxidative pentose phosphate cycle provides for the formation of two very important biosynthetic precursors, SH-7-P and E-4-P, which serve as precursors to the aromatic amino acids as discussed in Chapter 15. The pentose phosphate cycle operates to yield a fermentative end product only under certain conditions and in a few unusual organisms. *Acetobacter xylinum* uses a variation of this pathway to produce acetic acid (see Chapter 11).

The oxidative pentose cycle (Fig. 8-4) assumes significance in the energy production of some organisms. Although the cycle is operative in *E. coli*, yeasts, streptomycetes, and fungi, it has been difficult to assess the relative importance of this pathway as compared to the combined operation of the EMP pathway and TCA cycle. The aerobic organism, *Gluconobacter suboxydans*, cannot ferment glucose and does not appear to contain the enzymes of the TCA cycle. Studies on isotope distribution into various products indicate that this organism uses a modification of the pentose phosphate cycle as a pathway of carbohydrate utilization.

The isotope distribution label from glucose labeled with ^{14}C in the C-3 and C-4 positions is helpful in determining which major metabolic pathways are utilized by a given organism. As shown in Figure 11-2, utilization of the EMP pathway leads to distribution of ^{14}C in the carbon dioxide formed. No other products are labeled. If the HMP pathway is used, there is no labeling of carbon dioxide, and the ethanol and lactate show an equal label distribution. If the Entner-Doudoroff pathway is utilized, half the label will be found in carbon dioxide and half in ethanol. Although this method of determining the route of carbohydrate metabolism has some limitations, especially if more than one pathway is operative at the same time, it provides an important basis for later studies using a variety of other techniques.

GLUCONEOGENESIS

Growth of microorganisms on so-called poor carbon sources, such as L-malate, succinate, acetate, or glycerol, requires the ability to synthesize hexoses needed for the production of cell wall mucopeptides, storage glycogen, and other compounds derived from hexose, such as pentoses, for nucleic acid biosynthesis. Hexose synthesis involves a reversal of carbon flow from pyruvate (**gluconeogenesis**). This could potentially be achieved by the reversal of the enzymes in the EMP glycolytic pathway (Fig. 8-1). However, of the major enzymatic reactions involved in glycolysis, three are insufficiently reversible to allow carbon flow from pyruvate in the direction of hexose synthesis.

First, pyruvate kinase is not reversible because the free-energy requirement is too great. Instead, the formation of PEP is catalyzed by **PEP carboxykinase**, the first committed step in gluconeogenesis (reaction 15 in Fig. 8-1). This Mg^{++}-dependent enzyme requires GTP as the phosphate donor:

$$oxaloacetate + GTP \longleftrightarrow PEP + CO_2 + GDP$$

A second irreversible enzyme is phosphofructokinase (reaction 5 in Fig. 8-1). To overcome this deterrent to gluconeogenesis, fructose-1,6-bisphosphatase (fructose-1,6-bisphosphate 1-phosphohydrolase, reaction 16 in Fig. 8-1) dephosphorylates FBP to yield F-6-P and P_i: The relative rates of glycolytic phosphofructokinase and gluconeogenic fructose bisphosphatase determine the direction of net carbon flux in the EMP pathway. In *E. coli* and *S. enterica*, mutants lacking fructose-1,6-bisphosphatase (encoded by *fbp*) cannot grow on L-malate, succinate, glycerol, or acetate (so-called gluconeogenic substrates). Organisms grown on pentoses can make hexose phosphates from GA-3-P.

The third bypass reaction required for gluconeogenesis involves dephosphorylation of G-6-P (reaction 17 in Fig. 8-1). Glucose-6-phosphatase removes P_i from G-6-P to yield free glucose. Thus, the formation of glucose from pyruvate requires a considerable expenditure of energy:

$$2 \text{ pyruvate} + 4ATP + 2GTP + 2NADH + 2H^+ + 4H_2O$$
$$\longrightarrow glucose + 2NAD^+ + 4ADP + 2GDP + 6P_i$$

Regulation

A major regulatory step in gluconeogenesis is PEP carboxykinase, encoded by *pckA* in *E. coli*. This enzyme is regulated by catabolite repression, a process in which gluconeogenesis is inhibited when glucose or other carbohydrate carbon sources are available. Maximum levels of PEP carboxylase are also induced at the onset of the stationary phase of growth, presumably to ensure the synthesis of adequate carbohydrate storage reserves or to provide metabolites from the upper part of the EMP pathway as the organism converts proteins to gluconeogenic amino acids. The stationary phase induction of PEP carboxykinase requires cyclic AMP as well as a regulatory signal, the nature of which has not been fully elucidated.

In *B. subtilis*, the genes governing the reactions in the glycolytic cycle have been identified as shown in Figure 8-1.

Glycogen Synthesis

Many organisms store glycogen as an energy reserve. Bacteria form glycogen using the enzyme ADP glucose pyrophosphorylase (reaction 18 in Fig. 8-1) and the branching enzyme glycogen synthase (1,4-α-D-glucan:1,4-α-D-glucan 6-α-D-glucanotransferase; reaction 19 in Figure 8-1):

$$\text{G-1-P} + \text{ATP} \longleftrightarrow \text{ADP glucose} + \text{PP}_i + \text{ADP glucose} + \alpha\text{-1,4-glucan}$$

$$\longrightarrow \alpha\text{-1,4-glucosylglycan} + \text{ADP}$$

In *E. coli*, mutants defective either in glycogen synthase or in ADP glucose pyrophosphorylase are unable to accumulate glycogen. Glycogen synthesis in *E. coli* is regulated by both the *relA* gene, which mediates the stringent response to amino acid starvation when the cells are using glucose but not when the cells are using glycerol, and by cyclic AMP. These two regulatory controls are independent of each other in that each regulatory process can be expressed in the absence of the other.

Glycogen synthesis in *E. coli* is regulated at the level of ADP glucose pyrophosphorylase (encoded by *glgC*). Glucose ADP synthetase, the first unique step in glycogen synthesis in this organism, is activated by glycolytic intermediates with FBP as the activator and AMP, ADP, and P_i as inhibitors. The ADP glucose synthetases of *E. coli* and *S. enterica* show considerable similarity in that both consist of four identical subunits and have the same spectrum of activators and inhibitors. Genetically, the *glg* genes of both organisms are clustered at the same point (75 min) on their respective genetic maps. A number of genes encoding catabolic, biosynthetic, and amphibolic enzymes in enteric bacteria are transcriptionally regulated by a complex catabolite repression-activation mechanism that involves enzyme III of the phosphotransferase system as one of the regulatory components. Comparable systems have been described for the regulation of gluconeogenesis in a wide variety of microorganisms from yeast to symbiotic nitrogen-fixing bacteria.

TRICARBOXYLIC ACID CYCLE

Sir Hans Krebs and co-workers demonstrated that C_4 dicarboxylic acids, such as succinate and malate, the C_5 compound α-ketoglutarate, and the C_6 compound citrate, were all oxidized by pigeon breast muscle. Citrate was synthesized from added oxaloacetate. The low yield of citrate was explained by the fact that both α-ketoglutarate and citrate were formed. Demonstration of succinate formation from fumarate and oxaloacetate in the presence of malonate (an inhibitor of succinate dehydrogenase) led to the conclusion that succinate could be formed by either oxidative or reductive reactions:

α-Ketoglutarate dehydrogenase

$$HOOC-CH_2-CH_2-CO-COOH \longrightarrow HOOC-CH_2-CH_2-COOH + CO_2 + 2H$$

Fumarate reductase

$$HOOC-CH=CH-COOH + 2H \longrightarrow HOOC-CH_2-CH_2-COOH$$

These observations led Krebs to propose the cyclic mechanism for the oxidation of pyruvate shown in Figure 8-5. He theorized that oxaloacetate condensed with a C_3 compound (presumably pyruvate) derived from glycolysis to yield a C_7 intermediate that was converted to citrate via decarboxylation. It was subsequently shown that a C_2 intermediate (acetyl-CoA) condensed with oxaloacetate to form citrate. The presence of the condensing enzyme (citrate synthase) has been confirmed in mammalian systems as well as in yeast and a wide variety of bacteria.

Malate represents a pivotal point in the cycle. It participates in several alternative reactions. It may be oxidized to oxaloacetate via the NAD-linked malate dehydrogenase (Fig. 8-5) as observed in *E. coli* or via a direct cytochrome-linked dehydrogenase as seen in *Pseudomonas* and *Serratia*. In either case, energy is generated. Many organisms can decarboxylate malate to pyruvate (malic enzyme), and subsequently carboxylate pyruvate, to form oxaloacetate. These reactions provide a scavenger system for reclaiming carbon dioxide (anapleurotic reactions). A few organisms apparently utilize this system for completion of the TCA cycle when the malate dehydrogenase enzyme is absent. However, the malic enzyme is not a normal link in the cyclic operation of the TCA cycle of most organisms. The fact that it is linked to NADP rather than NAD is of major significance in that it provides an alternate route of generating reducing activity for biosynthetic reactions that require reduced NADP.

The α-ketoglutarate dehydrogenase system is a complex containing three enzyme components: a thiamine pyrophosphate–dependent α-ketoglutarate dehydrogenase (Enz_1), dihydrolipoyl *trans*-succinylase (Enz_2), and an FAD-dependent dihydrolipoyl dehydrogenase (Enz_3):

$$\text{α-ketoglutarate} + \text{lipoate-}S_2\text{-Enz}_2 \xrightarrow{\textit{TPP-Enz}_1} \text{succinyl-S-lipoate-SH-Enz}_2 + CO_2$$

$$\text{succinyl-S-lipoate-SH-Enz}_2 + \text{CoA} \xrightarrow{\textit{Enz}_2} \text{succinyl-CoA} + \text{lipoate-}(SH)_2\text{-Enz}_2$$

$$\text{lipoate-}(SH)_2\text{-Enz}_2 + NAD^+ \xrightarrow{\textit{FAD-Enz}_3} \text{lipoate-}S_2\text{-Enz}_2 + NADH + H^+$$

Net: α-ketoglutarate + CoA + NAD^+ → succinyl-CoA + CO_2 + NADH + H^+

This complex of enzymes is comparable to the pyruvate dehydrogenase complex that catalyzes the oxidative decarboxylation of pyruvate at the initial stage of the TCA cycle (Fig. 8-5). However, the individual components differ from each other in physicochemical properties and specificity except for the third enzyme in which the components are identical and functionally interchangeable. Studies with mutants deficient in various components of the α-ketoglutarate dehydrogenase complex indicate that dihydrolipoyl components of both systems are encoded by *lpd*.

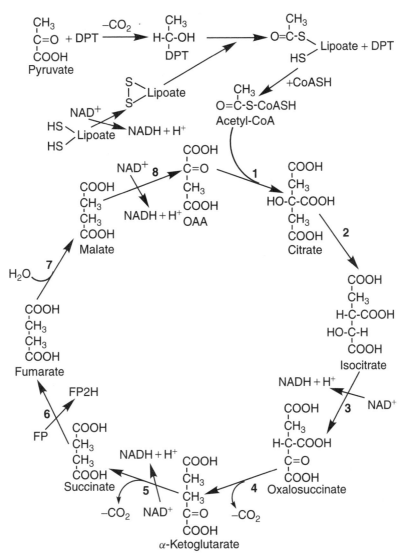

Fig. 8-5. The Krebs tricarboxylic (citric) acid cycle. OAA, oxaloacetate; DPT, diphosphoth-iamine; FP, flavoprotein. Structural genes for the enzymes in *Bacillus subtilis* are indicated by italicized three-letter designations: 1. Citrate synthase (*citZ*). 2. Aconitase (*citB*). 3 and 4. Isocitrate dehydrogenase (*citC*). 5. α-ketoglutarate dehydrogenase (*odhA, pdhA*). 6. Succinate dehydrogenase (*sdhA*). 7. Fumarase (*citG*). 8. Malate dehydrogenase (*citH*). The conversion of pyruvate to CO_2 and acetyl-CoA shown at the top of the diagram involves a group of enzymes known as the pyruvate dehydrogenase complex.

Under anaerobic conditions the TCA cycle no longer functions as such because the links to terminal respiration are required to maintain the activities and synthesis of succinate dehydrogenase and the α-ketoglutarate dehydrogenase complex. However, synthesis of these intermediates is required for biosynthetic reactions. Fumarate reductase activity is increased, providing a mechanism for continued succinate synthesis. Thus, the TCA cycle now functions as a branched biosynthetic pathway: one branch operating as a reductive pathway reversing the sequence from succinate to oxaloacetate and the other branch continuing to operate oxidatively to convert oxaloacetate to α-ketoglutarate as shown in Figure 8-6.

In actuality, the activity levels of a large number of enzymes that serve primarily aerobic functions are markedly reduced when *E. coli* is grown under anaerobic conditions. In this organism, a two-component signal transduction system, consisting of a transmembrane sensor protein, ArcB, and a cytoplasmic regulatory protein, ArcA, controls the expression of genes encoding enzymes involved in aerobic respiration. When oxygen is excluded, the Arc (aerobic respiratory control) system represses the expression of the structural genes for several flavoprotein-linked dehydrogenases, the cytochrome *o* complex, enzymes of the TCA cycle, glyoxylate shunt, and fatty acid degradation. Conversely, the oxygen-scavenging cytochrome *d* oxidase (encoded by the *cydAB* operon) is activated by the Arc system. The Fnr protein functions as an anaerobic repressor of both the cytochrome *o* oxidase complex (encoded by *cyoABCDE*) and cytochrome *d* oxidase (encoded by *cydAB*). Operation of this regulatory system is discussed in some detail in Chapter 5.

In Chapter 1 it is stated that all of the components of the cell are synthesized from only 12 precursor metabolites. Thus, the TCA cycle provides both reducing equivalents

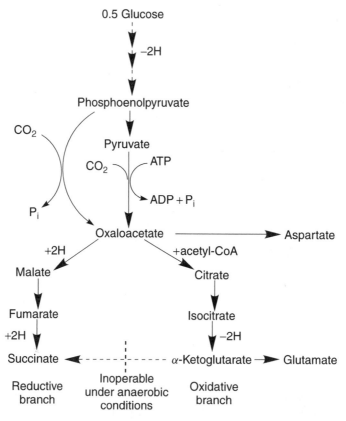

Fig. 8-6. Operation of the TCA cycle under anaerobic conditions.

to the terminal respiratory system and intermediates for the biosynthesis of amino acids and other vital cell constituents. Synthesis of these compounds at the expense of TCA cycle intermediates tends to diminish the activity of the cycle. However, these metabolites are replenished through carbon dioxide fixation by the following reactions:

Pyruvate carboxylase

$$\text{pyruvate} + CO_2 + ATP + H_2O + Mg_2^+ \longrightarrow \text{oxaloacetate} + ADP + P_i$$

Phosphoenolpyruvate carboxykinase

$$PEP + CO_2 + H_2O \longrightarrow \text{oxaloacetate} + P_i$$

Malic enzyme

$$\text{pyruvate} + CO_2 + NADPH + H^+ \longleftrightarrow \text{L-malate} + NADP^+$$

Heterotrophic organisms require at least one enzyme of this nature in order to grow aerobically on hexoses or glycolytic intermediates. The generation of oxaloacetate permits the continued flow of hexose carbon through the TCA cycle under conditions in which intermediates are being removed for biosynthesis. The same purpose is also served when oxaloacetate is used to initiate gluconeogenesis. Enzymes that serve in this capacity are termed **anapleurotic** (from the Greek, meaning "to fill up").

Transaminase reactions may also serve to generate oxaloacetate (from aspartate) or α-ketoglutarate (from glutamate). The aspartase reaction can provide oxaloacetate by producing fumarate that can be reduced to oxaloacetate. The glyoxylate cycle, discussed in the next section, can serve an anapleurotic role by allowing a bypass of carbon dioxide–releasing reactions. However, the glyoxylate cycle can function in this capacity only in organisms capable of utilizing acetate following induction on this substrate. Certain organisms deficient in pyruvate carboxylase activity, such as *Arthrobacter pyridinolis*, exhibit a nutritional requirement for malate in order to grow on glucose.

GLYOXYLATE CYCLE

Activation of acetate with CoA to form acetyl-CoA and the combined activities of the enzymes isocitritase (isocitrate lyase) and malate synthase provide for the operation of a C4 cycle called the glyoxylate cycle as shown in Figure 8-7. Isocitrate lyase converts isocitrate to succinate and glyoxylate. Malate synthase couples acetyl-CoA and glyoxylate to form malate. Operation of these enzymes results in the net formation of malate from 2 mol of acetate:

$$\text{isocitrate} \rightarrow \text{succinate} + \text{glyoxylate}$$
$$\text{glyoxylate} + \text{acetate-CoA} \rightarrow \text{malate} + \text{CoA}$$
$$\text{succinate} + \text{acetate} - 2(2H) \rightarrow\rightarrow\rightarrow \text{isocitrate}$$

Net: 2 acetate $- 2(2H) \rightarrow$ malate

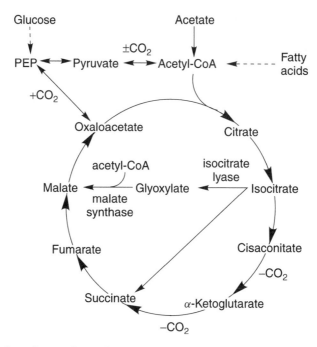

Fig. 8-7. The glyoxylate cycle or glyoxylate shunt. Note that the two CO_2-evolving steps of the TCA cycle are bypassed. Either glucose or fatty acids can serve as sources of acetate.

Activity of the glyoxylate bypass explains the ability of bacteria, yeast, and other microorganisms to utilize acetate as a sole source of carbon for growth.

The enzymes of the glyoxylate cycle are repressed by the presence of glucose or another more rapidly utilized substrate (see Chapter 5). As discussed previously, the Arc (anaerobic respiratory control) system and other regulatory factors repress the TCA cycle and the glyoxylate cycle under anaerobic conditions. Because of the very low redox potential of ferredoxin (see Fig. 9-6), many anaerobes and certain photosynthetic bacteria can form pyruvate, and in some cases, α-ketoglutarate, via reductive decarboxylation reactions. By this means, these organisms can circumvent the irreversibility of the oxidative decarboxylation of pyruvate to acetyl-CoA and CO_2 and of α-ketoglutarate to succinyl-CoA and CO_2 observed in other organisms. Thus, the utilization of acetate can occur in organisms that utilize ferredoxin in reductive metabolism.

The yeast *Saccharomyces cerevisiae* utilizes a reaction analogous to the conversion of isocitrate to glyoxylate and succinate in the metabolism of propionyl-CoA via a 2-methylcitrate cycle. This cycle is initiated by the formation of 2-methylcitrate from propionyl-CoA and oxaloacetate. 2-Methylcitrate is then converted into 2-methylisocitrate, which is subsequently cleaved to form pyruvate and succinate by a mitochondrial 2-methylisocitrate lyase encoded by *ICL2*. This reaction is very similar to the conversion of isocitrate to succinate and glyoxylate by the *ICL1*-encoded isocitrate lyase. The 2-methylcitrate cycle may be involved in the degradation of carbon skeletons of certain amino acids. For example, oxidative decarboxylation of 2-ketoisobutyrate, an intermediate in threonine catabolism, yields propionyl-CoA.

BIBLIOGRAPHY

Gombert, A. K., M. M. dos Santos, B. Christensen, and J. Nielsen. 2001. Network identification and flux quantification in the central metabolism of *Saccharomyces cerevisiae* under different conditions of glucose repression. *J. Bacteriol.* **83:**1441–51.

Hager, P. W., M. W. Calfee, and P. V. Phibbs. 2000. The *Pseudomonas aeruginosa* devB/SOL homolog, *pgl*, is a member of the *hex* regulon and encodes 6-phosphogluconolactonase. *J. Bacteriol.* **182:**3934–41.

Inui, M., K. Nakata, J. H. Roh, K. Zahn, and H. Yukawa. 1999. Molecular and functional characterization of the *Rhodopseudomonas palustris* no. 7 phosphoenolpyruvate carboxykinase gene. *J. Bacteriol.* **181:**2689–96.

LaPorte, D. C., S. P. Miller, and S. K. Singh. 2000. Glyoxylate bypass in *Escherichia coli.* In J. Lederberg (ed.), *Encyclopedia of Microbiology*, 2nd ed., Vol. 2. Academic Press, San Diego, pp. 556–61.

Luttik, M. A., P. Kötter, F. A. Salomons, I. J. van der Klei, J. P. van Dijken, and J. T. Pronk. 2000. The *Saccharomyces cerevisiae ICL2* gene encodes a mitochondrial 2-methylisocitrate lyase involved in propionyl-coenzyme A metabolism. *J. Bacteriol.* **182:**7007–13.

Peekhaus, N., and T. Conway. 1998. What's for dinner? Entner-Doudoroff metabolism in *Escherichia coli.* *J. Bacteriol.* **180:**3495–502.

Preiss, J. 2000. Glycogen biosynthesis. In J. Lederberg (ed.), *Encyclopedia of Microbiology*, 2nd ed., Vol. 2. Academic Press, San Diego, pp. 541–55.

Romano, A. H., and T. Conway. 1996. Evolution of carbohydrate metabolic pathways. *Res. Microbiol.* **147:**448–55.

Temple, L. M., A. E. Sage, H. P. Sweizer, and P. V. Phibbs, Jr. 1998. Carbohydrate metabolism in *Pseudomonas aeruginosa.* In T. C. Monti (ed.), *Pseudomonas.* Plenum Press, New York, pp. 35–72.

Tobish, S., D. Zühlke, J. Bernhardt, J. Stülke, and M. Hecker. 1999. Role of CcpA in regulation of the central pathways of carbon catabolism in *Bacillus subtilis.* *J. Bacteriol.* **181:**6996–7004.

Vinopal, R. T., and A. H. Romano. 2000. Carbohydrate synthesis and metabolism. In J. Lederberg (ed.), *Encyclopedia of Microbiology*, 2nd ed., Vol. 1. Academic Press, San Diego, pp. 647–68.

Walker, G. M. 2000. Yeasts. In J. Lederberg (ed.), *Encyclopedia of Microbiology*, 2nd ed., Vol. 4. Academic Press, San Diego, pp. 939–53.

CHAPTER 9

ENERGY PRODUCTION AND METABOLITE TRANSPORT

ENERGY PRODUCTION

The second law of thermodynamics states that systems will *spontaneously* change from states of higher order to lesser order. The cell, however, must do the opposite to survive. It must make order out of chaos. To accomplish this feat, cells first unleash and then harness the energy trapped within molecules encountered in the environment and use that energy to increase order (e.g., synthesize macromolecules or move chemicals into or out of the cell against a gradient). In this chapter, we discuss how cells steal energy from the environment.

Any discussion of energy must begin with a definition of **entropy**. Entropy (S), simply put, is the *measure* of disorder; the larger the value of S, the greater the disorder. Chemical reactions that cause a large *increase* in S [where the difference between product and precursor entropies (ΔS) is greater than zero] are favored and will occur spontaneously with the *release* of energy. Reactions that cause a *decrease* in S [ΔS < 0] increase order. These reactions require the input of energy and thus cannot occur spontaneously. An accounting device used to deduce the entropy change that occurs following a reaction is called the **Gibbs free energy (G).** G combines all the physical properties of a system including heat, pressure, volume, and energy. The free-energy *change* (ΔG) during a reaction (i.e., the G of the products minus the G of the starting materials) is a measure of the *disorder* created when the reaction occurs. Energetically favored reactions are those in which a large amount of free energy is released (i.e., they have a large *negative* ΔG), creating much disorder. In contrast, reactions with a large *positive* ΔG (such as would occur during peptide bond formation) create order and thus cannot occur spontaneously.

The goal of the cell is to harness the $-\Delta$G that results from converting glucose (a highly ordered compound) into CO_2 and H_2O and to use that energy to drive biosynthetic reactions that have a positive ΔG. Since ATP is the fuel used for most anabolic reactions, a major objective of the cell is to synthesize this high-energy nucleotide. The ΔG at equilibrium (ΔG^0) for ATP $\rightleftharpoons$ ADP + P$_i$ is -7.3 Kcal/mol. However, at normal cellular concentrations where there are 10 ATP molecules per ADP [(ATP) = 10(ADP)], ΔG^0 is even higher at -11 to -13 Kcal/mol. The cell must be able to *couple* the release of this much energy from glucose to the synthesis of ATP. This is accomplished by substrate-level phosphorylation and oxidative phosphorylation as discussed below.

The energy captured within ATP can then be harnessed to create order in the form of biosynthetic reactions. In a hypothetical enzyme reaction that converts substrates A$-$H and B$-$OH to A$-$B and H$_2$O, the energy from ATP hydrolysis is first used to convert B$-$OH to a higher-energy intermediate, B$-$O$-$PO$_4$. This compound is only transiently formed, with the energy released during its decay used by the enzyme to form A$-$B. Thus, the energy released from the ATP hydrolysis reaction (large $-\Delta G$) is coupled to the synthesis reaction (large $+\Delta G$). In this way, the cell can progressively create order.

Substrate-Level Phosphorylation

During glycolysis, energy in the form of ATP is generated by substrate-level phosphorylation. The oxidative steps in glycolysis or other comparable metabolic systems give rise to this energy. In the EMP pathway (Fig. 8-1) oxidation of glyceraldehyde-3-phosphate and the incorporation of P$_i$ result in the formation of 1,3-diphosphoglycerate. One of the phosphate bonds achieves a high-energy state. The energy inherent in this bond is used to transfer the P$_i$ to ADP to form ATP. Subsequent conversion of the 3-phosphoglycerate from this reaction to phosphoenol pyruvate again provides for the transfer of energy by the formation of ATP from ADP (Fig. 9-1).

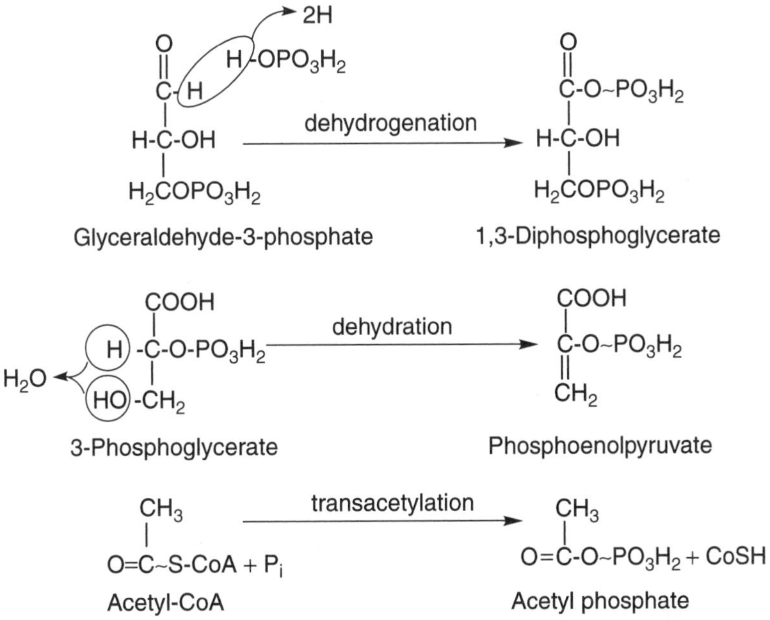

Fig. 9-1. Reactions involved in substrate-level formation of high-energy phosphate bonds.

The special physicochemical properties of certain compounds such as ATP enable them to transfer energy into an energy-requiring system. In this case, the energy transfer is associated with the transfer of a chemical unit — the phosphoryl group. In other systems, the energy transfer is associated with the transfer of electrons and protons in oxidation–reduction reactions. Although the energy of high-energy compounds is often said to reside in the P–O bond, it is more correct to refer to the **phosphoryl transfer potential** associated with phosphorylated compounds. The nucleoside triphosphates (ATP, GTP, etc.) are the most universal compounds with energy transfer potential in biological systems and serve as the immediate energy sources for biosynthesis as well as physicochemical activities, such as flagellar motor rotation or active transport. In the oxidation of triose phosphate, as shown in Figure 9-1, some of the details are not shown. In this reaction, the combined events of thiol addition, hydrogen transfer, and phosphorolysis by glyceraldehyde-3-phosphate dehydrogenase lead to formation of a phosphate bond that may be transferred to ADP:

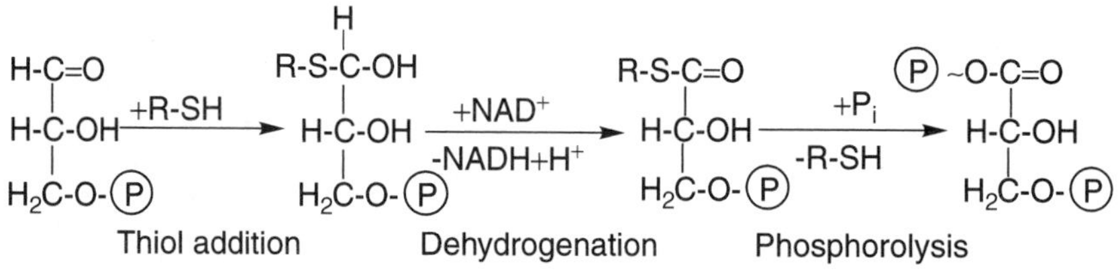

This reaction is initiated by condensation of the thiol (-SH) group at a specific cysteine residue at the catalytic site of the enzyme with the aldehyde carbonyl. Since thiol esters yield considerably more energy on hydrolysis than ordinary oxygen esters, this facilitates the addition of the phosphate group to the carbonyl group. Generation of high-energy phosphate bonds by reactions of this type are referred to as **substrate-level phosphorylations** to distinguish them from those generated by electron transport.

The number of moles of ATP generated will depend on the cleavage pathway utilized during the degradation of carbohydrates (Table 9-1). For example, pentose fermentation proceeds via C_2-C_3 cleavage to yield lactate and acetate as products:

$$\text{pentose} + 2P_i + 2ADP \longrightarrow \text{lactate} + \text{acetate} + 2ATP + 2H_2O$$

As shown in the last part of Figure 8-3, pentose fermentation will yield 2 mol of ATP per mol of pentose used if acetyl phosphate is formed as an intermediate. Phosphoketolase splits the pentose into acetyl phosphate and glyceraldehyde-3-phosphate. Acetyl phosphate can be converted to acetate by acetate kinase, yielding 1 mol of ATP. The glyceraldehyde-3-phosphate is utilized via the triose portion of the EMP pathway to yield 2 additional molecules of ATP. The phosphoketolase reaction has been shown to occur in *Acetobacter xylinum*, *Lactobacillus* (heterofermentative species), and *Leuconostoc*. In the genus *Bifidobacterium*, hexose phosphoketolase cleavage of F-6-P yields acetyl phosphate and erythrose-4-phosphate (E-4-P):

$$\text{F-6-P} + P_i \longrightarrow \text{acetyl-P} + \text{E-4-P}$$

Pentose phosphate is formed by the actions of transaldolase and transketolase. The pentose thus formed is then split by the phosphoketolase to form acetyl phosphate

TABLE 9-1. Energy Yields from Various Fermentation Pathways

Substrate	Cleavage Type	Products	Moles Triose-P Formed	Net ATP Yield (mol)
Hexose	FBP-aldolase[a]	2 lactate	2	2
Hexose	Phosphoketolase	1 lactate, 2 acetate, 1 CO_2	1	1
Pentose	Phosphoketolase	2 lactate, 1 acetate	1	2
Hexose	KDPG-aldolase[b]	2 ethanol, 2 CO_2	1	1
Aldonic	FBP-aldolase + KDPG-aldolase	1.83 lactate, 0.5	1	1.33[c]

[a] FBP-aldolase, fructose-1,6-bisphosphate aldolase.
[b] KDPG-aldolase, 2-keto-3-deoxy-6-phosphogluconate aldolase.
[c] Resting cells produce 1.5 lactate + 0.5 C (actually missing, but calculated as acetate via phosphoketolase. 1.83 lactate formed in the presence of arsenite.

Source: From Gunsalus, I. C., and C. W. Shuster, 1961. Energy-yielding metabolism in bacteria. In *The Bacteria*, Vol. II, Metabolism. I. C. Gunsalus and R. Y. Stanier (eds.). Academic Press, New York.

and glyceraldehyde-3-phosphate as shown previously. The hexose phosphoketolase reaction appears to be unique in the bifidobacteria and has not been found in other heterofermentative organisms. Anaerobic glycolytic organisms do not appear to contain these phosphoketolases.

Other energy-yielding reactions at the substrate level have been described. In *Clostridium cylindrosporum* the fermentation of purines yields ATP via the reaction:

$$HN{=}CH{-}NH{-}CH_2COOH + ADP + P_i + 2H_2O$$

$$\longrightarrow HCOOH + NH_3 + CH_2NH_2COOH + ATP$$

Reactions that lead to the formation of carbamoyl phosphate via energy coupling may eventually yield energy through the transfer of the phosphate group to ADP:

$$HN_2COOPO_3H_2 + ADP \longrightarrow NH_3 + CO_2 + ATP$$

The contribution of inorganic pyrophosphate (PP_i) and polyphosphate to the overall energy yield may be considerable. Although earlier workers had suspected that PP_i could serve as an energy source, most biochemists considered that hydrolysis of PP_i served to render irreversible reactions in which PP_i was formed. As discussed later in relation to the high-energy yields observed in propionic acid fermentations, PP_i can be used in place of ATP as a source of energy in certain fermentation reactions. Such reactions may be used more widely than is currently appreciated.

Oxidative Phosphorylation

When a carbohydrate is oxidized via a respiratory mechanism, energy is generated by passing electrons through a series of electron acceptors and donors until they ultimately reach a final e^- acceptor such as O_2 or nitrate. The energy inherent in the carbohydrate is gradually released during this series of coupled oxidation–reduction reactions and used to pump protons out of the cell via the membrane-bound cytochrome systems. Since membranes are impermeable to protons, this establishes an electrochemical gradient ($\Delta\mu H^+$) or proton motive force across the cell membrane. Proton motive force (Δp) is composed of electrical (charge) and chemical (pH) components according to the following relationship:

$$\Delta p = \frac{\Delta\mu H^+}{F} = \Delta\Psi - \frac{2.3RT}{F}\Delta pH$$

where $\Delta\Psi$ represents the transmembrane electrical potential and ΔpH is the pH difference across the membrane. R, T, and F are the gas constant, the absolute temperature, and the Faraday constant, respectively. Thus, (2.3RT/F) at $36\,^{\circ}C$ equals 60 mV, and the total driving force across the membrane generated by proton extrusion in millivolts (mV) is $\Delta\mu H^+ = \Delta\Psi - 60$ mV (ΔpH). The cell uses this energy to synthesize ATP. According to chemiosmotic theory, the two components of Δp are thermodynamically equivalent. A change of ΔpH by 1 unit results in the same change in the ATP/ADP equilibrium as a shift of $\Delta\Psi$ by 60 mV. PMF is also used directly to power flagellar movement and some transport systems.

Measurement of PMF

Electrical potential ($\Delta\Psi$) across a cell membrane can be measured using lipophilic cations that will freely pass through the membrane. Since the interior side of the membrane is negatively charged, lipophilic cations such as tetraphenyl phosphonium ion (TPP^+) will accumulate inside a cell depending on the extent of the negative charge. In a similar manner, ΔpH can be measured experimentally using radiolabeled weak acids such as benzoic acid. Weak acids disassociate based on their disassociation constants (pKa) as follows:

$$HA \rightleftharpoons H^+ + A^-$$

The protonated form (HA) passes freely back and forth across the membrane, whereas the deprotonated form (A^-) does not. At a pH value equal to the pKa, 50% of the acid will be disassociated. At 1 pH unit above the pKa, 90% will be disassociated, whereas at 1 pH unit below the pKa, only 10% will be disassociated.

So, 10% of a weak acid with a pKa of 5.5 will be protonated at pH 6.5. If cells are also present, the HA will pass into the cells, equilibrate with external HA, and disassociate intracellularly according to the intracellular pH. The disassociated form will become trapped and thus cannot equilibrate with the outside. This disassociation concomitantly lowers internal HA, which will reequilibrate with external HA, bringing more HA into the cell. This HA also disassociates according to the pH of the cell, continuing the cycle. Consequently, by measuring the accumulation of a radiolabeled weak acid, internal pH can be calculated according to the following formula:

$$pH_i = \log\left[\frac{\text{total acid concentration in}}{\text{total acid concentration out}}(10^{pK} + 10^{pH\ out}) - 10^{pK}\right]$$

Under aerobic conditions at pH 6, the PMF generated is approximately 190 mV with $\Delta\Psi$ and ΔpH contributing equally. However, what happens when $\Delta pH = O$ — that is, $pH_i = pH_o$? In this situation, all of PMF is derived from $\Delta\Psi$, the electrical component. In fact, $\Delta\Psi$ increases to compensate for the loss of ΔpH. This illustrates that the two components of PMF can be exchanged according to the needs of the cell. Ion circulations involving the exchange of protons for other cations such as Na^{2+} or K^+ are responsible for the interconversion between $\Delta\Psi$ and ΔpH (see chemiosmotic-driven transport in this chapter).

Electron Transport Systems

The cytoplasmic membranes of bacteria contain the electron transfer systems (ETS) that generate PMF by coupling the oxidation of NADH and other substrates to the expulsion of protons. The electron transport systems are composed of heme-containing components (cytochromes), iron-sulfur cluster enzymes, flavoproteins (containing FMN), and quinolones. A typical electron transport chain is depicted in Figure 9-2. Remember that different organisms, and even *E. coli* under different conditions, will use an ETS that employs a variety of substrates and terminal electron acceptors. Thus, the respiratory systems can be visualized as being branched at the dehydrogenase and terminal oxidase sites (Fig. 9-3). The electron transfer reactions are initiated by specific modular units. These are NADH, lactate, succinate, formate, or glycerol-3-phosphate dehydrogenases. Figure 9-4 illustrates how NAD-dependent dehydrogenases stereospecifically transfer a substrate-bound H (as deuterium) with two e^- to NAD^+, forming NADH. Some enzymes, such as succinate dehydrogenase, contain flavoproteins and are linked directly to the next step in the system, bypassing the need for NAD^+. FAD functions in flavoprotein-linked reactions as shown in Figure 9-5. Each oxidation–reduction system has a definite oxidation–reduction potential (E'_o) at which it functions. By determining the electrode potential for each system, the order in which each functions can be deduced (Fig. 9-6).

The quinone pool acting as a universal adapter collects the electrons from the different donor modules. The quinone pool passes the electrons on to the appropriate acceptor module depending on what terminal electron acceptor is being used (e.g., O_2, nitrate, fumarate). Quinones are mobile hydrogen carriers that shuttle between the large dehydrogenase and oxidase complexes. The predominant quinone, ubiquinone-8, functions in aerobic respiration while anaerobic respiratory chains may require menaquinone-8. The ETS system in Figure 9-2 begins with NADH dehydrogenase,

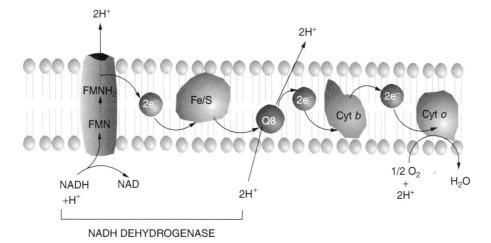

Fig. 9-2. An electron transport chain of *E. coli*.

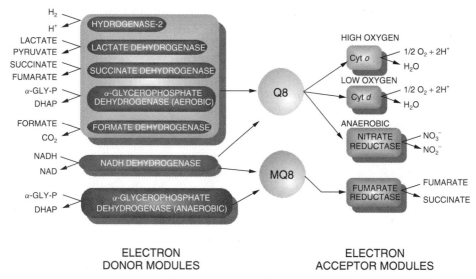

Fig. 9-3. Menu of electron donor and acceptor modules for *E. coli*. Selection (induction) of module combinations depends on the substrate available (left side of figure) and the redox potential of the final electron acceptor (right side of figure). Q8, ubiquinone; MQ8, menaquinone.

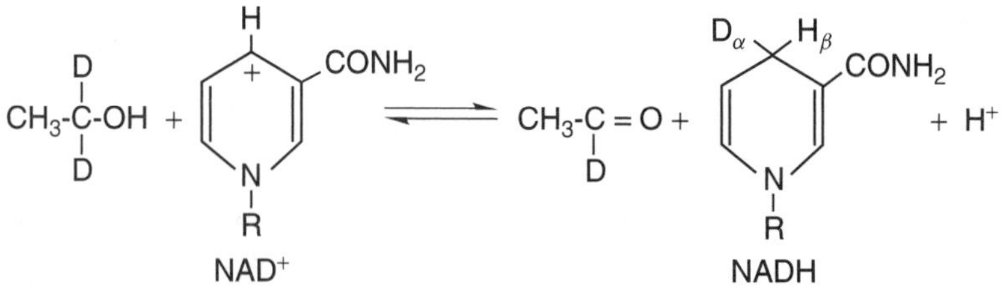

$$CH_3\text{-}\underset{\underset{D}{|}}{\overset{\overset{D}{|}}{C}}\text{-OH} + \quad \rightleftharpoons \quad CH_3\text{-}\underset{\underset{D}{|}}{C}=O + \quad + H^+$$

NAD$^+$ $\qquad\qquad\qquad\qquad\qquad$ NADH

Fig. 9-4. Deuterium labeling of NAD$^+$ during reduction by alcohol dehydrogenase. In the presence of alcohol dehydrogenase, deuterium (D) from deuterium-labeled alcohol is added stereospecifically to NAD$^+$. Under comparable circumstances NADP$^+$, which contains an additional phosphate group on the pentose of the adenylic acid moiety, reacts similarly.

a flavoprotein complex containing FMN and nonheme iron-sulfur clusters. NADH transfers two electrons and protons to FMN, forming FMNH$_2$. Transfer of the electrons to the Fe/S group causes the release of the 2 protons (H$^+$) to the outside, generating PMF. The electrons in the Fe/S group are then transferred to ubiquinone. Ubiquinone picks up the 2e$^-$, as well as 2H$^+$ from the cytosol, to form hydroquinone.

Under conditions of high oxygenation, hydroquinone will transfer its electrons to the cytochrome-o complex and release 2H$^+$ to the cell exterior, adding to PMF. The cytochrome-o complex is composed of four polypeptides that make up two spectral signals, termed cytochromes b_{555} and b_{562}. The subscript numbers correspond to maximum wavelength absorbance values. It may be difficult to visualize how proton translocation across the membrane can occur during the transfer of e$^-$ from hydroquinone to cytochrome o. It seems the ubiquinol oxidase site of the Cyt o complex faces the periplasm while the dioxygen reductase site faces the cytoplasm. Thus, the

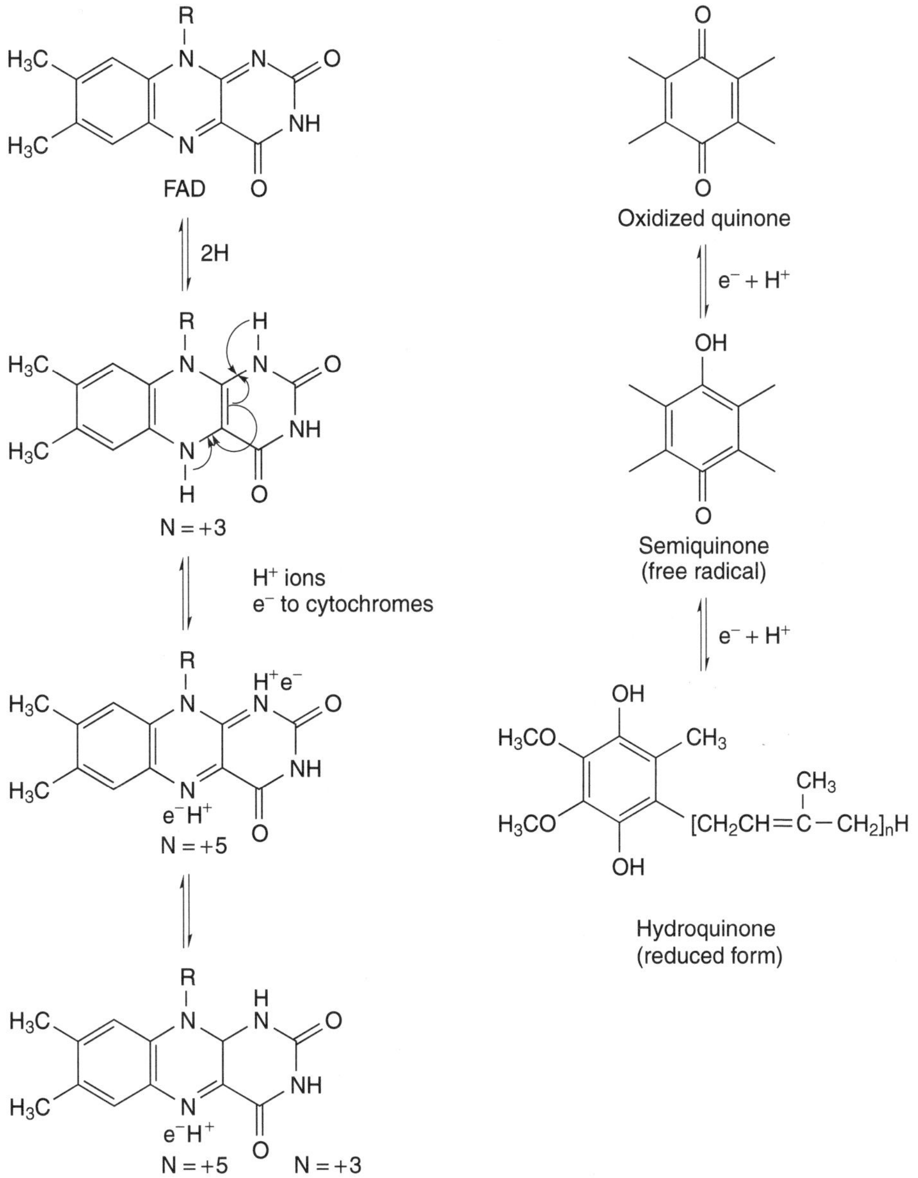

Fig. 9-5. Functions of flavin adenine dinucleotide (FAD) and ubiquinone in hydrogen (H$^+$) and electron (e$^-$)transfer reactions.

release of two protons to the periplasm is coupled to the consumption of two protons in the cytoplasm to convert $\frac{1}{2}O_2 + 2H^+ + 2e^-$ to H_2O.

When *E. coli* grows under low oxygen conditions, it requires a terminal oxidase with a higher affinity for oxygen. Thus, the high-affinity cytochrome *d* complex is induced and the lower-affinity Cyt *o* complex is repressed. A combination of regulatory proteins including Arc (see Chapter 18) are involved in this control. The Cyt *d* complex utilizes cytochrome b_{558} and cytochrome *d*.

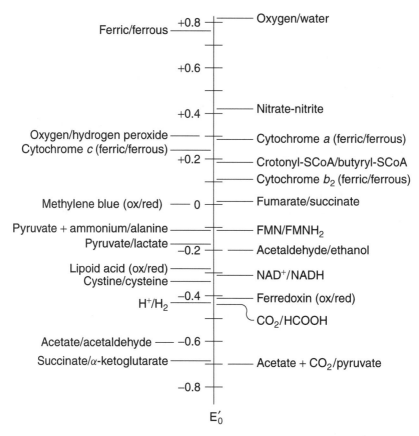

Fig. 9-6. Electrode potentials of some important biological systems. E_0, oxidation–reduction potential in volts.

Anaerobic Respiration

When placed under anaerobic conditions, enteric bacteria (such as *Escherichia coli*) cleave pyruvate to acetyl CoA and formate. If there are no appropriate alternative electron acceptors available, acetyl CoA and formate will be converted to fermentative end products (Fig. 9-7). But as noted above, many organisms such as *E. coli* can respire even under anaerobic conditions if alternative electron acceptors are present. Each system for accepting electrons can be viewed as a modular unit. Some of these acceptor modules are shown in Figure 9-3. They are complex membrane-associated enzymes. For example, fumarate reductase has an FAD-containing subunit, an iron-sulfur protein, and two membrane anchor proteins. Nitrate reductase is even more complex. A given modular system is induced when cells are grown in medium containing a specific acceptor molecule. There is, in fact, a hierarchy of transcriptional control corresponding to the potential energy that can be produced and the electrical potential of each acceptor.

The Fnr, NarL, and Arc systems described in Chapter 18 control the expression of the respiratory genes. Anaerobic growth on specific acceptor molecules also results in the synthesis of cytochromes specific for each acceptor module. Thus, the cytochrome *b* used for fumarate reductase is different from that used for nitrate reductase. As

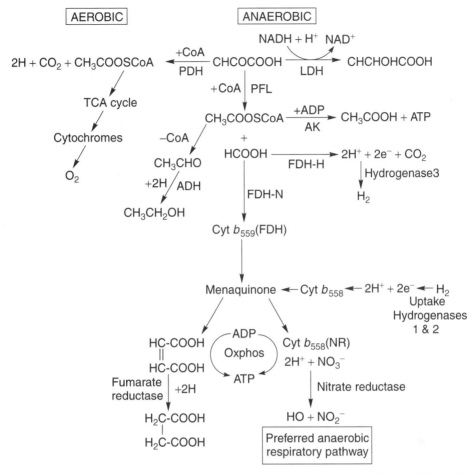

Fig. 9-7. Alternate pathways of pyruvate utilization under aerobic and anaerobic conditions. PDH, pyruvate dehydrogenase; PFL, pyruvate formate lyase; TCA, tricarboxylic acid cycle; FDH-N, formate dehydrogenase-N; FDH-H, formate dehydrogenase-H; Oxphos, oxidative phosphorylation; AK, acetate kinase; LDH, lactate dehydrogenase; ADH, alcohol dehydrogenase.

illustrated when oxygen was the terminal e^- acceptor, the transmission of electrons to nitrate or fumarate also translocates protons to the cell exterior, generating PMF. Figure 9-7 illustrates the alternative pathways for pyruvate that are possible depending on the type of terminal electron acceptor present.

Conversion of PMF to Energy

How PMF is converted into usable energy by the cell is certainly one of the keys to understanding life. There are several ways the cell can use PMF. The cell can directly generate ATP from PMF by reversing the action of the major H^+-translocating ATPase. These are the so-called $\mathbf{F_1F_0}$**-type ATP synthases** that are widely distributed in the bacterial kingdom. This type of ATPase is composed of two structurally and functionally distinct entities. The F_1 moiety lies extrinsic to the membrane in the cytoplasm and catalyzes the hydrolysis of ATP to ADP. F_1 is attached to the membrane by a pore complex termed F_0. F_0 serves as the conductor of H^+ across the membrane. Protons passing into the cell (i.e., down the electrochemical gradient) through F_1F_0 will drive F_1 to synthesize ATP (see Fig. 1-13 and Fig. 9-8). It takes three to four H^+ to generate one ATP.

PMF can also be used to drive the transport of some metabolites into the cell. Lactose, for instance, is transported into the cell by the integral membrane permease LacY. The energy required to pump lactose into the cell against an increasing lactose gradient comes from the symport of H^+ with lactose and the partial dissipation of the PMF. Additionally, the flagellar motor is driven by PMF. Each flagellar rotation requires the influx of $256H^+$!

Another consideration is how a facultative organism such as *E. coli* generates PMF under anaerobic conditions in the absence of any electron acceptor molecules — that is, under fermentative conditions. Even under these conditions the cell requires a PMF to

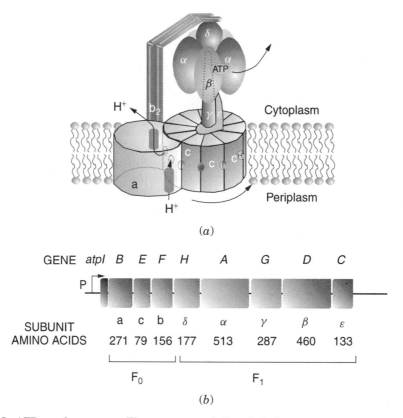

Fig. 9-8. ATP synthase rotor. The structure of *E. coli* F_1F_0 complex and genetic organization of the *atp* operon. (*a*) The ATP synthase shown with the cytoplasmic F_1 portion synthesizing ATP as H^+ is transferred through the F_0 pore and into the cell. (*b*) The *atp* operon is transcribed as a single polycistronic message from the promoter (P) region. The generic designations are shown above the operon; the corresponding protein products and the number of amino acids in each are shown below the operon.

transport carbohydrates and use its flagella. But how can an electrochemical gradient be formed if there are no active electron transport chains to pump H^+ out of the cell? Under these conditions, the F_1F_0 ATPase will use an ATP to pump H^+ out of the cell, thereby generating a PMF. Loss of the F_1F_0 ATPase, therefore, will prevent anaerobic growth in instances where the ATPase is required to generate PMF. In addition, these mutants will not grow aerobically on TCA cycle intermediates, since substrate-level phosphorylation is not possible. The H^+-translocating ATPase is the only way to synthesize ATP when growing on TCA cycle intermediates.

Structure of F_1F_0 and the *atp* Operon

As noted above, the ATP synthase consists of two functionally distinct entities, F_1 and F_0, which function in a rotary-like manner. The F_1 moiety retains ATPase function when stripped from the membrane. It consists of five subunits in an unusual stoichiometric ratio of $\alpha_3\beta_3\gamma\delta\varepsilon$ (aggregate molecular mass ca. 380 kDa). The F_0 pore consists of three subunits also in an unusual stoichiometry of $a_1b_2c_{10\pm1}$ (aggregate molecular mass ca. 150 kDa).

Figure 9-8a illustrates that F_1 is built as a hexamer of alternating α and β subunits that surround an asymmetric core consisting of the γ subunit. The interfaces between α and β subunits form three nucleotide-binding catalytic sites where ADP is converted to ATP (or vice versa). F_1 binds to F_0 through two distinct linkages: (1) a central stalk (or rotor) comprised of $\gamma\varepsilon$ anchored to the c-ring of the F_0 subunit and (2) a peripheral stalk (or stator) comprised of $b_2\delta$ that anchors the $\alpha_3\beta_3$ hexamer to the F_0a subunit. The a/c interface forms two unaligned partial channels for proton transport that open to opposite sides of the membrane. During ATP synthesis, each proton enters through the periplasmic channel in the a subunit and binds to a deprotonated c subunit. The c-ring then rotates one step to the right relative to the a subunit. This rotation moves a previously protonated c subunit into alignment with the cytoplasmic channel where the proton is lost and released to the cytoplasm.

The key to ATP synthesis is this c-ring rotation. Because the c-ring is anchored to the $\gamma\varepsilon$ pin, rotation of the c-ring rotor forces γ to rotate. The γ pin is forced to rotate within the catalytic $\alpha_3\beta_3$ hexamer because the stator moors the hexamer to the a subunit. Rotation of the γ pin within the hexamer drives the binding changes required to achieve net synthesis of ATP. One of the elegant proofs for subunit rotation involved attaching an actin filament to the c-ring and observing its rotation by fluorescence microscopy.

As noted earlier, when needed, the ATPase can work in reverse to pump protons out of the cell, generating the PMF needed to drive flagellar movement as well as a variety of transport systems. Thus, the cytoplasmic $\alpha_3\beta_3$ hexamer will also catalyze ATP hydrolysis, the energy from which rotates the γ pin in the opposite direction. This, in turn, reverses rotation of the c-ring, so cytoplasmic protons enter the cytoplasmic channel of the a subunit and are released on the periplasmic side, generating PMF.

It is worth discussing at this point how *E. coli* synthesizes the proper ratio of ATPase subunits. All of the genes encoding F_1F_0 reside in a single operon (Fig. 9-8b). RNA transcription of the entire operon occurs from a promoter region upstream of the first gene such that all of the genes are cotranscribed. With this in mind, it seems odd that with the unusual stoichiometries of the ATPase subunits, there is little if any excess a, b, γ, δ, or ε subunits in the cytoplasm. The proteins are all synthesized in the ratios as they occur in the assembled complex, but there is no differential transcriptional control that would explain why, for example, more c protein is produced than b or a. The answer appears to lie with the translational efficiencies of the various **translational initiation regions (TIR)** located at the beginning of each gene within the polycistronic message. Genes for subunits required at high stoichiometric ratios contain more efficient TIRs than do those required in smaller amounts. TIR efficiency can be correlated to the Shine-Dalgarno (SD) consensus sequence as well as to surrounding levels of secondary structure that could block ribosome access to the SD sites.

Energy Yield

The amount of energy derived from the oxidation of a given carbohydrate during respiration varies depending on several factors. These factors include the type of cytochrome system used to pump protons and generate PMF, the type of terminal electron acceptor and the efficiency of the ATP synthetase. For example, both $E.\ coli$ and yeast growing aerobically will produce $NADH + H^+$. Transfer of e^- from NADH to the cytochrome system of $E.\ coli$ will pump $4H^+$ out of the cell while the yeast cytochrome system will pump $6H^+$. Furthermore, the number of protons required to generate ATP via the $E.\ coli$ ATPase is perhaps two or three while the value for yeast mitochondrial ATP synthetase is closer to $2H^+$. So, 1 $NADH + H^+$ in $E.\ coli$ may yield between 1 and 2ATP molecules, whereas yeast can produce 3ATP molecules. Figure 9-9 illustrates the energy yield from glucose using yeast as the model system.

Generating ATP in Alkalophiles

Bacteria that live in very alkaline environments will have a difficult time generating a PMF, since there are usually more protons in the cell than outside (internal pH is more acidic than the external environment, e.g., pH_i 8 vs. pH_o 10). In these situations, the solution for generating ATP is to rely on **sodium motive force**. Again, an F_1F_0 ATPase is involved, but it couples ATP synthesis to movement of Na^+ rather than H^+.

Energetics of Chemolithotrophs

The chemolithotrophic bacteria derive energy from the oxidation of inorganic substrates. Best known are those oxidizing hydrogen to water, ammonia to nitrite and nitrite to nitrate, sulfur to sulfate and iron(II) to iron(III). Most chemolithotrophs contain all the components of a "normal" electron transport chain and therefore generate proton motive force in much the same manner as chemoorganotrophs (e.g., $Escherichia$ $coli$). The inorganic compounds serve as sources of energy in the form of electrons that can be used to pump H^+ (or other cations) out of the cell.

Hydrogen Oxidation. Aerobic hydrogen oxidation is well known in eubacteria such as $Escherichia\ coli$. Hydrogen, via hydrogenase 1 or 2 (hydrogenase 3 in $E.\ coli$ is involved with the production of H_2 as part of the formate-hydrogen lyase system), is introduced into the electron transport chain by transfer to NAD^+, forming NADH. Subsequent electron transfers are equivalent to what have already been discussed. Anaerobic hydrogen oxidation is known best in the methanogens in which CO_2 serves as the hydrogen acceptor:

$$CO_2 + 4H_2 \longrightarrow CH_4 + 2H_2O \quad (\Delta G'_0 = -130\ KJ/CH_4)$$

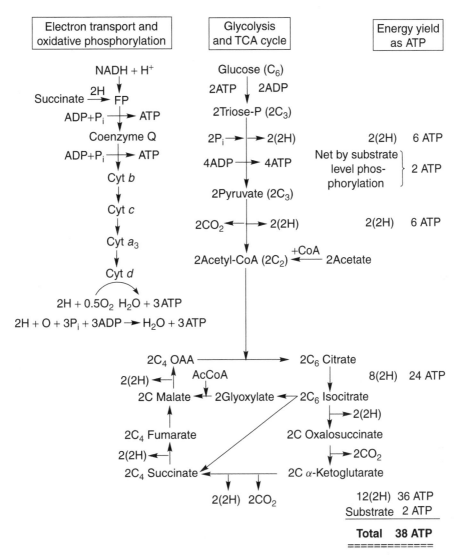

Fig. 9-9. Theoretical energy yield as ATP from glycolysis and the TCA cycle. The calculations shown here assume that each pair of hydrogen atoms (2H) released from the substrate yields 3 ATP. The reaction shown as $ADP + P_i \rightarrow ATP$ represents the action of ATP synthase. Two turns of the TCA cycle are required to completely oxidize the 2 acetyl-CoA derived from glucose. Each 2H generated by the system yields 1 molecule of water ($2H + 0.5O_2 \rightarrow H_2O$). Overall reaction: $C_6H_{12}O_6 + 38P_i + 38ADP + 6O_2 \rightarrow 6CO_2 + 6H_2O + 38ATP$. Total ATP from TCA cycle: $12(2H) + 6O_2 + 36P_i \rightarrow 6H_2O + 36ATP$.

Hydrogen can also be oxidized by some sulfate-reducing bacteria as:

$$4H_2 + SO_4^{2-} \longrightarrow H_2S + 2H_2O + 2OH^- \quad (\Delta G_0' = -152 \text{ KJ/mol})$$

Inorganic Nitrogen Compounds. Nitrifying bacteria use either ammonia or nitrite as their energy source:

$$NH_3 + 1\tfrac{1}{2}O_2 \longrightarrow HNO_2 + H_2O \quad (\Delta G = -272 \text{ KJ/mol})$$

$$NO_2^- + H_2O \longrightarrow NO_3^- + 2H^+ + 2e^- \quad (\Delta G = -73 \text{ KJ/mol})$$

Iron Oxidation. Organisms that can derive metabolic energy from Fe^{2+} include *Thiobacillus ferrooxidans* and *Leptospirillum ferrooxidans*. In all cases, oxidation takes place at acidic pH values, physiological optimum pH 2. This allows an increased solubility of the ferric ion product. These organisms, then, are acidophilic, which presents some interesting problems in terms of maintaining internal pH (see "pH Homeostasis").

$$4FeSO_4 + O_2 + 2H_2SO_4 \longrightarrow 2Fe_2(SO_4)_3 + 2H_2O \text{ (at pH2} \Delta G = -30KJ/mol)$$

These and other forms of chemolithotrophy are discussed in further detail below.

pH Homeostasis

Most enzymes and proteins have a rather narrow pH optima for activity or function. Nevertheless, bacteria have an amazing capacity to grow over a wide range of pH values. The bacteria can be divided into three groups depending on their preferred range of growth pH. Organisms that prefer acidic, neutral, or alkaline environments are called acidophiles, neutralophiles, and alkalophiles, respectively. *E. coli* is a neutralophile, but even it can grow over a 1,000-fold range of pH, from pH 5 to pH 8, and it can survive over a 10 million–fold range, from pH 2.0 to pH 9.0! Over external pH values from 4 to 9, the internal pH (pH_i) does not vary more than 10-fold and usually much less than that. Over growth pH values from 5.5 to 8, for example, pH_i only wavers between 7.2 and 7.7. How this extraordinary degree of homeostasis is accomplished remains one of the fundamental questions of bacterial physiology.

There are a number of factors that can influence pH_i. Among these are the buffer capacity of the cell, the outward transport of protons associated with respiration or ATP hydrolysis, and electroneutral transport systems that exchange protons for certain cations, particularly Na^+ and K^+. For example, as environmental pH becomes more acidic, any decrease in pH_i could trigger a K^+/H^+ antiport system that would bring K^+ into the cell while extruding a proton. This will maintain internal pH near 7.5 and create a larger ΔpH. Likewise, an increase in external pH will cause alkalinization of the cytoplasm, which could trigger Na^+/H^+ antiport systems that will extrude Na^+ and import protons to acidify the cytoplasm (see chemiosmotic-driven transport).

Aside from pH homeostasis, bacteria can induce acid protection systems called the acid tolerance response (ATR) and acid resistance, both of which increase survival during exposure to extremely low pH. The specifics of ATR systems differ among bacteria but are generally believed to include mechanisms designed to protect or repair macromolecules exposed to low pH. In organisms such as *Salmonella* and *E. coli*, they protect cells to about pH 3 but not below. *Shigella* and *E. coli* have additional acid resistance systems that are remarkable in their ability to protect cells under acid conditions as low as pH 2. For technical reasons, we do not know what happens to pH_i between pH 2 and 3, but details of the acid resistance systems have been exposed.

E. coli possess three acid resistance systems that are induced under different conditions. Two of these systems depend on the presence of particular amino acids at the acidic pH to protect the cell against acid stress: a glutamate-dependent system and an arginine-dependent system. Specific amino acid decarboxylases are involved with each system as well as membrane-bound antiporters that exchange substrate (e.g., glutamate) in the media for the decarboxylation product (e.g., gamma-amino butyric acid). How the systems function to protect the cell against protons is not clear, but mounting evidence suggests they may not directly enhance pH homeostasis, as originally proposed. They may be somehow linked to energy production at low pH. The importance of these AR systems is that they enable very low numbers of *Shigella* or *E. coli* to survive the gastric acidity barrier and cause disease.

METABOLITE TRANSPORT

The cytoplasmic membrane forms a hydrophobic barrier impermeable to most hydrophilic molecules. Consequently, mechanisms must exist that enable the cell to transport growth solutes into the cell and waste solutes out (useful transport websites include: http//www.biology.ucsd.edu/~msaier/transport/http//www-biology.ucsd.edu/~ipaulsen/transport/index.html). The outer membranes of gram-negative cells (e.g., *E. coli*) contain narrow channels that permit passive diffusion of hydrophilic compounds such as sugars, amino acids, and certain ions to the periplasm. Only compounds smaller than 600 to 700 daltons can pass. Hydrophobic compounds of any size are excluded. These pores are formed by porin proteins (e.g., OmpF or OmpC of *E. coli*). Chapter 7 discusses the structure of pores in detail and Chapter 18 provides details about their regulation.

A few compounds essential to bacterial growth, such as vitamin B_{12}, oligosaccharides, and iron chelates that are too large to cross the outer membrane via normal pores, have specific transport proteins or specific pores in the outer membrane that mediate their passage. For example, maltose traverses the outer membrane via a specific pore composed of maltoporin.

Oxygen, carbon dioxide, ammonia, and water all diffuse through both the outer and inner membranes without hindrance. But while most essential nutrients can diffuse through the outer membrane, all require one or more transport systems to cross the inner membrane and gain access to the cytoplasm. The goal of most of these transport systems is to concentrate nutrients inside the cell that may be in short supply externally. Consequently they work against a concentration gradient and require energy in some form to accomplish their tasks. There are several general transport mechanisms involved in membrane transport. These are diagrammed in Figure 9-10 and are divided into four groups: facilitated diffusion, chemiosmotic-driven transport, binding protein–dependent transport, and group translocation.

Facilitated Diffusion

Facilitated diffusion is the only transport system that does not require energy. As a result, it only allows for passage of a substrate *down* a concentration gradient. Consequently, the substrate will never achieve a concentration inside the cell greater than what exists outside. Inner membrane **channel proteins** are responsible for this type of transport. The glycerol transporter (GlpF), a mechanosensitive ion channel (MscL), and a voltage-sensitive ion channel are examples in *E. coli*.

Glycerol is one of the few compounds that enters prokaryotic cells by facilitated diffusion. The glycerol facilitator (GlpF) is a pore-type channel protein present in

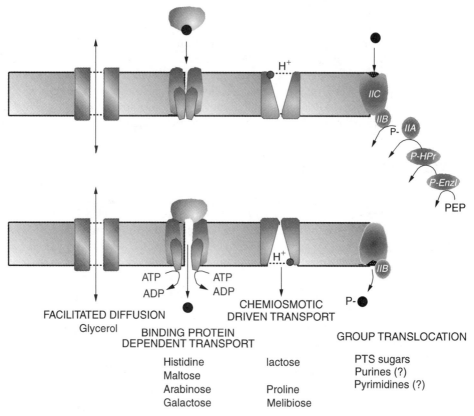

Fig. 9-10. Possible models for cytoplasmic membrane transport systems. Each of the systems probably has one or more membrane-spanning proteins that forms a specific channel in the cytoplasmic membrane. Facilitated diffusion and ion-driven transport systems require only one gene product, whereas binding protein-dependent transport and group translocation systems require several gene products. Facilitated diffusion allows entry and exit of substrate through a specific pore. Binding of substrate to specific sites accessible to the periplasmic side of the active transport systems coupled with an appropriate energy source allows a conformational change in the carrier proteins and substrate release inside the cell.

the inner membrane of *E. coli*. The *glpF* gene occurs in an operon with *glpK* (encoding glycerol kinase). This porin-type channel also allows passage of polyhydric alcohols and other unrelated small molecules such as urea and glycine but excludes charged molecules such as glyceraldehyde-3-phosphate (G3P) and dihydroxyacetone phosphate (DOHAP).

Aquaporin-Z (AqpZ). The entry and release of water from cells is a necessary feature of many physiological functions. However, the diffusion of water through a lipid bilayer is too low to explain the high osmotic permeability exhibited by cell membranes. The Aqp channel will mediate inwardly and outwardly directed osmotic fluxes of water triggered by abrupt changes in the extracellular osmolality. However, since *aqpZ* mutants remain viable, it is believed that other water channels are present that will compensate for the loss.

Mechanosensitive Channels

Turgor pressure (see Chapter 18) is defined as the outwardly directed pressure used by the cell to maintain shape and to allow for expansion of the cell membrane as the cell grows. The concentration of intracellular solutes (glutamate, potassium) determines the extent of turgor pressure (usually around 4 atm for *E. coli*). Upon transfer to medium that is *lower* in osmolarity, water will rapidly enter the cell. In the absence of mechanisms to reduce turgor, the pressure inside could rise to 11 atm, threatening the integrity of the cell. It has been proposed that catastrophic damage to the cell membrane is avoided by rapid activation of mechanosensitive channels (MscL, MscS). In contrast to aquaporin, mechanosensitive channels will permit inward or outward flow of ions. Transient opening of these channels allows K^+, glutamate, other compatible solutes, and ATP to exit and sodium and H^+ to enter. The result is a lowering of intracellular-compatible solute concentration and, thus, turgor.

ATP-Binding Cassette Transporter Family

Periplasmic proteins of gram-negative bacteria can be released into the surrounding medium if cells are subjected to osmotic shock (cells suspended in buffered 20% glucose containing EDTA are centrifuged and rapidly resuspended into cold $MgCl_2$). Among these periplasmic proteins are **binding proteins** for specific nutrients such as SO_4^{2-}, amino acids, sugars, and other compounds. In gram-positive bacteria, where no outer membrane is present to compartmentalize proteins, the binding proteins are present either as lipoproteins, tethered to the external surface of the cytoplasmic membrane, or as cell surface–associated proteins, bound to the external membrane surface via electrostatic interactions. The binding proteins function by transferring the bound substrate to a compatible membrane-bound complex of four proteins of which two are integral membrane proteins and two are peripheral, membrane-associated cytoplasmic proteins (Fig. 9-10). It is remarkable that these transport systems, sometimes called **traffic ATPases**, bear striking subsequence homologies to each other across phylogenetic domains. The cytoplasmic members of these systems contain two specific ATP-binding motifs: the Walker A box [GX(S/T)GXGK(S/T)(S/T)], which forms a so-called P-loop that interacts with the γ phosphate of ATP, and the Walker B box (ILLLD). As a result, these proteins are called ATP-binding cassette, or ABC proteins.

The transport process is energized by ATP or other high-energy phosphate compounds such as acetyl phosphate. Approximately 40% of the substrates transported by *E. coli* involve periplasmic-binding protein and ABC transporter mechanisms. It appears with these systems that transfer of the substrate from the periplasmic-binding protein to the membrane complex triggers ATP hydrolysis, which in turn leads to the opening of a pore that allows unidirectional diffusion of the substrate to the cytoplasm. The histidine permease of *E. coli* is one of the best-studied examples of a traffic ATPase. More information on ABC transporters can be found at http//ir2lcb.cnrs-mrs.fr/ABCdb/.

Chemiosmotic-Driven Transport

These systems accomplish movement of a molecule across the membrane at the expense of a previously established ion gradient such a as proton motive or a sodium motive force. About 40% of the substrates that enter *E. coli* involve ion-driven transport. There are three basic types: symport, antiport, and uniport. **Symport** involves the simultaneous transport of two substrates in the same direction by a single carrier. For example, a proton gradient can allow symport of an oppositely charged ion or a neutral molecule. Transport of lactose by the LacY permease is such an example. **Antiport** is the simultaneous transport of two like-charged compounds in opposite directions by a common carrier. The Na^+/H^+ antiporters of *E. coli* (NhaA and NhaB) are examples that are believed to be important for generating sodium motive force and maintaining neutral internal pH under alkaline growth conditions. **Uniport** occurs when movement of a substrate is independent of any coupled ion. Transport of glycerol is an example of uniport and was described above as facilitated diffusion. Table 9-2 lists examples of chemiosmotic transporters and their coupling ions.

Group translocation couples transport of a substrate to its chemical modification (e.g., by attaching a phosphate or coenzyme A group to the substrate). This traps the substrate within the cell in a form different from the exogenous substrate, so the concentration gradient of unmodified substrate never equilibrates. The phosphotransferase system (*pts*) involved in the transport of many carbohydrates utilizes this approach (see below).

TABLE 9-2. Representative Chemiosmotic Porters

Uniport	Symport	Antiport
Glycerol[a]	H^+/amino acid	H^+: cation
	H^+/glycine[a]	H^+:Ca^{2+a}
	H^+/histidine[a]	H^+:$CaHPO_4^a$
	H^+/lysine[a]	H^+:K^{+a}
	H^+/phenylalanine[a]	H^+:Na^{+a}
	H^+/organic acid	K^+: cation
	H^+/DL-lactate[a]	K^+:$CH_3NH_3^{+a}$
	H^+/pyruvate[a]	$H_2PO_4^-$: organic anion
	H^+/succinate[a]	$H_2PO_4^-$:hexose 6-phosphate[a]
	H^+/gluconate[a]	$H_2PO_4^-$:glycerol 3-phosphate[a]
	H^+/sugar	$H_2PO_4^-$:phosphoenolpyruvate[b]
	H^+/arabinose[a]	
	H^+/galactose[a]	
	H^+/lactose[a]	
	H^+/inorganic anion	
	H^+/phosphate[a]	
	Na^+/amino acid	
	Na^+, H^+/glutamate[a]	
	$Na^+ (H^+)$/proline[a]	
	Na^+/sugar	
	$Na^+ (H^+)$/melibiose[a]	
	Mg^{2+}/organic acid	
	Mg^{2+}, H^+/citrate[c]	

[a]*E. coli.*
[b]*Salmonella typhimurium.*
[c]*Bacillus subtilis.*

Establishing Ion Gradients

The establishment of ion gradients is of supreme importance to microorganisms. Proton and sodium gradients are important in various organisms for energy production (see above) and in transport. Some consideration should be given at this point to how the gradients are established and maintained. The cell membrane is impermeable, for the most part, to charged ions, so the cell has an opportunity to control the flow of ions across the membrane through ion-specific transport systems. As already detailed, proton gradients (known as the proton motive force) are established principally by the electron transport systems that pump protons out of the cell as electrons are transferred down the system. There are also specific membrane-bound ATPases that can couple the hydrolysis of ATP with the export of H^+, Na^+, and K^+. In addition, as shown in Table 9-2, there are a variety of antiport systems that can exchange ions such that one gradient can help build another (e.g., Na^+/H^+). So the picture that develops is of an interrelated series of ion circulations whose purpose is to provide energy to the cell.

SPECIFIC TRANSPORT SYSTEMS

ATP-Linked Ion Motive Pumps

The two major classes of these ion pumps are the F-type (F_1F_0) and P-type (E_1E_2) ATPases. The F-type ATPase is involved with pumping H^+ out of the cell or in coupling H^+ movement into the cell with the generation of ATP. The P-type ATPases are remarkably similar to each other despite the wide range of ions they transport. They include potassium, magnesium, calcium, cadmium (resistance), and arsenate (resistance). All members have the following properties: (1) a phosphorylated intermediate (aspartyl-phosphate in all ATPases where the phosphorylated amino acid has been determined); (2) two conformational forms of the phosphorylated intermediate, referred to as E_1 and E_2, which differ in reactivity to substrates and proteases; (3) a large (ca. 100 kDa) membrane-bound subunit with six to eight membrane-spanning regions and several regions of amino acid sequence homology. Most P-type ATPases only have this single large subunit. An exception is the Kdp ATPase.

The **Kdp ATPase** of *Escherichia coli* was first identified from the analysis of mutants that could not grow on low concentrations of potassium. It is a three-subunit enzyme whose subunit ratio appears to be $A_2B_2C_2$, although the exact stoichiometry is unknown (Fig. 9-11). The KdpB protein spans the membrane seven times, is homologous to other P-type ATPases (e.g., contains a conserved DKTGT motif), and is the subunit that is phosphorylated by ATP (putative phosphorylation site Asp307), which couples energy to transport. The KdpA subunit, which traverses the membrane 10 times, binds K^+ and probably forms most, if not all, of the membrane channel for K^+.

All other P-type ATPases couple the export of one ion from the cytoplasm to the import of a different ion from outside the cell (e.g., Na^+K^+ ATPase). The Kdp ATPase is different from all other P-type ATPases in that it transports ions exclusively to the intracellular compartment. The inset in Figure 9-11 illustrates the proposed kinetic model for K+ transport via the Kdp system. The low-energy state of the system (E_1) changes to high energy (E_2P) following phosphorylation. In this state, sites on the

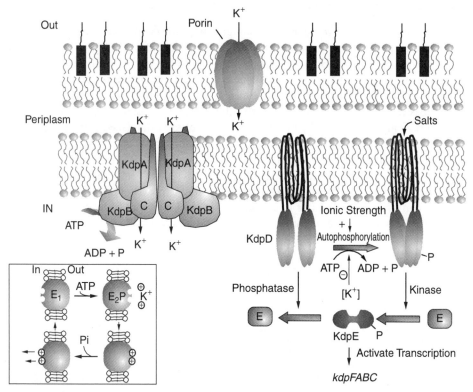

Fig. 9-11. The Kdp potassium transport ATPase of *E. coli* consists of three inner membrane proteins: KdpA, KdpB, and KpdC. KdpB is homologous to other E_1-E_2 ATPases. The outer membrane porin protein is not specified, and there is no periplasmic K^+-binding protein. The regulatory proteins KdpD and KdpE form a two-component signal transduction system that undergoes a cycle of auto- and *trans*-phosphorylation. Autophosphorylation is activated by intracellular ionic strength and an extracellular rise in osmolarity, and repressed by potassium. The phosphorylated form of KdpE activates transcription of the *kdpFABC*.

exterior of KdpA are exposed and K^+ can enter. Opening of the KdpA membrane channel and dephosphorylation accompany movement of K^+ to the inner surface and a return of the complex to E_1. The number of ions transported per ATP hydrolyzed is not known.

Transcriptional control of the *kdpFABC* operon is unusual in that the system is expressed under two seemingly contradictory conditions: when cells are unable to maintain the desired pool of K^+ and when experiencing osmotic upshift. Another, constitutively expressed, non-P-type system called Trk is involved with "housekeeping" potassium transport during normal growth conditions. The *kdp* operon is regulated by a typical two-component system involving KdpD, the sensor kinase, and KdpE, the response regulator (Fig. 9-11). However, the environmental signal recognized by KdpD has been unclear. It now appears that the sensing of K^+ and osmotic upshock by KdpD are mechanistically discriminated. Increased ionic strength triggers autophosphorylation of KdpD while high internal K^+ concentration inhibits autophosphorylation. Although this makes intuitive sense, osmotic upshift by salts can override the downregulating effect of high internal K^+ concentration and stimulate KdpD phosphorylation. Under any of these scenarios, phosphorylated KdpD transphosphorylates KdpE, which in turn activates transcription of the *kdpFABC* operon. The higher level of Kdp proteins increases K^+ transport. The reasons for the dual environmental control of *kdp* make perfect sense. Obviously, the cell needs to increase K^+ uptake when internal levels fall too low, but since K^+ is a compatible solute, the cell will also want to increase K^+ transport following osmotic shock (see Chapter 18).

The Histidine Permease

A member of the traffic ATPases, the histidine permease comprises the histidine-binding protein, HisJ, and three proteins, HisQ, HisM, and HisP, which form the membrane-bound complex. The membrane-bound complex consists of one copy each of HisQ and HisM and two copies of the ATP-binding subunit HisP. The receptor HisJ binds histidine and changes conformation, allowing it to interact with the membrane-bound complex. Translocation then occurs through a series of conformational changes concomitant with ATP hydrolysis and is hypothesized to take place through a pore formed by HisQ and HisM.

Iron

Iron is an essential nutrient for bacteria, serving as a cofactor for many enzymes and as a redox center in cytochromes and iron-sulfur proteins. However, iron is initially inaccessible to bacteria in the natural environment. Within animal fluids and tissues, proteins such as transferrin, lactoferrin, or ferritin sequester iron. Outside hosts, FeII rapidly oxidizes to FeIII and forms insoluble ferric hydroxide polymers. Bacteria have evolved elaborately clever mechanisms to wrest iron from the environment. A common strategy, used also by our reference organism *E. coli,* involves the synthesis and secretion of small, high-affinity, iron-binding chelates (called **siderophores**) that bind environmental iron. The iron chelate is then transported back into the cell by a specialized transport system and the iron is released.

A major siderophore of *E. coli* is **enterochelin** (Fig. 15-21). Its chemical synthesis and structure are described in Chapter 15. The mechanism by which desferrienterochelin is secreted remains a mystery. Once in the environment, it will bind to Fe^{3+}, after which the products of the *fep* genes enable movement of ferrienterochelin back into the cell (Fig. 9-12). The outer membrane FepA protein (79 kDa) functions as a monomer and enables transport of the complex across the outer membrane. It contains as many as 29 ß-sheet transmembrane spanning regions and an amino acid sequence called a **Ton box**. This region is predicted to contact the TonB inner membrane protein in a way that will "transduce" potential energy from the cytoplasmic membrane to FepA and drive transport of the siderophore across the outer membrane (Fig. 9-12).

Two other inner membrane proteins, ExbB and ExbD, form a heterohexamer with a central channel that accommodates the transmembrane segment of TonB. PMF is harnessed by the passage of protons through the ExbDB hexamer. This in some way energizes TonB to interact with FepA (and other TonB-dependent outer membrane proteins), triggering a conformational change in FepA that facilitates ferrienterochelin transport to the periplasm. Next, a periplasmic ferrienterochelin-binding protein (FebB) delivers the complex to the inner membrane transport system comprised of FepG, FepD,

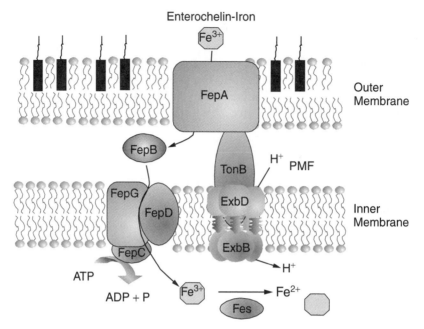

Fig. 9-12. The Ent enterochelin iron system transports the trimeric catechol enterochelin (hexagon). Enterochelin passes through the FepA protein of the outer membrane in a process requiring energy coupling through TonB. The periplasmic FepB protein passes the Fe^{3+} enterochelin to the inner membrane proteins FepG and FepD. Passage across the inner membrane requires ATP energy through FepC. In the cytoplasm, the Fes enterochelin esterase cleaves enterochelin, allowing the release and subsequent reduction of iron.

and FepC. FepC is a membrane-associated ATPase that provides the energy required to effect transport of the complex through FepG and FepD and into the cytoplasm. Once inside the cell, the enterochelin backbone is cleaved by Fes esterase, reducing the affinity for iron from 10^{-52} to 10^{-8}.

Regulation of this system is also interesting. The Fe^{3+} released in the cell is reduced by an unknown mechanism to Fe^{2+}. As ferrous iron accumulates in the cell, it is thought to bind to the regulatory protein **Fur** (ferric uptake regulator). Fur is actually a Mn^{2+}-binding protein with two binding sites for this metal. How and where (and even if) iron binds to Fur is as yet unclear. Nevertheless, as iron levels increase in the cell, metallated Fur binds to a 17 bp consensus sequence (the IRON or FUR box) located in front of the seven transcriptional units in *E. coli* that contain the *ent, fep,* and *fes* genes and repress their expression. Fur, however, has a role in the cell beyond that of controlling iron uptake. At least 50 *E. coli* or *S. enterica* genes are regulated by Fur, some in an iron-independent fashion.

Phosphotransferase System

The phosphotransferase system (PTS) is involved in both the transport and phosphorylation of a large number of carbohydrates, in movement toward these carbon sources (see "Chemotaxis" in Chapter 7) and in the regulation of several other metabolic pathways (see "catabolite control" in Chapter 5). In this group translocation transport system, carbohydrate phosphorylation is coupled to carbohydrate transport, the energy for which is provided by the EMP intermediate phosphoenolpyruvate (PEP). Figure 9-13 illustrates that there are two proteins common to all of the PTS carbohydrates. They are enzyme I (*ptsI* in *E. coli*) and the histidine protein HPr (*ptsH*). They are soluble, cytoplasmic proteins that participate in the phosphorylation of all PTS carbohydrates in a given organism and therefore are called the **general PTS proteins**.

In contrast, the enzyme IIs (EIIs) are carbohydrate specific. EIIs consist of three domains (A, B, and C) that may be combined in a single membrane-bound protein or split into two or more proteins (depending on the system) called IIA, IIB, and IIC. The B and C domains may form one protein called IICB. Likewise, the A and B subunits may form one protein called IIAB. IIC is an integral membrane protein (or domain). IIA (formerly called enzyme III) is soluble. In any scenario, the phospho group from PEP is transferred to the incoming carbohydrate via phospho intermediates of EI, HPr, EIIA, and EIIB. EIIC forms the translocation channel and at least part of the specific carbohydrate-binding site. On the basis of sequence homologies, EIIs may be grouped into four classes: the **mannitol class**, the **glucose class**, the **mannose class**,

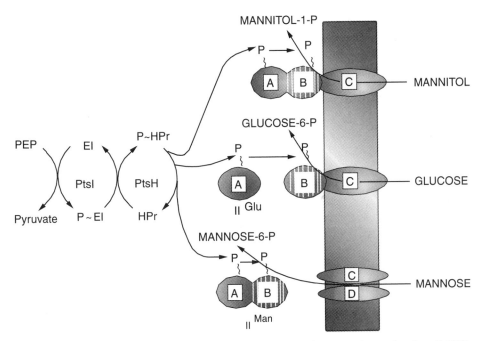

Fig. 9-13. Organization of PTS systems. EI and HPr are the general proteins for all PTSs. Of the many EIIs, only three are shown: those specific for mannitol (Mtl), glucose (Glc), and mannose (Man). Each contains two hydrophilic domains: IIA (formerly EIII or III) containing the first phosphorylation site (P-His), and IIB containing the second phosphorylation site (either a P-Cys or a P-His residue). The membrane-bound, hydrophobic domain IIC may be split into two domains (IIC and IID). II^{Mtl}, II^{Glc}, and II^{Glc}, and II^{Man} are specific for mannitol, glucose, and mannose, respectively. P indicates the phosphorylated forms of the various proteins. (*Source:* Derived from Postma, et al. *Microbiol Rev.* **57**:543–594, 1993.)

and in a few gram-positive organisms, galactose and the disaccharide lactose are PTS carbohydrates forming the **lactose class**.

A discussion of the PTS role in catabolite repression of gene expression can be found in Chapter 4. The PTS system is also involved in chemotaxis, a basic discussion of which is found in Chapter 7. The PTS allows chemotaxis toward PTS carbohydrates. Stimulation corresponds to uptake and phosphorylation of a substrate through EII. No MCP is involved in this process. CheB and CheR also are not involved. While the exact mechanism is not clear, it is thought that dephosphorylated EI or HPr that forms during carbohydrate transport could directly or indirectly decrease the level of P-CheA or P-CheY, resulting in increased smooth swimming and positive chemotaxis.

SUMMARY

The remarkable diversity of compounds tapped by the microbial world for energy underscores the basic evolutionary concept of "finding your own niche." While one energy source in a given environment (e.g., glucose) may be scarce, there will surely be an organism that has evolved to unleash energy intrinsic to a different compound plentiful to that environment (e.g., Fe^{2+}). So, too, the multitude of mechanisms used by various species to generate PMF belies the force of necessity to avoid competition with other species, if at all possible.

Microbiology students should stand in awe of the complexity of biochemical processes interwoven to make a living cell. Nowhere is this more evident than when considering all the ramifications involved in transporting an ion from one cell compartment to another. Moving just one Na^+ ion will have a ripple effect on many other ion movements — all so that the cell can maintain an overall homeostasis. In fact, the volume and diversity of ion traffic that occurs back and forth across the bacterial membrane makes the Los Angeles freeway look underutilized.

BIBLIOGRAPHY

Energy Production

Audia, J. P., C. C. Webb, and J. W. Foster. 2001. Breaking through the acid barrier: an orchestrated response to proton stress by enteric bacteria. *Int. J. Med. Microbiol.* **291**:97–106.

Boyer, P. D. 1997. The ATP synthase — a splendid molecular machine. *Annu. Rev. Biochem.* **66**:717–49.

Gunsalus, I. C., and C. W. Shuster. 1961. Energy-yielding metabolism in bacteria. In I. C. Gunsalus and R. Y. Stanier (eds.), *The Bacteria, Metabolism* Vol. 2. Academic Press, New York.

Hutcheon, M. L., T. M. Duncan, H. Ngai, and R. L. Cross. 2001. Energy-driven subunit rotation at the interface between subunit *a* and the *c* oligomer in the F(O) sector of *Escherichia coli* ATP synthase. *Proc. Natl. Acad. Sci. USA* **98**:8519–8524.

Richardson, D. J. 2000. Bacterial respiration: a flexible process for a changing environment. *Microbiology* **146**:551–71.

Saier, M. H., Jr. 1997. Peter Mitchell and his chemiosmotic theories. *ASM News* **63**:13–21.

Stock, D., A. G. Leslie, and J. E. Walker 1999. Molecular architecture of the rotary motor in ATP synthase. *Science* **286**:1700–5.

Metabolite Transport

Booth, I. R., and P. Luis. 1999. Managing hypoosmotic stress: aquaporins and mechanosensitive channels in *Escherichia coli*. *Curr. Opin. Microbiol.* **2:**166–9.

Calamita, G. 2000. The *Escherichia coli* aquaporin-Z water channel. *Mol. Microbiol.* **37:**254–62.

Gassel, M., and K. Altendorf. 2001. Analysis of KdpC of the K(+)-transporting KdpFABC complex of *Escherichia coli*. *Eur. J. Biochem.* **268:**1772–81.

Gassel, M., A. Siebers, W. Epstein, and K. Altendorf. 1998. Assembly of the Kdp complex, the multi-subunit K$^+$-transport ATPase of *Escherichia coli*. *Biochim. Biophys. Acta* **1415:**77–84.

Gatti, D., B. Mitra, and B. P. Rosen. 2000. *Escherichia coli* soft metal ion-translocating ATPases. *J. Biol. Chem.* **275:**34,009–12.

Genco, C. A., and D. White Dixon. 2001. Emerging strategies in microbial haem capture. *Mol. Microbiol.* **39:**1–11.

Higgs, P. I., P. S. Myers, and K. Postle. 1998. Interactions in the TonB-dependent energy transduction complex: ExbB and ExbD form homomultimers. *J. Bacteriol.* **180:**6031–38.

Jung, K., M. Veen, and K. Altendorf. 2000. K$^+$ and ionic strength directly influence the autophosphorylation activity of the putative turgor sensor KdpD of *Escherichia coli*. *J. Biol. Chem.* **275:**40,142–7.

Krulwich, T. A., M. Ito, and A. A. Guffanti. 2001. The Na(+)-dependence of alkaliphily in *Bacillus*. *Biochim. Biophys. Acta* **1505:**158–68.

Linton, K. J., and C. F. Higgins. 1998. The *Escherichia coli* ATP-binding cassette (ABC) proteins. *Mol. Microbiol.* **28:**5–13.

Poole, R. K., and G. M. Cook. 2000. Redundancy of aerobic respiratory chains in bacteria? Routes, reasons and regulation. *Adv. Microb. Physiol.* **43:**165–224.

Wilson, T. H., and P. Z. Ding. 2001. Sodium-substrate cotransport in bacteria. *Biochim. Biophys. Acta* **1505:**121–30.

_____

CHAPTER 10

METABOLISM OF SUBSTRATES OTHER THAN GLUCOSE

_____

UTILIZATION OF SUGARS OTHER THAN GLUCOSE

Glucose is the preferred carbon source for most of the common heterotrophic bacteria. Nevertheless, microorganisms possess remarkable versatility in their ability to use a wide range of other compounds as sources of carbon and energy. Most (if not all) of these pathways invariably lead to the production of intermediates that can enter one of the central pathways described in Chapter 8 (i.e., EMP, Entner-Doudoroff, pentose shunt, TCA cycle, or glyoxylate cycle). Many of these alternate carbon sources are utilized only after a period of **induction** of the enzymes required for their transport and metabolism. The presence of glucose in the growth medium generally inhibits the expression of catabolic enzymes for these substrates.

Monod first described this glucose effect in 1947 for the inhibition of β-galactosidase synthesis in *Escherichia coli*, and in 1961 Magasanik coined the term **catabolite repression** for this phenomenon. In media containing glucose and another carbohydrate, bacteria exhibit two complete growth cycles (diauxic growth) separated by a lag period. Glucose prevents entry of the second substrate by a process known as **inducer exclusion** and represses induction of the genes coding for the enzymes required for utilization of the second substrate. The details of the genetic regulation of carbohydrate metabolism are in Chapter 5.

Lactose

In *E. coli*, lactose utilization requires the induction, under the control of the *lac* operon, of a specific permease for its transport, and β-galactosidase, the enzyme that cleaves it to form D-galactose and D-glucose (Fig. 10-1). The D-galactose is phosphorylated to galactose-1-phosphate and metabolized via the Leloir pathway to yield fructose-6-phosphate (Fig. 10-2). Fructose-6-phosphate is ultimately utilized via the EMP pathway.

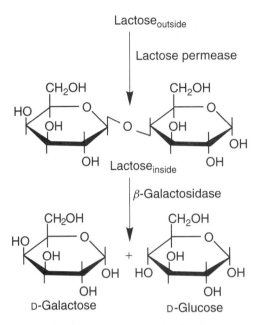

Fig. 10-1. Lactose utilization in *E. coli.*

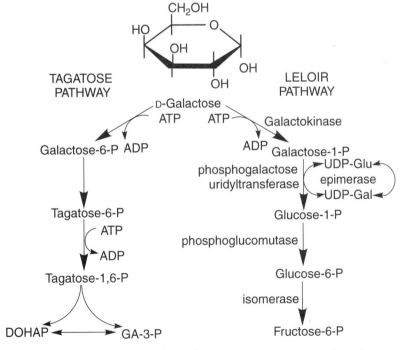

Fig. 10-2. The tagatose and Leloir pathways of galactose utilization.

Some organisms, such as *Lactobacillus casei*, do not contain the *lac* operon system but possess a PEP phosphotransferase system (PTS) that phosphorylates lactose via a specific enzyme II (see Chapter 5). The intracellular lactose is split by phospho-β-galactosidase to yield glucose and galactose-6-phosphate. Galactose-6-phosphate is metabolized via the tagatose-6-phosphate pathway to yield glyceraldehyde-3-phosphate and dihydroxyacetone phosphate (Fig. 10-2). In both cases, the D-glucose is metabolized via the EMP pathway. In *Staphylococcus aureus*, the tagatose-6-phosphate pathway is required for the utilization of galactose as well as lactose. Strains of *Klebsiella* lacking both *lac* and *gal* operons (Lac⁻Gal⁻) give rise to lactose-utilizing mutants that transport lactose via the PEP-PTS and metabolize lactose via phospho-β-galactosidase. The resultant galactose-6-phosphate is metabolized via the tagatose phosphate pathway.

Galactose

The inducible enzyme system required for galactose utilization in *S. cerevisiae* is under the control of a complex system of structural and regulatory genes that includes a galactose permease (encoded by *GAL2*) and three enzymes of the Leloir pathway: galactokinase, galactose-1-phosphate uridyltransferase, and uridine diphosphoglucose-4-epimerase encoded by the structural genes *GAL1*, *GAL7*, and *GAL10*. The enzyme phosphoglucomutase converts glucose-1-phosphate to glucose-6-phosphate, which then enters the glycolytic pathway. The galactose pathway enzymes are coordinately controlled by a positive factor required for the expression of structural genes and a negative factor that interacts with the inducer (galactose) to modulate the function of the positive factor. Expression of the structural genes is controlled by carbon catabolite repression.

Maltose

Utilization of maltose, a disaccharide of glucose, by *E. coli* requires the expression of genes concerned with maltose uptake (*malEGF*, *malK*, and *lamB*) and induction of amylomaltase and maltodextrin phosphorylase (encoded by the *malPG* operon). Expression of these genes is induced by maltose and is mediated indirectly by the *malT* activator. The cAMP receptor protein (CRP) binds to cAMP and positively regulates the *malT* gene and the *malEGF* operon. The MalT protein mediates the action of cAMP-CRP on malPQ genes. Amylomaltase hydrolyzes maltose with the production of D-glucose and the polysaccharide maltodextrin (Fig. 10-3). Maltodextrin is converted to G-1-P by maltodextrin phosphorylase. With the isomerization of G-1-P to G-6-P by phosphoglucomutase, both products are utilized via the EMP pathway. Similar systems for the utilization of maltose are active in a number of other organisms.

Mannitol

Catabolism of mannitol and other hexitols (D-glucitol or sorbitol, and D-galactitol) by *E. coli* involves transport by specific enzyme IIs of the PEP-PTS. The phosphorylated derivative is then oxidized to the corresponding sugar phosphate by a specific hexitol phosphate dehydrogenase (Fig. 10-4). The catabolic enzymes for mannitol, glucitol, and galactitol are encoded within three operons (*mtl, gut,* and *gat*) in *E. coli*. All

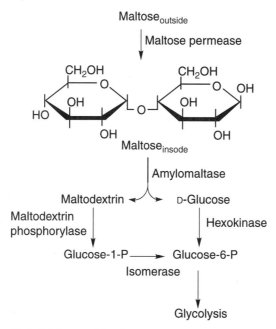

Fig. 10-3. Maltose utilization pathways in *Escherichia coli.*

three operons have the same gene order: a regulatory gene (C), the hexitol-specific enzyme II of the PEP-PTS (A), and the hexitol phosphate dehydrogenase (D). In the case of mannitol, mannitol-1-phosphate is oxidized by a specific NAD-dependent dehydrogenase to yield F-6-P, which enters the EMP pathway. *Bacillus subtilis* transports D-glucitol (sorbitol) through an inducible permease. The free hexitol is oxidized to D-fructose by an inducible glucitol dehydrogenase. D-Fructose then enters the EMP pathway via F-6-P.

Mannitol is also produced as an end product of glucose metabolism in *E. coli, Staphylococcus aureus, Lactobacillus brevis, Leuconostoc mesenteroides, Absidia glauca*, and *Rhodococcus erythropolis*. Many organisms utilize a mannitol cycle as a means of regenerating NADP through the NADP-dependent mannitol dehydrogenase that oxidizes mannitol to fructose (Fig. 10-4). The fruiting body of *Agaricus bisporous* accumulates large amounts of mannitol, whereas only low amounts are found in the mycelium. *E. coli* can also convert D-mannitol to D-ribose utilizing a pathway involving the synthesis of ribulose-5-phosphate (Fig. 10-4).

Fucose and Rhamnose

E. coli uses interesting parallel pathways for L-fucose and L-rhamnose metabolism. These pathways involve the induction of specific permeases, kinases, isomerases, and aldolases (Fig. 10-5). A cluster of genes at 60.2 minutes on the *E. coli* genetic map encodes the enzymes involved in fucose utilization. The *fuc* genes are organized into four operons under the influence of *fucR*, a positive regulatory protein. Fuculose-1-phosphate is a true inducer. The genes for the rhamnose system constitute a well-defined operon, whereas the fucose system maintains the gene for aldolase (*ald*) under separate

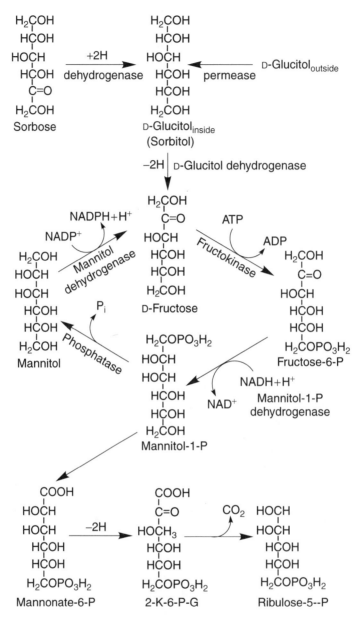

Fig. 10-4. Pathways of metabolism of sorbose and mannitol. In many organisms the mannitol cycle functions as a means of regulating reduced NADP levels.

control. The pathways converge with the formation of dihydroxyacetone phosphate and lactaldehyde. Dihydroxyacetone phosphate enters the central metabolic pathway at the triose level.

Under aerobic conditions an NAD-linked dehydrogenase oxidizes L-lactaldehyde to L-lactate and an FAD-linked dehydrogenase oxidizes L-lactate to pyruvate, which then enters the central pathway. Anaerobically, a single NAD-linked oxidoreductase

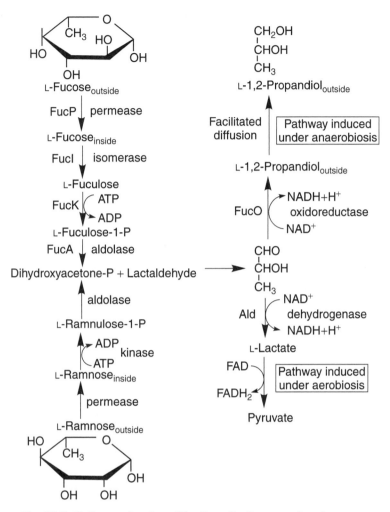

Fig. 10-5. Pathways for the utilization of L-fucose and L-rhamnose.

is induced by either L-fucose or L-rhamnose and serves both pathways. In either case, L-1,2-propanediol is formed and excreted. However, each methyl pentose exerts its influence at a different level. L-fucose is a transcriptional activator of *fucO*, whereas L-rhamnose functions posttranscriptionally. The L-fucose pathway can be recruited for the utilization of 6-deoxy-L-talitol and D-arabinose. Identical systems for the dissimilation of L-fucose and L-rhamnose are present in *S. enterica* and *K. pneumoniae*.

Mellibiose, Raffinose, Stachyose, and Guar Gum

These compounds all contain α-1,6-linked galactose residues, as shown in Figure 10-6. Many species of enteric bacteria can ferment mellibiose and raffinose, but relatively few can utilize galactomannans such as guar gum. *Bacteroides ovatus*, an obligately anaerobic resident of the intestinal tract of humans, develops α-galactosidase activity when grown on mellibiose, raffinose, or galactomannan. However, β-D-mannanases,

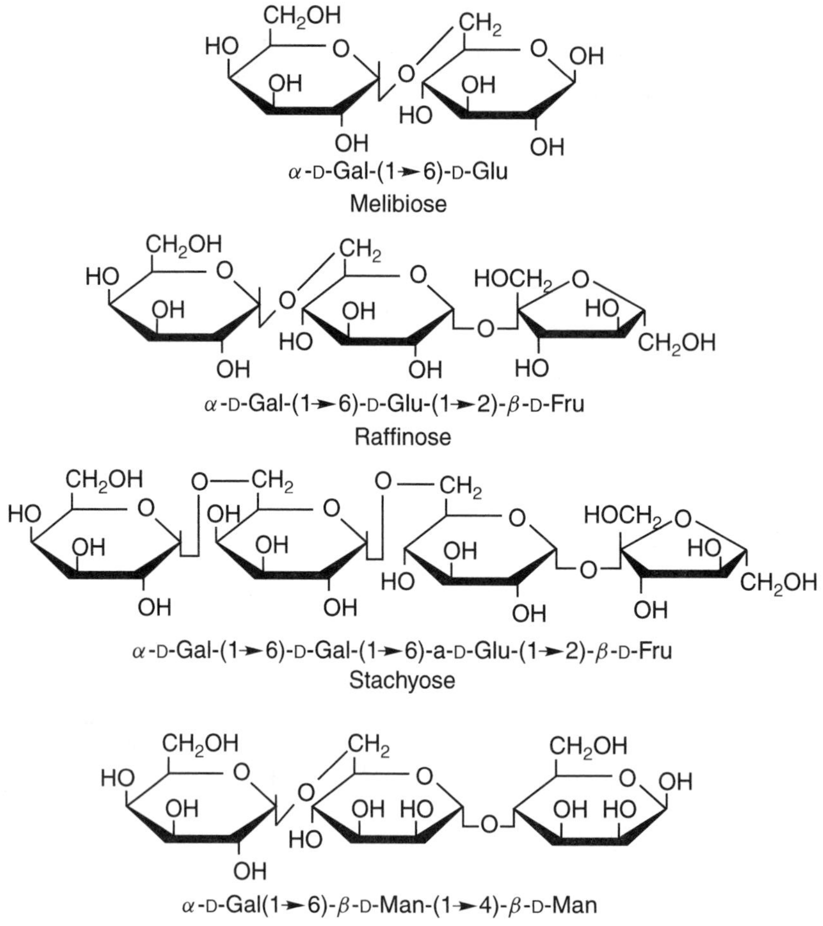

Fig. 10-6. Structures of mellibiose, raffinose, stachyose, and guar gum (galactomannan).

enzymes that degrade the mannose backbone of guar gum (galactomannan), are produced only during growth on this substrate. The α-galactosidase I differs from the α-galactosidase II produced during growth on mellibiose, raffinose, or stachyose.

PECTIN AND ALDOHEXURONATE PATHWAYS

Pectin, a highly methylated form of poly-β-1,4-D-galacturonic acid, is a major constituent of plant cell walls. The ability to degrade pectin contributes to the virulence of bacterial and fungal phytopathogens such as *Erwinia chrysanthemi* and *E. carotovora*, *Moloninia fructigena*, *Cladosporium cucumerinum*, and *Botrytis cinerea*. The rumen bacteria, *Butyrivibrio fibrisolvens* and *Lachnospira multiparus*, provide nutrients to the ruminant animal because of their ability to degrade pectin and cellulose.

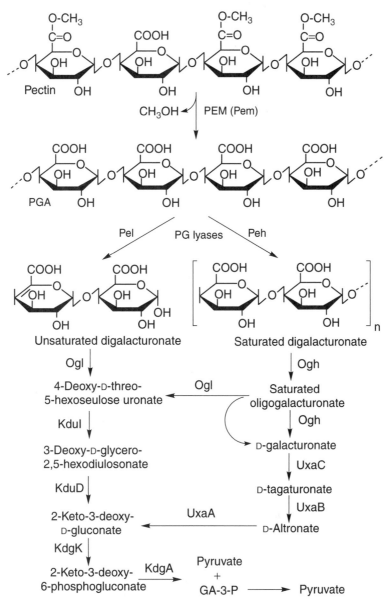

Fig. 10-7. Pathways of pectin catabolism in bacteria. The three-letter designation for the structural enzymes are: Pem, pectin methylesterase (PME); polygalacturonate (PGA) lyases (PelA to PelE) are extracellular enzymes. Further degradation of the breakdown products are conducted by: Ogh, oligogalacturonate hydrolase; Ogl, oligogalacturonate lyase; KduI, 4-deoxy-L-*threo*-5-hexoseulose uronate isomerase; KduD, keto-deoxyuronate dehydrogenase; KdgK, 2-keto-3-deoxygluconate kinase; KdgA, 2-keto-3-deoxy-6-phosphogluconate aldolase; UxaA, altronate hydrolase; UxaB, altronate oxidoreductase; UxaC, uronate isomerase.

Bacteroides thetaiotaomicron, a gram-negative anaerobe found in the human intestinal flora, can also degrade pectin and other plant polysaccharides.

E. chrysanthemi and *E. carotovora* initiate the degradation of pectin by means of a group of enzymes acting sequentially as shown in Figure 10-7. Pectin methylesterase (Pem) removes methoxyl groups linked to C-6 to yield polygalacturonate, which is degraded by pectate lyases (encoded by *pelABCDE*) to form unsaturated digalacturonate. Alternatively, an exopolygalacturonase produces digalacturonate. The polygalacturonate residues are then degraded via either of the two alternative pathways that lead to the formation of 2-keto-3-deoxy-6-phosphogluconate (KDPG) as shown in Figure 10-7. This metabolite then follows the Entner-Doudoroff pathway to pyruvate and triose phosphate.

CELLULOSE DEGRADATION

Cellulose is a β-1,4-linked glucose polymer that occurs in crystalline or amorphous forms and is usually found along with other oligosaccharides in the walls of plants and fungi. Cellulose-digesting microorganisms in the rumen of herbivorous animals are responsible for the ability of ruminants to use cellulose as a source of energy and building blocks for biosynthesis. A major obstacle to economical animal protein production has been the inefficient utilization of cellulose-containing materials by the rumen microflora. The ubiquitous distribution of cellulose in municipal, agricultural, and forestry waste emphasizes the potential use of cellulose for conversion to useful products such as single-cell protein or fermentation products such as methane or alcohol. As a consequence, the degradation of cellulose has been a continuing subject of intense study.

Cellulose degradation requires the combined activities of three basic types of enzymes (Fig. 10-8). Initially, an *endo-β*-1,4-glucanase cleaves cellulose to smaller oligosaccharides with free-chain ends. Then *exo-β*-1,4-glucanases remove disaccharide cellobiose units from either the reducing or nonreducing ends of the oligosaccharide chains. Cellobiose is then hydrolyzed to glucose by β-glucosidases.

These cellulolytic enzymes may be produced as extracellular proteins by organisms such as *Trichoderma* or *Phanaerochete* (filamentous fungi), or by *Cellulomonas, Microbispora,* or *Thermomonaspora* (Actinomycetes). Rumen bacteria such as *Ruminococcus flavofaciens* and *Fibrobacter succinogenes*, or gram-positive anaerobes such as *Clostridium thermocellum, C. cellulovorans,* or *C. cellulolyticum*, produce a cell-bound multienzyme complex called the **cellulosome.** With the aid of the electron microscope, cellulosomes can be seen as protuberances on the cell surface (Fig. 10-9). The cellulosome of *C. cellulovorans* contains three major subunits: a scaffolding protein, CipA; an exoglucanase, ExgS; and an endoglucanase, EngE. Also present are endoglucanases EngB, EngL, and EngY, and a mannanase, ManA. The scaffolding protein serves as a cellulose-binding factor. Another component, present in duplicate and referred to as dockerin, mediates the association of cellulose fibers with the scaffolding protein. Various models have been proposed to conceptualize the complex interaction of the cellulosome with cellulose fibers during the digestion process.

A wide diversity of actively cellulolytic organisms is important in industrial applications, in the rumen of animals, and in the digestive systems of arthropods that degrade wood. Termites and other arthropods that degrade wood owe their ability to digest cellulose to the presence of specific microbial symbionts in their digestive tract.

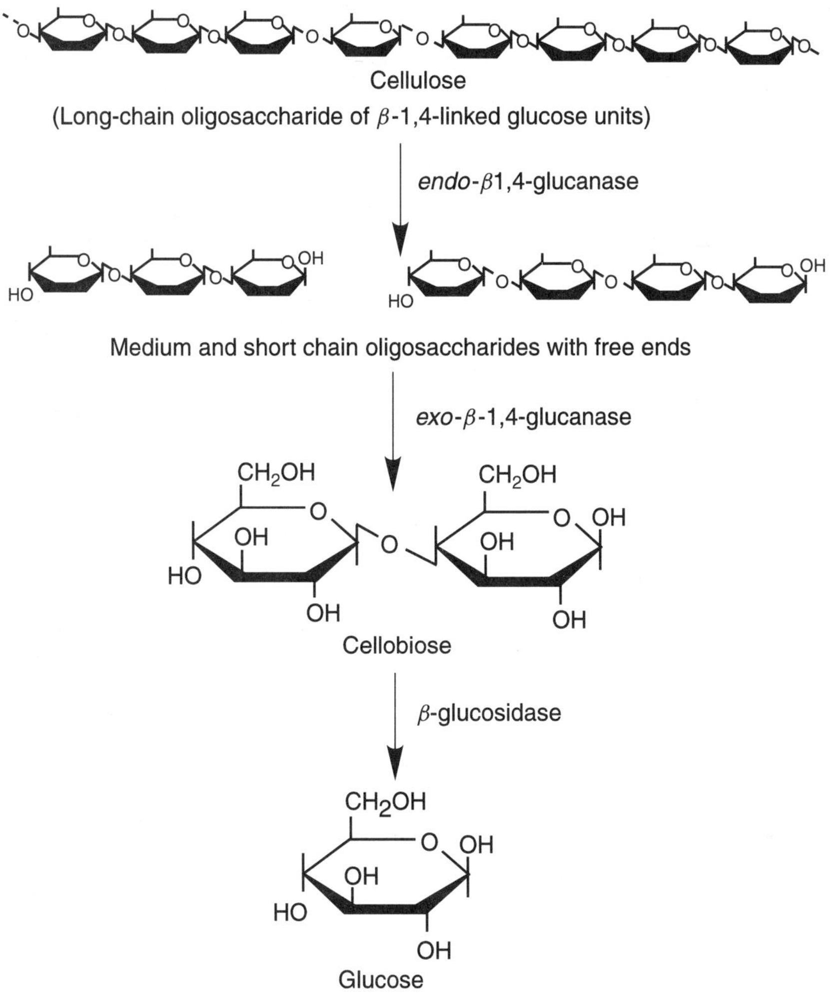

Cellulose

(Long-chain oligosaccharide of β-1,4-linked glucose units)

endo-β1,4-glucanase

Medium and short chain oligosaccharides with free ends

exo-β-1,4-glucanase

Cellobiose

β-glucosidase

Glucose

Fig. 10-8. Enzymatic degradation of cellulose.

STARCH, GLYCOGEN, AND RELATED COMPOUNDS

Starch, one of the most common storage compounds in plants, is a mixture of 25% amylose and 75% amylopectin. Amylose consists of long, unbranched chains of glucose in α-(1,4) linkage. Although not truly soluble in water, amylose forms hydrated micelles in which the polysaccharide chain forms a helical coil. By comparison, amylopectin is a highly branched form of starch in which the backbone consists of glucose chains in α-(1,4) linkage with α-(1,6) linkages at the branch points (Fig. 10-10). Glycogen, the main storage compound in animal cells, is comparable to amylopectin in that the main backbone also consists of glucose units in α-(1,4) linkage but with more frequent α-(1,6) branches. Pullulan, a starch-like polysaccharide, is a 1,4:1,6-α-D-glycan composed of maltotriose units in 1,6-α-linkage.

Amylose can be hydrolyzed by α-amylase, which cleaves the α-(1,4) linkages to yield a mixture of α-glucose and α-maltose. Amylose is also hydrolyzed by β-amylase,

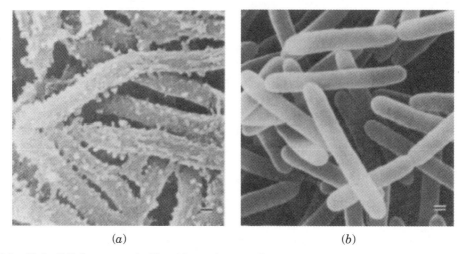

(a)   *(b)*

Fig. 10-9. Cellulosomes of *Clostridium thermocellum.* Scanning electron micrographs of ferritin-labeled cellobiose-grown cells. Prior to processing for electron microscopy, the cells were treated with cationized ferritin. Wild-type cellulose-digesting cells *(a)* are easily distinguishable from mutant cells *(b)* by the protuberances that cover the entire cell surface of the wild type. Bars = 200 nm. (source From Bayer, E. A. and R. Lamed, 1986. *J. Bacteriol.* **167**:828–836.)

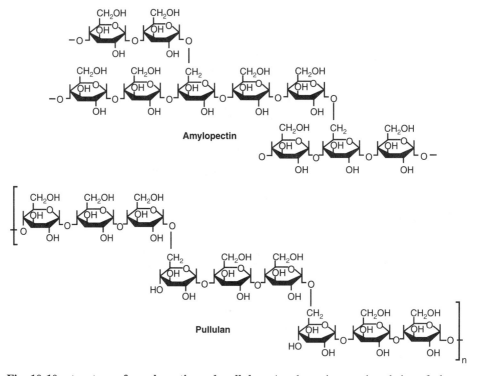

Fig. 10-10. structure of amylopectin and pullulan. Amylopectin contains chains of glucose molecules in α-(1,4) linkage with occasional α-(1,6)-linked branches. The structure of glycogen is similar except that there are more frequent α-(1,6)-linked branches. Pullulan is composed of maltotriose units of three glucose molecules in α-(1,4) linkage connected by α-(1,6) linkages.

liberating β-maltose in stepwise fashion from the nonreducing end of the molecule. Thus β-amylases invert the configuration at the C-1 position during cleavage of the α-(1,4)-glucosidic bond, whereas α-amylases retain the α configuration at the C-1 position. These enzymes also hydrolyze amylopectin or glycogen to yield glucose, maltose, and a highly branched core, limit dextrin. A debranching enzyme, α-(1,6)-glucosidase, is capable of hydrolyzing the α-(1,6) linkages in limit dextrin. The combined action of α-(1,6)-glucosidase and α-amylase is required to completely degrade amylopectin or glycogen to glucose and maltose. Thus, α-amylases are endoenzymes that can bypass the α-(1,6) branch points of amylopectin, whereas β-amylases are exoenzymes that cannot hydrolyze amylopectin internally to the α-(1,6) branch points. Both α- and β-amylases and debranching enzymes (also called pullulanases) are produced by a number of bacteria and fungi. *Aspergillus* species produce glucoamylases, enzymes that degrade starch to glucose. Extracellular β-amylases have been observed in *B. polymyxa, B. megaterium,* and other *Bacillus* species.

The amylases secreted by a variety of *Bacillus* species have been studied intensively because of their industrial application. The α-amylases from *B. subtilis* are saccharifying enzymes that produce mostly glucose and maltose from starch. On the other hand, the α-amylases produced by *B. licheniformis* and *B. amyloliquefaciens* are liquefying enzymes that yield mostly maltosaccharides. Under the conditions of nutrient deprivation, *B. subtilis* activates the structural gene for α-amylase, *amyE*. The synthesis of α-amylase is not inducible in the classic sense in that no compound can be shown to trigger the action of *amyE*. However, synthesis of the enzyme is repressed by glucose and other readily metabolizable carbon sources. Several genes appear to increase the formation of α-amylase and its secretion.

For ecological as well as industrial reasons, the thermostability of α-amylases has been of considerable interest. Thermophilic species, such as *B. acidocaldarius* and *B. stearothermophilus*, produce amylases that are stable at temperatures ranging from 58 °C to 80 °C. However, mesophilic species, such as *B. licheniformis* and *B. amyloliquefaciens*, also produce amylases that are active at temperatures in excess of 75 °C. Comparative studies of the enzymes and the structural genes of mesophilic and thermophilic species reveal considerable amino acid homology between the enzymes from the two groups, indicating that these proteins are related on an evolutionary basis.

Thermophilic anaerobes produce highly active thermostable starch-degrading enzymes. *Clostridium thermosulfurogenes* produces an extracellular, thermoactive, and thermostable β-amylase and a cell-bound glucoamylase. It does not produce a debranching enzyme. *Clostridium thermosulfuricum* produces a cell-bound glucoamylase and a debranching enzyme, pullulanase, which are thermoactive and thermostable. Mixed cultures of these organisms exhibit enhanced production of β-amylase, glucoamylase, and pullulanase, with concomitant increase in ethanol production from starch. The β-amylase of *C. thermosulfurogenes* is expressed at high levels only when the organism is grown on maltose or other carbohydrates containing maltose units. Glucose represses β-amylase synthesis, but cAMP does not eliminate the repressive effect.

Bacteroides thetaiotaomicron, a gram-negative anaerobe found in high numbers in the human colon, can ferment a variety of polysaccharides, including amylose, amylopectin, and pullulan. In this organism the degradative enzymes are not extracellular but are cell associated, indicating that the first step in starch utilization involves binding of starch to the bacterial surface. This action is followed by translocation through the outer membrane into the periplasm where the starch-degrading enzymes are located. *B. thetaiotaomicron* produces four maltose-inducible outer membrane proteins and one enzyme located in the cytoplasmic membrane.

A 115 kDa outer membrane protein is essential for maltoheptaose utilization, whereas the other outer membrane proteins are involved in starch utilization. Two constitutively produced outer membrane proteins are also involved in starch utilization. Various bacteria produce four types of pullulan-hydrolyzing enzymes. A glucoamylase hydrolyzes pullulan from the nonreducing ends to produce glucose. A pullulanase (type I) found in *K. pneumoniae* and *B. thetaiotaomicron* breaks α-(1,6) glucosidic linkages to produce maltotriose. An isopullulanase found in *Aspergillus niger* hydrolyzes α-(1,4) glucosidic linkages in pullulan to produce panose [a branched trisaccharide with two α-(1,4) linkages and one α-(1,6) branch]. An α-glucosidase is found in *B. thetaiotaomicron* in addition to pullulanases I and II.

Cyclodextrins are cyclic oligosaccharides containing from 6 to 12 glucopyranose units bonded through α-(1,4) linkages, as shown in Figure 10-11. These compounds are widely used in the food and pharmaceutical industry as solubilizing and stabilizing agents. They have also found application as the bonded phase for high-performance liquid chromatography (HPLC) used in the separation and identification of organic compounds. Certain organisms produce cyclodextrins in the process

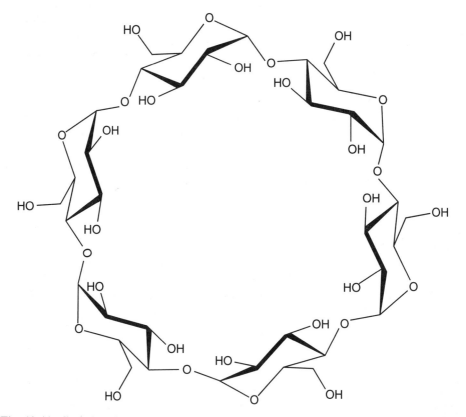

Fig. 10-11. Cyclodextrin structure. Cyclodextrins are cyclic oligosaccharides containing from 6 to 12 glucopyranose units bonded through α-(1,4) linkage.

of degrading starch. *Bacillus stearothermophilus* degrades starch by means of the enzyme cyclodextrin glucanotransferase. Cyclodextrins are resistant to hydrolysis by many starch-splitting enzymes. *Bacillus macerans, B. coagulans, B. sphaericus*, and *C. thermohydrosulfuricum* produce a cyclodextrinase that can hydrolyze linear maltodextrins as well as cyclodextrin.

METABOLISM OF AROMATIC COMPOUNDS

Many microorganisms can utilize aromatic compounds as their sole source of carbon. The aromatic ring structure is degraded and the products are converted to compounds that can enter central metabolic pathways to provide carbon sources and energy. Degradation of aromatic compounds is of major importance because of the widespread occurrence of these substances in natural as well as synthetic materials. Hydrocarbon-degrading microorganisms facilitate reentry of the carbon from these compounds into the natural biological cycles, preventing their accumulation in the environment. Accumulation of toxic aromatic hydrocarbons, particularly herbicides, pesticides, and other industrial waste products, can lead to serious contamination of ground water and cause serious ecological damage.

A fundamental aspect of the metabolism of aromatic ring compounds involves the conversion of the ring structure to a form that can be cleaved to yield an open chain, which can then be metabolized to intermediates in the central metabolic pathways. Under aerobic conditions, monooxygenases or dioxygenases introduce hydroxyl groups that facilitate ring cleavage. As shown in Figure 10-12, several pathways lead to the formation of catechol or similar compounds that can undergo either ortho- or meta-cleavage to yield intermediates leading into the general metabolic schemes. The meta-cleavage pathway results in the formation of pyruvate and acetaldehyde, whereas the ortho-pathway yields β-ketoadipate, a compound that is cleaved to form succinate and acetate.

Several ring oxidation mechanisms are known. *Pseudomonas putida* mt-2 oxidizes the methyl group of toluene to form benzyl alcohol. Subsequent reactions yield catechol, the substrate for the meta-cleavage pathway that leads to keto acid formation. *P. putida* PpF1 initiates the oxidation of toluene by incorporating oxygen into the aromatic nucleus to form *cis*-toluene dihydrodiol, which is converted to 3-methylcatechol. This compound is then degraded via the meta-cleavage pathway to yield an α-keto acid. *P. mendocina* oxidizes toluene to *p*-cresol, which undergoes ortho-cleavage to form β-ketoadipate. This compound is split by succinyl-CoA transferase to yield succinate and acetate as intermediates that can be used in the TCA cycle. The keto acid intermediates formed via the meta-cleavage route ultimately lead to acetaldehyde and pyruvate, compounds that can enter the central metabolic pathways.

The demonstration by Stanier and his colleagues in 1947 of the mechanism of the oxidation of mandelate to intermediates of the TCA cycle by *P. putida* represents the first example of sequential induction of enzymes of a catabolic pathway. **Sequential induction** refers to the fact that each of the enzymes in the pathway is inducible by its specific substrate.

Compounds with more complex ring structures, such as naphthalene, anthracine, or phenanthrine, undergo sequential ring fission to form ring structures that can

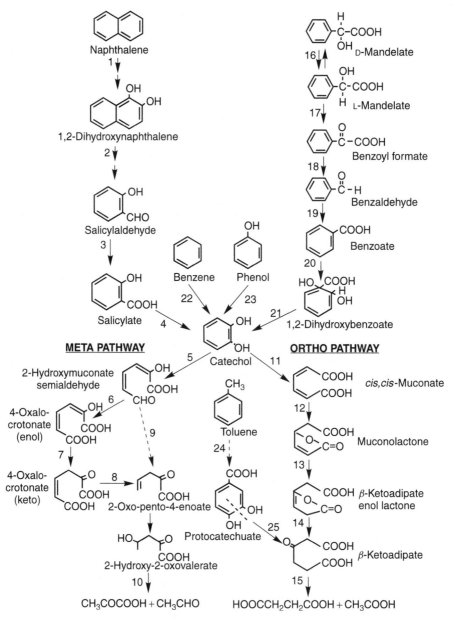

Fig. 10-12. Metabolism of aromatic hydrocarbons. The enzymes are: (1) naphthalene oxygenase, (2) 1,2-dihydroxynaphthalene oxygenase, (3) salicylaldehyde dehydrogenase, (4) salicylate hydroxylase, (5) catechol 2,3-oxygenase, (6) 2-hydroxymuconic semialdehyde dehydrogenase, (7) 4-oxalocrotonate tautomerase, (8) 4-oxalocrotonate decarboxylase, (9) 2-hydroxymuconic semialdehyde hydrolase, (10) 4-hydroxy-2-oxovalerate aldolase, (11) catechol 1,2-oxygenase, (12) *cis, cis*-muconate lactonizing enzyme, (13) (+)-muconolactone isomerase, (14) β-ketoadipate enol-lactone hydrolase, (15) succinyl CoA-β-ketoadipate thiolase, (16) mandelate racemase, (17) L-(+)-mandelate dehydrogenase, (18) benzoylformate decarboxylase, (19) NAD- and NADP-linked benzaldehyde dehydrogenases, (20, 21) benzoic acid oxidase, (22) benzene hydroxylase, (23) phenol hydroxylase, (24) toluene hydroxylase, (25) protocatechuate ortho-cleavage enzyme.

then be degraded via the ortho- or meta-cleavage pathways (Fig. 10-12). Chlorinated compounds, such as chlorobenzoate or polychlorinated biphenyls, often require dehalogenation before the aromatic ring can be hydroxylated to facilitate ring fission. In general, the more complex the ring structures, the more resistant polyaromatic hydrocarbons are to degradation. So far in this discussion, we have mentioned primarily pseudomonads that are well known for their ability to degrade aromatic hydrocarbons. Under natural conditions, a consortium of microorganisms may be

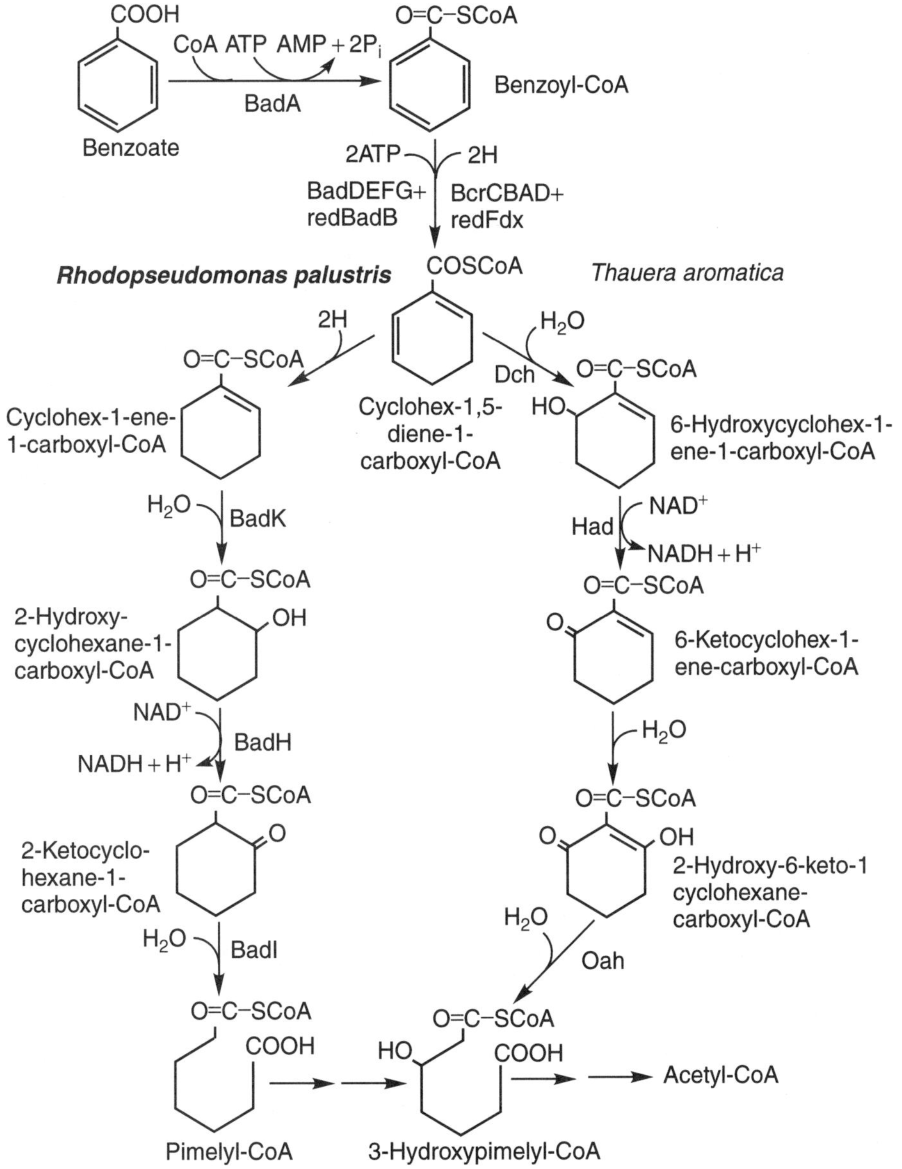

Fig. 10-13. Comparison of anaerobic benzoate degradation by *R. palustris* and *T. aromatica.*
(*Source*: Redrawn from C. S. Harwood, et al., FEMS Microbiol. Rev. 22, 439–458 1999.)

required to rapidly degrade a compound with four or more fused rings. For example, benzo[a]pyrene, a potent carcinogen that commonly occurs in complex mixtures such as diesel fuel, contains five fused rings. Rapid mineralization of this compound was best accomplished by a combination of organisms phylogenetically related to the bacterial genera *Sphingobacterium, Pseudomonas, Aquabacterium, Burkholderia, Ralstonia, Sphingomonas, Mycobacterium,* and *Alcaligenes.*

Even more complex organic materials such as lignin are resistant to microbial attack. However, many fungi and a few bacteria are capable of degrading lignin or its breakdown products. These organisms are essential for the return of the constituents of these heteropolymers to the natural carbon and nitrogen cycles. The wood-rotting fungus *Phanerochaete chrysosporium* is capable of degrading the pesticide DDT and other complex organic compounds because of their structural similarity to components of the lignin complex.

Studies on the degradation of aromatic compounds cited thus far have all involved the incorporation of molecular oxygen into the aromatic nucleus to form dihydroxylated intermediates. Under anaerobic conditions, the aromatic ring structures are attacked reductively. A comparison of anaerobic benzoate degradation by *Rhodopseudomonas palustris* and *Thauera aromatica* is shown in Figure 10-13. Subsequent metabolism involves the formation of intermediates leading to the formation of acetyl-CoA. Acetyl-CoA can then enter the central metabolic pathway. A wide assortment of aromatic compounds are degraded to benzoyl-CoA via several novel reactions including carboxylation of phenolic compounds, reductive elimination of ring substituents such as hydroxyl or amino groups, oxidation of methyl substituents, *O*-demethylation reactions, and shortening of aliphatic side chains.

BIBLIOGRAPHY

Pectin Utilization

Stutzenberger, F. 2000. Pectinases. In J. Lederberg (ed.), *Encyclopedia of Microbiology,* 2nd ed., Vol. 1. Academic Press, San Diego, pp. 562–79.

Cellulose Utilization

Béguin, P., and J. -P. Aubert. 2000. Cellulases. In J. Lederberg (ed.), *Encyclopedia of Microbiology,* 2nd ed., Vol. 1. Academic Press, San Diego, pp. 744–58.

Desvaux, M., E. Guedon, and H. Petitdemange. 2001. Carbon flux distribution and kinetics of cellulose fermentation in steady-state continuous cultures of *Clostridium cellulolyticum* on a chemically defined medium. *J. Bacteriol.* **183:**119–30.

Ding, S. -Y., et al. 2001. Cellulosomal scaffolding-like proteins from *Ruminococcus flavofaciens. J. Bacteriol.* **183:**1945–53.

Felix, C. R., and L. G. Ljungdahl. 1993. the cellulosome: the exocellular organelle of *Clostridium. Annu. Rev. Microbiol.* **47:**791–819.

Kataeva, I. A., R. D. Seidel III, X.-L. Li, and L. G. Ljungdahl. 2001. Properties and mutation analysis of the CelK cellulose-binding domain from the *Clostridium thermocellum* cellulosome. *J. Bacteriol.* **183:**1552–59.

Kruus, K., A. C. Lua, A. L. Demain, and J. H. D. Wu. 1995. The anchorage function of CipA (CelL), a scaffolding protein of the *Clostridium thermocellum* cellulosome. *Proc. Natl. Acad. Sci. USA* **92:**9254–58.

Tamaru, Y., and R. H. Doi. 1999. Three surface layer homology domains at the N terminus of *Clostridium cellulovorans* major cellulosomal subunit EngE. *J. Bacteriol.* **181**:3270–76.

Zhou, S., F. C. Davis, and L. O. Ingram. 2001. Gene integration and expression and extracellular secretion of *Erwinia chrysanthemi* endoglycanase CelY (*celY*) and CelZ (*celZ*) in ethanolic *Klebsiella oxytoca* P2. *Appl. Environ. Microbiol.* **67**:6–14.

Utilization of Starch, Glycogen, and Related Compounds

Delia, J. N., and A. A. Salyers. 1996. Contribution of a neopullulanase, a pullulanase, and an α-glucosidase to growth of *Bacteroides thetaiotaomicron* on starch. *J. Bacteriol.* **178**:7173–79.

Duffner, F., C. Bertoldo, J. T. Andersen, K. Wagner, and G. Antranikian. 2000. A new thermoactive pullulanase from *Desulfurococcus mucosus:* cloning, sequencing, purification, and characterization of the recombinant enzyme after expression in *Bacillus subtilis. J. Bacteriol.* **182**:6331–8.

Shipman, J. A., K. H. Cho, H. A. Siegel, and A. A. Salyers. 1999. Physiological characterization of SusG, an outer membrane protein essential for starch utilization by *Bacteroides thetaiotaomicron. J. Bacteriol.* **181**:7206–11.

Vieille, C., A. Savchenko, and J. G. Zeikus. 2000. Amylases, microbial. In J. Lederberg (ed.), *Encyclopedia of Microbiology*, 2nd ed., Vol. 1. Academic Press, San Diego, pp. 171–9.

Utilization of Aromatic Hydrocarbons

Cowles, C. E., N. N. Nichols, and C. S. Harwood. 2000. BenR, a XylS homologue, regulates three different pathways of aromatic acid degradation in *Pseudomonas putida. J. Bacteriol.* **182**:6339–46.

Harwood, C. S., G. Burchhardt, H. Herrmann, and G. Fuchs. 1999. Anaerobic metabolism of aromatic compounds via the benzoyl-CoA pathway. *FEMS Microbiol. Rev.* **22**:439–58.

Kanaly, R. A., and S. Harayama. 2000. Biodegradation of high molecular weight polycyclic aromatic hydrocarbons by bacteria. *J. Bacteriol.* **182**:2059–67.

Kanaly, R. A., R. Bartha, K. Watanabe, and S. Harayama. 2000. Rapid mineralization of benzo[a]pyrene by a microbial consortium growing on diesel fuel. *Appl. Environ. Microbiol.* **66**:4205–11.

Ou, L. -T.2000. Pesticide biodegradation. In J. Lederberg (ed.), *Encyclopedia of Microbiology*, 2nd ed., Vol. 3. Academic Press, San Diego, pp. 594–606.

Parke, D., D. A. D'Argenio, and L. N. Ornston. 2000. Bacteria are not what they eat: that is why they are so diverse. *J. Bacteriol.* **182**:257–63.

Young, D. M., D. Parke, D. A. D'Argenio, M. A. Smith, and L. N. Ornston. 2001. Evolution of a catabolic pathway. *ASM News* **67**:362–69.

CHAPTER 11

FERMENTATION PATHWAYS

The major pathways used by microorganisms for the dissimilation of carbohydrates and the generation of biologically useful energy (as PMF or ATP) and reducing power (as NADH or NADPH) are delineated in previous chapters. The introductory reactions whereby a variety of alternate sugars and other substrates are metabolized to intermediates that can enter these central metabolic routes are also described. Individual organisms may use the fructose bisphosphate aldolase, phosphoketolase, or ketogluconate aldolase pathways or variations of them. Additional fermentative products may be derived from pyruvate or other intermediates. The fermentative pathways occurring in some of the major groups of microorganisms are outlined in Figure 11-1 and are considered in detail in this chapter.

FERMENTATION BALANCES

A thorough evaluation of the pathways of carbohydrate fermentation requires qualitative identification of and quantitative accounting for the amount of products recovered. To assess the accuracy of the analytical determinations, a **carbon balance** or **carbon recovery** can be calculated, as shown in the sample fermentation in Table 11-1. Under a number of circumstances, the balance may not be ideal because some portion of the carbohydrate may be assimilated into cellular constituents or compounds other than the major substrate added may be utilized to form some portion of the recovered products.

Oxidation–reduction reactions play a major role in the fermentative metabolism of carbohydrates. It is, therefore, advisable to perform an **oxidation–reduction** or **O–R balance** on the products formed. The O–R balance provides an indication as to whether the formed products balance with regard to their oxidized or reduced states. It may not be possible to balance the hydrogen and oxygen of the substrate directly because hydrations or dehydrations may occur as intermediary steps in the fermentation pathway. However, a simple way to achieve the same end is to calculate an **oxidation**

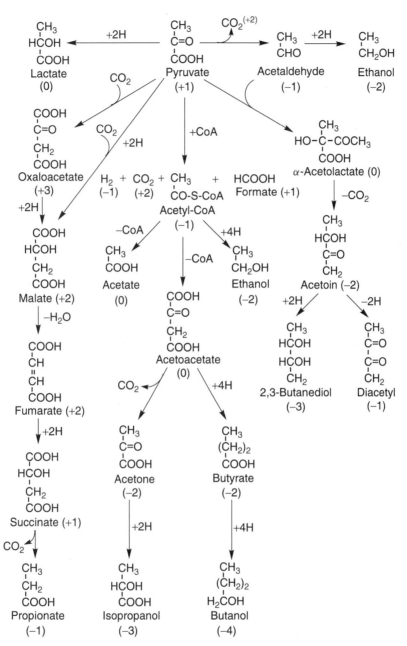

Fig. 11-1. Major pathways of fermentation product formation from pyruvate. Numbers in parentheses are the oxidation values calculated on the basis of the number of oxygen atoms less one-half the number of hydrogen atoms as shown in Table 11-1.

value for each compound and compare the total amount of oxidized and reduced products as shown in Table 11-1. If the ratio of oxidized products to reduced products is close to the theoretical value of 1.0, this provides further indication that the products are in balance.

TABLE 11-1. Fermentation Balance for *Lactobacillus pentoaceticus*

Compound	Mmol	Mmol Carbon[a]	Oxidation Value[b]	Oxidized Products	Reduced Products	C_1 Observed	C_1 Calculated[c]
Glucose	100	600	0				
Lactate	96	288	0				
Glycerol	7	21	−1		7		
Ethanol	86	172	−2		172		86
Acetate	7	14	0				7
CO_2	89	89	+2	178		89	
Totals		584		178	179	89	93

Carbon Recovery	Oxidation–Reduction	C_1 Balance
$\dfrac{\text{mmol product}}{\text{mmol substrate}} \times 100$	$\dfrac{\text{Oxidized products}}{\text{Reduced products}}$	$\dfrac{\text{Observed } C_1}{\text{Calculated } C_1}$
$\dfrac{584}{600} \times 100 = 97.4\%$	$\dfrac{178}{179} = 1.0$	$\dfrac{89}{93} = 0.96$

[a] mmol C = mm compound × number of C atoms.
[b] Oxidation value = (number of O atoms) − (number of H atoms/2).
[c] Calculated C_1 = expected amount for each C_2 observed. In the example above, for each mmole of pyruvate converted to ethanol, an equal amount of CO_2 is expected:

$$\text{Pyruvate } (C_3) \longrightarrow \text{ethanol } (C_2) + CO_2(C_1)$$

Similarly, for each mmol of pyruvate converted to acetate, an equal amount of CO_2 is expected:

$$\text{Pyruvate } (C_3) \longrightarrow \text{acetate } (C_2) + CO_2(C_1)$$

The **oxidation value** of a compound is determined from the number of oxygen atoms less one-half the number of hydrogen atoms. For glucose, which has 6 oxygen atoms and 12 hydrogen atoms, the oxidation value is 0 (Table 11-1). Thus, glucose is referred to as a **neutral compound**, and to be equally balanced the fermentation products should contain equivalent amounts of oxidized and reduced products. The carbon in the end products in the example shown in Table 11-1 is very close to the amount of fermented glucose (97.4%). In addition, the amount of oxidized product (carbon dioxide) is approximately equal to the amount of reduced products (glycerol and ethanol), so the O–R balance is very close to the theoretical value of 1.0. Figure 11-1 shows the general routes of formation and the oxidation values for a number of common fermentation products.

Another aspect of fermentation balances is the C_1 **balance**. The amount of expected C_1 product is calculated from the amounts of those products for which CO_2 or formate is expected as an accompanying product. For example, if a C_2 compound such as ethanol or acetate is among the final products, an equal amount of CO_2 will be expected, since ethanol is derived from pyruvate by decarboxylation. Note that in Table 11-1 the C_1 balance (0.96) is very close to the theoretical value of 1.0.

YEAST FERMENTATION

Although well over 800 species of yeasts are known, the paradigm for studies of physiology, intermediary metabolism, and genetics is *Saccharomyces cerevisiae*. This yeast species uses the EMP pathway of glucose metabolism under the conditions of neutral or slightly acid pH and an anaerobic environment. The major products formed under these conditions are carbon dioxide and ethanol. The sequence of reactions involved in this pathway is presented in Figure 8-1. However, certain facets of the alcoholic fermentation of yeast bear additional consideration.

When yeast cells are grown in high concentrations of glucose, the cells function under a regulatory control system called carbon catabolite repression or glucose repression. By using ^{13}C-labeled substrates, it is possible to make reliable estimates as to the metabolic fluxes occurring in the cell. In batch culture, 16.2 of every 100 molecules of glucose consumed by the cells enters the pentose phosphate pathway, whereas the same relative flux is 44.2 per 100 molecules in a chemostat. The TCA cycle does not operate as a cycle in batch cultures of growing yeast cells. In contrast, chemostat cultures of yeast use the TCA cycle in its intact form.

The ethanolic fermentation in yeast may be altered as a result of a number of changes in the culture medium. In the presence of sodium sulfite, acetaldehyde is trapped as a bisulfite addition complex, and glycerol is formed as a major product:

$$glucose + HSO_3^- \longrightarrow glycerol + acetaldehyde\text{-}HSO_3^- + CO_2$$

Under these conditions, acetaldehyde is unable to serve as a hydrogen acceptor. Dihydroxyacetone phosphate becomes the preferred hydrogen acceptor, yielding glycerol-3-phosphate, which is then hydrolyzed to glycerol and P_i. The fermentation is not shifted entirely in the direction of glycerol formation as it is not possible to add sufficient bisulfite to bind all of the acetaldehyde without incurring additional toxic effects. Thus, some ethanol will be found among the products.

Under alkaline conditions still another type of fermentation pattern is observed:

$$2 \; glucose \longrightarrow 2 \; glycerol + acetate + ethanol + 2CO_2$$

Under these conditions a dismutation reaction occurs in which 1 mol of acetaldehyde is oxidized to acetate and another is reduced to ethanol. This sequence is catalyzed by two NAD-linked dehydrogenases and requires a balance between the following reactions:

$$CH_3CHO + NAD^+ + H_2O \longrightarrow CH_3COOH + NADH + H^+$$

$$CH_3CHO + NADH + H^+ \longrightarrow CH_3CH_2OH + NAD^+$$

$$GA\text{-}3\text{-}P + P_i + NAD^+ \longrightarrow 1,3\text{-}diphosphoglycerate + NADH + H^+$$

$$dihydroxyacetone\text{-}P + NADH + H^+ \longrightarrow glycerol\text{-}3\text{-}P + NAD^+$$

This series of reactions is sometimes referred to as the glycerol-3-phosphate shuttle. The oxidation of acetaldehyde to acetate yields no ATP, since the reaction proceeds directly without the intermediary formation of acetyl-CoA.

The fermentation products formed by yeast can also be altered drastically through metabolic engineering. For example, introduction of a lactate dehydrogenase gene from bovine muscle (*LDH-A*) into *S. cerevisiae* engenders the production of lactic acid at levels rivaling those achieved by lactic acid bacteria.

In cell-free yeast extracts the addition of P_i results in a marked increase in the fermentation rate. The rate eventually subsides to that of the control without added P_i. Addition of a second quantity of P_i will again increase the fermentation rate. This phenomenon (sometimes referred to as the Harden-Young effect) results from incorporation of P_i into organic phosphate esters, particularly fructose-1,6-bisphosphate, F-6-P, G-6-P, and G-1-P. Inorganic phosphate is assimilated during the formation of 1,3-diphosphoglycerate. In subsequent steps, ADP is phosphorylated to ATP, which in turn permits the phosphorylation of additional glucose. Accumulation of phosphate esters limits the amount of P_i available for the metabolism of additional glucose. This effect can be counteracted by the addition of arsenate ion, which results in the formation of phosphoglyceryl arsenate. This compound hydrolyzes nonenzymatically, preventing the accumulation of phosphorylated intermediates. During the usual procedure of preparing cell-free extracts of yeast, the enzyme ATPase may be inactivated. Preparations in which ATPase is active, or to which exogenous ATPase is added, do not exhaust the P_i, and phosphate esters do not accumulate. Under these conditions, fermentation continues at a constant rate.

In a resting yeast cell suspension that is fermenting glucose, the introduction of oxygen results in the cessation of ethanol formation. This phenomenon, termed the **Pasteur effect**, has been studied extensively in attempts to elucidate the mechanism involved. Growing yeast cells do not show a noticeable Pasteur effect. Growing cells respire only 3 to 20% of the catabolized sugar and the rest is fermented. The mechanism involved in the cessation of fermentation in resting cells (i.e., under conditions of nitrogen starvation) is a progressive inactivation of the sugar transport systems. As a consequence, the contribution of respiration to glucose catabolism, which is small in growing cells, becomes quite significant under the conditions of nitrogen starvation.

Some of the difficulty in studying alcoholic fermentation in yeast results from the fact that several isozymes of alcohol dehydrogenase (ADH) are present, and the activities of these and other enzymes concerned with carbohydrate metabolism are partitioned between the cytoplasm, the mitochondrial membrane, and the mitochondrial cytosol. The key enzyme of fermentative alcohol production (ADH1) reduces acetaldehyde to ethanol in the presence of NADH and serves to maintain the redox balance of glucose fermentation occurring in the cytoplasm of yeast cells. During ethanol oxidation, ADH2 catalyzes the formation of acetaldehyde.

Oxidative utilization of ethanol in the mitochondrion involves ADH3. The inner mitochondrial membrane is impermeable to NAD^+ or NADH. Yeast mutants lacking ADH1 may regenerate cytoplasmic NAD^+ by reducing DHAP to glycerol-3-phosphate. Glycerol-3-phosphate can enter the mitochondrial membrane, where it is reoxidized to DHAP by an FAD-linked glycerol-3-phosphate dehydrogenase. The electrons generated in this oxidation are transferred to the respiratory chain. For this reason, ADH1-deficient cells are unable to grow anaerobically. As a result of the release of P_i from glycerol-3-phosphate, glycerol appears as an end product. However, yeast deficient in at least four of the known ADH isozymes still produces up to one-third of the theoretical maximum yield of ethanol from glucose. This may result from the presence of additional forms of

ADH. The published genome for *S. cerevisiae* identified at least seven genes that could potentially code for members of the ADH family. Glycerol is a major fermentation product, but acetaldehyde and acetate are also produced. Ethanol production in multiply ADH-deficient cells is dependent on mitochondrial electron transport associated with the inner mitochondrial membrane.

LACTIC ACID–PRODUCING FERMENTATIONS

In lactic acid–producing bacteria two major types of lactic acid fermentation occur. Those species that ferment glucose primarily to lactic acid are **homofermentative** and those that produce a mixture of other products along with lactic acid are **heterofermentative.** Some homofermentative species can form a wider variety of products if the conditions of the fermentation are altered. At alkaline pH the amounts of formate, acetate, and ethanol are increased at the expense of lactate (Table 11-2). As discussed earlier, isotope-labeling studies are helpful in determining the major pathway used in the formation of end products. *Lactobacillus casei, L. pentosus*, and *S. faecalis* were shown to produce ^{14}C-methyl-labeled lactate from glucose-1-^{14}C with 50% dilution and ^{14}C-carboxyl-labeled lactate from glucose-3,4-^{14}C without dilution of the specific activity. These labeling patterns fit the distribution of carbon atoms expected from utilization of the EMP pathway, as shown in Figure 11-2. In *L. lactis* an operon has been shown to code for phosphofructokinase, pyruvate kinase, and lactate dehydrogenase, three enzymes crucial to the operation of the EMP pathway. Distribution of ^{14}C-labeled substrates indicates that more than 90% of the carbon in fermented sugars is converted into metabolic end products, usually lactic acid (see Fig. 11-2). As little as 5% of the glucose carbon consumed is converted into biomass.

It has been proposed that the term **homolactic** be used to refer to lactic acid bacteria that contain aldolase but not transketolase. True homolactic acid bacteria metabolize glucose via the EMP pathway while those that are truly heterofermentative presumably follow a hexose monophosphate pathway. However, some data suggest that a combination of both pathways may be operative in certain organisms.

TABLE 11-2. Effect of pH on Glucose Fermentation by *S. faecalis*

Products Formed[a]	pH		
	5.0	7.0	9.0
Lactic acid	174.0	146.0	122.0
Acetic acid	12.2	18.8	31.2
Formic acid	15.4	33.6	52.8
Ethanol	7.0	14.6	22.4
Carbon recovery %	95.0	90.0	88.0
O-R index[b]	1.02	1.18	1.18

[a]Amounts of products shown as μmol/100 μmol of fermented glucose.
[b]See Table 7-5 for definition of O-R (oxidation–reduction) index and its calculation.
Source: From Wood, 1961. Fermentation of carbohydrates and related compounds. In *The Bacteria*, Vol. II, Metabolism. I. C. Gunsalus and R. Y. Stanier (Eds.). Academic Press, New York, p. 59.

Embden-Meyerhof-Parnas Pathway

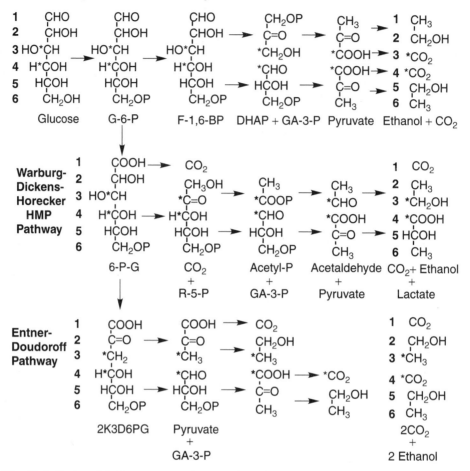

Fig. 11-2. Isotope-labeling patterns in central pathways of carbohydrate metabolism in microorganisms. $*C$ indicates ^{14}C-labeled carbon atoms.

One of the major factors controlling the type of product formed is the amount of reducing equivalents available. In the oxidative step in the triose phosphate pathway, NAD^+ is reduced. The electrode potential of this system ($NAD^+/NADH_2$) is -0.28 V. The electrode potential of the lactate dehydrogenase reaction is -0.18 V. Since this is at a higher potential, reduced NAD will donate its hydrogen atoms to pyruvate, reducing it to lactate. However, if some other hydrogen acceptor is available, reduced NAD will donate its hydrogen atoms to this acceptor rather than to pyruvate if the electrode potential of the alternative reaction is higher (see Fig. 9-6). For example, the potential for the reduction of acetaldehyde to ethanol is -0.07 V. If acetaldehyde is formed, then NADH will donate its hydrogen atoms to this system. Under the usual conditions of lactic acid fermentation, acetaldehyde does not appear to be formed in any appreciable quantity. However, by alteration of the pH, the fermentation may be diverted to a variety of products (Table 11-3). Since the streptococci appear to utilize the EMP pathway, the diversion of fermentation must occur at a step after pyruvate

TABLE 11-3. Examples of Acetone Butanol and Mixed-Solvent Fermentations

Products[a]	1[b]	2[c]	3[d]	4[e]	5[f]	6[g]
CO_2	195.5	220.0	207.0	174.0	215.0	195.0
H_2	233.0	165.9	111.1	229.4	137.0	54.0
Acetate	42.6	24.8	20.3	48.5	16.0	5.0
Butyrate	75.3	7.1	14.5	59.5		
Formate	trace				10.0	
Lactate				25.7		
Ethanol		4.9			122.0	95.0
Butanol		47.4	50.2			4.5
Acetone		22.3			28.0	6.0
Isopropanol			18.0			
Acetoin		5.7				
2,3-butanediol					12.0	39.0
C recovery (%)	97.0	98.0	93.5	98.0	105.0	100.0
O-R balance	1.02	1.01	1.05	0.99	0.90	1.13

[a] Amounts expressed as millimoles per 100 mmol of glucose used.
[b] *C. saccharobutyricum.* Donker, H. L. 1926. Ph.D. thesis, Technishe Hooge school, Delft, The Netherlands.
[c] *C. acetobutylicum.* Donker, 1926.
[d] *C. butylicum.* Sjolander, N. W. 1937. *J. Bacteriol.* **34**:419.
[e] *C. thermosaccharolyticum.* Osburn, G. L., R. W. Brown, and C. H. Werkman. 1937. *J. Biol. Chem.* **121**:685.
[f] *B. acetobutylicum.* Osburn, et al., 1937.
[g] *B. polymyxa.* Osburn, et al., 1937.

formation rather than by metabolizing hexose through an alternative pathway. Under these conditions, a dismutation of pyruvate occurs, giving rise to lactate, acetate, and formation through the following sequence:

$$CH_3COCOOH + CoASH \longrightarrow CH_3CO\text{-}SCoA + HCOOH$$

$$CH_3CO\text{-}ScoA + P_i \longrightarrow CH_3COOPO_3H_2 + CoASH$$

$$CH_3COCOOH + 2H \longrightarrow CH_3CHOHCOOH$$

$$CH_3COOPO_3H_2 + ADP \longrightarrow CH_3COOH + ATP$$

NET: 2 pyruvate $+ 2H + ADP + P_i \longrightarrow$ lactate $+$ acetate $+$ formate $+$ ATP

Ethanol, when produced, is formed via reduction of acetyl phosphate:

$$CH_3COOPO_3H_2 + 2(2H) \longrightarrow CH_3CH_2OH + P_i$$

Because of a difference in the pyruvate cleaving systems, formate, rather than carbon dioxide is the major one carbon product. Carbon dioxide is formed in small amounts by some homolactic organisms under conditions in which the fermentation is diverted from the production of lactate.

Homofermentative organisms, such as *S. faecalis*, have been shown to possess high levels of glucose-6-phosphate dehydrogenase and 6-phosphogluconate dehydrogenase.

However, a homofermentative pattern is maintained through regulatory interrelationships between the EMP and HMP pathways (see Fig. 11-3). The specific activation of lactate dehydrogenase (LDH) by FDP is characteristic of *S. bovis* and many other lactic acid–producing bacteria. The level of LDH, which is dependent on FDP for activity, decreases as fermentation becomes heterolactic. In *S. faecalis* a single NADP-linked 6PGD is observed when glucose is the primary energy source. Gluconate-adapted cells produce two 6-PGDs: one specific for NADP and the other specific for NAD. The NADP-linked enzyme is insensitive to FDP inhibition but is inhibited by ATP and other nucleotides. The G-6-P isomerase is inhibited by 6-PG.

When there is a cellular demand for ATP, these regulatory activities direct glucose carbon through the energy-producing EMP pathway. Inhibition of G-6-P isomerase by 6-PG prevents the accumulation of F-6-P. Alteration of the fermentative pattern of these organisms by change in pH or other environmental influences may result from the disruption of these regulatory functions. In *S. mutans*, triose phosphates (glyceraldehyde-3-phosphate or dihydroxyacetone phosphate) strongly inhibit pyruvate-formate lyase (PFL), the first step in the dismutation sequence. Inhibition by triose phosphates in cooperation with a reactivating effect of ferredoxin may regulate pyruvate-formate lyase activity.

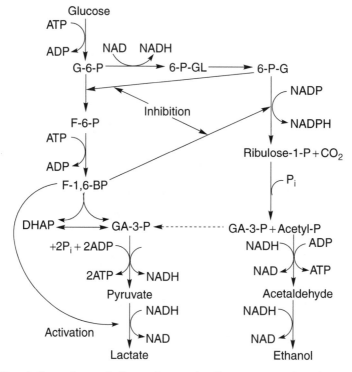

Fig. 11-3. Regulation of metabolic pathways in *Streptococcus faecalis*. 6-Phosphogluconate (6-P-G) inhibits the activity of hexose diphosphate isomerase. Fructose-1,6-bisphosphate (F-1,6-BP) inhibits the activity of 6-phosphogluconate dehydrogenase. Lactate dehydrogenase is activated by F-1,6-BP. (*Source*: Redrawn from Moat, A. G. 1985. Biology of lactic acetic, and propionic acid bacteria. In *Biology of Industrial Microorganisms*, A. L. Demain and N. A. Solomon (Eds.). Benjamin-Cummings, Menlo Park, CA. pp. 143–186.)

The genus *Lactococcus* includes *L. lactis*, *L. cremoris*, and *L. diacetylactis*. These species, as well as various species of *Leuconostoc* and *Lactobacillus*, produce acetoin, diacetyl, and 2,3-butanediol. These compounds are responsible for the characteristic and desirable flavor in certain dairy products such as butter. In certain other fermentations, the presence of these compounds is undesirable, and it is of importance to be able to minimize their production. As shown in Figure 11-4, there are several routes for the production of acetoin and diacetyl. With acetate as substrate, acetyl-CoA is formed directly from acetate without the intermediary formation of pyruvate. *Lactococcus diacetylactis*, *Lactobacillus casei*, and other lactate-producing organisms that require lipoic acid for growth activate acetate to acetyl phosphate and then form acetyl-CoA via acetate kinase and phosphotransacetylase. Therefore, in media devoid of lipoic acid, these organisms cannot form acetyl-CoA from pyruvate.

In a lipoate-free medium containing glucose and acetate, acetoin is formed via both acetyl-CoA-C_2-TPP condensation and C_2-C_3 condensation routes. In *L. lactis* the gene for α-acetolactate decarboxylase (*aldB*) is clustered with the genes for the branched-chain amino acids. This enzyme appears to play a dual role in this organism by regulating valine and leucine biosynthesis as well as catalyzing the second step of the pathway to acetoin and 2,3-butanediol. Control of the shift from homolactic to mixed-acid fermentation seems to be governed by the NADH/NAD$^+$ ratio related to the flux-controlling activity of glyyceraldehyde-3-phosphate dehydrogenase.

During growth in milk, PFL and lactate dehydrogenase are important to sustain growth of *L. lactis* and to regenerate reduced cofactors under anaerobic conditions. The gene encoding PFL has been cloned and characterized. The enzyme is activated

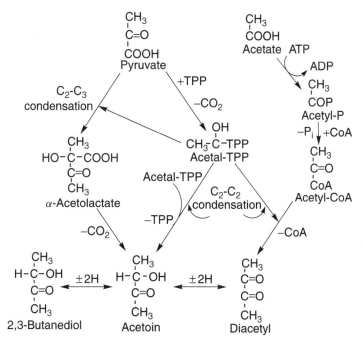

Fig. 11-4. Alternate routes to acetoin, diacetyl, and 2,3-butanediol. TPP, thiamine pyrophosphate.

via free-radical formation and is degraded in the presence of oxygen. Under anaerobic conditions, the carbon flux from pyruvate is distributed mainly between the two competing enzymes: lactate dehydrogenase (LDH) and pyruvate-formate lyase (PFL). Regulation of the shift from homofermentative to mixed-acid formation is associated primarily with the influence of allosteric effectors acting on the LDH and PFL enzymes.

Organisms that metabolize sugars via fermentative pathways and do not use molecular oxygen as a final hydrogen acceptor or generate ATP via oxidative phosphorylation are anaerobes. However, if such organisms grow in the presence of oxygen, they may be referred to as aerotolerant. In the presence of oxygen, toxic products such as hydrogen peroxide may be generated by the autooxidation of reduced FAD, and hydroxyl radicals may be formed by the interaction of this compound with superoxide. Anaerobic organisms that tolerate the presence of oxygen are usually able to do so by virtue of the presence of enzymes that protect them from these toxic compounds (see further discussions in Chapter 18). Some fermentative organisms such as *L. plantarum* are aerotolerant simply because they do not interact with molecular oxygen. High intracellular concentrations of Mn^{2+} also aid these organisms in scavenging superoxide anion.

Streptococcus pyogenes and other lactic acid bacteria produce NADH oxidase, a unique flavoprotein that catalyzes the four-electron reduction of oxygen to water. This enzyme may actually allow certain fermentative organisms to use molecular oxygen to advantage in producing energy. Under anaerobic conditions the conversion of pyruvate to lactate or a mixture of formate, acetate, and ethanol serves to oxidize NADH back to NAD^+. In the presence of oxygen, flavin-containing enzymes catalyze the oxidation of NADH by oxygen.

Although *S. mutans* depends on glycolysis and prefers an anaerobic environment, the presence of oxygen induces the formation of NADH oxidase and superoxide dismutase. Two types of NADH oxidase are induced in oxygen-tolerant strains of *S. mutans*. A peroxide-forming oxidase (Nox-1) catalyzes the reaction:

$$NADH + H^+ + O_2 \longrightarrow NAD^+ + H_2O_2$$

The second oxidase (Nox-2) catalyzes a four-electron reduction of oxygen by NADH:

$$2NADH + 2H^+ + O_2 \longrightarrow 2NAD^+ + 2H_2O$$

The presence of free FAD enhances the activity of Nox-1 but not that of Nox-2. *S. mutans* also produces a peroxidase (AhpC) encoded by *ahpC*, a gene located upstream of the *nox-1* gene. This peroxidase enzyme complements Nox-1 by reducing the peroxide to water and emulating the action of Nox-2. However, Nox-2 appears to play an important role in aerobic energy metabolism in oxygen-tolerant *S. mutans*, whereas Nox-1 serves a negligible role despite the fact that it is the only enzyme produced when the organism is grown under anaerobic conditions. This organism also produces another antioxidant system, encoded by the *dpr* gene that is independent of the Nox-1-AhpC system and can compensate for deletions of both *nox-1* and *ahpC* in mutant strains. The product of the *dpr* gene, Dpr, is an iron-binding protein that presumably contributes to oxygen tolerance by maintaining a low level of iron.

Most lactic acid bacteria do not produce cytochromes because they are unable to synthesize heme. However, when grown in the presence of hematin, *Enterococcus faecalis* produces functional cytochromes that yield ATP via oxidative phosphorylation. An *E. faecalis* gene cluster, *cydABCD*, encodes a functional cytochrome *bd*–terminal oxidase that is produced when the cells are grown in the presence of heme. The ability to form cytochromes seems to be limited to *E. faecalis* and a few strains of *L. lactis*. Other lactic acid bacteria produce menaquinones, which are isoprenoid quinones that function in electron transport and possibly in oxidative phosphorylation.

BUTYRIC ACID — AND SOLVENT-PRODUCING FERMENTATIONS

Members of the *Clostridium*, *Butyrivibrio*, *Bacillus*, and other less-well defined flora of marsh sediments, wetwood of living trees, and anaerobic sewage digestion systems produce butyric acid, butanol, acetone, isopropanol, or 2,3-butanediol. Hydrogen, carbon dioxide, acetate, and ethanol are commonly found among the fermentation products.

Clostridium acetobutylicum utilizes the EMP pathway for glucose catabolism with the formation of ethanol, carbon dioxide, hydrogen, acetone, isopropanol, butyrate, and butanol from pyruvate via acetyl-CoA as shown in Figure 11-1. During growth, the organism first forms acetate and butyrate (acidogenic phase). As the pH drops and the culture enters the stationary phase, there is a metabolic shift to solvent production (solvetogenic phase). The pivotal reaction in these fermentations is the formation of acetyl-CoA from pyruvate. Acetyl-CoA can then undergo condensation to form acetoacetate, which may be reduced to butyrate and butanol or cleaved via decarboxylation to acetone. Acetone may be further reduced to isopropanol. Acetyl-CoA may also be reduced to acetaldehyde and ethanol. With other species such as *B. acetoethylicum* or *B. polymyxa*, lactate or 2,3-butanediol may be observed as shown in Table 11-3.

In balancing the fermentation products formed in these complex fermentations, it is not difficult to account for the carbon in the products formed. However, formulating a theoretical scheme to account for a balance of oxidized and reduced products may be more difficult. Consider the following equation:

$$4 \text{ Glucose} \longrightarrow 2 \text{ acetate} + 3 \text{ butyrate} + 8CO_2 + 8H_2$$

$$\text{Carbon balance}: (4 \times 6) = (2 \times 2) + (3 \times 4) + (8 \times 1) = 24$$

$$\text{O–R balance}: 0 = 0 + 3(-2) + 8(+2) + 8(-1) = +16 - 14$$

The carbon atoms balance, but there is a deficit of some reduced product. In most clostridial fermentations there is a compensatory formation of small amounts of several reduced products as shown in Table 11-3. Production of these compounds appears to be necessary because the condensation of 2 mol of acetyl-CoA to form acetoacetyl-CoA and the subsequent conversion to acetone via acetoacetate decarboxylase results in a sharp decrease in the number of hydrogen acceptors available. Acetyl-CoA can act as a hydrogen acceptor, giving rise to ethanol, and acetone can be further reduced to isopropanol. In fermentations that show the best O–R balance, ethanol, acetoin, and 2,3-butanediol are found among the final end products.

The overall equation for mixed butyrate, acetone, and ethanol fermentation can account for both the carbon and the O–R balance:

$$C_6 \longrightarrow \text{butyrate } (C_4) + 2C_1 + 2H_2$$
$$C_6 \longrightarrow \text{acetone } (C_3) + 3C_1 + 2H_2$$
$$C_6 \longrightarrow 2 \text{ ethanol } (C_2) + 1C_1 + 2H_2$$

NET: $3C_6 \longrightarrow C_4 + C_3 + 2C_2 + 7C_1 + 6H_2$ (carbon balance: $18 = 18$)

O-R: $0 \longrightarrow (-1) + (-2) + 2(-2) + 7(+2) + 6(-1) = +14 - 14 = 0$

Generally, small amounts of either oxidized or reduced products will be formed to provide for a balance of the oxidized and reduced products. The oxidation values are given in Figure 11-1. Representative examples of acetone-butanol and mixed-solvent fermentations are shown in Table 11-3.

Formate is produced in relatively few of these fermentations. Although carbon dioxide and hydrogen gas are produced as in the fermentation of gram-negative organisms, in clostridia and other anaerobes hydrogen is formed through a pyruvate : ferredoxin (Fd) oxidoreductase without the intermediary production of formate. The reduced ferredoxin is converted to hydrogen by hydrogenase:

$$\text{pyruvate} + \text{Fd} \longrightarrow \text{acetyl-CoA} + CO_2 + \text{FdH} + H^+$$

$$\text{FdH} + H^+ \longrightarrow H_2 + \text{Fd}$$

The flow of electrons from NADH to FdH to H_2 explains why these organisms produce large quantities of H_2 and do not consume H_2.

An enzyme that contains two Fe_4S_4 clusters catalyzes the oxidation of pyruvate to acetyl-CoA, carbon dioxide, and reduced ferredoxin. Ferredoxin also contains two Fe_4S_4 clusters arranged in approximately two-fold symmetry within the ferredoxin protein. Hydrogenases that result in the production of hydrogen gas contain iron-only enzymes. The *C. pasteurianum* enzyme contains five distinct iron-sulfur clusters covalently attached to the protein. One Fe_2S_2 cluster and three Fe_4S_4 clusters transfer electrons to the hydrogen center, the presumed catalytic site of the enzyme. The hydrogen center is composed of a six-iron cluster and a two-iron subcluster bound to carbon monoxide and/or cyanide. The arrangement of the iron (Fe) and sulfur (S) atoms in the four-iron cluster is shown in Figure 11-5.

Production of acetic acid from glucose by *C. thermoaceticum* is efficient in that 3 molecules of acetate are produced per molecule of glucose. This is achieved by a novel series of reactions that involves the synthesis of acetate from carbon dioxide:

$$C_6H_{12}O_6 \longrightarrow \text{EMP} \longrightarrow 2CH_3COOH + 2CO_2 + 8H^+ + 8e^-$$
$$2CO_2 + 8H^+ + 8e^- \longrightarrow CH_3COOH$$

NET: $C_6H_{12}O_6 \longrightarrow 3CH_3COOH$

It is particularly significant that the synthesis of acetate from carbon dioxide by this organism was the first demonstration of total synthesis of an organic compound from

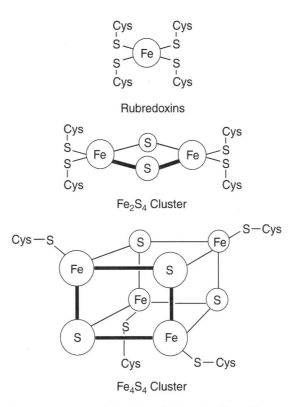

Fig. 11-5. proposed arrangements of the iron (Fe) and sulfur (S) atoms in rubredoxin, ferredoxin, and hydrogenase systems. Cys-S represents the linkage with the S in cysteine. In 8Fe/8S systems, the iron and sulfur atoms are arranged in two tetrameric Fe_4S_4 clusters.

carbon dioxide by a heterotrophic organism. This pioneering work by Harland G. Wood in the 1940s and 1950s conclusively demonstrated that all forms of life fix carbon dioxide. However, most heterotrophs can only add carbon dioxide to a preexisting compound and form a carboxyl group. Bacteria that can form an organic compound from carbon dioxide and hydrogen contain hydrogenase, an enzyme that converts hydrogen to two protons and two electrons. These electrons provide the necessary reductive capacity for the utilization of carbon dioxide. The overall reaction involves participation of ferredoxin as a reduced electron carrier and the enzymes hydrogenase, carbon monoxide dehydrogenase, and methylenetetrahydrofolate reductase.

FERMENTATIONS OF THE MIXED-ACID TYPE

Under anaerobic conditions and in the absence of alternate electron acceptors, members of the *Enterobacteriaceae* (*Escherichia*, *Enterobacter*, *Salmonella*, *Klebsiella*, and *Shigella*) ferment glucose to a mixture of acetic, formic, lactic, and succinic acids and ethanol (Table 11-4). The production of acetate and formate as major products is notable. As much as 85% of the glucose fermented by *E. coli* is metabolized via the EMP pathway. Other pathways must contribute to the products to some extent,

TABLE 11-4. Examples of Mixed-Acid Fermentations

Products[a]	*Escherichia Coli*	*Enterobacter Aerogenes*	*Salmonella Typhi*
Lactate	108.8	53.4	121.7
Ethanol	41.3	59.4	25.4
Acetate	32.0	10.1	25.6
Formate	1.6	5.5	39.33
CO_2	54.0	126.9	0
H_2	45.2	44.2	0
Succinate	18.0	6.0	10.8
Acetoin	0	0.4	0
2,3-Butanediol	0	34.6	0
Carbon recovery (%)	100.0	99.5	93.3
O-R balance	0.99	0.99	0.97

[a] Amounts are expressed as millimoles per 100 mmol glucose used.

however. As opposed to clostridial fermentations in which acetyl-CoA, carbon dioxide, and dihydrogen usually arise without formate as an intermediate, formate is consistently found as a product of sugar metabolism by enteric bacteria. A CoA-dependent pyruvate-formate lyase (Pfl) encoded by the *pfl* gene initiates the sequence

$$CH_3COCOOH + CoASH \longleftrightarrow CH_3COSCoA + HCOOH$$

Phosphotransacetylase converts acetyl-CoA to acetyl phosphate:

$$CH_3COSCoA + P_i \longleftrightarrow CH_3COOPO_3H_2 + CoASH$$

Acetate kinase (encoded by the *ackA* gene in *E. coli*) generates ATP from acetyl phosphate:

$$CH_3COOPO_3H_2 + P_i \longleftrightarrow CH_3COOH + ATP$$

This enzyme results in the production of acetate and generates a major portion of the ATP produced during anaerobic growth. The formate-hydrogen lyase (FHL) system converts formate to H_2 and CO_2. The FHL system actually consists of two enzymes: formate dehydrogenase-H (FDH-H), which yields carbon dioxide and a reduced carrier; and a hydrogenase-3, which produces H_2 from the reduced carrier:

$$HCOOH + carrier \longrightarrow CO_2 + carrier\text{-}2H$$
$$carrier\text{-}2H \longleftrightarrow H_2 + carrier$$
$$\overline{}$$
$$\textbf{Net: } HCOOH \longrightarrow CO_2 + H_2$$

Induction of FDH-H (encoded by *fdhF*) requires the presence of formate, molybdate, the absence of electron acceptors, such as oxygen or nitrate, and acidic pH. Expression of *fdhF* requires an alternate σ factor (NtrA) and an upstream activating sequence (UAS). Expression of *pfl* is dependent on phosphoglucoisomerase and phosphofructokinase activities.

Lactic acid arises from pyruvate through the activity of LdhA induced under the conditions of anaerobiosis and a low pH. Production of ethanol occurs via the reduction of acetyl-CoA to ethanol through a CoA-linked acetaldehyde dehydrogenase coupled to an NAD-linked alcohol dehydrogenase.

Acetoin is formed by *Enterobacter aerogenes* but not by *E. coli*. A variety of other organisms (e.g., *Serratia*, *Erwinia*) also form acetoin or its oxidation (diacetyl) or reduction (2,3-butanediol) products, as shown in Figures 11-1 and 11-4. In gram-negative organisms, thiamine pyrophosphate is involved as a cofactor and α-acetolactate is produced as an intermediate in acetoin formation. Acetate induces the production of the acetolactate-forming enzyme, acetolactate decarboxylase, and diacetyl reductase. Diacetyl reductase also functions as an acetoin dehydrogenase in *E. aerogenes* and serves as a regulator of the balance between acetoin and 2,3-butanediol.

In *E. coli* the direct formation of diacetyl from 2-hydroxyethyl-TPP and acetyl-CoA has been demonstrated. Diacetyl is reduced to acetoin by an $NADP^+$-specific diacetyl reductase that does not act on acetoin (Fig. 11-4). Acetoin is a major catabolic product of *B. subtilis* grown aerobically in glucose. Since acetoin is neutral, large amounts of glucose can be produced without excess acid production.

Two genes, *ilvBN* (coding for acetohydroxy acid synthase) and *alsS* (coding for α-acetolactate synthase), are involved in producing acetolactate from pyruvate in *B. subtilis*. Acetolactate can be converted to acetoin by spontaneous decarboxylation at low pH or via the action of acetolactate decarboxylase (encoded by *alsD*). Acetoin serves as a carbon storage compound. It is reimported from the growth medium and utilized during the stationary phase when other carbon sources have been depleted. There appears to be more than one pathway of acetoin utilization in *B. subtilis*. Three genes form an operon, *acuABC*, genes that encode acetoin utilization enzymes. Another system, the *acoABCL* operon, encoding the multicomponent acetoin dehydrogenase enzyme complex, appears to constitute the main route of acetoin catabolism in this organism. Expression of this operon is induced by acetoin and repressed by glucose. An *acoR* gene, located downstream from the *acoABCL* operon, encodes a positive regulator, which stimulates the transcription of the operon.

A survey of anaerobic fermentation balances with various substrates using a nonintrusive nuclear magnetic resonance (NMR) spectroscopy technique showed that substrates more reduced than glucose yield more reduced product as ethanol. More oxidized substrate resulted in the production of more neutral product in the form of acetate. The redox level of the substrate is important in governing the ratio of oxidized to reduced products. For example, sugar alcohols, which are more reduced than the corresponding hexoses, must yield a higher proportion of more reduced fermentation products to achieve hydrogen balance.

Fermentation of citrate (oxidation value: $+3$) yields sufficient oxidized products in the form of HCOOH (oxidation value: $+1$) and carbon dioxide (oxidation value: $+2$) to balance the oxidation/reduction equation. The fermentation of citrate by *Klebsiella pneumoniae* begins with cleavage by citrate lyase to form acetate and oxaloacetate. Oxaloacetate is decarboxylated by the oxaloacetate decarboxylase Na^+ pump. Part of the energy from this reaction is used for citrate transport. The formed pyruvate is degraded by pyruvate-formate lyase to acetyl-CoA and formate. Phosphotransacetylase forms acetyl phosphate from acetyl-CoA and P_i. Transfer of high-energy phosphate from acetyl phosphate to ADP generates ATP, leaving acetate as a final end product. A portion of the formate is acted upon by formate-hydrogen lyase, yielding CO_2 and H_2. An NADP-reducing hydrogenase converts H_2 to $NADPH + H^+$, which is available for biosynthetic reactions.

Utilization of glycerol (oxidation value: -1) involves transport into the cell via a facilitator encoded by *glpF*. Glycerol is phosphorylated by glycerol kinase (encoded by *glpK*) trapping it inside the cell as glycerol-3-phosphate (G-3-P). Under anaerobic conditions a dehydrogenase encoded by the *glpACB* operon yields dihydroxyacetone phosphate, which can then be metabolized via the triose phosphate pathway. In the presence of oxygen, an aerobic dehydrogenase encoded by the *glpD* operon accomplishes the same end. The *fnr* gene product FNR, a pleiotropic activator of genes involved in anaerobic respiration, regulates the utilization of glycerol, G-3-P, and glycerophosphate diesters. The Arc (aerobic respiratory control system) controls the expression of several genes encoding enzymes involved in aerobic respiration.

PROPIONIC ACID FERMENTATION

Propionate, acetate, and carbon dioxide are the major products of the fermentation of glucose, glycerol, and lactate by *Propionibacterium*, *Veillonella*, *Bacteroides*, and some species of clostridia. *Clostridium propionicum*, *Bacteroides ruminicola*, and *Peptostreptococcus* can remove water from lactate to form acrylate with subsequent reduction to propionate. Coenzyme A and lactyl CoA dehydrase mediate the reaction

$$
\begin{array}{cccc}
\text{CH}_3 & \text{CH}_2 & \text{CH}_3 & \text{CH}_3 \\
| & || & | & | \\
\text{HCOH} \xrightarrow{-\text{H}_2\text{O}} & \text{HC} \xrightarrow{+2\text{H}} & \text{CH}_2 \xrightarrow{-\text{COSCoA}} & \text{CH}_2 \\
| & | & | & | \\
\text{COSCoA} & \text{COSCoA} & \text{COSCoA} & \text{COOH} \\
\text{Lactyl-CoA} & \text{Acrylyl-CoA} & \text{Propionyl-CoA} & \text{Propionate}
\end{array}
$$

The two hydrogen atoms required for the reduction of acrylyl-CoA to propionate to propionyl-CoA arise through the formation of acetate, CO_2, and 2(2H) from a portion of the total lactate utilized:

$$\text{lactate} \longrightarrow \text{acetate} + CO_2 + 2(2H)$$

$$\underline{2 \text{ lactate} + 2(2H) \longrightarrow 2 \text{ propionate}}$$

Net: $3 \text{ lactate} \longrightarrow 2 \text{ propionate} + \text{acetate} + CO_2$

In *Propionibacterium* and *Veillonella*, propionate formation arises via a more complex series of reactions:

$$1.5 \text{ glucose} + 3\text{ADP} + 3\text{P}_i \longrightarrow 3 \text{ pyruvate} + 3\text{ATP} + 3(2H)$$

$$\text{pyruvate} + \text{ADP} + \text{P}_i \longrightarrow \text{acetate} + CO_2 \text{ ATP} + 2H$$

$$\underline{2 \text{ pyruvate} + 2\text{ADP} + 2\text{P}_i + 4(2H) \longrightarrow 2 \text{ propionate} + CO_2 + 6\text{ATP} + 2H_2O}$$

Net: $1.5 \text{ glucose} + 6 \text{ ADP} + 6\text{P}_i \longrightarrow \text{acetate} + 2 \text{ propionate} + CO_2 + 6\text{ATP} + 2H_2O$

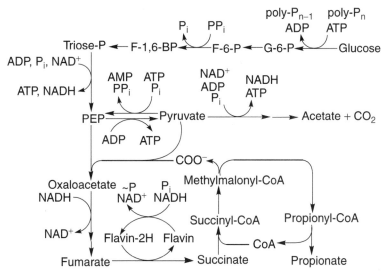

Fig. 11-6. formation of propionate, acetate, and CO$_2$ by *Propionibacterium*. (*Source*: From H. G. Wood, 1986, personal communication.)

The details of this pathway from glucose to propionate, acetate, and CO$_2$ in propionibacteria are shown in Figure 11-6.

A major factor in the elucidation of this route of propionate formation was the demonstration of transcarboxylation by methylmalonyl-oxaloacetate transcarboxylase. This multimeric transcarboxylase catalyzes the reversible transfer of a carboxyl group from methylmalonyl CoA to pyruvate to form propionyl-CoA and oxaloacetate. The transferred carboxyl group is never released or exchanged with the CO$_2$; nor is it exchanged with the CO$_2$ dissolved in the medium. In this reaction, biotin plays an important catalytic role. Cobalamin (vitamin B$_{12}$) also serves as a cofactor in the formation of methylmalonyl-CoA from succinyl-CoA in the reaction sequence

$$\text{enz-biotin-CO}_2 + \text{pyruvate} \longrightarrow \text{oxaloacetate} \longrightarrow \text{enz-biotin}$$

$$\text{oxaloacetate} + 4\text{H} \longrightarrow \text{succinate} + \text{H}_2\text{O}$$

$$\text{succinyl-CoA-B}_{12}\text{-enz} \longrightarrow \text{methylmalonyl-CoA} + \text{B}_{12}\text{-enz}$$

$$\text{methylmalonyl-CoA} + \text{enz-biotin} \longrightarrow \text{propionyl-CoA} + \text{enz-biotin-CO}_2$$

$$\text{propionyl-CoA} + \text{succinate} \longrightarrow \text{succinyl-CoA} + \text{propionate}$$

_____

NET: $\text{pyruvate} + 4\text{H} \longrightarrow \text{propionate} + \text{H}_2\text{O}$

As shown in Figure 11-6, the hydrogen atoms required for the reduction of oxaloacetate to succinate are obtained from reactions in the conversion of glucose to acetate and carbon dioxide via pyruvate. Under appropriate conditions and in the presence of the requisite enzymes, yields of 11 mol of ATP per 3 mol of glucose can be achieved:

$$3 \text{ glucose} + 4\text{PP}_i + 11\text{ADP} \longrightarrow 2 \text{ acetate} + 4 \text{ propionate} + 2\text{CO}_2 + 11\text{ATP} + 4\text{PP}_i$$

The participation of polyphosphate and PP$_i$ also explains the high cell yields observed in the growth of propionibacteria. Demonstration of reactions in which the energy inherent in polyphosphate and PP$_i$ may be utilized and not wasted through hydrolysis has far-reaching implications in other systems, since many biological reactions yield PP$_i$ or polyphosphate as products. Some examples are as follows:

$$\text{Glucokinase : glucose} + \text{polyP}_n \longrightarrow \text{G-6-P} + \text{polyP}_{n-1}$$

$$\text{Phosphofructokinase : F-6-P} + \text{PP}_i \longrightarrow \text{F-1,6-BP} + \text{P}_i$$

$$\text{Carboxytransphosphorylase : PEP} + \text{CO}_2 + \text{P}_i \longrightarrow \text{oxaloacetate} + \text{PP}_i$$

Net conversion of pyruvate to oxaloacetate:

$$\text{pyruvate} + \text{ATP} + \text{P}_i \longrightarrow \text{oxaloacetate} + \text{AMP} + 2\text{PP}_i$$

ACETIC ACID FERMENTATION

The acetic acid bacteria are divided into two genera: *Acetobacter* and *Gluconobacter*. Both species are obligate aerobes that oxidize sugars, sugar alcohols, and ethanol with the production of acetic acid as the major end product. Electrons from these oxidative reactions are transferred directly to the respiratory chain. The respiratory chain of *G. suboxydans* consists of cytochrome *c*, ubiquinone, and a terminal cytochrome-*o* ubiquinol oxidase. In *A. aceti* the composition of the respiratory chain varies according to the degree of aeration during growth.

Gluconobacter suboxydans lacks a functional TCA cycle, although all of the enzymes of this cycle except succinate dehydrogenase are present. This organism cannot ferment glucose or other carbohydrates since the EMP and Entner-Doudoroff pathways are not present. A modified pentose cycle is used for the metabolism of glucose under aerobic conditions (Fig. 11-7). *Acetobacter* species have a functional TCA cycle. The acetyl phosphate resulting from the C$_2$-C$_3$ cleavage of pentose is oxidized to CO$_2$ and energy is generated via oxidative phosphorylation.

A characteristic activity of *Acetobacter* and *Gluconobacter* is the oxidation of ethanol to acetic acid. Ethanol oxidation occurs via two membrane-associated dehydrogenases: alcohol dehydrogenase and acetaldehyde dehydrogenase:

$$\text{CH}_3\text{CH}_2\text{OH} \longrightarrow \text{CH}_3\text{CHO} + 2\text{H} \longrightarrow \text{CH}_3\text{COOH} + 2\text{H}$$

The electrons generated in ethanol oxidation are transferred directly to the respiratory chain.

Commercial production of acetic acid is carried out in two steps. Alcohol is first produced by yeast fermentation converting glucose to CO$_2$ and alcohol. The second stage uses *A. aceti* to convert the ethanol to acetic acid, with the reducing equivalents being transferred to oxygen via the respiratory chain as described above. The process requires carefully controlled conditions to achieve the maximum yield of acetate.

Acetate is a major product of carbohydrate metabolism in *B. subtilis*. Conversion of pyruvate to acetate occurs primarily during exponential growth. During the stationary phase, acetate is generated via butanediol. Overacidification of the medium

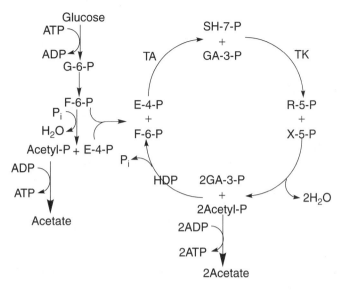

Fig. 11-7. Modified pentose cycle for *Acetobacter* and *Gluconobacter*. TA, transaldolase; TK, transketolase; E-4-P, erythrose-4-phosphate; SH-7-P, sedoheptulose-7-phosphate; R-5-P, ribose-5-phosphate; X-5-P, xylulose-5-phosphate; HDP, hexose diphosphate; GA-3-P, glyceraldehyde-3-phosphate. (*Source*: From A. G. Moat, 1985. Biology of lactic, acetic, and propionic acid bacteria. In *Biology of Industrial Microorganisms*. A. L. Demain and N. A. Solomon (Eds.). Benjamin Cummings, Menlo Park, CA, pp. 143–188.)

due to pyruvate and acetate accumulation during exponential growth is apparently alleviated by acetoin production under aerobic conditions or 2,3-butanediol under anaerobic conditions. These compounds can be reutilized in the stationary phase. The phosphotransacetylase (*pta*) gene, together with the acetate kinase (*ackA*) gene, encode the enzymes that catalyze the conversion of acetyl-CoA to acetate via acetyl~P. These two genes are organized into an operon and are involved in the maintenance of the intracellular acetyl-CoA and acetyl~P pools.

BIBLIOGRAPHY

Fermentation Pathways

Böck, A. 2000. Fermentation. In J. Lederberg (ed.), *Encyclopedia of Microbiology*, 2nd ed., Vol. 20. Academic Press, San Diego, pp. 343–9.

Yeast Fermentation

Gombert, A. K., M. M. DosSantos, B. Christensen, and J. Nielsen. 2001. Network identification and flux quantification in the central metabolism of *Saccharomyces cerevisiae* under different conditions of glucose repression. *J. Bacteriol.* **183**:1441–51.

Hanegraaf, P. P. F., A. H. Stouthamer, and S. A. L. M. Kooijman. 2000. A mathematical model for yeast respiro-fermentative physiology. *Yeast* **16**:423–37.

Ostergaard, S., L. Olsson, and J. Nielsen. 2000. Metabolic engineering of *Saccharomyces cerevisiae*. *Microbiol. Mol. Biol. Rev.* **64**:34–50.

Walker, G. M. 2000. Yeasts. In J. Lederberg (ed.), *Encyclopedia of Microbiology*, 2nd ed., Vol. 4. Academic Press, San Diego, pp. 939–53.

Walker, G. M. 1998. *Yeast Physiology and Biotechnology*. John Wiley & Sons, Chichester, UK.

Lactic Acid Fermentation

Arnau, J., et al. 1997. Cloning, expression, and characterization of the *Lactococcus lactis pfl* gene, encoding pyruvate formate-lyase. *J. Bacteriol.* **179:**5884–91.

Garrigues, C., P. Loubiere, N. D. Lindley, and M. Cocaign-Bosquet. 1997. Control of the shift from homolactic acid to mixed-acid fermentation in *Lactococcus lactis*: predominant role of the NADH/NAD$^+$ ratio. *J. Bacteriol.* **179:**5282–7.

Gibson, C. M., T. C. Mallett, A. Claiborne, and M. G. Caparon. 2000. Contribution of NADH oxidase to aerobic metabolism of *Streptococcus pyogenes*. *J. Bacteriol.* **182:**448–55.

Goupil-Feuillerat, N., et al. 2000. Dual role of α-acetolactate decarboxylase in *Lactococcus lactis* subsp. *lactis*. *J. Bacteriol.* **179:**6285–93.

Higuchi, M., Y. Yamamoto, L. B. Poole, et al. 1999. Functions of two types of NADH oxidases in energy metabolism and oxidative stress of *Streptococcus mutans*. *J. Bacteriol.* **181:**5940–7.

Litchfield, J. H. 2000. Lactic acid, microbially produced. In J. Lederberg (ed.), *Encyclopedia of Microbiology*, 2nd ed., Vol. 3. Academic Press, San Diego, pp. 9–17.

Llanos, R. M., C. J. Harris, A. J. Hillier, and B. E. Davidson. 1993. Identification of a novel operon in *Lactococcus lactis* encoding three enzymes for lactic acid synthesis: phosphofructokinase, pyruvate kinase, and lactate dehydrogenase. *J. Bacteriol.* **175:**2541–51.

Melchiorsen, C. R., et al. 2000. Synthesis and posttranslational regulation of pyruvate formate-lyase in *Lactococcus lactis*. *J. Bacteriol.* **182:**4783–8.

Moat, A. G. 1985. Biology of the lactic, acetic, and propionic acid bacteria. In A. L. Demain and N. A. Solomon (eds.), *Biology of Industrial Microorganisms*. Benjamin-Cummings, Menlo Park, CA, pp. 143–88.

Novák, L., and P. Loubiere. 2000. The metabolic network of *Lactococcus lactis*: distribution of ^{14}C-labeled substrates between catabolic and anabolic pathways. *J. Bacteriol.* **182:**1136–43.

Somkuti, G. A. 2000. Lactic acid bacteria. In J. Lederberg (ed.), *Encyclopedia of Microbiology*, 2nd ed., Vol. 3. Academic Press, San Diego, pp. 1–8.

Winstedt, L., L. Frankenberg, L. Hederstedt, and C. von Wachenfeldt. 2000. *Enterococcus faecalis* V583 contains a cytochrome *bd*-type respiratory oxidase. *J. Bacteriol.* **182:**3863–6.

Yamamoto, Y., M. Higuchi, L. B. Poole, and Y. Kamio. 2000. Role of the *dpr* product in oxygen tolerance in *Streptococcus mutans*. *J. Bacteriol.* **182:**3740–7.

Butyric Acid and Solvent-Producing Fermentations

Adams, M. W. W., and E. I. Stiefel. 1998. Biological hydrogen production: not so elementary. *Science* **282:**1842–3.

Cheryan, M. 2000. Acetic acid production. In J. Lederberg (ed.), *Encyclopedia of Microbiology*, 2nd ed., Vol. 1. Academic Press, San Diego, pp. 13–17.

Das, A., and L. G. Ljungdahl. 2000. Acetogenesis and acetogenic bacteria. In J. Lederberg (ed.), *Encyclopedia of Microbiology*, 2nd ed., Vol. 1. Academic Press, San Diego, pp. 18–27.

Ljungdahl, L. G. 1994. The acetyl-CoA pathway and the chemiosmotic generation of ATP during acetogenesis. In H. L. Drake (ed.), *Acetogenesis*. Chapman Hall, New York.

Mitchell, W. J. 1999. Physiology of carbohydrate to solvent conversion by clostridia. *Adv. Microb. Physiol.* **39:**31–130.

Nair, R. V., et al. 1999. Regulation of the *sol* locus genes for butanol and acetone formation in *Clostridium acetobutylicum* ATCC824 by a putative transcriptional repressor. *J. Bacteriol.* **181**:319–30.

Peters, J. W., W. N. Lanzilotta, B. J. Lemon, and L. C. Seefelt. 1998. X-ray crystal structure of the Fe-only hydrogenase (Cpl) from *Clostridium pasteurianum* to 1.8 angstrom resolution. *Science* **282**:1853–8.

Saint-Amans, S., et al. 2001. Regulation of carbon and electron flow in *Clostridium butyricum* VPI 3266 grown on glucose-glycerol mixtures. *J. Bacteriol.* **183**:1748–54.

Mixed-Acid Fermentations

Ali, N. O., J. Bignon, G. Rapoport, and M. Debarbouille. 2001. Regulation of the acetoin catabolic pathway is controlled by sigma L in *Bacillus subtilis*. *J. Bacteriol.* **183**:2497–504.

Kessler, D., and J. Knappe 1996. Anaerobic dissimilation of pyruvate. In F. C. Neidhardt, et al. (eds.), *Escherichia coli and Salmonella: Cellular and Molecular Biology*, 2nd ed. ASM Press, Washington, DC, pp. 199–205.

Steuber, J., W. Krebs, M. Bott, and P. Dimroth. 1999. A membrane-bound NAD(P)$^+$-reducing hydrogenase provides reduced pyridine nucleotides during citrate fermentation by *Klebsiella pneumoniae*. *J. Bacteriol.* **181**:241–5.

Propionic Acid Fermentation

Deborde, C., and P. Boyaval. 2000. Interactions between pyruvate and lactate metabolism in *Propionibacterium freudenreichii* subsp. *shermanii*: in vivo ^{13}C nuclear magnetic resonance studies. *Appl. Environ. Microbiol.* **66**:2012–20.

Acetic Acid Fermentation

Cheryan, M. 2000. Acetic acid production. In J. Lederberg (ed.), *Encyclopedia of Microbiology*, 2nd ed., Vol. 1. Academic Press, San Diego, pp. 13–7.

Presecan-Siedel, E., A. Galinier, R. Longin, et al. 1999. Catabolite regulation of the *pta* gene as part of carbon flow pathways in Bacillus subtilis. *J. Bacteriol.* **181**: 6889–97.

CHAPTER 12

PHOTOSYNTHESIS AND INORGANIC METABOLISM

Organisms that use C_1 compounds (e.g., CO_2 or CH_4) as their major or sole source of carbon and energy are called **autotrophs**. **Methylotrophs** use methane (CH_4) or methanol (CH_3OH) as their sole source of carbon. Autotrophic organisms that use light as a source of energy are **photoautotrophs.** The source of utilized energy serves as a physiological distinction, as shown in Table 12-1.

CHARACTERISTICS AND METABOLISM OF AUTOTROPHS

Photosynthetic Bacteria and Cyanobacteria

Most living forms on earth are ultimately dependent on the process of photosynthesis. This process occurs in green plants, algae, cyanobacteria, and photosynthetic bacteria. A large community of marine microorganisms, generally referred to as phytoplankton, contains many species of cyanobacteria (representative examples: *Prochlorococcus, Synechococcus*, and *Anabaena*) that comprise the largest population of photosynthetic organisms on the planet. Many plants and microorganisms also conduct nitrogen fixation (see Chapter 14), providing a basis of continuity for all other life. Some reactions in the photosynthetic process are quite slow and inefficient. Therefore, one major aspect of the study of photosynthetic organisms is the improvement of the efficiency of the process through genetic engineering. Photosynthetic bacteria are found in the deeper waters of permanently stratified (**meromictic**) lakes where the conditions are anaerobic, but light is available.

Differentiation between the photosynthetic bacteria and the cyanobacteria (sometimes referred to in the past as blue-green algae) is based on the type of photosensitive pigments produced. Prokaryotes such as the cyanobacteria (*Anabaena, Synechococcus, Prochlorococcus*) that conduct true photosynthesis contain **chlorophyll a**, which is common to the eukaryotic algae and green plants. Water serves as the electron

TABLE 12-1. Principal Groups of Autotrophs

Energy Source	Group	Genera
H_2	Hydrogen bacteria	*Ralstonia* (formerly *Alcaligenes*), *Nocardia, Xanthobacter, Pseudomonas derxia*
NH_3	Nitrifying bacteria	*Nitrosolobus, Nitrosomonas, Nitrocystis*
NO_2	Nitrifying bacteria	*Nitrobacter, Nitrospina, Nitrosococcus*
N_2	Nitrogen-fixing bacteria	*Azotobacter, Anabaena, Prochlorococcus, Rhizobium*
H_2S, S, $S_2O_3^{2-}$	Sulfur bacteria	*Thiobacillus, Sulfolobus, Desulfotomaculum, Wolinella, Desulfovibrio, Beggiatoa*
Fe^{2+}	Iron bacteria	*Gallionella, Sphaerotilus, Thiobacillus ferrooxidans, Leptothrix, Shewanella oneidensis*
CH_4, CH_3OH	Methylotrophs	*Hyphomicrobium, Methylomonas, Methylobacterium, Methylosinus, Paracoccus, Pseudomonas*
H_2, CO_2, Formate Methylamine Trimethylamine Acetate	Methanogens	*Methanobacterium, Methanobrevibacter, Methanococcus, Methanomicrobium, Methogenium, Methanospirillum, Methanosarcina*
Light	Phototrophs	*Rhodobacter, Anabaena, Prochlorococcus, Synechococcus,* algae

donor and oxygen is generated by photolysis. The purple bacteria (*Thiorhodaceae*) contain **bacteriochlorophyll a** or **b**. The *Thiorhodaceae* utilize H_2S and/or inorganic compounds as electron donors, and their metabolism does not involve molecular oxygen (i.e., it is anaerobic). Green bacteria (*Chlorobacteriaceae*) contain **bacteriochlorophyll c** or **d** and small amounts of **bacteriochlorophyll a**, using H_2S and/or organic compounds as electron donors and following anaerobic metabolic pathways. The structure and biosynthesis of bacteriochlorophyll has been studied in detail and is discussed in Chapter 15.

Production of light-absorbing carotenoid pigments also represents a differentiating characteristic. All algae and green plants contain β-carotene. The purple sulfur and nonsulfur bacteria contain a variety of carotenoid pigments of both aliphatic and aryl types, whereas the green bacteria contain only aryl carotenoids (Fig. 12-1). The carotenoid pigments absorb light energy and transfer it to the chlorophyll molecules of the antenna.

Algae and the cells of higher plants contain chloroplasts. Comparable structures (**chromatophores**) are observed in the photosynthetic bacteria. The photosynthetic apparatus of *Rhodococcus sphaeroides* consists of a series of **intracytoplasmic membranes** (ICMs) that appear as vesicular invaginations originating from the cytoplasmic membrane. *R. sphaeroides* carries out anoxigenic photosynthesis but is also capable of both aerobic and anaerobic respiration as well as fermentation.

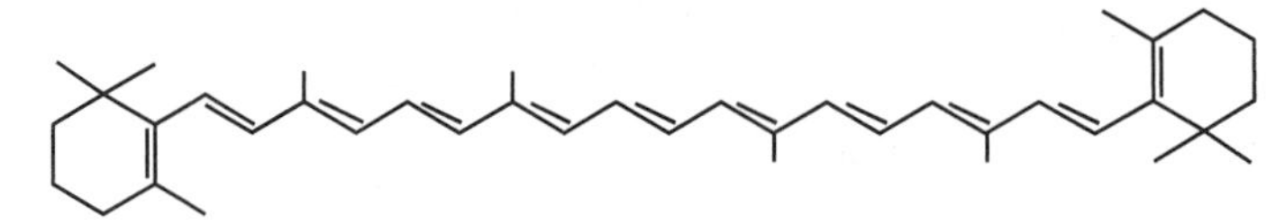

Alicyclic, β-carotene, algae, green plants

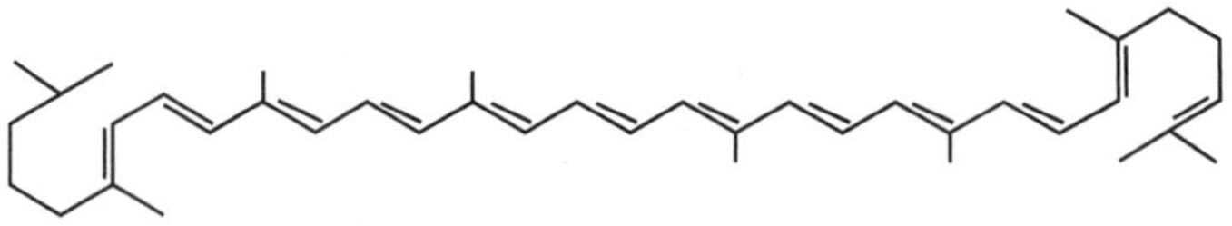

Aliphatic, lycopene, purple bacteria

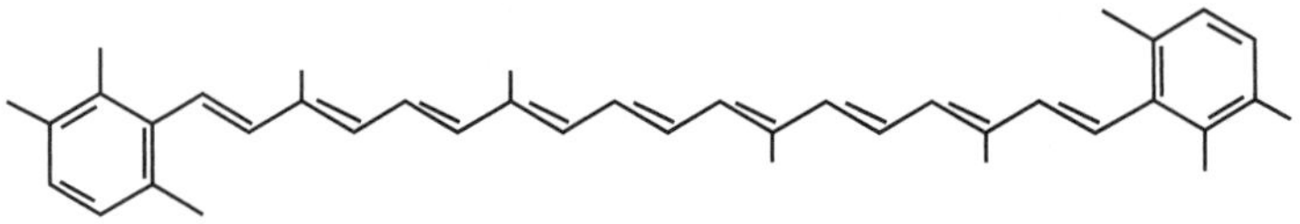

Aryl, isorenieratene, green bacteria

Fig. 12-1. Examples of carotenoid pigments produced by plants, algae, and photosynthetic bacteria. Although there are many variations, all carotenoids are of one of these three basic types.

The *Chlorobacteriaceae* (green bacteria) contain vesicles enclosed within a thin nonunit membrane that is not directly associated with the cell membrane. Metabolically, the green bacteria are strict anaerobic organisms that are obligately photosynthetic. They utilize H_2S, thiosulfate, or H_2 as an electron donor and CO_2 as the carbon source:

$$CO_2 + 2H_2S + light \longrightarrow (CH_2O) + H_2O + 2S$$

$$2CO_2 + 2Na_2S_2O_3 + 3H_2O + light \longrightarrow 2(CH_2O) + 2NaHSO_4$$

$$CO_2 + 2H_2 + light \longrightarrow (CH_2O) + H_2O$$

The purple bacteria contain two groups: the purple sulfur bacteria (*Thiorhodaceae*) that use H_2S as an electron donor and the purple nonsulfur bacteria (*Athiorhodaceae*) that depend on organic compounds such as short-chain fatty acids for photosynthetic metabolism. Poly-β-hydroxybutyrate is the end product:

$$CO_2 + 2CH_3CHOHCH_3 + light \longrightarrow (CH_2O) + H_2O + 2CH_3COCH_3$$

$$2CH_3COOH + 2CoASH \longrightarrow 2CH_3COSCoA$$

$$2CH_3COSCoA \longrightarrow CH_3COCH_2COSCoA + CoASH$$

$$nCH_3CHOHCH_2COSCoA \longrightarrow (CH_3CHOHCH_2COOH)_n + CoASH$$

Poly-β-hydroxybutyrate serves as a major storage reserve material in these organisms. It is also an important reserve energy source in many other organisms. The cyanobacteria are considered to be very early evolutionary forms because of their lack of dependence on oxygen and on the basis of molecular evidence derived from 16S rRNA sequencing. Phylogenetic analysis of *c*-type cytochromes and rRNA sequences has established a relationship between cyanobacteria and the chloroplasts of green algae and higher plants. These lines of evidence provide support for the concept of prokaryotic origins of chloroplasts along similar lines of development attributed to mitochondria.

Autotrophic CO_2 Fixation and Mechanisms of Photosynthesis

Photoautotrophs and chemoautotrophs, in which CO_2 serves as the sole or principal source of cellular carbohydrate, fix CO_2 via either the reductive pentose phosphate pathway (Calvin) cycle or the reductive C_4-dicarboxylic acid pathway. These systems were first discovered in green plants. Originally, all green plants were thought to assimilate atmospheric CO_2 via the reductive pentose pathway (Fig. 12-2) in which phosphoglyceric acid (PGA) is the first stable product (hence the designation C_3 plants). Subsequently, an alternative pathway of CO_2 fixation was discovered in which C_4 dicarboxylic acids (oxaloacetate and malate) were found as the primary products of photosynthesis. Within a taxonomic category, plants with C_3 photosynthesis are considered to be ancestral to those with C_4 primary photosynthetic products.

In photosynthetic and autotrophic bacteria, CO_2 fixation occurs primarily via the reductive pentose phosphate pathway (Fig. 12-2). In this system reduction of 1 mol of CO_2 to the oxidation level of carbohydrate involves the oxidation of 2 mol of NADPH and the hydrolysis of 3 mol of ATP. Only two of the reactions, phosphoribulokinase and ribulose bisphosphate carboxylase (RuBisCO), are specific to photosynthetic or chemoautotrophic organisms. The other reactions are held in common with the carbohydrate metabolism of nonphotosynthetic organisms. The reductive pentose cycle constitutes the **dark reaction** of photosynthesis. Six turns of the cycle result in the synthesis of 1 mol of hexose (F-6-P):

$$6CO_2 + 6H_2O + 18ATP + 12NADPH + 12H^+$$
$$\longrightarrow F\text{-}6\text{-}P + 18ADP + 12NADP^+ + 17P_i$$

The remainder is recycled through the reductive pathway as shown in Figure 12-2.

The reductive C_4-dicarboxylic acid pathway (Fig. 12-3) is present in a number of photosynthetic bacteria. In some organisms, such as the *Chlorobium*, it is the only cyclic pathway for CO_2 assimilation. Organisms that use the C_4 pathway possess the enzyme pyruvate-orthophosphate dikinase, which synthesizes phosphoenolpyruvate (PEP):

$$pyruvate + ATP + P_i + Mg^{++} \longrightarrow PEP + AMP + PP_i$$

This enzyme differs from the PEP synthase of *E. coli* and other bacteria that can utilize C_4 acids in that it produces orthophosphate rather than monophosphate. *Chlorobium thiosulfatophilum*, a member of the green sulfur bacteria, requires P_i in addition to Mg^{++} and ATP for the formation of PEP from pyruvate, supporting the fact that in photosynthetic bacteria, such as C_4 plants, pyruvate-orthophosphate dikinase rather than PEP synthase is used to form PEP in the photosynthetic assimilation of CO_2. The reductive carboxylic acid cycle is essentially a reverse of the TCA cycle in which pyruvate oxidase and α-ketoglutarate oxidase systems are replaced by ferredoxin-dependent pyruvate synthetase and α-ketoglutarate synthetase. This system is also of major importance in the metabolism of anaerobic bacteria.

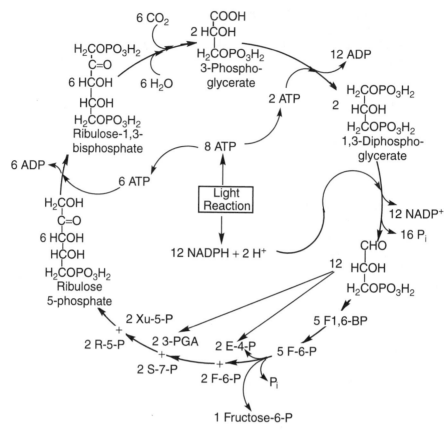

Fig. 12-2. The reductive pentose phosphate pathway. Since 3-phosphoglycerate is the first stable product of atmospheric CO_2 fixation, this pathway is sometimes referred to as the C_3 pathway. This cycle of reactions constitutes the dark reaction of photosynthesis because the energy required in the form of ATP has already been generated during photophosphorylation.

Photosynthesis, whether in green plants, algae, cyanobacteria, or photosynthetic bacteria, begins with the absorption of light by a pigment molecule and the delivery of the absorbed light energy to electron carriers that can transduce the energy into chemical form. The function of the light-harvesting pigments or **antennae** is to capture photons. Energy in the excited pigments is channeled into a complex called the **reaction center**. The reaction center functions as a battery that transfers electrons across the photosynthetic membrane and provides the energy for the fixation of carbon dioxide. The components of the bacterial photosynthetic reaction center have been studied in considerable detail. A somewhat simplistic diagram of the reaction center is shown in Figure 12-4.

Within the reaction center there are four bacteriochlorophyll molecules. Two of these are referred to as a **special pair** because they absorb light and transfer it to an electron. The two additional bacteriochlorophyll molecules appear to be inactive and are referred to as "voyeur chlorophyll." Once the energy of the photon has been transferred to an

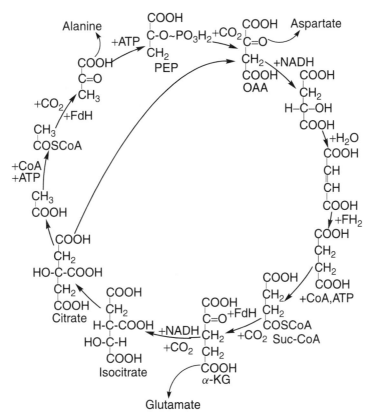

Fig. 12-3. The reductive C_4-carboxylic acid cycle. This is the only cyclic pathway of CO_2 assimilation in certain photosynthetic bacteria such as *Chlorobium*. OAA, oxaloacetate; PEP, phosphoenolpyruvate; Suc-CoA, succinyl-coenzyme A; FH_2, reduced ferredoxin derived from photosynthesis; FH_2, reduced flavin.

electron, the electron moves to a bacteriopheophytin molecule, creating a positive charge on the special pair of chlorophyll molecules. The electron then travels to a quinone. A soluble cytochrome molecule transfers its electron to the special pair. The cytochrome acquires a positive charge and the special pair of bacteriochlorophyll is neutralized. The excited electron is then passed to the second quinone.

The terminal steps resulting in the phosphorylation of ADP represent an additional series of electron transfer reactions involving ferredoxin, $NADP^+$, and cytochromes. The generation of ATP in photosynthesis is a process comparable to that utilized in coupling phosphorylation to electron transport during respiration (i.e., chemiosmotic coupling to an ATPase). In noncyclic photophosphorylation (Fig. 12-5) electrons are transferred from chlorophyll to ferredoxin, flavoprotein, and then to $NADP^+$. An electron donor (water in plants and algae; H_2, H_2S, or various organic compounds in photosynthetic bacteria) transfers electrons to cytochrome, producing the chemical energy needed to phosphorylate ADP. In cyclic photophosphorylation, ATP is generated from ADP and P_i with no other net chemical change (Fig. 12-5). Since cyclic

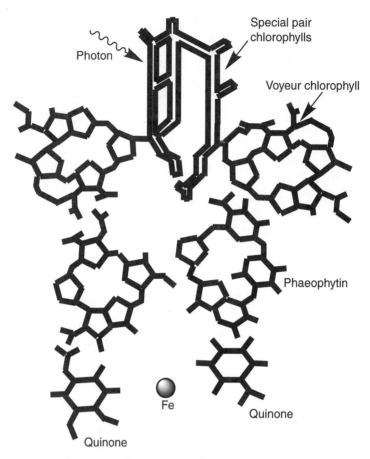

Fig. 12-4. Conceptual drawing of the bacterial photosynthetic reaction center.

photophosphorylation does not generate $NADH + H^+$, compounds such as H_2, H_2S, or other available compounds provide reducing power.

In cyanobacteria and the eukaryotic red algae, phycobiliproteins are the most prominent light-harvesting polypeptides of the cell. These polypeptides are highly pigmented, water-soluble proteins that make up a major portion of the soluble cell protein. The major phycobiliproteins are phycoerythrin, phycocyanin, and allophycocyanin. Phycobilisome complexes appear as rows of closely spaced granules at the outer surface of the photosynthetic (thylakoid) membranes of red algae and cyanobacteria. The composition of the phycobilisome complex of the filamentous cyanobacterium *Fremyella diplosiphon* is altered by growth under red light as compared to green light. The differences in composition of the phycobilisome structure are the result of altered expression of the genes coding for phycobiliproteins.

Hydrogen Bacteria

Members of this group include the examples given in Table 12-1. These organisms utilize H_2 to provide energy and reducing power for growth and CO_2 fixation. Most of these bacteria are facultatively autotrophic and grow readily on organic substrates.

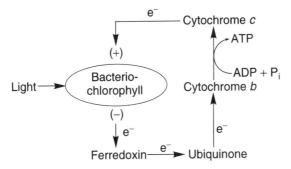

Cyclic photophosphorylation–photosynthetic bacteria

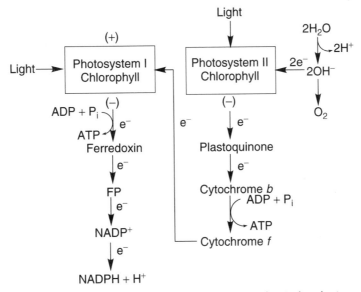

Noncyclic photophosphorylation–algae, cyanobacteria, plants

Fig. 12-5. Comparison of cyclic and noncyclic photophosphorylation. In cyclic photophosphorylation ATP is produced, but no reducing equivalents are generated. In the noncyclic pathway, two molecules of ATP are produced, reduced NADP is generated, and oxygen is produced by photolysis of water.

Many heterotrophic bacteria are capable of using H_2 to provide reducing power and energy for metabolic purposes but cannot support CO_2 fixation.

The facultative chemolithoautotrophic proteobacterium, *Ralstonia eutropha* (formerly *Alcaligenes eutrophus*), can utilize H_2 as a sole source of energy. Two energy-generating (NiFe) hydrogenases are present. One, a membrane-bound hydrogenase, is primarily involved in electron transport–coupled phosphorylation, whereas the other is a cytoplasmic enzyme that reduces NAD to provide reducing equivalents.

Reduction of CO_2 by H_2 can be shown as

$$2H_2 + CO_2 \longrightarrow (CH_2O) + H_2O$$

Utilizable energy in the form of ADP is generated from the oxidation of H_2 by hydrogenase:

$$2H_2 + 0.5O_2 + NAD^+ + ADP + P_i \longrightarrow H_2O + NADH + ATP$$

Carbon dioxide is assimilated autotrophically through the essential reactions of phosphoribulokinase and ribulose-1,5-bisphosphate carboxylase and the Calvin cycle (Fig. 12-3). In *Ralstonia eutropha* phosphoribulokinase is partially inactivated when an autotrophic culture is shifted to heterotrophic growth with pyruvate as the source of carbon and energy. Reactivation of phosphoribulokinase occurs after exhaustion of pyruvate from the medium. The hydrogen autotroph, *Xanthobacter*, can grow autotrophically with either hydrogen or methanol as an energy source. Hydrogen is oxidized by a membrane-bound hydrogenase. Methanol is oxidized to formaldehyde, formate, and then to CO_2 by the sequential action of methanol dehydrogenase, formaldehyde dehydrogenase, and formate dehydrogenase:

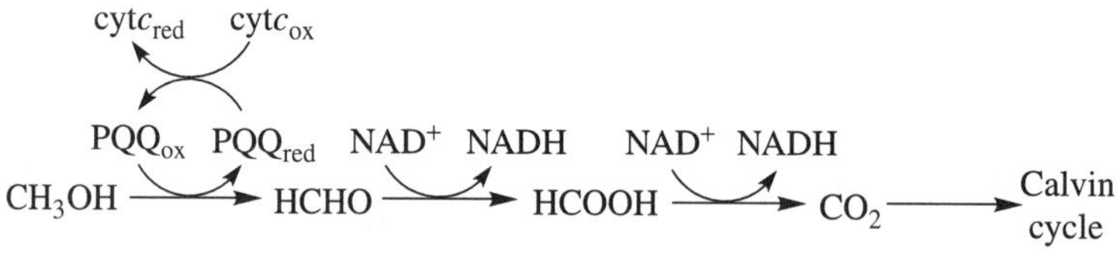

Nitrifying Bacteria

These organisms are important participants in the nitrogen cycle; hence their activities and relationship to nitrogen metabolism are discussed in greater detail in Chapter 14. *Nitrosomonas* is the most common organism found in the soil that oxidizes ammonia to nitrite:

$$2NH_3 + 3O_2 \longrightarrow 2NO_2^- + 2H^+ + 2H_2O$$

Nitrobacter is the most common organism in soil that oxidizes nitrite to nitrate:

$$2NO_2^- + 3O_2 \longrightarrow 2NO_3^-$$

Both *Nitrosomonas* and *Nitrobacter* have been shown to possess a specialized mechanism for ATP production, and reduced NAD is required for the assimilation of carbon dioxide. A portion of the ATP derived from oxidative phosphorylation at the cytochrome level is used to reduce pyridine nucleotides.

Sulfur Bacteria

All members of this group, *Thiorhodaceae*, are capable of growth on elemental sulfur. Many can utilize thiosulfate ($S_2O_3^-$) as well. The biochemistry of two of the sulfate-reducing bacteria, *Desulfovibrio* and *Desulfotomaculum*, are fundamentally different. In the case of *Desulfovibrio*, the PP_i produced during the formation of adenylyl sulfate

(APS) from ATP and sulfate in the first step of sulfate reduction is hydrolyzed to P_i by inorganic pyrophosphatase:

$$ATP + SO_4^{2-} \longrightarrow APS + PP_i$$

$$PP_i + H_2O \longrightarrow 2P_i$$

By this process, the energy in PP_i is dissipated by hydrolysis, and to obtain a net yield of ATP during growth on lactate plus sulfate, *Desulfovibrio* must carry out electron transfer–coupled phosphorylation.

By comparison, *Desulfotomaculum* conserves the bond energy in PP_i by means of the enzyme acetate:PP_i phosphotransferase and the subsequent formation of ATP by acetate kinase:

$$acetate + PP_i \longrightarrow acetyl\ phosphate + P_i$$

$$acetyl\ phosphate + ADP \longrightarrow acetate + ATP$$

These reactions allow *Desulfotomaculum* to use PP_i as a source of energy for growth with acetate and sulfate. The conversion of APS (adenosine phosphosulfate) to sulfite by APS reductase requires the addition of two electrons:

$$APS + 2e^- \longrightarrow AMP + SO_3^{2-}$$

The further reduction of sulfite to sulfide requires the action of sulfite reductase (a), trithionate reductase (b), and thiosulfate reductase (c), and the recycling of sulfite:

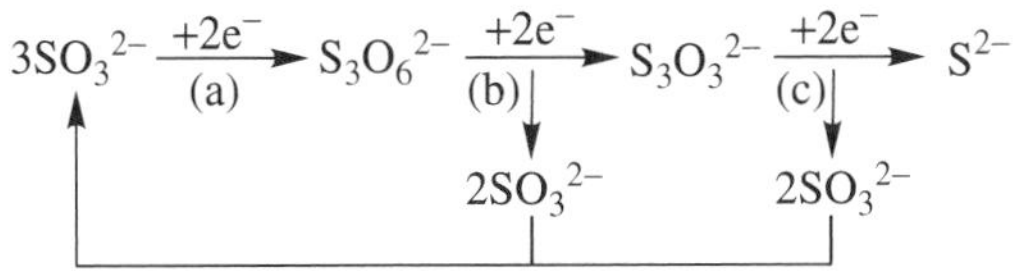

The sulfur-dependent archaea found in the vicinity of hot springs are able to grow chemoautotrophically using CO_2 as the sole carbon source and the oxidation of elemental sulfur with oxygen yielding sulfuric acid:

$$2S^0 + 3O_2 + 2H_2O \longrightarrow 2H_2SO_4$$

However, *Sulfolobus ambivalens* is able to live by an anaerobic mode of chemoautotorophy using CO_2 as the sole carbon source but using H_2 for the reduction of sulfur to H_2S:

$$S^0 + H_2 \longrightarrow H_2S$$

Iron Bacteria

Thiobacillus ferrooxidans, Gallionella, Leptothrix, Sulfolobus, Sphaerotilus, and *Shewanella oneidensis* are capable of oxidizing ferrous iron to ferric iron as a means of generating biologically useful energy:

$$Fe^{2+} + H^+ + 0.25O_2 \longrightarrow Fe^{3+} + 0.5H_2O + 40\ kcal$$

Thiobacillus ferrooxidans is an obligate autotroph. While it can be grown heterotrophically in the absence of an oxidizable iron source, continued cultivation on an organic substrate renders it incapable of growth with ferrous iron as the sole energy source. *T. ferrooxidans* differs from other autotrophic organisms in that it cannot revert to an autotrophic mode of life after prolonged cultivation on organic substrates. The transition of *T. ferrooxidans* to obligate organotrophy is governed by a number of factors including the pH of the medium, the incubation temperature, the availability of oxygen, the age of the cells at the time of transition, and the type of energy and carbon source available. Conversion to organotrophy results in a gradual loss of the ability to oxidize Fe^{2+} and cessation of CO_2 fixation. *Gallionella, Sphaerotilus*, and other iron-oxidizing organisms appear to be facultative and can be readily grown as heterotrophs and then returned to growth on iron.

Shewanella oneidensis, a metal-reducing bacterium found in soils, can use ferric iron as a terminal electron acceptor. This organism is able to reductively dissolve Fe^{3+}-containing minerals such as goethite (α-FeOOH) or hematite (α-Fe$_2$O$_3$). Under anaerobic conditions *S. oneidensis* can apparently generate two energized membranes using a system of proteins that shuttle electrons from an energy source in the cytoplasm, across the plasma membrane and periplasmic space, to the outer membrane. Once in the outer membrane, iron reductases appear to transfer electrons directly to Fe^{3+} in the crystal structure of minerals, causing a weakening of the iron-oxygen bond and reductive dissolution of the mineral. Using atomic force microscopy it has been possible to show that the affinity between *S. oneidensis* and goethite increases by two to five times under anaerobic conditions. An iron reductase within the outer membrane is apparently mobilized and specifically interacts with the goethite surface to facilitate the electron transfer process.

METHYLOTROPHS

Methylotrophic bacteria are able to utilize methane, methanol, methylamine, or formate as the sole source of carbon and energy. The term **methanotroph** designates methylotrophs that can use methane for carbon and energy. There are also several species of yeasts and molds that can use methane or methanol. Most methylotrophs are obligate in that they can only use C_1 compounds. The general pathway of oxidative reactions is shown in Figure 12-6.

Two types of methylotrophic bacteria have been identified on the basis of the mode of assimilation of formaldehyde. **Type I methylotrophs** use the ribulose monophosphate pathway for formaldehyde assimilation:

$$3HCHO + 3 \text{ ribulose monophosphate} \longrightarrow 3 \text{ hexulose-6-phosphate}$$

The hexulose-6-phosphate is metabolized via the central pathway to form glyceraldehyde-3-phosphate. The overall reaction is

$$3 \text{ HCHO} + ATP \longrightarrow \text{glyceraldehyde-3-phosphate} + ADP$$

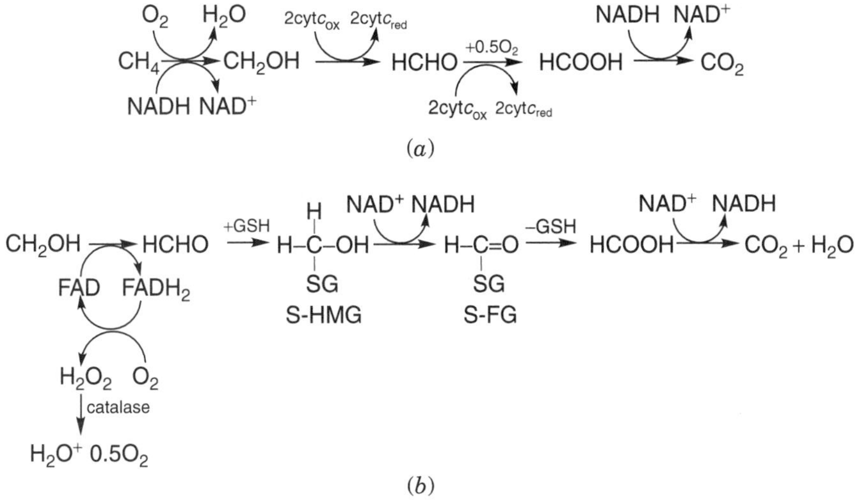

Fig. 12-6. (a) **General pathway of oxidative reactions in methylotrophs. (b) Pathway for the conversion of methanol to CO_2 and H_2 in a methylotrophic yeast.** GSH, glutathione; S-HMG, *S*-hydroxymethyl glutathione; S-FG, *S*-formylglutathione.

Type II methylotrophs use the serine pathway for formaldehyde assimilation:

$$2 \text{ HCHO} + 2 \text{ glycine} \longrightarrow 2 \text{ serine}$$

$$2 \text{ serine} \longrightarrow 2 \text{ glycerate} \longrightarrow 2 \text{ phosphoglycerate}$$

The overall reaction is

$$2 \text{ HCHO} + CO_2 + 3 \text{ ATP} + 2 \text{ NADH}$$

$$\longrightarrow 2 \text{ phosphoglycerate} + 2 \text{ ADP} + P_i + \text{NAD}^+$$

Bacterial methylotrophs include *Paracoccus denitrificans* and several species of *Pseudomonas, Hyphomicrobium,* and *Xanthobacter.* However, most of the obligate methylotrophs belong to the genera *Methylophilus, Methylobacterium, Methylococcus, Methylosinus,* or *Methylomonas.*

Methylotrophic yeast includes *Hansenula, Candida, Torulopsis,* and *Pichia.* The metabolic pathway for the conversion of methanol to CO_2 and H_2O appears to be similar for several of these yeasts. The pathway of *Pichia pastoris* involves alcohol oxidase, catalase, formaldehyde dehydrogenase, *S*-formylglutathione, and formate dehydrogenase in the sequence of reactions shown in Figure 12-6. In yeasts, the alcohol dehydrogenase and catalase reactions take place in peroxisomes, membranous organelles containing flavin-linked oxidases that regenerate oxidized flavin by reaction with O_2. Synthesis of some of these enzymes is tightly regulated and several of the genes involved in methanol utilization appear to be controlled at one level by a glucose catabolite repression–depression mechanism. The structural genes for alcohol dehydrogenase and two other enzymes in the sequence are regulated by methanol at the level of transcription.

Methane is produced in anaerobic environments such as natural wetlands but is also a major agricultural and industrial by-product. As the most abundant organic gas in the atmosphere it absorbs terrestrial radiation (infrared radiation) more effectively than does CO_2. As a result, methane contributes more heavily to global warming. Methanotrophic bacteria are distributed widely and play a significant role in moderating global warming by oxidizing most of the methane before it reaches the atmosphere. Methanol-oxidizing organisms are useful for the production of single-cell protein; microbial cells are used as animal feed supplements. Growing such organisms on materials that would otherwise be disposed of as waste provides an important means of recycling these materials into useful products. Some of the methylotrophs display a wide range of biotransformations of potential commercial importance.

Methanogens

Methanogenic organisms gain energy by using H_2 to reduce CO_2 to CH_4. These organisms can also decarboxylate acetate to CO_2 and CH_4. Methane formation represents the terminal portion of a complex series of anaerobic reactions that occur in nature and involve a number of organisms that degrade biopolymers such as cellulose, starch, or proteins to acetate, H_2, and CO_2. Conversion of complex organic material to these simple products requires the action of both primary and secondary fermenters from the clostridia and other anaerobic organisms. Primary fermenters can yield acetate, H_2, and CO_2. Other products require additional degradation by the secondary fermenters. Methanogenic microorganisms conduct the last portion of the conversion sequence to yield CH_4 as a final product.

Methanogens belong to the archaea (archaebacteria). The major genera are listed in Table 12-1. Most methanogens can produce CH_4 from H_2 and CO_2 as shown in the first equation below. Only the Methanosarcinales (e.g., *Methanosarcina barkeri*) can reduce other substrates to CH_4 according to the following equations:

$$4H_2 + CO_2 \longrightarrow CH_4 + 2H_2O$$

$$4HCOOH \longrightarrow CH_4 + 3CO_2 + 2H_2O$$

$$4CH_3NH_2Cl + 2H_2O \longrightarrow 3CH_4 + CO_2 + 4NH_4Cl$$

$$2(CH_3)_2NHCl + 2H_2O \longrightarrow 3CH_4 + CO_2 + 2NH_4Cl$$

$$4(CH_3)_2NCl + 6H_2O \longrightarrow 9CH_4 + 3CO_2 + 4NH_4Cl$$

$$CH_3COOH \longrightarrow CH_4 + CO_2$$

Conversion of acetate to CH_4 and CO_2 involves the following intermediary steps:

$$CH_3COOH + ATP \longrightarrow CH_3COOPO_3H_2 + ADP$$

$$CH_3COOPO_3H_2 + CoASH \longrightarrow CH_3COSCoA + P_i$$

$$CH_3COSCoA + THSPt \longrightarrow CH_3THSPt + CoASH + CO_2$$

$$THSPt + HSCoM \longrightarrow CH_3SCoM + THSPt$$

$$CH_3SCoM + HSCoB \longrightarrow CoM\text{-}S\text{-}S\text{-}CoB + CH_4$$

$$CoM\text{-}S\text{-}S\text{-}CoB \longrightarrow HSCoM + HSCoB$$

Reduction of CO_2 to CH_4 follows the pathway shown in Figure 12-7 and involves the function of several unique coenzymes: methanofuran (MFR); tetrahydromethanopterin (H_4MPT); deazaflavin F_{420} as an electron donor; coenzyme M (HS-$CH_2CH_2SO_3^-$, or HSCoM); and coenzyme B (HSCoB, 7-mercaptoheptanoylthreonine phosphate). Several of these coenzymes were once thought to be present only in methanogenic archaea. However, it has now been shown that the C_1 transfer enzymes and their cofactors, CoM and CoB, function in methylotrophic bacteria as well. The chemical structures of these cofactors are shown in Figure 12-8. All methanogens use the major energy-yielding step associated with the reduction of a methyl group to CH_4, although different species may obtain electrons for the reductive step from the oxidation of a variety of substrates.

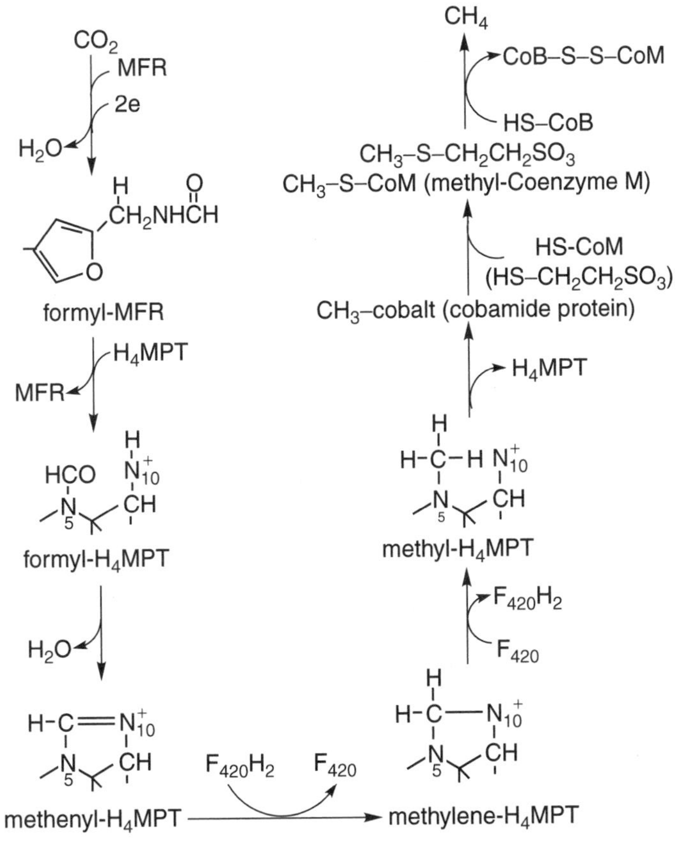

Fig. 12-7. Pathway for the reduction of CO_2 to CH_4 by methanogens. MFR, methanofuran; H_4MPT, tetrahydromethanopterin; $F_{420}H_2$, deazaflavin F_{420}, electron donor; HS-CoB, coenzyme B (7-mercaptoheptanoylthreonine phosphate), electron donor. Complete structures of the unusual cofactors that participate in methanogenesis are shown in Figure 12-8. (*Source*: From R. S. Wolfe, *ASM News* **62:**529–534, 1996.)

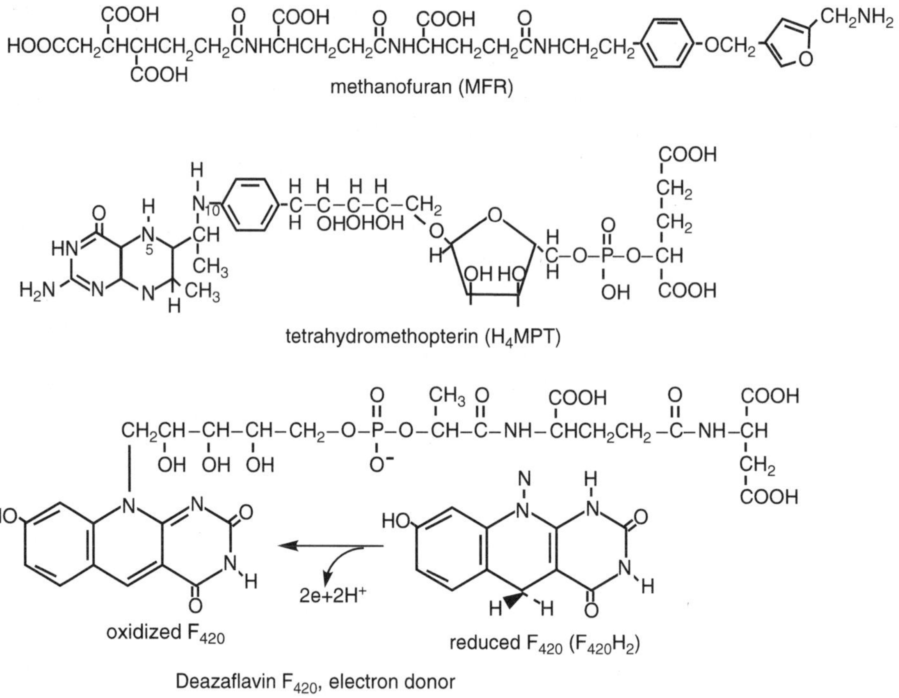

Fig. 12-8. Structures of the unusual coenzymes that participate in methanogenesis. The pathway of reduction of CO_2 to CH_4 is shown in Fig. 12-7. (*Source*: From R. S. Wolfe, *ASM News* **62**:529–534, 1996.)

BIBLIOGRAPHY

Burke, S. A., S. L. Lo, and J. S. Krzycki. 1998. Clustered genes encoding the methyltransferases of methanogenesis from monomethylamine. *J. Bacteriol.* **180**:3432–40.

Campbell, D., V. Hurry, A. K. Clarke, P. Gustafsson, and G. Öquist. 1998. Chlorophyll fluorescence analysis of cyanobacterial photosynthesis and acclimation. *Microbiol. Mol. Biol. Rev.* **62**:667–83.

Chistoserdova, L., J. A. Vorholt, R. K. Thauer, and M. E. Lidstrom. 1998. C1 transfer enzymes and coenzymes linking methylotrophic bacteria and methanogenic archaea. *Science* **281**:99–102.

Cooper, A. J. L. 1998. Advances in enzymology of the biogeochemical sulfur cycle. *Chemtracts—Biochem. Mol. Biol.* **11**:729–47.

de Wit, R. 2000. Sulfide-containing environments. In J. Lederberg (ed.), *Encyclopedia of Microbiology*, 2nd ed., Vol. 4. Academic Press, San Diego, pp. 478–94.

Grahame, D. Q., and S. Gencic. 2000. Methane biochemistry. In J. Lederberg (ed.), *Encyclopedia of Microbiology*, 2nd ed., Vol. 3. Academic Press, San Diego, pp. 188–98.

Hanson, R. S., and T. E. Hanson. 1996. Methanotrophic bacteria. *Microbiol. Rev.* **60**:439–71.

Hatchikian, E. C., V. Magro, N. Forget, Y. Nicolet, and J. C. Fontecilla-Camps. 1999. Carboxy-terminal processing of the large subunit of [Fe] hydrogenase from *Desulfovibrio desulfuricans* ATCC 7757. *J. Bacteriol.* **181**:2947–52.

Kleihues, L., O. Lenz, M. Bernhard, T. Buhrke, and B. Friedrich. 2000. The H2 sensor of *Ralstonia eutropha* is a member of the subclass of regulatory [NiFe] hydrogenases. *J. Bacteriol.* **182:**2716–24.

Lens, P., and L. H. Pol. 2000. Sulfur cycle. In J. Lederberg (ed.), *Encyclopedia of Microbiology*, 2nd ed., Vol. 4. Academic Press, San Diego, pp. 495–505.

Lower, S. K., M. F. Hochella, Jr., and T. J. Beveridge. 2001. Bacterial recognition of mineral surfaces: nanoscale interactions between *Shewanella* and α-FeOOH. *Science* **292:**1360–3.

Murrell, J. C., and I. R. McDonald. 2000. Methylotrophy.. In J. Lederberg (ed.), *Encyclopedia of Microbiology*, 2nd ed., Vol. 3. Academic Press, San Diego, pp. 245–55.

Partensky, F., W. R. Hess, and D. Vaulot. 1999. *Prochlorococcus*, a marine photosynthetic prokaryote of global significance. *Microbiol. Mol. Biol. Rev.* **63:**106–27.

Pieulle, L., V. Magro, and C. Hatchikian. 1997. Isolation and analysis of the gene encoding the pyruvate-ferredoxin oxidoreductase of *Desulfovibrio africanus*, production of the recombinant enzyme in *Escherichia coli*, and effect of carboxyterminal deletions on its stability. *J. Bacteriol.* **179:**5684–92.

Saier, M. H., Jr. 2000. Bacterial diversity and the evolution of differentiation. *ASM News* **66:** 337–43.

Schink, B. 1997. Energetics of syntrophic cooperation in methanogenic degradation. *Microbiol. Mol. Biol. Rev.* **61:**262–80.

Schwartz, E., T. Buhrke, U. Gerisher, and B. Friedrich. 1999. Positive transcriptional feedback controls hydrogenase expression in *Alcaligenes eutrophus* H16. *J. Bacteriol.* **181:**5684–92.

Sowers, K. R. 2000. Methanogenesis. In J. Lederberg (ed.), *Encyclopedia of Microbiology*, 2nd ed., Vol. 3. Academic Press, San Diego, pp. 204–26.

Wolfe, R. S. 1996. 1776–1996: Alessandro Volta's combustible air. *ASM News* **62:**529–34.

Xiong, J., W. M. Fischer, K. Inoue, M. Nakahara, and C. E. Bauer. 2000. Molecular evidence for the early evolution of photosynthesis. *Science* **289:**1724–30.

Yoon, H. -S., and J. W. Golden. 1998. Heterocyst pattern formation controlled by a diffusible peptide. *Science* **283:**935–8.

Yoon, K. -S., T. E. Hanson, J. L. Gibson, and F. R. Tabita. 2000. Autotrophic CO2 metabolism. In J. Lederberg (ed.), *Encyclopedia of Microbiology*, 2nd ed., Vol. 1. Academic Press, San Diego, pp. 349–58.

Zeilstra-Ryalls, J., M. Gomelsky, J. M. Eraso, A. Yeliseev, J. O'Gara, and S. Kaplan. 1998. Control of photosystem formation in *Rhodobacter sphaeroides*. *J. Bacteriol.* **180:**2801–9.

CHAPTER 13

LIPIDS AND STEROLS

The major lipid-containing components of microbial cells are membranous structures such as the cytoplasmic membrane, the outer membrane of gram-negative bacteria, the intracytoplasmic membranes of photosynthetic bacteria, and the nuclear and mitochondrial membranes of eukaryotic cells. Other lipids play important roles as electron carriers, enzyme cofactors, transport agents in cell wall synthesis, and light-absorbing pigments.

A common feature of all lipids is their insolubility in water. Membrane lipids are **amphipathic** — that is, they have **hydrophilic** (water-soluble) and **hydrophobic** (water-insoluble) regions that cause them to orient into bilayers. This characteristic provides a solubility barrier to polar solutes in an aqueous environment. Phospholipids comprise the bimolecular leaflet structure of the cytoplasmic membrane, which is both flexible and self-sealing. Monounsaturated and ring-containing fatty acids contribute to the flexibility and fluidity that enables membranes to undergo changes in shape that accompany cell growth or movement.

In the fungi, sterols are associated with the cytoplasmic membrane and are involved with permeability functions as indicated by the leakage of metabolites following the binding of polyene antibiotics, such as nystatin or amphotericin B, to the sterol groups. Certain lipids, for example, glycosyldiglycerides and lipoteichoic acids, are found in the cell surface of gram-positive but not gram-negative bacteria. Conversely, gram-negative bacteria display an outer membrane whose outer face is composed largely of lipopolysaccharides (LPS) while gram-positive bacteria are devoid of these structures. Glycolipids are present in non-sterol-requiring *Mycoplasma* but not in the sterol-requiring members of this group, suggesting replacement of sterols by glycolipids in the membrane structures of the first group. In fungi there is an association of lipids and sterols with the development of respiratory activity.

Although there is still a great deal to be determined with regard to the precise role of lipids in various cellular activities, our knowledge of the biosynthesis and metabolism of lipids and the regulation of these activities is well developed.

LIPID COMPOSITION OF MICROORGANISMS

Representative examples of the major types of naturally occurring fatty acids are shown in Table 13-1. Of the examples shown, not all are found in microbial cells. In contrast to eukaryotic organisms, bacterial cells contain few fatty acids with greater than 19 carbon atoms, the shorter chain-length acids being primarily of importance as metabolic intermediates in fatty acid biosynthesis. The predominant saturated fatty acid in microorganisms is palmitic acid (C_{18}). Lesser quantities of stearic (C_{18}), myristic (C_{14}), and lauric (C_{12}) acids are observed. The principal unsaturated fatty acids are monoenoic acids (one unsaturated position). Di-, tri-, or polyenoic acids are not found in bacteria. Many unusual fatty acids are present in smaller quantities. Branched, hydroxylated, or methylated fatty acids and cyclopropane ring-containing fatty acids are present in many microorganisms. Corynolic acid (a branched, dihydroxy fatty acid with 52 carbon atoms) and mycolic acid (a branched, dihydroxy fatty acid with 87 to 88 carbon atoms) are examples of fatty acids unique to the *Corynebacterium* and *Mycobacterium*, suggesting a close taxonomic relationship between these two genera.

Yeasts and molds have a fatty acid composition more closely related to that of plants and their seeds. The common brewer's and baker's yeasts contain about 5 to 8% lipids (dry weight), with palmitoleic acid being a major constituent. However, when baker's yeast is grown under anaerobic conditions, palmitic acid synthesis increases at the expense of palmitoleic acid. Many yeasts and molds accumulate lipids, usually as triglycerides containing a high percentage of palmitic acid, particularly near the end of the growth cycle.

Archaea produce a number of unusual lipid structures that contain ether linkages. These organisms are divided into three major groups: halophiles, methanogens, and thermophiles. Thermophiles thrive in extreme environments that are uninhabitable by any other living forms because their unusual structural and biochemical factors aid in their survival and functioning under these harsh conditions.

Sterols are commonly present in fungi but are rarely observed in bacteria. Exceptions are found: *Mycoplasma* incorporates sterols from the growth medium into cell membranes; *Streptococcus pneumoniae* incorporates cholesterol into cytoplasmic membranes; and methanotropic bacteria synthesize sterols in significant quantities.

By means of high-resolution gas chromatography, an accurate system of species identification has been developed on the basis of fatty acid composition.

Straight-Chain Fatty Acids

Straight-chain fatty acids are the major constituents of membrane phospholipids. Palmitic, stearic, hexadecanoic, octadecenoic, cyclopropanic, 10-methylhexadecanoic, 2- or 3-hydroxyl fatty acids, and a few others are commonly observed in bacterial cells. Saturated and unsaturated straight-chain fatty acids with less than 10 or 12 carbon atoms are usually intermediates of degradative pathways or biosynthesis of longer-chain fatty acids.

The conditions under which microbial cells are grown can markedly influence their fatty acid constituents. Composition of the growth medium, availability of oxygen, temperature, pH, and the age of the culture can each affect the distribution of fatty acids. Young cultures of *E. coli* contain large quantities of C_{16}- and C_{18}-monoenoic acids and only small amounts of C_{17}- and C_{19}- cyclopropane

TABLE 13-1. Representative Examples of Naturally Occurring Fatty Acids

Number of C atoms	Generic Name	Common Name	Composition
Saturated Fatty Acids ($-C-C-C-$)			
1	Methanoic	Formic	HCOOH
2	Ethanoic	Acetic	CH_3COOH
3	Propanoic	Propionic	C_2H_5COOH
4	Butanoic	Butyric	C_3H_7COOH
6	Hexanoic	Caproic	$C_5H_{11}COOH$
16	Hexadecanoic	Palmitic	$C_{15}H_{31}COOH$
18	Octadecanoic	Stearic	$C_{17}H_{35}COOH$
Unsaturated Fatty Acids (*Monoenoic*) ($-C-C=C-C-$)			
4	*trans*-2-Butenoic	Crotonic	$C_4H_6O_2$
16	*cis*-9-Hexadecenoic	Palmitoleic	$C_{16}H_{30}O_2$
18	*cis*-9-Octadecenoic	Oleic	$C_{18}H_{34}O_2$
18	*trans*-11-Octadecenoic	Vaccenic	$C_{18}H_{34}O_2$
Unsaturated Fatty Acids (*Dienoic*) ($-C-C=C-C=C-$)			
6	2,4-Hexadienoic	Sorbic	$C_6H_8O_2$
18	*cis*-9, *cis*-12-Octadecadienoic	α-Linoleic	$C_{18}H_{32}$
Unsaturated Fatty Acids (*Trienoic*) ($-C=C-C=C-C=C-$)			
18	*cis*-6, *cis*-9, *cis*-12-Octadecatrienoic	γ-Linoleic	$C_{18}H_{30}O_2$
18	*cis*-9, *cis*-12, *cis*-15-Octadecatrienoic	α-Linoleic	$C_{18}H_{30}O_2$
Unsaturated Fatty Acids (*Tetraenoic*) ($-C=C-C=C-C=C-C=C-$)			
20	5,8,11,14-Eicosatetraenoic	Arachidonic	$C_{20}H_{32}O_2$
Hydroxyalkanoic Acids ($-CH_2-CHOH-COOH$)			
12	2-Hydroxydodecanoic	2-Hydroxylauric	$C_{12}H_{24}O_2$
18	2-Hydroxyoctadecanoic	2-Hydroxystearic	$C_{18}H_{36}O_2$
Keto, Epoxy, and Cyclo Fatty Acids			
5	4-ketopentanoic ($-C-CO-C-$)	Levulinic	$C_5H_8O_2$
19	ω-(2-*n*-Octylcycloprop-1-enyl)-octanoic (9,10-methyleneoctadec-9-enoic)	Sterculic	$C_{19}H_{34}O_2$

$$CH_3-(CH_2)_7-C\underset{\diagdown \diagup}{=}C-(CH_2)_7-COOH$$
$$CH_2$$

TABLE 13-1. (*continued*)

Hydroxy Unsaturated Fatty Acids

18	D-12-Hydroxy-*cis*-9-octadecenoic	Ricinoleic	$C_{19}H_{34}O_2$

Branched-Chain Fatty Acids

5	3-Methylbutanoic	Isovaleric	$C_5H_{10}O_2$
16	14-Methylpentadecanoic	*iso*-Palmitic	$C_{18}H_{32}O_2$
19	1-D-10-Methyloctadecanoic	Tuberculostearic	$C_{19}H_{38}O_2$
20	18-Methylnonadecanoic	*iso*-Arachidic	$C_{20}H_{40}O_2$

fatty acids. In the late exponential phase of growth, the cyclopropane fatty acids are greatly increased at the expense of the unsaturated fatty acids. In *S. aureus* about 5% of the total fatty acids are unsaturated when the organism is grown on a minimal defined medium. When human serum is added to the medium, the proportion of unsaturated fatty acid increases to 27%. Under aeration *S. aureus* displays a decrease in the percentage of saturated fatty acids. In anaerobically grown staphylococci, C_{18} and C_{20} acids comprise 59% of the total fatty acids. Unsaturated fatty acids are not present in *B. acidocaldarius*, an organism characterized by its tolerance of high temperature (50–70 °C) and acidity (pH 2–5). This organism notably produces a high proportion of unusual lipids. The ability of the organism to grow under these extreme conditions is attributed, in part, to the unusual fatty acid composition.

Fungi contain the same major classes of lipids and fatty acids as bacteria, but the relative amounts of each class may vary. In *Phycomyces*, yeasts, *Euascomycetes*, and *Basidiomycetes* the C_{16} and C_{18} fatty acids predominate. The fatty acids of *S. cerevisiae* are also in the C_{16} and C_{18} class, with only 1 to 2% of the total fatty acids having chain lengths greater than 18 carbon atoms. These range from C_{19} to C_{34}, with C_{26}-unsaturated acids constituting 18% of the fatty acids with greater than 18 carbon atoms at the midexponential phase of growth and 64% at the late exponential phase.

Branched-Chain Fatty Acids

Among bacterial genera, relatively high concentrations of branched-chain fatty acids have been found in *Bacillus*, *Staphylococcus*, *Corynebacterium*, *Mycobacterium*, *Pseudomonas*, and *Spirochaeta*. The most common types of branched-chain fatty acids observed are of the iso- and anteiso-configuration:

$$
\begin{array}{cc}
\begin{array}{c}
CH_3CH_2 \\
\diagdown \\
CH_2-(CH_2)_n-COOH \\
\diagup \\
CH_3
\end{array}
&
\begin{array}{c}
CH_3 \\
\diagdown \\
CH_2-(CH_2)_n-COOH \\
\diagup \\
CH_3
\end{array}
\\
\text{Anteiso-} & \text{Iso-}
\end{array}
$$

Iso- and anteiso-fatty acids are characteristic constituents of lipids in many different organisms.

Bacillus subtilis was one of the first bacterial species in which branched-chain fatty acids were identified. At least six different branched-chain fatty acids representing 60% of the total fatty acids have been found. Corynemycolenic acid, produced by *Corynebacterium*, has the following structure:

$$CH_3-(CH_2)_3-CH=CH-(CH_2)_7-\underset{\underset{OH}{|}}{CH}-\underset{\underset{C_{14}H_{29}}{|}}{CH}-COOH$$

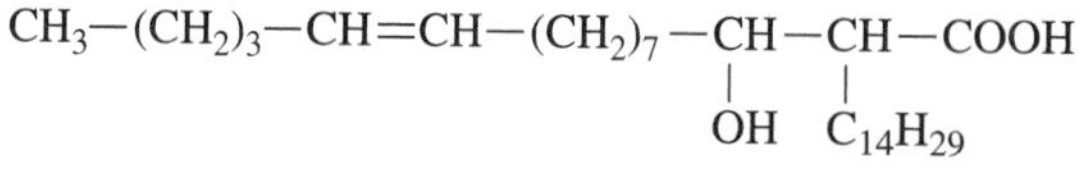

Three branched-chain hydroxy acids, 2-hydroxy-9-methyl-decanoic acid, 3-hydroxy-9-methyl-decenoic acid, and 3-hydroxy-11-methyl-dodecanoic acid, have been identified in *Pseudomonas maltophila*. Several species of *Mycobacterium* contain 2-methyl-3-hydroxypentanoic acid:

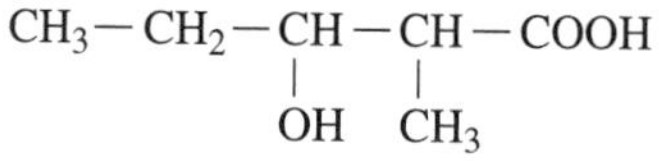

$$CH_3-CH_2-\underset{\underset{OH}{|}}{CH}-\underset{\underset{CH_3}{|}}{CH}-COOH$$

Many other 3-hydroxy fatty acids have been found in microorganisms. Members of the genus *Bacteroides* contain hydroxy fatty acids as major constituents. The LPS from *B. fragilis* contains 13-methyltetradecanoic, D-3-hydroxypentadecanoic, D-3-hydroxyhexadecanoic, D-3-hydroxy-15-methylhexadecanoic, and D-3-hydroxyheptadecanoic acids. The D-3-hydroxyheptadecanoic acids predominate. Lesser amounts of the iso-branched C_{15}, straight-chain C_{16}, and anteiso-branched C_{17} acids are present.

The relatively high level of hydroxy acid is too great to be accounted for simply as cell-envelope-bound LPS fatty acid. Many of these hydroxy acids may be associated with the sphingolipids present in a high level in *Bacteroides*. Poly-β-hydroxybutyrate is present in prominent intracellular granules and serves as an energy reserve polymer in bacteria, cyanobacteria, and eukaryotic organisms. 3-Hydroxy fatty acids can also serve as precursors to the monoenoic fatty acids. See further discussion on biosynthesis of the branched chain fatty acids in the "Biosynthesis of Fatty Acids" section.

Ring-Containing Fatty Acids

In 1962, Hoffman discovered the first ring-containing fatty acid, lactobacillic acid, in *L. arabinosus*. It was shown to be *cis*-11,12-methyleneoctadecanoic acid:

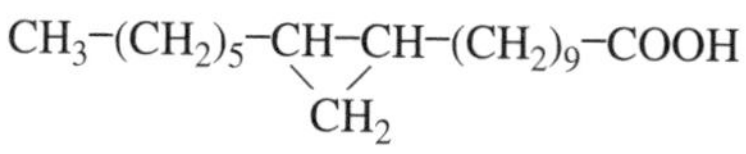

$$CH_3-(CH_2)_5-\underset{\underset{\diagdown\;\diagup}{}}{CH-CH}-(CH_2)_9-COOH$$
$$CH_2$$

Greater than 15% of the fatty acids of most lactobacilli are ring-containing fatty acids, whereas less than 5% of the fatty acids of *Bifidobacterium* are ring-containing fatty acids. Along with lactobacillic acid, two other cyclopropane fatty acids are commonly found in bacteria. A C_{17}-cyclopropane fatty acid, *cis*-9,10-methylenehexadecanoic acid, is present in *E. coli* and other gram-negative organisms and has the following structure:

$$CH_3-(CH_2)_5-\underset{\underset{\diagdown\;\diagup}{}}{CH-CH}-(CH_2)_7-COOH$$
$$CH_2$$

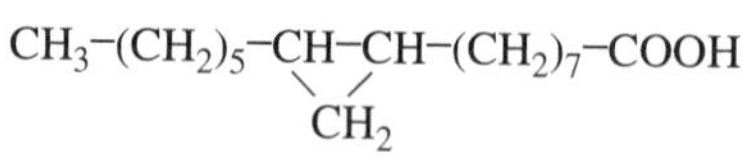

The marked inhibition of cyclopropane fatty acid synthesis, which accompanies the induction of filamentous forms of *E. coli*, suggests that ring-containing fatty acids play an important structural role in the cell envelope. Another commonly occurring cyclopropane fatty acid, 9,10-methyleneoctadec-9-enoic acid, is shown in Table 13-1.

Cyclopropane fatty acids are formed from unsaturated fatty acids present in phospholipids. They are formed only in organisms capable of making the corresponding unsaturated fatty acids (palmitoleic, *cis*-vaccenic, and oleic acids). Once formed, the unsaturated position is converted to a cyclopropane ring by the transfer of a methyl group from *S*-adenosylmethione catalyzed by the cyclopropane fatty acid synthase. This enzyme binds to bilayers of unsaturated fatty acid–containing phospholipids and forms cyclopropane rings on both faces of the membrane bilayer. Modification of the unsaturated positions in the fatty acids of the membrane occurs at the onset of the stationary phase of growth. Protection of the fatty acid from oxidation or from chemical degradation is offered as one explanation for this change in structure. Cyclopropane fatty acid synthase is induced by rapid limitation of oxygen, by initiation of respiration, or by the addition of nitrate or thiosulfate to the culture medium. Although the precise role of cyclopropane fatty acids remains elusive, there does seem to be considerable evidence to suggest that they play a role in the pathogenicity of *M. tuberculosis* and *Helicobacter pylori*.

Alk-1-enyl Ethers (Plasmalogens)

Phosphoglycerides that yield aldehydes upon hydrolysis contain long-chain fatty acids in ether linkage with the C-1 of glycerol and are called **plasmalogens.** These long-chain aldehydes are otherwise similar in composition to long-chain fatty acids. For example, the long-chain aldehydes derived from plasmalogens of *C. butyricum* are comparable to the long-chain fatty acids of this organism. The other carbon atoms of the glycerol backbone are linked to a fatty acid at C-2 and a phosphate ester at C-3 (Fig. 13-1). Although plasmalogens are known to occur in higher organisms, surveys of a wide variety of microorganisms have failed to reveal their presence in aerobic, facultative, or microaerophilic bacteria. In anaerobic genera, the plasmalogens occur in addition to, rather than in place of, the usual phospholipids. Strict anaerobes from the genera *Bacteroides, Clostridium, Desulfovibrio, Peptostreptococcus, Propionibacterium, Ruminococcus, Selenomonas, Sphaerophorus, Treponema*, and *Viellonella* all contain plasmalogens, the molar ratio of aldehydes/phosphorus varying from 0.004 to 1.04.

The major components of the plasmalogens of *C. butyricum* are 16:0 (saturated C_{16} fatty acid), 16:1 (monounsaturated C_{16} fatty acid), 17:cyc (C_{17} cyclopropane fatty acid), 18:0 (saturated C_{18} fatty acid), 18:1 (monounsaturated C_{18} fatty acid), and 19:cyc (C_{19} cyclopropane fatty acid). The fatty aldehydes obtained from the plasmalogens of *Selenomonas ruminantium* consist of normal saturated and monounsaturated fatty aldehydes, with 12 to 18 carbon-chain lengths predominating. A marine bacterium contains phosphatidic acid mixed with its plasmalogen analogs. Phosphatidylserine and its plasmalogen analog are major phosphoglycerides of the strictly anaerobic rumen bacterium *Megasphaera elsdenii*. Phosphoglycerides of the anaerobic lactate-fermenting *Veillonella parvula* and *Anaerovibrio lipolytica* also contain a high proportion of heptadecenoic acyl and alk-1-enyl ether moieties.

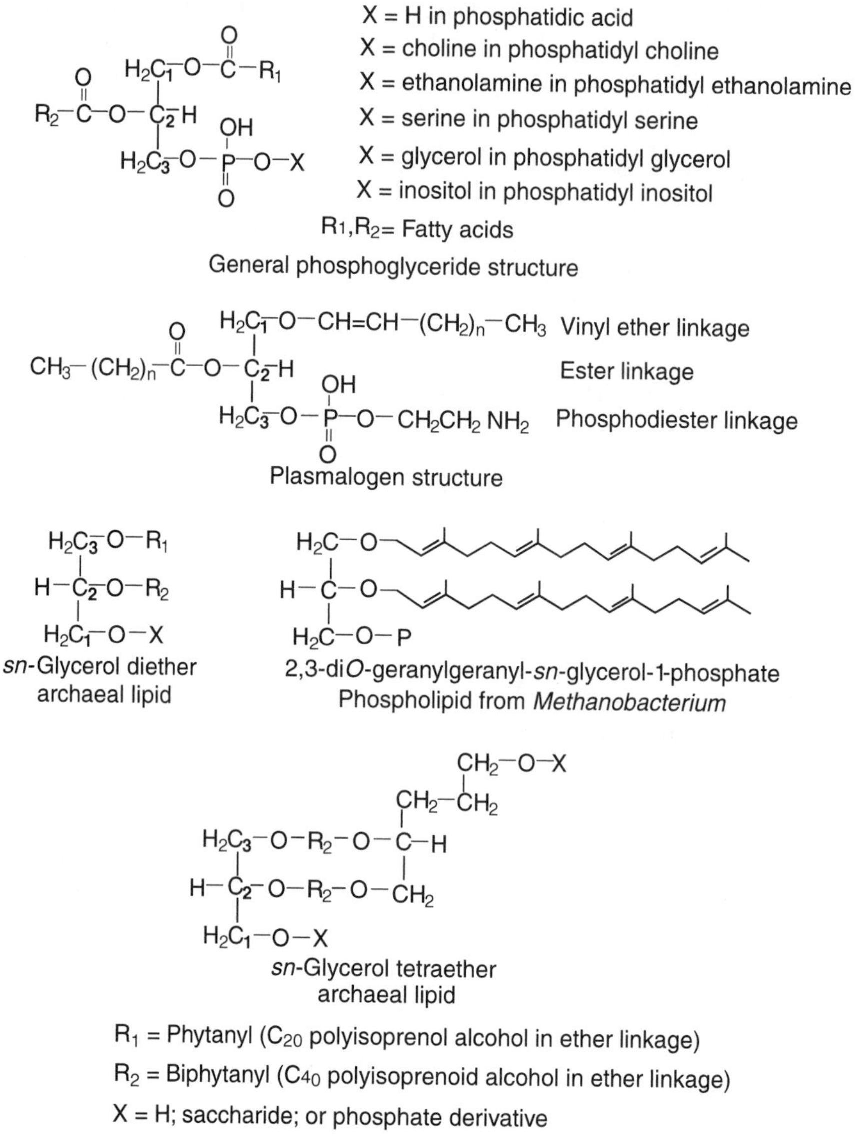

X = H in phosphatidic acid
X = choline in phosphatidyl choline
X = ethanolamine in phosphatidyl ethanolamine
X = serine in phosphatidyl serine
X = glycerol in phosphatidyl glycerol
X = inositol in phosphatidyl inositol

R_1, R_2 = Fatty acids

General phosphoglyceride structure

Vinyl ether linkage

Ester linkage

Phosphodiester linkage

Plasmalogen structure

sn-Glycerol diether
archaeal lipid

2,3-di$O$-geranylgeranyl-sn-glycerol-1-phosphate
Phospholipid from *Methanobacterium*

sn-Glycerol tetraether
archaeal lipid

R_1 = Phytanyl (C_{20} polyisoprenol alcohol in ether linkage)

R_2 = Biphytanyl (C_{40} polyisoprenoid alcohol in ether linkage)

X = H; saccharide; or phosphate derivative

Fig. 13-1. Structure of phospholipids, plasmalogens, and archaeal membrane lipids. A fundamental feature of the Archaea is that the glycerophosphate backbone is sn-glycerol-1-phosphate, the enantiomer of that found in bacteria and eukaryotes.

Alkyl Ethers

Long-chain alcohols bound to glycerol in alkyl ether linkage occur in the Archaea, a group of bacteria that exist primarily in extremely harsh environments. Their ability to thrive in these extreme conditions is considered to be due, at least in part, to these unusual lipid structures. The Archaea contain predominantly di-O-alkyl analogs of phosphatidylglycerophosphate. The lipids of these organisms are formed by condensation of glycerol or more complex polyols with isoprenoid alcohols containing

20, 25, or 40 carbon atoms. A cardinal feature of all archaeal phospholipids is the presence of 2,3-di-O-sn-glycerol as the backbone structure as opposed to the sn-1,2-glycerol found in most other naturally occurring glycerophosphates or diacylglycerols (see Fig. 13-1). To date, no exceptions have been found to this fundamental difference in stereoconfiguration.

Phospholipids (Phosphoglycerides)

Mono-, di-, and triglycerides are the major lipid components of plants and animals. In contrast, bacteria contain relatively low concentrations of glycerides, suggesting that they play a minor role in microbial structure or function. In mammals, a high percentage of stored lipids are in the form of glycerides, while in microorganisms the phospholipids (phosphoglycerides) are the predominant type (Fig. 13-1). Mammalian cells contain phosphatidylcholine as the most common phospholipid. Bacteria usually contain phosphatidylethanolamine, phosphatidylserine, phosphatidylglycerol, and phosphatidic acid. Lysophosphatidyl compounds, which lack one of the three fatty acid chains, are sometimes found in microorganisms. Phosphatidylinositol is low in content in most bacteria but is found in higher concentration in mycobacteria, fungi, and protozoa.

Gram-negative bacteria contain large amounts of phospholipids. Together with LPS, they constitute up to 20 to 40% of the dry weight of the cell envelope. Some of these lipids, especially C_{55}-isoprenols, serve as lipid carriers in the transferase enzymes involved in peptidoglycan synthesis (see Chapter 7). The phospholipids of the cell membrane are considered to exist as bimolecular leaflets. Many marine and estuarine bacteria contain significant quantities of lysophosphatidylethanolamine. Cardiolipin (diphosphatidylglycerol) and cardiolipin synthetase are absent in many nonfermentative isolates of marine bacteria. There are two metabolic pools of phosphatidylglycerol and cardiolipin in *B. stearothermophilus*, *E. coli*, *S. aureus*, and *H. parainfluenzae*. Cardiolipin is an important constituent of the membrane of *H. parainfluenzae*.

As mentioned previously, the archaea produce a variety of unusual lipid structures that contain ether linkages. Although quite different in their chemical structure, most of these lipids are comparable to bacterial phospholipids in that they contain single or bipolar head groups with hydrophobic alkyl side chains. They form membrane bilayers with the polar head group toward the aqueous phase that are indistinguishable from those of bacteria. Neutrophilic halophiles contain C_{20} phytanyl substituents (Fig. 13-1). However, in alkaliphilic halophiles, C_{25} sesterpanylic compounds may predominate. The membranes of methanogens contain diphytanyl-glycerol diether and dibibiphytanyl-diglycerol tetraether. The presence of tetraether lipids in methanogens correlates with the ease of demonstrating freeze-fracture planes in these organisms. Above the 45 to 50% level of tetraether core lipids, the frequency of membrane fracture is markedly reduced. The membrane structures of thermophiles show considerable variation, but C_{20} phytanyl and C_{40} biphytanyl chains are found in *Desulfurococcus*, *Thermoproteus*, *Thermofilum*, and *Pyrodictum*. *Sulfolobus* contains a variety of tetraethers of two classes. The first class is comprised of glycerol-dialkylglycerol-tetraethers that contain two sn-2,3-glycerol components bridged by two isoprenoid C_{40} diols through ether linkages. In the second class, referred to as glycerol-dialkylnonitol tetraethers, a branched nonitol replaces one of the glycerols.

Glycolipids

Glycosyldiglycerides may contain glucose, galactose, or mannose. The structure of the glycolipid α-D-glucopyranosyl-(1,2)-α-D-glucopyranosyl-(2,3)-diglyceride is

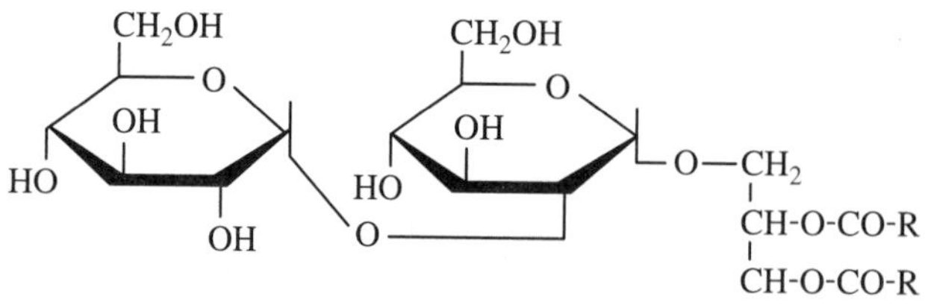

Monoglycosyl derivatives apparently represent biosynthetic precursors of the diglycosyl derivatives. Glycosyldiglycerides are widely distributed in gram-positive bacilli, but thus far have not been found in gram-negative bacilli. They have either α,α or β,β configurations and do not seem to occur in the α,β form. The nonsterol-requiring *Mycoplasma* contains glycolipids, suggesting that glycolipids replace sterols in either the structure and/or the function of the membrane.

In most glycolipids the carbohydrate moiety is linked to the glycerol component. However, glycolipid structures have been observed in which the carbohydrate portion is attached directly to the fatty acid. A rhamnolipid found in *P. aeruginosa* contains two molecules of rhamnose and two molecules of decanoic acid:

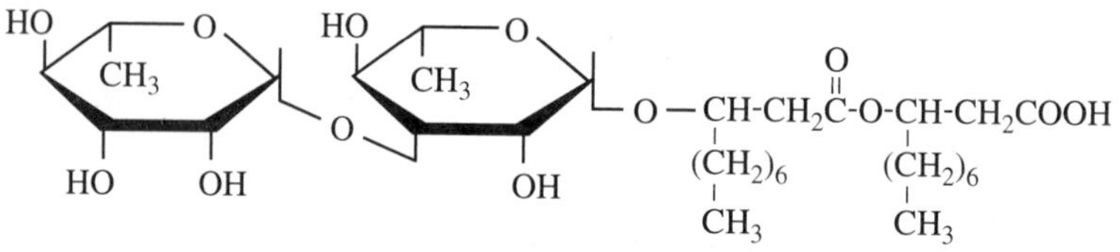

A glycolipid of major importance in *M. tuberculosis*, the **cord factor**, is associated with the virulence of this organism as well as with the characteristic serpentine growth pattern. Cord factor has been shown to contain two molecules of mycolic acid attached to trehalose. In human and bovine strains of *M. tuberculosis*, the substituent side chain is $C_{24}H_{49}$, as shown in the following structure:

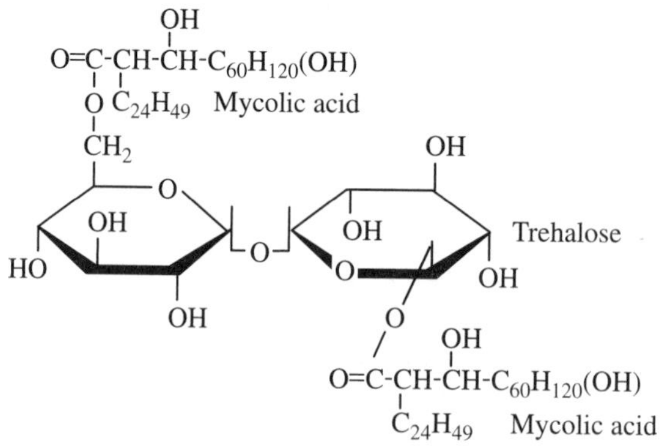

The side chain in the cord factor in avian and saprophytic mycobacteria is $C_{22}H_{45}$. Other variations in the chemical structure of mycolic acids have been found as a larger number of microorganisms have been examined.

A unique phenolic glycolipid in *M. leprae* contains 3,6-di-*O*-methyl-glucose, 2,3-di-*O*-methylrhamnose, and 3-*O*-methylrhamnose linked to phenol-dimycoserosyl phthiocerol:

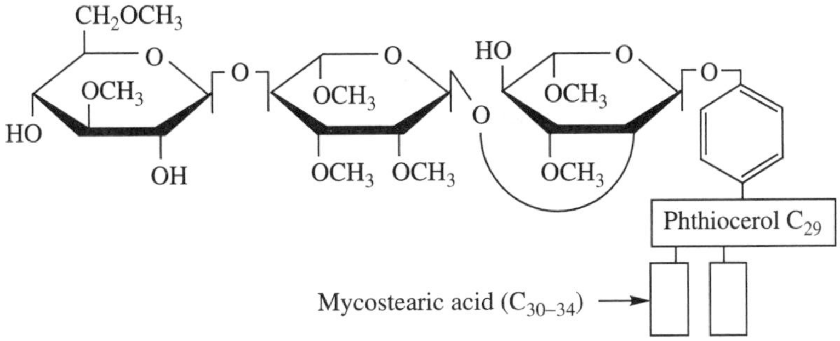

This unusual antigen of *M. leprae* suppresses the in vitro mitogenic response of lymphocytes from lepromatous patients. The suppressor T cells of these patients recognize the specific terminal trisaccharide moiety that triggers the suppression. Lymphocytes from patients with tuberculoid leprosy, lepromin-positive contacts, or normal donors show no comparable suppression of lymphocyte proliferation. Removal of the mycoserosic acid side chain has no effect on in vitro suppression. However, absence of the 3′ terminal methyl group or removal of the terminal sugar abolishes or significantly reduces the suppressive effect. Comparable glycolipids from *M. kansasii* and *M. bovis* have no effect on the mitogenic response.

BIOSYNTHESIS OF FATTY ACIDS

Fatty acid biosynthesis has been studied extensively in bacteria and yeast. Acetyl-CoA is the ultimate precursor of fatty acid carbons. Acetyl-CoA carboxylase, a biotin-containing enzyme that catalyzes the ATP-dependent fixation of CO_2 into acetyl-, propionyl-, and butyryl-CoA, catalyzes the first committed reaction in the pathway (Fig. 13-2). The reaction occurs in two stages: the carboxylation of biotin with bicarbonate is catalyzed by biotin carboxylase, and transfer of the carboxyl group from carboxybiotin to acetyl-CoA to form malonyl-CoA is mediated by carboxyltransferase (Fig. 13-3). The first stage is followed by a series of reactions that ultimately lead to the synthesis of long-chain fatty acids at the C_{16} (palmitic) and C_{18} (stearic) chain length (Fig. 13-2). Initially, it was puzzling to find that CO_2, an essential reactant in the system, was not incorporated into the fatty acids. This occurs because the CO_2 used to form malonyl-CoA is subsequently removed in the condensation step. It is actually the liberation of CO_2 that shifts the equilibrium in the direction of synthesis.

Because of its crucial role as the first committed step in fatty acid biosynthesis, the acetyl-CoA carboxylase system has been considered as one of the most likely sites of regulation of fatty acid biosynthesis. In mammalian systems, several mechanisms of regulation of acetyl-CoA carboxylase have been proposed. Allosteric regulation by

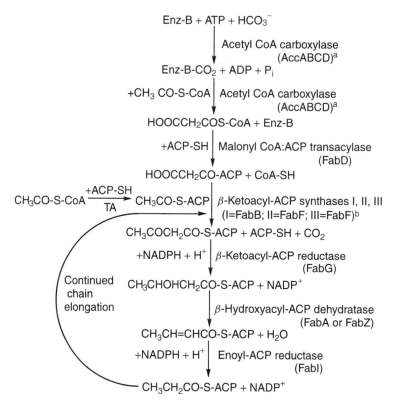

Fig. 13-2. Reactions in fatty acid biosynthesis. Enz-B, enzyme-biotin; CoA, coenzyme A; ACP, acyl carrier protein; TA, acetyl-CoA-ACP transacylase. The three-letter designations are for the structural proteins. [a]See Figure 13-3 for details of acetyl-CoA carboxylase system involving enzymes AccABCD. [b]The FabH enzyme, β-ketoacyl-ACP synthase III, catalyzes the condensation of acetyl-CoA and malonyl-ACP, whereas FabB and FabF, β-ketoacyl-ACP synthases I and II, are active with C_2-C_{14}-ACP but inactive with C_{16}-ACP, and are inactive with CoA derivatives.

citrate and isocitrate, feedback inhibition by the end product (a long-chain acyl-CoA), covalent modification by phosphorylation and dephosphorylation, and regulation at the level of gene expression have all been implicated as possible mechanisms. In *E. coli* the four components of the acetyl-CoA carboxylase system function as an enzyme complex. The genes that encode the biotin carboxyl carrier protein (BCCP), *accB*, and biotin carboxylase, *accC*, are cotranscribed (*accBC*) and map at 72 minutes. The genes encoding the carboxyltransferase α (*accA*) and β (*accD*) subunits map at different regions of the chromosome. The rates of transcription of the genes encoding all four subunits of acetyl-CoA carboxylase are directly related to the rate of cell growth. The promoter sequences of these genes and certain features of their respective promoter regions indicate a role in regulation of the complex.

Fatty acid biosynthesis in bacteria is conducted by a soluble, disassociated system designated as **type II**. At least eight individual enzyme components are readily separated and purified from *E. coli* and *P. shermanii*. The system requires acetyl-CoA, malonyl-CoA, NADH, NADPH, and ACP for fatty acid synthesis. In *E. coli* and other

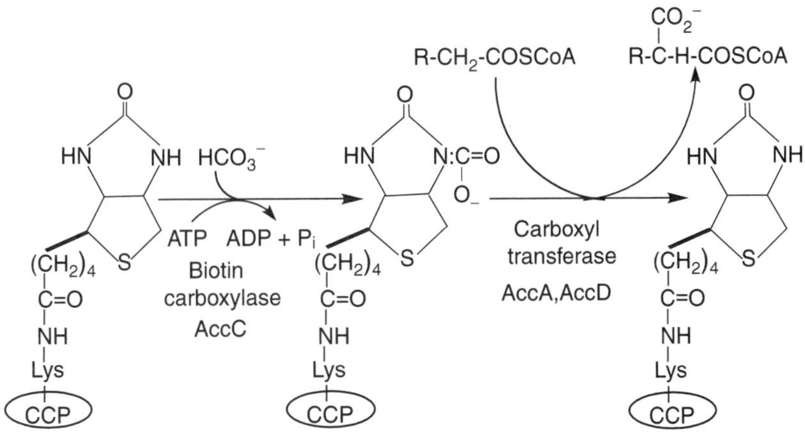

Fig. 13-3. The role of biotin in the acetyl-CoA carboxylase system. This is the first committed step in fatty acid biosynthesis. The *E. coli* complex consists of four subunits. The three-letter designations are the products of the corresponding structural genes in *E. coli*.

bacteria, 12 or more genes code for the enzymes depicted in Figure 13-2. The genes *acpP*, *fabD*, *fabF*, *fabG*, and *fabH* are encoded in a *fab* gene cluster that has been termed the *fab* operon The enzyme β-ketoacyl-ACP synthase I (FabB) is required for the elongation of unsaturated fatty acids. Mutants lacking FabB activity are unable to synthesize either palmitoleic or *cis*-vaccenic acids and require fatty acids for growth.

β-Ketoacyl-ACP synthase II (FabF) is responsible for the temperature-dependent regulation of fatty acid composition. Mutants lacking FabF are deficient in *cis*-vaccenic acid but grow normally. β-Ketoacyl-ACP synthase III (FabH) catalyzes the formation of acetoacyl-ACP. The *fabA* gene encodes 3-dehydrodecanoyl-ACP dehydratase. Malonyl-CoA-ACP transacylase (FabD) catalyzes the transacylation of ACP with malonate. Enoyl-ACP reductase (FabI) conducts the final step in the elongation process. The product is recycled through the last four steps to yield long-chain fatty acids. In *E. coli*, FabI is the target of triclosan, a broad-spectrum antibiotic active against many bacteria and fungi. Mutations in the *fabI* gene confer triclosan resistance. Triclosan is widely used as an ingredient in antiseptic soaps, toothpastes, and other materials. The development of triclosan resistance may select for multiple resistance to other agents, including many clinically useful antibiotics. Triclosan (also known as irgasan) provides for the selective growth of *P. aeruginosa* on pseudomonas isolation agar.

The ACP that functions in fatty acid synthesis is very stable to heat and acid treatment. The structure of the prosthetic group is

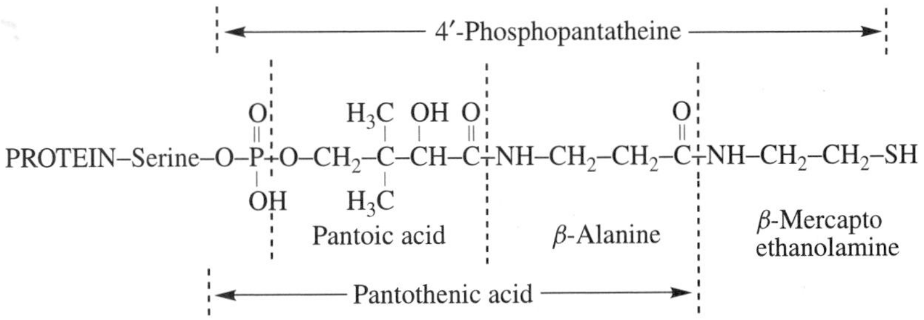

The composition of this prosthetic group is very similar to that of coenzyme A. The ACP is specific to the fatty acid synthase system and cannot replace coenzyme A.

In *E. coli* unsaturated fatty acids are synthesized simultaneously with saturated fatty acids. The divergence occurs at the C_{10} stage (β-hydroxy-decanoate). At this step there is apparently a competition between β-hydroxyacyl-ACP dehydrase, which forms α,β,γ-*cis* double bonds, as shown in Figure 13-4. Continued elongation leads to the formation of unsaturated fatty acids. As a result of this mode of unsaturated fatty acid synthesis, only monounsaturated fatty acids are formed, and *cis*-vaccenic acid ($18:1^{\Delta 11}$) is the major unsaturated fatty acid in organisms that use the anaerobic pathway (see Fig. 13-4 and Table 13-2).

In yeasts, fungi, and other eukaryotes, a multienzyme complex called **type I** carries out fatty acid synthesis. Mycobacteria also contain a **type I** fatty acid synthase system. The **type I** fatty acid synthase in yeast consists of two multifunctional peptide chains each having six subunits ($\alpha_6\beta_6$) that carry out the same succession of reactions shown in Figure 13-2. An acyl carrier protein component serves as the carrier to move the growing fatty acid chain to the active sites. In yeast the elongation is terminated at

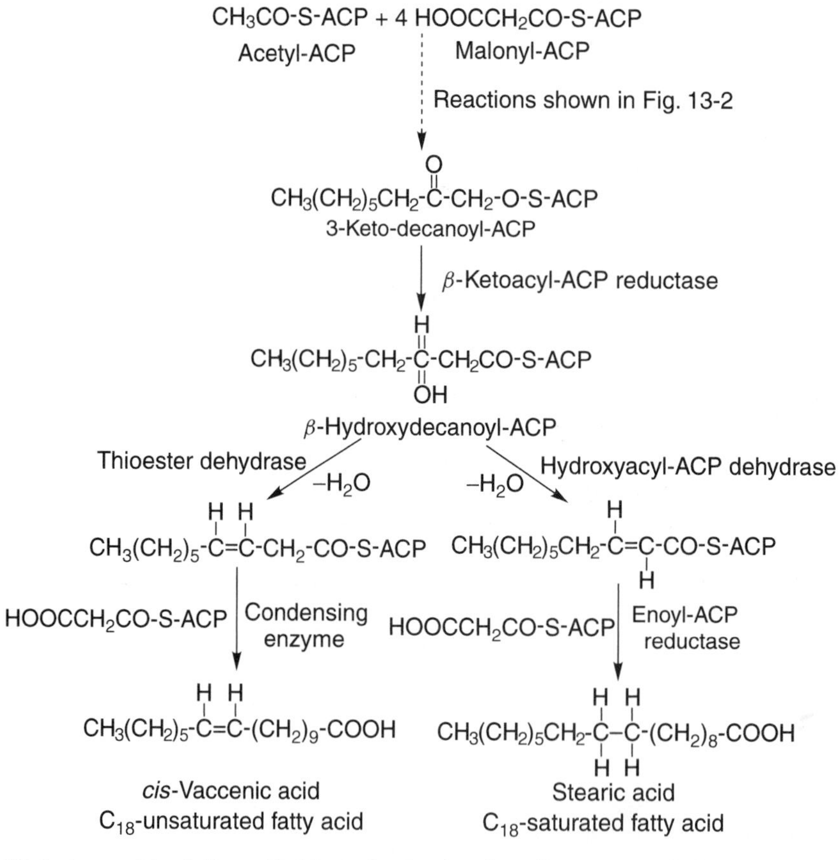

$$CH_3CO\text{-}S\text{-}ACP + 4\ HOOCCH_2CO\text{-}S\text{-}ACP$$

Acetyl-ACP   Malonyl-ACP

Reactions shown in Fig. 13-2

$$CH_3(CH_2)_5CH_2\overset{\overset{\displaystyle O}{\|}}{\text{-C-}}CH_2\text{-O-S-ACP}$$

3-Keto-decanoyl-ACP

β-Ketoacyl-ACP reductase

$$CH_3(CH_2)_5\text{-}CH_2\overset{\overset{\displaystyle H}{|}}{\underset{\underset{\displaystyle OH}{|}}{\text{-C-}}}CH_2CO\text{-S-ACP}$$

β-Hydroxydecanoyl-ACP

Thioester dehydrase   Hydroxyacyl-ACP dehydrase

$-H_2O$ $-H_2O$

$$CH_3(CH_2)_5\overset{\overset{\displaystyle H\ \ H}{|\ \ \ |}}{\text{-C=C-}}CH_2\text{-CO-S-ACP}$$ $$CH_3(CH_2)_5CH_2\overset{\overset{\displaystyle H}{|}}{\underset{\underset{\displaystyle H}{|}}{\text{-C=C-}}}CO\text{-S-ACP}$$

HOOCCH₂CO-S-ACP | Condensing enzyme   HOOCCH₂CO-S-ACP | Enoyl-ACP reductase

$$CH_3(CH_2)_5\overset{\overset{\displaystyle H\ \ H}{|\ \ \ |}}{\text{-C=C-}}(CH_2)_9\text{-COOH}$$ $$CH_3(CH_2)_5CH_2\overset{\overset{\displaystyle H\ \ H}{|\ \ \ |}}{\underset{\underset{\displaystyle H\ \ H}{|\ \ \ |}}{\text{-C-C-}}}(CH_2)_8\text{-COOH}$$

cis-Vaccenic acid   Stearic acid
C_{18}-unsaturated fatty acid C_{18}-saturated fatty acid

Fig. 13-4. Anaerobic fatty acid biosynthesis showing divergence of the saturated and unsaturated pathways. The pathway to saturated fatty acids shown in Figure 13-2 is followed until the C_{10} level (β-hydroxydecanoyl-ACP), at which point there is a competition between the enzymes involved in forming saturated and unsaturated fatty acids.

TABLE 13-2. Distribution of Pathways for Biosynthesis of Long-Chain Fatty Acids

Anaerobic Pathway	Aerobic Pathway
Escherichia coli	*Alcaligenes faecalis*
Salmonella enterica	*Corynebacterium diphtheriae*
Serratia marcescens	*Mycobacterium phlei*
Azotobacter agilis	*Bacillus* (several species)
Agrobacterium tumefaciens	*Micrococcus luteus*
Lactobacillus plantarum	*Beggiatoa* spp.
Staphylococcus haemolyticus	*Myxobacter* spp.
Clostridium pasteurianum	*Leptospira canicola*
Clostridium butyricum	*Saccharomyces cerevisiae*
Caulobacter crescentus	*Candida lipolytica*
Propionibacterium	*Neurospora crassa*
Cloroflexus auranticus	*Stigmatella auranticus*
Chlorobium limicola	

the C_{16} (palmitic acid) to C_{18} (stearic) level, and a transferase forms a coenzyme A derivative. Mammals, yeast, algae, protozoa, and some highly aerobic bacteria (*Alcaligenes*, *Mycobacterium*, *Corynebacterium*, *Bacillus*, and others as shown in Table 13-2) use the **aerobic pathway** to synthesize long-chain unsaturated fatty acids. In *S. cerevisiae* oxidative desaturation by a specific desaturase, Ole1p, introduces a double bond between positions 9 and 10 after the long-chain saturated fatty acid has been synthesized:

$$CH_2(CH_2)_7-\overset{\overset{\displaystyle H}{|}}{\underset{\underset{\displaystyle H}{|}}{C}}-\overset{\overset{\displaystyle H}{|}}{\underset{\underset{\displaystyle H}{|}}{C}}-(CH_2)_7-CO-SCoA \xrightarrow[]{O_2 \quad 2H_2O} CH_2(CH_2)_7-\overset{\overset{\displaystyle H}{|}}{C}=\overset{\overset{\displaystyle H}{|}}{C}-(CH_2)_7-CO-SCoA$$

Stearoyl-CoA (18:0)   Oleoyl-CoA ($18:1^{\Delta 11}$)

Approximately 70 to 80% of the total fatty acids in membrane lipids of the yeast *Saccharomyces cerevisiae* consist of the unsaturated fatty acids palmitoleic acid (16 : 1) and oleic acid (18 : 1). The remaining fatty acids are saturated, consisting primarily of palmitic acid (16 : 0) and lesser amounts of stearic acid (18.0) and myristic acid (14 : 0).

Most other fungi contain the di- and trienoic acids linoleic acid (18 : 2) and α-linolenic acid (18 : 3). Yeast utilizes fatty acid desaturases for the formation of unsaturated fatty acids. The Δ-9 desaturase catalyzes the formation of the double bond between the ninth and tenth carbon atoms of both palmitoyl (16 : 0) and stearoyl (18 : 0) CoA substrates to form 16 : 1 and 18 : 1 unsaturated fatty acids. The activity of the Δ-9 desaturase is markedly reduced by the addition of unsaturated fatty acids, presumably because the level of the desaturase mRNA is repressed under these conditions. Longer-chain fatty acids of 24 to 26 carbon atoms are synthesized by a membrane-bound elongase containing two components: Elo2p and Elo3p. These very long-chain fatty acids comprise only 1% of the cellular fatty acid pool, but they are essential for the ceramide found in sphingolipids and the lipid portion of glycosylphosphatidylinositol-anchored membrane proteins.

Under anaerobic conditions yeast exhibit a requirement for unsaturated fatty acids that stems from their requirement for coupling oxidative phosphorylation to ATP synthesis in the yeast mitochondrion. This effect is obvious since yeast use the aerobic pathway (desaturase enzyme) for the formation of unsaturated fatty acids. However, studies with mutant strains show that *S. cerevisiae* also requires unsaturated fatty acids for aerobic growth. The required amount is 4-fold higher than that required for anaerobic growth.

BIOSYNTHESIS OF PHOSPHOLIPIDS

Phospholipids (mainly phosphatidylcholine, phosphatidylethanolamine, phosphatidylinositol, and phosphatidylserine) are the major structural components of cellular membranes. Biosynthesis of these important compounds, more correctly referred to as phosphoglycerides, begins with the formation of an *sn*-glycerol-3-phosphate (G-3-P). The G-3-P may be formed from dihydroxyacetone phosphate (DOHAP) derived from hexose cleavage by the action of the biosynthetic G-3-P dehydrogenase encoded by *gpsA*:

$$DOHAP + NAD(P)H + H^+ \leftrightarrow G\text{-}3\text{-}P + NAD(P)^+$$

sn-Glycerol-3-phosphate may also be produced from glycerol via the action of glycerol kinase encoded by *glpK*:

$$glycerol + ATP \rightarrow G\text{-}3\text{-}P + ADP$$

The first committed steps in phosphoglyceride biosynthesis involve coupling two molecules of fatty acid to G-3-P (Fig. 13-5). This reaction is catalyzed by two acyl transferases. Glycerol-3-phosphate acyltransferase (PlsB) adds one unsaturated fatty acid to form 1-acyl-glycerol-3-phosphate. An unsaturated fatty acid is added to the C-2 position by 1-acylglycerophosphate acyltransferase (PlsK) to form phosphatidic acid. In bacteria, a fatty acid-ACP is the donor rather than a fatty acid-CoA. The next step involves formation of cytidine diphosphoglyceride (CDP-diglyceride) by phosphatidate cytidyltransferase, as shown in Figure 13-5.

In mammalian cells, the seeds of plants, and in lower eukaryotes, triglycerides are formed in considerable quantities and serve as storage compounds. In microorganisms, particularly bacteria, phospholipids are more commonly observed. *E. coli* normally contains approximately 70% phosphatidylethanolamine, 25% phosphatidylglycerol, and 5% cardiolipin.

At this point the pathway branches. Addition of L-serine by phosphatidylserine synthase (encoded by *pssA*) yields phosphatidyl serine. Decarboxylation of phosphatidylserine by phosphatidylserine decarboxylase (encoded by *psd*) yields phosphatidylethanolamine, one of the major phospholipids found in bacteria. In the other branch of the pathway, addition of a second molecule of G-3-P to the CDP-diglyceride by phosphatidylglycerophosphate synthetase (encoded by *pgsA*) results in the release of CMP and the formation of phosphatidylglycerolphosphate. Removal of P_i by phosphatidylglycerolphosphate phosphatase (encoded by *pgsA*, *pgsB*) yields phosphatidyl glycerol. In bacteria, two molecules of phosphatidylglycerol are condensed by

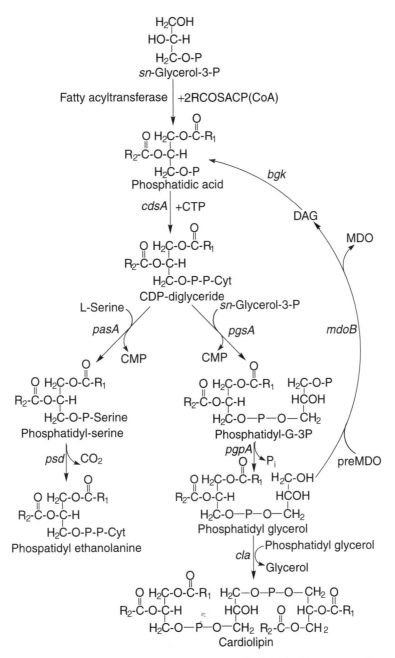

Fig. 13-5. Biosynthesis of phosphoglycerides (phospholipids). The structural genes for the *E. coli* enzymes are: *plsB*, *sn*-glycerol-3-phosphate acyltransferase; *plsK*, 1-acylglyc-erol-3-phosphate acyltransferase; *cdsA*, phosphatidate cytidyltransferase; *pssA*, phosphatidyl-serine synthase; *psd*, phosphatidylserine decarboxylase; *pgsA*, phosphatidylglycerophosphate synthase; *pgpA*, phosphatidylglycerophosphate phosphatase; *cls*, cardiolipin (diphosphatidylglyc-erol) synthase.

cardiolipin synthase (encoded by *cls*), releasing glycerol to form cardiolipin (diphosphatidylglycerol). In mammalian systems, the formation of cardiolipin involves the formation of a CDP diglyceride as an intermediate.

In addition to their role as structural components of the cell membrane, another significant aspect of phospholipids is the production of derivatives that serve as messengers in cellular regulation. Phospholipids are acted upon by several classes of phospholipases (A_2, C, and D) to produce a variety of intracellular and intercellular messengers:

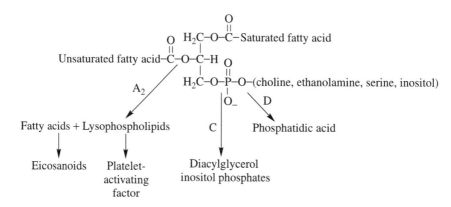

Many of these regulatory compounds are of major significance in mammalian systems. However, the formation of phosphatidylinositol and its phosphorylated derivatives in yeast is an important aspect of yeast signal transduction. Phospholipid breakdown by various classes of phospholipases gives rise to compounds that serve as messengers in the utilization of signal transduction cascades in the regulation of the cell cycle, response to environmental stress, and the process of sporulation in yeast.

DEGRADATION OF FATTY ACIDS

Most bacteria do not readily degrade fatty acids. However, once induced, they may become very active fatty acid oxidizers. For example, *E. coli* will grow on long-chain fatty acids only after a prolonged lag phase. Cells induced to grow on a fatty acid, such as palmitate, will grow immediately on fatty acids without any appreciable lag. The activities of the enzymes of fatty acid oxidation shown in Figure 13-6 increase or decrease coordinately, depending on the conditions under which the cells were previous grown. Reversal of the strong glucose inhibition of fatty acid utilization by cAMP indicates that catabolite repression is the mechanism involved.

The structural genes (*fad* genes) encoding enzymes of fatty acid degradation are observed at four distinct loci on the *E. coli* gene map and encode at least five enzyme activities involved in the transport, acylation, and β-oxidation of medium-chain (C_6-C_{10}) and long-chain (C_{12}-C_{18}) fatty acids. Long-chain fatty acids can induce the *fad* genes, whereas medium-chain fatty acids cannot. Therefore, the wild-type *E. coli* can utilize long-chain fatty acids, such as oleate ($C_{18}:1$), but not medium-chain length, such as decanoate (C_{10} phenotype), as the sole carbon and energy source. The *fadR* gene is a multifunctional regulatory gene mapping at 25.5 minutes that exerts negative control over the *fad* regulon. The product of the *fadR* gene (FadR) is a diffusible protein that exerts control over fatty acid degradation by decreasing the transcription of the *fad* structural genes.

In addition to the *fad* enzymes, expression of the glyoxylate shunt enzymes is also required for the growth of *E. coli* on acetate or fatty acids as a sole carbon source. In wild-type *E. coli*, repression of the acetate (*ace*) operon is under the control of *fadR* and *iclR* regulatory genes for the *aceBA* operon). Both the *fadR* and *iclR* genes regulate the glyoxylate shunt at the level of transcription. The *fadR* gene seems to play a role in the regulation of unsaturated fatty acid biosynthesis as evidenced by the fact that *fadR* mutants synthesize significantly less unsaturated fatty acids than wild-type cells. A functional *fadR* gene is required for optimal synthesis of unsaturated fatty acids.

The *fadAB* and *fadR* genes have been cloned and characterized with regard to their specific functions. In *E. coli* a multifunctional enzyme complex is encoded by the *fadAB* operon. Two protein subunits (42,000 and 78,000 Da) in this enzyme complex exhibit 3-ketoacyl-CoA thiolase, 3-hydroxyacyl-CoA epimerase, and enoyl-CoA hydratase. Two identical small subunits are encoded within the *fadA* gene, which contains thiolase activity. The other four activities are associated with the large subunit encoded by *fadB*. The intact complex is required for function. Both *fadA* and *fadB* are transcribed as a single transcriptional unit with the direction of transcription being from *fadA* to *fadB*. Plasmids carrying the *fadR* gene suppress the expression of the *fadR* operon. A 19,000 Da protein, FadR, exerts control over the *fad*, *ace*, and *fab* structural genes.

Neurospora crassa also possesses an inducible β-oxidation system. Activities of the enzymes in this system are enhanced after a shift from a sucrose to an acetate growth medium. The induction is even more pronounced after growth in the presence of oleate as the sole carbon and energy source. The enzymes of this pathway are localized in the glyoxysome microbodies, which are distinct from the peroxisomes (subcellular organelles whose main function is to process substances for elimination). The β-oxidation system in *N. crassa* relies on acyl-CoA dehydrogenase rather than acyl-CoA oxidase in the first step. The system does not seem to be under the same type of coordinate regulation as the fatty acid oxidation system in *E. coli*. By comparison, β-oxidation in yeasts takes place in peroxisomes. Peroxisomal β-oxidation involves catalase in the reoxidation of $FADH_2$. These organelles have a novel long-chain acyl-CoA synthetase whose product is exclusively used for β-oxidation and not for lipid synthesis. Lipolytic yeasts, such as *Candida rugosa*, secrete lipases that initiate the process by degrading triacylglycerol to fatty acids and glycerol as described previously.

Some of the earliest investigations into the metabolism of fatty acids and triglycerides dealt primarily with the degradation of triglycerides by lipases and further oxidation of the liberated fatty acids. Lipases attack triglycerides in the following manner:

Triglyceride   Fatty acids   Glycerol

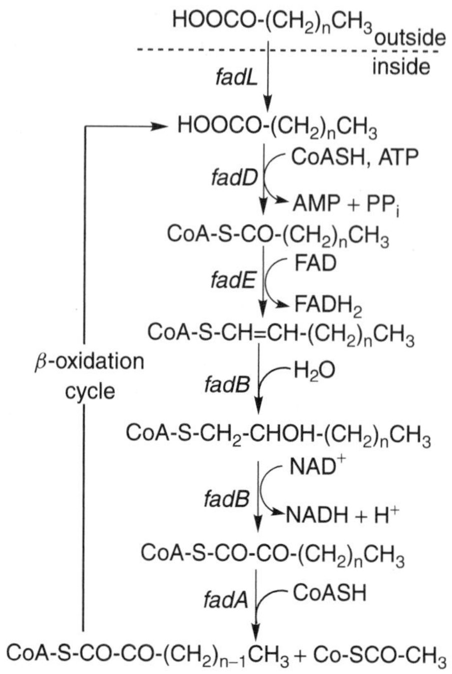

$$HOOCO\text{-}(CH_2)_nCH_3 \text{ outside}$$

Fig. 13-6. Degradation of fatty acids (β-oxidation cycle). The *E. coli* structural genes and enzymes are: *fadL*, fatty acid transport protein; *fadD*, acyl-CoA synthetase; *fadE*, acyl-CoA dehydrogenase; *fadB*, 3-hydroxyacyl-CoA dehydrogenase; *fadA*, β-ketoacyl-CoA thiolase.

The liberated glycerol is oxidized to glycerol-3-phosphate, which is then metabolized via the triose phosphate portion of the EMP pathway. The free fatty acids are converted to acyl-CoA derivatives by acyl-CoA synthetase and subsequently oxidized via the sequence of reactions shown in Figure 13-6. These reactions are repeated until the fatty acids have been reduced to acetyl-CoA. The two-carbon fragments are then metabolized by the glyoxylate cycle. In organisms that do not have a functional glyoxylate cycle, acetoacetate, β-hydroxybutyrate, and acetone may be produced.

BIOSYNTHESIS OF ISOPRENOIDS

Isoprenoids are compounds composed of repeating five carbon isopentenyl diphosphate (IPP) units. In bacteria, polyprenyl phosphates such as undecaprenol are involved in cell wall biosynthesis (Chapter 7). The electron transport agents menaquinone and ubiquinone (Chapters 9 and 15) contain isoprenols. Carotenoids (Chapter 12) are composed of isoprenoid units. The sterols, chlorophylls, and plastoquinones found in eukaryotic organisms also contain isoprenoids. All isoprenoids are synthesized by consecutive condensations of the five-carbon monomer isopentenyl diphosphate (IPP):

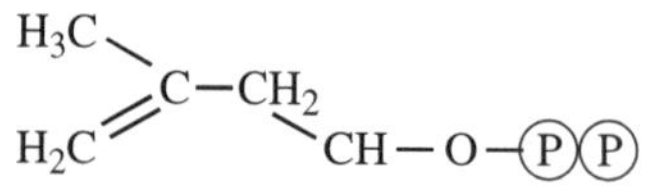

In bacteria, the C_{55}-isoprenol, bactoprenol, which participates in cell wall biosynthesis, is composed of repeating isoprenyl subunits and has the following structure:

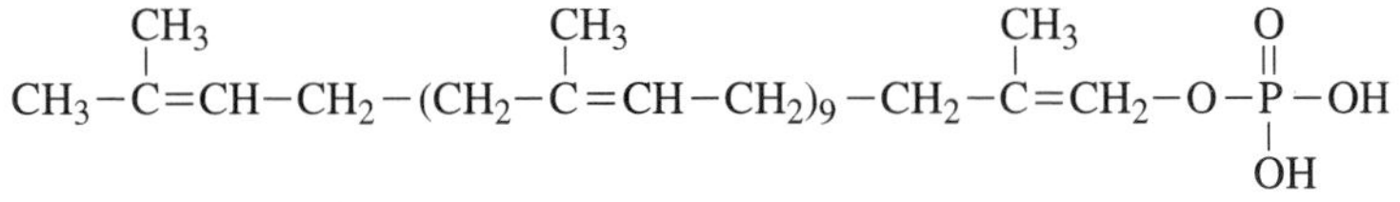

$$CH_3-\overset{\overset{\displaystyle CH_3}{|}}{C}=CH-CH_2-(CH_2-\overset{\overset{\displaystyle CH_3}{|}}{C}=CH-CH_2)_9-CH_2-\overset{\overset{\displaystyle CH_3}{|}}{C}=CH_2-O-\overset{\overset{\displaystyle O}{||}}{\underset{\underset{\displaystyle OH}{|}}{P}}-OH$$

The central isoprenoid precursor, isopentenyl diphosphate (IPP), can be synthesized via the mevalonate pathway as shown in Figure 13-7. In staphylococci, streptococci, and enterococci, genes are present that encode all of the enzymes of the mevalonate pathway, but none of the genes coding for enzymes of the GAP-pyruvate pathway (see below) are found. The mevalonate pathway has been demonstrated in *Myxococcus, Lactobacillus, Methanococcus, Methanobacterium, Pyrococcus, Archeoglobus,*

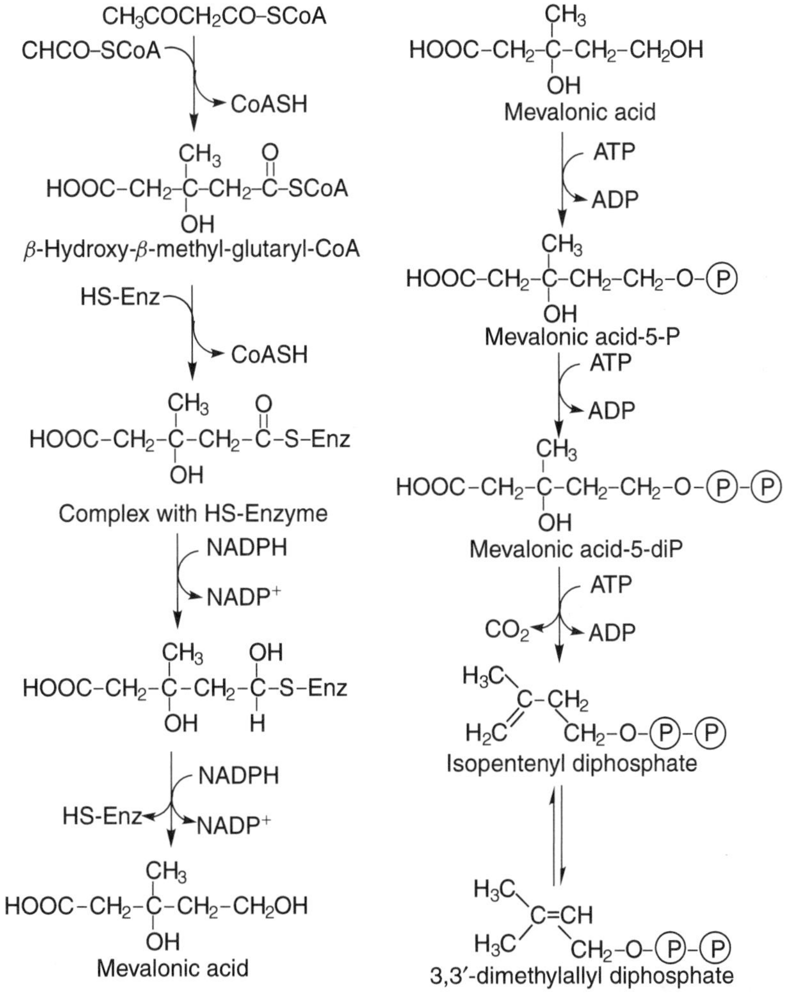

Fig. 13-7. Biosynthesis of mevalonic acid and its conversion to isoprenoid structural form.

and *Streptomyces* species. In all species using the mevalonate pathway, hydroxymethylglutaryl coenzyme A (HMG-CoA) synthetase and HMG-CoA reductase genes are closely linked. Genes encoding mevalonate kinase, mevalonate decarboxylase, and phosphomevalonate kinase apparently constitute an operon. These enzymes are essential for the growth of *S. pneumoniae* and presumably other gram-positive cocci. Phylogenetic analysis provides evidence to suggest that gram-positive cocci obtained the genes for the mevalonate pathway through one or more horizontal gene transfer events involving primitive eukaryotes.

Some organisms use an alternative pathway to IPP, the pyruvate-GAP pathway. This pathway involves condensation of pyruvate and glyceraldehyde-3-phosphate (GAP) to form 1-deoxy-D-xylulose 5-phosphate (DXP). Subsequent conversion of DXP to 2-*C*-methyl-D-erythritol 4-phosphate (MEP) leads to the formation of isopentenyl diphosphate:

Alternate pathway to isopentenylphosphate (IPP)

Recent studies reveal that almost all of the eubacteria except for *S. aureus* and other gram-positive cocci and *B. burgdorferi* utilize only the nonmevalonate pathway for isoprenoid biosynthesis. Most gram-negative bacteria and *B. subtilis* possess only components of the GAP-pyruvate pathway. The nonmevalonate pathway has also been demonstrated in *Chlamydia trachomatis*, *Aquifex aeolicus*, *Synecocystis*, *Scendesmus*, and *M. tuberculosis*.

Yeasts and molds contain sterols in their cytoplasmic membranes. A few prokaryotic organisms such as *Mycoplasma* have sterols in their membranes. Sterols function in membrane permeability. In fungi, a branch of the pathway of fatty acid synthesis leads to the formation of mevalonic acid, the first intermediate in sterol biosynthesis. Through a complex series of reactions, mevalonic acid is converted to farnesyl pyrophosphate and then to squalene. Squalene then undergoes several cyclization reactions to ultimately give rise to lanosterol, the precursor of many sterol derivatives. The pathways leading to the formation of sterols are shown in Figures 13-7 and 13-8.

The pathway to mevalonic acid (Fig. 13-7) has been demonstrated in yeast as well as in mammalian liver. The condensing enzyme responsible for the coupling of acetoacetyl-CoA and acetyl-CoA has been isolated and purified from yeasts. Hydroxymethylglutaryl-CoA can also arise from malonyl-CoA via reactions that are identical to those responsible for the initiation of fatty acid biosynthesis. Free hydroxymethylglutarate is not readily utilized as a precursor of steroids by yeast or mammalian liver, presumably because of the absence of an activating enzyme. On the other hand, *Phycomyces blakesleeanus* can readily incorporate this compound since it has the necessary activating enzyme.

The *Mycoplasmataceae* are separated into two groups on the basis of their nutritional requirements for lipid components. The genus *Mycoplasma* requires sterols for growth,

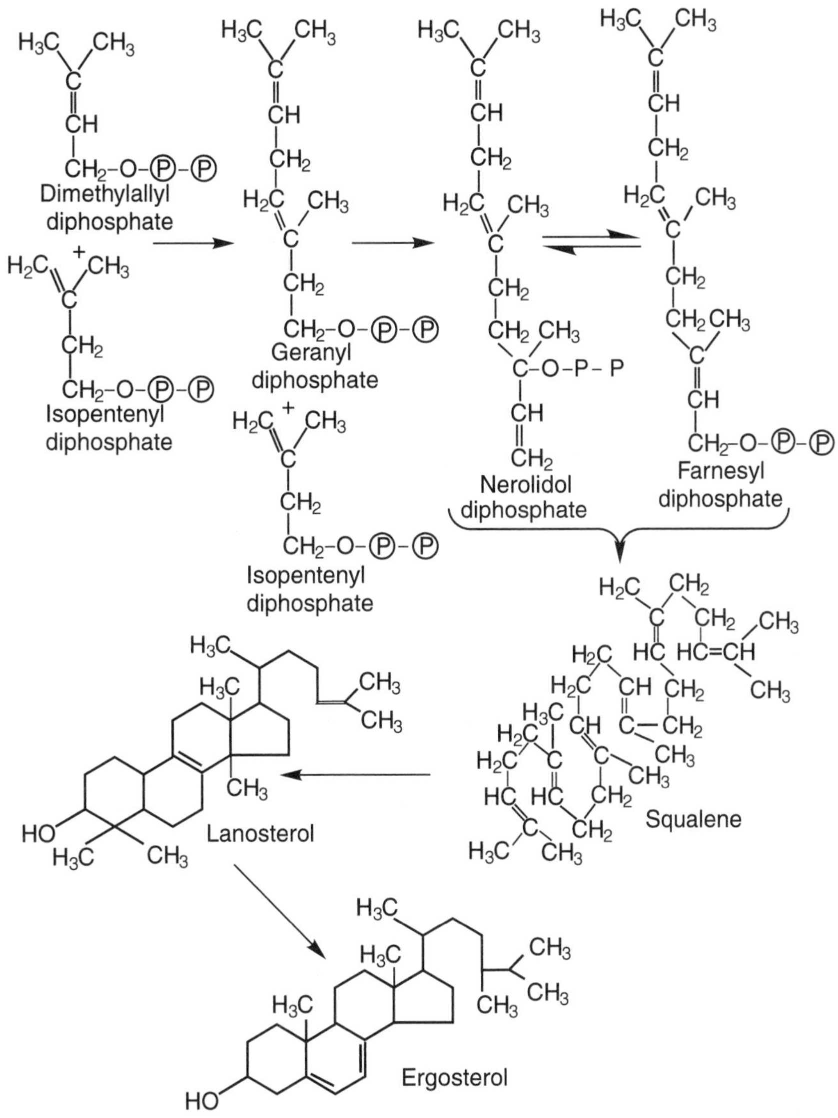

Fig. 13-8. Biosynthesis of nerolidol and farnesyl pyrophosphates and their conversion to squalene and ergosterol. It should be noted that there are several intermediate steps between squalene and lanosterol and between lanosterol and ergosterol.

whereas *Acholeplasma* synthesize carotenoid compounds that assume the functional capacity of sterols. Sterol and carotenol can apparently substitute for one another. Cholesterol added to the culture medium is used in place of carotenoids whose synthesis is repressed under these growth conditions (for examples of carotenoid structures, see Chapter 12).

Acholeplasma laidlawii can synthesize carotenoids from acetate. The organism contains a specific acetokinase and phosphotransacetylase, both of which are necessary for the synthesis of acetyl-CoA, and the β-ketothiolase and CoA transferase enzymes required to make acetoacetyl-CoA. The presence of β-hydroxymethylglutaryl-CoA–condensing enzyme and reductase activity in *Acholeplasma,* together with the ability to incorporate ^{14}C-labeled acetate into mevalonate, indicates that these organisms use the same pathway as yeast for the synthesis of mevalonate. The absence of these enzyme activities in *Mycoplasma hominis* explains its growth requirement for sterol. In other *Mycoplasma* species (e.g., *M. gallisepticum*), the metabolic block occurs subsequent to the mevalonate step in the biosynthetic pathway to terpenoids.

In order to participate in the biosynthesis of sterols, mevalonic acid must first be converted into the isoprenoid structural form, as shown in Figure 13-7. This sequence of reactions is present in yeast. There is some evidence that 5-diphospho-3-phosphomevalonic acid, a very unstable intermediate, is formed. This would account for the arrangement of the double bonds in the isopentenyl structure. Isopentenyl diphosphate and dimethylallyl diphosphate are condensed with the elimination of pyrophosphate to form geranyl pyrophosphate. This compound condenses with another molecule of isopentenyl diphosphate, yielding nerolidol diphosphate and, ultimately, *trans, trans*-farnesyl diphosphate (Fig. 13-8). The reaction sequence leading to the formation of nerolidol and farnesyl diphosphate has been demonstrated in yeast.

Squalene is formed by the condensation of one molecule of farnesyl diphosphate and one molecule of nerolidol diphosphate (Fig. 13-8). The conversion of squalene to ergosterol in yeast or to cholesterol in mammals involves a number of intermediary steps. In yeast, an epoxide intermediate precedes the formation of lanosterol. At least eight additional steps are required to convert lanosterol to ergosterol. *S*-adenosylmethionine donates the additional carbon (C_{28}) at the lanosterol level. In mammalian systems, the conversion of squalene to cholesterol requires at least two sterol carrier proteins (SCP_1 and SCP_2) to bind the substrate and make its reaction to the sterol-synthesizing enzymes present in the microsomes. These sterol carrier proteins differ in several respects from the acyl carrier protein involved in fatty acid biosynthesis. However, both of these compounds serve the analogous function of maintaining the solubility and reactivity of the substrate.

Yeasts, including *S. cerevisiae*, require sterols for growth. Under aerobic conditions, most yeasts are able to synthesize the required level of ergosterol. Under anaerobic conditions, yeast cannot synthesize sterols or unsaturated fatty acids and require their addition to the growth medium. Yeast mutants auxotrophic for sterols and unsaturated fatty acids are altered in the fatty acid composition of their mitochondrial phospholipids. Incorporation of fatty acids into phospholipids varies with the sterol and unsaturated fatty acids supplied. Ergosterol, in the presence of linoleic or linolenic acids or a mixture of palmitoleic and oleic acids, permits excellent growth. Substitution of other sterols, such as cholesterol, or addition of oleic acid as the sole fatty acid, results in poor growth.

The genetic basis for limiting sterol synthesis in *S. cerevisiae* resides in the regulation of hydroxymethylglutaryl-CoA reductase activity. A decrease in the specific activity of this enzyme correlates with accumulation of squalene during supplementation with ergosterol or mevalonolactone. The addition of ergosterol results in feedback inhibition of hydroxymethylglutaryl-CoA reductase.

BIBLIOGRAPHY

Altincicek, B., A.-K. Kollas, S. Sanderbrand, et al. 2001. GcpE is involved in the 2-*C*-methyl-D-erythritol 4-phosphate pathway of isoprenoid biosynthesis in *Escherichia coli. J. Bacteriol.* **183**:2411–6.

Choi, K.-H., R. J. Heath, and C. O. Rock. 2000. β-Ketoacyl carrier protein synthase III (FabH) is a determining factor in branched-chain fatty acid biosynthesis. *J. Bacteriol.* **182**:365–70.

Crick, D. C., M. C. Schulbach, E. E. Zink, et al. 2000. Polyprenyl phosphate biosynthesis in *Mycobacterium tuberculosis* and *Mycobacterium smegmatis. J. Bacteriol.* **182**:5771–8.

Cunningham, F. X., T. P. LaFond, and E. Gantt. 2000. Evidence of a role for LytB in the nonmevalonate pathway of isoprenoid biosynthesis. *J. Bacteriol.* **182**:5841–8.

Davis, M. S., and J. E. Cronan, Jr. 2001. Inhibition of *Escherichia coli* acetyl coenzyme A carboxylase by acyl-acyl carrier protein. *J. Bacteriol.* **183**:1499–1503.

Dowhan, W. 1997. Molecular basis for membrane phospholipid diversity: Why are there so many lipids? *Annu. Rev. Biochem.* **66**:199–232.

Exton, J. H. 1997. New developments in phospholipase D. *J. Biol. Chem.* **272**:15,579–82.

Greenberg, M. L., and J. M. Lopes. 1996. Genetic regulation of phospholipid biosynthesis in *Saccharomyces cerevisiae. Microbiol. Rev.* **60**:1–20.

Grogan, D. W., and J. E. Cronan, Jr. 1997. Cyclopropane ring formation in membrane lipids of bacteria. *Microbiol. Mol. Biol. Rev.* **61**:429–41.

Howland, J. L. 2000. *The Surprising Archaea.* Oxford University Press, New York.

Hrafnsdóttir, S., and A. K. Menon. 2000. Reconstitution and partial characterization of phospholipid flippase activity from detergent extracts of the *Bacillus subtilis* cell membrane. *J. Bacteriol.* **182**:4198–206.

Jackson, M., D. C. Crick, and P. J. Brennan. 2000. Phosphatidylinositol is an essential phospholipid of mycobacteria. *J. Biol. Chem.* **275**:30,092–9.

Kennedy, J., et al. 1999. Modulation of polyketide synthase activity by accessory proteins during lovastatin biosynthesis. *Science* **284**:1368–72.

Kikuchi, S., I. Shibuya, and K. Matsumoto. 2000. Viability of an *Escherichia coli pgsA* null mutant lacking detectable phosphatidylglycerol and cardiolipin. *J. Bacteriol.* **182**:371–6.

Kutchma, A., T. T. Hoang, and H. P. Schweizer. 1999. Characterization of a *Pseudomonas aeruginosa* fatty acid biosynthetic gene cluster: purification of acyl carrier protein (ACP) and malonyl-coenyzme A:ACP transacylase (FabD). *J. Bacteriol.* **181**:5498–504.

Kuzayama, T., M. Takagi, S. Takahashi, and H. Seto. 2000. Cloning and characterization of 1-deoxy-D-xylulose 5-phosphate synthase from *Streptomyces* sp. strain CL190, which uses both the mevalonate and nonmevalonate pathways for isopentenyl diphosphate biosynthesis. *J. Bacteriol.* **182**:891–7.

Leman, J. 2000. Lipids, microbially produced. In J. Lederberg (ed.), *Encyclopedia of Microbiology* 2nd ed., Vol. 3. Academic Press, New York, pp. 62–70.

Mdluli, K., et al. 1998. Inhibition of a *Mycobacterium tuberculosis* β-ketoacyl ACP synthase by isoniazid. *Science* **280**:1607–10.

Nishihara, M., T. Yamazaki, T. Oshima, and Y. Koga. 1999. *sn*-Glycerol-1-phosphate-forming activities in *Archaea:* separation of archaeal phospholipid biosynthesis and glycerol catabolism by glycerophosphate enantiomers. *J. Bacteriol.* **181**:1330–3.

Parks, L. W., and S. M. Casey. 1995. Physiological implications of sterol biosynthesis in yeast. *Annu. Rev. Microbiol.* **49**:95–116.

Rock, C. O. 2000. Lipid biosynthesis. In J. Lederberg (ed.), *Encyclopedia of Microbiology*, 2nd ed., vol. 3. Academic Press, New York, pp. 55–61.

Schneiter, R., et al. 2000. Elo1p-dependent carboxy-terminal elongation of $C14:1\Delta^9$ to $C16:1\Delta^{11}$ fatty acids in *Saccharomyces cerevisiae*. *J. Bacteriol.* **182:**3655–60.

Smirnova, N., and K. A. Reynolds. 2001. Engineered fatty acid biosynthesis in *Streptomyces* by altered catalytic function of β-ketoacyl-acyl carrier protein synthase III. *J. Bacteriol.* **183:**2335–42.

Takagi, M., T. Kuzuyama, S. Takahashi, and H. Seto. 2000. A gene cluster for the mevalonate pathway from *Streptomyces* sp. strain CL190. *J. Bacteriol.* **182:**4153–7.

Vilchèze, C., et al. 2000. Inactivation of the *inhA*-encoded fatty acid synthase II (FASII) enoyl-acyl carrier protein reductase induces accumulation of the FASI end products and cell lysis of *Mycobacterium smegmatis*. *J. Bacteriol.* **182:**4059–67.

Walker, G. M. 1998. *Yeast Physiology and Biotechnology*. John Wiley & Sons, Chichester, UK, 350p.

Wilding, E. I., J. R. Brown, A. P. Bryant, et al. 2000. Identification, evolution, and essentiality of the mevalonate pathway for isopentenyl diphosphate biosynthesis in gram-positive cocci. *J. Bacteriol.* **182:**4319–27.

Wilding, E. I., D.-Y. Kim., A. P. Bryant, et al. 2000. Essentiality, expression and characterization of the class II 3-hydroxy-3-methylglutaryl coenzyme A reductase of *Staphylococcus aureus*. *J. Bacteriol.* **182:**5147–52.

CHAPTER 14

NITROGEN METABOLISM

Microorganisms play a major role in the nitrogen cycle. A unique group of bacteria fix atmospheric nitrogen (dinitrogen) into ammonia and assimilate the ammonia into amino acids. Certain nitrogen-fixing bacteria have established a symbiotic relationship with plants where they fix atmospheric nitrogen into forms that can be utilized by the host plants. Degradation of nitrogenous compounds by microorganisms is also an important aspect of the nitrogen cycle. Without this degradation and subsequent return of nitrogen from a wide variety of complex natural and artificial compounds to the nitrogen cycle, higher forms of life could not exist. An in-depth knowledge of these processes may aid materially in overcoming the imbalances created in the nitrogen cycle by overloading it with waste and excretory materials and improper use of agricultural lands by overabundant applications of fertilizers, herbicides, and pesticides.

In this chapter, we consider the contributions of microorganisms to the nitrogen cycle and the underlying mechanisms of the processes of nitrogen fixation, metabolism of inorganic nitrogen compounds, and assimilation of inorganic nitrogen into amino acids. While urea is not an inorganic compound, it is widely used as a fertilizer because it is readily broken down into carbon dioxide and ammonia. For this reason, urea metabolism is also described in this chapter. Reactions involved in the interconversion of the amino acids, especially amino group transfer (transamination), and other important reactions of amino acids are discussed. The biosynthesis of amino acids, purines, pyrimidines, and other nitrogen-containing compounds is covered in Chapters 15 and 16.

BIOLOGICAL NITROGEN FIXATION

Dutch farmers associated the establishment of rich stands of clover with the improvement of subsequent crops grown in the same fields. Although they were unaware of the precise nature of the relationship, they noted the development of nodules on the root system of clover. These observations provided the foundations for systematic crop rotation implemented by Townsend in England during the agricultural revolution of the eighteenth century. The photograph in Figure 14-1 shows extensive nodule development on the root system of a soybean plant, which is a legume. A legume is a plant that bears seed pods, such as pea or soybean. These plants often form a symbiotic association with bacteria that form prominent nodules on the roots. These nodules in the root system harbor millions of bacteria that convert atmospheric nitrogen (dinitrogen) to ammonia (NH_3).

The works of Atwater and of Hellriegel and Wilfarth, published in the 1880s, were among the first to provide scientific data supporting the importance of the root nodules and the bacteria within them in the process of nitrogen fixation by clover and other leguminous plants. Despite criticisms by scientists of the time, their reports initiated a concerted effort to determine the nature of the symbiotic organisms associated with the

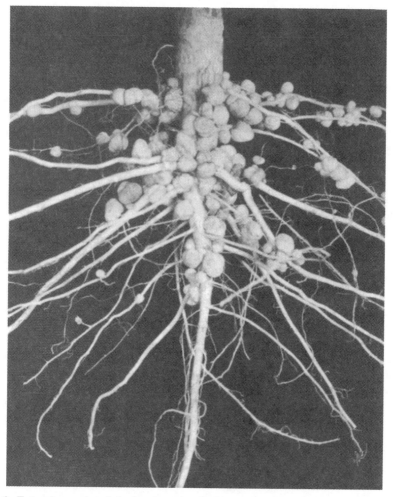

Fig. 14-1. Extensive root nodule development in the root system of soybean (*Glycine max*), a leguminous plant. (Photo courtesy of R. S. Smith, Milwaukee, WI.)

root nodules of legumes (and other plants), as well as characterization of the numerous free-living microorganisms capable of nitrogen fixation.

For many years, it has been considered that the ammonia produced by the bacteria in the nodule was transferred to the plant where it was then assimilated into amino acids and other nitrogenous compounds. More recently, it has been determined that under certain conditions the bacteria assimilate the ammonia into the amino acid alanine, which is then transported into the plant system. A number of examples of nitrogen-fixing organisms and their associated plant partners are listed in Table 14-1.

Symbiotic nitrogen-fixing organisms and the photosynthetic nitrogen fixers appear to account for most of the atmospheric nitrogen assimilated into organic forms in nature. Under conditions in which fixed nitrogen is low, root-nodulated angiosperms and gymnosperms and the cyanobacteria may be especially valuable in improving soil fertility. The estimated amount of nitrogen fixed by symbiotic bacterial-inoculated legumes in the United States approximately equals the amount of nitrogen supplied by farmers as nitrogen fertilizer. In the past, the contribution of free-living forms was probably underestimated. With the introduction of the acetylene reduction technique* for assessing the nitrogen-fixing potential of organisms in natural environments, it has been possible to show the presence of nitrogen-fixing organisms in a number of settings, including the intestinal tract of nonruminant mammals. Although free-living organisms, in general, appear less efficient in their ability to fix nitrogen, their number, variety, and ubiquitous distribution suggest that they are of major ecological importance. The cyanobacteria have a distinct advantage in that the energy (as ATP) and the reducing power (as NADH) required for nitrogen fixation can be supplied by photosynthesis, a process that makes it possible for them to become established in environments unfavorable for the development of other nitrogen-fixing organisms.

Rapidly rising energy and labor costs have made it less economical to increase plant growth by the use of ammonia fertilizer. Commercial production of ammonia by the Haber-Bosch process (catalyst-mediated reduction of hydrogen with nitrogen under high pressure and temperature) is expensive. Furthermore, continued application of chemical fertilizers at a high rate has threatened water supplies and the ecological balance of rivers and streams. Thus, attention has been focused on the improvement of plant yields through the development of new associations between nitrogen-fixing bacteria and plants. In the pursuit of this goal, intensive efforts have been directed toward studies of the genetics, biochemistry, and ecology of both free-living and symbiotic nitrogen-fixing organisms. It has already become apparent that the photosynthetic capacity of plants is one limiting factor in nitrogen fixation. Work on the selection and breeding of plant strains with greater photosynthetic efficiency has also been intensified. Another long-range goal has been to attempt development of an association between nitrogen-fixing bacteria and the root system of highly efficient photosynthetic plants such as corn.

---

* Originally, mass spectrometry was used to measure the amount of ^{15}N-labeled nitrogen gas reduced to ammonia by nitrogenase. The discovery that nitrogenase can reduce acetylene to ethylene proved very useful, since this reaction can be measured by the somewhat simpler technique of gas chromatography.

TABLE 14-1. Examples of Nitrogen-Fixing Organisms

Symbiotic Association of Various Genera with Leguminous Plants

Azorhizobium caulindans, tropical legume (*Sesbania rostrata*)
Allorhizobium undicola, lotus (*Lotus albicus*)
Bradyrhizobium japonicum, soybean (*Glycine max*)
Mesorhizobium amorphae, false indigo (*Amorpha fruticosa*)
Rhizobium trifolii, clover (*Trifolium, Crotolaria*)
Sinorhizobium meliloti, alfalfa (*Medicago sativa*)

Symbiotic Association of Actinomycetes with Angiosperms

Frankia sp., alder (*Alnus*)
Frankia sp., bog myrtle or sweet gale (*Myrica*)
Frankia sp., oleasters (*Shepherdia, Eleagnus, Hippophae*)
Frankia sp., New Jersey tea (*Ceanothus*)

Symbiotic Association with Leaf-Nodulating Plants

Klebsiella aerogenes

Symbiotic Association of Marine Bacteria with Bivalves

Aerobic, chemoheterotrophic sp., bivalves (*Teredinidae*)

Symbiotic Association with Marine Diatoms

Richelia intracellularis (cyanobacterium), *Rhizoselenia*

Associative Interaction with Grasses

Azospirillum brasiliense, tropical grasses
Azospirillum lipoferum, tropical grasses, maize
Azotobacter paspali, tropical grass (*Paspalum notatum*)

Free-Living Bacteria and Cyanobacteria

Aerobic, heterotrophic
 Azotobacter, Derxia, Azomonas, Biejerinkia, Nocardia, Pseudomonas
Aerobic, phototrophic
 *Anabaena, Calothrix, Nostoc, Gleotheca, Cylindrospermum,
 Aphanocapsa*
Facultative, heterotrophic
 *Enterobacter cloacae, Klebsiella pneumoniae, Bacillus polymyxa,
 Desulfovibrio desulfuricans, D. gigas, Achromobacter*
Anaerobic, heterotrophic
 Clostridium pasteurianum, C. butyricum, Propionispira arboris
Anaerobic, phototrophic
 *Chromatium vinosum, Rhodospirillum rubrum, Rhodopseudomonas
 sphaeroides, R. capsulata, Rhodomicrobium vernielli, Rhodocyclus,
 Chlorobium limocola*
Nonphotosynthetic, autotrophic
 *Methanobacterium, Methylococcus, Methylosinus, Methanococcus,
 Methanococcus, Methanosarcina*

THE NITROGEN FIXATION PROCESS

Fixation of atmospheric dinitrogen (N_2 or $N{\equiv}N$) is accomplished by a variety of bacteria and cyanobacteria utilizing a multicomponent **nitrogenase system**. Despite the variety of organisms capable of fixing nitrogen, the nitrogenase complex appears to be remarkably similar in most organisms (Fig. 14-2). Nitrogenase consists of two oxygen-sensitive proteins. **Component I (dinitrogenase)** is a molybdenum-iron protein containing two subunits. **Component II (dinitrogenase reductase)** is an iron-sulfur protein that transfers electrons to dinitrogenase. These proteins, together with ATP, Mg^{2+}, and a source of electrons, are essential for nitrogen-fixing activity.

The overall process of nitrogen fixation is accomplished at considerable expense of energy, requiring from 12 to 16 molecules of ATP and 6 to 8 electrons, depending on the manner in which the equation is viewed. In one form, the equation may be written

$$N_2 + 6\ H^+ + 6\ e^- + 12\ ATP + 12\ H_2O \longrightarrow 2\ NH_3 + 12\ ADP + 12\ P_i$$

This form of the equation does not take into account the fact that dihydrogen (H_2) is an obligate product of the nitrogenase reaction. If H_2 is considered, then the equation

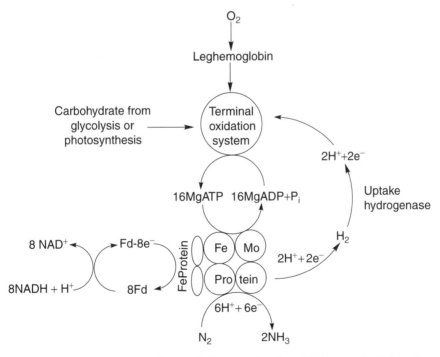

Fig. 14-2. The nitrogenase complex and the associated activities required for nitrogen fixation. The FeMo protein, dinitrogenase, is also referred to as component I. The Fe protein, dinitrogenase reductase (component II), contains a 4Fe-4S cluster that is not shown in the diagram. Fd, ferredoxin. The overall reaction requires 8 NADH + H$^+$. Six of these are used to reduce N_2 to NH_3 and two are used to form H_2. The uptake hydrogenase returns H to the system, thus conserving energy.

can be shown as

$$N_2 + 8\ H^+ + 8\ e^- + 16\ ATP + 12\ H_2O \longrightarrow 2\ NH_3 + H_2 + 16\ ADP + 16\ P_i$$

Thus, the theoretical stoichiometric relationship between N_2 fixation and H_2 production is related by the equation

$$N_2 + 8\ H^+ + 8\ e^- \longrightarrow 2\ NH_3 + H_2$$

The production of hydrogen during nitrogen fixation is an energy-expensive process. In actuality, most aerobic nitrogen fixers rarely evolve H_2 because a membrane-bound uptake hydrogenase recycles the H_2 and produces ATP through respiration to help support the ATP requirements of the system.

An anaerobic environment is essential for nitrogenase activity because of the oxygen lability of both proteins in the complex. An uptake hydrogenase coupled to a dioxygen-consuming pathway helps to maintain an anaerobic environment. Anaerobiosis is essential because oxygen represses the hydrogen uptake system. In the cyanobacteria, the problem of conducting oxygen-evolving photosynthesis and nitrogen fixation simultaneously is circumvented by sequestration of the nitrogenase system in specialized cells called heterocysts. Otherwise, they can only fix N_2 in the dark under anoxic conditions.

Components of the Nitrogenase System

The nitrogen fixation (*nif*) genes in *K. pneumoniae* are listed in Table 14-2 along with their known product or function. The nitrogenase complex of this organism contains two separable proteins. Component I (dinitrogenase) is a α_2,β_2 tetramer of 240 kDa encoded by *nifK* and *nifD*. The iron-molybdenum cofactor (FeMo-co) of the dinitrogenase is synthesized under the direction of six *nif* genes (*nifQ, B, V, N, E, H*). Mutants of *nifB* and *nifNE* accumulate an inactive apo-component I (ApoI). ApoI is an oligomer containing an additional protein, the product of *nifY*, which disassociates from the complex upon activation by the addition of FeMo-co with restoration of the ability to

TABLE 14-2. Nitrogen Fixation (*nif*) Genes and Their Products in *Klebsiella pneumoniae*[a]

Gene	Product	Gene	Product
Q	FeMoCo, Mo uptake	U	Protein associated with FeS center
B	FeMoCo synthesis	X	Regulatory protein
A	Positive regulator	N	FeMoCo
L	Negative regulator	E	FeMoCo synthesis
F	Flavodoxin, electron acceptor	Y	Unknown function
M	Fe protein activation	T	Unknown function
Z	Insertion of FeMoCo protein	K	MoFe protein β-subunits
W	Activation of dinitrogenase	D	MoFe protein α-subunits
V	Homocitrate synthesis	H	Dinitrogenase reductase
S	Protein associated with FeS center	J	Pyruvate oxidoreductase

[a]The genes are listed in the order in which they appear on the detailed chromosome map of *Klebsiella pneumoniae*.

fix N$_2$. Component II (dinitrogenase reductase, encoded by *nifH*) is an α_2 protein (ca. 60 kDa) containing a single four-iron four-sulfur (Fe$_4$S$_4$) center. This protein binds and hydrolyzes Mg ATP when an electron is transferred from reduced ferredoxin to dinitrogenase. Ferredoxin and/or flavodoxin can serve as electron donors.

Regulation of the 17 *nif* genes in *K. pneumoniae* is under the direction of the *nifLA* operon. The NtrC protein activates the transcription of the *nifLA* operon under the conditions of nitrogen limitation. The NifA protein is a positive regulatory factor required for *nif* gene transcription (Table 14-2). The *nifL* gene product interacts with the NifA protein to prevent NifA activation in the presence of fixed nitrogen (ammonia or amino acids) or oxygen. The functions of *nifZ, W, U, S, X,* and *T* are still not well defined. However, there is evidence that the products of the *nifW* and *nifZ* genes may be involved in processing one of the structural components of nitrogenase and that the product of the *nifX* genes is a positive regulator of the *nif* regulon in response to ammonia and oxygen (Fig. 14-3).

The nitrogen-fixing systems of *Azotobacter* species have been well characterized. Three different nitrogenases found in *A. vinelandii* and *A. chroococcum* are regulated by the Mo or V content of the culture medium. Nitrogenase 1 is produced by both organisms in the presence of Mo. Dinitrogenase reductase is composed of two identical subunits encoded by *nifH*. Dinitrogenase 1 is a tetramer of two pairs of nonidentical subunits encoded by *nifD* and *nifK*. The three structural genes appear in the order *nifHDK* and comprise an operon. Nitrogenase 2 is produced by both *A. vinelandii* and *A. chroococcum* grown in a nitrogen-free medium in the presence of V. Dinitrogenase reductase 2 is a dimer encoded by *vnfH*. Dinitrogenase 2, encoded by *vnfD* and

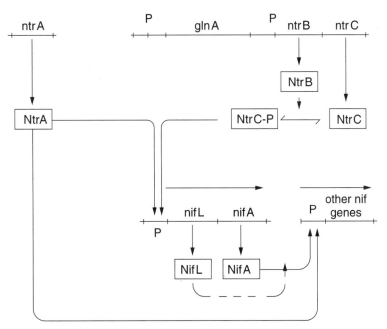

Fig. 14-3. Model for *nif* regulation in *Klebsiella pneumoniae*. The thin arrows indicate regulatory functions and the thick horizontal arrows represent transcripts. (*Source*: From Dixon, R. A., *J. Gen. Microbiol.* **130**:2745–2755, 1984.)

vnfK, is a tetramer composed of two pairs of subunits containing Fe and V. A third nitrogenase, nitrogenase 3, is encoded by alternate nitrogen fixation (*anfHDK*) genes, which are expressed in *A. vinelandii* only in the absence of Mo and V. Dinitrogenase reductase 3 contains two identical subunits and dinitrogenase 3 is present in two active configurations, $\alpha_2\beta_2$ and $\alpha_1\beta_2$. The regulatory genes *nifA, vnfA,* and *anfA* are required for the expression of nitrogenases 1, 2, and 3, respectively. An additional regulatory gene, *nfrX*, is required for growth on N_2 (diazotrophic growth) in the presence or absence of Mo.

The nitrogenase system from *Clostridium pasteurianum* is similar to that found in *K. pneumoniae* and *A. vinelandii*. Two proteins are required for nitrogen fixation and ATP-dependent H_2 evolution. One protein, component I, is an FeMo protein having dinitrogenase activity. A second protein, the Fe protein, or component II, has dinitrogenase reductase activity. Although there are structural similarities between the nitrogenase complexes of various organisms, the primary structure of those from *C. pasteurianum* is quite different from those of *K. pneumoniae* and *A. vinelandii*. The nitrogenase components from a wide variety of bacteria can interact successfully with one another. The FeMo and Fe proteins from different organisms show a remarkably high degree of successful formation of active hybrid nitrogenases. However, the components of *C. pasteurianum* nitrogenase are much less effective in forming heterologous complexes than are mixtures of the components from gram-negative organisms.

A number of phototrophic bacteria including strict anaerobes such as *Chromatium, Chlorobium,* and *Rhodospirillum rubrum* can fix nitrogen anaerobically in the light because their photosynthetic systems do not evolve O_2. However, the most important phototrophs from an ecological viewpoint are the cyanobacteria. They are capable of performing oxygenic photosynthesis and fixing CO_2 through the reductive pentose phosphate pathway (see Chapter 12). Cyanobacteria lack α-ketoglutarate dehydrogenase but utilize this intermediate as a substrate for nitrogen assimilation. Nitrate, ammonia, urea, and N_2 are used as nitrogen sources. In the absence of fixed nitrogen, most cyanobacteria are able to fix N_2 using a nitrogenase system similar to that described in other diazotrophic organisms. However, they face a dilemma in that nitrogenase is extremely sensitive to oxygen. *Anabaena* and *Nostoc* form heterocysts, which are specialized cells that contain most of the nitrogenase activity. In heterocysts, the photosynthetic system is effectively inactivated and any residual O_2 is eliminated by highly active respiration.

Some cyanobacteria such as *Anabaena variabilis* produce two Mo-dependent nitrogenases: Nif1 functions in heterocysts while Nif2 functions under anoxic conditions in vegetative cells. A number of cyanobacteria that do not form heterocysts are also known. They apparently operate under microaerophilic conditions when fixing nitrogen. Among the free-living nitrogen-fixing bacteria, the cyanobacteria contribute much more significantly to the soil nitrogen because they obtain their energy from photosynthesis and are therefore capable of colonizing soil or other habitats where conditions are sparse. They also function symbiotically with primitive plants such as lichens, liverworts, and water ferns.

Of the many genes required for heterocyst function, only *ntcA, hanA,* and *hetR* appear to be required for the initiation of heterocyst development. NtcA functions as a nitrogen-dependent global transcriptional regulator in all cyanobacteria. In *Anabaena*, *ntcA* mutants fail to grow on nitrate, do not form heterocysts, or undergo nitrogenase

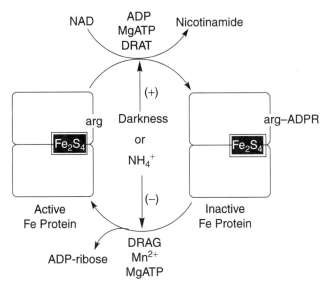

Fig. 14-4. Diagrammatic view of regulation of nitrogenase by ADP ribosylation of the Fe protein in *Rhodospirillum rubrum*. DRAG, dinitrogenase reductase–activating glycohydrolase; DRAT, dinitrogenase reductase ADP ribosyltransferase; Fe_2S_4, the iron-sulfur in the Fe protein; arg, arginine 101 position in the Fe protein. (Simplified version of diagram in C. M., Halbleib et al., *J. Biol. Chem.* **275:**3493–3500, 2000.)

synthesis. The *hetF* gene product is essential for heterocyst development in the filamentous cyanobacterium *Nostoc punctiforme*. HetF appears to cooperate with HetR in a positive regulatory pathway in differentiating heterocysts.

Nitrogenase synthesis in a variety of organisms is generally subject to the close regulatory controls at the level of *nif* gene transcription. However, many of these organisms also display a posttranslational regulation of nitrogenase activity effected by small extracellular concentrations of ammonia. In some organisms this rapid and reversible inhibition of nitrogenase, termed **ammonia switch-off**, is the result of a covalent modification of the Fe protein (dinitrogenase reductase) in response to the addition of ammonia. Formation of an inactive form of nitrogenase in the photosynthetic bacterium *R. rubrum* results from ADP ribosylation of an arginine residue in the Fe-protein component (Fig. 14-4). This activity is catalyzed by dinitrogenase reductase ADP ribosyltransferase (DRAT). Activation of the modified protein by removal of the ADP-ribose moiety is catalyzed by dinitrogenase reductase activating glycohydrolase (DRAG). Other systems in which covalent modification of the Fe protein occurs include *A. brasiliense*, *A. lipoferrum*, and *A. vinelandii*. In *A. amazonense* a different, noncovalent inhibitory mechanism results in only a partial inhibition of nitrogenase activity by ammonia. In *Rhodospirillum rubrum* nitrogen fixation is also tightly regulated by transcriptional regulation of *nif* gene expression.

SYMBIOTIC NITROGEN FIXATION

The successful establishment of a symbiotic relationship between a bacterium such as *Sinorhizobium meliloti* and its host plant (*Medicago sativa*, alfalfa), culminating in the formation of nodules that conduct active nitrogen fixation, is a complex process that occurs in several stages: attachment of the bacteria to root hairs, root hair curling, formation of a "shepherd's crook," development of infection threads within root hairs, growth of the threads toward the inner cortex of the root, and formation of a nodule meristem in the inner root cortex (Fig. 14-5). These cytological changes in the root hair provide a means of passage of the organism into the internal root system where it infects a root cell. This infection causes the cell to swell and divide, forming a thick mass of cells called the root nodule. In the root nodule a differentiated form of the bacterium, termed a **bacteroid**, is capable of nitrogen fixation.

A portion of the genomes of both the plant and the bacteria are expressed only in the symbiotic state. Considering the complexity of the interaction between the bacterium and the host plant in the development of effective symbiosis, it is not surprising that attempts to develop strains that are more efficient in their nitrogen-fixing capability have been hampered by the fact that indigenous strains in the soil are more competitive than laboratory-developed strains in their ability to initiate nodule development. Nevertheless, continued efforts to engineer more effective strains of nitrogen-fixing bacteria are important because of the increase in the cost of nitrogen fertilizer and the undesirable contamination of rivers and streams that occurs through the heavy application of it.

The nodulation process requires the exchange of a series of signals between the plant and bacteria. Initially, plant genes coding for the production of flavenoids are expressed, and these compounds are excreted from the roots. Flavenoids activate NodD1 and start

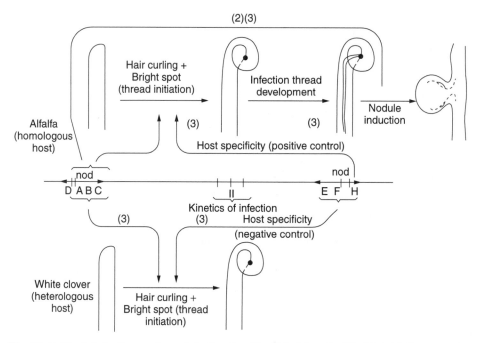

Fig. 14-5. Root infection and modulation by *Sinorhizobium meliloti* in *Medicago sativa* (alfalfa). Assignment of development step control in common and specific *nod* loci of *S. meliloti*. Thick arrows represent development steps. Thin arrows represent genetic control. (*Source*: From Debellé, F., et al., *J. Bacteriol* **168**:1075–1086, 1986.)

the sequence of Nod (nodulation factor) synthesis. Nod factors induce root hair curling and initiate nodule development (Fig. 14-6). Invasion of the root nodule by rhizobia is dependent on the production of at least one of several complex polysaccharides: succinoglycan, exopolysaccharide II, or a capsular polysaccharide (CPS or KPS) containing 3-deoxy-D-manno-2-octulosonic acid (see Fig. 14-7).

The early events in root nodule formation by rhizobia in leguminous plants are under the control of genes located on a large plasmid called **symbiosis** or **Sym plasmid**. Plant

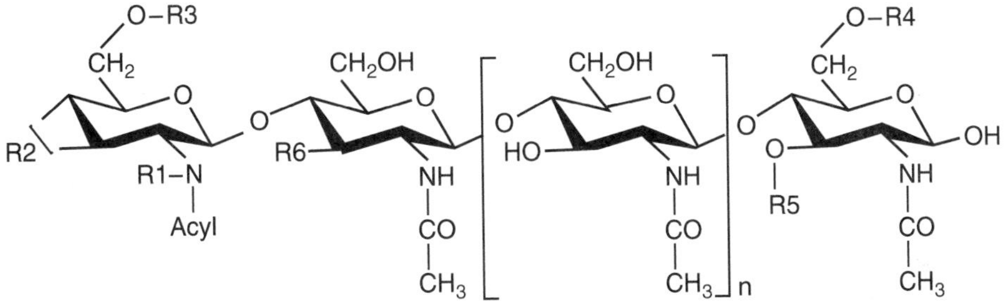

Fig. 14-6. General structure of the Nod factors produced by Rhizobia. Acyl substituents are most often C_{16}, C_{18}, or C_{20} saturated or unsaturated fatty acids; R1 is usually a methyl group or H; R2 is frequently a carbamoyl group or OH; R3 may be H, a carbamoyl group, or an acetyl group; R4 may be fucose, methylated fucose, acetylated fucose, sulfate, or H; R5 is usually H, but may occasionally be an arabinosyl substituent; R6 is usually OH, but may occasionally be fucose or an acetyl group; n usually varies between 1 and 2, but may occasionally be 3 or 0.

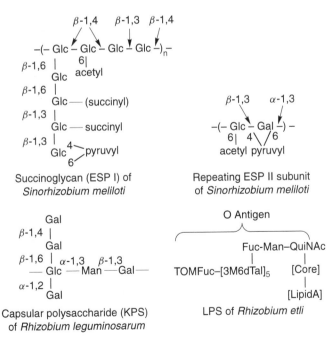

Fig. 14-7. Polysaccharides involved in nodule development by rhizobia. Glc, glucose; Gal, galactose; Man, mannose; TOMFuc, tri-O-methylated fucose; 3M6dTal, 3-O-methyl-6-deoxytalose; Fuc, fucose; QuiNAc, *N*-acetyl quinovosamine (2-amino-2,6-dideoxyglucose.)

TABLE 14-3. Genes Involved in Nodulation and Their Known Function or Activity

Designation	Function or Activity
nodD	Positive transcription activator; binds to the *nod* box upstream of all inducible operons
nod box	Highly conserved DNA sequence; *cis*-acting regulatory element of *nod* gene expression
nodA	*N*-Acyltransferase
nodB	Oligochitin deacetylase
nodC	Oligochitin synthase
nodFEL	Operon involved in addition of acyl substituents to core lipooligosaccharide
nodH	Addition of substituents to core lipooligosaccharide
nodI	Involved in efficiency of nodulation and development of infectious threads
nodJ	Involved in efficiency of nodulation and development of infectious threads
nodM	Fructose-6-phosphate:glutamine amidotransferase
nodP	Sulfate + APS (adenylyl sulfate) → PAPS (3′-phosphoadenylyl sulfate)
nodQ	Addition of substituents to core lipooligosaccharide
nodS	*N*-methyltransferase
nodU	6-O-Carbamoyltransferase
nodX	Addition of acetyl substituent to core lipooligosaccharide
nodZ	6-O-Fucosyltransferase
noeA	Sulfuryltransferase
noeC	Addition of substituent to core lipooligosaccharide
noeI	2-O-Methyltransferase
noeJ	Mannose-1-phosphate guanyltransferase
noeL	GDP-Mannose-4,6-dehydratase
noeK	Phosphomannose mutase
nolL	O-Acetyltransferase
nolO	Carbamoyltransferase
nolK	Fucose synthase

flavenoids, in concert with NodD, induce the expression of *nod* genes. The NodD protein serves as a positive transcription activator. It binds to the *nod* box, located upstream of all inducible *nod* genes and operons. The *nod* box, a highly conserved DNA sequence, appears to function as a *cis*-acting regulatory element of *nod* gene expression. The *nodABC* genes common to all rhizobia are required for the synthesis of a lipooligosaccharide that triggers root nodule formation. Addition of various substituents to the core compound imparts host specificity to the lipooligosaccharide. Addition of these components occurs under the influence of *nodH* and *nodQ* and the *nodFEL* operon (see Table 14-3). The NodI and NodJ proteins are involved in efficiency of nodulation and play a role in the normal development of infection threads. Once bacteroid development is complete, the inducible *nod* genes are no longer transcribed. This transcription switch-off prior to the release of the bacteria from the infection thread is a general phenomenon observed in all rhizobia and is the result of a negative regulatory control mechanism.

Under the conditions of symbiosis, the host plant provides the bacteroid with reduced carbon in the form of C_4-dicarboxylic acids (succinate, malate, and fumarate). These compounds serve as energy sources for the fixation of nitrogen to ammonia. Dicarboxylic acids are present in high concentration in the nodule and are the most effective substrates for respiration and subsequent ATP-utilizing nitrogen fixation in the bacteroid. The dicarboxylic acid transport genes (*dct*) are located on a megaplasmid. Mutants defective in dicarboxylate transport form ineffective nodules. The products of *dctB* and *dctD* regulate the expression of *dctA*, which encodes a transport protein. DctB is a sensor protein that activates DctD by phosphorylation. The DctD protein activates transcription at the σ^{54}-dependent *dctA* promoter. In addition to activating *dctA* transcription, DctD can repress the expression of *dctA*. In uninfected cells, inactive DctD binds to the *dctA* promoter and prevents its activation by NtrC.

For many years, the paradigm of symbiotic nitrogen fixation has been the release of ammonia directly from the bacteroid to the plant. Active ammonia transport systems have been described in some symbiotic microorganisms. However, recent studies reveal that alanine, rather than ammonia, is the form of nitrogen transported to the plant. It has been shown that rhizobia can assimilate ammonia into pyruvate, forming alanine via alanine dehydrogenase. The equilibrium of the alanine dehydrogenase reaction favors ammonia assimilation.

Labeling experiments with $^{15}N_2$ using highly purified nodule bacteroids of *B. japonicum* show that recently formed ammonia from nitrogenase was incorporated into alanine and then transported to the plant from the bacteroid. Secretion of labeled alanine was dependent on nitrogenase activity as bacteroids exposed to inactivating levels of oxygen or to the energy uncoupling agent CCCP (carbonyl cyanide *m*-chlorophenylhydrazone) did not excrete alanine. The peribacteroid membrane is inverted with respect to the bacteroid, so alanine secretion by the bacteroid to the plant cytoplasm can occur quickly, but it cannot be readily reabsorbed. It is considered that alanine serves primarily as a transport mechanism for fixed nitrogen, since most organisms use glutamate as the central compound in nitrogen metabolism (see "General Reactions of Amino Acids".)

INORGANIC NITROGEN METABOLISM

The assimilation of inorganic nitrogen ends with the incorporation of ammonia into organic compounds. Since ammonia is the only form of inorganic nitrogen that can be directly assimilated into amino acids, the ability of an organism to utilize other forms of inorganic nitrogen depends on the presence of enzymes or enzyme systems that are able to convert these compounds to ammonia. This process is referred to as **denitrification**. The reverse process, converting ammonia to nitrate and nitrite, is called **nitrification** (Fig. 14-8). Table 14-4 provides examples of organisms known to carry out reactions of inorganic nitrogen. Although it is not an inorganic compound, the metabolism of urea is discussed in this section because it is so readily degraded to ammonia and carbon dioxide by ureases found in many genera of bacteria and fungi.

Many of the organisms that conduct the reactions shown in Table 14-4 do not actually assimilate the nitrogen. In some organisms, nitrate may be used as a terminal electron acceptor in place of oxygen. The end product is nitrite or N_2. Nitrate respiration yields biologically useful energy under anaerobic conditions. Nitrate assimilation occurs by sequential reduction to nitrite, hydroxylamine, and, finally, ammonia. Organisms that

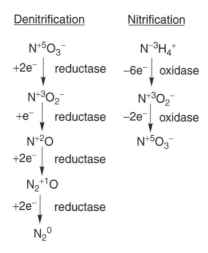

Fig. 14-8. Pathways of inorganic nitrogen metabolism. Denitrification and nitrification. The superscript numbers on the nitrogen indicate the valence state or oxidation level of the nitrogen in the compounds.

conduct both nitrate respiration and nitrate assimilation contain two nitrate reductases. In *E. coli* and *Neurospora* the assimilatory enzyme is a soluble cytoplasmic enzyme, whereas the respiratory enzyme is membrane-bound and sensitive to oxygen. Both enzymes contain flavin as the prosthetic group. In most organisms the assimilatory enzyme is repressed by ammonia and induced in the presence of either nitrate or nitrite. Algae readily utilize nitrate as a source of inorganic nitrogen, but the route of assimilation is less well characterized than in bacteria and fungi. Ammonia is considered to be the final product of nitrate reduction by algae.

Denitrification by members of the obligately chemolithotrophic *Nitrobacteriaceae* is considered to be the major source of assimilable nitrogen (ammonia) in soils. The ammonia oxidizer, *Nitrosomas europaea*, and the nitrite oxidizer, *Nitrobacter winogradskyi*, have been widely studied. *Nitrosomonas europaea* obtains all of its energy for growth from the oxidation of ammonia to nitrite. The oxidation of ammonia to hydroxylamine is an oxygen-dependent reaction catalyzed by ammonia monooxygenases:

$$NH_3 + O_2 + 2\ e^- + 2\ H^+ \longrightarrow NH_2OH + H_2O$$

Hydroxylamine is further oxidized to nitrite by hydroxylamine oxidoreductase:

$$NH_2OH + H_2O \longrightarrow NO_2^- + 5\ H^+ + 4\ e^-$$

Two of the four electrons generated from hydroxylamine oxidation are used to support the oxidation of additional ammonia molecules, while the other two electrons enter the electron transfer chain and are used to support ATP synthesis via oxidative phosphorylation and CO_2 reduction. The oxidative chain from NO_2^- to O_2 in *N. winogradskyi* consists of membrane-bound cytochrome *c* oxidase. *Nitrobacter* is obligately dependent on nitrite as the sole source of nitrogen and energy and requires carbon dioxide for growth. *Nitrobacter* is, therefore, dependent on *Nitrosomonas* or other organisms in the natural environment for its nitrogen supply.

TABLE 14-4. Known Biological Reactions of Inorganic Nitrogen

Reaction	Electrons Required	Representative Organisms Performing Reaction	Requirements (comments)
$NO_3^- \rightarrow NO_2^-$	2	*Escherichia coli, Micrococcus denitrificans, Bacillus subtilis, Haemophilus influenzae, Pseudomonas aeruginosa, Neurospora crassa, Achromobacter fischeri*	Varies among organisms NAD, NADP, FAD, FMN, Mo, cytochromes *c, b*
$NO_2^- \rightarrow NH_4^+$	6	*Bacillus pyocyaneus, Neurospora crassa, Bacillus pumilis, Clostridium pasteurianum, Desulfovibrio desulfuricans*	NADPH, NADH, FAD, Cu, Fe
NO_3^- or $NO_2^- \rightarrow$ NO, N_2O, and N_2 Denitrification	Varies from 1 to 10	*Micrococcus denitrificans, Denitrobacillus, Pseudomonas, Spirillum itersonii, Bacillus licheniformis, Achromobacter, T. thioparus Thiobacillus denitrificans*	See requirements for individual reactions
$NO_2^- \rightarrow$ NO Nitrite reduction	1	*Pseudomonas stutzeri, E. coli, Bacillus subtilis, P. aeruginosa*	NAD, NADP, FMN, FAD, Fe^{2+}
$NO \rightarrow N_2$ or N_2O Nitric oxide reductase	2 or 4	*Pseudomonas stutzeri, E. coli, Bacillus subtilis, Pseudomonas aeruginosa, Clostridium pasteurianum, Pseudomonas aeruginosa*	NAD, NADP, FAD, FMN, Fe^{2+}
$N_2O \rightarrow N_2$ Nitrous oxide reductase	2	*Pseudomonas stutzeri*	Cu
$H_2N_2O_2 \rightarrow 2NH_4^+$ Hyponitrite assimilation	8	*Neurospora crassa, E. coli*	NADPH, Fe^{2+}

(continued overleaf)

TABLE 14-4. (*continued*)

Reaction	Electrons required	Representative Organisms Performing Reaction	Requirements (comments)
$N_2 \rightarrow 2NH_4^+$ Nitrogen fixation	6	See Table 14-1 for representative organisms	Mo, Fe, V
$NH_2OH \rightarrow NH_4^+$ Hydroxylamine reductase	2	*Neurospora crassa, Azotobacter vinelandii, Desulfovibrio desulfuricans, Bacillus pumilis, Clostridium pasteurianum, Pseudomonas aeruginosa*	Varies—NADH, FAD, Mn^{2+}
$NH_2NH_2 \rightarrow 2NH_4^+$ Hydrazine reductase	2	*Micrococcus lactilyticus*	Measured hydrogen uptake with hydrazine added
$NH_4^+ \rightarrow$ organic compounds	In some cases, 2	Large group	Varies
$NH_4^+ \rightarrow NO_2^-$ or $NH_2OH \rightarrow NO_2^-$ Nitrification	−6 or −4	*Nitrosomonas* spp.	Acceptor—cytochrome c or phenazine methosulfate
$NO_2^- \rightarrow NO_3^-$ Nitrite reductase	−2	*Nitrobacter* spp.	Cytochrome, Fe^{3+}

Nitrite formation by heterotrophic soil organisms appears to be rather limited. Nitrate formation, on the other hand, is accomplished by a significant number and variety of heterotrophic species present in the soil microflora. *Aspergillus flavus* is particularly active and must be considered as a major source of nitrate in soils.

Denitrifying bacteria, such as *Pseudomonas stutzeri*, use the reduction of nitrous oxide (N_2O) to dinitrogen (N_2) for the generation of energy. Reduction of N_2O is usually the terminal step of bacterial denitrification, proceeding from nitrate to nitrite to nitric oxide to nitrous oxide and the final reduction of nitrous oxide to N_2 by N_2O reductase. A gene cluster containing the genes necessary for the reduction of nitrite (*nir*), nitric oxide (*nor*), and nitrous oxide (*nos*) has been identified in *P. stutzeri*. There are at least 15 genes in this cluster arranged in the order *nos-nir-nor*. The structural gene *nosZ* codes for the copper-containing enzyme N_2O reductase. Three other genes, *nosDFY*, are involved in the formation of the copper chromophore of the enzyme. A regulatory gene, *nosR*, is present within the *nos* region of the denitrification gene cluster of *P. stutzeri*. It is essential for the expression of the *nosZ* gene, and there is indirect evidence that the transcriptional regulator Fnr may also be involved in *nosZ* expression (see Chapter 5 for a discussion of regulatory mechanisms).

Urea is a simple organic compound with the structure H_2NCONH_2. It is widely used as an inexpensive nitrogen fertilizer because it is readily degraded to ammonia and carbamate by the enzyme urease:

$$H_2NCONH_2 + H_2O \longrightarrow NH_3 + H_2NCOOH$$

Carbamate spontaneously decomposes to yield ammonia and carbonic acid:

$$H_2NCOOH + H_2O \longrightarrow NH_3 + H_2CO_3$$

Complete urease gene clusters have been found in *Bacillus* spp, *Helicobacter pylori*, *Klebsiella aerogenes*, *Proteus mirabilis*, and *Yersinia enterocolitica*. In addition, a number of other bacteria, including soil bacteria such as *Sinorhizobium meliloti*, have been studied in some detail with regard to the genetics of urease production and its regulation. The enzyme has been identified as a virulence factor in *H. pylori* and other organisms that cause infections in humans and animals. Urease activity can serve as a diagnostic test for infection by *H. pylori*. Urease is notable because it was the first enzyme to be crystallized — a feat accomplished by Dr. James B. Sumner of Cornell University in 1926 and for which he later received the Nobel Prize.

In enteric bacteria urease expression is positively regulated and transcription is activated either in the absence of an assimilable nitrogen source or in the presence of urea. On the other hand, in *Streptococcus salivarius*, urease expression is derepressed at low pH and is enhanced in the presence of excess carbohydrate. Ureolysis by *S. salivarius* appears to provide protection of the organisms against acid damage and may also enable the organisms to acquire nitrogen when carbohydrates are present in excess. Wild-type strains of *S. salivarius* are protected against acid killing through physiologically relevant concentrations of urea, whereas a urease-deficient mutant is not. This organism can use urea as a source of nitrogen for growth exclusively through a urease-dependent pathway.

The bacterial urease from *K. pneumoniae* is composed of three basic subunits, UreA, UreB, and UreC, which appear as a trimer in the intact enzyme. Each UreABC unit contains two coordinated nickel ions, separated from one another by 3.6 Å and bridged by a carbamylated lysine residue. Within this center are three metal-coordinated water molecules, four histidine ligands, one aspartate ligand, and two noncoordinated histidines that function in the conversion of urea into ammonia and carbon dioxide.

ASSIMILATION OF INORGANIC NITROGEN

Many microorganisms synthesize all of their amino acids and other nitrogenous compounds using ammonia and carbon chains derived from carbohydrate metabolism. The available pathways for ammonia assimilation are quite limited, however. Of all the amino acids found in proteins and other cellular constituents, ammonia can be directly assimilated into only a few. These amino acids then serve as donors of their amino nitrogen via transamination to keto acid precursors to form all of the other amino acids. The potential routes of ammonia assimilation are through the synthesis of glutamate, alanine, or aspartate. The major enzymes involved in ammonia assimilation are the glutamate dehydrogenases (GDH) and two enzymes that operate in tandem, glutamine synthetase (GS) and glutamate synthase, as follows:

Glutamate dehydrogenases (GDH)

α-ketoglutarate + + NH_4^+ + NADH + H^+ $\longleftrightarrow$ L-glutamate + NAD^+

α-ketoglutarate + NH_4^+ + NADPH + H^+ $\longleftrightarrow$ L-glutamate + $NADP^+$

Glutamine synthase (GS)–glutamate synthase (GOGAT)

L-glutamate + NH_4^+ + ATP $\longrightarrow$ L-glutamine + ADP + P_i

α-ketoglutarate + L-glutamine + NADPH + H^+ $\longrightarrow$ 2 L-glutamate + $NADP^+$

Net: α-ketoglutarate + NH_4^+ + ATP + NADPH + H^+

$\longrightarrow$ L-glutamate + ADP + P_i + $NADP^+$

The abbreviation GOGAT frequently used for this enzyme system is derived from its previous trivial name, glutamine amide-2-oxoglutarate amino transferase.

Other enzymes that may play a role in ammonia assimilation in some organisms are alanine dehydrogenases and aspartase:

Alanine dehydrogenases

pyruvate + NH_4^+ + NADH + H^+ $\longleftrightarrow$ L-alanine + NAD^+

pyruvate + NH_4^+ + NADPH + H^+ $\longleftrightarrow$ L-alanine + $NADP^+$

Aspartase

fumarate + NH_4^+ $\longleftrightarrow$ L-aspartate

The route(s) of ammonia assimilation varies from one organism to another depending on the ammonia assimilation enzymes present. In the majority of organisms that have been studied, glutamate is the most widely used route for ammonia assimilation. In organisms capable of synthesizing both the NADP-linked glutamate dehydrogenase and the GS–GOGAT system, the NADP–GDH is functional at high concentrations of ammonia while the GS–GOGAT system is most active at concentrations of ammonia below 1 mM. These pathways are highly regulated as the concentration of ammonia, glutamate, and glutamine are key sensors that relate the nitrogen status of the organism. The yeast, *Saccharomyces cerevisiae*, assimilates at least two-thirds of its amino nitrogen requirement via glutamate, and, when the ammonia concentration in the culture medium is high, utilizes the NADP–GDH for glutamate synthesis. At low concentrations of ammonia, the GS–GOGAT system is operative. In nitrogen-fixing species of *Bacillus* (*B. polymyxa, B. macerans*), NADP–GDH activity is several-fold higher than that of GS and is the predominant pathway for ammonia assimilation.

A few organisms appear to be incapable of forming the GS–GOGAT system, and, therefore utilize one or more of the alternative routes for ammonia assimilation. *Rhodospirillum purpureus,* which does not fix N_2, uses exogenously supplied ammonia via the NADP-linked GDH. The only ammonia assimilation pathway in *Streptococcus sanguis, S. bovis, S. mutans,* and *S. salivarius* is NADP–GDH regardless of the external ammonia concentration. In some *Bacillus* species, the NADP–ADH and NADP–GDH enzymes are highest in activity when ammonia is in high concentration. Under N_2-fixing conditions rhizobia appear to form alanine via ADH and export alanine rather than ammonia to the plant host.

Aspartase is sufficiently active in *Klebsiella aerogenes* to function as an important assimilatory enzyme, particularly after growth on C_4-dicarboxylic acids. Under most conditions of growth, however, aspartase appears to serve in a dissimilatory (ammonia-releasing) role.

As described in Chapter 5, the GS–GOGAT system and related enzymes are highly regulated by the Ntr (nitrogen regulation) system, particularly in *E. coli* and *K. aerogenes*. On the other hand, in *Bacillus* spp. there is no evidence for a global regulatory system analogous to the Ntr system. Activity of the GS–GOGAT system in *B. subtilis* and *B. licheniformis* is regulated by the available nitrogen source (feedback inhibition). The GS regulatory protein GlnR negatively regulates transcription of the *B. subtilis* GS structural gene, while GltC stimulates the expression of GS. In *B. licheniformis*, the GOGAT enzyme (GltS) consists of two unequal subunits. The larger subunit catalyzes the glutaminase reaction:

$$\text{glutamine} + H_2O \longrightarrow \text{glutamate} + NH_3$$

An ammonia transfer reaction is catalyzed by the small subunit:

$$NH_3 + \alpha\text{-ketoglutarate} + NADPH + H^+ \longrightarrow \text{glutamate} + NADP^+$$

In *B. subtilis*, GS (GlnA) and a regulatory protein, GlnR, are encoded in an operon. Regulation of the *glnRA* operon involves the action of both GlnR and GS. Here GlnR is a repressor that interferes with transcription under conditions of nitrogen excess.

In the nitrogen-fixing anaerobe *Clostridium kluyveri*, the NADP–GDH pathway plays an important role in ammonia assimilation in ammonia-grown cells but plays only a minor role to that of the GS–GOGAT pathway in nitrogen-fixing cells — conditions in which the intracellular ammonia concentration is low. In *C. butyricum* the GS–GOGAT system is the predominant pathway for ammonia assimilation with either ammonia or N_2 as the source of nitrogen. In *C. acetobutylicum* there is no evidence for a global Ntr system and the GS enzyme is not regulated by adenylylation. Instead, a promoter, P3, which controls the transcription of an antisense RNA, is present in the downstream region of *glnA* DNA.

In phototrophic bacteria, such as the nonsulfur purple bacterium *Rhodobacter capsulatum*, the NADP–ADH aminating activity can function as an alternative route for ammonia assimilation when GS is inactive. The ADH is induced in cells grown on pyruvate plus nitrate, pyruvate plus ammonia, or L-alanine under both light-anaerobic and dark-heterotrophic conditions. Aminating activity is strictly NADPH dependent, whereas deaminating activity is strictly NAD dependent.

GENERAL REACTIONS OF AMINO ACIDS

Amino Acid Decarboxylases

Microorganisms exhibit decarboxylase activity for many amino acids including aspartate, glutamate, ornithine, lysine, arginine, tyrosine, phenylalanine, cysteic acid, diaminopimelic acid, hydroxyphenyl serine, histidine, tryptophan, 5-hydroxytryptophan, and possibly others. The general reaction for all of these enzymes is

$$R\text{-}CHNH_2COOH \longrightarrow R\text{-}CH_2NH_2 + CO_2$$

As far as is known, pyridoxal phosphate is the coenyzme of all the amino acid decarboxylases, with the exception of histidine decarboxylase, for which the cofactor is pyruvate. All of the amino acid decarboxylases are essentially irreversible and therefore are not of importance in the biosynthesis of most amino acids. However, in the case of diaminopimelic acid (DAP) decarboxylase, lysine is the final product. Amino acid decarboxylases are produced at low pH and their range of optimal activity is pH 3–5. Thus, excess acidity resulting from the production of acid end products may be regulated by amino acid decarboxylase activity, particularly in anaerobic proteolytic organisms such as the clostridia.

The diamine putrescine, the decarboxylation product of ornithine, is an essential growth factor for several organisms and is a biosynthetic precursor of both spermidine and spermine, polyamines found in a wide variety of microorganisms (see Chapter 15). Spermine is present in eukaryotic organisms but is found in only a few bacteria (e.g., *Pseudomonas aeruginosa* and *Bacillus stearothermophilus*). Spermidine is more widely distributed, being found in bacteria and fungi as well as in higher organisms.

Germinating conidia of *N. crassa* produce an active glutamate decarboxylase. This is the first step in a pathway that leads to a rapid increase in aspartate in the amino acid pool of this organism.

Amino Acid Deaminases

Deamination of amino acids occurs via several quite different reactions:

1. Oxidative deamination
   a. NAD^+- or $NADP^+$-linked deamination
   b. FAD- or FMN-linked deamination
2. Nonoxidative deamination

The oxidative deamination of glutamic acid by the reversible glutamate dehydrogenases has been discussed earlier in relation to ammonia assimilation. In most organisms the NAD-specific enzyme operates catabolically, whereas the NADP-linked enzyme functions in glutamate synthesis.

Alanine dehydrogenase (ADH) is present in members of the genus *Bacillus*, the photosynthetic nonpurple sulfur bacteria *Rhodobacter,* and others. The reaction proceeds via an α-imino intermediate:

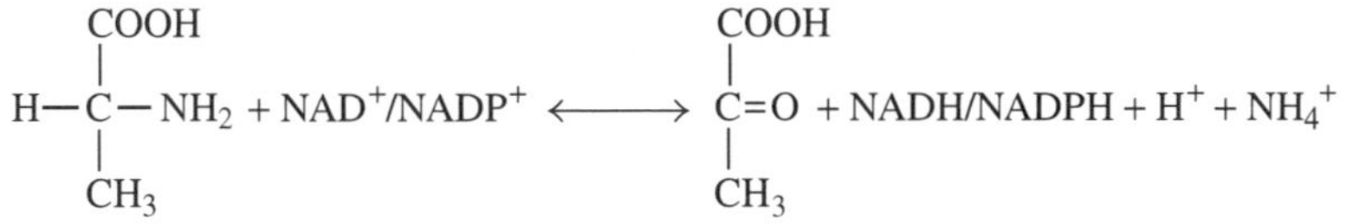

$$H-\underset{\underset{CH_3}{|}}{\overset{\overset{COOH}{|}}{C}}-NH_2 + NAD^+/NADP^+ \longleftrightarrow \underset{\underset{CH_3}{|}}{\overset{\overset{COOH}{|}}{C}}=O + NADH/NADPH + H^+ + NH_4^+$$

The ubiquitous distribution and high activity of ADH in the aerobic bacilli has been considered as evidence that the reverse reaction catalyzed by this enzyme may serve as a major route of ammonia assimilation in these organisms. This conclusion has also been drawn in the case of the symbiotic rhizobia.

Amino acid oxidases, sometimes called aerobic or oxidative deaminases, involve reactions catalyzed by enzymes containing FAD (flavin adenine dinucleotide) or FMN (flavin mononucleotide). The reaction proceeds in two stages:

$$R\text{-}CHNH_2COOH + Enz\text{-}FAD \longrightarrow \alpha\text{-keto acid} + NH_3 + CO_2 + Enz\text{-}FADH_2$$

$$Enz\text{-}FADH_2 + O_2 \overset{nonenzymatic}{\longrightarrow} Enz\text{-}FAD + H_2O_2$$

Reduced FAD (or reduced FMN) may react nonenzymatically with molecular oxygen as shown above or it may transfer the hydrogen to other hydrogen acceptors. The amino acid oxidases are nonspecific in that a single enzyme may catalyze the oxidation of a variety of amino acids, for example, methionine, phenylalanine, tyrosine, leucine, isoleucine, valine, norvaline, alanine, tryptophan, and cysteine in decreasing order of activity. Another may oxidize proline, hydroxyproline, citrulline, histidine, and arginine. The rate of oxidation may differ for each amino acid, and the order of activity may differ with the source of the enzyme. Both D- and L-amino acid oxidases are known, but a single enzyme is usually specific for one configuration. These reactions yield no useful energy if the reduced FAD is oxidized nonenzymatically by molecular oxygen. It is highly unlikely that these enzymes play any significant role in the assimilation of ammonia.

Nonoxidative deaminases, such as aspartic acid deaminase (aspartase), serine, and threonine deaminases (dehydratases), and cysteine desulfhydrase, are specific in their substrate requirements.

Serine and **threonine deaminases (dehydratases)** catalyze the following type of reaction:

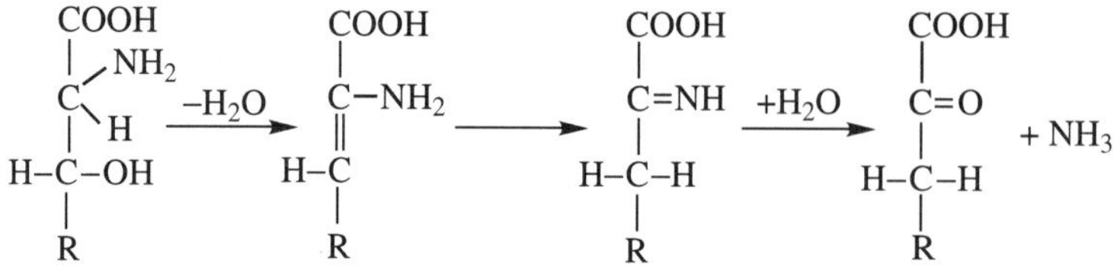

R = H in serine; CH$_3$ in threonine

imino acid

pyruvate or α-ketobutyrate

An intramolecular transfer of hydrogen atoms occurs and water is removed (dehydration) to produce an imino acid via a β-elimination reaction. In the second stage of the reaction, water reacts nonenzymatically with the imino acid to release ammonia. Pyridoxal-5′-phosphate is the cofactor in this reaction. The two deaminating enzymes have been shown to be distinct. The level of L-serine deaminase varies as a function of nitrogen nutrition, carbon source, and the supply of glycine and leucine. Glycine and leucine induce the formation of serine deaminase. The enzyme seems to play a role in a number of pathways in which serine is generated and further metabolized as part of the main carbon pathway. Both *E. coli* and *S. enterica* produce biodegradative as well as biosynthetic threonine deaminases. The latter provides α-ketobutyrate as an intermediate in the formation of isoleucine. The biodegradative enzyme is induced under anaerobic conditions in amino acid–rich medium, requires cAMP for its synthesis, and is sensitive to catabolite repression by glucose.

Aspartase is present in a number of organisms. It converts aspartate to fumarate and ammonia:

$$HOOCCH_2CHNH_2COOH \longleftrightarrow HOOCCH{=}CHCOOH + NH_3$$

There is no hydrogen exchange in this reaction. Instead, an intramolecular transfer of hydrogen occurs, and pyridoxal-5-phosphate is not involved as a cofactor in the reaction. The reaction is reversible, providing a potential route for ammonia assimilation. Although the equilibrium of the reaction is such that it could serve as an ammonia assimilation pathway, it is considered to function primarily in a catabolic capacity.

Cysteine desulfhydrase (dehydratase) has a reaction mechanism similar to that of serine and threonine deaminases except that hydrogen sulfide, rather than water, is removed via a β-elimination reaction:

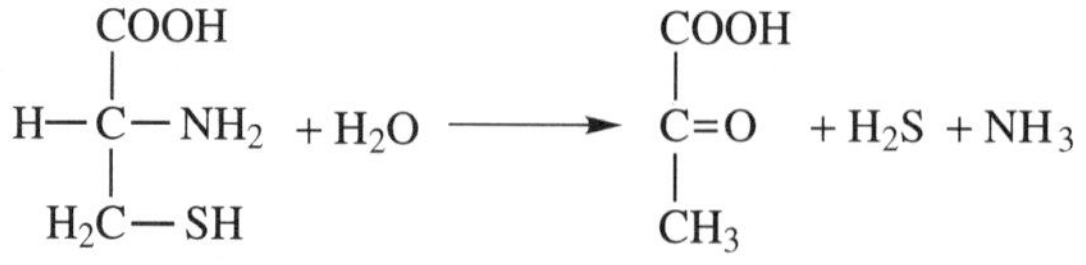

Dehydratases and desulfhydrases are essentially irreversible reactions and are not considered reactions that could readily participate in ammonia assimilation.

Phenylalanine deaminase (ammonia lyase) is produced by yeasts, molds, and bacteria and catalyzes the nonoxidative deamination of L-phenylalanine to *trans*-cinnamic acid:

$$\text{〈} \rangle\text{-CH}_2\text{CHNH}_2\text{COOH} \longrightarrow \text{〈} \rangle\text{-CH=CHCOOH} + \text{NH}_3$$

A dehydroalanine group serves as a cofactor in the reaction. The enzyme is of interest because of its potential for use in the treatment and diagnosis of phenylketonuria and has industrial applications in the synthesis of L-phenylalanine from *trans*-cinnamic acid. In *Rhodotorula glutinis* this enzyme serves as the initial step in a metabolic pathway, leading to the formation of benzoate and 4-hydroxybenzoate.

Amino Acid Transaminases (Aminotransferases)

In 1945, Lichstein and Cohen first demonstrated transaminase activity in bacteria. Transfer of the α-amino nitrogen between glutamate, aspartate, alanine, and their corresponding α-keto acids was shown to be similar to reactions observed in mammalian tissues. The two major transaminases demonstrated were aspartate amino transferase and alanine aminotransferase:

$$\text{aspartate} + \alpha\text{-ketoglutarate} \longleftrightarrow \text{oxaloacetate} + \text{glutamate}$$

$$\text{alanine} + \alpha\text{-ketoglutarate} \longleftrightarrow \text{pyruvate} + \text{glutamate}$$

Pyridoxal phosphate is the coenzyme for all known transaminase reactions.

Transamination was first thought to be a relatively limited activity confined to these three amino acids and their keto analogs. However, in the early 1950s, Feldman and Gunsalus, and Rudman and Meister, showed that a number of amino acids would undergo transamination with α-ketoglutarate to form the corresponding amino acids. Transaminases are fully reversible reactions, so glutamate is a key donor for the synthesis of most other amino acids. This result led to a more generalized view of transamination in which any amino acid and any keto acid could exchange the amino group:

$$\text{amino acid}_1 + \text{keto acid}_2 \longleftrightarrow \text{keto acid}_1 + \text{amino acid}_2$$

Transaminase activity is ubiquitous as evidenced by the fact that keto analogs can replace many of the amino acids for the growth of amino acid–requiring mutants. The reaction is not totally universal, however. Four major transaminases have been identified in *E. coli* as shown in Table 14-5. By these criteria, strains lacking transaminase C (*avtA* mutants) should have no nutritional requirement, whereas strains lacking transaminase B (*ilvE* mutants) should require isoleucine. In practice, most *ilvE* mutants require only isoleucine, but some show additional requirements for valine or leucine as a result of reduced expression of either *avtA* or other genes distal to *ilvE*. These transaminases have not, in general, been well identified in other organisms.

TABLE 14-5. Major Transaminases in *Escherichia coli*

Transaminase	Interacts with	Gene
Aromatic amino acids	Tyrosine Phenylalanine Glutamate Leucine Aspartate Methionine	*tyrB*
Aspartate (transaminase A)	Aspartate Glutamate Tyrosine Phenylalanine	*aspC*
Transaminase B (branched-chain aminotransferase)	Glutamate Leucine Isoleucine Valine Phenylalanine Methionine	*ilvE*
Transaminase C (alanine-valine aminotransferase)	Alanine Valine α-Aminobutyrate	*avtA*

However, most of these activities appear to be present in organisms that display a general ability to synthesize the common amino acids.

The aspartate aminotransferase (L-aspartate:2-oxoglutarate aminotransferase) is ubiquitous. However, the enzyme has been studied in detail only in *E. coli, P. putida, S. cerevisiae,* and the archeon *Sulfolobus solfataricus.* A thermophilic *Bacillus* species produces an aspartate aminotransferase that shows some sequence similarity in the N-terminal region between the eubacterial and archaeal enzymes.

Other transaminases are known. For example, *S. enterica* displays a glutamine amidotransferase encoded within the *trpD* gene. It serves a dual role in transferring the amino group of glutamine to chorismic acid in the synthesis of anthranilic acid. Glutamine amidotransferase activity is also present in a tryptophan gene (*trypE*) in *Bacillus pumilis.*

Amino Acid Racemases

A number of microorganisms contain enzymes that catalyze the conversion of D-amino acids to L-amino acids via the general reaction

$$
\begin{array}{ccc}
\text{COOH} & & \text{COOH} \\
| & & | \\
\text{H} - \text{C} - \text{NH}_2 & \longleftrightarrow & \text{H}_2\text{N} - \text{C} - \text{H} \\
| & & | \\
\text{R} & & \text{R}
\end{array}
$$

Most biochemical compounds are asymmetric and there is a tendency for one form to predominate over the other. Amino acids in naturally occurring proteins are usually in

L-configuration. However, the cell walls and polypeptide capsules of many organisms contain D-amino acids. Since most biosynthetic reactions lead to the synthesis of the L-amino acids, racemases are necessary for the conversion of certain amino acids to the D configuration for the synthesis of these specialized cell structures. As an example, D-alanine is a structural component of the cell wall of *Enterococcus faecalis* as well as several other gram-positive organisms. When grown in a medium lacking pyridoxal, D-alanine becomes a specific growth requirement for *E. faecalis* because alanine racemase is inactive under the conditions of pyridoxal deficiency. D-Amino acids are found in the polypeptide capsules of members of the genus *Bacillus* and in peptide antibiotics, providing other indications of the importance of racemases in microbial metabolism.

Role of Pyridoxal-5′-Phosphate in Enzymatic Reactions with Amino Acids

In the foregoing discussion of the various reactions in which amino acids participate, it was mentioned that pyridoxal-5′-phosphate (PLP), the coenzyme form of vitamin B_6, functions in many of these reactions. PLP is the most versatile of all enzyme cofactors in that it can catalyze several distinct chemical reactions. Such versatility seems to be due, in part, to the PLP cofactor acting as an electron sink. For example, a proton may be removed from the α-carbon of the amino acid substrate, with the resultant stabilization of the carbanion at the C_α or C'-4 position; or the electrons may flow into the ring, neutralizing a positively charged pyridine nitrogen (i.e., a quinoid structure may be one of the intermediates). It is assumed that appropriate groups on the apoenzyme hold the PLP in the precise alignment necessary for rupture of the $-C-X$ bond (Fig. 14-9).

Depending on the type of electron shifts that take place, nine main types of reactions are recognized: transamination, β decarboxylation, α decarboxylation, aldol cleavage,

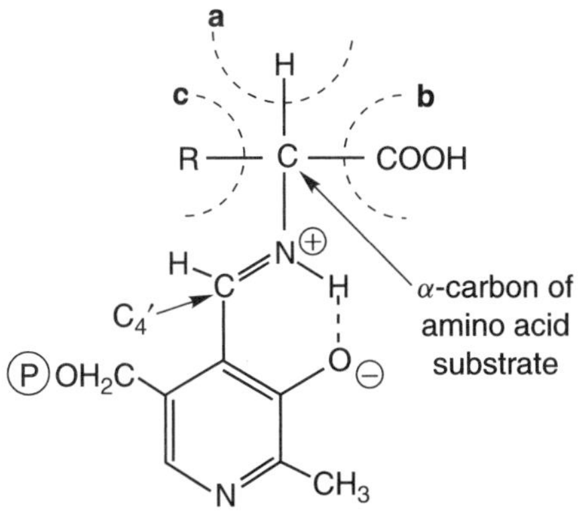

Fig. 14-9. Functions of pyridoxal-5′-phosphate (vitamin B_6) in reactions of amino acids. The product of the enzyme-catalyzed reaction will depend on which of the four bonds projecting from the α-carbon atom is split. Reaction (**a**) occurs in racemization, transamination, β elimination, γ elimination, β replacement, γ replacement, and β decarboxylation. Reaction (**b**) occurs in α-decarboxylation. Reaction (**c**) occurs in aldol cleavage. Although all three of these types of reactions are well known, almost invariably each enzyme is quite specific as to which bond it will break. This remarkable specificity occurs because the enzyme can "hold" the substrate in such a way that only the required bonds can be broken. This alignment is probably due to the strategic positioning of certain groups within the active site of the enzyme.

cleavage, γ elimination, β elimination, γ displacement, or β displacement may occur. In addition, various PLP enzymes carry out unique reactions that do not fall into any of these categories — for example, dialkyl amino acid transaminase, tryptophan synthase, threonine synthase, and δ-amino-levulinate synthase. Three enzymes are necessary for the reduction of cytidine diphosphate (CDP) sugars to 1,3-dideoxy sugars. One of these enzymes contains pyridoxamine. Pyridoxamine, not pyridoxal, is a required growth factor for certain organisms. These are the only known cases where the cofactor requirement is for the amine rather than the pyridoxal form. The 1,3-dideoxy sugars are important components of certain bacterial cell walls.

THE STICKLAND REACTION

Some members of the genus *Clostridium* can utilize amino acids as a source of energy by means of coupled oxidation–reduction reactions involving certain amino acids as hydrogen donors and others as receptors:

$$R_1\text{-}CH_2NH_2\text{–}COOH + R_2\text{-}CH_2NH_2\text{–}COOH + H_2O$$
$$\longrightarrow R_1\text{-}CH_2\text{–}COOH + R_2\text{-}COCH_2\text{–}COOH + 2NH_3$$

A specific example of such a reaction involving glycine and alanine takes place according to the following reaction scheme:

$$CH_2NH_2COOH + CH_3CHNH_2COOH + H_2O$$
$$\longrightarrow CH_3COOH + CH_3COCOOH + 2NH_3$$

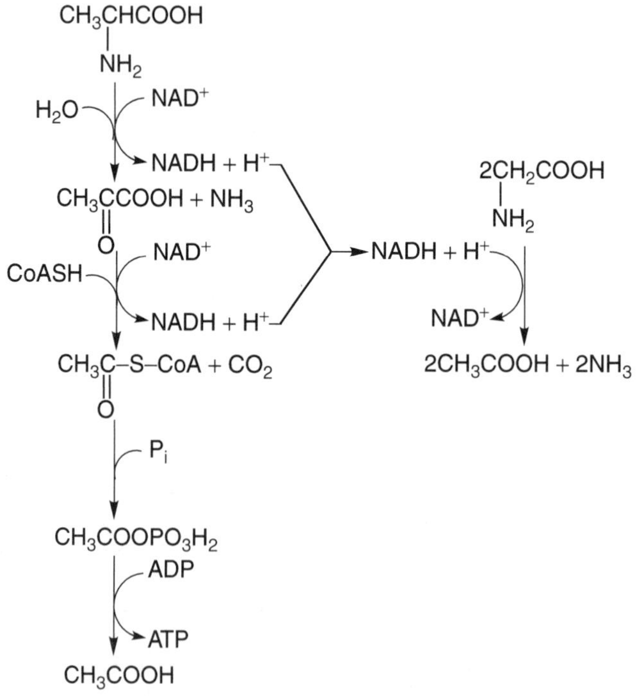

Fig. 14-10. Mechanism of the Stickland reaction.

This type of reaction leads to the formation of short-chain fatty acids and keto acids. The Stickland reaction involves several steps in which NAD first accepts a hydrogen atom from the amino acid donor and transfers it to the acceptor amino acids. The latter reaction is catalyzed by an amino acid reductase (Fig. 14-10). Certain amino acids serve preferentially as hydrogen donors and others as hydrogen acceptors. The rate of the reaction is quite rapid, and this system is used by proteolytic clostridia to gain energy via substrate-level phosphorylation.

BIBLIOGRAPHY

Nitrogen Fixation

Broughton, W. J., S. Jabouri, and X. Perret. 2000. Keys to symbiotic harmony. *J. Bacteriol.* **182:**5641–2.

Debellé, F., C. Rosenberg, J. Vasse, F. Maillet, E. Martinez, J. Dénarié, and G. Truchet, 1986. Assignment of symbiotic developmental phenotypes to common and specific nodulation (*nod*) genetic loci of *Rhizobium meliloti*. *J. Bacteriol.* **168:**1075–86.

Fischer, H.-M. 1994. Genetic regulation of nitrogen fixation in rhizobia. *Microbiol. Rev.* **58:**352–86.

Gage, D. J., R. Bobo, and S. R. Long. 1996. Use of green fluorescent protein to visualize the early events of symbiosis between *Rhizobium meliloti* and alfalfa (*Medicago sativa*). *J. Bacteriol.* **178:**7159–66.

Graham, P. H. 2000. Nodule formation in legumes. In J. Lederberg (ed.), *Encyclopedia of Microbiology*, 2nd ed., Vol. 3. Academic Press, San Diego, pp. 407–17.

Green, L. S., Y. Li, D. W. Emerich, F. J. Bergersen, and D. A. Day. 2000. Catabolism of α-ketoglutarate by a *sucA* mutant of *Bradyrhizobium japonicum*: evidence for an alternative tricarboxylic acid cycle. *J. Bacteriol.* **182:**2838–44.

Halbleib, C. M., Y. Zhang, and P. W. Ludden. 2000. Regulation of dinitrogenase reductase ADP-ribosyltransferase and dinitrogenase reductase-activating glycohydrolase by a redox-dependent conformational change of nitrogenase Fe protein. *J. Biol. Chem.* **275:**3493–500.

Herrero, A., A. M. Muro-Pastor, and E. Flores. 2001. Nitrogen control in cyanobacteria. *J. Bacteriol.* **183:**411–25.

Hill, S., and G. Sawyers. 2000. Azotobacter. In J. Lederberg (ed.), *Encyclopedia of Microbiology*, 2nd ed., Vol. 1. Academic Press, San Diego, pp. 359–71.

Kaiser, B. N., et al. 1998. Characterization of an ammonium transport protein from the peribacteroid membrane of soybean nodules. *Science* **281:**1202–5.

Kuykendall, L. D., F. M. Hashem, R. B. Dadson, and G. H. Elkan. 2000. Nitrogen fixation. In J. Lederberg (ed.), *Encyclopedia of Microbiology*, 2nd ed., Vol. 3. Academic Press, San Diego, pp. 379–91.

Ma, Y., and P. W. Ludden. 2001. Role of the dinitrogenase reductase arginine 101 residue in dinitrogenase reductase ADP-ribosyltransferase binding, NAD binding, and cleavage. *J. Bacteriol.* **183:**250–6.

Noel, K. D., L. S. Forsberg, and R. W. Carlson. 2000. Varying the abundance of O antigen in *Rhizobium etli* and its effect on symbiosis with *Phaseolus vulgaris* . *J. Bacteriol.* **182:**5317–24.

Pellock, B., et al. 1999. Biosynthesis, regulation, and control of molecular weight, distribution of symbiotically important *Rhizobium meliloti* exopolysaccharides. In A. Steinbüchel (ed.), *Biochemical Principles and Mechanisms of Biosynthesis and Degradation of Polymers*. Wiley-VCH, Weinheim, New York, p. 104–12.

Pellock, B. J., H.-P. Cheng, and G. C. Walker. 2000. Alfalfa root nodule invasion efficiency is dependent on *Sinorhizobium meliloti* polysaccharides. *J. Bacteriol.* **182:**4310–8.

Perret, X., C. Staehelin, and W. J. Broughton. 2000. Molecular basis of symbiotic promiscuity. *Microbiol. Mol. Biol. Rev.* **64:**180–201.

Poole, P., and D. Allaway. 2000. Carbon and nitrogen metabolism in *Rhizobium. Adv. Microb. Physiol.* **43:**117–63.

Postgate, J. R. 1998. *Nitrogen Fixation,* 3rd ed. Cambridge University Press, New York.

Rudnick, P. A., T. Arcondéguy, C. K. Kennedy, and D. Kahn. 2001. *glnD* and *mviN* are genes of an essential operon in *Sinorhizobium meliloti. J. Bacteriol.* **183:**2682–5.

Thiel, T., and B. Pratte. 2001. Effect on heterocyst differentiation of nitrogen fixation in vegetative cells of the cyanobacterium *Anabaena variabilis* ATCC 29413. *J. Bacteriol.* **183:**280–6.

van Rhijn, P., and J. Vanderleyden. 1995. The *Rhizobium*-plant symbiosis. *Microbiol. Rev.* **59:**124–42.

Waters, J. K., B. L. Hughs II, L. C. Purcell, K. O. Gerhardt, T. P. Mawhinney, and D. W. Emerich. 2000. Alanine, not ammonia, is excreted from N_2-fixing soybean nodule bacteroids. *Proc. Natl. Acad. Sci. USA* **95:**12038–42.

Wong, F. C. Y., and J. C. Meeks. 2001. The *hetF* gene product is essential to heterocyst differentiation and affects HetR function in the cyanobacterium *Nostoc punctiforme. J. Bacteriol.* **183:**2654–61.

Zhang, Y., E. L. Pohlmann, C. M. Halbleib, P. W. Ludden, and G. P. Roberts. 2001. Effect of P_{II} and its homolog GlnK on reversible ADP-ribosylation of dinitrogenase reductase by heterologous expression of the *Rhodospirillum rubrum* dinitrogenase reductase ADP-ribosyl transferase-dinitrogenase reductase-activating glycohydrolase regulatory system in *Klebsiella pneumoniae. J. Bacteriol.* **183:**1610–20.

Inorganic Nitrogen

Knowles, R. 2000. Nitrogen cycle. In J. Lederberg (ed.), *Encyclopedia of Microbiology*, 2nd ed., Vol. 3. Academic Press, San Diego, pp. 379–91.

Lin, J. T., and V. Stewart. 1998. Nitrate assimilation by bacteria. *Adv. Microb. Physiol.* **39:**1–30.

Urease

Chen, Y.-Y., C. A. Weaver, and R. A. Burne. 2000. Dual functions of *Streptococcus salivarius* urease. *J. Bacteriol.* **182:**4667–9.

Hausinger, R. P., G. J. Colpas, and A. Soriano. 2001. Urease: a paradigm for protein-assisted metallocenter assembly. *ASM News* **67:**78–84.

Mobley, H. L., M. D. Island, and R. P. Hausinger. 1995. Molecular biology of ureases. *Microbiol. Rev.* **59:**451–80.

Assimilation of Inorganic Nitrogen

Hu, P., T. Leighton, G. Ishkhanova, and S. Kustu. 1999. Sensing of nitrogen limitation by *Bacillus subtilis*: comparison to enteric bacteria. *J. Bacteriol.* **181:**5042–50.

Reitzer, L. 2000. Amino acid function and synthesis. In J. Lederberg (ed.), *Encyclopedia of Microbiology*, 2nd ed., Vol. 1. Academic Press, San Diego, pp. 134–51.

CHAPTER 15

BIOSYNTHESIS AND METABOLISM OF AMINO ACIDS

Amino acid biosynthesis is discussed most conveniently in the context of families of amino acids that originate from a common precursor:

1. **Glutamate or α-Ketoglutarate Family.** Glutamate, glutamine, glutathione, proline, arginine, putrescine, spermine, spermidine, and in yeasts and molds, lysine. A tetrapyrrole (heme) precursor, δ-aminolevulinate, arises from glutamate in some organisms.

2. **Aspartate Family.** Aspartate, asparagine, threonine, methionine, isoleucine, and, in bacteria, lysine.

3. **Pyruvate Family.** Alanine, valine, leucine, and isoleucine.

4. **Serine-Glycine or Triose Family.** Serine, glycine cysteine, and cystine. In yeasts and molds, mammals, and some bacteria, δ-aminolevulinate is formed by the condensation of glycine and succinate.

5. **Aromatic Amino Acid Family.** Phenylalanine, tyrosine, and tryptophan. Additional compounds that can originate from the common aromatic pathway include enterochelin, p-aminobenzoate, ubiquinone, menaquinone, and NAD.

6. **Histidine.** This amino acid originates as an offshoot of the purine pathway in that a portion of the histidine molecule is derived from the intact purine ring.

In some instances, it is convenient to discuss the catabolic pathways of the amino acids in conjunction with the discussion of their biosynthesis.

THE GLUTAMATE OR α-KETOGLUTARATE FAMILY

Glutamine and Glutathione Synthesis

The importance of glutamate as one of the primary amino acids involved in the assimilation of ammonia is discussed in Chapter 14. Glutamate and glutamine play a central role in amino acid biosynthesis by the ready transfer of amino or amide groups, respectively, in the synthesis of other amino acids by transamination or transamidation reactions. Glutamine is synthesized from glutamate with the participation of ammonia and ATP. Glutathione, a disulfide-containing amino acid whose functions have only recently begun to be understood in detail, is synthesized in two steps. The coupling of L-glutamate and L-cysteine in the presence of ATP to form γ-glutamylcysteine is catalyzed by a specific synthase. In the presence of glycine and ATP, glutathione synthase forms glutathione (Fig. 15-1).

The Proline Pathway

The pathway to proline (Fig. 15-1) involves formation of γ-glutamylphosphate from L-glutamate and ATP by γ-glutamyl kinase (ProB). In the presence of NADPH, γ-glutamylphosphate is reduced to glutamate γ-semialdehyde by ProA. Glutamate γ-semialdehyde cyclizes spontaneously to form 1-pyrroline-5-carboxylate. Pyrroline-5-carboxylate is converted to proline by a specific reductase (ProC).

Aminolevulinate Synthesis

In *S. enterica* and *E. coli*, δ-aminolevulinic acid (ALA), the first committed precursor to tetrapyrroles, arises from glutamate. This C_5 pathway, which was originally thought to occur primarily in plants and algae, is now firmly established as a major route to ALA in several bacterial species. Prior to this discovery, the condensation of glycine and succinyl-CoA by the enzyme ALA synthase (the C_4 pathway) was considered to be the only route of ALA formation. It is still the major route of ALA formation in mammals, fungi, and certain bacteria, such as *Rhodopseudomonas sphaeroides, R. capsulatus*, and *Bradyrhizobium japonicum*. The C_5 pathway to ALA involves conversion of glutamate to glutamyl-tRNA^{Glu} by glutamyl-tRNA synthetase, reduction to glutamate γ-semialdehyde (GSA) by an NADPH-dependent glutamyl-tRNA reductase (HemA), and transamination by glutamate γ-semialdehyde aminomutase (HemL) to form ALA:

$$
\begin{array}{ccccccc}
\text{COOH} & & \text{COOH} & & \text{COOH} & & \text{COOH} \\
| & & | & & | & & | \\
\text{HCH} & & \text{HCH} & & \text{HCH} & & \text{HCH} \\
| & \xrightarrow[]{\text{tRNA}^{Glu},\ \text{ATP}} & | & \xrightarrow[\text{HemA}]{\text{NADPH}} & | & \xrightarrow[\text{HemL}]{\text{PALP}} & | \\
\text{HCH} & & \text{HCH} & & \text{HCH} & & \text{HCH} \\
| & & | & & | & & | \\
\text{HCNH}_2 & & \text{HCNH}_2 & & \text{HCNH}_2 & & \text{HCNH}_2 \\
| & & | & & | & & | \\
\text{COOH} & & \text{C=O} & & \text{C=O} & & \text{C=O} \\
& & | & & | & & | \\
& & \text{tRNA}^{Glu} & & \text{H} & & \text{CH}_2\text{NH}_2 \\
\text{Glutamate} & & \text{Glutamyl-tRNA} & & \text{GSA} & & \text{ALA}
\end{array}
$$

The details of the pathway of tetrapyrrole (heme) biosynthesis are considered in the section on tetrapyrrole pathways.

The Arginine Pathway

Bacteria and fungi synthesize ornithine via a series of *N*-acetyl derivatives (Fig. 15-2). The function of the *N*-acetyl group is to prevent the premature cyclization of 1-pyrroline-5-carboxylate to proline. There is a divergence in the pathway in different

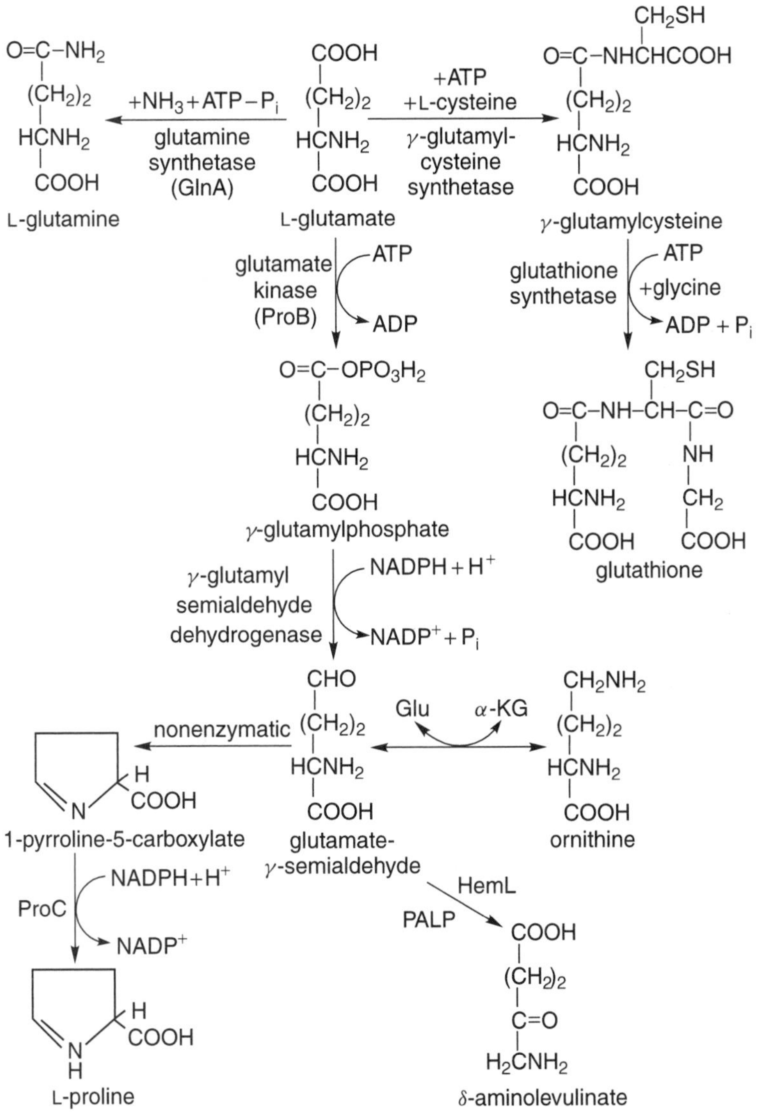

Fig. 15-1. Pathways from glutamate to glutamine, glutathione, proline, ornithine, and ALA. The transaminase that converts glutamate γ-semialdehyde to ornithine is reversible. However, it is generally considered that this reaction is primarily involved in ornithine degradation. Ornithine is normally synthesized via the pathway shown in Figure 10-2. ProC, 1-pyrroline-5-carboxylate reductase; PALP, pyridoxal phosphate; HemL, glutamate γ-semialdehyde aminomutase.

organisms depending on the manner in which the acetyl group is removed. In *Enterobacteriaceae* and *Bacillaceae*, *N*-acetylornithine is deacylated via acetylornithine deacetylase (ArgE). In *N. gonorrhoeae, Pseudomonadaceae*, cyanobacteria, photosynthetic bacteria, and yeasts and molds, the acetyl group of *N*-acetylornithine is recycled by ornithine acetyltransferase (ArgJ).

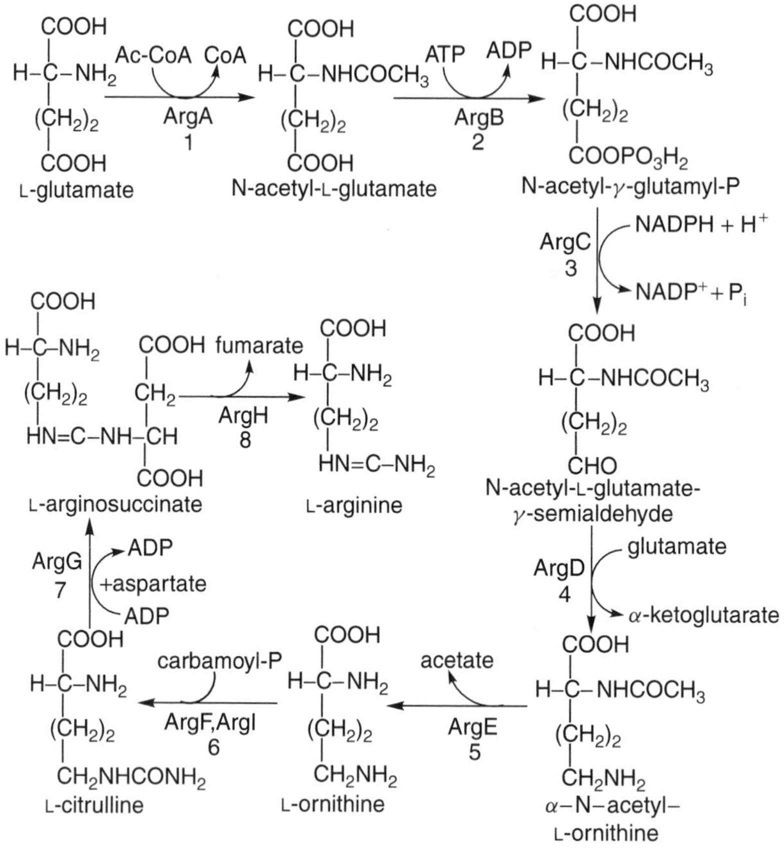

Fig. 15-2. Pathway of arginine biosynthesis. The enzyme designations for *Escherichia coli* are: ArgA, acetylglutamate synthetase; ArgB, acetylglutamate kinase; ArgC, acetylglutamate γ-semialdehyde dehydrogenase; ArgD, acetylornithine-δ-transaminase; ArgF, ArgI, ornithine transcarbamylase; ArgG, arginosuccinate synthase; ArgH, arginosuccinase. In yeasts, molds, *E. coli, P. mirabilis, S. marcescens,* and some other enterobacteria, step 5 occurs as shown. In *Pseudomonas fluorescens, Micrococcus glutamicus, Anabaena variabilis,* and several other bacteria, step 5 involves transacylation between glutamate and acetylornithine.

The eight enzymes involved in the arginine pathway (Fig. 15-2) are found in *E. coli, S. enterica, Proteus, Pseudomonas, B. licheniformis, B. subtilis, B. sphaericus, S. bovis, N. gonorrhoeae, N. crassa, A. niger, S. cerevisiae,* and *C. albicans.*

Carbamoyl phosphate is a common precursor in the biosynthesis of arginine and pyrimidines. In *E. coli* and *S. enterica* a single carbamoyl phosphate synthetase catalyzes the reaction

$$2ATP + HCO_3^- + \text{L-glutamine} + H_2O$$
$$\longrightarrow NH_2COOPO_3H_2 + 2P_i + 2ADP + \text{glutamate}$$

Ammonia can replace glutamine as a nitrogen donor in vitro, but glutamine is the physiologically preferred substrate, and K^+ and Mg^{2+} are required participants. The enzyme is composed of two nonidentical subunits. The smaller subunit, encoded by *carA*, acts as a glutamine amidotransferase. The larger subunit, encoded by *carB*, carries out the remaining functions. Since carbamoyl phosphate is involved in two major biosynthetic pathways, expression of the *carAB* operon is regulated by both arginine and pyrimidines. The enzyme is also subject to allosteric control by intermediates in both pathways. Ornithine stimulates enzyme activity, whereas UMP is inhibitory. The enzyme is activated by IMP and PRPP, coordinating its activity with purine biosynthesis as well. In some earlier studies, certain organisms appeared to depend on the nonenzymatic formation of carboxylamine and its conversion by carbamoyl phosphokinase:

Nonenzymatic:

$$NH_4^+ + HCO_3^- \longrightarrow NH_2COO^- + H_3O^+$$

Carbamoyl phosphokinase:

$$NH_2COO^- + ATP + Mg^{2+} \longrightarrow NH_2COOPO_3H_2 + ADP$$

The relative importance of this reaction in ammonia assimilation as compared to the GDH or GS–GOGAT systems has never been adequately assessed.

In *S. cerevisiae* and *N. crassa* there are two carbamoyl phosphate synthetases. One enzyme, carbamoyl phosphate synthetase A, is linked to the arginine pathway and is subject to repression by arginine. The other enzyme, carbamoyl phosphate synthetase P, is linked to the pyrimidine biosynthesis pathway and is subject to both repression and feedback inhibition by pyrimidines. Localization of carbamoyl phosphate synthetase A within mitochondria seems to play a major role in the channeling of this precursor. In yeast there seems to be little channeling of carbamoyl phosphate, since both carbamoyl phosphate synthetases are in the cytoplasm and contribute to a common pool of carbamoyl phosphate.

Some microorganisms are able to generate usable energy as ATP by the anaerobic degradation of arginine using the arginine deiminase (ADI) pathway (Fig. 15-3). In *Bacillus licheniformis* a complex of three enzymes (ArcA, ArcB, and ArcC) and an arginine-ornithine antiporter (ArcD) are involved in arginine degradation. The ArcD protein mediates proton motive force–driven uptake of arginine and ornithine and the stoichiometric exchange between arginine and ornithine. The *arc* genes of *B. licheniformis* appear in the same order (*arcABDC*) as in *C. perfringens*. In *P. aeruginosa* these genes appear in the order *arcDABC*. *Lactobacillus sake* contains genes for the arginine deiminase pathway that include *arcD*, an arginine-ornithine antiporter, and a transaminase-encoding gene, *arcT*. The ADI pathway is active in a number of other aerobic spore-forming bacilli as well as in *Mycoplasma, Streptococcus, Lactococcus, Lactobacillus, Clostridium, Pseudomonas*, and the cyanobacterium *Synechocystis*.

Arginine degradation in *E. coli, K. pneumoniae*, and *P. aeruginosa* may follow yet another catabolic pathway: the arginine succinyltransferase (AST) pathway. In *E. coli* the enzymes in this pathway are encoded by genes in the *astCADBE* operon representing the structural genes for arginine succinyltransferase, succinylarginine dihydrolase, succinylornithine transaminase, succinylglutamic semialdehyde dehydrogenase,

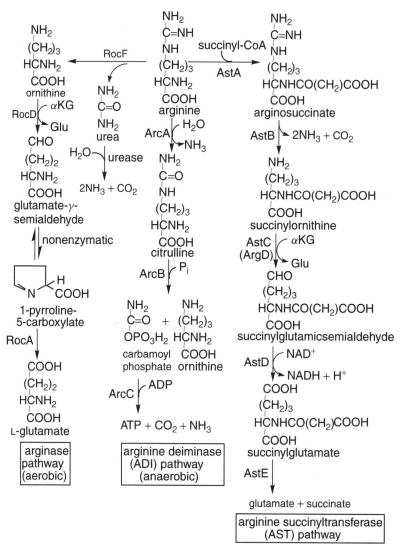

Fig. 15.3. Arginine catabolic pathways. Enzyme designations are: RocF, arginase; RocD, ornithine aminotransferase; RocA, pyrroline-5-carboxylate dehydrogenase; ArcA, arginine deiminase; ArcB, catabolic ornithine carbamoyltransferase; ArcC, carbamate kinase; AstA, arginine succinyltransferase; AstB, succinylarginine dihydrolase; AstC, succinylornithine transaminase; AstD, succinylglutamate semialdehyde dehydrogenase; AstE, succinylglutamate desuccinylase. Note that in *P. aeruginosa* the enzymes of the arginine succinyltransferase pathway are designated Aru (for arginine utilization) and are represented by the designations AruA, AruB, AruC, AruD, and AruE, respectively.

and succinylglutamate desuccinylase (Fig. 15-3). The final products are glutamate and succinate. Disruption of any of the genes in this pathway prevents arginine catabolism and impairs ornithine utilization. In *P. aeruginosa* these same genes are designated *aru* (for arginine utilization). Arginine metabolism is of considerable significance in *P. aeruginosa* as evidenced by the strong chemotactic activity for this amino acid and the fact that there are four different catabolic pathways for arginine utilization: arginine deiminase, arginine succinyltransferase (AST), arginine dehydrogenase, and arginine decarboxylase.

In *S. cerevisiae, N. crassa*, and in mammalian cells, arginine is degraded via the arginase pathway. In fungi, the arginine catabolic enzymes arginase (encoded by *CAR1*) and ornithine transaminase (encoded by *CAR1*) work in tandem to degrade arginine:

$$\text{L-arginine} \longrightarrow \text{urea} + \text{L-ornithine}$$

$$\text{ornithine} + \alpha\text{-ketoglutarate} \longrightarrow \text{glutamate} + \text{glutamate } \gamma\text{-semialdehyde}$$

Arginase activity in mammals is so rapid that insufficient amounts of arginine are available for protein synthesis and arginine is required as one of the essential amino acid nutrients. In yeasts and molds, the presence of an alternative route of arginine synthesis and regulatory controls over arginase activity prevent the occurrence of arginine deprivation.

In *N. crassa*, arginine metabolism is compartmentalized between the cytoplasm, the mitochondrion, and vacuoles. In addition, arginase activity is inhibited both in vitro and in vivo by citrulline and ornithine, eliminating a potentially wasteful catabolism of endogenous arginine. In *S. cerevisiae*, which degrades arginine and proline via the common intermediate 1-pyrroline-5-carboxylate, there is also a compartmentalized system for the metabolism of these amino acids. The proline biosynthetic enzyme 1-pyrroline-5-carboxylate reductase is located in the cytoplasm, whereas the proline-degrading enzyme 1-pyrroline-5-carboxylate dehydrogenase is in a particulate fraction of the cell, presumably in the mitochondrion. Arginase is encoded by *CAR1*, which is inducible by arginine. Regulated expression of the *CAR1* gene requires three upstream activation sequences and an upstream repression sequence.

Polyamine Biosynthesis

Polyamines are widely distributed in bacteria, yeasts, and molds as well as in higher forms. A number of growth processes are affected by polyamines. However, their precise role in governing cell growth and differentiation has not been established. The major pathway to polyamines in yeast is via the decarboxylation of ornithine to putrescine (Fig. 15-4). Ornithine decarboxylase is the rate-limiting step and appears to be a common intermediate in most organisms.

In bacteria and plants an alternate pathway to putrescine proceeds via decarboxylation of arginine to agmatine by a biosynthetic arginine decarboxylase (encoded by *speA* in *E. coli*). Agmatine ureohydrolase removes urea from agmatine to yield putrescine:

Arginine   Agmatine   Putrescine

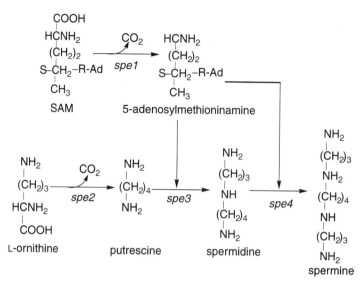

Fig. 15.4. Polyamine biosynthesis in *Saccharomyces cerevisiae*. Genes coding for the enzymes in the pathway are: *spe1*, ornithine decarboxylase; *spe2*, S-adenosylmethionine (SAM) decarboxylase; *spe3*, aminopropyl transferase (spermidine synthase); *spe4*, spermine synthase. (*Source*: Redrawn from Tabor, C. W., et al., *Fed. Proc.* **41**:3084–3088, 1982.)

Exogenous arginine acts as a signal for the selective utilization of this pathway in *E. coli*. This organism lacks arginase and cannot convert arginine to ornithine. When arginine is added to the growth medium, ornithine levels decline due to inhibition of arginine biosynthesis.

The α-Ketoadipate Pathway to Lysine

In yeasts and molds, and in a rare thermophilic species of bacteria, the biosynthetic pathway to lysine emanates from α-ketoglutarate (Fig. 15-5). The bacterial pathway to lysine, which is initiated by the condensation of pyruvate and aspartate-β-semialdehyde, is part of the aspartate family of amino acid biosynthetic routes and is shown in Figure 15-6. These completely divergent routes of lysine biosynthesis represent a major phylogenetic difference between bacteria and fungi. Only recently has any exception to this difference been demonstrated. So far, the only bacterium that has been shown to utilize the aminoadipate pathway to lysine is an extreme thermophile, *Thermus thermophilus*. This represents the first reported incidence where lysine is not synthesized via diaminopimelate in bacteria.

The series of reactions from homocitrate to α-ketoadipate are analogous to the reactions involved in the conversion of citrate to α-ketoglutarate in the citric acid cycle. The formation of homocitrate from acetyl-CoA and α-ketoglutarate is inhibited by lysine, indicating a feedback control mechanism for the pathway. The initial step in the pathway is also subject to repression by lysine.

Biosynthesis of the β-lactam antibiotics (penicillins and cephalosporins and their derivatives) occurs by a branch from the lysine pathway in *Penicillium chrysogenum*, *Cephalosporium acremonium*, *Streptomyces clavuligerus*, and related organisms. In virtually all organisms the pathway is initiated by the condensation of

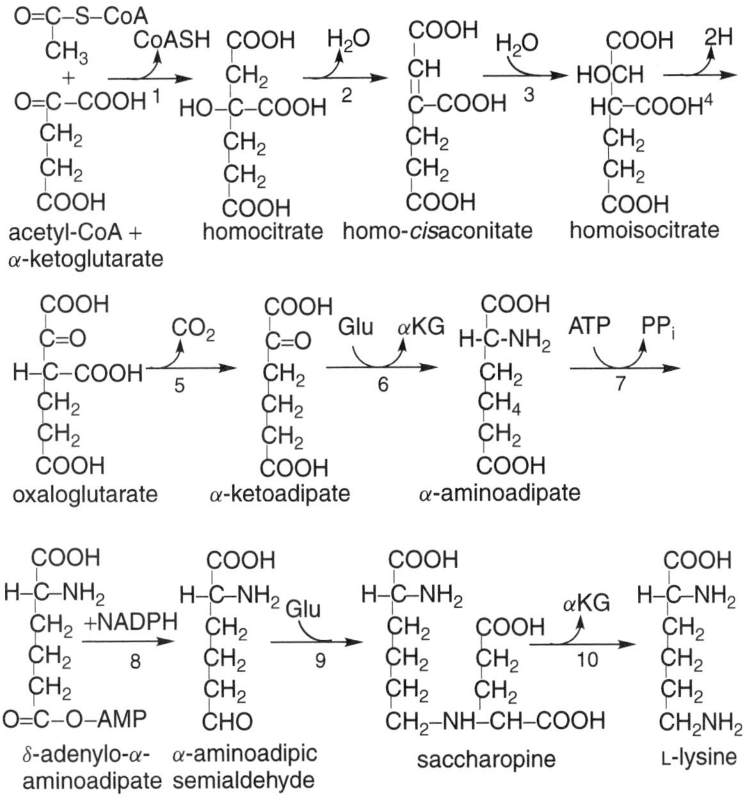

Fig. 15-5. Biosynthesis of lysine in yeasts and molds. The enzymes involved are: 1, homocitrate synthase; 2, homoaconitate dehydratase (*LYS7*); 3, homoaconitate hydratase (*LYS4*); 4 and 5, NAD^+-mediated homoisocitrate dehydrogenase; 6, α-aminoadipate transaminase; 7, α-aminoadipate semialdehyde dehydrogenase; 8, NADPH-mediated reduction of δ-adenylo-aminodipate to form aminoadipate semialdehyde; 9, saccharopine dehydrogenase ($NADP^+$, L-glutamate forming); 10, saccharopine dehydrogenase (NAD^+, lysine forming). The gene designations are for *S. cerevisiae*. To date only one bacterium, an extremely thermophilic organism, *Thermus thermophilus*, has been shown to follow this pathway.

L-α-aminoadipate, L-cysteine, and L-valine to form the tripeptide δ-(L-α-aminoadipyl)-L-cysteinyl-L-valine (ACV) mediated by the ATP-dependent ACV synthetase:

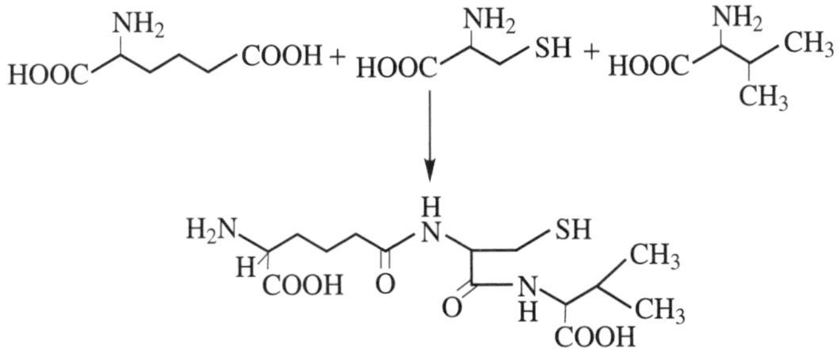

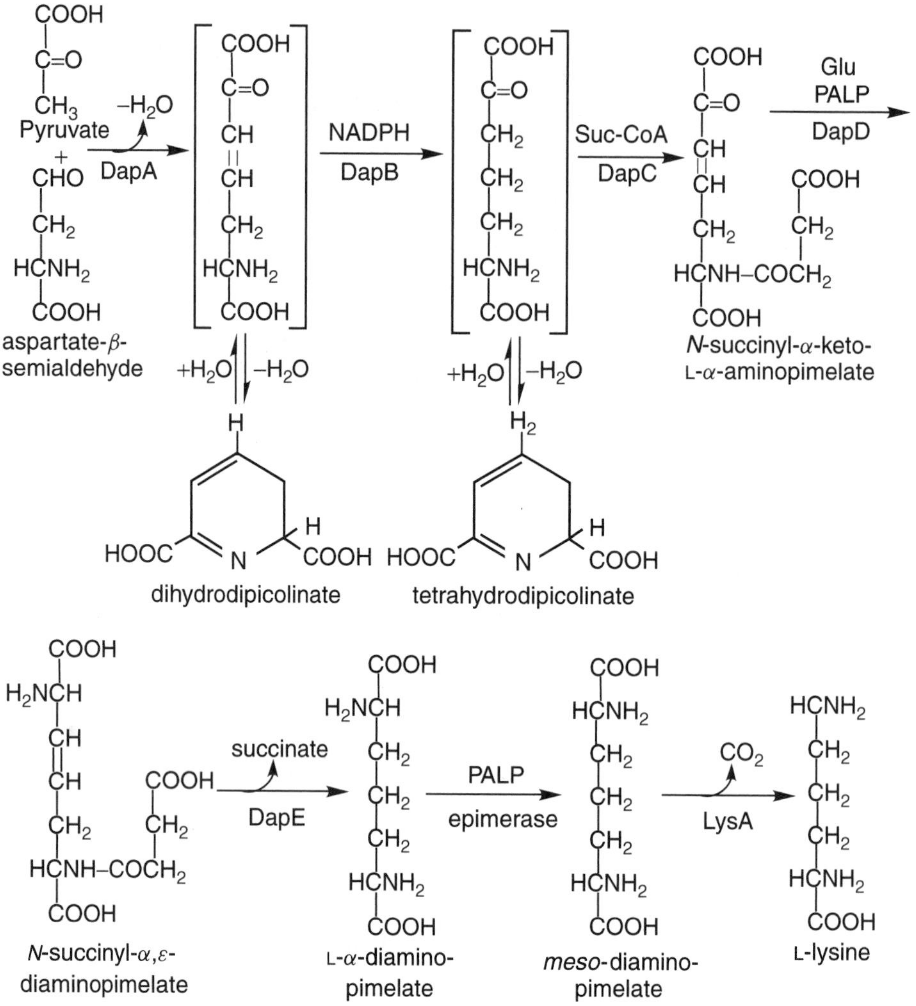

Fig. 15-6. Biosynthesis of lysine in bacteria. The enzyme designations are for *Escherichia coli.* DapA, dihydrodipicolinate synthase; DapB, dihydrodipicolinate reductase; DapC, tetra-hydro-dipicolinate succinylase; DapD, succinyl diaminopimelate aminotransferase; DapE, succinyl diaminopimelate desuccinylase; LysA, diaminopimelate decarboxylase; PALP, pyridoxal phosphate.

Subsequent ring closures form the β-lactam and thiazolidine ring systems to yield the basic penicillin double-ring structure:

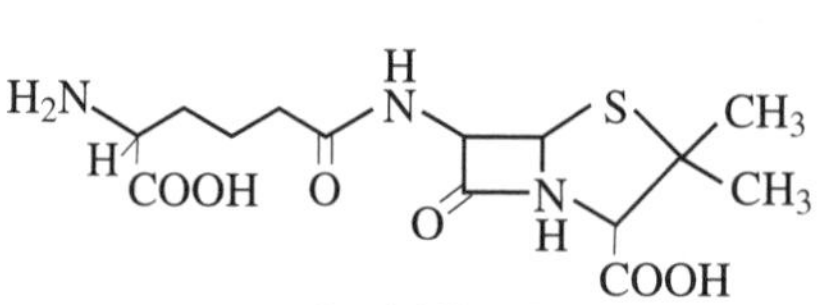

Penicillin N

Expansion of the ring to add an additional carbon atom gives rise to the cephalosporin series of derivatives.

THE ASPARTATE AND PYRUVATE FAMILIES

The aspartate and pyruvate families of amino acids are discussed together since there is a distinct overlap in the enzymes involved in the terminal steps of the biosynthesis of the branched-chain amino acids as shown in Figure 15-7. Valine and leucine carbon chains are derived from pyruvate. In bacteria, threonine, isoleucine, methionine, and lysine all emanate from aspartate. Fungi utilize similar pathways with the exception of lysine, which is synthesized via a completely different pathway as discussed previously.

Asparagine Synthesis

Asparagine is synthesized from aspartate, ammonia (or glutamine), and ATP by the enzyme asparagine synthetase. Two types of asparagine synthetase are known. One is

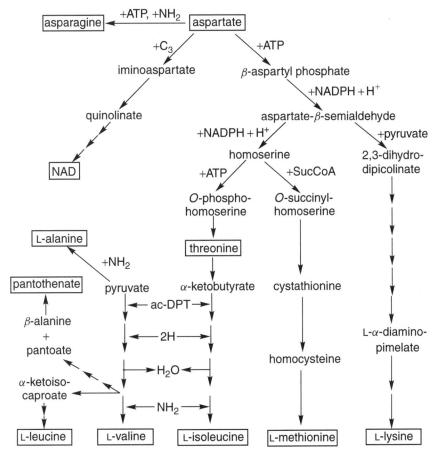

Fig. 15-7. Outline of the pyruvate and aspartate families of amino acids. Compounds enclosed in boxes represent final end products. SucCoA, succinyl-coenzyme A; ac-DPT, acetal diphosphothiamine.

AsnA, which can use only ammonia as the amino donor:

$$\text{L-aspartate} + NH_3 + ATP \longrightarrow \text{L-asparagine} + AMP + PP_i$$

The other is AsnB, which can use either glutamine or ammonia as the nitrogen donor, although glutamine is the preferred substrate:

$$\text{L-aspartate} + ATP + \text{glutamine} \longrightarrow \text{L-asparagine} + \text{L-glutamate} + AMP + PP_i$$

Both AsnA and AsnB are produced by *E. coli* and *K. aerogenes*, and activity of either enzyme is sufficient to provide an adequate supply of asparagine. A deficiency of both enzymes is required for asparagine auxotrophy. Eukaryotes apparently produce only the glutamine-dependent type of asparagine synthetase. *Bacillus subtilis* produces three genes coding for the AsnB-type asparagine synthetase (*asnB, asnH,* and *asnO* or *yisO*), but no gene coding for the ammonia-dependent AsnA-type enzyme has been found. The three genes are expressed differently during growth. In a medium favoring sporulation, expression of *asnB* was found only during exponential growth, that of *asnH* was elevated at the transition between exponential growth and stationary phase, and that of *asnO* occurred only during sporulation. Strains lacking *asnO* failed to sporulate, indicating a specific involvement of this gene in the sporulation process.

The Aspartate Pathway

In both bacteria and fungi, the enzymes aspartokinase and aspartate semialdehyde dehydrogenase initiate the aspartate pathway of amino acid biosynthesis (Fig. 15-8).

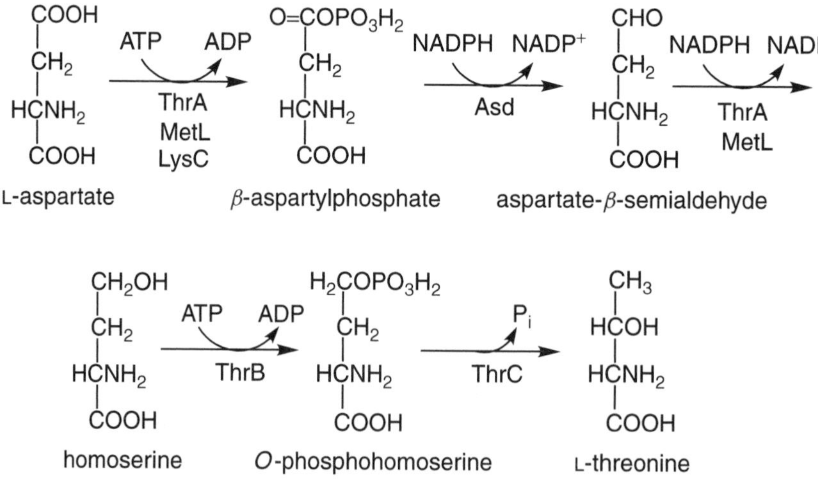

Fig. 15-8. Initial steps in the aspartate pathway of amino acid biosynthesis. The branch leading to threonine is shown here. Enzyme designations are for *E. coli*. ThrA, threonine-specific aspartokinase I–homoserine dehydrogenase I; MetL, methionine-specific aspartokinase II–homoserine dehydrogenase II; LysC, lysine-specific aspartokinase III; Asd, aspartate semialdehyde dehydrogenase; ThrB, homoserine kinase; ThrC, threonine synthetase.

In *E. coli* there are three aspartokinases, each of which is specific to the end product of the pathway involved in its synthesis. Aspartokinase I (ThrA) is specific for the threonine branch, aspartokinase II (MetL) is specific for the methionine branch, and aspartokinase III (LysC) is specific for the lysine branch. Aspartate semialdehyde dehydrogenase (Asd) forms aspartate-β-semialdehyde by removal of phosphate from aspartyl phosphate and reduction using NADPH. Aspartate β-semialdehyde stands at the first branch point in the multibranched pathway. Reduction of aspartate-β-semialdehyde to homoserine is catalyzed by homoserine dehydrogenase I (ThrA), which is specific for the threonine branch of the pathway, or homoserine dehydrogenase II (MetL), which is specific for the methionine branch. O-phosphohomoserine is converted to threonine by threonine synthetase (ThrC).

The Bacterial Pathway to Lysine

Condensation of aspartate β-semialdehyde with pyruvate yields dihydrodipicolinate, the first intermediate in the bacterial pathway to lysine (Fig. 15-6). This pathway is of special interest because dipicolinic acid is produced during sporulation in *Bacillus* species and either diaminopimelic acid or lysine is present in the peptidoglycan structures of all prokaryotes that produce a rigid cell wall. This pathway to lysine has been shown to occur in *E. coli*, *S. enterica*, *S. aureus*, *E. faecalis*, *B. subtilis*, and the cyanobacteria, and, with the exception of one highly thermophilic species mentioned earlier, seems to occur only in bacteria.

The biosynthetic pathway depicted in Figure 15-6 involves the formation of succinylated derivatives of α-keto-L-α-diaminopimelate and diaminopimelate as intermediates. *E. coli*, *S. enterica*, *S. aureus*, and a number of other bacteria utilize this pathway. Certain *Bacillus* species form acetylated derivatives exclusively. A few organisms (e.g., *B. sphaericus*, *C. glutamicum*) convert tetrahydrodipicolinate (piperideine-2,6-dicarboxylate) to D,L-diaminopimelate in a single step via diaminopimelate dehydrogenase. However, in *C. glutamicum* the pathway using succinylated intermediates appears to function along with the direct dehydrogenase pathway. This organism is of interest because it is used for the commercial production of lysine. The dehydrogenase pathway is apparently a prerequisite for handling increased flow of metabolites to D,L-diaminopimelate and lysine.

Because of the insolubility of iron at physiological pH, microorganisms have evolved a variety of systems (siderophores) to facilitate the acquisition of iron (see Chapter 7). *E. coli*, *Shigella*, and *Salmonella* synthesize an iron-chelating siderophore, aerobactin, from lysine. The iron-regulated aerobactin operon is found on a ColV-K30 plasmid. This operon consists of at least five genes for synthesis (*iuc*, iron uptake chelate) and transport (*iut*, iron uptake transport) of aerobactin. The biosynthetic pathway starts with oxidation of lysine to N^ε-hydroxylysine by N^ε-lysine monooxygenase (LucD). As shown in Figure 15-9, the hydroxyl derivative is acetylated by LucB (N-acetyltransferase) to form N^ε-acetyl-N^ε-hydroxylysine. Citrate addition to form aerobactin is catalyzed by aerobactin synthetase (LucC). Similar siderophores called rhizobactins are produced by rhizobia. In *Sinorhizobium meliloti* the process begins with the conversion of L-aspartic-β-semialdehyde to α-ketoglutarate, conversion of L-glutamate to L-2,4-diaminobutyric acid, and subsequent condensation with 1,3-diaminopropane and citrate to yield rhizobactin 1021, a structure comparable to aerobactin.

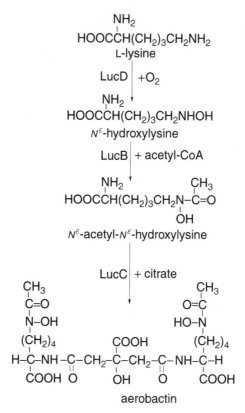

Fig. 15-9. Aerobactin synthesis from lysine. The enzyme designations for *E. coli* are: LucD, N^ε-lysine monooxygenase; LucB, N^ε-acetyltransferase; LucC, aerobactin synthase.

Threonine, Isoleucine, and Methionine Formation

Another branch point in the aspartate pathway occurs at homeserine. One branch leads to the formation of threonine (Fig. 15-8) and ultimately isoleucine. Homoserine can also be converted to methionine via several alternate routes. In *E. coli, S. enterica*, and other enteric organisms, conversion of homoserine to homocysteine involves either of two alternative pathways (Fig. 15-10). The enzyme homoserine succinyltransferase (MetA) condenses succinyl-CoA and homoserine to yield *O*-succinylhomoserine. Cystathionine γ-synthase is apparently capable of catalyzing the direct reaction of hydrogen sulfide (H₂S) or methylsulfide with *O*-succinylhomoserine to form homocysteine.

The methylation of homocysteine to form methionine may occur via either of two enzymes. One enzyme (MetH) is vitamin B_{12} dependent and requires NADH, FAD, SAM, and either 5-methyltetrahydrofolate or its triglutamyl derivative. A vitamin B_{12}-independent enzyme (MetE) requires only Mg^{2+} and 5-methyltetrahydropteroyltriglutamate. The 5-methyltetrahydrofolate is formed from serine and tetrahydrofolate (THF) through the action of serine hydroxymethyltransferase (GlyA), which converts THF to 5,10-methylene-THF. The enzyme 5,10-methylenetetrahydrofolate reductase (MetF) uses reduced FAD (FADH₂) to convert 5,10-methylene-THF to 5-methyl-THF.

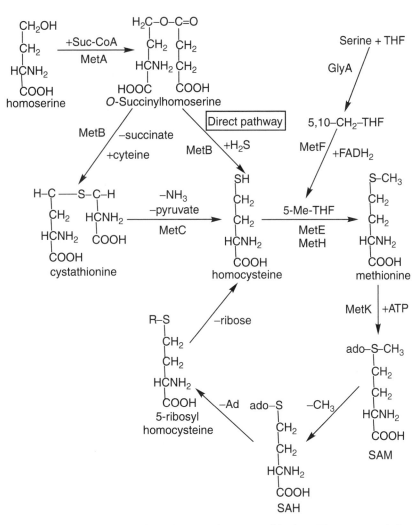

Fig. 15-10. Alternate pathways from homoserine to methionine. The enzyme designations are for *E. coli*. MetA, homoserine acyltransferase; MetB, cystathionine-γ-synthase; MetC, cystathionase; 5-Me-THF, 5-methyltetrahydrofolate; 5,10-CH$_2$-THF, 5,10-methylenetetrahydrofolate; MetF, 5,10-methylenetetrahydrofolate reductase; GlyA, serine hydroxymethyltransferase; MetE, tetrahydropteroyltriglutamate methyltransferase; MetH, vitamin B$_{12}$–dependent homocysteine-N^5-methylenetetrahydrofolate transmethylase (requires catalytic amounts of SAM and FADH); MetK, s-adenosylmethionine synthetase.

In *Neurospora*, the conversion of homoserine to homocysteine occurs via an analogous sequence of reactions in which homoserine and acetyl-CoA are condensed to form *O*-acetylhomoserine. The reaction is catalyzed by homoserine acetyltransferase. In this sequence, cystathionine γ-synthase exchanges cysteine for the acetyl group in *O*-acetylhomoserine to release acetate and form cystathionine. This pathway is also followed in *Aspergillus nidulans*.

An alternate route of homocysteine formation involving the direct sulfhydrylation of *O*-acetylhomoserine, *O*-succinylhomoserine, or *O*-phosphohomoserine is mediated by homocysteine synthase:

$$O\text{-acetylhomoserine} + H_2S \longrightarrow \text{L-homocysteine} + \text{acetate}$$

$$O\text{-succinylhomoserine} + H_2S \longrightarrow \text{L-homocysteine} + \text{succinate}$$

$$O\text{-phosphohomoserine} + H_2S \longrightarrow \text{L-homocysteine} + P_i$$

One or both of these activities has been reported in *N. crassa, S. cerevisiae*, and *E. coli*. The first of these reactions appears to be the main pathway in yeast; however, alternative routes for methionine biosynthesis are also available.

Isoleucine, Valine, and Leucine Biosynthesis

The isoleucine carbon skeleton is derived, in part, from aspartate via the deamination of threonine. It is convenient to discuss the biosynthesis of all three branched-chain amino acids at the same time because of the close interrelationship of the pathways. In the isoleucine-valine pathway, the same enzymes catalyze four of the steps in both sequences as shown in Figure 15-11. The immediate precursor of valine, α-ketoisovalerate, represents another branch point. Condensation with acetyl-CoA initiates the series of reactions leading to leucine synthesis, as shown in Figure 15-11. Via another series of reactions, α-ketoisovalerate is converted to pantoic acid, a precursor of pantothenic acid. The basic series of reactions leading to the formation of the branched-chain amino acids appears to be quite similar in most microorganisms. For example, the isopropylmalate pathway to leucine is widespread among diverse organisms capable of leucine biosynthesis.

Occasionally, organisms are found with novel routes of biosynthesis. In *Leptospira*, there is evidence that isoleucine carbon arises via a pathway involving citramalate as an intermediate. The origins of the carbon skeletons of the other amino acids are the same as in other prokaryotes.

Regulation of the Aspartate Family

Regulation of the biosynthesis of the amino acids of the aspartate family is complex because of the multiple branches and, to some extent, the interrelationships between the aspartate and pyruvate families. In *E. coli* K-12, primary regulation is exerted at two points. The major one is at the aspartokinase reaction, which regulates the flow of carbon to all of the amino acids formed (Fig. 15-12). The second site is the conversion of aspartate-β-semialdehyde to homoserine, a reaction catalyzed by homoserine dehydrogenases I and II. Aspartokinase I and homoserine dehydrogenase I activities are associated with a single multifunctional enzyme, ThrA, which is subject to allosteric inhibition by threonine. The synthesis of ThrA is repressed by a combination of threonine and isoleucine.

The structural genes for aspartokinase I–homoserine dehydrogenase I (*thrA*), homoserine kinase (*thrB*), and threonine synthase (*thrC*) lie in close proximity to one another on the *E. coli* map and constitute an operon. Aspartokinase II and homoserine dehydrogenase II activities are also catalyzed by a bifunctional protein (MetL).

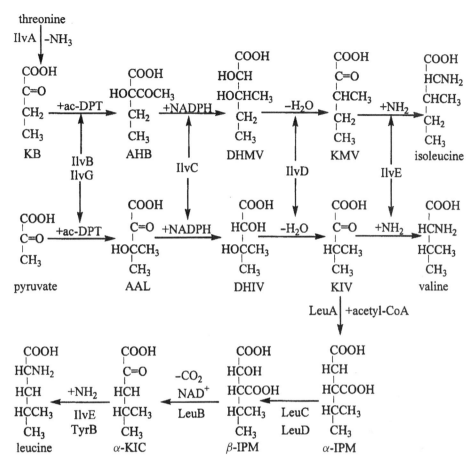

Fig. 15-11. Biosynthesis of the branched-chain amino acids, isoleucine, valine, and leucine. Enzyme designations are for *E. coli*. IlvA, threonine deaminase; IlvB, valine-sensitive acetohydroxy acid synthase I; IlvG, acetohydroxy acid synthase II; IlvC, acetohydroxy acid isomero-reductase; LeuA, α-isopropylmalate synthase; LeuC, LeuD, isopropylmalate dehydratase; LeuB, β-isopropylmalate dehydrogenase; TyrB, aromatic amino acid aminotransferase. KB, α-keto-butyrate; AHB, α-aceto-α-hydroxybutyrate; DHMV, α,β-dihydroxy β-methylisovalerate; KMV, α-keto-β-methylvalerate; α-IPM, α-isopropylmalate; β-IPM, β-isopropylmalate; α-KIC, α-keto-isocaproate.

Transcription of *metL* is regulated by methionine. However, neither of the two activities is feedback inhibited by methionine, threonine, SAM, or by combinations of these compounds. Aspartokinase III (LysC) is regulated by lysine through both feedback and repression mechanisms. Comparable regulatory systems for the aspartate pathway are present in other members of the *Enterobacteriaceae*, particularly *Salmonella, Enterobacter, Edwardsiella, Serratia*, and *Proteus*.

In *E. coli*, feedback inhibition plays a prominent role in regulation of the lysine pathway. A constitutive mutant for the synthesis of the lysine-sensitive aspartokinase

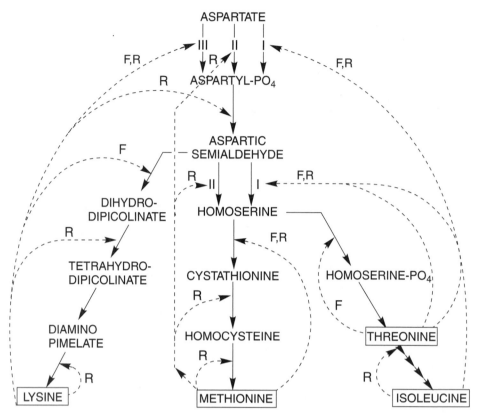

Fig. 15-12. End-product regulation of the aspartate family of amino acids in *E. coli.*
F, feedback inhibition; R, repression of enzyme synthesis. Aspartokinase I–homoserine
dehydrogenase I is a bifunctional enzyme encoded by *thrA*. Aspartokinase II–homoserine
dehydrogenase II is a bifunctional enzyme encoded by *metL*. Aspartokinase III is encoded
by *lysC*.

III (*lysC*) has been isolated. The aspartokinase III protein may serve as an inducer
of diaminopimelate decarboxylase synthesis and thus regulate lysine synthesis in this
manner. In some organisms, **concerted feedback** or **multivalent inhibition** is a more
common type of regulation of the enzymes of the aspartate family. In this type of
inhibition, all of the products of the pathway act together to effect the inhibition.
Single aspartokinase enzymes subject to inhibition by lysine and threonine have been
reported.

THE SERINE-GLYCINE FAMILY

In bacterial, fungal, and mammalian systems, metabolic relationships between glycine
and serine may be described by the general scheme shown in Figure 15-13. Serine is
a precursor of L-cysteine (Fig. 15-14). Glycine contributes carbon and nitrogen in the
biosynthesis of purines, porphyrins, and other metabolites. As shown in Figure 15-13,
serine hydroxymethyltransferase (SHMT), the *glyA* gene product, converts serine to

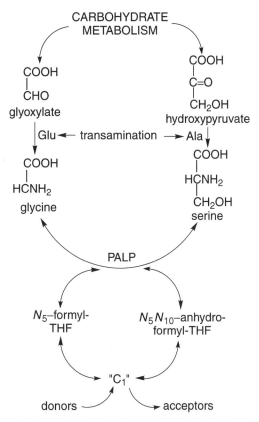

Fig. 15-13. Glycine-serine interrelationships. THF, tetrahydrofolate; PALP, pyridoxal phosphate.

glycine and 5,10 methylenetetrahydrofolate, a major contributor of one-carbon units, to the formation of methionine, purines, and thymine. Oxidative cleavage of glycine by the glycine cleavage (GCV) enzyme system provides a second source of one-carbon units:

$$CH_2NH_2COOH + THF \longrightarrow NH_3 + CO_2 + 5,10\text{-methylenetetrahydrofolate}$$

The GCV system has been described in several bacterial species including *E. coli, Peptococcus glycinophilus*, and *Arthrobacter globiformis*. It consists of four proteins: the P protein, a pyridoxal phosphate enzyme; the H protein, a lipoate-containing hydrogen carrier; the T protein that transfers the methylene carbon to THF; and the L-protein, a lipoamide dehydrogenase encoded by *lpd* that is common to the pyruvate and α-ketoglutarate dehydrogenase complexes. The genes encoding these enzymes form an operon that maps at 65.2 minutes on the *E. coli* chromosome. The regulatory proteins Lrp, GcvA, and PurR are involved in the regulation of the glycine-inducible GCV system.

Serine, glycine, and cysteine are synthesized via a pathway emanating from 3-phosphoglycerate and proceeding through a series of phosphorylated intermediates (Fig. 15-14). Serine may be derived from glycine formed from glyoxylate by

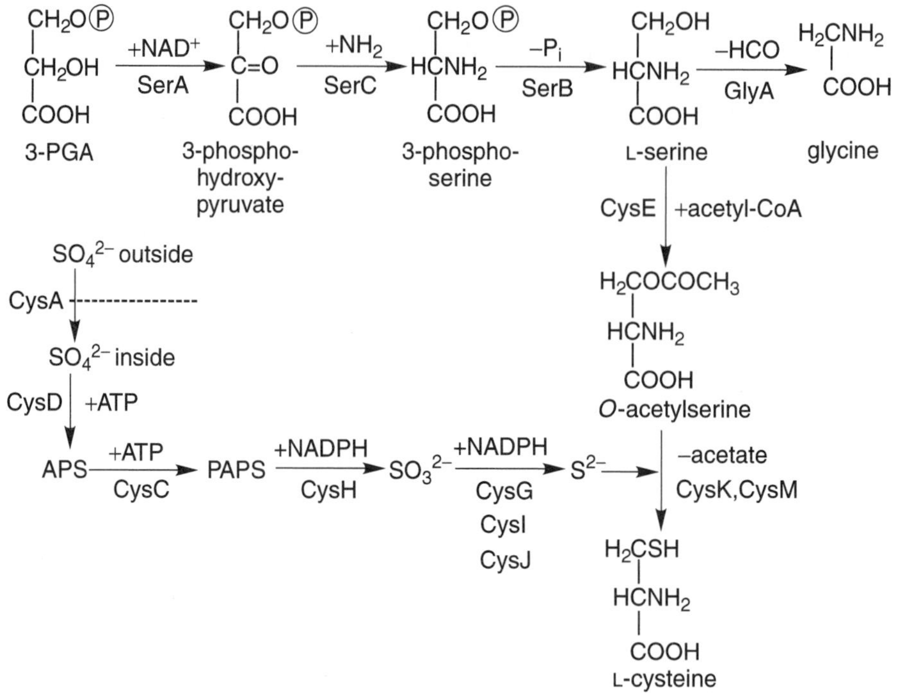

Fig. 15-14. Biosynthesis of serine, glycine, and cysteine. The enzyme designations for *E. coli* are: SerA, 3-phosphoglycerate (3-PGA) dehydrogenase; SerC, phosphoserine aminotransferase; SerB, phosphoserine phosphatase; GlyA, serine hydroxymethyltransferase; CysA, sulfate permease; CysD, sulfate adenylyltransferase; CysC, adenylylsulfate (APS) kinase; CysH, 3′-phosphoadenylyl sulfate (PAPS) reductase; CysG, CysI, CysJ, sulfite reductase; CysE, serine acetyl-transferase; CysK, acetylserine sulfhydrylase A; CysM, acetylserine sulfhydrylase B. The *cys* operon is regulated by *cysB*.

transamination. Deamination of threonine to α-ketobutyrate with cleavage to acetyl-CoA and glycine can also give rise to serine.

Bakers yeast possesses the phosphorylated pathway to serine from 3-phosphoglycerate as well as the glyoxylate pathway. Regulation of the phosphorylated pathway by serine feedback inhibition of 3-phosphoglycerate dehydrogenase and regulation of glyoxylate transaminase has been demonstrated in *S. cerevisiae*. In bacteria the phosphorylated pathway appears to be the main route of serine formation. The bacterial pathway is also regulated by feedback inhibition of 3-phosphoglycerate dehydrogenase.

Sulfur is a constituent of several important biological components including cysteine, methionine, thiamine, biotin, lipoic acid, and coenzyme A. Sulfur usually occurs as sulfate and is transported into the cell via a sulfate permease. In most bacteria, sulfate is incorporated into organic compounds via the formation of adenylylsulfate (APS) and phosphoadenylyl sulfate (PAPS) with reduction to sulfide and incorporation into L-cysteine. Cysteine provides the sulfur atom for other sulfur-containing compounds, including methionine. The converging pathways of sulfate reduction and the formation of *O*-acetylserine shown in Figure 15-14 appears to be the most common route for the synthesis of cysteine. However, cysteine may also arise from serine and hydrogen sulfide by the action of serine sulfhydrase (cysteine synthase). Cysteine is also formed by transsulfuration between homocysteine and serine (Fig. 15-10). In *K. aerogenes* the reduced sulfur of methionine can be recycled into cysteine. Under conditions where methionine serves as the sole source of sulfur, a pathway involving SAM forms homocysteine (Fig. 15-10). The yeast, *S. cerevisiae*, and some bacteria assimilate sulfide into homoserine to form homocysteine (Fig. 15-10). The homocysteine can then be converted to methionine.

Cysteine biosynthesis is regulated by the genes that code for the enzymes in the pathways of sulfate reduction and the formation of *O*-acetylserine from serine (Fig. 15-14). *O*-Acetylserine induces sulfate permease activity and also induces the rest of the enzymes in the sulfate reduction sequence. In *S. enterica* each of the enzymes in the sequence is repressed by cysteine. Sulfate permease (CysA) and serine acetyltransferase (CysE) are also inhibited by cysteine via a feedback mechanism. Although the biosyntheses of cysteine and methionine are obviously interrelated, it is the sulfate transport system and the pathway involved in the activation of sulfur that govern the synthesis of cysteine through positive control mechanisms, whereas methionine biosynthesis is under negative control. However, it has been proposed that the *O*-acetyl derivatives play a parallel role in the regulation of the synthesis of methionine and cysteine by inducing the initial steps in the pathway. Three of the genes controlling the pathway of sulfate activation (*cysIJH*) are arranged in a cluster forming an operon. Other regulatory interrelationships also exist between the cysteine and methionine pathways.

Aminolevulinate and the Pathway to Tetrapyrroles

As discussed earlier, aminolevulinate (ALA) can be synthesized by condensation of glycine and succinyl-CoA (C$_4$ pathway) or from glutamate (C$_5$ pathway). In either case, ALA is the first committed step in the formation of tetrapyrroles, as shown in Figure 15-15. The pathway diverges at uroporphyrinogen III (UroIII), one branch leading to the synthesis of siroheme and vitamin B$_{12}$ (cobalamin) and the other to the formation of protoporphyrin IX. At this point, the pathway branches again, one branch leading to heme and the other to bacteriochlorophyll.

THE AROMATIC AMINO ACID PATHWAY

Phenylalanine, Tyrosine, and Tryptophan

Demonstration of shikimate, anthranilate, and indole as precursors of tryptophan and other aromatic amino acids in fungi and bacteria provided the earliest clues as to the intermediates in the aromatic amino acid pathway. Inhibition by analogs of the aromatic amino acids, isolation of mutants with multiple requirements for aromatic compounds, and other nutritional and biochemical findings indicated that these compounds shared a common origin. Shikimate substituted for the aromatic amino acid requirements of many auxotrophs, indicating that it was the common precursor of at least five different aromatic compounds. As a result of these findings, shikimate was considered to be the major branch point compound in the aromatic pathway. However, in both bacteria and fungi it was ultimately shown that chorismate (from the Greek meaning "to branch") serves in this capacity (Fig. 15-16).

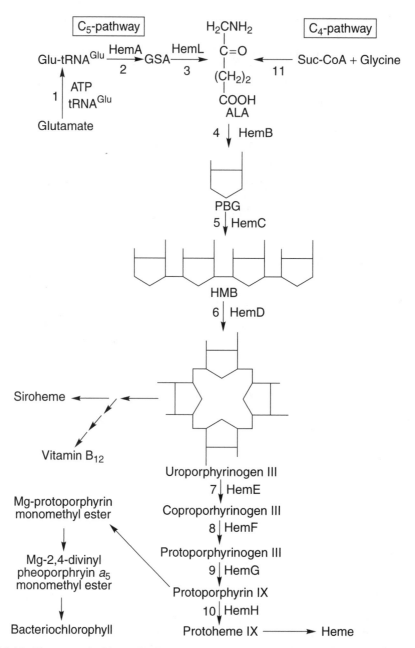

Fig. 15-15. Tetrapyrrole biosynthesis. The enzymes in the pathway are (1) glutamyl-tRNA synthase, (2) NAD(P)H glutamyl-tRNA reductase, (3) GSA (glutamate-γ-semialdehyde) 2,1-aminotransferase, (4) PBG (porphobilinogen) synthase, (5) HMB (hydroxymethylbilane) synthase, (6) UroIII (uroporphyrinogen) synthase, (7) uroporphyrinogen III decarboxylase, (8) coproporphyrinogen III oxidase, (9) protoporphyrinogen IX oxidase, (10) ferrochetolase, (11) ALA synthase.

The Common Aromatic Amino Acid Pathway

The aromatic amino acids phenylalanine, tyrosine, tryptophan, and several other related aromatic compounds are produced via a common pathway that begins with the condensation of erythrose-4-phosphate (E-4-P) and phosphoenolpyruvate (PEP) to form 3-deoxy-D-arabino-heptulosonate 7-phosphate (DAHP). The enzymes participating in the common aromatic pathway and its branches to tyrosine and phenylalanine are shown in Figure 15-16. DAHP is converted to shikimate and then to chorismate.

Chorismate stands at a branch point leading to the formation of several other amino acids and related metabolites. The formation of many additional aromatic compounds via branches of this pathway emphasizes the need for a complex regulatory system. Furthermore, a mutant blocked in the common pathway prior to a branch point may exhibit multiple nutritional requirements. Although there are many similarities in the aromatic amino acid pathway in all bacteria and fungi, there are several differences in detail regarding the manner in which tryptophan, tyrosine, and phenylalanine are formed. Even more apparent are the number of variations in the ways in which other related compounds are formed through branches or extensions from the common pathway.

In *E. coli* and *S. enterica*, regulation of the common aromatic pathway is modulated through three unlinked genes, *aroF*, *aroG*, and *aroH*, which encode isozymes of the first enzyme, DAHP synthetase, sensitive to tyrosine, phenylalanine, and tryptophan, respectively. Although all three DAHP synthetases are regulated transcriptionally, feedback inhibition is quantitatively the major control mechanism in vivo. In *S. enterica* the structural gene for the phenylalanine-inhibitable DAHP synthetase (*aroH*) is linked to *gal*. The structural gene for the tryptophan-associated isozyme (*aroF*) is linked to *aroE*, whereas that for the tyrosine-inhibitable isozyme (*aroG*) is linked to *pheA* and *tyrA*. The *pheA* and *tyrA* gene products are the respective branch-point enzymes leading to the phenylalanine and tyrosine pathways. Only the tyrosine-related isozyme (*aroG*) is completely repressed, as well as feedback inhibited, by low levels of tryrosine.

In *E. coli* there is a similar pattern of isozymic control of DAHP synthetase and control of the genes in the aromatic amino acid pathway. The Tyr repressor, the tyrR gene product, mediated by tyrosine and phenylalanine, respectively, represses expression of aroF. The TyrR protein also regulates the expression of several other genes concerned with aromatic amino acid synthesis or transport. These include *aroL*, *aroP*, *tyrB*, *tyrP*, and *mtr* (resistance to the tryptophan analog 5-methyltryptophan). Expression of *aroH* is controlled by the Trp repressor (encoded by *trpR*), which also regulates the expression of the *trp* operon, the *mir* gene, as well as the expression of *trpR* itself.

In a wide variety of fungi, the genes coding for enzymes 2 to 6 in the prechorismate pathway (Fig. 15-16) are arranged in a cluster, designated the *arom* gene cluster. In most fungi these enzymes sediment in a complex aggregate on centrifugation in sucrose density gradients. The sedimentation coefficients for them are very similar in representative examples of Basidiomycetes (*Coprinus lagopus, Ustilago maydis*), Ascomycetes (*N. crassa, S. cerevisiae, A. nidulans*), and Phycomycetes (*Rhizopus stolonifer, Phycomyces nitens, Absidia glauca*). By comparison, the five enzymes catalyzing the conversion of DAHP to chorismate in bacteria are physically separable in *E. coli, S. enterica, E. aerogenes, Anabaena variabilis*, and *Chlamydomonas reinhardtii*. In *Euglena gracilis* an enzyme aggregate with five activities has been found. This aggregate can be disassociated into smaller components.

In some bacteria, the genes coding for enzymes of the chorismate pathway appear to be scattered on the chromosome. However, in *B. subtilis*, a cluster of contiguous genes for tryptophan biosynthesis represents a polycistronic operon. The *trpEDCFBA* genes form an operon similar to operons found in enteric bacteria. However, in *B. subtilis*, the *trp* genes are part of a **supraoperon** containing *trpEDCFBA-hisH-tyrA-aroE*, as shown in Figure 15-17a. The *aroFBH* genes may also be present at the 5′ end of this supraoperon. All *aroF* mutants (chorismate synthetase) also lack dehydroquinate synthase (AroB) activity. The gene that specifies AroB is closely linked to *aroF*. Both genes are part of the *aro* gene cluster. Mutants lacking chorismate mutase activity also lack DAHP synthetase and shikimate kinase activity, presumably as a result of their aggregation in a multienzyme complex. As an indication of the complexity of the supraoperon, the *mtrAB* operon of *B. subtilis* encodes GTP cyclohydrolase I (MtrA), an enzyme involved in folate biosynthesis. The *mtrB* gene is a *trans*-acting RNA-binding regulatory protein activated by tryptophan. Transcription termination at the attenuator preceding the *trp* gene cluster presumably occurs as a consequence of binding of the activated MtrB protein to the nascent transcript (see later discussion on interpathway regulation in Chapter 16).

The genes controlling tryptophan synthesis in enteric bacteria are also arranged in an operon. The *E. coli trp* operon and the enzymes under its control are shown in Figure 15-17b. There are five structural genes for the enzymes in the operon and they are induced coordinately in response to the availability of tryptophan. In *E. coli* the activities of the first two enzymes are catalyzed by an aggregate formed from the products of the first two genes in the operon. A single protein catalyzes steps 3 and 4. Tryptophan synthase converts indoleglycerol phosphate to indol and then couples indol to serine to form tryptophan. The enzyme is a complex formed from the products of the last two genes (*trpA* and *tryB*). A regulatory site preceding the structural genes regulates the synthesis of mRNA^{Trp} and the enzymes of the tryptophan pathway, thus reducing operon expression. This **attenuator** function apparently occurs at the level of transcription by providing a region in which transcription is terminated (see Chapter 5 for additional details concerning regulation of gene expression).

Pathways to Tyrosine and Phenylalanine

The biosynthesis of L-tyrosine and L-phenylalanine from chorismate can occur via alternate routes in many organisms (Fig. 15-18). In *E. coli*, TyrA (chorismate mutase

---

Fig. 15-16. Aromatic amino acid pathways. The enzyme designations are for *E. coli*. AroF, AroG, AroH, 3-hydroxy-L-arabino-heptulosonate 7-phosphate (DHAP) synthetases; AroB, 3-dehydroquinate synthase; AroD, 3-dehydroquinate dehydratase; AroE, dehydroshikimate reductase; AroL, shikimate kinase II; AroA, 5-enolpyruvylshikimate 5-phosphate synthase; AroC, chorismate synthase; PheA, bifunctional enzyme, chorismate mutase P-prephenate dehydrogenase; TyrA, bifunctional enzyme, chorismate mutase T-prephenate dehydrogenase; TyrB, tyrosine aminotransferase; TrpE, anthranilate synthase; TrpD, anthranilate phosphoribosyl transferase; trpC, bifunctional enzyme, $N-$(5-phosphoribosyl) anthranilate isomerase-indole-3-glycerophosphate synthetase; TrpA,TrpB, tryptophan synthase; Tna, tryptophanase. PEP, phosphoenolpyruvate; E-4-P, erythrose-4-phosphate; EPPK, 3-enoyl-pyruvyl-3-phosphoshikimate; CDRP, 1-(O-carboxy-phenylamino)-1-deoxyribulose-5-phosphate; 4-HPP, 4-hydroxyphenylpyruvate.

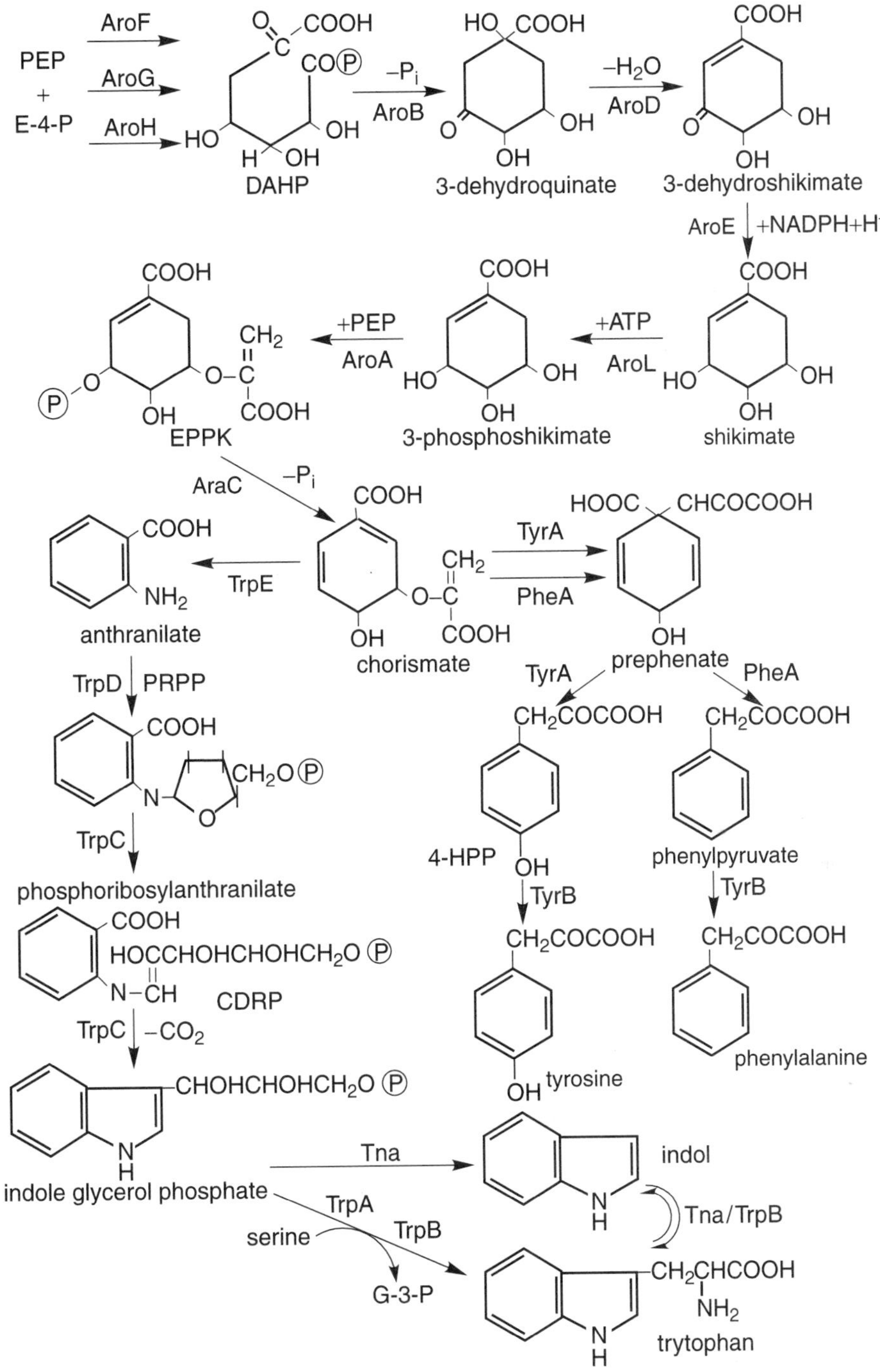

AroF
AroG
AroH
PEP
+
E-4-P
DAHP
$-P_i$
AroB
3-dehydroquinate
$-H_2O$
AroD
3-dehydroshikimate
AroE
$+NADPH+H^+$
shikimate
+ATP
AroL
3-phosphoshikimate
+PEP
AroA
EPPK
AraC
$-P_i$
chorismate
TrpE
anthranilate
TyrA
PheA
prephenate
TrpD
PRPP
TrpC
phosphoribosylanthranilate
CDRP
TrpC
$-CO_2$
4-HPP
phenylpyruvate
TyrB
tyrosine
phenylalanine
indole glycerol phosphate
Tna
indol
Tna/TrpB
TrpA
TrpB
serine
G-3-P
trytophan

Bacillus subtilis

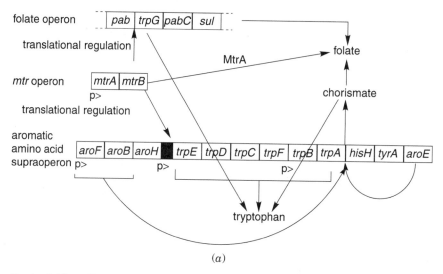

(a)

Escherichia coli

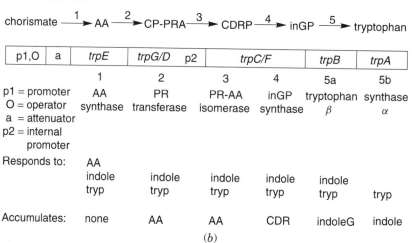

(b)

Fig. 15-17. Organization of the genes for tryptophan biosynthesis. (A) The aromatic amino acid supraoperon of *B. subtilis* showing interrelationships with the genes from the folate operon and the *mtr* operon. The MtrA protein (GTP cyclohydrolase I) catalyzes formation of the initial pteridine ring (dihydroneopterin triphosphates) of folate. The proteins AroF, AroB, and AroE are required for chorismate synthesis. The proteins Pab, TrpG, and PabC are involved in *p*-aminobenzoate PAB synthesis from chorismate. Sul catalyzes condensation of PAB and the pteridine ring. MtrB is considered to bind to a target sequence that overlaps the *trpG* ribosome-binding site, resulting in transcriptional regulation of the *trp* gene cluster. The *trpEDCFBA* and *trpG* gene products are necessary for tryptophan biosynthesis from chorismate in some strains of *Bacillus*. p> indicates positions of known promoters; ■ denotes position of the *trp* attenuator; = = = indicates open reading frames upstream of *pab* and downstream of *sul*. (*Source*: Redrawn from Babitzke, P. et al., *J. Bacteriol.* **174**:2059-2064, 1992). (B). Organization of the genes of the tryptophan operon of *E. coli*. AA, anthranilate; CP-PRA, *N*−(o-carboxyphenyl)-phosphoribosylamine; CDRP, 1-(o-carboxyphenylamino)-1-deoxyribose 5-phosphate; inGP, indoleglycerolphosphate.

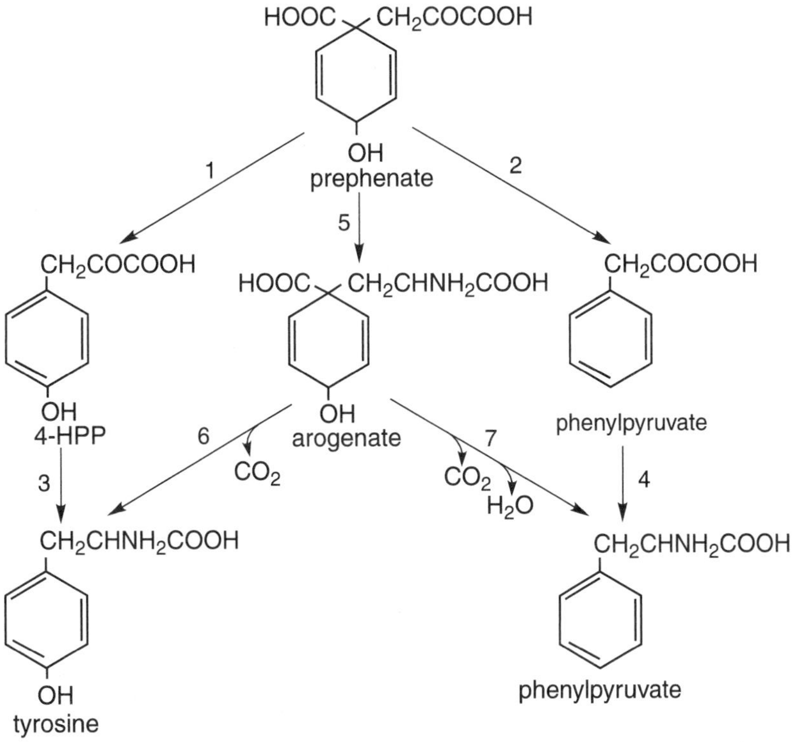

Fig. 15-18. Alternate pathways to phenylalanine and tyrosine. The pathways shown are present in *Pseudomonas aeruginosa*. Various organisms may express any combination of these pathways. The dehydrogenases may be either NAD^+ or $NADP^+$ linked. The enzymes are (1) prephenate dehydrogenase, (2) prephenate dehydratase, (3) 4-hydroxyphenylpyruvate (4-HPP) aminotransferase, (4) phenylpyruvate aminotransferase, (5) prephenate aminotransferase, (6) arogenate dehydratase, and (7) arogenate dehydrogenase.

T-prephenate dehydrogenase) is a bifunctional enzyme that can convert chorismate to prephenate and then to phenylpyruvate. Alternatively, prephenate may be converted to arogenate by transamination, with subsequent conversion of arogenate to either tyrosine (by arogenate dehydrogenase) or phenylalanine (by arogenate dehydratase). Utilization of these alternate pathways varies greatly from one organism to another. In *E. coli, S. enterica, B. subtilis*, and other common bacteria, the prephenate pathway to phenylpyruvate and phenylalanine and to 4-hydroxyphenylpyruvate and tyrosine is the predominant route as shown in Figure 15-15.

By comparison, *Euglena gracilis* uses arogenate as the sole precursor of both tyrosine and phenylalanine via a pathway in which arogenate rather than prephenate is the branch point. The arogenate pathway is common in the cyanobacteria. However, the degree to which it is used and the manner in which the alternative pathways are regulated differ widely from one species to another. In *P. aeruginosa* and other pseudomonads, these alternate pathways coexist. Their presence accounts for the unusual resistance of these organisms to analogs of the aromatic amino acids. Under certain circumstances, the arogenate pathway appears to function as an unregulated overflow route for tyrosine production. The arogenate pathway is a obligatory route to tyrosine in *Corynebacterium glutamicum, Brevibacterium flavum,* and *B. ammoniogenes.* By comparison, arogenate appears to be a deadend metabolite in *N. crassa,* the prephenate pathway being used as the major route to tyrosine and phenylalanine.

Salmonella enterica and *E. coli* synthesize tyrosine and phenylalanine by virtually separate pathways in that there are separate DAHP synthetases and chorismate mutases coded for by genes specific to either the tyrosine or phenylalanine pathways. The chorismate mutase T and prephenate dehydrogenase coded by *tyrA* are specific for tyrosine formation while chorismate mutase P and prephenate dehydratase coded by *pheA* are specific for phenylalanine biosynthesis. In *S. enterica* the genes *aroF* and *tyrA,* which specify the tyrosine-repressible DAHP synthase (AroF) and chorismate mutase T-prephenate dehydrogenase, comprise an operon. The genes *aroG* and *pheA,* which specify phenylalanine-repressible DAHP synthase (AroG) and chorismate mutase P-prephenate dehydratase (PheA), regulate the synthesis of phenylalanine as shown in Figure 15-19. These bifunctional enzymes are found in *S. enterica, E. coli,* and *E. aerogenes.* They are not found as bifunctional enzymes in *N. crassa, S. cerevisiae,* or *B. subtilis.*

Within the genera *Pseudomonas, Xanthomonas,* and *Alcaligenes,* variation in the enzymes of tyrosine biosynthesis comprises five groups that compare favorably with rRNA–DNA hybridization groups. The rRNA homology groups I, IV, and V all lack activity for arogenate/NADP dehydrogenase. Group II species possess arogenate dehydrogenase (with lack of specificity for NAD or NADP) and are sensitive to feedback inhibition by tyrosine. The arogenate dehydrogenase of species in group III is insensitive to feedback inhibition. Group IV displays prephenate/NADP dehydrogenase activity, whereas groups I and V lack this activity.

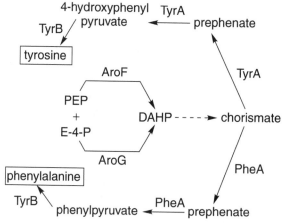

Fig. 15-19. Genes and enzymes of phenylalanine and tyrosine biosynthesis in *Escherichia coli* and *Salmonella enterica*. PEP, phosphoenolpyruvate; E-4-P, erythrose-4-phosphate; DAHP, 3-deoxy-D-arabinoheptulonate 7-phosphate. Enzyme designations are: AroF, tyrosine-repressible DAHP synthase; AroG, phenylalanine-repressible DAHP synthase; TyrA, chorismate mutase T-prephenate dehydrogenase; PheA, chorismate mutase P-prephenate dehydratase; TyrB, tyrosine aminotransferase.

Mammalian species appear to have the ability to convert phenylalanine to tyrosine through the action of phenylalanine hydroxylase:

$$\text{phenylalanine} + 0.5O_2 \longrightarrow \text{tyrosine} + H_2O$$

In microorganisms this activity is uncommon. Dual requirements for tyrosine and phenylalanine generally accompany mutational loss of enzyme activity at a point prior to the branch point in the synthesis of these amino acids. *Pseudomonas* forms an inducible phenylalanine hydroxylase. Since hydroxylation reactions often represent the initiating step in the degradation of aromatic compounds, as discussed in Chapter 10, it seems plausible that a highly degradative organism, such as *Pseudomonas*, would possess such activity for degradative purposes and only secondarily to provide tyrosine from phenylalanine.

p-Aminobenzoate and Folate Biosynthesis

Many bacteria and fungi, including *E. coli, Streptomyces griseus*, and *Pseudomonas acidovorans*, convert chorismate to *p*-aminobenzoate (PAB) via a two-step process (Fig. 15-20). In the first step, chorismate is converted to 4-amino-4-deoxychorismate (ADC) by ADC synthase (PabA, PabB) in the presence of glutamine. In the absence of PabA and glutamine, PabB can convert chorismate to ADC if an excess of ammonia is provided. ADC is converted to PAB by removal of the pyruvyl group and concomitant aromatization of the ring by a PALP-containing enzyme ADC lyase (PabC).

PAB is then condensed with 6-hydroxymethyl-7,8-pterin pyrophosphate (DHPS) by dihydropteroate synthase (SulA). The resultant dihydropteroate is then converted to 7,8-dihydrofolate by the addition of glutamate by dihydrofolate synthetase (FolC). After 7,8-dihydrofolate is reduced to tetrahydrofolate by dihydrofolate reductase, the synthase enzyme then serves to add additional glutamate units to form folylpoly-glutamate.

The pteridine component of folate is derived from guanosine triphosphate (GTP). GTP cyclohydrolase I (FolE; in *B. subtilis* this enzyme (mtrA) is the product of the *mtrA* gene) catalyzes the first committed step in the biosynthesis of the pterin ring of folate. In this reaction the C-8 carbon of the GTP is removed as formate and the remaining enzyme-bound intermediate is converted to 7.8-dihydro-D-*erythro*-neopterin (H$_2$-neopterin) 3'-phosphate. Removal of the phosphate residues yields 7,8-dihydroneopterin. 7,8-Dihydroneopterin is then converted to 6-hydroxymethyl-7,8-dihydropterin by the action of dihydrohydroxylmethylpterin pyrophosphokinase (FolK in *E. coli*; SulD in *Streptococcus pneumoniae*). In *S. pneumoniae*, SulD is a bifunctional enzyme with both 6-hydroxymethyl-7,8-dihydropterin kinase and 7,8-dihydroneopterin aldolase activities.

In the archaea *Methanococcus thermophila* and *Methanobacterium thermoautotrophicum*, ΔH GTP cyclohydrolase I is absent and GTP is transformed via a multistep process into H-neopterin 2' : 3'-cyclic phosphate. This compound is then hydrolyzed to H-pterin via a phosphorylated intermediate. Formation of the pterin ring is catalyzed by two or more enzymes rather than the GTP cyclohydrolase I enzyme present in eukaryotes and bacteria.

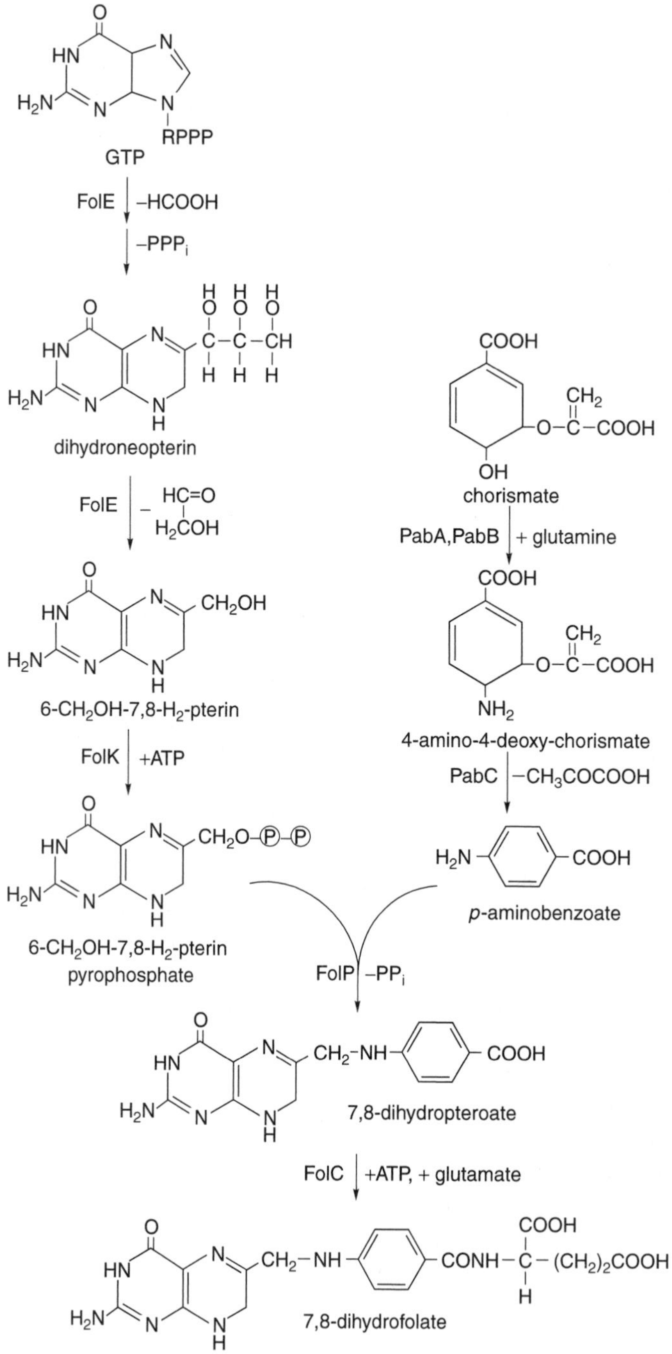

Fig. 15-20. PABA and folate biosynthesis. The enzyme designations for *Escherichia coli* are: PabA/PabB, 4-amino-4-deoxychorismate (ADC) synthase; PabC, ADC lyase; FolE, GTP cyclohydrolase; FolK, dihydro-hydroxymethylpterin pyrophosphokinase; FolP, dihydropteroate synthetase; FolC, dihydrofolate:folylpolyglutamate synthetase. [Pterin is the trivial name for 2-amino-4-hydroxypteridine; neopterin is the trivial name for 6(D-*erythro*-1′,2′,3′-trihydroxypropyl) pterin.]

Enterobactin Biosynthesis

The catechol siderophore enterobactin (enterochelin) is an iron-chelating compound that facilitates the transport of iron (see Chapter 7). It is a cyclic trimer of *N*-2,3-dihydroxybenzoylserine, which is synthesized from chorismate. The first stage of this pathway is the formation of 2,3-dihydroxybenzoate via the intermediates isochorismate and 2,3-dihydro-2,3-dihydrobenzoate as shown in Figure 15-21. The

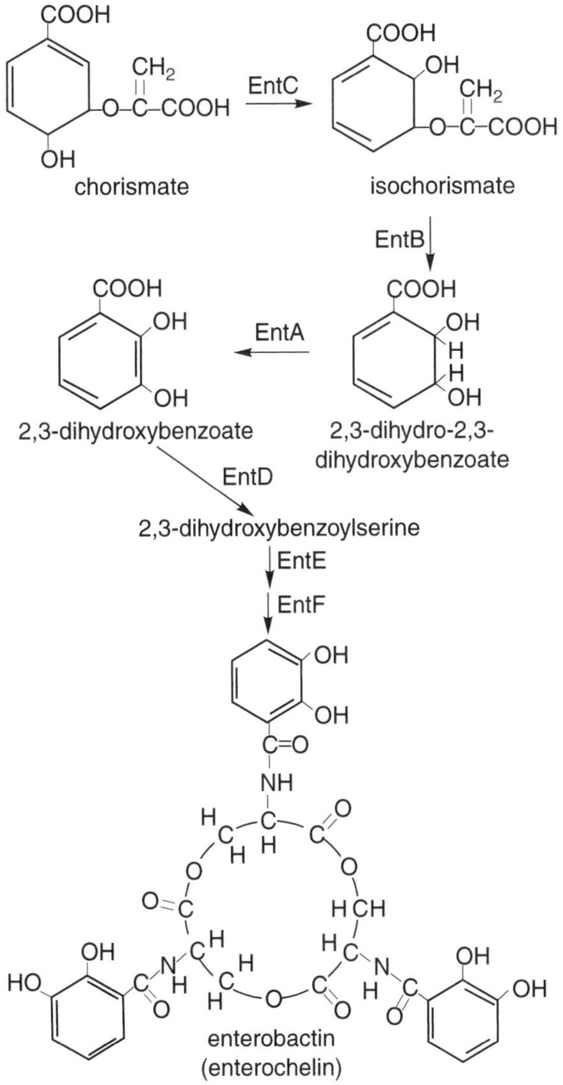

Fig. 15-21. Biosynthesis of enterobactin (enterochelin). The enzyme designations for *E. coli* are: EntC, isochorismate synthase; EntB, 2,3-dihydroxybenzoate synthetase; EntA, 2,3-dihydro-2,3-dihydroxybenzoate dehydrogenase; EntD, 2,3-dihydroxybenzoylserine synthetase; EntE, EntF serve to condense three molecules of 2,3-dihydroxylbenzoylserine into the cyclic enterobactin structure.

enzymes catalyzing these steps are isochorismate synthetase (EntC), 2,3-dihydro-2,3-dihydroxybenzoate synthetase (EntB), and 2,3-dihydro-2,3-dihydroxybenzoate dehydrogenase (EntA). In *E. coli* eight genes govern the biosynthesis of enterobactin and its function. These genes occur in a cluster at 13 minutes on the *E. coli* linkage map. The second stage of enterobactin synthesis is catalyzed by a sequence of enzymes (EntD, EntE, EntF, and EntG) present in a multienzyme complex. EntD couples 2,3-dihydroxybenzoate and serine to form 2,3-dihydroxybenzoylserine. The remaining enzymes catalyze the coupling of three molecules of 2,3-dihydroxybenzoylserine to form the cyclical enterobactin structure.

The Pathway to Ubiquinone

Ubiquinone (coenzyme Q), the only nonprotein component of the electron transport chain, functions in terminal respiration (Chapter 9). Ubiquinone is synthesized from chorismate via the pathway shown in Figure 15-22. The first intermediate specific to the ubiquinone pathway, 4-hydroxybenzoate, is formed by the action of chorismate lyase (UbiC). In some organisms tyrosine can serve as a source of 4-hydroxybenzoate through the intermediary formation of 4-hydroxyphenylpyruvate. An octaprenyl group (R_8) is added by the action of UbiA (4-hydroxybenzoate octaprenyltransferase). Via decarboxylation, 3-octaprenyl-4-hydroxybenzoate is converted to 2-octaprenylphenol. UbiB, UbiH, and UbiF catalyze the three monooxygenase steps in the pathway. UbiE, an S-adenosylmethionine:2-demethylquinol methyltransferase, catalyzes the methylation steps in both the ubiquinone and menaquinone pathways.

Menaquinone (Vitamin K) Biosynthesis

Menaquinone (vitamin K_2) is a methylnaphthoquinone with the structure shown in Figure 15-23. Biosynthesis of menaquinone originates with the addition of α-ketoglutarate to the ring of chorismate and concomitant removal of pyruvate and CO_2. The details of the pathway are shown in Figure 15-23. All naturally occurring menaquinones have *trans*-configurations in the double bonds of the prenyl side chain. Menaquinone is essential for electron transfer to fumarate during anaerobic growth of *E. coli* (see Chapter 9).

Biosynthesis of Nicotinamide Adenine Dinucleotide (NAD)

Both NAD and NADP, the functional forms of nicotinic acid, are synthesized from tryptophan in mammals, *N. crassa*, and *S. cerevisiae*, as shown in Figure 15-24. The bacterium *Xanthomonas pruni* also synthesizes NAD from tryptophan. However, most bacteria use an entirely different pathway to NAD that involves the condensation of aspartate and dihydroxyacetone phosphate (DOHAP) and the subsequent conversion of the condensation product (iminoaspartate) to QA as shown in Figure 15-24. The pathway from QA to NAD appears to be identical in all organisms regardless of the mode of QA formation. The enteric bacteria (*E. coli* and *S. enterica*) and most other common bacteria (including *M. tuberculosis*) use the aspartate-DOHAP route to QA. Under anaerobic conditions, *S. cerevisiae* uses the aspartate-DOHAP pathway. This change most likely occurs because several oxidative steps in the pathway from tryptophan to quinolinate require molecular oxygen. For this reason, the bacterial

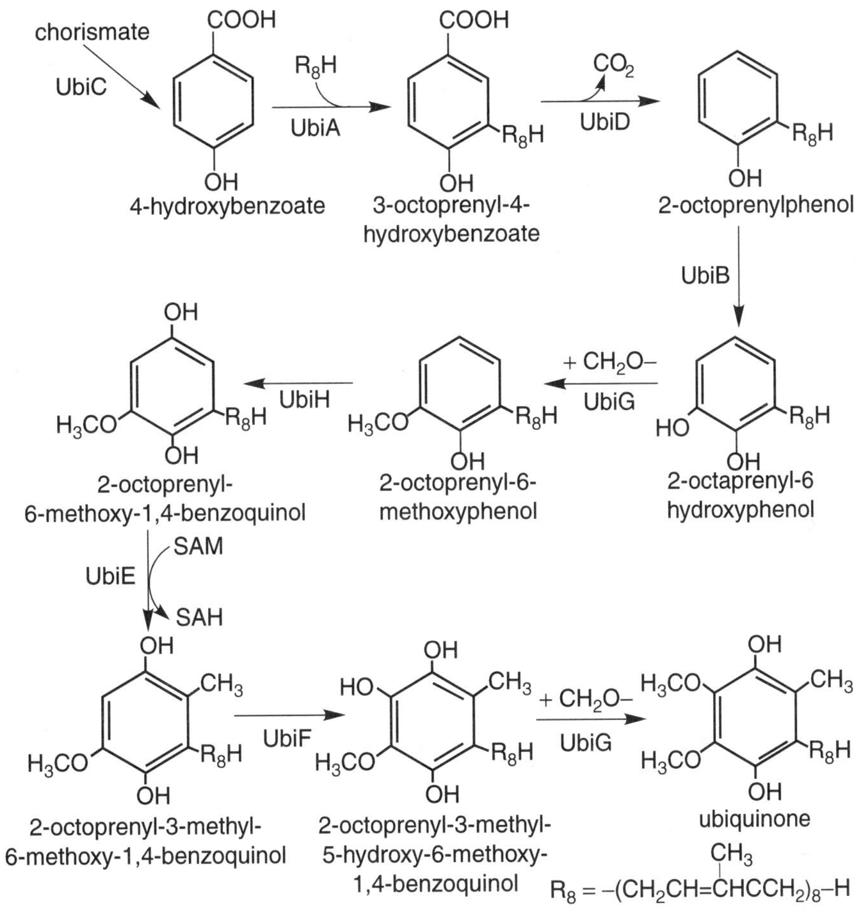

Fig. 15-22. Biosynthesis of ubiquinone (coenzyme Q). The enzyme designations are for *E. coli*. UbiC, chorismate lyase; UbiA, 4-hydroxybenzoate polyprenyl (R) transferase; UbiD, decarboxylation of 3-octoprenyl-4-hydroxybenzoate to 2-octaprenylphenol; UbiB, conversion of 2-octoprenylphenol to 2-octaprenyl-6-hydroxyphenol (flavin reductase); UbiG, o-methylation of 2-octaprenyl-6-hydroxyphenol to 2-otaprenyl-6-methoxyphenol; UbiH, conversion of 2-octaprenyl-6-methoxyphenol to 2-octaprenyl-6-methoxy-1,4-benzoquinol; UbiE, S-adenosyl-L-methionine (SAM)-dependent methyltransferase converts 2-octaprenyl-6-methoxy-1,4-benzoquinol to 2-octaprenyl-3-methyl-6-methoxy-1,4-benzoquinol; UbiF, converts 2-octaprenyl-3-methyl-6-methoxy-1,4-benzoquinol to 2-octaprenyl-3-methyl-5-hydroxy-6-methoxy-1,4-benzoquinol; UbiG, o-methylation of 2-octaprenyl-3-methyl-5-hydroxy-6-methoxy-1,4-benzoquinol to ubiquinone 8.

pathway is referred to as anaerobic. The anaerobe, *C. butylicum*, follows a unique pathway via condensation of aspartate and formate to yield the intermediate *N*-formylaspartate. The addition of acetate and subsequent ring closure yields quinolinate (Figure 15-24).

Organisms that demonstrate a specific nutritional requirement for nicotinamide utilize a unique route of NAD biosynthesis that involves conversion of nicotinamide

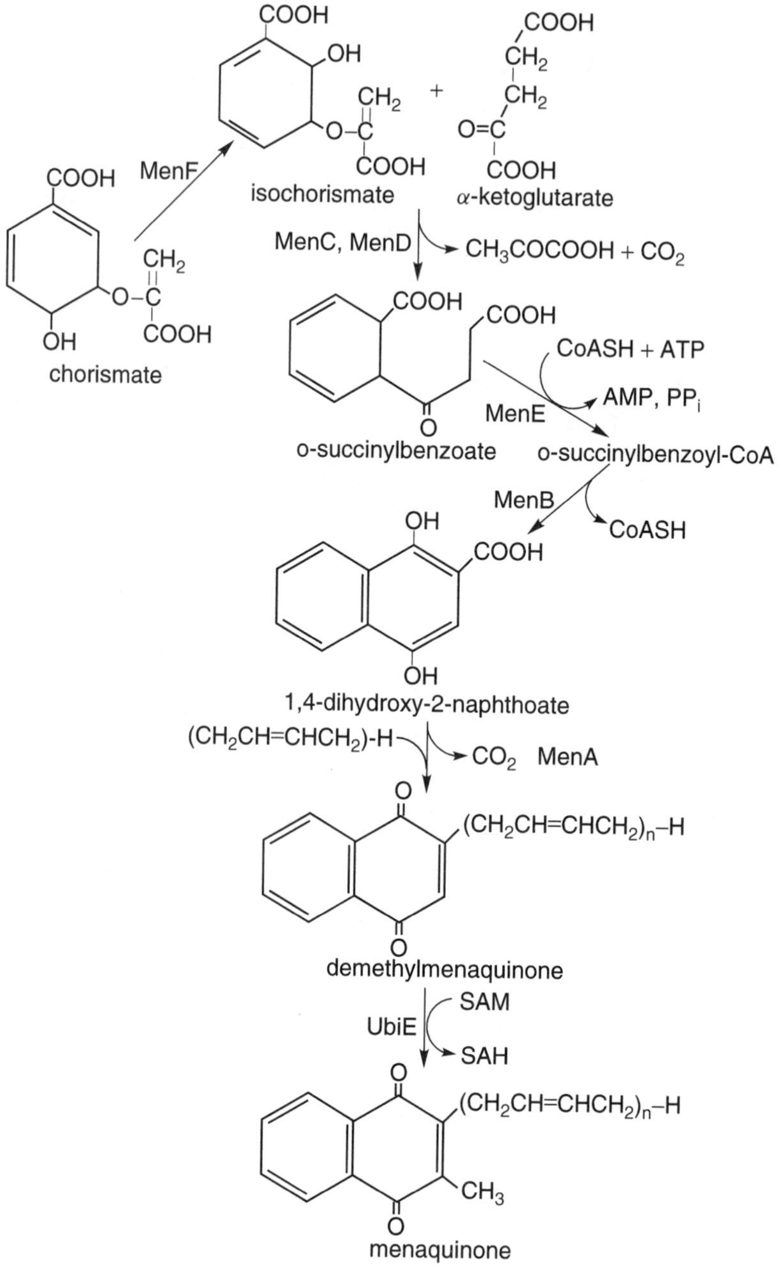

Fig. 15-23. Menaquinone biosynthesis. The enzyme designations are for *Escherichia coli*. MenF, menaquinone-specific isochorismate synthase; MenC, o-succinylbenzoate synthase II; MenD, o-succinylbenzoate synthase I [o-succinylbenzoate is the trivial name for 4-(2-carboxy-phenyl)-4-oxybutyrate]; MenE, o-succinylbenoate-CoA synthase; MenB, 1,4-dihydroxy-2-naphthoate synthase; MenA, 1,4-dihydroxy-L-napthoate octaprenyltransferase; UbiE, S-adenosylmethionine (SAM):2-demethylmenaquinone methyltransferase. SAH, S-adenosylhomocysteine.

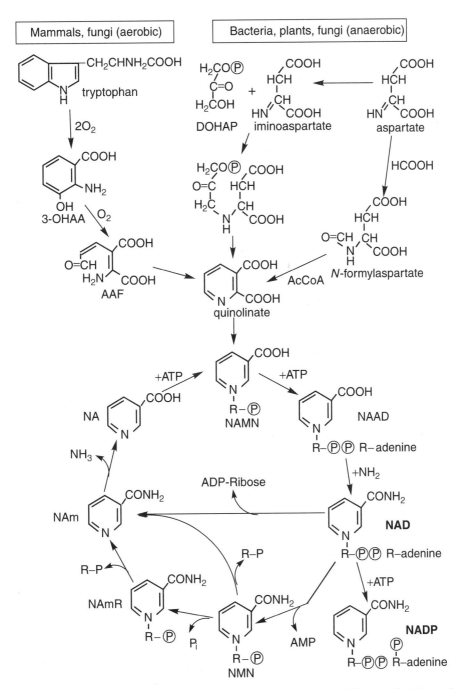

Fig. 15-24. Alternate pathways for NAD biosynthesis and the pyridine nucleotide cycles.
3-OHAA, 3-hydroxyanthranilate; AAF, 2-acroleyl-3-aminofumarate; DOHAP, dihydroxyacetone
phosphate; AcCoA, acetyl coenzyme A; NAMN, nicotinic acid mononucleotide; NAAD, nicotinic acid adenine dinucleotide; NAD, nicotinamide adenine dinucleotide; NADP, nicotinamide
adenine dinucleotide phosphate; NMN, nicotinamide mononucleotide; NamR, nicotinamide ribo-
side; NA, nicotinic acid; Nam, nicotinamide.

to NAD without prior deamidation:

$$\text{nicotinamide} + \text{PRPP} + \text{ATP} \longrightarrow \text{nicotinamide mononucleotide (NMN)}$$

$$\text{NMN} + \text{ATP} \longrightarrow \text{NAD}$$

Lactobacillus fructosus and *Haemophilus influenzae* use this route for NAD biosynthesis. It is used for the same purpose in the nucleus of mammalian cells. The second reaction, catalyzed by NAD pyrophosphorylase (ATP:NMN adenylyltransferase), is considered to be located exclusively in the nucleus of mammalian cells and can be used as an identifying marker for testing the purity of isolated nuclei.

Bacteria use NAD as a substrate in several important reactions including DNA repair. DNA ligases catalyze the formation of phosphodiester bonds at single-strand breaks between adjacent $3'$-OH and $5'$-phosphate termini in double-stranded DNA. DNA ligases that require ATP for adenylation are found in eukaryotic cells and in bacteriophages. All known bacterial DNA ligases require NAD rather than ATP. The overall reaction for the NAD^+-dependent DNA ligase is

$$\text{nicked DNA} + NAD^+ \longrightarrow \text{repaired DNA} + \text{NMN} + \text{adenosine}$$

A comparison of the amino acid composition and DNA sequences indicate that the NAD-dependent ligases are phylogenetically unrelated to the ATP-dependent DNA ligases. The first step of DNA ligation in bacteria requires adenylation of the ε-amino group of lysine 112 at the reactive site of the enzyme by NAD creating an adenylated enzyme intermediate with AMP covalently bound to the enzyme and releasing NMN. In the second step of the reaction, the adenylate group is transferred from Lys-112 to the terminal $5'$-phosphate at the DNA nick, resulting in the formation of a phosphodiester bond and sealing the DNA strand. The fact that NAD-dependent ligases are unique to bacteria suggests that this enzyme may be a selective target for new antibacterial agents.

Enzymes referred to as ADP ribosylases or NAD glycohydrolases catalyze reactions in which the ADP-ribosyl moiety of NAD is transferred to an acceptor to yield mono- or poly-ADP-ribosylated derivatives:

$$\text{NAD} + \text{acceptor} \longrightarrow \text{acceptor-(ADP-ribose)} + \text{nicotinamide}$$

Cholera toxin is an ADP ribosylase. In Chapter 14 the role of NAD glycohydrolase in regulating nitrogen fixation is discussed. In the light organ of the Hawaiian squid, the symbiont *Vibrio fischeri* uses an NAD glycohydrolase to regulate its activity. Streptococci produce NADase, a characteristic used as an identifying marker.

Utilization of NAD as a substrate in a number of reactions emphasizes the need for rapid recycling of NAD by means of salvage pathways or pyridine nucleotide cycles (PNCs). A number of alternative PNCs have been shown to operate in microorganisms as shown in Figure 15-24.

Some organisms that degrade NAD to nicotinamide and nicotinic acid cannot recycle these products and accumulate them in the culture medium. The human strains, but not bovine or avian strains of *M. tuberculosis*, can be identified on the basis of this characteristic. The anaerobe *Clostridium butylicum* also accumulates nicotinic acid because it is unable to recycle it. Although these organisms can synthesize NAD, they lack the requisite enzymes to form nucleotide derivatives of nicotinic acid.

HISTIDINE BIOSYNTHESIS

The biosynthesis of histidine occurs via a unique pathway that is more closely linked to the metabolism of pentoses and purines than to any of the other amino acid families. The pathway of histidine biosynthesis, shown in Figure 15-25, is initiated by the coupling of phosphoribosyl pyrophosphate (PRPP) with ATP at N-1 of the purine ring. This is followed by opening of the ring. The N-1 and N-2 of the imidazole ring of histidine are thus derived from the N-1 and N-2 of the adenine ring. Aminoimidazolecarboxamide ribonucleotide (AICRP) derivatives are normally bound to the enzymes involved in their formation and rearrangement. Cleavage of the open-ring structures gives rise to imidazole-glycerol phosphate (IGP) and AICRP, which is recycled via the purine biosynthetic pathway to reform the purine nucleotide. There do not appear to be any branches from this pathway to other end products. The chemical intermediates and the enzymes involved in histidine biosynthesis are virtually identical in a variety of bacteria and fungi, but the arrangement of the genes controlling the synthesis and operation of the pathway differs markedly from one organism to another.

In *S. enterica*, 8 genes arranged in a single operon control 10 steps in the synthesis of histidine (Fig. 15-26). The genes in *Staphylococcus aureus* are similarly arranged. In yeast, the *his4* region specifies three of the enzyme activities of the histidine pathway, namely, the second, third, and tenth steps. The enzymes responsible for these actions are tightly associated in a multifunctional complex. There is no evidence for aggregation of the remainder of the enzymes, nor are the genes specifying the other enzymes contiguous with these genes or with each other. This contrasts with the close linkage of the 10 *his* genes in *S. enterica* but is comparable to many other gene–enzyme relationships in fungi. The equivalent genes coding for the second, third, and tenth steps in *Neurospora* are also arranged in a cluster. They code for a complex composed of nonidentical isomers. Purified preparations of the complex contain all of these enzyme activities.

The arrangement of the genes in the *his* operon of *S. cerevisiae* is shown in Figure 15-26. The genes in the *his* operon do not correspond to the order of the reactions in the pathway. Mechanisms for regulation of histidine biosynthesis include (1) operon repression mediated at an attenuated site in response to the availability of histidyl-tRNA^{his}, (2) transcriptional control responsive to the general availability of amino acids involving the general signal molecule guanosine tetraphosphate, and (3) feedback inhibition of the first enzyme of the pathway (ATP–phosphoribosyltransferase, ATP–PRT). Details of the various mechanisms of regulation are presented in Chapter 5.

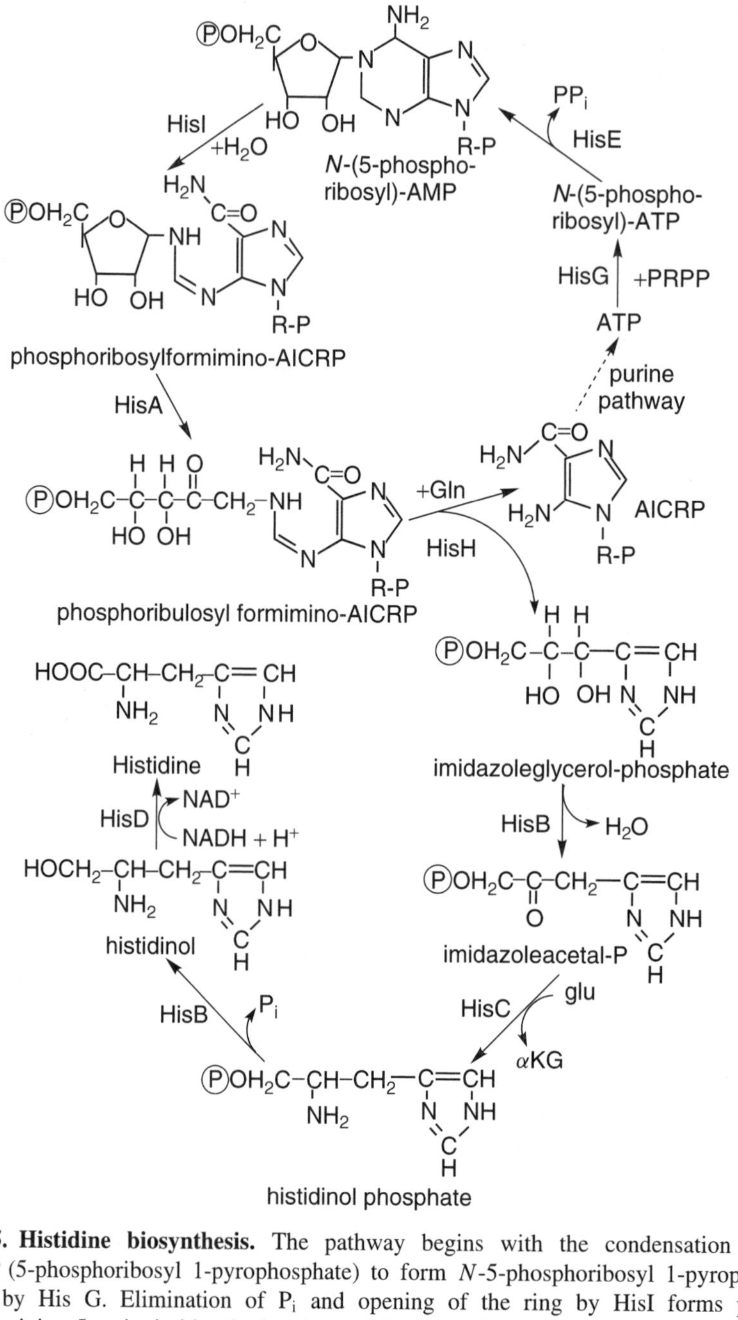

Fig. 15-25. Histidine biosynthesis. The pathway begins with the condensation of ATP and PRPP (5-phosphoribosyl 1-pyrophosphate) to form N-5-phosphoribosyl 1-pyrophosphate catalyzed by His G. Elimination of P_i and opening of the ring by HisI forms phosphoribosyl-formimino-5-aminoimidazole-4-carboxamide ribonucleotide. HisA mediates conversion to 5-phosphoribulosyl-formimino-5-aminoimidazole-4-carboxamide ribonucleotide. Cleavage of this intermediate by HisH results in the formation of imidazole glycerol-phosphate and AICRP (aminoimidazolecarboxamide ribonucleotide). AICRP is recycled through the purine pathway. HisB converts imidazole glycerol-phosphate to imidazole acetal-phosphate. Transamination by HisC yields histidinol phosphate. Removal of P_i by HisB yields histidinol. Via an enzyme-bound intermediate (histidinal), HisD converts histidinol to histidine.

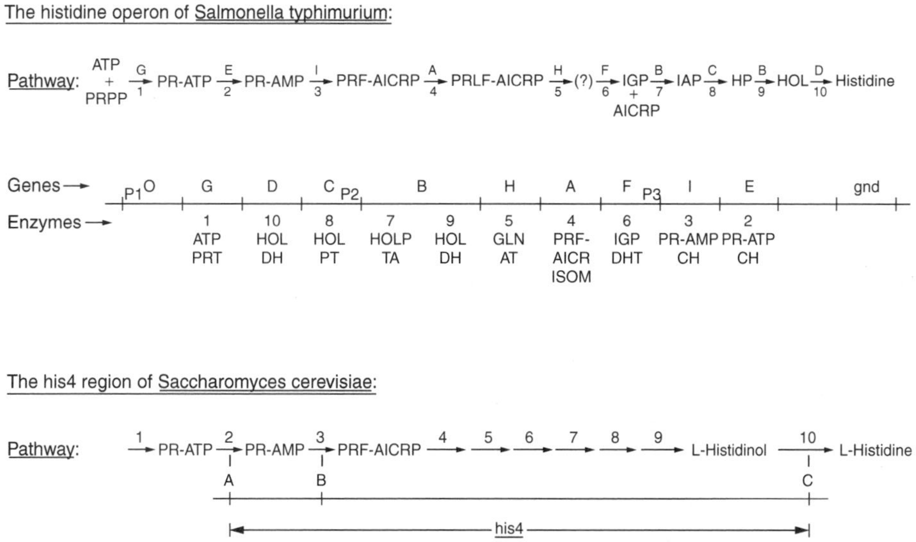

The histidine operon of <u>Salmonella typhimurium</u>:

The his4 region of <u>Saccharomyces cerevisiae</u>:

Fig. 15-26. **Arrangement of the genes in the histidine operon of *Salmonella enterica* and the *his4* region of *Saccharomyces cerevisiae*.** The equivalent genes for the second, third, and tenth steps in *Neurospora crassa* are clustered in a manner similar to those in yeast. Abbreviations: (1) ATP-PRT, ATP-phosphoribosyltransferase; (10) HOL-DH, histidinol dehydrogenase; (8) HOL-PT, histidinol phosphatase; (7) HOLP-TA, histidinol phosphate transaminase; (9) histidinol dehydrogenase; (5) GLN-AT, glutamine amidotransferase; (4) PRF-AICR ISOM, phosphoribosyl-formimino-5-aminoimidazolecarboxamide ribotide isomerase; (6) IGP-DHT, imidazoleglycerolposphate dehydrogenase; (3) PR-AMP CH, phosphoribosyl-AMP cyclohydrolase; PR ATP CH, phosphoribosyl-ATP-cyclohydrolase. P1, P2, and P3 are promoters. In *S. enterica*, the genes are transcribed in a clockwise direction for O through E.

BIBLIOGRAPHY

Amino Acids

Michal, G. 2000. *Biochemical Pathways*, M. Gerhard (ed.). John Wiley & Sons, New York.

Reitzer, L. 2000. Amino acid function and synthesis. In J. Lederberg (ed.), *Encyclopedia of Microbiology*, 2nd ed., Vol. 1. Academic Press, New York, pp. 134–51.

Sahm, H., and L. Eggeling. 2000. Amino acid production. In J. Lederberg (ed.), *Encyclopedia of Microbiology*, 2nd ed., Vol. 1. Academic Press, New York, 152–61.

Glutamate (α-Ketoglutarate) Family

Brakhage, A. A. 1998. Molecular regulation of beta-lactam biosynthesis in filamentous fungi. *Microbiol. Mol. Biol. Rev.* **62**:547–5845.

Casqueiro, J., S. Gutiérrez, O. Bañuelos, M. J. Hijarrubia, and J. F. Martín. 1999. Gene targeting in *Penicillium chrysogenum*: disruption of the *lys2* gene leads to penicillin overproduction. *J. Bacteriol.* **181**:1181–8.

Hochuli, M., H. Patzelt, D. Oesterhelt, K. Wüthrich, and T. Szyperski. 1999. Amino acid biosynthesis in the halophilic archeon *Haloarcula hispanica*. *J. Bacteriol.* **181**:3226–37.

Jensen, S. E., and A. L. Demain. 1995. Beta-lactams. In L. C. Vining and C. Stuttard (eds.), *Genetics and Biochemistry of Antibiotic Production*. Butterworth-Heinemann, Boston, pp. 239–68.

Kobashi, N., M. Nishiyama, and M. Tanokura. 1998. Aspartate kinase-independent lysine synthesis in an extremely thermophilic bacterium, *Thermus thermophilus*: lysine is synthesized via α-aminoadipic acid not via diaminopimelic acid. *J. Bacteriol.* **181**:1713–8.

Maghnouj, A. 1998. The arcABDC gene cluster, encoding the arginine deiminase pathway of *Bacillus licheniformis*, and its activation by the arginine repressor ArgR. *J. Bacteriol.* **180**:6468–75.

Martín, J. F. 1998. New aspects of genes and enzymes for β-lactam antibiotic biosynthesis. *Appl. Microbiol. Biotechnol.* **50**:1–15.

Messenguy, F., F. Vierendeels, B. Scherens, and E. Dubois. 2000. In *Saccharomyces cerevisiae*, expression of arginine catabolic genes *CAR1* and *CAR2* in response to exogenous nitrogen availability is mediated by the Ume6(CargR1)-Sin3 (CargRII)-Rpd3 (CargRIII) complex. *J. Bacteriol.* **182**:3158–64.

Nicoloff, H., J.-C. Hubert, and F. Bringel. 2000. In *Lactobacillus plantarum*, carbamoyl phosphate is synthesized by two carbamoyl-phosphate synthetases (CPS): carbon dioxide differentiates the arginine-repressed from the pyrimidine-regulated CPS. *J. Bacteriol.* **182**:3416–22.

Park, S.-M., C.-D. Lu, and A. T. Abdelal. 1997. Cloning and characterization of *argR*, a gene that participates in regulation of arginine biosynthesis and catabolism in *Pseudomonas aeruginosa* PAO1. *J. Bacteriol.* **179**:5300–8.

Quintero, M. J., A. M. Muro-Pastor, A. Herrero, and E. Flores. 2000. Arginine catabolism in the cyanobacterium *Synechocystis* sp. strain PCC 6803 involves the urea cycle and arginase pathway. *J. Bacteriol.* **182**:1008–15.

Rius, N., and A. L. Demain. 1997. Lysine ε-aminotransferase, the initial enzyme of cephalosporin biosynthesis in actinomycetes. *J. Microbiol. Biotech.* **7**:95–100.

Schneider, B. L., A. K. Kiupakis, and L. J. Reitzer. 1998. Arginine catabolism and the arginine succinyltransferase pathway in *Escherichia coli*. *J. Bacteriol.* **180**:4278–86.

Zéñiga, M., M. Champomier-Verges, M. Zagoree, and G. Pérez-Martinez. 1998. Structural and functional analysis of the gene cluster encoding the enzymes of the arginine deiminase pathway of *Lactobacillus sake*. *J. Bacteriol.* **180**:4154–9.

Aspartate and Pyruvate Families

Blanco, J., J. J. R. Coque, and J. F. Martin. 1998. The folate branch of the methionine biosynthesis pathway in *Streptomyces lividans*: disruption of the 5,10-methylenetetrahydrofolate reductase gene leads to methionine auxotrophy. *J. Bacteriol.* **180**:1586–91.

Epelbaum, S., R. A. LaRossa, T. K. VanDyk, T. Elkayam, D. M. Chipman, and Z. Barak. 1998. Branched-chain amino acid biosynthesis in *Salmonella typhimurium*: a quantitative analysis. *J. Bacteriol.* **180**:4056–67.

Liu, G., J. Casqueiro, O. Bañuelos, R. A. Cardoza, S. Gutiérrez, and J. F. Martin. 2001. Targeted inactivation of the *mecB* gene, encoding cystathionine-lyase, shows that the reverse transsulfuration pathway is required for high-level cephalosporin biosynthesis in *Acremonium chrysogenum* C10 but not for methionine induction of the cephalosporin genes. *J. Bacteriol.* **183**:1765–72.

Lynch, D., J. O'Brien, T. Welch, P. Clarke, P. O. Cuév, J. H. Crosa, and M. O'Connell. 2001. Genetic organization of the region encoding regulation, biosynthesis, and transport of rhizobactin 1021, a siderophore produced by *Sinorhizobium meliloti*. *J. Bacteriol.* **183**:2576–85.

VanDyk, T. K., B. L. Ayers, R. W. Morgan, and R. A. LaRossa. 1998. Constricted flux through the branched-chain amino acid biosynthetic enzyme acetolactate synthase triggers elevated expression of genes regulated by *rpoS* and internal acidification. *J. Bacteriol.* **180:**785–92.

Vermeij, P., and M. A. Kertesz. 1999. Pathways of assimilative sulfur metabolism in *Pseudomonas putida*. *J. Bacteriol.* **181:**5833–7.

Wehrmann, A., B. Phillip, H. Sahm, and L. Eggeling. 1998. Different modes of diaminopimelate synthesis and their role in cell wall integrity: a study with *Corynebacterium glutamicum. J. Bacteriol.* **180:**3159–65.

Yoshida, K. I., Y. Fujita, and S. D. Ehrlich. 1999. Three asparagine synthetase genes of *Bacillus subtilis. J. Bacteriol.* **181:**6081–91.

Serine-Glycine Family

Cooper, A. J. L., and A. D. Hanson. 1998. Advances in enzymology of the biogeochemical sulfur cycle. *Chemtracts — Biochem. Mol. Biol.* **11:**729–47.

Seiflin, T. A., and J. G. Lawrence. 2001. Methionine-to-cysteine recycling in *Klebsiella aerogenes. J. Bacteriol.* **183:**336–46.

Thomas, D., and Y. Surdin-Kerjan. 1997. Metabolism of sulfur amino acids in *Saccharomyces cerevisiae. Microbiol. Mol. Biol. Rev.* **61:**503–32.

Aromatic Amino Acid Family

Babitzke, P., P. Gollnick, and C. Yanofsky. 1992. The *mtrAB* operon of *Bacillus subtilis* encodes GTP cyclohydrolyse I (MtrA), an enzyme involved in folic acid biosynthesis, and MtrB, a regulator of tryptophan biosynthesis. *J. Bacteriol* **174:**2059–64.

Bhattacharyya, D. K., O. Kwon, and R. Meganathan. 1997. Vitamin K_2 (menaquinone) biosynthesis in *Escherichia coli*: evidence for the presence of an essential histidine residue in o-succinylbenzoyl coenzyme A synthetase. *J. Bacteriol.* **179:**6061–5.

Cantoni, R., M. Branzoni, M. Labò, M. Rizzi, and G. Riccardi. 1998. The MTCY428.08 gene of *Mycobacterium tuberculosis* codes for NAD^+ biosynthesis. *J. Bacteriol.* **180:**3218–21.

Crowley, P. J., J. A. Gutierrez, J. D. Hillman, and A. S. Bleiweis. 1997. Genetic and physiologic analysis of a formyl-tetrahydrofolate synthetase mutant of *Streptococcus mutans. J. Bacteriol.* **179:**1563–72.

Howell, D. M., and R. H. White. 1997. D-erythro-Neopterin biosynthesis in the methanogenic archea *Methanococcus themophila* and *Methanobacterium thermoautotrophicum* ΔH. *J. Bacteriol.* **179:**5165–70.

Kaczmarek, F. S., R. P. Zaniewski, T. D. Gootz, et al. 2001. Cloning and functional characterization of an NAD^+-dependent DNA ligase from *Staphylococcus aureus. J. Bacteriol.* **183:**3016–24.

Kiswasnathan, V. K., J. M. Green, and B. P. Nichols. 1995. Kinetic characterization of 4-amino-4-deoxychorismate synthase from *Escherichia coli. J. Bacteriol.* **177:**5918–23.

Kwon, O., M. E. S. Hudspeth, and R. Meganathan. 1996. Anaerobic biosynthesis of enterobactin in *Escherichia coli*: regulation of *entC* expression and evidence against its involvement in menaquinone (vitamin K_2) biosynthesis. *J. Bacteriol.* **178:**3252–9.

Lee, P. T., A. Y. Hsu, H. T. Ha, and C. F. Clarke. 1997. A C-methyltransferase involved in both ubiquinone and menaquinone biosynthesis: isolation and identification of the *Escherichia coli ubiE* gene. *J. Bacteriol.* **179:**1748–54.

Poon, W. W., D. E. Davis, H. T. Ha, T. Jonassen, P. N. Rather, and C. F. Clarke. 2000. Identification of *Escherichia coli ubiB*, a gene required for the first monooxygenase step in ubiquinone biosynthesis. *J. Bacteriol.* **182:**5129–246.

Rowland, B. M., and H. W. Taber. 1996. Duplicate isochorismate synthase genes of *Bacillus subtilis*: regulation and involvement in the biosyntheses of menaquinone and 2,3-dihydroxybenzoate. *J. Bacteriol.* **178:**854–61.

Song, J., T. Xia, and R. A. Jensen. 1999. Phhb, a *Pseudomonas aeruginosa* homolog of mammalian pterin 43a-carbinolamine dehydratase/DcoH, does not regulate expression of phenylalanine hydroxylase at the transcriptional level. *J. Bacteriol.* **181:**2789–96.

Stabb, E. V., K. A. Reich, and E. G. Ruby. 2001. *Vibrio fischeri* genes *hvnA* and *hvnB* encode secreted NAD$^+$-glycolydrolases. *J. Bacteriol.* **183:**309–17.

Suvarna, K., D. Stevenson, R. Meganathan, and M. E. S. Hudspeth. 1998. Menaquinone (vitamin K_2) biosynthesis: localization and characterization of the *menA* gene from *Escherichia coli*. *J. Bacteriol.* **180:**2782–7.

Viswanathan, V. K., J. M. Green, and B. P. Nichols. 1995. Kinetic characterization of 4-amino-4-deoxychorismate synthase from *Escherichia coli*. *J. Bacteriol.* **177:**5918–23.

Yang, Y., G. Zhao, T. Man, and M. E. Winkler. 1998. Involvement of the *gapA*- and *epd (gapB)*-encoded dehydrogenases in pyridoxal 5-phosphate coenzyme biosynthesis in *Escherichia coli* K-12. *J. Bacteriol.* **180:**4294–9.

Histidine

P. Alifano, R. Fani, P. Lio, A. Lazcano, M. Bazzicalupo, M. S. Carlomagno, and C. B. Bruni. 1996. Histidine biosynthetic pathways and genes: structure, regulation, and evolution. *Microbiol. Rev.* **60:**44–69.

CHAPTER 16

PURINES AND PYRIMIDINES

Our knowledge of the structure of nucleic acids and their subunits, the purines and pyrimidines, was gained by the isolation and chemical characterization of these structures from a variety of biological materials. The basic purine and pyrimidine structures and the numbering system recommended by *Chemical Abstracts* are shown in Table 16-1. The most commonly encountered purines are **adenine** (6-aminopurine) and **guanine** (2-amino-6-oxopurine). A number of methylated purines have been found in a variety of sources. Their significance is discussed in conjunction with their role in specialized functions.

Several pyrimidine derivatives have been characterized. **Uracil** (2,4-dioxopyrimidine), **cytosine** (4-amino-2-oxopyrimidine), and **thymine** (5-methyl-2,4-dioxopyrimidine or 5-methyluracil) are most common. Thymine is found only in DNA; uracil is present only, in RNA; 5-hydroxymethylcytosine is found in the DNA of T-even (T2, T4, T6, etc.) bacteriophages of *E. coli* and not in the DNA of the host cell (see Chapter 6). This most significant discovery by Wyatt and Cohen in 1953 proved to be of inestimable value in studying the synthesis of bacteriophage DNA, as it provided the distinguishing feature necessary to show that the synthesis of host cell DNA is terminated when bacteriophage DNA synthesis is initiated (see Chapter 6).

BIOSYNTHESIS OF PURINES

Evidence as to the origins of the purine molecule was obtained from studies with isotopically labeled precursors. Birds were used in the original experiments because of the high percentage of nitrogen excreted as uric acid or its derivatives ($>80\%$). From these studies it was possible to determine that glycine was incorporated intact into the 4, 5, and 7 positions of the purine ring. Formate was consistently incorporated into the 2 and 8 positions and carbon dioxide into position 6 (Fig. 16-1). Pigeon liver homogenates were found to synthesize purines de novo from these same precursors,

TABLE 16-1. Basic Purine and Pyramidine Structures and Numbering System [a]

Chemical Structure	Recommended IUPAC nName	Common Name Synomyms
	1*H*-Purin-6-amine	Adenine 6-AminoPurine
	2-Amino-1,7-dihydro-6-*H*-purin-6-one	Guanine 2-Amino-6-oxopurine (2-amino-hypoxanthine)
	2,4(1*H*,3*H*)-Pyrimidinedione	Uracil 2,4-Dioxopyramidine
	5-Methyl-2-4-(1*H*3*H*)-Pyrimidinedione	Thymine 5-Methyl-2,4-dioxopyramidine (5-Methyl uracil)
	4-Amino-2(1*H*)-pyrimidinedione	Cytosine 4-Amino-2-oxopyrimidine

[a] The chemical structures of these compounds are often shown as the tautomeric form with the maximum number of double bonds within the ring(s). However this is true only for adenine. The most commonly encountered tautomeric form is reflected in the name assigned by the IUPAC commission and give in the table. The 1977 *Chemical Abstracts* listing of chemical names gives the IPPAC recommendations for purine and pyrimedine nomenclature.

and the basic scheme for the formation of purines was considered to be

$$CO_2 + NH_3 + glycine + HCOOH + R\text{-}1\text{-}P \longrightarrow ribose\ intermediates$$

$$ribose\ intermediates + HCOOH + P_i \longrightarrow inosine\text{-}5'\text{-}phosphate$$

$$inosine\text{-}5'\text{-}phosphate \longrightarrow inosine + P_i \longrightarrow hypoxanthine + R\text{-}1\text{-}P$$

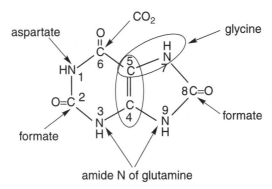

Fig. 16-1. Precursors of the purine molecule. In avian systems, the contribution of the amino nitrogen of aspartate to N-1 and the amide nitrogen of glutamine to N-3 and N-9 has been verified by ^{15}N-labeling studies. Comparable isotope labeling studies have not been performed with microorganisms. However, other evidence indicates that these same donors provide the nitrogen for positions 1, 3, and 9 in the purines of microbial systems. The C-4, C-5, and N-7 are derived intact from glycine.

One of the first compounds suggested as a potential intermediate in the biosynthesis of purines by microorganisms was aminoimidazolecarboxamide (AICA). Somewhat ironically, AICA did not serve as an intermediate in the synthesis of IMP by the avian system. In microorganisms, considerable variation was observed in the ability of AICA to support the growth of purine-requiring auxotrophs. Eventually, it was demonstrated that the true intermediates in the pathway were the ribonucleotides rather than the free bases. Aminoimidazole and AICA were among the first compounds to be identified as degradative metabolites of purines by *C. cylindrosporum*. A purine-requiring mutant of *E. coli* accumulated an arylamine that differed from AICA in several respects and was converted to AICA by an AICA-accumulating mutant. It was found to be aminoimidazole or its ribonucleotide (AIR). Adenine auxotrophs of yeasts or yeasts grown under the condition of biotin deficiency accumulated aminoimidazole derivatives. Further work in a variety of systems eventually confirmed the participation of AIR, AICAR, and FAICAR as intermediates in the purine pathway (Fig. 16-2).

In the pigeon liver system, the formation of AIR from FGAM and the intermediates involved in the pathway to IMP and AMP were identified. Several microbial systems were also shown to follow the same pathway shown in Figure 16-2. The enzyme responsible for the conversion of AMP to AMP-S was shown to be the same enzyme that catalyzed the conversion of SAICAR to AICAR. The formation of GMP from IMP was also elucidated in microbial systems. Inosine-5-monophosphate (IMP) first undergoes a dehydrogenation to form XMP. Then, via an amido-transferase reaction, the amido-nitrogen of glutamine is transferred to the 2 position to form GMP.

The amino acid histidine and the essential B vitamins, folate, riboflavin, and thiamine, may be considered products of the purine biosynthetic pathway in that a portion of each of their structures is derived from purines. The regulatory systems required for coordinating the flow of metabolites through these multiple branches of the purine pathway are discussed later.

In histidine biosynthesis, as described in Chapter 15, ATP is the initial substrate. Coupling of PRPP at N-1 of the purine ring of ATP and ring cleavage results in

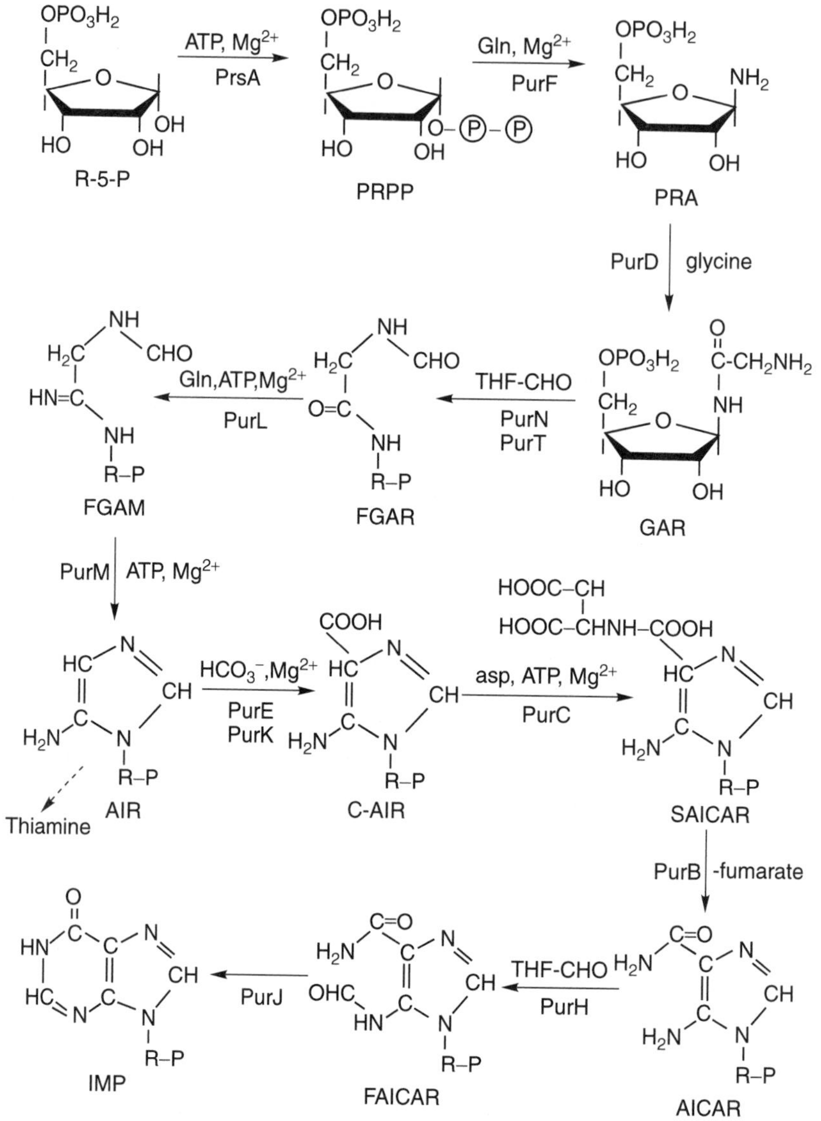

Fig. 16-2. Purine biosynthesis. R-5-P, ribose 5-phosphate; PRPP, 5-phosphoribosyl-1-pyro-phosphate; PRA, 5-phosphoribosylamine; GAR, glycinamide ribonucleotide; FGAR, 1-N-formylglycinamide ribonucleotide; FGAM, α-N-formylglycinamidine ribonucleotide; AIR, aminoimidazole ribonucleotide; C-AIR, 5-amino-4-carboxyimidazole ribonucleotide; SAICAR, 5-amino-4-imidazole-(N-succinylo-)-carboxamide ribonucleotide; AICAR, 5-aminoimidazole-4-carboxamide ribonucleotide; FAICAR, 5-formamidoimidazole-4-carboxamide ribonucleotide; IMP, inosine 5′-monophosphate. Enzyme designations for *E. coli* and *S. enterica* are: Prs, ribose phosphate pyrophosphokinase or PRPP synthetase; PurF, aminophosphoribosyltrans-ferase; PurD, phosphoribosylglycinamide synthetase; PurI, PurL, phosphoribosylglycinamide formyltransferase; PurG, phosphoribosylformylglycinamidine synthetase; PurI, PurM, phosphori-bosylaminoimidazole synthetase; PurE, phosphoribosylaminoimidazole carboxylase; PurC, phos-phoribosylaminoimidazole succinocarboxamide synthetase; PurB, adenylosuccinate lyase; PurH, phosphoribosylaminoimidazolecarboxamide formyltransferase; PurJ, IMP cyclohydrolase.

formation of the histidine precursor imidazoleglycerol phosphate and AICAR. The AICAR is recycled through the purine biosynthetic pathway.

Synthesis of the pterin moiety of folate starts with GTP. Details of folate synthesis are discussed in Chapter 15 in conjunction with the derivation of PAB from chorismate as one of the several branches of the common aromatic amino acid biosynthetic pathway. The GTP cyclohydrolase I initiates the pathway, leading to the formation of a pterin derivative which, when coupled to PAB, forms 7,8-dihydropteroate, the precursor of 7,8-dihydrofolate (see Fig. 15-20).

The pyrimidine ring and the ribityl side chain of riboflavin are derived from GTP, as shown in Figure 16-3. The GTP cyclohydrolase II, encoded by *ribA*, cleaves the imidazole ring of GTP to form 2,5-diamino-6-ribosylamino-4(3H)-pyrimidinone

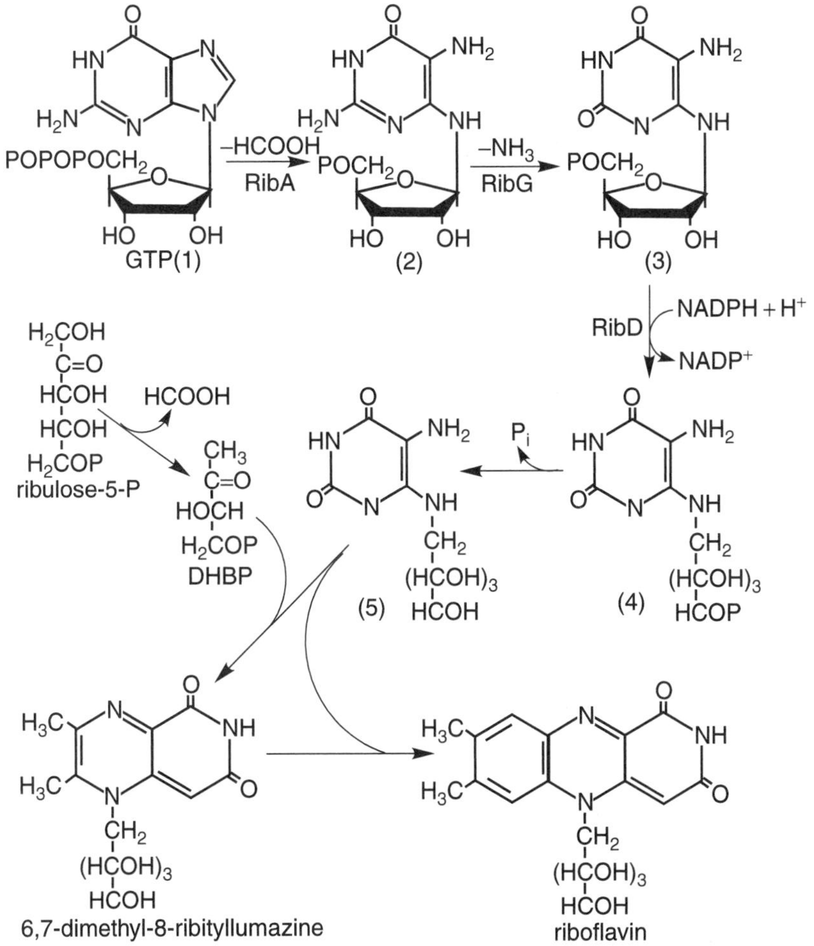

Fig. 16-3. Riboflavin biosynthesis. Enzyme designations are for *E. coli*. RibA, GTP cyclohydrolase II; RibD, pyrimidine deaminase; RibG, reductase; (2) 2-diamino-6-ribosylamino-4 (3H)-pyrimidinedione 5′-phosphate; (3) 5′-amino-6-ribosylamino-2,4(1H,3H)-pyrimidinedione 5′-phosphate; (4) 5′-amino-6-ribitylamino-2,4 (1H,3H)-pyrimidinedione 5′-phosphate; (5) 5-amino-6-ribitylamino-2,4 (1H,3H)-pyrimidinedione.

5'-phosphate (phosphoribosylamino-pyrimidine or PRP) with the release of formate and inorganic pyrophosphate. Ultimately, condensation of 5-amino-ribitylamino-2,4-($1H$,$3H$)-pyrimidinedione with 3,4-dihydroxy-2-butanone 4-phosphate yields 6,7-dimethyl-8-ribityllumazine, the direct precursor of riboflavin (Fig. 16-3). Ribulose 5-phosphate is the precursor of 3,4-dihydroxy-2-butanone 4-phosphate, which forms the xylene ring of the riboflavin molecule.

The pyrimidine ring of thiamine (vitamin B_1) is synthesized from AIR as a branch of the purine biosynthetic pathway. Thiamine contains a pyrimidine ring and a thiazole ring. Two precursors, 4-amino-5-hydroxymethyl-2-methylpyrimidine pyrophosphate and 4-methyl-5-(β-hydroxyethyl)thiazole monophosphate, are synthesized separately and coupled to form thiamine monophosphate (TMP). This compound is then phosphorylated to form thiamine pyrophosphate (TPP), as shown in Figure 16-4. *Salmonella enterica* mutants blocked prior to the AIR step require both purine and thiamine for growth. Isotope labeling studies confirmed that AIR is the source of the carbon and nitrogen of the pyrimidine ring of thiamine. The genes coding for enzymes in the TPP pathway comprise a tightly linked cluster, *thiCEFGH*, at 90 minutes on the *E. coli* genetic map. The *thiC* gene product is required for the synthesis of the hydroxymethyl-pyrimidine precursor of TPP. The remaining enzymes (ThiEFGH) catalyze the synthesis of the thiazole precursor.

Under anaerobic conditions, *S. enterica* can form the pyrimidine moiety of thiamine independently of the *purF* locus. In the absence of oxygen, exogenous pantothenate satisfies the thiamine requirement of *purF* mutants. Only the PurF enzyme is bypassed, however, and PurD, PurG, and PurI are still required for the formation of AIR and its conversion to the pyrimidine ring of thiamine.

BIOSYNTHESIS OF PYRIMIDINES

Pyrimidine biosynthesis is initiated by the formation of carbamoyl phosphate (Fig. 16-5). In view of the importance of carbamoyl phosphate in both arginine and pyrimidine synthesis, coordinating the channeling of this intermediate into the two pathways is critical. In *Neurospora*, two carbamoyl phosphate gradients are maintained by two separate carbamoyl phosphate synthetases. Carbamoyl phosphate synthetase A provides a pool of carbamoyl phosphate specifically for the arginine pathway, as discussed in Chapter 15. Carbamoyl phosphate synthetase P provides a carbamoyl phosphate pool specifically for the pyrimidine pathway. Other organisms, for example, yeast, apparently achieve by regulation what *Neurospora* accomplishes by compartmentation.

Aspartate and carbamoyl phosphate are coupled to form carbamoyl aspartate (ureidosuccinate), which is cyclized by a separate enzyme to form dihydroorotate (Fig. 16-5). After oxidation to orotate, the nucleotide is formed by phosphoribosyl transferase. Note that, in comparison to purine biosynthesis, the nucleotide stage is established *after* completion of ring formation. Uridine-5'-phosphate (UMP) is converted to UTP. Then UTP is aminated to form cytidine triphosphate (CTP). There is no known enzyme that converts cytosine to CMP. For cytosine to serve as a nutritional source of pyrimidines, it must be deaminated to uracil.

The deoxynucleoside diphosphates are formed from the corresponding ribonucleoside diphosphate (NDP) by the action of thioredoxin, a sulfhydryl-containing protein

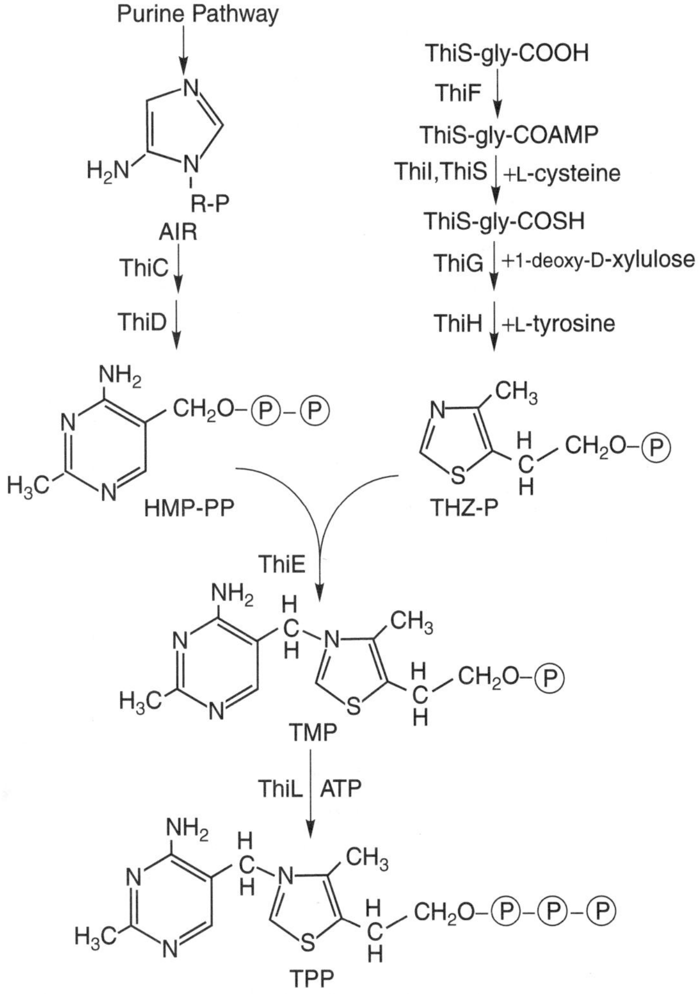

Fig. 16-4. Thiamine (vitamin B$_1$) biosynthesis. The enzyme designations are for *S. enterica*. TPP, thiamine pyrophosphate; TMP, thiamine monophosphate; HMP-PP, hydroxymethyl-pyrimidine pyrophosphate; THZ-P, 4-methyl-5-β-hydroxyethyl-thiamzole monophosphate.

cofactor, and thioredoxin reductase, a system that undergoes cyclic reduction and reoxidation of FAD and NADPH:

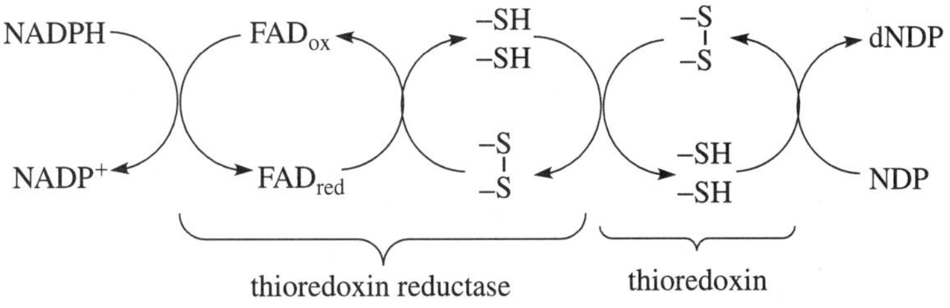

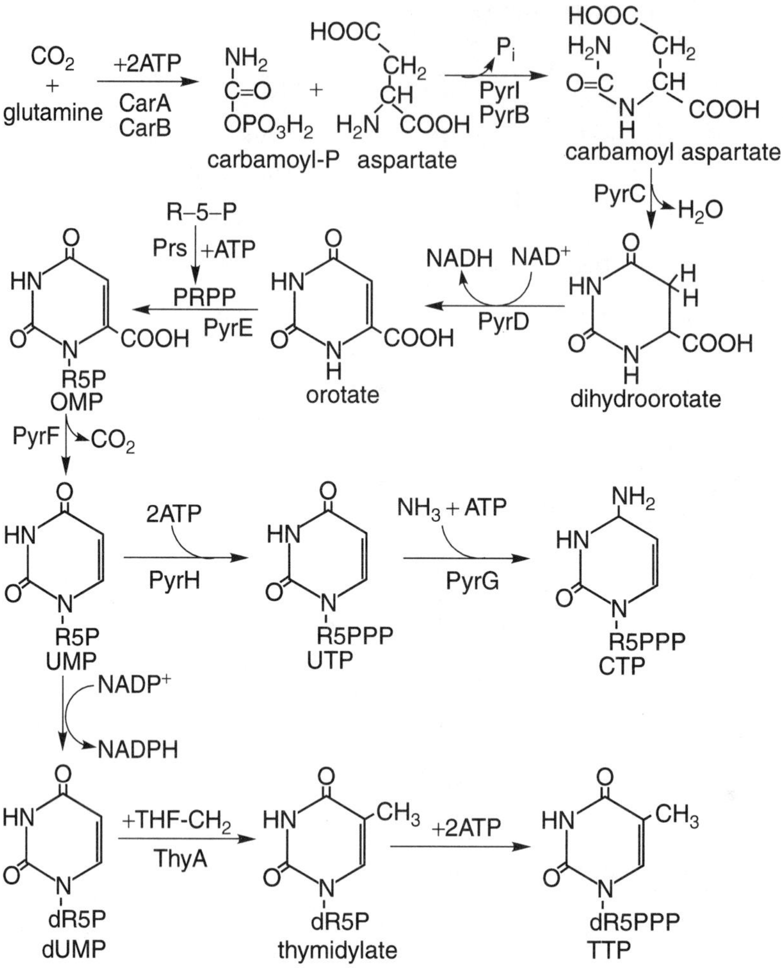

Fig. 16-5. Pyrimidine biosynthesis. Enzyme designations are for *E. coli.* CarA, CarB, carbamoylphosphate synthetases A and B; PyrI, PyrB, aspartate carbamoyltransferase; PyrC, dihydroorotase; PyrD, dihydroorotate dehydrogenase; Prs, PRPP synthetase; PyrE, orotate phosphoribosyltransferase; PyrF, orotidine-5'-phosphate (OMP) decarboxylase; PyrH, pyridine nucleotide phosphorylase; PyrG, cytidine triphosphate (CTP) synthetase; ThyA, thymidylate synthase. UMP, uridylate; UTP, uridine triphosphate; CTP, cytidine triphosphate; dUMP, deoxyuridine 5'-phosphate; TTP, thymidine triphosphate.

Thymidylate (TMP) is formed from dUMP by transfer of a methyl group from N_5,N_{10}-methylene tetrahydrofolate. In the formation of hydroxymethyl-deoxycytidylate, apparently no reduction accompanies the C-1 transfer from the N_5,N_{10}-methylene tetrahydrofolate, so the complete hydroxymethyl group is added. Both the purine and pyrimidine derivatives must be converted to the trinucleotide stage before they can be incorporated into nucleic acid.

In *E. coli* and *S. enterica*, dCTP deaminase forms most of the dUTP, the precursor of thymidylate. The dUTP is degraded by dUTPase (Dut), yielding PP_i and dUMP, the substrate for thymidylate synthase. The breakdown of dUTP prevents its incorporation into DNA by DNA polymerase in place of dTTP. Accumulation of dUTP or incorporation of uracil into DNA is probably not lethal until at least 10% of DNA thymine is replaced by uracil. Above this level, excess uracil in DNA may result in degradation by repair enzymes, causing lethal double-strand breaks. Uracil-containing DNA may not be recognized by DNA-binding enzymes or may turn off protein synthesis by a regulatory system. Note the existence of a DNA repair system that deals with the problem (see Chapters 2 and 3).

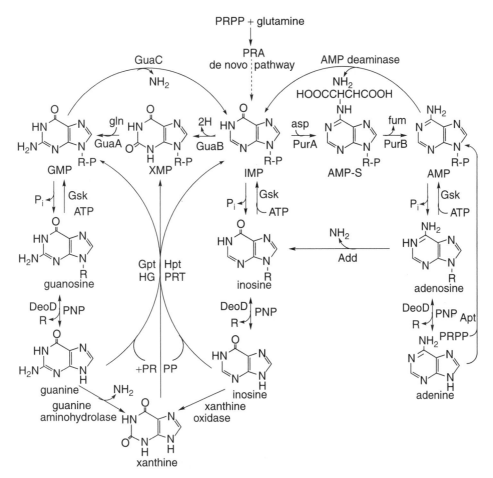

Fig. 16-6. Interconversion of purines and their derivatives. Enzyme designations are for *E. coli*. GuaA, GMP synthase; GuaB, IMP dehydrogenase; GuaC, GM reductase; PurA, adenylosuccinate synthetase; PurB, adenylosuccinate lyase; Hpt, hypoxanthine phosphoribosyltransferase (H-PRT); Gpt, guanosine/xanthine phosphoribosyltransferase (G-PRT); A-PRT, adenine phosphoribosyltransferase (Apt); Add, adenosine deaminase; gln, glutamine; asp, aspartate; fum, fumarate; PNP, purine nucleotide phosphorylase (DeoG); Ask, adenosine kinase; Gsk, guanosine-inosine kinase; PRPP, 5-phosphoribosylpyrophosphate; PRA, phosphoribosylamine; GMP, guanosine monophosphate; XMP, xanthine monophosphate; IMP, inosine monophosphate; AMP-S, succinyloadenosine monophosphate; AMP, adenosine monophosphate.

INTERCONVERSION OF NUCLEOTIDES, NUCLEOSIDES, AND FREE BASES: SALVAGE PATHWAYS

It is often assumed that adenine and guanine or their ribonucleotides are freely interconverted by microorganisms. This is not the case. There is a limited ability to interconvert one base with another and an even more limited ability to interconvert nucleotides and nucleosides as shown in Figures 16-6 and 16-7. Adenine and guanine cannot be directly converted from one to the other. The free purine bases are converted directly to their nucleotides by pyrophosphorylases (phosphoribosyltransferases) by reaction with PRPP. Two distinct purine-utilizing enzymes, one specific for the conversion of adenine to AMP and the other specific for the conversion of hypoxanthine and guanine to IMP and GMP, respectively, are present in a variety of bacteria. These enzymes are subject to feedback inhibition and may play a role in the regulation of purine transport and incorporation into nucleic acids. Nucleotidases and nucleosidases are involved in the sequential degradation of the nucleosides to the free bases and are primarily catabolic enzymes. Glycohydrolases may degrade the nucleotides to the free bases and ribose phosphate.

As shown in Figure 16-6, the interconversion of the purine bases may be quite complex and often occurs only through a virtual maze of enzyme reactions. As a specific example, consider the pathway by which *S. enterica* converts exogenous adenine to guanine nucleotide (GMP). Adenine is converted to adenosine, followed by deamination to inosine and subsequent phosphorolysis to hypoxanthine. Hypoxanthine

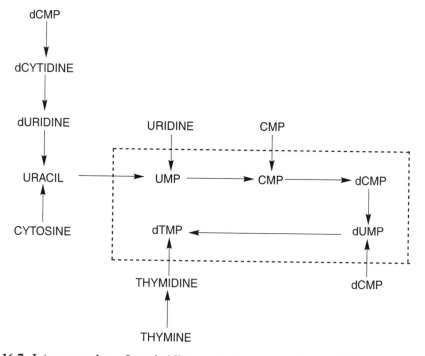

Fig. 16-7. Interconversion of pyrimidines and their derivatives. dCMP, deoxycytidine monophosphate; CMP, cytosine monophosphate; UMP, uridine monophosphate; dTMP, thymidine monophosphate.

is then converted to inosine monophosphate (IMP) by inosine monophosphate pyrophosphorylase. The IMP is then converted sequentially to XMP and GMP.

At the nutritional level, the ability of a given organism to interconvert adenine and guanine is reflected in the ability of the individual purines to promote the growth of purine-requiring strains. Adenine serves quite readily as the sole purine nutrient for many purine-requiring organisms while guanine shows a limited capability in this regard. The capacity of organisms to interconvert purines and their nucleotides also influences the action of a variety of purine antagonists. Regulation of the *gua* operon (which controls the conversion of IMP to XMP and GMP) is effected by adenine and guanine nucleotides rather than the free bases.

Some microorganisms, such as the lactobacilli, possess nucleoside-N-glycosyl transferases, which catalyze the transfer of the ribosyl or deoxyribosyl moieties from nucleosides to free bases. However, a wide variety of bacteria, including many of the enteric bacteria, appear to interconvert nucleosides and free bases through the coupling of nucleoside phosphorylases and do not possess nucleoside-N-glycosyl transferases. Both purine and pyrimidine bases may be converted to the nucleoside and nucleotide stages by the combined action of nucleoside phosphorylases and nucleoside kinases:

$$\text{pyrimidine} + \text{R-5-P} \longleftrightarrow \text{pyrimidine-ribose} + \text{P}_i$$

$$\text{pyrimidine-ribose} + \text{ATP} \longleftrightarrow \text{pyrimidine-ribose-P} + \text{ADP}$$

REGULATION OF PURINE AND PYRIMIDINE BIOSYNTHESIS

Both feedback inhibition and repression-depression control systems regulate the activities of the purine and pyridine pathways. As shown in Figure 16-8, the steps in the purine pathway subject to regulation are the first step, encoded by *purF*, and the enzymes immediately after the branch point at IMP (encoded by *purA, purB, guaB*, and *guaA*). The activity of PurF is inhibited by either AMP or GMP. Purine-requiring mutants that accumulate intermediates in the culture medium no longer do so when the purines or the nucleotides are added. The enzymes at the IMP branch point are affected by the specific nucleotide being synthesized. The IMP dehydrogenase is inhibited by feedback and its synthesis is repressed by GMP. The two enzymes that convert IMP to GMP are encoded by the *guaBA* operon and are coordinately controlled. The two genes encoding the enzymes responsible for the conversion of IMP to AMP, *purA* and *purB*, are subject to end product repression. Synthesis of PRA is inhibited by AMP through a feedback mechanism. Regulation of the enzymatic steps leading from IMP to AMP and GMP forms a metabolic "figure 8" insofar as the end products of the two branches loop back to inhibit the formation of the XMP or AMP-S intermediates (Fig. 16-8).

The purine pathway genes are distributed throughout the *E. coli* chromosome both as single genes and small operons. They are negatively regulated at the transcriptional level by the PurR protein and its corepressors. The *purR* gene encodes an aporepressor that combines with the purine corepressors hypoxanthine and guanine, resulting in an increased affinity for a 16 bp palindromic operator in each of the eight operons of the purine regulon. Other genes that supply intermediates for purine nucleotide synthesis or are involved in the synthesis or salvage of pyrimidine nucleotides are also part of

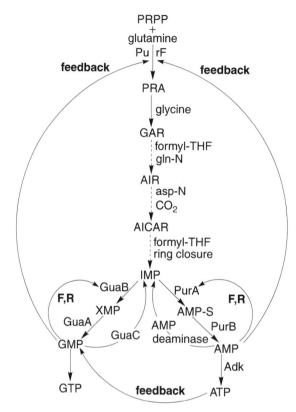

Fig. 16-8. Regulatory mechanisms in purine biosynthesis. PRPP, phosphoribosylpyrophosphate; PRA, phosphoribosylamine; GAR, glycinamide ribonucleotide; AIR, aminoimidazoleribonucleotide; AICAR, 5-aminoimidazole-4-carboxamide ribonucleotide; IMP, inosine monophosphate; XMP, xanthosine monophosphate; GMP, guanosine monophosphate; GTP, guanosine triphosphate; AMP-S, succinyloadenosine monophosphate, AMP, adenosine monophosphate. Enzyme designations for *E. coli* are: PurF, aminophosphoribosyl transferase; GuaB, IMP dehydrogenase; GuaA, GMP synthase; PurA, adenylosuccinate synthase; PurB, adenylosuccinate lyase; Adk, adenylate kinase. AMP represses each of the individual steps.

the PurR-regulated regulon. This cross-pathway regulation is necessary to assure the proper supply of precursors for each of the pathways involved. These genes include *glyA* (encoding serine hydroxymethyltransferase, a major contributor of C-1 units), *codA* (encoding cytosine deaminase, a pyrimidine salvage enzyme), *pyrC* (encoding dihydroorotase), *pyrD* (encoding dihydroorotate dehydrogenase), *prsA* (encoding PRPP synthase), *speA* (encoding arginine decarboxylase), and *glnB* (encoding PII protein in nitrogen regulation). Each of these enzymes contains a site to which PurR binds in vitro and is coregulated in vivo by *purR*.

The expression of *E. coli purR* is autoregulated. Autoregulation at the level of transcription requires two operator sites, designated $purR_{O1}$ (O_1) and $purR_{O2}$ (O_2). Operator O_1 is in the region of DNA between the transcription start site and the site for translation initiation, and O_2 is in the protein-coding region. Operator site O_2, located within the *purR* coding sequence, binds the repressor with 6-fold lower affinity than O_1 but still appears to make an important contribution to the in vivo autoregulation.

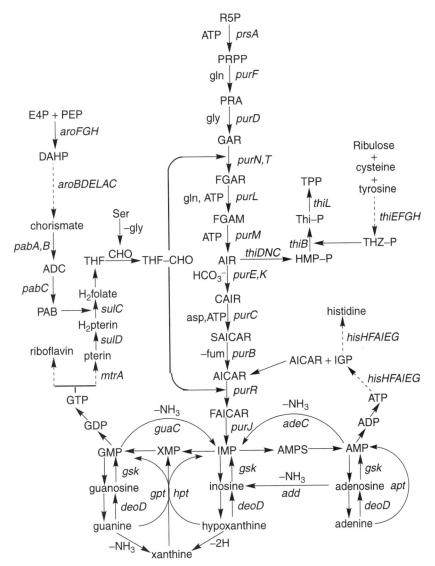

Fig. 16-9. Biosynthetic pathways leading from the de novo purine biosynthetic pathway.
R5P, ribose 5-phosphate; PRPP, 5-phosphoribosylpyrophosphate; PRA, 5-phosphoribosylamine;
GAR, glycinamide ribonucleotide; FGAR, 1-N-formylglycinamide ribonucleotide; FGAM,
α-N-formylglycinamidine ribonucleotide; SAICAR, 5-amino-4-imidazole-(-N-succinylo-)-car-
boxamide ribonucleotide; AICAR, 5-aminoimidazole-4-carboxamide ribonucleotide; FAICAR,
5-formamidoimidazole-4-carboxamide ribonucleotide; IMP, inosine 5′-monophosphate; AMP-S,
N-succinylo-AMP; AMP, adenosine 5′-monophosphate; XMP, xanthosine 5′-monophosphate;
GMP, guanosine 5′-monophosphate; GTP, guanosine triphosphate; ATP, adenosine triphosphate; THF, tetrahydrofolate; THF-CHO, N_5, N_{10}-tetrahydrofolate; E4P, erythrose 4-phos-
phate; PEP, phosphoenolpyruvate; DAHP, 3-hydroxy-D-arabino-heptulosonate 7-phosphate;
ADC, 4-amino-4-deoxychorismate; PAB, p-aminobenzoate; H$_2$pterin, 6-CH$_2$OH-7.8-dihydroxy-
pteroate; H$_2$folate, 7,8-dihydofolate; IGP, imidazoleglycerol phosphate; HMP-P, 4-amino-5-
hydroxymethylpyrimidine phosphate; THZ-P, 4-methyl-5-(β-hydroxyethyl) thiazole phosphate;
Thi-P, thiamine monophosphate; TPP, thiamine pyrophosphate.

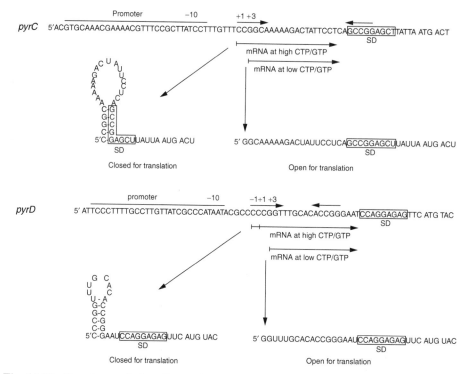

Fig. 16-10. Structure of the *S. enterica* serovar Typhimurium *pyrC* and *pyrD* promoter-leader regions and the proposed regulatory mechanism. Nucleotides are numbered relative to the promoter −10 region (overlined) where +1 at the consensus spacing of 7 bp downstream of the −10 element defines the in vivo transcriptional initiation point in repressing conditions. The Shine-Dalgarno (SD) regions are boxed, and the regions of hyphenated dyad symmetry are indicated by arrows above the sequence. Arrows below the sequence represent the transcripts arising in conditions of different CTP/GTP pool ratios. The putative secondary structures formed at the 5′ ends of the transcripts are shown. (*Source*: from Sørensen, K. I., et al. *J. Bacteriol.* **175:**4137–4144, 1993.)

In *B. subtilis* the genes encoding enzymes for the 10 steps to IMP and the 4 genes encoding the conversion of IMP to AMP and GMP form a 12-gene *pur* operon. The *pur* operon is subject to dual control by adenine and guanine compounds. An adenine compound represses transcription initiation and a guanine compound regulates transcription termination-antitermination in a mRNA leader region preceding the first structural gene.

In considering the regulation of purine biosynthesis, complexities arise as a result of the branch points leading to the pyrimidine ring of thiamine, the pterin moiety of folate, the ribitylaminopyrimidine of riboflavin, and a portion of the ring structure of histidine (Fig. 16-9). The enzyme PRPP synthetase, encoded by *prsA*, participates in several biosynthetic pathways. It is involved in the biosynthesis of purine and pyrimidine nucleotides, tryptophan, histidine, and pyridine nucleotides [NAD(P)]. The PRPP synthetase enzyme is subject to repression by pyrimidine nucleotides and is also under the regulatory control of PurR. In view of the increasing number of genes that are subject to cross-pathway regulation, it seems likely that an even greater number of genes containing sequences related to PurR-binding sites will be revealed through further research in this area.

Aspartate transcarbamylase (ATCase), the first enzyme specific to the pyrimidine pathway, represents one major site of regulatory control. In *E. coli*, the *pyrBI* operon encodes the catalytic (*pyrB*) and regulatory (*pyrI*) subunits of ATCase. The activity of ATCase is subject to allosteric regulation by the activator ATP and the inhibitors CTP and UTP, which bind to the regulatory subunit and alter substrate binding at the catalytic site. This dual control by purines and pyrimidines provides the cell with an efficient mechanism for maintaining the proper ratio of these compounds. The system serves as a prototype for allosteric regulatory systems. The level of pyrBI expression, which is negatively regulated by pyrimidine availability, also controls ATCase activity. This regulation occurs primarily through UTP-sensitive attenuation control, with additional pyrimidine-mediated regulation occurring independently at the level of transcription initiation (see Chapter 5). Of two promoters, P_1 and P_2, located upstream of *pyrB*, greater than 95% of the *pyrBI* transcripts are initiated at promoter P_2, with only a small portion of the level of ATCase influenced by the P_1 promoter.

In *S. enterica* and *E. coli*, expression of *pyrC* (encoding dihydroorotase) and *pyrD* (encoding dihydroorotase dehydrogenase) is regulated in response to fluctuations in the intracellular CTP/GTP pool ratio. High CTP/GTP pool ratios repress expression by production of an mRNA initiated with a CTP downstream of the leader region. As shown in Figure 16-10, this transcript is inefficiently translated because of its capacity to form a stable secondary structure (hairpin) at the 5′ ends of the transcripts, thus sequestering sequences required for ribosomal binding. The potential for hairpin formation is controlled through CTP/GTP-modulated selection of the transcriptional start site. See Chapter 5 for an additional discussion of regulatory mechanisms.

BIBLIOGRAPHY

Purines and Pyrimidines

Bera, A. K., S. Chen, J. L. Smith, and H. Zalkin. 2000. Temperature-dependent function of the glutamine phosphoribosylpyrophosphate amidotransferase ammonia channel and coupling with glycinamide ribonucleotide synthetase in a hyperthermophile. *J. Bacteriol.* **182**:3734–9.

Craig, S. P. III, and A. E. Eakin. 2000. Purine phosphoribosyltransferases. *J. Biol. Chem.* **275**:20,231–4.

Guetsova, M. L., T. R. Crother, M. W. Taylor, and B. Daignan-Fornier. 1999. Isolation and characterization of the *Saccharomyces cerevisiae XPT1* gene encoding xanthine phosphoribosyl transferase. *J. Bacteriol.* **181**:2984–6.

Lennon, B. W., C. H. Williams, Jr., and M. L Ludwig. 2000. Twists in catalysis: Alternating conformations of *Escherichia coli* thioredoxin reductase. *Science* **289**:1190–4.

Martinussen, J., J. Schallert, B. Andersen, and K. Hammer. 2001. The pyrimidine operon *pyrRPB-carA* from *Lactococcus lactis*. *J. Bacteriol.* **183**:2785–94.

Nygaard, P., and H. H. Saxlid. 2000. Nucleotide metabolism. In J. Lederberg (ed.), *Encyclopedia of Microbiology*, 2nd ed., Vol. 3. Academic Press, New York, pp. 418–30.

Turner, R. J., E. R. Bonner, G. K. Grabner, and R. L. Switzer. 1998. Purification and characterization of *Bacillus subtilis* PyrR, a bifunctional pyr mRNA-binding attenuation protein/uracil phosphoribosyltransferase. *J. Biol. Chem.* **273**:5932–8.

Riboflavin Biosynthesis

Eberhardt, S., S. Korn, F. Lottspeich, and A. Bacher. 1997. Biosynthesis of riboflavin: an unusual riboflavin synthase of *Methanobacterium thermoautotrophicum*. *J. Bacteriol.* **179**:2938–43.

Mack, M., A. P. G. M. van Loon, and H.-P. Hohmann. 1998. Regulation of riboflavin biosynthesis in *Bacillus subtilis* is affected by the activity of the flavokinase/flavin adenine dinucleotide synthetase encoded by *ribC*. *J. Bacteriol.* **180**:950–5.

Richter, G., M. Fischer, C. Kreiger, et al. 1997. Biosynthesis of riboflavin: characterization of the bifunctional deaminase-reductase of *Escherichia coli* and *Bacillus subtilis*. *J. Bacteriol.* **179**:2022–8.

Thiamine Biosynthesis

Claas, K., S. Weber, and D. M. Downs. 2000. Lesions in the *nuo* operon, encoding NADH dehydrogenase complex I, prevent PurF-independent thiamine synthesis and reduce flux through the oxidative pentose phosphate pathway in *Salmonella enterica* serovar typhimurium. *J. Bacteriol.* **182**:228–32.

Frodyma, M., A. Rubio, and D. M. Downs. 2000. Reduced flux through the purine biosynthetic pathway results in an increased requirement for coenzyme A in thiamine synthesis in *Salmonella enterica* serovar typhimurium. *J. Bacteriol.* **182**:236–40.

Gralnick, J., E. Webb, B. Beck, and D. Downs. 2000. Lesions in *gshA* (encoding γ-L-glutamyl-L-cysteine synthetase) prevent aerobic synthesis of thiamine in *Salmonella enterica* serovar typhimurium LT2. *J. Bacteriol.* **182**:5180–7.

Zilles, J. I., T. J. Kappock, J. Stubbe, and D. M. Downs. 2001. Altered pathway routing in a class of *Salmonella enterica* serovar typhimurium mutants defective in aminoimidazole ribonucleotide synthetase. *J. Bacteriol.* **183**:2234–40.

CHAPTER 17

BACTERIAL CELL DIVISION

Previous chapters discuss the processes of energy production and biosynthesis of amino acids, purines, pyrimidines, and other building blocks for macromolecular synthesis including chromosome replication, cell wall biosynthesis, and the formation of other vital cellular components such as the cell membrane. The flow of these metabolites and timing of the individual activities must be intricately coordinated so that they lead to the orderly process of cell duplication. In this chapter, emphasis is placed on the processes that take place during the cell cycle: the coordination of chromosome replication with formation of a division septum, the production and activity of various proteins essential to the cell division process, and the regulation of these events. Because more has come to be known about the genetics and regulation of cell division in *Escherichia coli*, this organism has become the paradigm for studies on bacterial cell division. Therefore, we consider cell division in gram-negative rods first.

CELL DIVISION IN GRAM-NEGATIVE RODS

As discussed in Chapter 7, the cell wall peptidoglycan of *E. coli* and other gram-negative bacteria is extended by a process of intercalation of new wall material orchestrated by a large number of enzymes. At some point, a series of events results in the assembly of a large group of proteins into a ring where a constriction occurs between two daughter cells. This ring structure and its component parts are called a **divisome**. Table 17-1 lists many of the genes that encode proteins contributing to peptidoglycan synthesis and cell division.

Under a variety of circumstances, cell division can be inhibited without disrupting the continued linear elongation of the cell. The action of certain β-lactam antibiotics, a variety of chemical and physical agents, alterations in the nutritional environment, and mutations that directly affect the process of cell division can result in the formation of filaments lacking septation. One very fruitful approach has been the use of mutants

TABLE 17-1.   Function of *E. coli* Genes Involved in Cell Wall Synthesis and Cell Division

Gene	Map Location	Reported Function
P_{mra}	2	Promoter for first 9 genes (*mraA* to *ftsW*) in 2 min *mra* cluster[a]
mraA	2	D-Alanine carboxypeptidase
mraB	2	Peptidoglycan-synthesizing enzyme
mrbA	2	UDP-*N*-acetylglucosaminyl-3-enolpyruvate reductase
mrbB	2	Peptidoglycan (PG)-synthesizing enzyme
mrbC	2	PG-synthesizing enzyme
mrcA	2	PBP1A (*ponA*); PG-synthesizing enzyme
mrcB	2	PBP1B (*ponB*); PG-synthesizing enzyme
mraY	2	UDP-*N*-acetylmuramyoylpentapeptide:undecaprenol-PO_4 phosphatase
ftsL	2	Essential cytoplasmic membrane protein involved in cell division; depletion of FtsL results in Y shapes, etc.
fts36	2	(*lts33*), filamentous; maps upstream of *ftsI*
ftsI	2	PBP3 (*pbpB; sep*), membrane protein, septum PG formation
murE	2	*meso*-DAP-adding enzyme
murF	2	D-Alanyl-D-alanine adding enzyme
murD	2	D-Glutamic acid adding enzyme
ftsW	2	Septum PG formation; may pair with PBP-3 like RodA/PBP-2
murG	2	*N*-acetylglucosaminyl transferase; final step in lipid cycle of PG synthesis
murC	2	L-Alanine adding enzyme
ddlB	2	D-Alanyl-D-alanine ligase
dacA	2	PBP5; D-alanine carboxypeptidase
dacB	2	PBP4; D-alanine carboxypeptidase
dacC	2	PBP6: D-alanine carboxypeptidase
ftsA	2	Forms filaments with constrictions at septal sites
ftsZ	2	Mg^{2+}-dependent GTPase; FtsZ forms ring structure at cell division site; facilitates integration of PBP3 into CM; FtsZ overproduction induces minicell formation; forms filaments with no sign of septa; reduced D-alanine carboxypeptidase activity; interacts with SOS-response-associated cell division; inhibits SulB (SfiB) action
sulB	2	(Suppressor of *lon*, or *sfiB*, suppressor of filamentation); allelic to *ftsZ*; functions in septation process
envA	2	(*divC*); forms chains of unseparated cells; reduced levels of NAc muramyl L-alanine amidase
ftsM	2	(*supU*;*serU*); encodes $tRNA_2^{Ser}$; regulation of cell division
secA	2	(*azi*; *pea*); secretion of envelope proteins; terminal end of 2 min region
mrcB	3.6	(*pbpF, ponB*); peptidoglycan synthetase; PBP1Bs
rodA	14.3	(*mrdB*); affects cell shape, mecillinam sensitivity
rodB	14.3	(*mrdA,pbpA*), PBP2; mecillinam resistance
mukB	21.0	Required for chromosome partitioning; DNA binding; *mukB* mutants are anucleate
minB	26.4	*min* operon; formation of minicells containing no DNA, positioning division septum
minC	26.4	Inhibition of FtsZ ring at division site
minD	26.4	Affects cell division and growth; membrane ATPase that activates MinC
minE	26.4	Reverses inhibition by MinC of FtsZ ring

(continued overleaf)

TABLE 17-1. (*continued*)

Gene	Map Location	Reported Function
dicB	35.5	Control of cell division
xerD	41.7	*cer*-specific recombination
pbpG	47.9	PBP7
zipA	54.5	FtsZ interacting protein; essential gene affects cell division and growth; septal ring structural protein
parC	68.2	Cell partitioning; topoisomerase IV subunit A
parE	68.4	nfxD; cell partitioning; topoisomerase IV subunit B
murA	71.8	*mrbA, murZ*; UDP-N-enoylpyruvoyl transferase; phosphomycin resistance
xerC	86.1	*cer*-specific recombination
murI	89.7	*mbrC, dga, glr;* glutamate racemase; D-glutamate synthesis; essential for peptidoglycan synthesis
murB	89.9	UDP-N-acetylglucosaminyl-3-enolpyruvate reductase
murH	99.3	Terminal stage in peptidoglycan synthesis, incorporating disaccharide peptide units into wall

[a]Individual promoters are distributed throughout the 2 min region as well.

defective in cell division. Temperature-sensitive mutants that continue to elongate at 42 °C and form long aseptate filaments containing nucleoids at regular intervals but divide normally at the permissive temperature (28 °C) are designated *fts* for filamentation temperature sensitive. The *fts* genes can be designated as functioning early or late in the division process on the basis of their morphology.

Mutants showing no signs of septation, such as *ftsZ*, apparently act early in the sequence, whereas mutants in *ftsA* and *ftsI* give rise to filaments with indications of an attempt to form septa. By combining a *ftsZ* mutant with morphology mutants such as *rodA* or *pbpA* that govern the cylindrical shape of the cell, it was possible to discern that FtsZ acts at an early stage. Combination with *ftsA, ftsQ,* and *ftsI* mutations resulted in elongated morphology with indications of initiation of septation, suggesting that these genes act at a later stage. Mutation in a cell separation gene (*envA*) causes *E. coli* cells to form chains of cells during fast growth in rich medium. The mutation is associated with low levels of N-acetylglucosaminyl-L-alanine amidase, an enzyme involved in splitting peptidoglycan molecules between the N-acetylmuramic acid residue and the pentapeptide side chain.

The proposed steps in the cell division process of *E. coli* are shown in Table 17-2. Three main groups of enzymes are involved. The first group contains enzymes that form the peptidoglycan sacculus around the growing cell. In the absence of modifying activities, this set of enzymes produces the minimum amount of peptidoglycan needed to completely surround the cell—that is, it produces spherical cells that enlarge without division. The second groups modifying reactions—contains for example, the RodA-PBP2 system—that results in the development of a cylindrical shape to the peptidoglycan sacculus. These functions must be active over the entire surface (except at the poles) because preexisting cylindrical surfaces are converted to spherical surfaces after inactivation of either of these two gene products. Also, a spherical surface formed

TABLE 17-2. Genes Involved in Morphogenesis and Cell Division in *Escherichia coli*

Genes for Peptidoglycan Synthesis	Genes for Rod Cell Shape	Early Acting Cell Division Genes		Late Acting Cell Division Genes	Cell Division Genes Present at Constriction
mraA	*rodA*	*ftsZ*	*zipA*	*ftsZ*	*ftsZ*
mraB	*pbpA*	*minE*	*ftsA*	*zipA*	*minB*
mrbA		*ftsW*		*ftsA*	*ftsA*
mrbB				*ftsW*	*ftsW*
mrbC				*ftsI*	*ftsI*
mrcA				*ftsL*	*ftsL*
mrcB				*ftsN*	*ftsN*
mraY				*ftsQ*	*ftsQ*
murE				*ftsK*	*ftsK*
murF				*envA*	
murG					
murC					
ddl					
dacA					
dacB					
dacC					

in the absence of these products is replaced by or converted to a cylindrical surface after reactivation of this system.

The third group, responsible for cell division through the formation of new septation sites, acts periodically and is confined to the site of septation at the cell surface. A switch to formation of the type of peptidoglycan (murein) required for septation occurs late in the *E. coli* cell cycle. Inhibition by β-lactam agents gives rise to long, multinucleate nonseptate filaments. Benzylpenicillin and cephalexin selectively bind to PBP3, a high-molecular-weight protein specifically involved in the synthesis of peptidoglycan at the septum. The PBP3 protein is encoded by *ftsI* (*pbpB*) and displays both transpeptidation and transglycosylation activity. The FtsI protein interacts with the other gene products (FtsA, FtsQ, FtsW, and RodA) to maintain the rod shape. FtsW pairs with PBP3 in the same manner that RodA and PBP2 interact to regulate peptidoglycan synthesis. The PBP2–RodA and PBP3–FtsW pairs of membrane proteins carry out alternating morphogenetic modifications of the growing murein sacculus.

Switching from cell elongation to septum formation may result from the change in the relative availability of the different peptide chains required by the two competing systems. Using the technique of fusing a green fluorescent protein (GFP) derived from *Aequorea victoria*, a marine jellyfish, to the FtsI protein, it has been possible to show that this protein forms a septal ring. This penicillin-binding protein (FtsI) is a late recruit to the division site, and its septal location is dependent on interaction with FtsZ, FtsA, FtsQ, and FtsL as well as an intact membrane anchor.

The FtsZ protein acts at an early stage of the septation process (Table 17-2 and Fig. 17-1). FtsZ is a Mg-dependent GTPase dispersed throughout the cell. Just prior to the onset of septal invagination, FtsZ is recruited from the cytoplasm to the division site,

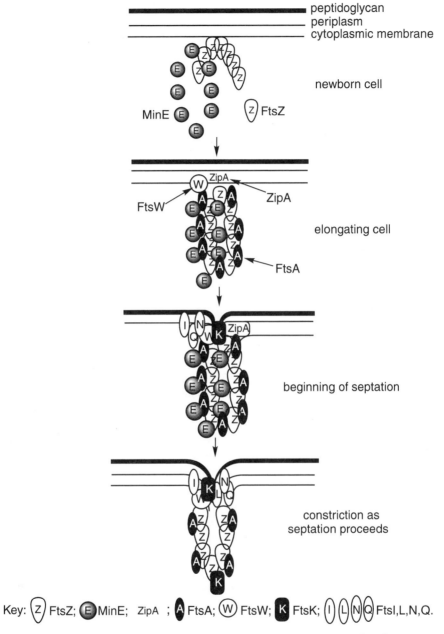

Key: $\widehat{Z}$ FtsZ; ⬤MinE; ZipA ; **A** FtsA; Ⓦ FtsW; ◼ FtsK; ⬮⬮⬮⬮FtsI,L,N,Q.

Fig. 17-1. Conceptual diagram of the role of cell division components in *E. coli.* The process begins with the assembly of FtsZ at the midpoint of the cell and targets MinE to the same location. In the initial stage of septation, ZipA, FtsA, and FtsW associate with the FtsZ ring followed shortly by FtsK, FtsI, FtsL, FtsN, and FtsQ. As septation proceeds, the ring contracts and a few of the FtsZ components begin leaving the ring. FtsK may aid in binding the FtsZ ring to the membrane. See Table 17-1 for a description of the function of the individual components.

where it forms a ring that remains associated with the leading edge of the invaginating septum until septation is completed.

The FtsZ ring system that forms the cytoskeletal framework of the division process is homologous to the tubulin of eukaryotic cells. FtsZ shows sequence homology with tubulin, and the atomic structure reveals virtually identical folds. Both FtsZ and tubulin assemble into protofilaments that can switch from a straight conformation to a curved one. As the cytokinetic ring constricts, the protofilaments curve to accommodate the decreasing diameter and may actually provide the force required for constriction. The shift from a straight to a curved conformation is powered by GTP hydrolysis. It is assumed that the carboxy-terminal peptide of the FtsZ ring is attached to the inner face or the inside of the ring.

ZipA, the integral membrane protein that is an essential component for cell division in *E. coli*, contains an amino-terminal transmembrane anchor and a three-part cytoplasmic domain. ZipA presumably serves as a membrane anchor for FtsZ but may also stabilize and cross-link FtsZ polymers. The carboxy-terminal domain of ZipA binds to a short conserved peptide at the carboxyl terminus of FtsZ. The carboxy-terminal peptide is not required for polymerization of FtsZ or for formation of the Z ring, but it is essential for binding of FtsA and ZipA to FtsZ. FtsA is an actin homolog necessary for the cell division process. The structures of FtsA and ZipA appear to have the flexibility needed to permit the membrane-tethered ZipA to reach around the protofilament to bind the carboxy-terminal peptide on the inner face of FtsZ.

As discussed in Chapter 2, cell division does not take place until a round of replication has been completed. This observation suggests that cell division is triggered by events that occur at or near the termination of replication. The most important of these is decatenation of the replicated chromosomes. The separated chromosomes are then partitioned into the two daughter cells. Division occurs subsequent to or simultaneously with partitioning.

In *E. coli* several enzymes appear to participate in resolving chromosome dimers formed during replication. Two site-specific recombinases, XerC and XerD, act in concert at a site near the terminus of chromosome replication (*dif*) to separate chromosome dimers into monomers during the process of cell division. Mutations in *xerC* and *xerD* result in the development of filamentous cells containing abnormally partitioned cells. It has also been shown that the FtsK protein must be in place at the septal region for the Xer-mediated resolution of chromosomes to occur. Topoisomerase IV, encoded by *parE* and *parC*, is also responsible for decatenation of newly formed chromosomes. In *parE* and *parC* mutant cells, newly replicated chromosomes fail to be decatenated. The *mukB* gene encodes a protein (MukB) involved in chromosome partitioning in *E. coli*. A suppressor gene *smbA* (suppressor of *mukB*) is essential for cell proliferation in the temperature range of 22–42 °C. Cells lacking the SmbA protein cease macromolecular synthesis.

One concept of the placement of the division plane in *E. coli*, nucleoid occlusion, is based on the observation that cytokinesis takes place only between duplicated nucleoids that have become fully separated. The location of the division plane is directly related to the cellular position of the replicating chromosome. Recent evidence that the positioning of the FtsZ ring is affected by the positioning of the segregating nucleoid lends credence to this view. The positioning of the FtsZ ring is inhibited by unsegregated nucleoids in an *E. coli* mutant defective in topoisomerase IV at 42 °C (*parC*). At the permissive temperature (28 °C), the nucleoids separate normally and the

FtsZ ring is assembled at the center of the cell, and division occurs in a normal manner. At the higher temperature, separation of the nucleoids is incomplete, and FtsZ rings are formed on either side of the unsegregated nucleoids, resulting in either anucleate cells or cells containing the incompletely separated chromosomes. FtsZ rings form in many of the anucleate cells and initiate constriction, indicating that the division site is not specified by the nucleoid. If the mutant cells are shifted back to the permissive temperature, the nucleoids begin to separate and the FtsZ ring is assembled in the spaces between the nucleoids.

The *min* operon contains *minB, minC, minD,* and *minE* genes. The *minB* mutant is the original minicell-forming mutant isolated by Howard Adler and colleagues in 1967. Under normal circumstances, the septum in an *E. coli* cell is placed accurately at the midpoint. Potential division sites (PDSs) are also present at the poles, but the products of the *min* operon, MinC, D, and E, specifically repress usage of these sites. The MinC protein can inhibit FtsZ ring assembly at all PDSs, but requires the ATPase MinD for this function. Mutants lacking MinC or MinD produce numerous minicells, reflecting frequent aberrant septation near the cell poles. The MinE protein suppresses the MinCD-mediated division block specifically at midcell, ensuring proper placement of the septum.

Using fusion with a green fluorescent protein (GPF), it has been shown that the MinE protein accumulates in a ring structure at or near the midpoint of the cell. Assembly of the MinE ring also requires MinD, which serves to direct MinE to the cell membrane. MinD displays dynamic properties and accumulates alternately in either one of the cell halves in what appears to be a rapidly oscillating membrane association-disassociation cycle imposed by MinE (Fig. 17-2).

The DicB protein, encoded by *dicB* (see Table 17-1), is another cell division inhibitor unrelated to MinC, MinD, or MinE. It interacts with MinC to cause division inhibition by a MinD-independent mechanism.

As indicated at the outset of the discussion on cell division, the growth and division of the murein sacculus must be an orderly and coordinated process. Because of the intricacies involved, it has been suggested that a multienzyme complex orchestrates the growth of the murein layer. Experimental evidence supports the concept that murein hydrolases and synthases interact with each other. Lytic transglycosylases immobilized to a matrix specifically retained the bifunctional transglycosylase/transpeptidases PBP1A, PBP1B, and PBP1C, the transpeptidases PBP2 and PBP3, and the endopeptidases PBP4 and PBP7. This group of enzymes is needed to enlarge murein according to the three-for-one growth mechanism shown in Figure 17-3. This strategy has been named the "make-before-break" system by Arthur Koch.

In Figure 17-3, a dimer of a bifunctional PBP (PBP1A or PBP1B) could supplement a preexisting primer murein strand to form a murein triplet, which in turn is attached to the sacculus by transpeptidation catalyzed by a dimer of a transpeptidase such as PBP2 or PBP3. The docking strand could be removed specifically by the action of a dimer of an endopeptidase (PBP4 or PBP7) splitting the cross bridges to the left and right of the old strand, and a lytic transglycosylase, that depolymerize the polysaccharide strand. The model shown in Figure 17-3 suggests that a murein triplet may be synthesized in one step. As shown in Table 17-3, the division system is postulated to contain an additional monofunctional transglycosylase that synthesizes the middle strand of the triplet. The cell division process also differs from that of

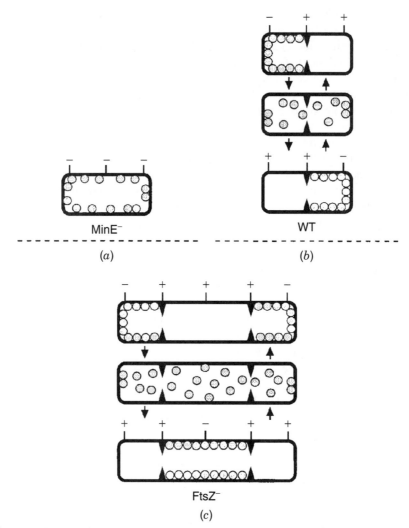

Fig. 17-2. Model for the MinD and MinE action in preventing aberrant septation events. MinD is represented by gray spheres, the MinE ring by filled triangles, and PDSs by either a minus sign (blocked by MinC/MinD action) or a plus sign (not blocked by MinC/MinD, available for assembly of FtsZ ring) sign. (*a*) In the absence of MinE, MinD is distributed evenly over the membrane. Provided MinC is present, this prevents septal ring formation at all PDSs, resulting in the formation of nonseptate filaments. (*b*) In wild-type (WT) cells, MinD oscillates from one side of the MinE ring to the other, alternately blocking division at each of the polar PDSs. For simplicity it is assumed that the MinC/MinD division block is relieved as soon as MinD leaves a PDS, although it may well remain refractive to FtsZ assembly for some period afterward. (*c*) In the absence of FtsZ, multiple MinE rings define three or more cell segments. As in wild-type cells, MinD oscillates between the segments flanking each MinE ring. (*Source*: From D. M. Raskin and P. A. J. de Boer, *Proc. Natl. Acad. Sci. USA* **96**:4971–4976, 1999.)

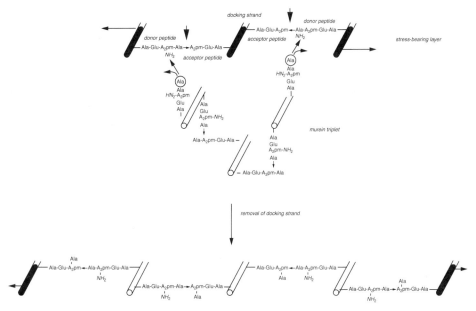

Fig. 17-3. Model for the growth of murein following a make-before-break strategy (three-for-one growth model). Three cross-linked glycan strands from the monolayered, stress-bearing murein are shown in gray. A triple pack of newly synthesized, cross-linked glycans still in a relaxed state is shown in outline. The murein triplet is covalently attached to the free amino groups present in the donor peptides of the cross-links on both sides of a strand, called the docking strand, which is substituted by acceptor peptides. Specific cleavage of the preexisting cross-links (arrowheads) results in the replacement of the docking strand by the murein triplet. The rods represent the glycan strands. A_2pm, diaminopimelic acid. (From J.-V. Höltje, *Microbiol. Mol. Biol. Rev.* **62:**181–203, 1998.)

TABLE 17-3. Elongation and Restriction Complexes in *E. coli* Morphogenesis

Elongation Complex	Constriction Complex
Lytic transglycolase	Lytic transglycolase
Endopeptidase	Endopeptidase
Transpeptidase	Transpeptidase
Transpeptidase-transglycolase	Transpeptidase-transglycolase
Penicillin-binding protein 2 (PBP2)	Penicillin-binding protein 3 (PBP3)
	Transglycosylase

elongation by the presence of the transpeptidase PBP3. The PBP2 transpeptidase, known to be responsible for the rod shape, would be a specific component of the elongation complex.

A comprehensive study of *E. coli* mutants lacking all possible combinations of 8 of the 12 penicillin-binding proteins (PBPs) revealed some interesting information (for a listing of the PBPs and their functions, see Table 7-4). Through genetic engineering it was possible to construct a set of 192 *E. coli* strains containing every possible combination of deletions of PBPs 1A, 1B, 4, 5, 6, and 7, AmpC, and AmpH. In addition, the *dacD* gene was deleted from two septuple mutants, creating strains lacking 8 genes. The only deletion combinations that could not be produced were those lacking both PBP1A and PBP1B because this combination is lethal. Surprisingly, all other deletion mutants were viable even though, at the extreme, 8 of the 12 known PBPs had been eliminated.

Furthermore, when both PBP2 and PBP3 were inactivated by the β-lactam antibiotics mecillinam and aztreonam, respectively, several mutants did not lyse but continued to grow as enlarged spheres, so one mutant synthesized osmotically resistant peptidoglycan when only 2 of 12 PBPs—PBP1B and PBP1C—remained active. The studies show that from among these proteins, PBP2 and PBP3, plus either PBP1A or PBP1B, are the only PBPs required for laboratory growth of regularly dividing rod-shaped cells. This set of *E. coli* strains supplies genetic backgrounds in which the physiological roles of nonessential PBPs and other peptidoglycan-reactive enzymes may be observed and investigated in greater detail.

CELL DIVISION IN GRAM-POSITIVE COCCI

In its simplest aspect, the cell cycle of a bacterial cell can be envisioned as occurring in phases that include B, the interval between cell division and the initiation of DNA synthesis; C, the period of DNA synthesis (chromosome replication); and D, the period between termination of chromosome replication and the end of cell division. T, the total cell cycle, is equal to $B + C + D = 1$.

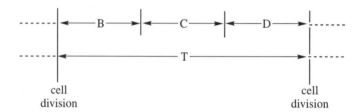

In dividing cells of *Enterococcus hirae*, an equatorial band develops at the initiation of cell division (Fig. 17-4). This fortuitous occurrence permits visual description of the cell cycle in this organism on rather precise terms. A cross-wall is assembled at each of these sites. As a constricting division furrow bilaterally splits this cross-wall, the two cleaved cross-wall layers separate and expand to form two new polar caps. As the formation of new poles nears completion, new sites must be initiated for continued cell growth. These nascent septa appear early in the cell cycle. The absence of significant turnover of peptidoglycan in *E. hirae* aids in distinguishing the preexisting wall surface from the newly extended areas.

The raised wall bands, at relatively fixed distances from the pole, are observed at all stages of the division cycle. With cells of increased length, an increased area of wall is observed between the bands and an equatorially located nascent septum. The wall bands mark the meeting of polar caps made during a previous cell cycle, with equatorially located walls synthesized during the most recent cell cycle. The events that occur during cell division are more or less continuous. However, for the

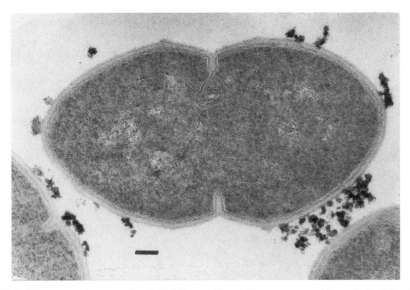

Fig. 17-4. Electron micrograph of dividing cells of *Enterococcus hirae*. Raised bands are observed around the circumference of the cell wall. Nascent septa appear at these sites early in the cell cycle. The wall bands mark the meeting of the polar cap, made during a previous cell cycle, with the equatorially located wall synthesized during the most recent cell cycle. Bar equals 0.1 μm. (*Source*: From Higgins and L. Daneo-Moore, *J. Cell Biol.* **61**:288–300, 1974.)

convenience of description they can be divided into stages as outlined here and in Figure 17-5:

Stage 1: Centripetal penetration of the wall is initiated under an equatorial band. A notch is formed in the external surface of the wall directly above the nascent cross-wall, and two external wall bands appear.

Stage 2: A thin cross-wall penetrates a short distance (70–80 nm) into the cell. The new wall bands are separated by the insertion of a newly synthesized peripheral wall.

Stage 3: The nascent cross-wall thickens somewhat as the new wall bands become separated still further.

Stage 4: When the two new hemispheres have reached adequate size and the wall bands are near the equators of the new daughter cells, the penetration of the equatorial cross-wall resumes. As it penetrates into the cytoplasm, the cross-wall continues to peel apart into two layers of wall that become the poles of the newly formed cells. In rapidly growing cells, the entire process begins again at the two new subequatorial bands.

Stage 5: Final separation of the two new daughter cells occurs. Although the separation may begin as early as the beginning of stage 4, the event that can be quantitated as a doubling in cell number is not completed until the end of stage 5.

In a rich growth medium, the mass doubling time of *E. hirae* is about 30 minutes. Since DNA synthesis takes longer (C = 50–52 min), a new round of chromosome

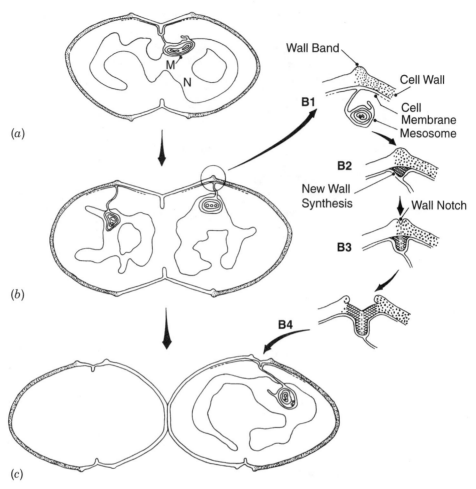

Fig. 17-5. Stages in the cell division cycle of *Enterococcus hirae*. The diplococcus in (*a*) is in the process of growing a new wall at its cross-wall and segregating its nuclear material (N) to the two nascent daughter cocci. In rapidly growing exponential phase cultures before completion of the central cross-wall, new sites of wall elongation are established at the equators of each of the daughter cells at the junction of the old polar wall (stippled) and the new equatorial wall beneath a band of wall material that encircles the equator (*b*). Beneath each band a mesosome (M) is formed while the nucleoids separate and the mesosomes at the central site are lost. The mesosomes appear to be attached to the plasma membrane by a thin membranous stalk (**B1**). Invagination of the septal membrane appears to be accompanied by centripetal cross-wall penetration (**B2**). A notch is then formed at the base of the nascent cross-wall, creating two new wall bands (**B3**). Wall elongation at the base of the cross-wall pushes the newly made wall outward. At the base of the cross-wall, the new wall peels apart into the peripheral wall, pushing the wall bands apart (**B4**). When a sufficient new wall is made so that the wall bands are pushed to a subequatorial position [e.g., from (C) to (A) to (B)], a new cross-wall cycle is initiated. Meanwhile, the initial cross-wall centripetally penetrates into the cell, dividing it into two daughter cocci. At all times the body of the mesosomes appears to be associated with the nucleoid. Doubling of the number of mesosomes seems to precede completion of the cross-wall by a significant interval. Nucleoid shapes and the position of mesosomes are based on projections of reconstructions of serially sectioned cells. (*Source*: From M. L. Higgins and G. D. Shockman, *Crit. Rev. Microbiol.* **1**:29–72, 1971.)

replication must be initiated before the old round of synthesis is completed (**dichoto-mous replication**). It follows, therefore, that wall band splitting and initiation of chromosome replication do not occur simultaneously. The rounds of replication must have been completed early enough so that segregation can take place. Consequently, wall band splitting indirectly controls both cell division and chromosome replication. Thus, it may be concluded that growing streptococci respond to the alteration of nutritional conditions mainly by altering the rate at which they fashion the cell wall. Only to a small extent do they respond by shifting the time of wall band splitting to earlier portions of the cycle.

By comparing the cell cycle events of band splitting (BS) and cell division (CD), as shown in Figure 17-6, it can be concluded that the cell initiates wall band splitting when the growth zones cannot function rapidly enough to allow the increase of surface area required to accommodate continuing production of cytoplasm. Although initiation of new growth sites seems to be independent of normal chromosome replication, it has been shown that chromosome replication is necessary for the terminal events of growth site development that result in the division of a site into two separate poles. Mitomycin C, which inhibits DNA synthesis rapidly and with a minimal effect on the synthesis of other macromolecules, causes an eventual cessation of cell division. However, the growth sites enlarge until they reach about 0.25 μm^3 of cell volume and do not increase further in size. This occurs regardless of whether they are formed before or after the inhibition of DNA synthesis. When those sites approach this 0.25 μm^3 limit, new sites are initiated. These observations suggest that regardless of whether chromosome replication is inhibited, sites have the same finite capacity to enlarge to this size, and when this capacity is reached, new sites are initiated.

The genes governing peptidoglycan biosynthesis and cell division in *Enterococcus faecalis* and *Staphylococcus aureus*, corresponding to those in the 2 min region of *E. coli*, have been identified and sequenced. As shown in Figure 17-7, the gene clusters from *E. coli*, *B. subtilis*, *S. aureus*, and *E. faecalis* show some similarities

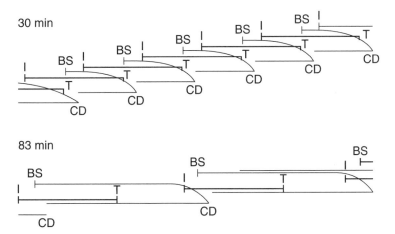

Fig. 17-6. Representation of the cell cycle of *Enterococcus hirae* at two growth rates. I, initiation of chromosome replication; T, termination of chromosome replication; BS, band-splitting events; CD, cell division events. (*Source*: From A. L. Koch and M. L. Higgins, *J. Gen. Microbiol.* **130:**725–745, 1984.)

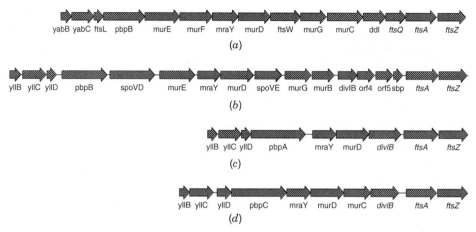

Fig. 17-7. Comparison of the clusters of genes coding for peptidoglycan synthesis and cell division in four bacterial species. (*a*) *E. coli*; (*b*) *B. subtilis*; (*c*) *S. aureus*; (*d*) *E. faecalis*. The arrows indicate open reading frames (ORFs) and directions of transcription. Gene designations are under the ORFs. Unidentified ORFs are designated as orfs. (*Source*: From M. J. Pucci, et al., *J. Bacteriol.* **179**:5632–5635, 1997.)

of organization including the presence of penicillin-binding proteins at the left ends and *ftsA* and *ftsZ* cell division genes at the right ends of the clusters. However, there are also some important differences, including the absence of several genes and the comparative sizes of the *divIB* and *ftsQ* genes. There is also a wide range of amino acid sequence similarities between gram-positive cocci and the other organisms. It should be emphasized that this comparison has been made only for genes present in the 2 min region of the *E. coli* chromosome map.

As indicated in Table 17-1, a substantial number of genes involved in peptidoglycan synthesis and cell division in *E. coli* map to other locations on the chromosome. Also, many of the *E. coli* genes encoding important PBPs map elsewhere on the *E. coli* chromosome map (see Table 7-4). Clearly, several genes essential to the synthesis of peptidoglycan and cell division in gram-positive cocci as well as in *E. coli* must be located at other sites on the chromosome.

A 50 k Da protein, PcsB, has been characterized and found to be present in all of 27 clinical isolates of group B streptococci tested. This protein shows significant similarity to a murein hydrolase from *Listeria monocytogenes*. Insertional inactivation of *pcsB* resulted in mutant strain Sep1, which exhibited a drastically reduced growth rate compared to the parental strain and displayed increased susceptibility to osmotic pressure and to various antibiotics. Electron microscopy revealed growth in clumps, cell separation in several planes, and multiple-division septa within single sells. The data suggest a pivotal role of PcsB for cell division and antibiotic tolerance in group B streptococci.

Resistance to ampicillin in *E. faecium* results from increased production of a low-affinity PBP, PBP5, thought to be intrinsic to all strains of this organism. In *E. hirae*, a close relative of *E. faecium*, increased production of PBP5 has been attributed to deletion of *psr*, a presumed repressor of *pbp5* expression. Mutations in *pbp5* of *E. faecium* result in a decrease in the penicillin-binding capability of PBP5.

Vancomycin is also an inhibitor of peptidoglycan synthesis. It binds to the terminal D-alanyl-D-alanine of pentapeptide precursors, preventing polymerization and cross-linking. Vancomycin resistance is due to the acquisition and expression of resistance operons resulting in the synthesis of precursors that terminate in D-alanine-D-lactic acid. The precursors bind glycopeptides with low affinity and cause destruction of normal pentapeptide precursors. Demonstration of the presence of a *vanB*-containing transposon integrated within a larger transferable element that also contains a mutated *pbp5* gene encoding high-level resistance to ampicillin provides a plausible explanation for the emergence of a number of high-level ampicillin-resistant strains of *E. faecium*.

In *Streptococcus pneumoniae* the class A high-molecular-weight PBPs are bimodular enzymes. In addition to a central penicillin-binding transpeptidase domain, they contain an N-terminal glycosyltransferase domain. Mutations in the *S. pneumoniae* genes for PBP1A, PBP1B, and PBP2A isolated by insertion duplication mutagenesis within the glycosyltransferase domain provide evidence that their function is not essential for cellular growth. PBP1B PBP2A and PBP1A PBP1B double mutants could be isolated but showed defects in positioning of the septum. Attempts to obtain a PBP2A PBP1A double mutant failed. All mutants with a disrupted *pbp2a* gene showed higher sensitivity to moenomycin, an antibiotic known to inhibit PBP-associated glycosyltransferase activity. This finding indicates that PBP2A is the primary target for glycosyltransferase inhibitors in *S. pneumoniae*. Since this organism is a major human pathogen causing pneumonia, meningitis, and ear infections, it has been the target of several studies directed at PBPs.

In all, six PBPs have been identified in *S. pneumoniae*. PBPs 1A, 2B, and 2A belong to class A while PBPs 2x and 2B are monofunctional class B proteins. Deletion of *pbp2x* and *pbp2b* is lethal for *S. pneumoniae,* while deletion of *pbp1a* is tolerated, probably due to compensation by PBP1B. The specific resistance to ceftriaxone and cefotaxime of strains of *S. pneumoniae* isolated from the hospital environment is mediated by modification of PBP2x and PBP1B. Gene transfer of *pbp1a, pbp2x,* and *pbp2b* from resistant strains conferred penicillin resistance on sensitive *S. pneumoniae* strains under laboratory conditions. Protection by moenomycin of the glycosyltransferase domain against trypsin degradation has been interpreted as an interaction between the glycosyltransferase domain and moenomycin. Studies on disruption of the activities of PBPs indicate that individually the *pbp1a, pbp1b*, and *pbp2a* genes are dispensable but that either *pbp1a* or *pbp2a* is required for growth in vitro. The results of many of these studies indicate the potential for acquisition by *S. pneumoniae* of high-level β-lactam resistance by interspecies gene transfer.

CELL DIVISION IN GRAM-POSITIVE BACILLI

The morphogenetic approach to the investigation of cell division that proved so useful in the study in gram-negative organisms has also been useful for studies with gram-positive bacilli. As show in Figure 17-8, there is considerable similarity between the genes in the 133° region of the genetic map of *B. subtilis* and the genes in the *mra* (2 min) region of the *E. coli* genetic map that are involved in the synthesis of cell wall peptidoglycans and cell division. Mutants of *B. subtilis* have been found that form filaments at high temperature (40–50 °C) but grow normally at the permissive temperature (20–35 °C).

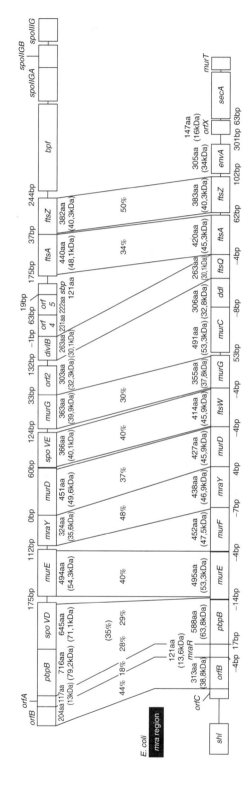

Fig. 17-8. Comparison of the 133° region of the _B. subtilis_ genetic map with the _mra_ (2 min) region of the _E. coli_ genetic map. Similar pairs of genes are joined by dotted lines and similarities in predicting amino acid sequences are indicated as percent (%) identity. The gaps between adjacent coding regions are indicated in base pairs (bp). (From C. E. Buchanan, Ghuysen and Hackenbeck, A. O. Henriques, and P. J. Piggot 1994. Cell wall changes during bacterial endospore formation. In Ghuysen, J.-M. and R. Hackenbeck (Eds.). _Bacterial Cell Wall_, Elsevier, Amsterdam, The Netherlands, pp. 167–186.)

576

There are a number of similarities in the proteins involved in cell division in *B. subtilis* as compared to *E. coli*. However, in *B. subtilis* there are no homologs of ZipA or FtsN. FtsZ and three membrane-bound division proteins — DivIB (a homolog of FtsQ), DivIC (comparable to FtsL), and PBP2B (a homolog of PBP3) — localize at the division site. As in *E. coli*, *FtsZ* appears to assemble early prior to the appearance of FtsL, DivIC, DivIB, or PBP2B. However, there are a number of differences in the *B. subtilis* cell division system. Depletion of FtsL results in rapid degradation of DivIC, while DivIB and PBP2B are undetectable, suggesting that these proteins assemble at the division site after FtsL or at the same time. This instability of one division protein in the absence of another is not observed in *E. coli*.

When PBP2B is depleted at 37 °C, DivIB and DivIC localizations are significantly decreased and occur only at midfilament positions. PBP2B also fails to localize in *divIB* and *divIC* temperature-sensitive mutants at 48 °C. The interdependency of DivIB, DivIC, and PBP2B suggests that these three proteins localize together or that at the nonpermissive temperature the other two proteins are required to maintain PBP2B at the septal site. Immunofluorescence microscopy studies show that DivIC is required for DivIB localization. However, DivIC localization is dependent on DivIB only at high growth temperatures, at which DivIB is essential for division. FtsZ localization is required for septal recruitment of DivIB and DivIC, but FtsZ can be recruited independently of DivIB.

The high-molecular-weight PBP1 may also play a role in cell division in *B. subtilis*. Cells lacking PBP1 sometimes grow as filaments. PBP1 (encoded by *ponA*) is expressed predominantly during vegetative growth. Vegetative *B. subtilis* cells also contain the high-molecular-weight PBP2C and PBP4. However, PBP1 appears to be functionally more important. Cells lacking PBP1 have a reduced growth rate in a variety of growth media and are longer, thinner, and more bent than wild-type cells. PBP1-deficient cells require increased levels of Mg^{2+}, and under conditions of deficiency of this cation, they are either unable to grow or they grow as filaments. Using epitope tagging and immunofluorescence microscopy, PBP1 has been shown to be localized at division sites in vegetative *B. subtilis*. Growing *ponA* mutant cells display a significant septation defect. Also, while FtsZ localizes normally in most *ponA* mutant cells, a significant portion of these cells display FtsZ rings with aberrant structure or improper localization, suggesting that the lack of PBP1 affects FtsZ ring stability or assembly. This appears to be the first example of a high-molecular-weight PBP localized to the bacterial division septum.

Several genes that play roles in chromosome partitioning in *B. subtilis* include *smc*, *spoOJ*, and *spoIIIE*. Proteins involved in structural maintenance of chromosomes (SMC proteins) are found in eukaryotes, archaea, and prokaryotes. The SMC protein of *B. subtilis* may have a function analogous to that of MukB in *E. coli*. SMC protein has the ability to aggregate and/or renature single-stranded DNA in ATP-dependent reactions in vitro. A technique for visualizing specific regions on the chromosome in living cells that uses green fluorescent protein in combination with time-lapse microscopy has shown that the SMC protein is required for proper arrangement of the chromosome and for efficient segregation of replication termini but not for bipolar movement of the chromosome.

The SpoOJ protein is a member of the ParB family of partition proteins. It binds to multiple sites in the origin region of the chromosome and forms a large nucleoprotein complex that is visible by immunofluorescence microscopy or green fluorescent protein fusion. This complex may play a role in pairing newly replicated sister origin regions or in structurally organizing the origin region. A double mutant, *smc spoOJ*, has a synthetic lethal phenotype in rich medium. SpoIIIE is involved in postseptational chromosome segregation during sporulation. However, it is also required for efficient segregation during vegetative growth when normal cell division or chromosome partitioning has been perturbed. During vegetative growth, SpoIIIE may provide a backup mechanism for chromosome partitioning when normal partitioning is defective. Mutants defective in nucleoid structure are unable to move chromosomes out of the way of the invaginating septum, and SpoIIIE is involved in repositioning these bisected chromosomes.

In *B. subtilis* the CodV and RipX proteins show 35% and 44% identity with the XerC and XerD proteins, respectively. The XerC and XerD proteins of *E. coli* are considered to act at the *dif* site to resolve dimeric chromosomes formed by recombination during replication. The *dif* gene, for deletion-induced filamentation, is a *recA*-independent recombination site at the terminus of the chromosome. In *B. subtilis,* genetic and biochemical evidence reveals that a 28 bp sequence of DNA (Bs*dif*), lying 6° counterclockwise from the terminus of replication (172°), is the site at which RipX and CodV catalyze site-specific recombination reactions required for normal chromosome partitioning. Cultures of *ripX* mutants contain a subpopulation of unequal-size cells held together in long chains. These chains include anucleate cells and cells with aberrantly dense or diffuse nucleoids, indicating a chromosome partitioning failure. In *B. subtilis* catenation nodes are removed from replicated chromosomes by ParC and ParE (subunits of topoisomerase IV). Both CodV and RipX are required for efficient recombination of substrates in vitro.

A number of genes involved in the cell division process in *B. subtilis* are primarily concerned with the unequal division of cells during the sporulation process. The role of these genes in sporulation is discussed in Chapter 19.

BIBLIOGRAPHY

General Reviews

Harry, E. J. 2001. Bacterial cell division: regulating Z-ring formation. *Mol. Microbiol.* **40:**795–803.

Höltje, J.-V. 1998. Growth of the stress-bearing and shape-maintaining murein sacculus of *Escherichia coli. Microbiol. Mol. Biol. Rev.* **62:**181–a203.

Lutkenhaus, J., and S. G. Addinall. 1997. Bacterial cell division and the Z ring. *Annu. Rev. Biochem.* **66:**93–116.

Lutkenhaus, J., and A. Mukherjee. 1996. Cell division. In F. C. Neidhardt, et al. (ed.), *Escherichia coli and Salmonella: Cellular and Molecular Biology,* 2nd ed. American Society for Microbiology, Washington, DC, pp. 1615–36.

Margolin, W. 1999. The bacterial cell division machine. *ASM News* **65:**137–43.

Naninga, N. 1998. Morphogenesis in *Escherichia coli. Microbiol. Mol. Biol. Rev.* **62:**110–29.

Naninga, N. 2000. Cell division, prokaryotes. In J. Lederberg (ed.), *Encyclopedia of Microbiology,* 2nd ed., Vol. 1. Academic Press, San Diego, pp. 704–9.

Naninga, N. 2001. Cytokinesis in prokaryotes and eukaryotes: common principles and different solutions. *Microbiol. Mol. Biol. Rev.* **65:**319–33.

Rothfield, I., S. Justice, and J. Garcia-Lara. 1999. Bacterial cell division. *Annu. Rev. Genet.* **33:**423–48.

Sawitzke, J., and S. Austin. 2001. An analysis of the factory model for chromosome replication and segregation in bacteria. *Mol. Microbiol.* **40:**786–94.

Cell Division in Gram-Negative Rods

Begg, K., Y. Nikolaichik, N. Crossland, and W. D. Donachie. 1998. Roles of FtsA and FtsZ in activation of division sites. *J. Bacteriol.* **180:**881–4.

Chen, J. C., D. S. Weiss, J.-M. Ghigo, and J. Beckwith. 1999. Septal localization of FtsQ, an essential cell division protein in *Escherichia coli. J. Bacteriol.* **181:**521–30.

Den Blaauwen, T., N. Buddelmeijer, M. E. G. Aarsman, C. M. Hameete, and N. Nanninga. 1999. Timing of FtsZ assembly in *Escherichia coli. J. Bacteriol.* **181:**5167–75.

Denome, S. A., P. K. Elf, T. A. Henderson, D. E. Nelson, and K. D. Young. 1999. *Escherichia coli* mutants lacking all possible combinations for eight penicillin binding proteins: viability, characteristics, and implications for peptidoglycan synthesis. *J. Bacteriol.* **181:**3981–93.

Donachie, W. D. 2001. Co-ordinate regulation of the *Escherichia coli* cell cycle or *the cloud of unknowing. Mol. Microbiol.* **40:**779–85.

Erickson, H. P. 2000. Dynamin and FtsZ: missing links in mitochondrial and bacterial division. *J. Cell. Biol.* **148:**1103–5.

Erickson, H. P. 2001. The FtsZ protofilament and attachment of ZipA — structural constraints on the FtsZ power stroke. *Curr. Opin. Cell Biol.* **13:**55–60.

Ghigo, J.-M., and J. Beckwith. 2000. Cell division in *Escherichia coli*: role of FtsL domains in septal localization, function, and oligomerization. *J. Bacteriol.* **182:**116–29.

Hale, C. A., and P. A. J. de Boer. 1997. Direct binding of FtsZ to ZipA, an essential component of the septal ring structure that mediates cell division in *E. coli. Cell* **88:**175–85.

Hale, C. A., and P. A. J. de Boer. 1999. Recruitment of ZipA to the septal ring of *Escherichia coli* is dependent on FtsZ and independent of FtsA. *J. Bacteriol.* **181:**167–76.

Hale, C. A., and P. A. J. de Boer. 2000. ZipA-induced bundling of FtsZ polymers mediated by an interaction between C-terminal domains. *J. Bacteriol.* **182:**5153–66.

Hara, H., S. Yasuda, K. Horiuchi, and J. T. Park. 1997. A promoter for the first nine genes of the *Escherichia coli mra* cluster of cell division and cell envelope biosynthesis genes, including *ftsI* and *ftsW. J. Bacteriol.* **179:**5802–11.

Levine, C., and K. J. Marians. 1998. Identification of *dnaX* as a high-copy suppressor of the conditional lethal and partition phenotypes of the *parE10* allele. *J. Bacteriol.* **180:**1232–40.

Lu, C., M. Reedy, and H. P. Erickson. 2000. Straight and curved conformations of FtsZ are regulated by GTP hydrolysis. *J. Bacteriol.* **182:**164–70.

Ma, X., D. W. Earhardt, and W. Margolin. 1996. Colocalization of cell division proteins FtsZ and FtsA to cytoskeletal structures in living *Escherichia coli* cells by using green fluorescent protein. *Proc. Natl. Acad. Sci. USA* **93:**12,998–13,003.

Ma, X., Q. Sun, R. Wang, et al. 1997. Interactions between heterologous FtsA and FtsZ proteins at the FtsZ ring. *J. Bacteriol.* **179:**6788–97.

Powell, B. S., and D. L. Court. 1998. Control of *ftsZ* expression, cell division, and glutamine metabolism in Luria-Bertani medium by the alarmone ppGpp in *Escherichia coli. J. Bacteriol.* **180:**1053–62.

Raskin, D. M., and P. A. J. de Boer. 1997. The MinE ring: an FtsZ-independent cell structure required for selection of the correct division site in *E. coli. Cell* **91:**685–94.

Raskin, D. M., and P. A. J. de Boer. 1999. Rapid pole-to-pole oscillation of a protein required for directing division to the middle of *Escherichia coli. Proc. Natl. Acad. Sci. USA* **96:**4971–6.

Rowland, S. L., X. Fu, M. A. Sayed, Y. Zhang, W. R. Cook, and L. I. Rothfield. 2000. Membrane redistribution of the *Escherichia coli* MinD protein induced by MinE. *J. Bacteriol.* **182**:613–9.

Sun, Q., X. C. Yu, and W. Margolin. 1998. Assembly of the FtsZ ring at the central division site in the absence of the chromosome. *Mol. Microbiol.* **29**:491–503.

Wang, L., M. K. Khattar, W. D. Donachie, and J. Lutkenhaus. 1998. FtsI and FtsW are localized to the septum in *Escherichia coli. J. Bacteriol.* **180**:2810–6.

Weiss, D. S., J. C. Chen, J.-M. Ghigo, D. Boyd, and J. Beckwith. 1999. Localization of FtsI (PBP3) to the septal ring requires its membrane anchor, the Z ring, FtsA, FtsQ, and FtsL. *J. Bacteriol.* **181**:508–20.

Yu, X.-C., A. H. Tran, Q. Sun, and W. Margolin. 1998. Localization of cell division protein FtsK to the *Escherichia coli* septum and identification of a potential N-terminal targeting domain. *J. Bacteriol.* **180**:1296–1304.

Zhou, P., Bogan, J. A., Welch, K., et al. 1997. Gene transcription and chromosome replication in *Escherichia coli. J. Bacteriol.* **179**:163–9.

Cell Division in Gram-Positive Cocci

Buist, G., J. Kok, K. J. Leenhouts, et al. 1995. Molecular cloning and nucleotide sequence of the gene encoding the major peptidoglycan hydrolase of *Lactococcus lactis*, a muramidase needed for cell separation. *J. Bacteriol.* **177**:1554–63.

Carias, L. L., S. D. Rudin, C. J. Donskey, and L. B. Rice. 1998. Genetic linkage and cotransfer of a novel, *vanB*-containing transposon (Tn5382) and a low-affinity penicillin-binding protein 5 gene in a clinical vancomycin-resistant *Enterococcus faecium* isolate. *J. Bacteriol.* **180**:4426–34.

Di Guilmi, A. M., N. Mouz, L. Martin, et al. 1999. Glycosyltransferase domain of penicillin-binding protein 2a from *Streptococcus pneumoniae* is membrane associated. *J. Bacteriol.* **181**:2773–81.

Filipe, S. R., E. Severina, and A. Tomasz. 2000. Distribution of the mosaic structured *murM* genes among natural populations of *Streptococcus pneumoniae. J. Bacteriol.* **182**:6798–805.

Grohs, P., L. Gutmann, R. Legrand, et al. 2000. Vancomycin resistance is associated with serine-containing peptidoglycan in *Enterococcus gallinarum. J. Bacteriol.* **182**:6228–32.

Groicher, K. H., B. A. Firek, D. Fujimoto, and K. W. Bayles. 2000. The *Staphylococcus aureus lrgAB* operon modulates murein hydrolase activity and penicillin tolerance. *J. Bacteriol.* **182**:1794–801.

Mengin-Lecreulx, D., T. Falla, D. Blanot, et al. 1999. Expression of the *Staphylococcus aureus* UDP-*N*-acetylmuramoyl-L-alanyl-D-glutamate: lysine ligase in *Escherichia coli* and effects on peptidoglycan biosynthesis and cell growth. *J. Bacteriol.* **181**:5909–14.

Paik, J., I. Kern, R. Lurz, and R. Hakenbeck. 1999. Mutational analysis of the *Streptococcus pneumoniae* bimodular class A penicillin-binding proteins. *J. Bacteriol.* **181**:3852–6.

Pinho, M. G., H. de Lencastre, and A. Tomasz. 2000. Cloning, characterization, and inactivation of the genes *pbpC*, encoding penicillin-binding protein 3 of *Staphylococcus aureus. J. Bacteriol.* **182**:1074–9.

Pucci, M. J., J. A. Thanassi, H.-T. Ho, P. J. Falk, and T. J. Dougherty. 1995. *Staphylococcus haemolyticus* contains two D-glutamic acid biosynthetic activities, a glutamate racemase and a D-amino acid transaminase. *J. Bacteriol.* **177**:336–42.

Pucci, M. J., J. A. Thanassi, L. F. Discotto, R. E. Kessler, and T. I. Dougherty. 1997. Identification and characterization of cell wall–cell division gene clusters in pathogenic gram-positive cocci. *J. Bacteriol.* **179**:5632–5.

Reinscheid, D. J., B. Gottschalk, A. Schubert, et al. 2001. Identification and molecular analysis of PcsB, a protein required for cell wall separation of group B streptococci. *J. Bacteriol.* **183:**1175–83.

Sieradzki, K., and A. Tomasz. 1999. Gradual alterations in cell wall structure and metabolism in vancomycin-resistant mutants of *Staphylococcus aureus*. *J. Bacteriol.* **181:**7566–70.

Sugai, M., T. Fujuwara, K. Ohta, et al. 1997. *epr*, Which encodes glycylglycine endopeptidase resistance, is homologous to *femAB* and affects serine content of peptidoglycan cross bridges in *Staphylococcus capitis* and *Staphylococcus aureus*. *J. Bacteriol.* **179:**4311–8.

Wada, A., and H. Watanabe. 1998. Penicillin-binding protein 1 of *Staphylococcus aureus* is essential for growth. *J. Bacteriol.* **180:**2759–65.

Walsh, A. W., P. J. Falk, J. Thanassi, et al. 1999. Comparison of the D-glutamate adding enzymes from selected gram-positive and gram-negative bacteria. *J. Bacteriol.* **181:**5395–401.

Zorzi, W., X. Y. Zhou, O. Dardenne, et al. 1996. Structure of the low-affinity penicillin-binding protein 5 PBP5fm in wild-type and highly penicillin-resistant strains of *Enterococcus faecium*. *J. Bacteriol.* **178:**4948–57.

Cell Division in Gram-Positive Bacilli

Graumann, P. L. 2000. *Bacillus subtilis* SMA is required for proper arrangement of the chromosome and for efficient segregation of replication termini but not for bipolar movement of newly duplicated origin regions. *J. Bacteriol.* **182:**6463–71.

Katis, V. L., and R. G. Wake. 1999. Membrane-bound division proteins DivIB and DivIC of *Bacillus subtilis* function solely through their external domains in both vegetative and sporulation division. *J. Bacteriol.* **181:**2710–8.

Levin, P. A., J. J. Shim, and A. D. Grossman. 1998. Effect of *minCD* on FtsZ ring position and septation in *Bacillus subtilis*. *J. Bacteriol.* **180:**6048–51.

Pedersen, L. B., E. R. Angert, and P. Setlow. 1999. Septal localization of penicillin-binding protein 1 in *Bacillus subtilis*. *J. Bacteriol.* **181:**3201–11.

Regamey, A., E. J. Harry, and R. G. Wake. 2000. Mid-cell Z ring assembly in the absence of entry into the elongation phase of the round of replication in bacteria: coordinating chromosome replication with cell division. *Mol. Microbiol.* **38:**423–34.

Sciochetti, S. A., P. J. Piggot, D. J. Sherratt, and G. Blakely. 1999. The *ripX* locus of *Bacillus subtilis* encodes a site-specific recombinase involved in proper chromosome partitioning. *J. Bacteriol.* **181:**6053–62.

Sciochetti, S. A., P. J. Piggot, and G. W. Blakely. 2001. Identification and characterization of the *dif* site from *Bacillus subtilis*. *J. Bacteriol.* **183:**1058–68.

Sharpe, M. E., and J. Errington. 1998. A fixed distance for separation of newly replicated copies of *oriC* in *Bacillus subtilis*: implications for coordination of chromosome segregation and cell division. *Mol. Microbiol.* **28:**981–90.

Sharpe, M. E., P. M. Hauser, R. G. Sharpe, and J. Errington. 1998. *Bacillus subtilis* cell cycle as studied by fluorescence microscopy: constancy of cell length initiation of DNA replication and evidence for active nucleoid partitioning. *J. Bacteriol.* **180:**547–55.

Vincente, M., and J. Errington. 1996. Structure, function and controls in microbial division. *Mol. Microbiol.* **20:**1–7.

Wang, X., and J. Lutkenhaus. 1993. The FtsZ protein of *Bacillus subtilis* is localized at the division site and has GTPase activity that is dependent upon FtsZ concentration. *Mol. Microbiol.* **9:**435–42.

CHAPTER 18

MICROBIAL STRESS RESPONSES

When supplied with sufficient nutrients and optimal growth temperature, pH, oxygen levels, and solute concentrations, microbes will grow at a maximum growth rate characteristic for the organism. Variations in any of these parameters can affect the maximum growth rate and, thus, can represent an **environmental stress** for the microbe. The ability of microbes to sense and respond (correctly) to impromptu alterations in the environment is crucial to their survival. In reality, conditions that allow for maximal growth rates outside the laboratory are few and far between. As a result, most bacteria live in a constant state of stress. This chapter covers the physiologic and genetic responses of certain well-characterized bacteria to various stress conditions.

OSMOTIC STRESS AND OSMOREGULATION

The concentration of solutes (e.g., salts, ions, metabolites) plays a critical role in microbial growth. In the laboratory, most microbes exhibit optimal growth in culture media of relatively low osmolarity. For many bacteria, **hypertonic** or **hyperosmotic** conditions result in water loss from the cytoplasm, causing the cell to shrink (**plasmolysis**). **Hypotonic** or **hypoosmotic** conditions result in an influx of water into the cytoplasm, which causes the cell to swell (**plasmoptysis**) and perhaps burst in a process referred to as **osmotic lysis**. Microbial membranes are freely permeable to water and, therefore, water inside the cell is essentially at equilibrium with the outside of the cell. Microbes with cell walls can and do maintain a high cytoplasmic solute concentration relative to the outside. This translates into lower **water activity (A$_w$)** inside the cell.

$$A_w = x \div (x + c)$$

[c = osmolality of solute(s); x = moles of water per liter (55.6)]

The lower cytoplasmic A_w causes water to at least try and flow into the cell. This places significant pressure on the microbial cell wall as the cell volume attempts to increase at the same time the cell wall prevents the cell from swelling. This pressure placed on the cell wall by the cytoplasmic membrane is referred to as **turgor pressure** or **turgor**. Turgor is opposed by the tension of the cell envelope. The function of osmoregulatory mechanisms or osmotic stress responses is to maintain turgor within limits, allowing for maintenance of cell viability.

Movement of water occurs by diffusion and, in a much more rapid process, through water-selective channels called **aquaporins**. The AqpZ channel of *E. coli* has been shown to mediate rapid and large water fluxes in both directions in response to sudden osmotic upshifts or downshifts, although its role in the cell is not essential. Turgor is maintained by regulating the total osmotic solute pool in the cytoplasm and the relative level of solutes in the periplasm (in Gram-negative bacteria) immediately outside the cytoplasmic membrane. In low-osmolality media, cytosolic osmolality is largely due to ionic solutes (e.g., K^+ ions); in high-osmolality media it largely involves neutral solutes (e.g., trehalose).

Some of the mechanisms described below are also responsible for maintaining osmotic homeostasis. For this reason, mechanisms of osmoregulation such as K^+ ion influx/glutamate biosynthesis–coupled systems are present all the time in the cell.

High Osmolality

As the osmolality of the surrounding environment increases, turgor pressure drops and growth slows or halts. Macromolecular biosynthesis is inhibited and respiration rates decline. The most rapid response to this osmotic upshock is an **increase in K^+ ion influx** through, at least, three uptake systems in *E. coli*: **Trk**, **Kdp**, and **Kup**. Homologous systems to Trk and Kdp have been identified in *Salmonella enterica* and other bacteria. The Trk and Kdp systems appear to be the major systems for K^+ uptake under these conditions, since they can achieve sufficiently high rates of uptake. The **Trk system** is composed of three components: **TrkA** (peripheral membrane protein), **TrkE** (membrane associated), and either **TrkH** (in *E. coli*; membrane-spanning protein) or **TrkG** (*E. coli* and other bacteria; membrane-spanning protein). The Trk system binds NAD(H) via TrkA and may regulate K^+ ion uptake. The **Kdp system** (see Chapter 9) is also a three-component system composed of **KdpA** (membrane-spanning protein), **KdpB** (integral membrane protein), and **KdpC** (peripheral membrane protein). KdpB is a **P-type ATPase** and likely provides the energy to drive K^+ ion influx through this system. Kup is a single, large membrane-spanning protein possessing a significant cytoplasmic tail domain. In addition to influx, K^+ ion accumulation results from plasmolysis and the closing of stretch-sensitive K^+ ion efflux channels.

Concurrent with the increase in K^+ ion influx, as a result of high external osmolality, is a decrease in intracellular putrescine levels due to increased excretion. Putrescine is a divalent cationic polyamine; thus, a divalent putrescine is replaced by the monovalent K^+ ion, allowing cytoplasmic osmolality to increase with minimal effect on the intracellular ionic strength.

The major anionic compound involved in osmoregulation (**osmolyte**) is **glutamate**. Glutamate is synthesized and accumulates quickly following osmotic upshock and is dependent on K^+ ion uptake. In *E. coli* and other enteric bacteria, glutamate is synthesized by two enzymes: **glutamate dehydrogenase (GDH)** and **glutamate synthase (GS)**.

Many microbes also accumulate the **disaccharide trehalose** (composed of two glucose residues) as a **compatible solute** in response to osmotic stress as well as starvation, thermal stress, and desiccation stress. Trehalose is synthesized by the products of the *otsAB* **operon**. **OtsA** is a trehalose-6-phosphate synthase and **OtsB** is a trehalose-6-phosphate phosphatase. In the absence of exogenous osmoprotectants (see below), *E. coli* can accumulate trehalose to levels that may comprise as much as 20% of the cytoplasmic osmolality under conditions of high osmolality.

Several compounds, when present externally, can stimulate bacterial growth rates under hyperosmotic conditions. These compounds are **osmoprotectants**. They are zwitterionic in nature and resemble glycine betaine and proline.

$(CH_3)_3\text{-}N^+\text{-}CH_2COO^-$

Betaine (glycinebetaine)

L-proline

Many potential osmoprotectants are found in plants and animals: betaine, betaine aldehyde, proline betaine, choline, choline-O-sulfate, stachydrine, and β-dimethyl-sulfonopropionate are present in plants, and carnitine is plentiful in animals. In *E. coli*, choline is converted to betaine aldehyde by the membrane-associated O_2-dependent BetA choline dehydrogenase, and then to glycine betaine by the soluble NADP-dependent betaine aldehyde dehydrogenase. Ecotoine produced by many halophilic (salt-loving; marine) bacteria is an excellent osmoprotectant for *E. coli* and other bacteria. These compounds are taken up into the cell by two osmotically regulated permeases, ProP and ProU, in *E. coli* and *S. enterica* and homologs are found in other bacteria. ProP has relatively low affinity for proline and appears to function primarily in the uptake of betaine. The ProU permease also has relatively low affinity for proline but is a high-affinity betaine transporter. The major proline transporter is PutP, but it mainly functions in proline utilization and is not involved in osmoprotection (see Chapter 5 for its regulation).

Interestingly, these mechanisms of osmoprotection exhibit a hierarchy of function. Initially, the intracellular level of K^+ ions rapidly rises, maximizing by 20 minutes. Glutamate accumulation follows somewhat more slowly. The cell recovers from plasmolysis as turgor is restored. Trehalose levels rise and K^+/glutamate levels decline after about 60 minutes. If an osmoprotectant such as betaine is added externally, trehalose concentrations decline and K^+ ion levels decrease even further. Restarting of macromolecular biosynthesis corresponds with a rise in trehalose levels.

Low Osmolality

Immediately following a significant decrease in osmolality (**hypoosmotic shock**), there is a rapid flux of water into the cell, increasing turgor. This results in stretching of the cell envelope and may lead to "cracks" in the membrane or may stretch existing pores or activate stretch-activated channels. The permeability of the membrane will be increased but only temporarily. Downshifts in osmolality can result in the extrusion of osmolytes and ions as well as the loss of amino acids, nucleotides, and other solutes from the cytoplasm.

Complex sugars, known as **membrane-derived oligosaccharides (MDOs)** that contain 6 to 12 glucose residues joined via $\beta(1 \rightarrow 2)$ and $\beta(1 \rightarrow 6)$ linkages, forming a branched structure, are found in the periplasm of Gram-negative bacteria. MDOs are substituted with *sn*-1-phosphoglycerol and phosphoethanolamine derived from the membrane phospholipids and also with *O*-succinyl ester residues. Synthesis of these compounds is induced by growth in conditions of low osmolality.

Osmotic Control of Gene Expression

EnvZ/OmpR is a paradigm of two-component systems (see Chapter 5). The outer membrane of *E. coli* K-12 contains several major outer membrane proteins. The relative amounts of two of them, coded by the *ompC* (47 min) and *ompF* (21 min) loci, are mediated by medium osmolarity. **OmpC** (mwt 36,500 mw) and **OmpF** (mwt 37,000) are porin proteins that form aqueous pores in the outer membrane, allowing polar molecules (<600 da) to cross the outer membrane barrier. In media of low osmolarity, OmpF protein is present in greater quantities than OmpC protein. In media of high osmolarity, the OmpC porin predominates over OmpF. While the quantitative ratios of these proteins vary, their combined levels remain fairly constant.

Regulation of these genes (Fig. 18-1) is mediated by the regulatory *ompB* locus (21 min). The *ompB* region actually is comprised of two genes, *ompR* and *envZ*, arranged in an operon. **EnvZ** (50,300 Da) is a transmembrane sensor protein that undergoes **autophosphorylation** at histidine 243 under conditions of high osmolarity. How EnvZ senses environmental change is unclear, since the periplasmic portion of the protein is apparently not required. Once phosphorylated, EnvZ will then transphosphorylate aspartate 55 in the receiver module of the true **DNA-binding regulator, OmpR** (27,400 Da), thereby increasing the level of OmpR~P in the cell. Phosphorylation of OmpR is required for the expression of both *ompC* and *ompF*. Presumably, conformational reshaping of OmpR by phosphorylation may increase DNA-binding affinity to the *ompC* and *ompF* promoters and enables interaction of OmpR with the α subunit of RNA polymerase, thereby activating transcription.

If OmpR~P is required for the transcription of both *ompF* and *ompC*, what could account for their reciprocal control? As shown in Figure 18-1, there are three OmpR-binding sites in front of *ompC* and *ompF*. At low osmolarity, the OmpR~P present in the cell binds to *ompF* sites F1, F2, and F3, activating *ompF* transcription, but only binds to *ompC* site C1. At high osmolarity, the higher level of OmpR~P resulting from EnvZ transphosphorylation now binds to C1, C2, and C3, activating *ompC* transcription. It is strange that this level of OmpR~P still binds at F1, F2, F3 but now represses *ompF*. It is proposed that when OmpR~P is bound to DNA, changes in osmolarity affect OmpR~P-OmpR~P protein interactions, which in turn alter the ability of OmpR to activate transcription at *ompF* (Fig. 18-1). Osmotic downshift — that is, when osmolarity shifts from high back to low — activates an EnvZ phosphatase that removes P from both OmpR and EnvZ, reversing the OmpC/OmpF expression ratio. A second phosphatase, SixA, may also be involved in this process.

A variety of studies indicate that alternative **phosphodonors** are capable of phosphorylating OmpR in the absence of EnvZ. One proven alternative phosphodonor is **acetyl phosphate**, although there may be additional cross-talk with other histidine kinases.

An additional control mechanism that appears to regulate the relative amounts of OmpC and OmpF involves a gene (*micF*) adjacent to *ompC*, which is transcribed

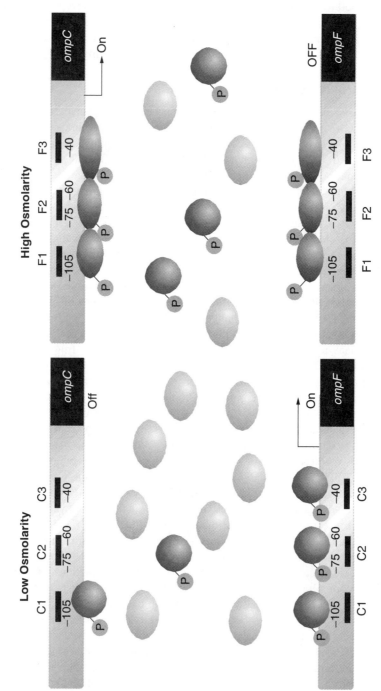

Fig. 18-1. Osmoregulation of major outer membrane proteins OmpF and OmpC.

in the opposite direction from *ompC*. The *micF* transcript (**micRNA**, 93 nucleotides) is complementary to the 5′ end region of *ompF* mRNA. Under conditions of high osmolarity, the *micF-ompC* region is induced and transcribed divergently from a central promoter. The resulting micRNA (also referred to as **antisense RNA**) inhibits translation of *ompF* mRNA by hybridizing to it. This RNA–RNA interaction is proposed to cause premature termination of *ompF* transcription and/or destabilization of *ompF* mRNA. The end result is that when more OmpC protein is produced, less OmpF protein is synthesized.

The role of the EnvZ-OmpR system in cellular physiology extends well beyond its role in governing porin expression. Studies have connected OmpR with flagellar expression, cell division, fatty acid transport, microcin synthesis, curli fibers, *Salmonella* virulence, and acid tolerance. One mechanism by which OmpR may affect virulence is through its involvement in controlling cytotoxicity toward infected macrophages. In addition, OmpR controls the expression of another two-component regulatory system, SsrAB, which controls expression of genes in the *Salmonella* pathogenicity island SPI-2. Recent evidence indicates that OmpR autoregulates its own expression in response to acid pH, providing a partial explanation for why the SPI-2 genes are induced by low pH.

AEROBIC TO ANAEROBIC TRANSITIONS

Facultative microorganisms such as *E. coli* and *S. enterica* are capable of modifying their metabolism to accommodate growth under either aerobic or anaerobic conditions. The transition between aerobic and anaerobic metabolism is accompanied by alterations in the rate, route, and efficiency of pathways of electron flow. Figure 18-2 illustrates the basic pathways utilized by *E. coli* for aerobic versus anaerobic electron flow. Under anaerobic conditions without alternate electron acceptors, pyruvate is converted to formate, acetate, or ethanol, CO_2 and H_2 gas (mixed acid fermentation). However, the choices and energy yield become more plentiful when alternate electron acceptors are available. *E. coli*, even under aerobic conditions, synthesizes two distinct **cytochrome oxidases** — **cytochrome** *o* (***cyo*** **operon**) and **cytochrome** *d* (***cyd*** **operon**) — produced under high O_2 and low O_2 conditions, respectively. Under anaerobic conditions, at least five more terminal oxidoreductases can be produced (Table 18-1).

TABLE 18-1. Characteristics of Electron Transport Systems in *E. coli*

Electron Acceptor	δG^{Oa} (kJ/mol)	Terminal Respiratory Enzyme	Operon	Chromosomal Location (min)
O_2	−233	Cytochrome *o* oxidase	*cyoABCDE*	10
O_2	−233	Cytochrome *d* oxidase	*cydAB*	17
NO_3^-	−144	Nitrate reductase	*narGHJI*	27
NO_3^-	−144	Nitrate reductase	*narZYWV*	33
DMSO	−92	DMSO/TMAO reductase	*dmsABC*	20
TMAO	−87	TMAO reductase	*torA*	28
Fumarate	−67	Fumarate reductase	*frdABCD*	94

[a]Free energy calculated by using NADH as an electron donor to the indicated electron acceptor.

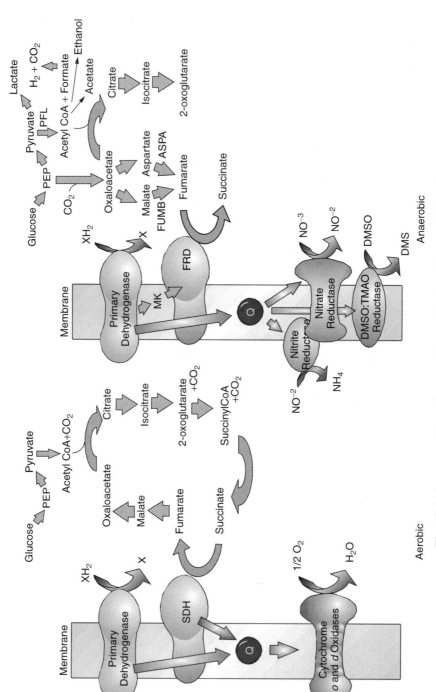

Fig. 18-2. Basic aerobic and anaerobic metabolic pathways of *E. coli*.

E. coli controls the production of the various respiratory pathway enzymes, first, in response to aerobic and anaerobic growth conditions and, second, to the availability of alternate electron acceptors. Clearly there is a hierarchy or preference for substrate use with the following order: oxygen > nitrate > DMSO > TMAO > fumarate. When several e^- acceptors are present simultaneously, the more energetically favored acceptor will be used first. An interesting question is how does the cell regulate such a complex system. There are three basic regulators described that sense changes in oxygen level or redox conditions. Their overlapping control circuits are illustrated in Figure 18-3.

Formate Nitrate Regulation

The **Fnr protein** (formate nitrate regulation) regulates over a hundred genes in response to the presence or absence of oxygen. The manner in which Fnr senses oxygen appears to involve an **oxygen-sensitive (4Fe-4S) center** (Cys-X_3-Cys-X_2-Cys-X_5-Cys). Active Fnr is a dimer and, therefore, contains two such centers. Exposure to oxygen causes partial disassembly of the two (4Fe-4S) centers to form two (2Fe-2S) centers. This disassembly converts Fnr dimers, which bind DNA, to monomers that are not active (Fig. 18-4). Under anaerobic conditions, Fnr represses *cyoABCDE*, *cydAD* and *narZYWV* (Fig. 18-3) but transciptionally activates *sdhCDAB* (**succinate dehydrogenase**), *frdABCD* (**fumarate reductase**), *dmsABC* (**DMSO reductase**), and *narGHJI* (**nitrate reductase**). Fnr bears significant sequence homology to the CRP protein and, like CRP, alters transcription through DNA bending and direct interactions with the α-subunit of RNA polymerase. In contrast to CRP, there is no evidence for cAMP involvement with Fnr.

Nitrate Response

Nitrate is the preferred **electron acceptor** for anaerobic cells because of the high midpoint potential of the nitrate/nitrite couple. In addition to *fnr*, the *narXL* **operon** and *narQ narP* are required to regulate respiratory gene expression in response to nitrate availability. DNA sequence analysis indicates *narX narL* and *narQ narP* are two-component regulatory systems that mediate a series of complex transcriptional adjustments in response to a dynamic ratio of two alternate electron acceptors: nitrate, and its reduction product, nitrite. **NarQ** and **NarX** are **histidine kinases** that sense nitrate and transmit a signal to their cognate response regulators **NarP** and **NarL**, respectively. This system contributes to the induction of *frd, dms*, and *narGHIJ* **operons** but has no effect on *cyo*, *cyd*, or *narZYWV*.

The control of the nitrite reductase operons *nir* and *nrf* by NarP and L offers a glimpse of the complex adjustments that occur in response to the cell's needs. Nitrate, at low concentration, induces nitrate reductase and the high-affinity Nrf nitrite reductase via the NarXL and NarQP systems. Thus, the cell will derive maximum energy yield from the nitrate present. However, at high nitrate concentrations, nitrate reductase is still expressed, but NarL will actually repress the **Nrf nitrite reductase** while inducing the low-affinity **NirB nitrate reductase** system. This differential regulation avoids generating too much proton motive force and energy that would result from using the high-affinity nitrite reductase under high nitrate conditions when plenty of energy is derived from nitrate reductase. Nevertheless, the low-affinity nitrite reductase is still necessary to detoxify the nitrite produced.

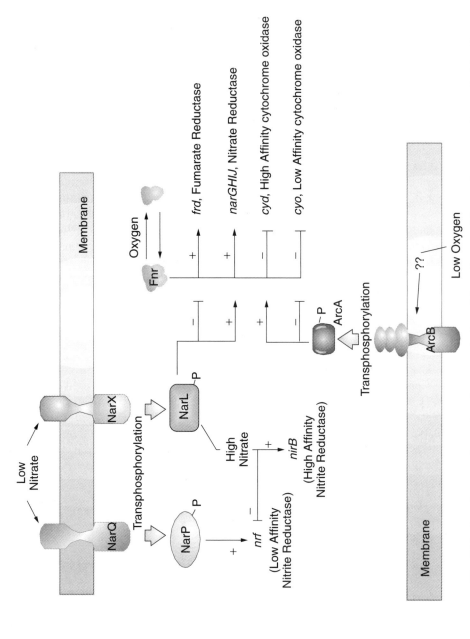

Fig. 18-3. Regulatory scheme for the control of aerobic and anaerobic respiratory pathways in *E. coli*.

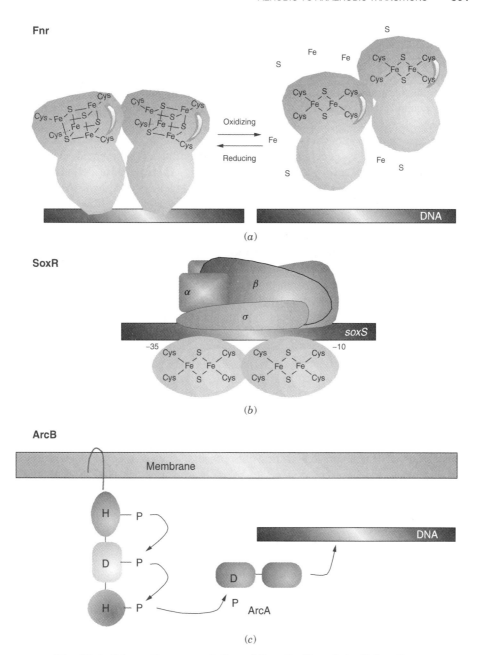

Fig. 18-4. Schematic representation of Fnr, SoxR, and ArcB function.

ArcAB System

The system that controls *cyo* and *cyd* includes the ***arcA*** and ***arcB*** products (*arc*, meaning anoxic redox control) comprising a two-component regulatory system. The system actually regulates a total of 30 genes. **ArcB** functions as the membrane-bound sensor/transmitter that communicates with **ArcA**, the receiver/regulator. ArcB is an unorthodox histidine kinase in that it contains three phosphotransfer signaling domains (Figs. 18-3 and 18-4). The first domain, the **histidine-kinase** module, phosphorylates itself at a conserved histidine (H292). The phosphate is subsequently transferred, first to an aspartate residue (D576) in the receiver module and then to a second histidine (H717) in the phosphotransfer domain located at the carboxyl end of ArcB. The phosphate is subsequently transferred to an aspartate residue (D54) within the amino terminus of ArcA.

The Arc system regulates the coordinate synthesis of tricarboxylic acid cycle enzymes. ArcA~P is generally a repressor of aerobically expressed target genes [eg., *acn* (**aconitase**), *cyo*, *gltA* (**citrate synthase**), *sdh* (**succinate dehydrogenase**)] and an activator of anaerobically expressed target genes [*cyd*, *hya* (**hydrogenase 1**), *pdu* (**propanediol degradation**)]. The **SixA phosphohistidine phosphatase** modulates the system by removing phosphate from ArcB residue H717, downregulating phosphotransfer. The nature of the sensed low oxygen signal, how it is sensed, and how the signal is transduced to activate ArcB kinase are unclear. A model in which ArcB senses proton motive force has been refuted, since the only amino acid (H47) with a physiologically relevant pK within the transmembrane region has proven dispensable. Other models invoke the ArcB-sensing redox state of the cell and/or metabolites (e.g., lactate, acetate) generated by anaerobic metabolism. Figure 18-3 presents a model depicting the regulatory interactions of Nar, Fnr, and Arc that govern selected oxygen-regulated genes.

OXIDATIVE STRESS

Generation of energy by the electron transport chain is dependent on the catalytic spin pairing of triplet oxygen. During this process, oxygen species that are toxic to DNA, protein, and lipid components of the cell are formed through both enzymatic and spontaneous chemical reactions. The reaction of oxidative enzymes with molecular oxygen can generate superoxide (superoxide anion):

$$O_2 + e^- + \text{oxidative enzymes} \longrightarrow O_2^{-\bullet}$$

Superoxide is relatively unreactive with DNA and proteins. However, it may interact in a number of enzymatic as well as spontaneous chemical reactions to produce more highly reactive oxygen derivatives such as **hydrogen peroxide** and **hydroxyl radicals**:

$$O_2^{-\bullet} + H_2O_2 \longrightarrow OH^- + OH^\bullet + O_2$$

Autooxidation of reduced FAD or reduced flavoprotein gives rise to hydrogen peroxide:

$$FADH_2 + O_2 \longrightarrow FAD + H_2O_2$$

The enzyme **NADPH oxidase** can generate superoxide anion and hydrogen peroxide:

$$NADPH + H^+ + 2O_2 \longrightarrow NADP^+ + O_2^{-\bullet} + H_2O_2$$

Superoxide can release Fe^{2+} and Fe^{3+} from various cellular compounds and ultimately give rise to hydroxyl radicals through the **Fenton reaction**:

$$Fe^{2+} + H_2O_2 \longrightarrow Fe^{3+} + OH^{\bullet} + OH^-$$

Reaction of superoxide anion with **nitric oxide** forms **peroxynitrite anion**:

$$O_2^{-\bullet} + NO^{\bullet} \longrightarrow ONOO^-$$

Peroxynitrite is highly reactive with various proteins. Methionine, cysteine, tyrosine, and tryptophan residues of proteins are especially vulnerable. Peroxynitrite can also give rise to other toxic derivatives that are highly reactive with biological compounds.

Most aerobic organisms are protected from the toxicity of superoxide and hydrogen peroxide by the enzymes **superoxide dismutase (SOD)** and **catalase**:

$$O_2^{-\bullet} + O_2^{-\bullet} + 2H^+ \longrightarrow O_2 + H_2O_2 \text{ (SOD)}$$
$$2H_2O_2 \longrightarrow 2H_2O + O_2 \text{ (catalase)}$$

Peroxidases can also catalyze the reduction of hydrogen peroxide by organic reductants such as **glutathione** or **ascorbic acid**:

$$H_2O_2 + 2RH \longrightarrow 2\ H_2O + 2\ R_{ox}$$

E. coli produces a cytoplasmic **Mn-SOD (SodA)** and **Fe-SOD (SodB)** that protect DNA and proteins from oxidation. Mutants deficient in these enzymes display enzyme inactivation, growth deficiencies, and DNA damage. A periplasmic **Cu/Zn-SOD (SodC)** protects the periplasmic and membrane constituents from exogenous superoxide. Several DNA repair systems also aid in recovery from oxidative damage (see Chapter 2).

The presence of SOD and catalase in anaerobic organisms would appear to be self-defeating since oxygen is a product of both reactions. Anaerobes must rely on other mechanisms to eliminate oxygen and prevent the formation of toxic oxygen species. One method used by organisms that do not carry out oxygen-dependent respiration involves a unique flavoprotein, **NADH oxidase**, which catalyzes the direct four-electron reduction of oxygen to water:

$$NADH + H^+ + 0.5\ O_2 \longrightarrow NAD^+ + H_2O$$

This enzyme actually allows streptococci and other lactic acid bacteria to use oxygen directly in the metabolism of carbohydrates without the inherent problems of oxygen toxicity caused by the formation of reactive oxygen derivatives.

A superoxide reductase system that provides protection against oxidative stress is present in the anaerobic sulfate-reducing bacterium *Desulfovibrio vulgaris* and in the hyperthermophilic anaerobe *Pyrococcus furiosus*. It has the advantage of eliminating superoxide without the formation of molecular oxygen. This system consists of the nonheme iron-containing proteins **rubrerythrin (Rbr)** and **rubredoxin oxidoreductase (Rbo)** that reduce superoxide to hydrogen peroxide without dismutation:

$$O_2^{-} \bullet + e^- + 2H^+ \longrightarrow H_2O_2$$

This system functions in conjunction with an **NADH peroxidase** that reduces hydrogen peroxide to water:

$$NADH + H^+ + H_2O_2 \longrightarrow NAD^+ + 2\ H_2O$$

These and other mechanisms used by anaerobic or facultative organisms to protect against oxygen damage are discussed later in reference to organisms that utilize primarily fermentation pathways in their metabolism.

Regulation of the Oxidative Stress Response

The natural by-products of aerobic metabolism are the reactive compounds superoxide (O_2^-) and hydrogen peroxide. These two species can lead to the generation of hydroxyl radicals (OH$^\bullet$), which can damage biological macromolecules. The oxidation stress modulon includes at least 80 proteins induced during exposure to superoxide. About half are also induced by H_2O_2. Research efforts have uncovered two regulons within this stress response, although there are certainly more. The two known systems, **OxyR** and **SoxRS**, used by *E. coli*, are outlined in Figure 18-5.

The **OxyR regulon** comprises nine of the proteins induced by H_2O_2. All are controlled by the positive regulator OxyR. OxyR is a class I type activator that, once bound to a target DNA sequence, interacts with the α C-terminal domain of RNA polymerase (see Chapter 5). The protein is activated by oxidation, which results in the formation of a disulfide bond between cysteine residues C199 and C208. Under noninducing conditions these cysteines are reduced by glutaredoxin I. For most OxyR-dependent genes, the oxidized but not the reduced form of OxyR binds target DNA sequences. However, the reduced form of this regulator can bind to the *oxyR-oxyS* target operator/promoter region but will not activate transcription until OxyR itself is oxidized. Some of the enzymes whose expression is regulated by OxyR include **catalase (*katG*), glutathione reductase (*gorA*), glutaredoxin I (*grxA*)**, and **alkyl hydroperoxide reductase (*ahpC* and *ahpF*)**. The likely in vivo function for alkyl hydroperoxide reductase would be the detoxification of lipid and other hydroperoxides produced during oxidative stress.

In *Salmonella*, reduced-OxyR was found to act as a repressor of the ***narZYWV* locus**, which encodes a stress-inducible aerobically expressed nitrate unresponsive **nitrate reductase (NR-Z)**. In the presence of H_2O_2, the oxidized-OxyR no longer represses *narZYWV* transcription and the operon becomes derepressed to about one-third of its maximal induced level of expression. Other loci have also been shown to be repressed by OxyR. Thus, OxyR, like many other response regulators, can function as both a transcriptional activator and repressor protein, depending on the target gene.

A second oxidative stress regulon comprises nine proteins induced by superoxide but not hydrogen peroxide. This regulon is under positive transcriptional control by the *soxRS* loci. Genes under *soxR* control include those responsible for **Mn^{2+}-containing superoxide dismutase (*sodA*)**, the DNA repair enzyme **endonuclease**

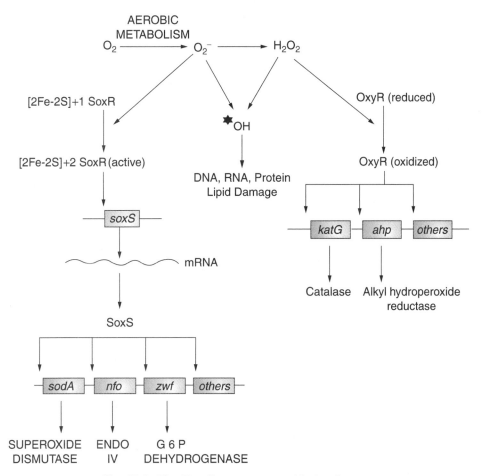

Fig. 18-5. The *E. coli* response to oxidative damage.

IV (*nfo*), **glucose-6-phosphate dehydrogenase** (*zwf*, production of NADPH), *fldA* **(flavoredoxin 1), NADPH:flavodoxin oxidoreductase** (*fpr*), and **fumarase C** (*fumC*). The current model for this system is that preexisting SoxR protein senses oxidative stress and then triggers expression of the *soxS* gene. The SoxS product then activates transcription from the promoters of the other members of the regulon.

As with Fnr, SoxR utilizes an Fe sulfur center to sense oxidative stress conditions in the cytoplasm. SoxR contains two stable (2Fe-2S) centers anchored to four cysteine residues near its carboxy terminus (Fig. 18-4). These (2Fe-2S) centers remain reduced (+1 state) under normal physiological conditions but rapidly oxidize (+2 state) when challenged with oxidative stress. It is believed that iron-sulfur centers are especially sensitive to redox reactions with superoxide. The change in redox state of the protein induces an active conformation of SoxR but does not change binding affinity.

Both oxidized and reduced forms of SoxR bind to the *soxS* promoter. Thus, oxidation, while not affecting binding, must enable SoxR to interact with RNA polymerase, stimulating open complex formation. There is some evidence that oxidized SoxR is reduced by the NADPH:flavoredoxin oxidoreductase/flavoredoxin couple with the oxidation of NADPH. The genes encoding these enzymes are themselves induced by SoxR, which suggests a mechanism of SoxRS autoregulation — that is, as the system is induced by oxidized SoxR (e.g., aerobic growth), the levels of the oxidoreductase/flavoredoxin couple increase, which will reduce SoxR and downregulate the system.

pH STRESS AND ACID TOLERANCE

Microbes can grow over a wide range of hydrogen ion concentrations (pH). For example, **acidophilic bacteria** will grow in acidic sulfur springs where a pH of 1 is common. In contrast, **alkalophilic bacteria** prefer to grow in environments such as soda lakes where pH conditions rise as high as pH 11. Most bacteria in the human sphere, however, prefer to grow at pH values closer to neutral and are called **neutralophilic**. They generally grow over a range of pH 5 to pH 9 and include *E. coli*, *S. enterica*, *Streptococcus lactis*, *Bacillus subtilis*, and many others. Although the neutralophiles do not generally grow under conditions of extreme acid or base, they can survive these exposures to various degrees if they are allowed to go through an adaptive transition whereby pH gradually changes.

It is well established that one way microbes respond to an acidifying pH is by producing enzymes that can convert acidic metabolites to neutral ones or neutral metabolites to alkaline products. Good examples of these types of enzymes are glutamate decarboxylase, lysine decarboxylase, and arginine decarboxylase of *E. coli*, all of which exhibit increased expression at external acidic pH. The concept of **pH homeostasis** or the ability of the microbe to control its internal pH is discussed in Chapter 9. It has been demonstrated that the major mechanisms typically used by Gram-negative organisms to control internal pH during growth involves the modulation of the primary proton pumps as well as the K^+/H^+ and Na^+/H^+ antiporters. However, additional adaptive mechanisms are engaged to survive under pH conditions outside the growth range.

Exposures of *S. enterica* and *E. coli* to a nonlethal acidic pH between pH 5 and 6 results in the induction of sets of genes whose products can protect the cell when exposed to potentially lethal pH conditions from pH 4 to pH 2. These inducible systems protect the cell at external pH values where the nonadaptive pH homeostasis mechanisms fail. The genetic and physiologic changes that occur in the cell are referred to as the **acid tolerance response** (ATR). Full induction of the ATR results in the increased expression of at least 50 newly synthesized or existing proteins called **acid shock proteins**, some of which are also heat shock induced.

One system important to acid tolerance in some organisms is the Mg^{+2}-dependent proton translocating ATPase (see "Energy Production" in Chapter 9). In Gram-positive organisms such as *Enterococcus faecalis*, the ATPase, which normally harnesses proton motive force to generate ATP, can work in reverse, hydrolyzing ATP to extrude a proton from the cell. As opposed to Gram-negative organisms that struggle to keep internal pH near 7.8, the streptococci are more flexible. Their only requirement is to maintain a pH of 0.5 to 1 unit — a goal the ATPase alone could accomplish.

The superior acid resistance of *E. coli* and *Shigella* over that of *S. enterica* is of particular interest from a medical view, since it helps explain the differences in infectious dose required by these organisms to overcome the **gastric acid barrier** and cause disease (about 10 organisms for *E. coli* and *Shigella* vs 10,000 for *Salmonella*). The primary reason for this difference is the production of **glutamate decarboxylase**, an enzyme missing from *S. enterica*. Besides the enzyme proper, this system includes a specific antiporter that links export of γ-amino butyric acid (the decarboxylation product of glutamate) to the import of more glutamtate. The pH optimum for this enzyme is around pH 5 and thus it does not play a role until internal pH falls to that level. Although initially thought to have a part in consuming and extruding intracellular protons, how this system really helps the cell survive pH 2 environments is not clear. Another organism that uses this system is the Gram-positive *Lactococcus lactis*.

In addition to glutamate decarboxylase, *Lactococcus* uses an **arginine deiminase** (ADI) pathway to generate ATP during acid stress. ADI converts arginine to citrulline and ammonia, ornithine transcarbamylase converts the citrulline to ornithine and carbamyl phosphate, and carbamate kinase cleaves carbamyl phosphate to ammonia and CO_2, generating an ATP in the process. The system also requires an arginine/ornithine antiporter to rid the cell of the end-product ornithine while importing new arginine substrate. The ammonia generated can alkalinize the environment, but since this system also helps survival of dilute cultures, the ATP generated must substantially aid in survival.

Helicobacter pylori, an important cause of gastric ulcers and cancer, utilizes another potent acid resistance system to survive in the stomach. The organism constitutively produces a powerful **urease**, a nickel-containing metalloenzyme that converts urea to carbon dioxide and ammonia. Contrary to what would be expected, the pH optimum of urease is around 7.5, raising the question of why cells grown at pH 7.5 have very little urease activity. There are two reasons. Although the urease is constitutively produced, the degradation of transcripts encoding nickel-incorporating enzymes appear subject to pH control. In addition, UreI is a membrane transporter of urea that only becomes activated at acid pH. Therefore, urea cannot gain access to intracellular urease except at low pH. Like the glutamate decarboxylase system, it is not entirely clear how urease provides acid resistance. The ammonia produced does not seem to be sufficient protection, since the system works even at low cell density — a condition where the small amount of ammonia produced will not change external pH. *Yersinia enterocolitica* is another Gram-negative enteropathogen that uses urease to gain safe passage through the stomach.

Much remains to be learned regarding how different bacteria cope with acid or alkaline threats to survival. What is clear is that the microbial world has taken full advantage of what is available in the environment to counter this form of stress and that successful strategies have been shared among species.

THERMAL STRESS AND THE HEAT SHOCK RESPONSE

As with pH, microbes exhibit a wide range of temperatures at which they can grow (Table 1-4). For example, bacteria can be isolated from hot springs where temperatures can reach as high of 90 °C or so (**thermophilic bacteria**) and from frozen tundra or polar caps where temperatures are below 0 °C (**psychrophilic bacteria**). Most bacteria, however, prefer to grow at milder temperatures such as 20 ° to 40 °C (**mesophilic bacteria**). This latter group includes *E. coli* and *S. enterica* serovars as well as other animal and human pathogens.

Upon a shift from 30° to 42 °C, *E. coli* and other bacteria transiently increase the rate of synthesis of a set of proteins called **heat shock proteins (HSPs)**. Many of these HSPs are required for cell growth or survival at more elevated temperatures (**thermotolerance**). Among the induced proteins are **DnaJ** and **DnaK**, the **RNA polymerase** σ^{70} **subunit** (*rpoD*), **GroES**, **GroEL** (see "Protein folding and Chaperones" in Chapter 2), **Lon protease** (see "Proteolytic Control" in Chapter 2), and **LysU**. There are nearly 50 heat shock–inducible proteins identified in *E. coli*. They can be subdivided into those regulated by alternative σ factors and two component regulatory systems.

The σ^H **regulon** provides protection against cytoplasmic thermal stress. The *E. coli* *rpoH* **locus** (formerly called *htpR*) encodes a 32 kDa σ factor, alternatively called σ^H or σ^{32}, which redirects promoter specificity of RNA polymerase. The σ^H protein regulates the expression of 34 heat shock genes. The simple explanation for how heat shock increases expression of the σ^H regulon is that heat shock first causes an elevation in σ^H levels, which in turn increases expression of the σ^H target genes.

Although the principle is simple, the controls governing σ^H production are complex. First, a temperature upshift from 30° to 42 °C results in the increased translation of *rpoH* message. *Cis*-acting mRNA sites within the 5′ region of *rpoH* message form temperature-sensitive secondary structures that sequester the ribosome-binding site. At higher temperatures, these secondary structures melt, thereby enabling more efficient translation of the *rpoH* message. In addition to the increased translation of *rpoH* message, the σ^H protein itself becomes more stable, at least transiently. The mechanism regulating proteolysis centers on whether σ^H associates with RNA polymerase. During growth at 30 °C, σ^H can be degraded by several proteases including **FtsH**, **HslVU**, and **ClpAP**. However, if σ^H is bound to RNAP, σ^H is protected from degradation.

The cell uses the **DnaK-DnaJ-GrpE chaperone team** to interact with σ^H at low temperature, sequestering σ^H from RNA polymerase (Fig. 18-6). Failure to bind RNAP facilitates degradation of the σ factor. Upon heat shock, there is an increase in the number of other unfolded or denatured proteins that can bind to DnaK or DnaJ. This reduces the level of free DnaK/DnaJ molecules available to bind σ^H, allowing σ^H to bind RNAP, which protects σ^H from degradation. As the cell reaches the adaptation phase following heat shock, the levels of DnaK and DnaJ rise (both are induced by σ^H) and can again bind σ^H, redirecting it toward degradation. Nevertheless, even though σ^H degradation resumes, translation of *rpoH* remains high at the elevated temperature and σ^H continues to accumulate, although at a slower rate.

In addition to translational and proteolytic controls, production of σ^H is regulated at the transcriptional level via a feedback mechanism. There are four promoters driving *rpoH* expression, three of which are dependent on σ^{70}, the **housekeeping σ factor**. The gene encoding σ^{70}, *rpoD*, is also a heat shock gene induced by σ^H. So increased production of σ^H increases σ^{70}, which increases transcription of *rpoH*. The fourth *rpoH* promoter is recognized by another σ factor, σ^E, encoded by *rpoE*. The heat shock response is also triggered by a variety of environmental agents such as ethanol, UV irradiation, and agents that inhibit DNA gyrase. Induction by all of these stimuli occurs through σ^H. How can all of these seemingly diverse stresses activate *rpoH*? The only explanation that appears reasonable is the accumulation of denatured or incomplete peptides. There is a potential **alarmone** that has been implicated in signaling expression of this global network. The molecule is **diadenosine 5′, 5′′′-P¹, P⁴-tetraphosphate**

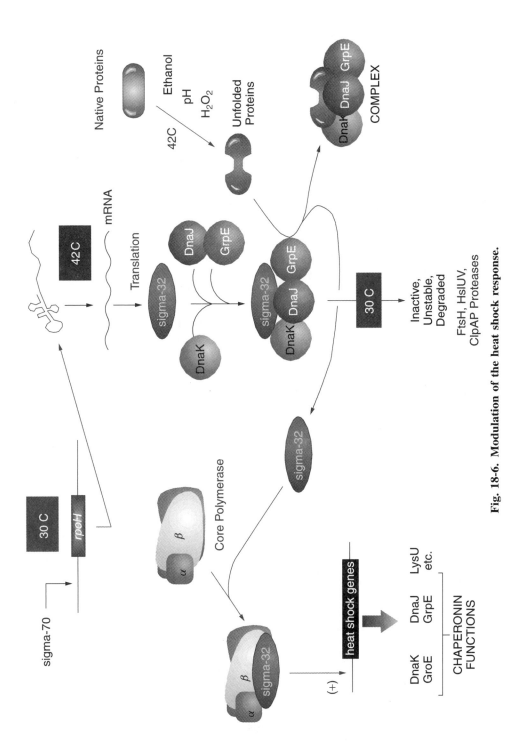

Fig. 18-6. Modulation of the heat shock response.

(AppppA), which is made by some aminoacyl–tRNA synthetases (e.g., *lysU*) at low tRNA concentrations. How this may influence the response is not known.

The σ^E **regulon** provides protection against extracytoplasmic stress. The hallmark of the Gram-negative cell is the existence of two membrane-bound subcellular compartments: the cytoplasm and the periplasm. Conditions in each of these compartments differ markedly. The cytoplasm is energy rich, reducing (low redox potential), and osmotically stable, whereas the periplasm lacks ATP, is oxidizing, and is in contact with the external milieu. Since optimal cell growth requires that the cell senses and responds to changes in these disparate subcellular compartments, it is not surprising that the stress responses in *E. coli* and *S. typhimurium* are compartmentalized into cytoplasmic and extracytoplasmic responses. The extracytoplasmic response pathways involve two partially overlapping signal transduction cascades: the σ^E and **Cpx systems**. These pathways are induced following the accumulation of misfolded proteins in the periplasm as a result of stresses such as high temperature, pH extremes, or carbon/energy starvation.

σ^E is a member of the **extracytoplasmic function (ECF) subfamily** of σ factors. In *E. coli*, σ^E is responsible for the transcription of up to genes, including *rpoH* (σ^H), *degP (htrA)* encoding a periplasmic protease for the degradation of misfolded proteins, *fkpA* encoding a periplasmic peptidyl prolyl isomerase, and *rpoE rseABC* operon (see below).

The gene encoding σ^E, *rpoE*, is the first member of an operon followed by the genes *rseA*, *B*, and *C* (*rse*, meaning regulators of σ^E). **RseA** is a transmembrane protein whose cytoplasmic C-terminal domain interacts with σ^E, acting as an anti-σ factor. The periplasmic face of RseA binds to the **periplasmic RseB**. Extracellular stress in some way signals increased proteolysis of RseA by the **periplasmic protease DegS**, thus relieving the anti-σ effect of RseA on σ^E. It has been proposed that RseB and perhaps other periplasmic proteins involved in protein folding protect RseA from degradation by binding to the RseA periplasmic domain, capping the target site of DegS. Stress-induced misfolding of periplasmic proteins would titrate the RseA cap proteins off of RseA, rendering the anti-σ factor vulnerable to attack by DegS. The result would be increased activity of σ^E leading to increased levels of σ^E protein and RseA anti-σ (since they form an operon). The increased amount of σ^E will drive further expression of genes whose products handle the periplasmic damage while the increased level of RseA will enable the cell to downregulate the system once the capping proteins are again free to bind and protect RseA from DegS degradation.

A second system dedicated to protecting the periplasmic perimeter of the cell is the CpxRA two-component system (see Chapter 5) with **CpxA** playing the role of membrane-localized sensor histidine kinase and **CpxR** as the cytoplasmic response regulator. CpxA responds to envelope stress by autophosphorylation followed by phosphotransfer to CpxR. CpxR~P activates expression of *dsbA* **(disulfide oxidoreductase)**, *ppiA* and *ppiD* encoding **peptidyl-prolyl-isomerases**, and, in *E. coli*, in conjunction with σ^E, *degP (htrA)*. In addition, CpxR~P activates transcription of *cpxP* encoding a small protein that negatively regulates the CpxAR regulon, probably by binding to a periplasmic domain of CpxA. The ability to autoactivate and then repress (via CpxP) enables a temporary amplification of the Cpx response that may be important to rescue cells from transitory stresses. **PrpA** and **PrpB** are **type I serine/threonine phosphatases** that also participate in the ECF pathway at least in part by affecting the phosphorylation level of CpxR.

B. subtilis has four classes of heat shock genes. All of them are not described here, but one is particularly interesting because of the mechanism used to control the genes. **Class I heat shock genes** include the major **chaperones DnaK-DnaJ-GrpE and GroEL-GroES**. Their transcription requires the housekeeping σ factor, σ^A (σ^{70}) and is negatively controlled by a repressor called **HrcA**. HrcA binds to a class I gene operator, a well-conserved 9 bp inverted repeat with a 9 bp spacer, called **CIRCE**. The **CIRCE/HrcA regulon** is normally repressed by the **HrcA repressor** but can be heat induced by inactivating the repressor. The molecular switch involves the **chaperone GroE**. Unlike σ^{32} in *E. coli*, the GroE chaperones bind to and facilitate folding of HrcA and thereby modulate repressor function. Titration of GroE by stress-induced misfolded proteins results in lower HrcA repressor activity.

NUTRIENT STRESS AND THE STARVATION — STRESS RESPONSE

Nutrient starvation and other environmental stresses are routine occurrences for most bacteria. Situations of true "feast" or nonstress conditions are few and far between for microbes outside the laboratory. Thus, microbes are most frequently found in a state of nutrient starvation or stress-induced slow growth or nongrowth. Fortunately this is the case, because given bacterial growth rates achieved under nonlimiting conditions in the laboratory, if microbial growth were nonlimiting in natural or animal host environments, the consequences would be deadly.

Starvation-Stress Response

When *E. coli*, *Salmonella*, and many other nondifferentiating microbes are starved for an essential nutrient such as a carbon-energy (C) source, they respond by inducing the expression of up to 50 or so new proteins or preexisting proteins. The genetic and physiologic reprogramming that occurs is the **starvation-stress response (SSR)**. The function of the SSR is to allow for the long-term starvation survival of the bacteria and to provide a general cross-resistance to a variety of other environmental stresses including extremes in temperature, pH, and osmolarity as well as exposure to reactive oxygen and nitrogen species and antimicrobial peptides/proteins.

It should be noted that a distinction must be made between starved cells and so-called stationary-phase cells. Typically, **stationary-phase cells** populate cultures that have stopped growing following exponential growth in rich or nonlimiting media, in contrast to **starved cells**, which populate cultures that have ceased growing in response to exhaustion of one or more defined nutrients. For stationary-phase cells, the condition that limits growth is not necessarily defined nor is it typically limited to a single stress; for starved cells, the limitation/stress that restricts growth is defined. Another key difference is that stationary-phase cultures normally achieve a much higher cell density compared with starved cultures, which can have a significant effect on overall cellular responses and long-term survival. Furthermore the genes/proteins expressed in stationary-phase cells may or may not overlap with those expressed in starved cells. The SSR refers specifically to the response of starved cells.

In general, for Gram-negative bacteria, starved cells are morphologically and physiologically very different from log-phase cells. The initial response to carbon-energy source limitation is to try and avoid the stress by increasing expression or expressing new uptake or scavenging systems to be able to utilize any nutrients that may become available. Persistence of the starvation-stress eventually results in a cell that is smaller, much more hardy, and metabolically efficient. This is mediated by the accumulation of at least two cellular nucleotides: **cyclic 3′, 5′ adenosine monophosphate (cAMP**; see "Catabolite Control" in Chapter 5) and **guanosine 3′, 5′-bis(diphosphate) (ppGpp**; see "Stringent Control" in this chapter). In addition, at least two alternative σ factors, σ^S (or σ^{38}; see Chapters 2 and 5) and σ^E (see above under heat shock response), encoded by the ***rpoS*** and ***rpoE*** genes, respectively, are key SSR regulators. These regulators are general factors controlling responses to starvation stress in a more global manner. However, this is not sufficient for the regulation of some stress-responsive genes. Thus, additional regulators may be involved in the regulation of specific genes, indicating a discriminating complexity for the SSR and other stress responses.

Examples of some regulatory proteins include **Fis (factor for inversion stimulation** first characterized for its role in flagellar phase variation; see "Phase Variation" in Chapter 5), **FadR (regulator of fatty acid metabolism**, which binds to medium/long-chain fatty acyl-CoA molecules and represses fatty acid biosynthetic genes and activates fatty acid degradation genes; see Chapter 13), **Lrp (leucine-responsive protein**, which controls certain aspects of amino acid metabolism; see "Serine/Glycine Family" in Chapter 15), **OxyR** (see above), **SpvA/SpvR (regulators of the *Salmonella* plasmid virulence** or ***spv* genes**; see "Paradigms of Bacterial Pathogenesis" in Chapter 20), **PhoP (response regulator that controls virulence factors**; see "Paradigms of Bacterial Pathogenesis" in Chapter 20), **IHF (integration host factor**; see Chapter 5), and **H-NS** ("Nucleoid Structure" in Chapter 2).

Some of the physiologic changes that occur during SSR include the expression of new or higher-affinity nutrient utilization systems to scavenge the environment for carbon-energy sources and other nutrients; the degradation of cellular RNA, proteins, and fatty acids; the reduction in the number of ribosomes; altering of the amounts and types of lipid components in the cytoplasmic membrane; an increase in the relative amounts of lipopolysaccharide or LPS in the outer membrane of Gram-negative bacteria; and the condensation of chromosomal DNA in order to protect it from damage.

Stringent Control

When bacteria experience conditions that limit the availability of one or more amino acids (shift from a rich medium to a minimal medium) or exhaust their primary carbon source, growth stops temporarily and rapid adjustments in metabolism are made. These include decreasing the rates of RNA accumulation (particularly stable RNAs such as rRNA and tRNA) and DNA replication, as well as reducing the biosynthesis of carbohydrates, lipids, nucleotides, peptidoglycan, and glycolytic intermediates. The transport of many macromolecular precursors into the cell is also shut down. This set of responses, characterized best as a response to amino acid starvation, is referred to as the **stringent response** or **stringent control**. The stringent response collectively enhances cellular viability during periods of amino acid or energy limitation and allows rapid recovery and reinitiation of growth when conditions improve.

For rapidly growing cells, a major amount of the available energy is used for ribosome synthesis. Therefore, blocking ribosome synthesis under amino acid starvation conditions is a major mode of energy conservation. Conditions causing a stringent response lead to an abrupt change in the rate of ribosome synthesis caused by inhibiting the transcription of genes encoding ribosome-associated components such as rRNA, tRNA, and mRNA for ribosomal proteins. When starved for amino acids, bacterial cells rapidly accumulate millimolar concentrations of two unusual nucleotides: guanosine pentaphosphate (guanosine 5′-triphosphate-3′-diphosphate, pppGpp) and guanosine tetraphosphate (guanosine 5′-diphosphate-3′-diphosphate, ppGpp). These nucleotides, first referred to as Magic Spot II and Magic Spot I, respectively, accumulate in *E. coli* and *Salmonella* as well as other bacteria during amino acid limitation.

Starvation for a specific amino acid will lead to an increase in the corresponding uncharged tRNA species. As the ratio of charged to uncharged tRNA falls, ribosomes stall on mRNAs when encountering the codon for that amino acid. Synthesis of pppGpp is triggered by the repeated binding of the uncharged tRNA molecules to the stalled ribosome. On the ribosome, the product of the *relA* gene (RelA, stringent factor), a ribosome-bound pyrophosphotransferase (pppGpp synthetase I) present on about 1% of ribosomes, catalyzes the formation of pppGpp from GTP and ATP, as shown in Figure 18-7. A second, ribosome-independent route to pppGpp also exists involving the *spoT* product (pppGpp synthetase II). SpoT is responsible for basal levels of (p)ppGpp. The enzyme pppGpp phosphohydrolase (*gpp*) is the major pppGpp hydrolase, degrading pppGpp to ppGpp. SpoT appears to have two functions: one in the synthesis of pppGpp and one in the degradation of ppGpp to GDP plus pyrophosphate.

A major result of the stringent response is a reduction in the rate of stable RNA accumulation — a response strongly correlated with the rise in (p)ppGpp concentration in the cell. Several studies suggest that RNA polymerase is a target of ppGpp action and that the regulated process during stringent control is transcription itself. The probable target is the ß-subunit of RNA polymerase (see Fig. 2-16c in Chapter 2). Both transcription initiation and polymerase pausing (important in certain regulatory mechanisms and transcription termination) are postulated as the major targets of ppGpp action. Thus, ppGpp would reduce the affinity of RNA polymerase for rRNA promoters or inhibit elongation of polymerase. Either effect will obviously lower the amount of rRNA available for ribosome synthesis, which can subsequently affect the synthesis of ribosomal proteins. On the other hand, ppGpp has also been found to stimulate transcription of several amino acid biosynthetic operons (e.g., *his*) and possibly other genes. Thus, it appears that ppGpp can act either as a negative effector or a positive effector depending on the target gene.

Fig. 18-7. Stringent control. (*a*) Ribosome-dependent and -independent pathways for the biosynthesis of guanosine 5′-diphosphate-3′-diphosphate. ppG, guanosine 5′-diphosphate; pppG, guanosine 5′-triphosphate; pppA, adenosine 5′-triphosphate;, adenosine 5′-monophosphate; ppGpp, guanosine 5′-diphosphate-3′-diphosphate; pppGpp, guanosine 5′-triphosphate-3′-diphosphate; pp$_i$, inorganic pyrophosphate. Ndk, nucleoside 5′-diphosphate kinase; Gpp, pppGpp 5′-phosphohydrolase. (*b*) Role of (p)ppGpp in maintaining an efficiently balanced pool of amino acids. In this example, histidine levels have fallen below optimum. The ensuing accumulation of ppGpp stimulates expression of the *his* operon while slowing translation and ribosome synthesis. As growth slows, the other amino acids accumulate and repress their own synthesis. Restoration of histidine levels shut off (p)ppGpp synthesis, allowing the resumption of normal ribosome synthesis rates.

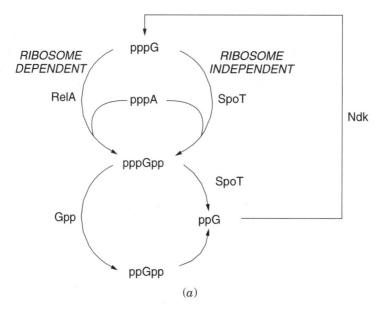

(a)

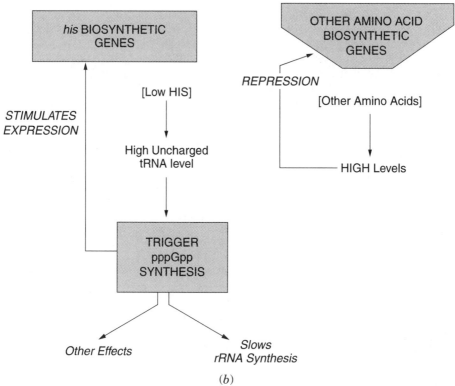

(b)

In vivo, the synthesis of (p)ppGpp on the ribosome is controlled by the charging of total tRNA species as a function of intracellular concentrations of the 20 amino acids. The ratio of charged to uncharged tRNA appears to be of central importance rather than the overall level of charged tRNA. It would appear that ppGpp is a component of a sensing mechanism that functions to adjust the synthesis, for example, of histidine biosynthetic enzymes with respect to the need for histidine relative to the total amino acid concentration of the cell as well as in the growth medium. Thus, along with the operon-specific attenuator mechanism that responds to the need for histidine, specifically altering the level of ppGpp enables the organism to sense how the supply of histidine, in this example, compares with the availability of all the amino acids in the cell — a kind of fine-tuning mechanism that maintains the correct relative levels of each amino acid.

EXTREMOPHILES

Although, the stress responses described above are important for the survival of most bacteria during exposure to the various stresses, some bacteria view stresses such as extremes in pH or temperature as a lifestyle choice. These bacteria survive and thrive in conditions that would send other microbes to their maker. **Acidophiles** typically survive and grow at an external pH (pH$_o$) below 4.0; **alkalophiles** grow only at a pH$_o$ of 8.0 or above. Similarly, **thermophiles** grow at temperatures generally around 50° to 65 °C and **psychrophiles** grow at temperatures as low as 5° to 10 °C.

Despite growing at pH 2–4, **acidophilic bacteria** maintain their cytoplasmic pH at 6.0 or higher. Consequently, there has been considerable interest in determining the mechanism whereby this large pH differential (ΔpH) is maintained. It is generally agreed that maintenance of this large transmembrane ΔpH is energy dependent and that its maintenance requires a transmembrane electrical potential ($\Delta\Psi$) that is positive inside (the reverse of that found in neutralophilic bacteria). A reverse transmembrane potential has been observed in both *Bacillus acidocaldarius* and *Thiobacillus acidophilis*. If this transmembrane potential is abolished, the ΔpH collapses. Despite the large ΔpH, the cytoplasmic pH is extremely stable. The cytoplasmic buffering capacity of *T. acidophilis* is responsible for the cytoplasmic pH homeostasis in metabolically comprised cells. When a large influx of H$^+$ occurs, the cytoplasmic buffering capacity prevents drastic changes in pH. In addition, the resultant increase in positive membrane potential due to this influx of H$^+$ eventually leads to cessation of further H$^+$ influx.

Studies with a heterotrophic, mesophilic, obligate acidophile provide additional insights into the nature of changes in membrane potential under active and inactive conditions (Fig. 18-8). Starving cells of this acidophile continue to show a ΔpH of about 1.7 but exhibit changes in membrane potential ($\Delta\Psi$) and proton motive force ($\Delta\rho$) that are just the opposite of those seen under conditions of optimal nutrition. The linkage of the transient H$^+$ influx with the rise of $\Delta\Psi$ and the cytoplasmic buffering capacity play central roles in acidophilism. It is considered that the same impermeant cellular macromolecules can account for both. Thus, the $\Delta\Psi$ represents a Donnan potential (electrochemical potential across a membrane at electrochemical steady state) that is offset in active cells by energy-dependent H$^+$ extrusion.

Many bacteria display optimal growth in alkaline media. Although a number of these are *Bacillus* species, alkalophilic strains of *Micrococcus*, *Pseudomonas*, *Clostridium*,

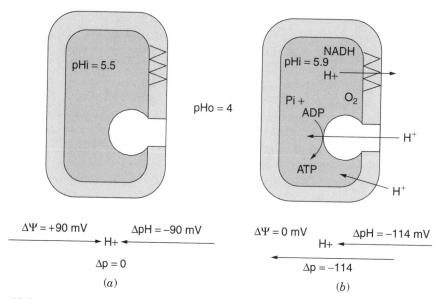

Fig. 18-8. Model for bioenergetic parameters in an obligately acidophilic bacterium at an outside pH (pH$_o$) of 4.0 in inactive (a) and respiring (b) cells. Left and right arrows indicate the forces impelling the H$^+$ ion flux into and out of the cells and are drawn to scale. There is no H$^+$ ion flux in inactive cells. In active cells H$^+$ ion influx equals H$^+$ ion efflux. (*Source:* From Goulborne et al., 1986.)

the photosynthetic bacterium *Ectothiorhodospiras,* and many others have been reported. These **alkalophiles** can grow only at pH levels of 8.0 to 11.5. A true alkalophile maintains a cytoplasmic pH of 9.0 or lower even at an external pH of 11.0, thus growing optimally under conditions in which the cytoplasm is more acidic than the external medium. The alkalophiles *B. firmus* and *B. alcalophilus* both depend on the activity of a Na$^+$/H$^+$ antiporter to achieve acidification of the cytoplasm relative to the exterior. A nonalkalophilic strain of *B. firmus* isolated after mutagenesis of the alkalophilic strain lacks the Na$^+$/H$^+$ antiporter. The inability of *B. firmus* to grow at neutral pH is not due to excessive acidification but is related to a failure of respiratory activity to generate a transmembrane electrical potential that is high enough to maintain certain cellular functions such as Na$^+$/solute symport and motility. Nonalkalophic mutant strains of *B. alcalophilus* exhibit loss of Na$^+$/H$^+$ antiport activity and Na$^+$ coupling of solute transport as well as lower amounts of membrane-bound cytochromes and a membrane-bound chromophore.

    Thermophilic organisms have been studied intensively in an effort to determine the mechanisms whereby these organisms not only survive but also prefer higher temperatures for growth. Chemical differences in the lipids found in thermophiles, higher metabolic rates facilitating rapid resynthesis of heat-denatured cellular components, and macromolecules with higher thermostability have all been suggested as mechanisms for greater heat tolerance in thermophiles. Thermophilic organisms contain lipids with higher melting points than those found in mesophiles, suggesting that the temperature at which the major lipid components of the cell melt may establish the upper limit for growth. In yeast, a direct correlation has been established between the growth temperature limits and the degree of unsaturation in mitochondrial lipids.

The lower the temperature limit of growth, the greater the degree of lipid unsaturation. The membrane lipid compositions of thermophilic yeasts are distinguished by the higher percentage (30–40%) of saturated fatty acids as compared with mesophilic and psychrophilic species. Psychrophilic yeasts contain approximately 90% unsaturated fatty acids, 55% of which is linolenic acid ($C_{\alpha-18:3}$).

Certain lines of evidence indicate that thermophilic organisms possess an intrinsic thermostability that is independent of any transferable, stabilizing factors. Membranes with greater heat stability, more rapid metabolic and growth rates, and factors that impart structural stability and greater inherent heat stability to individual protein macromolecules have been implicated as contributing factors. The greater heat stability of protein molecules from thermophilic organisms appears to reside in their ability to bind certain ions more tightly, thereby enhancing the establishment of more stable conformations. Proteins from thermophiles also contain increased levels of hydrophobic amino acids as compared to those from mesophiles. Although macromolecules from thermophiles are more thermostable, they are, for the most part, physicochemically similar to their mesophilic counterparts with regard to molecular weight, subunit composition, allosteric effectors, amino acid composition, and major amino acid sequences.

Certain attributes of protein thermostability are due to subtle changes in structure and to alterations in hydrogen bonding, hydrophobic interactions, and other noncovalent activities. For example, the thermostability of the tryptophan synthetase of *E. coli* is increased by amino acid substitutions that alter the hydrophobicity of the molecule in the absence of gross changes in conformation. In the thermophile *B. stearothermophilus*, a marked increase in thermostability is engendered by the presence of lysine in place of threonine in the plasmid-encoded enzyme that inactivates kanamycin. The nucleotide sequence of the plasmid encoding this enzyme in *B. stearothermophilus* differs by only one base from the plasmid coding for the identical enzyme in the mesophilic *S. aureus*. The lysine substitution permits increased electrostatic bridging with little significant change in the three-dimensional structure.

Isolation of thermophilic mutants of *B. subtilis* and *B. pumilus* and transformation of the thermophilic trait to mesophilic strains provides direct genetic evidence that the thermophilic trait is a phenotypic consequence of at least two unlinked genes. Additional studies of this type should provide information that could shed light on the specific nature of thermophily.

Psychrophilic organisms — those that exhibit an optimal temperature for growth at 15 °C or lower and a maximum temperature for growth at 20 °C — are defined as **obligate psychrophiles** to differentiate them from organisms that can grow at low temperatures but are actually mesophiles in terms of their optimum growth temperature. Although some confusion still exists with regard to the proper nomenclature and cardinal temperatures that differentiate psychrophiles from organisms that are **psychrotolerant** or **psychrotropic**, those organisms that grow well at temperatures at or below the freezing point of water are considered to be true psychrophiles.

The psychrophilic yeast *Leucosporidium* grows well at −1 °C. In this organism there is a positive correlation between the growth temperature and the unsaturated fatty acid composition of the cell lipids. At subzero temperatures (−1 °C) with ethanol as the substrate, 90% of the total fatty acid is unsaturated, with linolenic acid (35–50%) and linoleic acid (25–30%) predominating. At temperatures close to the maximum for growth, linolenic acid accounted for less than 20% of the total fatty acid, and oleic acid

(20–40%) and linolenic acid (30–50%) were the major components. Marked changes also occurred in the cytochrome composition of the cell, $a + a_3 : b : c$ were $1 : 1 : 2.9$ at 8 °C with glucose as the substrate, whereas at 19 °C they were $1 : 2.3 : 16.7$. Thus, it can be concluded that changes in membrane structure and composition are fundamental to temperature adaptation in psychrophilic yeasts.

The psychrophilic bacterium *Micrococcus cryophilus* also undergoes alteration in its lipid composition with changes in growth temperature. Cultures of this organism continue to grow without a lag following a sudden increase in temperature from 0° to 20 °C (upshift) or a reciprocal decrease (downshift). The growth rate changes gradually to that typical of cultures grown isothermally at the final temperature. After a temperature downshift, the phospholipid acyl chain length begins to change immediately, whereas there is a delay following upshift. However, the final fatty acid composition is attained within the same number of cell division times after an upshift or downshift. It appears, therefore, that this psychrophile is more stressed by a sudden increase in growth temperature than by a sudden decrease.

Studies on the viscosity and phase transition temperatures of lipids isolated from psychrophilic, psychrotropic, and mesophilic organisms are able to adjust their lipid-phase transition temperature to the growth temperature. By comparison, a psychrophilic *Clostridium* synthesizes lipids that have the same phase transition temperature after growth at different temperatures. This lack of growth temperature–inducible regulation of lipid-phase transition temperatures appears to be a molecular determinant for psychrophily in this organism. Comparisons of the properties of triosephosphate isomerase purified from psychrophilic, mesophilic, and thermophilic clostridia indicate that the purified enzymes have the same molecular weight, subunit molecular weight, and susceptibility to the active site-directed inhibitor, glycidol phosphate. However, their temperature and pH optima, as well as stabilities to heat, urea, and sodium dodecyl sulfate (SDS), differ markedly.

SUMMARY

Our understanding of stress responses has made great strides over the last several years, especially in the responses of microbes other than *E. coli* and *Salmonella*. However, the ever-increasing realization of the complexity and interrelationships of stress responses clearly indicates that we still have a great deal to uncover. Since most microbes spend the vast majority of their life under environmental stress, further knowledge of stress responses is critical to the complete understanding of microbial physiology. Information obtained may contribute to the development of new vaccines, new disease therapies, new agricultural strategies, new food safety procedures, and new antimicrobial agents. Thus, the study of stress responses continues to be an exciting area of basic and applied biomedical and biological research.

BIBLIOGRAPHY

Osmotic Stress and Osmoregulation

Booth, I. R., and P. Louis. 1999. Managing hypoosmotic stress: aquaporins and mechanosensitive channels in *Escherichia coli. Curr. Opin. Microbiol.* **2:**166–9.

Botsford, J. L., M. Alvarez, R. Hernandez, and R. Nichols. 1994. Accumulation of glutamate by *Salmonella typhimurium* in response to osmotic stress. *Appl. Environ. Microbiol.* **60**:2568–74.

Bremer, E. and R. Krämer. 2000. Coping with osmotic challenges: osmoregulation through accumulation and release of compatible solutes in bacteria. In G. Storz, and R. Hengge-Aronis (eds.), *Bacterial Stress Responses*. American Society for Microbiology Press, Washington, DC, pp. 79–97.

Head, C. G., A. Tardy, and L. J. Kenney. 1998. Relative binding affinities of OmpR and OmpR-phosphate at the *ompF* and *ompC* regulatory sites. *J. Mol. Biol.* **281**:857–70.

Leonardo, M. R., and S. Forst. 1996. Re-examination of the role of the periplasmic domain of EnvZ in sensing of osmolarity signals in *Escherichia coli. Mol. Microbiol.* **22**:405–13.

Matsubara, M., and T. Mizuno. 1999. EnvZ-independent phosphotransfer signaling pathway of the OmpR-mediated osmoregulatory expression of OmpC and OmpF in *Escherichia coli. Biosci. Biotechnol. Biochem.* **63**:408–14.

Aerobic to Anaerobic Transitions

Kwon, O., D. Georgellis, A. S. Lynch, D. Boyd, and E. C. Lin. 2000. The ArcB sensor kinase of *Escherichia coli*: genetic exploration of the transmembrane region. *J. Bacteriol.* **182**:2960–6.

Li, B., H. Wing, D. Lee, H. C. Wu, and S. Busby. 1998. Transcription activation by *Escherichia coli* FNR protein: similarities to, and differences from, the CRP paradigm. *Nucleic Acids Res.* **26**:2075–81.

Matsubara, M., and T. Mizuno. 2000. The SixA phospho-histidine phosphatase modulates the ArcB phosphorelay signal transduction in *Escherichia coli. FEBS Lett.* **470**:118–24.

Matsushika, A., and T. Mizuno. 2000. Characterization of three putative subdomains in the signal-input domain of the ArcB hybrid sensor in *Escherichia coli. J. Biochem. (Tokyo)* **127**:855–60.

Unden, G., S. Becker, J. Bongaerts, G. Holighaus, J. Schirawski, and S. Six. 1995. O_2-sensing and O_2-dependent gene regulation in facultatively anaerobic bacteria. *Arch. Microbiol.* **164**:81–90.

Unden, G., and J. Schirawski. 1997. The oxygen-responsive transcriptional regulator FNR of *Escherichia coli*: the search for signals and reactions. *Mol. Microbiol.* **25**:205–10.

Wang, H., and R. P. Gunsalus. 2000. The *nrfA* and *nirB* nitrite reductase operons in *Escherichia coli* are expressed differently in response to nitrate than to nitrite. *J. Bacteriol.* **182**:5813–22.

Oxidative Stress

Bauer, C. E., S. Elsen, and T. H. Bird. 1999. Mechanisms for redox control of gene expression. *Annu. Rev. Microbiol.* **53**:447–94.

Berlett, B. S., and E. R. Stadtman. 1997. Protein oxidation in aging, disease, and oxidative stress. *J. Biol. Chem.* **272**:20,313–6.

Clements, M. O., S. P. Watson, and S. J. Foster. 1999. Characterization of the major superoxide dismutase of *Staphylococcus aureus* and its role in starvation survival, stress resistance, and pathogenicity. *J. Bacteriol.* **181**:3898–903.

Fridovich, I. 1997. Superoxide anion radical ($O_2^{-\bullet}$), superoxide dismutases, and related matters. *J. Biol. Chem.* **272**:18,515–7.

Gibson, C. M., T. C. Mallett, A. Claiborne, and M. G. Caparon. 2000. Contribution of NADH oxidase to aerobic metabolism of *Streptococcus pyogenes. J. Bacteriol.* **182**:448–55.

Gort, A. S., and J. A. Imlay. 1998. Balance between endogenous superoxide stress and antioxidant defenses. *J. Bacteriol.* **180**:1402–10.

Henle, E. S., and S. Linn. 1997. Formation, prevention, and repair of DNA damage by iron/hydrogen peroxide. *J. Biol. Chem.* **272**:19,095–8.

Higuchi, M., et al. 1999. Functions of two types of NADH oxidases in energy metabolism and oxidative stress of *Streptococcus mutans*. *J. Bacteriol.* **181**:5940–7.

Jenney, F. E., Jr., M. F. J. M. Verhagen, X. Cui, and M. W. W. Adams. 1999. Anaerobic microbes: oxygen detoxification without superoxide dismutase. *Science* **286**:306–9.

Loewen, P. C., M. G. Klotz, and D. J. Hassett. 2000. Catalase — an "old" enzyme that continues to surprise us. *ASM News* **66**:76–82.

Lumppio, H. W., N. V. Shenvi, A. O. Summers, G. Voordouw, and D. M. Kurtz, Jr. 2001. Rubrerythrin and rubredoxin oxidoreductase in *Desulfovibrio vulgaris*: a novel oxidative stress protection system. *J. Bacteriol.* **183**:101–8.

McCormick, M. L., G. R. Buettner, and B. E. Britigan. 1998. Endogenous superoxide dismutase levels regulate iron-dependent hydroxyl radical formation in *Escherichia coli* exposed to hydrogen peroxide. *J. Bacteriol.* **180**:622–5.

Pomposiello, P. J., and B. Demple. 2000. Identification of SoxS-regulated genes in *Salmonella enterica* serovar Typhimurium. *J. Bacteriol.* **182**:23–9.

Pomposiello, P. J., and B. Demple. 2000. Oxidative stress. In J. Lederberg (ed.), *Encyclopedia of Microbiology*, 2nd ed. Academic Press, San Diego, pp. 526–32.

Zheng, M., and G. Storz. 2000. Redox sensing by prokaryotic transcription factors. *Biochem. Pharmacol.* **59**:1–6.

pH Stress and Acid Tolerance

Audia, J. P., C. C. Webb, and J. W. Foster. 2001. Breaking through the acid barrier: an orchestrated response to proton stress by enteric bacteria. *Int. J. Med. Microbiol.* **291**:97–106.

Castanie-Cornet, M. P., and J. W. Foster. 2001. *Escherichia coli* acid resistance: cAMP receptor protein and a 20 bp *cis*-acting sequence control pH and stationary phase expression of the *gadA* and *gadBC* glutamate decarboxylase genes. *Microbiology* **147**:709–15.

Foster, J. W. 2000. Microbial responses to acid stress. In G. Storz and R. Hengge-Aronis (eds.), *Bacterial Stress Responses*. American Society for Microbiology Press, Washington, DC, pp. 99–116.

Quivey, R. G., Jr., W. L. Kuhnert, and K. Hahn. 2000. Adaptation of oral streptococci to low pH. *Adv. Microb. Physiol.* **42**:239–74.

Weeks, D. L., S. Eskandari, D. R. Scott, and G. Sachs. 2000. A H+-gated urea channel: the link between *Helicobacter pylori* urease and gastric colonization. *Science* **287**:482–5.

Thermal Stress and the Heat Shock Response

Ades, S. E., L. E. Connolly, B. M. Alba, and C. A. Gross. 1999. The *Escherichia coli* sigma(E)-dependent extracytoplasmic stress response is controlled by the regulated proteolysis of an anti-sigma factor. *Genes Dev.* **13**:2449–61.

Missiakas, D., and S. Raina. 1997. Signal transduction pathways in response to protein misfolding in the extracytoplasmic compartments of *E. coli*: role of two new phosphoprotein phosphatases PrpA and PrpB. *EMBO J.* **16**:1670–85.

Missiakas, D., M. P. Mayer, M. Lemaire, C. Georgopoulos, and S. Raina. 1997. Modulation of the *Escherichia coli* sigmaE (RpoE) heat-shock transcription-factor activity by the RseA, RseB and RseC proteins. *Mol. Microbiol.* **24**:355–71.

Ravio, T. L., and T. J. Shihavy. 1999. The σ^E and Cpx regulatory pathways: overlapping but distinct envelope stress responses. *Curr. Opin. in Microbiol.* **2**:159–65.

Nutrient Stress and the Starvation Stress Response

Hengge-Aronis, R. 2000. The general stress response in *Escherichia coli*. In G. Storz and R. Hengge-Aronis (eds.), *Bacterial Stress Responses*. American Society for Microbiology Press, Washington, DC, pp. 161–78.

Jørgensen, F., M. Bally, V. Chapon-Herve, G. Michel, A. Lazdunski, P. Williams and G. S. A. B. Stewart. 1999. RpoS-dependent stress tolerance in *Pseudomonas aeruginosa*. *Microbiology* **145**:835–44.

Kvint, K., C. Hosbond, A. Farewell, O. Nybroe, and T. Nyström. 2000. Emergency derepression: stringency allows RNA polymerase to override negative control by an active repressor. *Mol. Microbiol.* **35**:435–43.

Sanders, J. W., G. Venema, and J. Kok. 1999. Environmental stress responses in *Lactococcus lactis*. *FEMS Microbiol. Rev.* **23**:483–501.

Spector, M. P. 1998. The starvation-stress response (SSR) of *Salmonella*. *Adv. Microb. Physiol.* **40**:233–79.

Trainor, V. C., R. K. Udy, P. J. Bremer, and G. M. Cook. 1999. Survival of *Streptococcus pyogenes* under stress and starvation. *FEMS Microbiol. Lett.* **176**:421–8.

Stringent Control

Barker, M. M., T. Gaal, C. A. Josaitis, and R. L. Gourse. 2001. Mechanism of regulation of transcription initiation by ppGpp. I. Effects of ppGpp on transcription initiation in vivo and in vitro. *J. Mol. Biol.* **305**:673–88.

Barker, M. M., T. Gaal, and R. L. Gourse 2001. Mechanism of regulation of transcription initiation by ppGpp. II. Models for positive control based on properties of RNAP mutants and competition for RNAP. *J. Mol. Biol.* **305**:689–702.

Chatterji, D., and A. Kumar Ojha. 2001. Revisiting the stringent response, ppGpp and starvation signaling. *Curr. Opin. Microbiol.* **4**:160–5.

Chatterji, D., N. Fujita, and A. Ishihama. 1998. The mediator for stringent control, ppGpp, binds to the beta-subunit of *Escherichia coli* RNA polymerase. *Genes Cells* **3**:279–87.

Eichel, J., Y. Y. Chang, D. Riesenberg, and J. E. Cronan, Jr. 1999. Effect of ppGpp on *Escherichia coli* cyclopropane fatty acid synthesis is mediated through the RpoS sigma factor (sigma S). *J. Bacteriol.* **181**:572–6.

Sorensen, M. A. 2001. Charging levels of four tRNA species in *Escherichia coli* Rel(+) and Rel(−) strains during amino acid starvation: a simple model for the effect of ppGpp on translational accuracy. *J. Mol. Biol.* **307**:785–98.

Extremophiles

Edwards, K. J., P. L. Bond, T. M. Gihring, and J. F. Banfield. 2000. An archaeal iron-oxidizing extreme acidophile important in acid mine drainage. *Science* **287**:1796–9.

Goulborne, E. Jr., M. Matin, E. Zychlinsky and A. Matin. 1986. Mechanism of ΔpH maintenance in active and inactive cells of an obligately acidophilic bacterium. *J. Bacteriol.* **166**:59–65.

Krulwich, T. A., M. Ito, D. B. Hicks, R. Gilmour, and A. A. Guffanti. 1998. pH homeostasis and ATP synthesis: studies of two processes that necessitate inward proton translocation in extremely alkaliphilic Bacillus species. *Extremophiles* **2**:217–22.

Matin, A. 1999. pH homeostatis in acidophiles. In J. C. Derek, and G. Gardew (eds.), *Bacterial Responses to pH*. John Wiley & Sons, Ltd., Chichester, England, pp. 152–65.

van de Vossenberg, J. L., A. J. Driessen, and W. N. Konings. 1998. The essence of being extremophilic: the role of the unique archaeal membrane lipids. *Extremophiles* **2**:163–70.

CHAPTER 19

BACTERIAL DIFFERENTIATION

Development or differentiation among eukaryotic microbes is well known and diverse, especially in the production of asexual conidia and sexual spores, but is not discussed here. In comparison, prokaryotic differentiation, although not as widespread, is beginning to show interesting diversity. The reasons for differentiation of developmental cycles in prokaryotes are reflected in the surprising number of examples including the differentiation into (1) a variety of "resting" cell forms that are more resistant to environmental stresses such as endospores (*Bacillus*) and myxospores (myxobacteria), (2) a number of cell forms specifically "designed" for dispersal of the bacteria within certain niches such as swarmer cells (*Caulobacter*), (3) cell forms performing specific functions such as heterocysts (*Anabaena*), and (4) cell forms designed to establish a symbiotic relationship with another organism for their mutual benefit such as the nitrogen-fixing nodules associated with legumes (*Rhizobium*) (see Chapter 14). This chapter highlights three interesting paradigms of bacterial differentiation: *Bacillus* endospore formation, myxospore/fruiting body formation in myxobacteria, and stalked and swarmer cell formation in *Caulobacter*.

BACILLUS ENDOSPORE FORMATION

As shown in Table 12-1, six separate genera of bacteria produce a developmental form called an **endospore**. All of these spores are resistant to heat, cold, radiation, and other adverse environmental conditions. However, the endospores produced by members of the *Bacillaceae* are probably the most resistant to both natural and artificially imposed hostile environmental conditions. The primary function of endospore formation appears to be the survival and dissemination of the species. Since endospore formation in bacteria is triggered by starvation, this phenomenon represents a special example of the stress responses discussed in Chapter 18. Although the production of endospores is displayed by several genera of bacteria (see Table 19-1), a much greater amount of

TABLE 19-1. Characteristics of Endospore-Forming Bacteria[a]

Genus	G + C (mol%)	Shape	Distinguishing Metabolic Traits
Bacillus	33–66	Rods	**Catalase positive**; most strict aerobe
Sporolactobacillus	38–40	Rods	**homolactic fermentation**; facultative anaerobe or microaerophilic
Clostridium	24–54	Rods or filaments	**Strict anaerobe**
Desulfotomaculum	37–50	Rods or filaments	**Sulfate reduction** and strict anaerobe
Sporosarcina	40–42	**Cocci in tetrads or packets**	Strict aerobe
Thermoactinomycetes	52–55	**Branched filaments**	Strict aerobe

[a]The major distinguishing factor that separates each spore former from the others is in bold type.

Source: From Slepecky, 1993, p. 1–21. In Biology of Bacilli: Applications to Industry. R. H. Doi and M. McGloughlin (Eds.). Butterworth-Heinemann, Boston, MA.

time and effort has been expended in the study of this process in the genus *Bacillus*, particularly *Bacillus subtilis*. For this reason, we emphasize the comparative biology, physiology, and genetics of bacterial endospore formation by this organism.

On a practical basis, the ability to produce resistant spores enables many organisms to survive autoclaving, radiation, or chemical processes for the preservation of foods or sterilization of material for medical procedures. Great care must be taken to establish the conditions required for their elimination. Concerted efforts to understand the developmental and regulatory factors governing sporulation may provide insights into other developmental processes in higher forms.

Life Cycle of *Bacillus*

All endospore-forming bacteria undergo a life cycle that includes vegetative growth in the presence of adequate nutrition and favorable environmental conditions. At the beginning of the stationary phase, when nutrients become limiting and pH and other conditions in the culture medium change, the cells have a variety of overlapping genetic networks available with which they can respond to this changing environment. Among the processes that can be activated under growth-restricting conditions are establishment of motility and competence for transformation. The decision to undergo sporulation appears to be a response of last resort in an effort to provide for survival.

A series of molecular switches appears to control the decision-making process to determine which of these responses will be followed. As the cells enter the stationary phase of growth, nutritional deprivation can trigger entry into sporulation. This process can be visualized by light and electron microscopy as a series of complex morphological changes that result in the formation of a highly resistant dormant form called an endospore. When conditions become favorable again, the spore can undergo activation, germination, and outgrowth into a metabolically active cell capable of entering the vegetative growth cycle, or, if conditions are unfavorable, the spore can proceed directly into another round of sporulation (the **microcycle**). The response of spores to activation by environmental factors that lead to germination and outgrowth need to be considered along with spore formation, since the genetic and regulatory machinery that governs these processes must be built into the spore as it is being formed.

Stages of Sporulation

The morphological events in the sporulation process as they occur in *Bacillus subtilis* are shown in Figure 19-1 and are as follows:

Stage 0 (vegetative cells): Cells undergoing normal vegetative growth are defined as being in stage 0 with regard to the sporulation cycle.

Stage I (axial filament): This stage has not been recognized universally, since mutants arrested in stage I were not found until recently. But since mutants blocked at stage I have been isolated, it is useful to consider this as a valid stage.

Stage II (septum formation): The first sign of sporulation is the formation of an asymmetrically sited division septum resulting in two distinct cells: the mother cell and the prespore. Each of the two cells receives a portion of the nuclear DNA. Intermediary stages designated IIi, IIii, and IIiii are based on the isolation of mutants blocked at these points between stages II and III.

Stages IIi, IIii, IIiii (prespore engulfment): The prespore is engulfed by the mother cell to form a protoplast.

Stage III (forespore development): The engulfed protoplast is now referred to as the forespore. In most electron micrographs the forespore is rather amorphous in appearance, presumably because no peptidoglycan layer has, as yet, been formed to provide a defined shape.

Stage IV (cortex formation): Shortly after forespore engulfment, a dense, narrow band of peptidoglycan, the cortex layer, is formed between the inner and outer membranes of the forespore. At this point, the beginnings of the spore coat can barely be discerned. As time progresses, the cortex becomes striated and multilayered.

Stage V (coat formation): Overlapping the cortex development process is the deposition of the multilayered spore coat. Discontinuous segments of coat material are observed. These segments coalesce into a continuous dense layer. Deposition of coat material at discontinuous points is more characteristic of coat formation in *Clostridium*. In *Bacillus*, deposition of coat material is a continuous process. However, there does not appear to be any fundamental difference in synthesis of the coat material in the two species.

Stage VI (maturation): Following the deposition of the cortex layers and formation of the spore coat, the final stage of maturation occurs. Observed as a whitening under phase-contrast microscopy, this process is associated with the synthesis of dipicolinic acid and calcium uptake into the mature spore. The characteristic properties of resistance, dormancy and germinability, appear at this stage.

Stage VII (release of the mature spore): The mother cell is lysed and the mature spore is released as a dormant form.

It should be emphasized that the cytological events outlined here and depicted in Figure 19-1 occur as an overlapping continuum rather than as separate and distinct events. These morphological stages serve as useful reference points. Mutants that permit the sporulation process to proceed to one of these stages but prevent any further development beyond this point are designated spo0, spoII, spoIII, spoIV, and so on. A temporally and spatially controlled program of gene expression drives the

Stage	Description	Morphology	Genes or enzymes involved; other properties
0	Vegetative cell	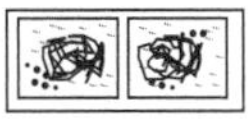	Non-refractile; normal vegetative growth stage
I $\sigma^H + \sigma^A$	Axial filament prespore	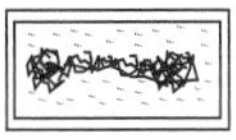	CitC⁻, isocitrate dehydrogenase deficient mutants, blocked at stage I; DNA replic.; FtsZ ring formed
Iii $\sigma^E + \sigma^F$	Asymmetric Septation		Forespore develops; alanine dehydrogenase; *spoOA*, *spoOK*, *spoOH*; chromosome pumped into forespore.
Iiii σ^E	Septal membrane Rearrangement		*spoIIA, spoIIP, spoIIG, spoIIB* + *spoVG*
Iiiii σ^E	Forespore engulfment		*spoIID, spoIIA(P); spoIVCA* + *spoVCB* + *spoIIIC* required for generation of σ^K
III $\sigma^K + \sigma^G$	Forespore development	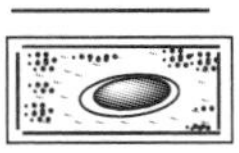	*spoIIIA, spoIIIE, spoIIIJ, spoIIID, spoIIIG, spoIVB, spoIVF, spoVB, spoVD, spoVE;* alkaline phosphatase; glucose dehydrogenase; aconitase, HR catalase
IV σ^K	PGCW and cortex synth.	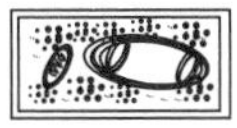	*cotD, cotT, gerE;* refractility, ribosidase, adenosine deaminase, dipicolinic acid
V σ^K	Coat synthesis		*cotE, gerE, cotA, cotB, cotC;* cysteine incorporation, chemical & UV resistance develops
VI	Maturation		Alanine racemase, heat resistance
VII	Mother cell lysis	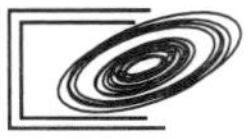	Mature spore is released from mother cell.
VIII	Free spore		Refractile; resistant to heat, radiation, etc.
	Germination		Non-refractile; sensitive to environmental factors

Fig. 19-1. Diagrammatic representation of the life cycle of *Bacillus subtilis*. As discussed in the text, the stages are designated by Roman numerals. The stage originally designated as I has not been universally recognized because mutants arrested at this stage were not found. However, mutants blocked at this stage have recently been described, so it will be considered as a valid stage. The cycle proceeds as a continuum and the stages are primarily for convenient reference to indicate a stage at which mutants block the progress of sporulation. The diagram provides a representative sample of mutants blocked at the stage indicated. The approximate stages at which various enzymatic activities and other properties develop are also shown. The diagram represents a composite of several diagrams from a number of the reviews cited at the end of this chapter.

sporulation process. Many of the genes required for the eventual response to activation, germination, and outgrowth must be built into the spore during the sporulation process, since they must be available for action within minutes of their activation. The program that regulates and orchestrates these events is discussed in further detail below.

PHYSIOLOGICAL AND GENETIC ASPECTS OF SPORULATION

Sporulation Genes

Over 125 gene products govern the complex morphological and biochemical changes that take place during sporulation. The genes involved in sporulation are scattered around the chromosome and are interspersed with many other genes that have no known role in sporulation. More than 60 genetic loci, designated *spo*, have been identified and named according to the morphological stage at which the sporulation is blocked. A partial list of these genes is shown in Table 19-2. Other genes that participate in sporulation include those coding for a number of small acid-soluble proteins (SASPs) of the spore core, or for the proteins in the spore coat, designated *cot*. Some of the genes involved in normal cell division, such as *ftsA* and *ftsZ*, are also involved in the asymmetric division process that initiates sporulation. Finally, those genes regulating

TABLE 19-2. A Partial List of Sporulation Genes and Their Function in *Bacillus subtilis*

Stage	Gene Designation	Function
I	*citC*	Isocitrate dehydrogenase
IIi	*spoOA*	Response regulator; phosphorylation initiates sporulation via phosphorelay system
	spoOK	Regulates phosphorelay system
	spoOH	Encodes σ^H
IIii	*spoIIG*	Pro σ^E and activating protease
	spoIIB + *spoVG*	Prespore engulfment; septum formation
IIiii	*spoIID*	Septal peptidoglycan hydrolysis
	spoIIA(P)	Processing of pro σ^E
III	*spoIIIA*	Prespore engulfment
	SpoIIIE	Controls σ^F expression
	spoIIiJ (*kinA*)	Histidine protein kinase; activates SpoOF
	spoIIID	Controls σ^E-dependent genes
	spoIIIG	Structural gene for σ^G
	spoIVB	Production of σ^K; signal peptide
	spoIVF (*spoIIIF*)	Processing of pro-σ^K
	spoVB	Cortex synthesis
	spoVD	Cortex synthesis
	spoVE	Cortex synthesis
IV	*cotD*	Coat synthesis
	cotT	Coat synthesis
	cotA	Coat synthesis
	cotB	Coat synthesis
	cotC	Coat synthesis
	gerE	Germination

spore germination (*ger*) are included, since they must be incorporated into the spore genome during the sporulation process.

Initiation

Endospore formation is initiated in response to starvation for carbon, nitrogen, or phosphorous. The very early stages of sporulation are reversible and transfer to fresh culture medium results in resumption of vegetative growth. Thus, sporulation seems to be initiated by the accumulation of factors that inhibit vegetative growth and derepress the spore genome. Sporulation is more efficient in a dense population of cells. At low cell density, sporulation efficiency is improved if the cells are suspended in medium previously conditioned by growth of cells to high density. This finding indicates that a substance produced by cells at high density is necessary for efficient sporulation. Production of this extracellular differentiation factor (EDF-1) or pheromone is governed by *spoOA, spoOB*, and *spoOH*.

Even if starvation conditions are imposed, sporulation will occur only if the cell has reached a certain stage in the vegetative growth cycle, the tricarboxylic acid (TCA) cycle is functional, and at least one external pheromone is present in sufficient concentration. Mutants blocked in the TCA cycle — for example, a mutant with a deletion of *citC*, the gene encoding isocitrate dehydrogenase — cause a block at stage I of sporulation. For this reason, it is useful to retain stage I as a valid part of the sporulation cycle. Once these conditions are met, a controlled program of sporulation gene expression largely replaces the pattern of vegetative gene expression. The temporal changes in gene expression that occur during sporulation are controlled by the sequential appearance or activation of sporulation-specific transcription factors (designated sigma, σ) that bind to the core RNA polymerase and direct it to transcribe only from specific promoters. Sigma factor binding also confers the capacity to recognize new classes of these factors as they appear.

The first visual recognition of the initiation of sporulation involves a special asymmetric cell division. Asymmetric placement of the septum separates the cell into the mother cell and the forespore. The forespore goes on to become the spore while the mother cell serves to nurture the cell until the spore is fully matured.

At this point, the sporulation gene program splits and two distinct programs become active: one in the mother cell and the other in the forespore. The mother cell produces certain proteins that are incorporated into the developing spore from the outside, and the forespore generates other proteins that complete the process by being added from the inside. The transcriptional regulator, SpoOA, plays a key role during the initiation stage. The SpoOA protein contains an aspartyl residue in the N-terminal region that is the target of phosphorylation by a histidine protein kinase, a component of the phosphorelay system (Fig. 19-2).

This complex system of regulation of stationary-phase gene expression involves at least two histidine protein kinases: KinA and KinB. They initiate the phosphorelay system by phosphorylating SpoOF. Other genes involved in this system include the *spoOB* operon, which contains *obg*, a GTP-binding protein essential for vegetative growth. It has been suggested that the *obg* gene product may be the protein that senses the decrease in GTP levels that occurs following nutrient deprivation and provides the link to sporulation events via the phosphorelay system. The experimental results suggest that a threshold level of activated SpoOA (SpoOA~P) induces sporulation gene

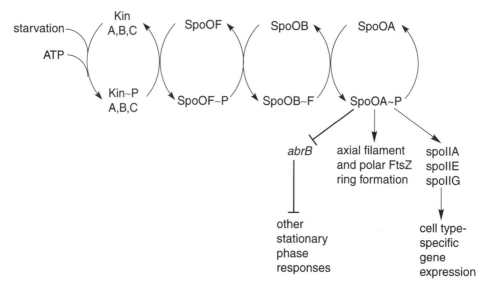

Fig. 19-2. Production of SpoOA~P by the phosphorelay system. The sensor kinases KinA, KinB, and KinC autophosphorylate on a histidine. Phosphate is transferred to SpoOF, then to SpoOB, and finally to SpoOA. Low levels of SpoOA~P are sufficient to repress transcription of *abrB*, derepressing expression of many of the stationary-phase response pathways negatively regulated by AbrB. Higher levels of SpoOA~P stimulate axial filament formation, polar septation, and transcription of genes (e.g., *spoIIA, spoIIE*, and *spoIIG*) required for cell type–specific gene expression. (*Source*: From Burkholder and Grossman, 2000.)

expression and that those cells able to induce the expression of early genes directly activated by SpoOA~P go on to produce mature spores. Genes that are under the apparent control of SpoOA cause the site of formation of the FtsZ ring to shift from midcell to polar sites as discussed further below.

Many stationary-phase genes are under the control of σ^H (SpoOH). This transcription factor is weakly expressed from a σ^A promoter in midlogarithmic phase but is greatly increased after the initiation of sporulation. Among the various genes controlled by σ^H is the sporulation-specific operon *spoIIA*, which produces the important σ^F.

As discussed in Chapter 17, FtsZ and FtsA are cell division proteins in *E. coli* that are recruited to the division site where they are involved in the formation of a ring that remains associated with the invaginating septum until septation is completed. Homologs of the *E. coli* cell division genes *ftsA* and *ftsZ* have been identified in *B. subtilis*. In *B. subtilis*, the *ftsA* gene was originally designated *spoIIG* or *spoIIN*. Both *ftsA* and *ftsZ* are required for vegetative septation as well as for the asymmetric septation that occurs in the initial stage of sporulation. The *ftsA* and *ftsZ* genes of *B. subtilis* constitute a simple operon expressed from promoter sequences immediately upstream of *ftsA*.

Both *ftsA* and *ftsZ* are transcribed from a distinct promoter, p2 (*ftsAp₂*), during sporulation but not during exponential growth. Transcription from p2 is dependent on RNA polymerase containing σ^H but does not require expression of other sporulation loci (*spoOA, spoOB, spoOE, spoOF,* or *spoOK*). Introduction of a *cat* cassette in the middle of promoter p2 does not affect vegetative growth but prevents postexponential symmetrical division and spore formation. A bifunctional protein, SpoIIE, interacts with FtsZ and is involved in the SpoOA-dependent switch in the positioning of FtsZ rings from midcell to polar locations.

Two proteins involved in chromosome partitioning, Soj (ParA) and SpoOJ (ParB), also appear to be involved in regulating the initiation of sporulation in response to chromosome structure or partitioning. Soj affects sporulation and SpoOA~P-dependent gene expression by decreasing the level of SpoOA protein. Soj negatively regulates transcription of *spoOA* and associates with the *spoOA* promoter region in vivo. Expression of *spoOA* from a heterologous promoter restores SpoOA levels and partly bypasses the sporulation and gene expression defects of a *spoOJ* null mutant.

Proteins homologous to the MinC and MinD proteins known to exist in *E. coli* have been identified in *B. subtilis*. The last two open-reading frames of the *divIVB* locus in *B. subtilis* are homologous to *minC* and *minD* of *E. coli*. However, the presence of an equivalent to MinE is less clear. The DivIVA protein displays some characteristics of a MinE-like protein and appears to have a function analogous to that of MinE in *E. coli*. The MinCD complex appears to play a role in asymmetric septum formation during the sporulation process.

Transition from Stage II to Stage III

Initiation of the prespore and mother cell programs of gene expression in the correct compartments is crucial to the success of sporulation. A sensitive and accurate signal from prespore to mother cell after septation is important in orchestrating further development of the spore. Two sporulation-specific transcription factors, σ^F in the prespore and σ^E in the mother cell, specify different programs of gene expression in the two compartments. Formation of the asymmetrically located septum during the initial stage of sporulation results in enclosure of the origin-proximal 30% of the chromosome in the prespore compartment. One copy of the chromosome is translocated from the mother cell into the prespore by an active process requiring SpoIIIE.

The SpoIIIE protein of *B. subtilis* is a DNA-dependent ATPase capable of tracking along DNA in the presence of ATP. The amino-terminal part of the protein mediates its location to the division septum. SpoIIIE acts as a DNA pump that actively moves one of the replicated pair of chromosomes into the prespore. A second copy of the chromosome remains in the mother cell.

The SpoIIR locus is near the origin of the chromosome and is transcribed only by RNA polymerase containing σ^F. The SpoIIR protein is the only σ^F-directed gene required for the activation of the transcription program directed by σ^E in the mother cell and ensures that σ^E is not activated until the septum is formed. Rapid activation of σ^E following septation may be important in preventing further septation. Normally, the *spoIIR* locus is situated at 324°, near the origin of replication (0/360°). Relocation of *spoIIR* to regions not in the origin-proximal part of the chromosome substantially reduces sporulation efficiency. This finding suggests that it is extremely important that the chromosomal location of *spoIIR* and septum positioning coincide with the partitioning of the chromosome during the asymmetric cell division.

The two unequal-size compartments, the mother cell and the forespore, are under the control of compartment-specific transcription factors. In the mother cell, σ^E is the first to be synthesized at the onset of sporulation. At the same time, σ^F is produced in the prespore. Neither σ^E nor σ^F become active until compartmentalization is complete. The inhibitor, SpoIIAB, keeps σ^F in an inactive state while σ^E is formed as an inactive proprotein (pro-σ^E). The binding of SpoIIAA to SpoIIAB instead of σ^F relieves the inhibition of σ^F by SpoIIAB. In the mother cell, SpoIIAA is kept inactive by phosphorylation. However, in the prespore, SpoIIE, a membrane-bound phosphatase, activates SpoIIAA.

Activation of pro-σ^E is accomplished by cleavage of a 27 amino acid sequence from the amino terminus catalyzed by the protease SpoIIGA. This protein is present in the preseptal cell but is maintained in an inactive state until septum formation is complete. SpoIIR triggers removal of the amino acids from pro-σ^E by SpoIIGA. Binding of SpoIIE to the forespore face of the septum has been suggested as an explanation for how the activation of σ^F is limited to the forespore. As the sporulation process continues, the transcription factor σ^K is activated in the mother cell and σ^G is activated in the forespore.

Shortly after the onset of differential gene expression in the mother cell and forespore compartments, the septum between them begins to migrate around the forespore until the leading edges of the membrane meet on the distal side of the forespore and fuse, releasing the forespore into the cytoplasm of the mother cell (see Fig. 19-1). This engulfment process results in the forespore being entirely enclosed within the mother cell and being bounded by two membranes: its own original cytoplasmic membrane and a membrane derived from the engulfing mother cell membrane. It is between these two membranes that the spore cell wall (the cortex) is synthesized.

The multilayed spore coat is assembled around the forespore within the mother cell. It is considered that thinning or removal of the peptidoglycan between the septal membranes allows the mother cell membrane to move around the forespore and eventually engulf it. Several gene products (SpoIIM, SpoIIP, SpoIID, SpoIIB, and SpoVG) have been implicated in the engulfment process. However, a spoIIB spoVG double mutant sporulates poorly and is blocked at an early stage of engulfment, with little or no thinning of septal peptidoglycan.

In SpoIIB mutants the septal peptidoglycan is incompletely degraded throughout the septum. When the forespore grows, it breaks the weakened septal peptidoglycan, resulting in broad bulges of the forespore into the mother cell. Engulfment is eventually completed, but the process proceeds more slowly. These observations suggest that SpoIIB facilitates the rapid and spatially regulated breakdown of septal peptidoglycan in the normal sporulation process.

Mutations in the σ^E-dependent *spoIID* gene cause a morphological block at intermediate stage IIii. In stage IIiii the edges of the septum begin to migrate toward the proximal pole of the cell. The edges of the septal membranes meet at the pole of the cell and fuse, completing the engulfment of the prespore within the mother cell cytoplasm (stage III). This final step is prevented by the *spoIIAC* (P) mutations, indicating that a σ^F-dependent gene is necessary for completion of engulfment.

Forespore Development

Once engulfment of the prespore is complete, it is referred to as a **forespore**. At this point, synthesis of σ^G, the product of the *spoIIIG* gene, begins with the resultant activation of the σ^G regulon in the forespore. The σ^G regulon contains a large number of genes that orchestrate many changes in the properties of the developing spore. Regulation of σ^G occurs at the transcriptional and posttranscriptional level. However, the genes involved in this stage vary considerably with regard to their regulation.

Mutations in *spoIIIA* and *spoIIIJ* allow engulfment to be completed but prevent further progression beyond stage III. The onset of the late stages of forespore development are dependent on events occurring in the mother cell at the level of *spoIIIG* transcription and σ^G activity. Specific intercommunication between the mother cell and the forespore appears to be necessary to couple the activation of σ^G in the forespore to the program of gene expression in the mother cell under the control of σ^K.

Three groups of σ^G-dependent genes are of major importance in the later stages of forespore development. These include the *ssp* genes encoding the small acid-soluble proteins (SASPs) in the spore core, the genes encoding sporulation-specific penicillin-binding proteins (PBPs), and the germination (*ger*) genes.

The structures of the nucleoids in the mother cell and forespore are quite different. Throughout the later stages of sporulation, the mother cell nucleoid retains the diffuse lobular structure of the vegetative cell nucleoid, while the forespore nucleoid is initially rather condensed and then assumes a ring-like structure. The development of the ring-like appearance of the forespore nucleoid is the result of the synthesis of a group of forespore-specific DNA-binding proteins termed the α/β-type SASPs, which saturate the forespore chromosome. These proteins also saturate the dormant spore chromosome. Sporulation of a *B. subtilis* strain lacking the majority of the α/β-type SASPs (termed $\alpha^-\beta^-$) exhibit a number of differences from that of wild-type strains, including delayed forespore accumulation of dipicolinate, overexpression of forespore-specific genes, and delayed expression of at least one mother cell–specific gene activated late in sporulation. These α/β-type SASPs appear to have global effects on gene expression during sporulation and spore outgrowth.

In addition to the role in the sporulation and germination processes, binding of the α/β-type SASPs to the outside of the DNA helix appears to straighten and stiffen the DNA while changing the DNA to an A-like helix. The properties of DNA containing bound SASPs change dramatically. DNA containing bound SASPs displays increased resistance to a variety of chemicals and appears to be inert to the action of wet heat and potentially mutagenic agents. Spore killing by dry heat and radiation occurs largely through DNA damage, and any deficiency of α/β-type SASPs renders spores more susceptible to their effects than wild-type spores. SASP binding to DNA suppresses formation of all cyclobutane pyrimidine dimers following UV irradiation and permits only the formation of thyminyl-thymine adducts whose repair is apparently much more error-free than repair of the cyclobutane dimers normally produced by UV irradiation.

Final Stages of Sporulation

The final stages of sporulation are controlled by the sporulation-specific transcription factor σ^K in the mother cell. The synthesis of σ^K results in activation of genes that participate in the formation of the cortex and the spore coat and other factors involved in maturation, as well as release of the free spore. Regulation of σ^K synthesis must be timed so that it occurs only after the appearance of σ^E and sealing off the forespore from the mother cell. A site-specific recombinase in the mother cell, encoded by the *spoIVCA* gene, brings together two partial coding sequences to form a circular DNA molecule. Another level of regulation involves processing of the inactive precursor, pro-σ^K, by proteolysis to form σ^K.

Spore Cortex Synthesis

The peptidoglycan cortex of bacterial endospores is required for maintenance of spore core dehydration, heat resistance, and dormancy. The spore cortex peptidoglycan differs in structure from that of the vegetative cell walls in that many of the peptide side chains are removed either partially, leaving single L-alanine side chains, or completely, with the formation of muramic-δ-lactam structures. The decrease in the number of side chains results in a very low degree of cross-linking of the peptidoglycan strands in the spore cortex, a property that may contribute to its flexibility and ability to undergo dehydration during the sporulation process.

Synthesis of the spore cortex involves the activity of several genes. The *spoVD* and *spoVE* genes map in a region of the chromosome containing a cluster of genes involved in cell wall synthesis and cell division. Both SpoVD and SpoVE proteins show amino acid sequence homology to PbpB (FtsI) and FtsW of *E. coli*. The *spoVD* gene is regulated independently of other genes in the cluster, is sporulation specific, and the gene product is involved in cross-wall synthesis, suggesting that its action may result in the production of a unique spore peptidoglycan. The presence of a σ^E-dependent promoter immediately upstream of *spoVE* allows expression during sporulation. The gene product appears to be similar to that of FtsW, suggesting a role in peptidoglycan synthesis. Several other genes, including *spoVB (spoIIIF), spoVG, gerJ,* and *gerM*, seem to be involved in cortex synthesis, but their role is less well defined.

Penicillin-binding proteins PBP2B and PBP3 increase during sporulation. Their map location immediately upstream of *spoVD* suggests that the gene for PBP2B may be a functional homolog of *pbpB* of *E. coli* and indicates a role in septum formation. A *B. subtilis* mutant, *dacB*, lacks the sporulation-specific D,D-carboxypeptidase penicillin-binding protein PBP5* and shows a significant increase in spore peptidoglycan cross-linking as compared to the wild type, but displays normal spore coat dehydration. A *cwlD* mutant produces spore peptidoglycan with no muramic-δ-lactam and shows a 2-fold increase in cross-linking, but has near-normal spore core dehydration and normal spore heat resistance. A *cwlD dacB* double mutant produces spore peptidoglycan with increased cross-linking but with little change in spore core hydration as compared to the wild type. Double mutants also produce novel muropeptides containing glycine with no significant changes in spore resistance or core hydration. Structural analysis of *B. subtilis* spore peptidoglycan during sporulation reveals that there is a gradient of cross-linking that spans the spore cortex peptidoglycan. However, the loss of this gradient in some mutant strains does not alter spore core dehydration, indicating that the gradient is not required for core dehydration.

Spore Coat Protein Synthesis

The multilayered spore coat is composed of at least 15 polypeptides plus an insoluble protein fraction. This coat provides a high degree of resistance to a variety of environmental factors, including mechanical stress, and degradative proteins, such as lysozyme. Coat formation involves the orderly assembly and cross-linking of upward of 20 coat proteins.

Genes for at least 7 coat proteins, designated *cotA* through *cotF* and *cotT*, have been identified. Most of the well-characterized *cot* genes have σ^K-dependent promoters even though they show considerable variation in their time of expression. The *cotE* gene has a σ^E-dependent promoter that permits earlier expression, but it also has a σ^K-dependent promoter allowing expression to continue once σ^K appears. The CotA, CotB, and CotC proteins are responsible for the formation of outer coat proteins. The CotD, CotE, and CotT proteins are involved in the formation of inner coat proteins. Mutations at *cotE* are deficient in proteins encoded by *cotA, cotB*, and *cotC* genes, indicating that CotE is deposited on the outside surface of the inner coat, where it serves as a basal protein on which the other proteins assemble.

Mutations in the *spoIVA* locus abolish cortex synthesis and interfere with the synthesis and assembly of the spore coat. The phenotypic properties of *spoIVA* mutants suggest a role for SpoIVA at an early stage in the morphogenesis of the spore outer layers. In the absence of SpoIVA, cortex synthesis is virtually absent and coat proteins accumulate in swirls in the mother cell cytoplasm rather than on the surface of the outer prespore membrane. The dual defects in both coat and cortex synthesis could be explained if coat deposition was dependent on completion of the cortex and SpoIVA was essential for cortex synthesis. The *B. subtilis* genome sequencing project revealed six open-reading frames — *yabD, yabE, yabF, yabG, yabH*, and *yabJ*, in the region between *dnaA* and *abrB* on the *B. subtilis* chromosome. The σ^K RNA polymerase transcribes the yabG gene during sporulation and mutant spores display altered-coat protein composition. YrbA (also called SafA, for SpoVID-associated factor A) associates with SpoVID to form a complex during the early stages of assembly of the spore coat. YrbA is one of several coat proteins that engender resistance of spores to lysozyme and may also play a role in germination.

The spores of *Bacillus* and *Clostridium* species contain as much as 10% of their dry weight as dipicolinic acid (DPA; pyridine-2,6-dicarboxylic acid). This compound is synthesized in the mother cell compartment during the later stages of sporulation but accumulates only in the developing forespore. Most of the DPA is in the core where it is chelated with Ca^{2+} and other ions. Since DPA is found only in dormant spores of *Bacillus* and *Clostridium* species, there has been considerable conjecture as to the role of DPA in the unique properties of spores.

Spores of a strain of *B. subtilis* containing a double mutation, $\Delta ger3spoVF$, lack a major germinant receptor and cannot synthesize DPA. The DPA-less spores have normal cortical and coat layers, as observed by electron microscopy, but their core region appears to be more hydrated than the spores with DPA. Analysis of the resistance to various agents of spores produced by wild-type, $\Delta ger3$, and $\Delta ger3spoVF$ revealed that (1) DPA and core water content play no role in spore resistance to dry heat, desiccation, or glutaraldehyde; (2) an elevated core water content is associated with decreased spore resistance to wet heat, hydrogen peroxide, formaldehyde, and the iodine-based disinfectant Betadine; (3) the absence of DPA increases spore resistance to UV radiation; and (4) wild-type spores are more resistant than $\Delta ger3$ spores to Betadine and glutaraldehyde.

The final step in sporulation is lysis of the mother cell and release of the fully formed spore (Fig. 19-1). If conditions are suitable, the spore can germinate and undergo outgrowth, converting back to a growing vegetative cell.

ACTIVATION, GERMINATION, AND OUTGROWTH OF BACTERIAL ENDOSPORES

The bacterial spore evolved as a means of surviving periods of low nutrient supply and other harsh environmental conditions. At the same time, the dormant spore must possess the ability to respond quickly and specifically to conditions favorable to reentry into the vegetative growth cycle. Inherent in the sporulation process, then, is the incorporation of an efficient mechanism for germination and outgrowth. This section examines the environmental factors that influence activation of the spore coat and its entry into the vegetative growth cycle.

Activation

The process of spore activation in natural environments is unknown. Under laboratory conditions, activation is most commonly achieved by heating (usually at $65-70\,°C$ for $30-45$ min). Exposure at low pH or low temperature, reducing agents, or a number of chemical agents may be equally effective in activating bacterial endospores. Activation can be distinguished from germination by a number of criteria. Activated spores still retain their resistance to heat, are refractile, are resistant to staining, and still contain large amounts of dipicolinic acid. Activation is generally considered nonessential. However, the percentage of germinating spores is greatly increased by activation. Indeed, it has been found that the number of germinating spores is often very low unless some form of activation is employed. The mechanism of the activation process is not known with certainty. Some activating agents appear to increase the permeability of the spore to germinants. Activation is reversible, whereas spores are committed to germinate following brief exposure to an effective germinant.

Germination

Germination occurs in response to specific germinants that act as triggers, leading to rapid changes in the structure and physiology of the spore. As already stated, the spore is committed to germinate even if the germinant is removed. Commitment precedes the detectable changes in the spore: loss of heat resistance, ion fluxes, release of Ca^{2+} and dipicolinic acid, hydrolysis of cortex peptidoglycan, rehydration of the core protoplast, and resumption of metabolic activity. During the early stages of germination, the light-scattering ability changes rapidly (within $6-8$ min) from phase-bright to phase-gray to phase-dark.

Effective germinants include (1) L-alanine and alanine analogs; (2) L-alanine plus inosine or other ribonucleotides; (3) sugars plus inorganic ions; (4) inorganic ions alone; or (5) asparagine, glucose, fructose, and KCl (AGFK). Spores of different species of bacilli may respond to one or all of these combinations. D-Alanine competitively inhibits the germinant action of L-alanine. Alcohols and methyl anthranilate inhibit germination triggering. Some inhibitors of germination are selective. For example, azide inhibits AGFK germination, whereas phenyl methyl sulfonyl fluoride inhibits only L-alanine germination. Inhibition of germination by protease inhibitors, such as tosyl arginine methyl ester, suggests a role for proteolytic activity in the germination process. Other inhibitors appear to be somewhat nonspecific in their action.

Germination Monitoring. Qualitative tests for monitoring spore germination include loss of optical density, loss of heat resistance, and changes in appearance under the phase-contrast microscope. A plate test that is widely used to distinguish Ger$^+$ and Ger$^-$ colonies involves detection of dehydrogenase-linked metabolism that is resumed upon germination by reduction of 2,3,5-triphenyltetrazolium chloride (a tetrazolium dye) added to the plating medium. Reduction of the tetrazolium dye stains the germinated colonies red, whereas ungerminated spores do not stain (appearing white). A high correlation between the absence of or marked reduction in tetrazolium reduction and a measurable defect in spore germination aids in the selection of germination-defective mutants.

Germination Loci. Genetic analysis of spore germination in *B. subtilis* has identified a number of germination (ger) loci. Germination mutants are grouped according to their phenotype. Spores of *gerA* and *gerC* mutants fail to germinate in response to L-alanine but are able to respond to AGFK. Mutants that respond normally to L-alanine but are not stimulated in response to AGFK are designated *gerB*, *gerK*, and *fruB*. Mutant spores that fail to germinate in response to either of these germinants have been designated *gerD* and *gerF*. The *gerE* gene may code for or regulate a protease required for the processing of polypeptides in the spore coat, during germination. These differences in response to germinants suggest the presence of at least two alternate systems for their detection and response. Mutants in a number of other loci (*gerE*, *gerJ*, *gerM*, *spoVIA*, *spoVIB*, *spoVIC*, *cotD*) affect both the germination properties of the spore and the overall structure of the cortex or coat. One novel gene, *yaaH*, encodes a spore protein produced in the mother cell compartment and is required for the L-alanine-stimulated germination pathway. Germination of *yaaH* mutant spores in AGFK is almost the same as that of wild-type spores.

Models of Spore Germination. Several models for the mode of action of triggering germinants have been proposed. Models that include metabolism of the germinants infer that the germinants are converted to some intracellular metabolites. Models based on the premise that the germinants are not metabolized propose that the germinants act allosterically on a receptor protein. Proponents of this latter model suggest that the germinants may alter membrane permeability or trigger proteolytic or cortex lytic enzyme activity.

Metabolic Changes During Germination. Dormant spores contain reduced levels of a number of important metabolites, particularly high-energy compounds such as ATP, sugar phosphates, reduced pyridine nucleotides, ribonucleotide triphosphates, and tRNA. Many other compounds, such as adenine nucleotides, ribonucleotides, pyridine nucleotides, and other RNA species, are present at levels approximating those in growing cells. One potential high-energy phosphate donor, 3-phosphoglycerate (PGA), is present in the dormant spores of some, but not all, strains at very high concentrations. Accumulation of ATP occurs within a few minutes after germination begins (150-fold increase in the first 5 min). This initial increase in ATP occurs at the expense of endogenous energy sources, particularly the large PGA pool.

Dormant spores contain PGA mutase, enolase, and pyruvate kinase, the three enzymes required for conversion of PGA to ATP. The dormant spore also contains sufficient enzymes of the glycolytic pathway and the hexose monophosphate shunt, since neither protein nor RNA syntheses are required for high-energy phosphate production through at least 40 minutes of germination. The mechanisms whereby these enzymes remain inactive in the dormant spore and are activated on germination are not known with certainty. After the exhaustion of the PGA pool, energy metabolism apparently occurs primarily via the hexose monophosphate shunt, since many enzymes of the TCA cycle are absent in germinating spores. The glucose dehydrogenase gene (*gdh*) is expressed only during sporulation.

The synthesis of RNA begins by the second minute of germination in *B. megaterium* and constitutes a major use for the ATP generated at this stage. Synthesis of RNA during the first 20 minutes of germination can be completely accomplished from nucleotides stored in the dormant spore. Approximately 85% of the ribonucleotides required for RNA synthesis during the first 25 to 50 minutes of germination are derived from RNA degradation. This hydrolytic activity is necessary for rapid RNA synthesis early in germination, since nucleotide biosynthesis is not demonstrable at this time.

Chromosome replication (DNA synthesis) does not begin until late in spore germination. Degradation of RNA supplies the nucleotides for DNA synthesis as well as for RNA synthesis. There is a temporal relationship between the increase in deoxyribonucleotide levels and the increased rate of DNA synthesis.

Germination-specific enzymes that hydrolyze spore-specific cortex peptidoglycans have been identified and studied in *B. subtilis* and *C. perfringens*. Cortex hydrolysis during germination is initiated by attack of spore cortex–lytic enzyme on intact spore peptidoglycan and leads to removal of cross–linkages. This step is followed by further degradation of the polysaccharide portion of the modified peptidoglycan by cortical fragment–lytic enzyme. The spore peptidoglycan hydrolase, SleC, which is active during germination, apparently exists in an inactive form (pro-SleC) on the outside of the cortex layer in the dormant spore. During germination, a germination-specific protease (CspC) cleaves amino acid residues from the pro-SleC, converting it to its active form. Apparently several serine proteases (CspA, CspB, CspC) are the products of genes positioned just upstream of the 5′ end of the *sleC* gene in *C. perfringens*.

A defect in the *cwlD* gene of *B. subtilis* causes a block in the formation of muramic acid lactam structure and leads to a lack of germination. Another cell wall hydrolase, CwlJ, also has an effect on germination. Spores deficient in *cwlJ* respond to both L-alanine and AGFK, but the refractility of spore suspensions decreased more slowly than in the case of the wild-type strain, and the mutant spores released less dipicolinic acid than the wild-type strain during germination.

A germination-specific *N*-acetylmuramoyl-L-alanine amidase is encoded by *sleB* in *B. subtilis* and *B. cereus*. Immunoelectron microscopy with anti-SleB antiserum and a colloidal gold–immunoglobulin G complex revealed that the enzyme is located just inside the spore coat layer in the dormant spore. The amidase appears to exist in a mature form lacking a signal sequence, indicating that *SleB* is translocated across the inner membrane of the forespore by a secretion signal peptide and is deposited in the cortex layer between the forespore inner and outer membranes. This peripheral location of the spore-lytic enzyme in the dormant spore suggests that spore germination may be initiated at the exterior of the cortex. Spores deficient in *sleB* do not release hexosamine at a significant level. However, a doubly deficient mutant, *cwlJ sleB*, produces spores that are unable to germinate but exhibit initial germination reactions such as partial decrease in refractility and slow release of dipicolinic acid.

Protein synthesis begins within the first 2 or 3 minutes of germination and can be accomplished solely from stored nitrogen reserves through the first 60 minutes of germination. The free amino acid content of dormant spores is limited but increases rapidly in germinating spores as a result of hydrolysis of spore protein. The proteins degraded in this process are the SASPs formed during sporulation (see earlier discussion). In *B. megaterium* and *B. subtilis*, a number of SASPs have been identified. Proteolytic activity is necessary for protein synthesis early in germination, since de novo biosynthesis of most amino acids is not demonstrable at this time.

Metabolism of small molecules during spore germination can be divided into two general categories or stages: stage I, or the turnover stage, from 0 to 70 minutes of germination; and stage II, or the synthesis stage, from 70 minutes and beyond. By the onset of stage II, the germinating spore gains the capacity for de novo biosynthesis of all small molecules required for the formation of DNA, RNA, and protein. Endogenous reserves of energy, amino acids, and nucleotides are exhausted by this time. During stage I, small molecules are derived primarily from stored reserves. Stage I consists of three overlapping substages — Ia (0–15 min), Ib (0–20 min), and Ic (0–70 min) — each representing the period when endogenous energy, nucleotide, and amino acid reserves, respectively, are sufficient for most of the metabolic and biosynthetic requirements.

Outgrowth

Although germinating spores lose the characteristics of the intact spore rather quickly, the germinants are by no means comparable to vegetative cells. Germinated spores have cytological structures typical of vegetative cells, are sensitive to environmental conditions, and are metabolically active. However, they lack many of the activities and properties of vegetative cells. The transition from germinated spore to vegetative cell has been termed **outgrowth** and involves a number of metabolic changes. Generally, if a suitable nutritional environment is provided, the germinated spore will continue rapidly into the outgrowth process. If nutritional requirements are lacking, the process may stop altogether, or the cells may proceed to sporulate and enter into **microcycle sporulation**. In the microcycle, asymmetric division, rather than symmetrical cell division, begins immediately after the elongation of the cell is initiated and the cells proceed to sporulate. However, one round of DNA replication must occur prior to reentry into the sporulation cycle.

The SASPs that saturate the spore chromosome and protect the spore from damage by UV radiation, heat, and peroxides are cleaved during spore germination by a germination protease and provide the amino acids for protein synthesis. This rapid degradation of SASPs is essential to allow for DNA transcription and DNA replication during spore outgrowth. If SASPs bind too tightly to DNA or their degradation is impaired, then spore outgrowth is inhibited. A mutant SASP, SspC, has a very high affinity for DNA and confers UV resistance on spores lacking α/β-type SASP. However, outgrowth of the germinated spores was blocked by 90% compared to wild-type spores, and the mutant form of SspC persisted in the germinated spores while wild-type SspC was almost completely degraded (see ref. by Moir and Smith, 1990). The persistence of the mutant form of SspC on DNA appears to interfere with transcription during spore outgrowth.

Isolation of mutants of *B. subtilis* that are temperature sensitive only during spore outgrowth suggests that there are functions that are specific to this stage of the life cycle. Studies of these mutants show that unique RNAs are synthesized during spore outgrowth. It has been possible to isolate and determine the chromosomal location of these genes from a cloned library of *B. subtilis* DNA using hybridization with labeled RNA prepared from outgrowing spores in the presence of a large excess of competing vegetative RNA.

MYXOBACTERIAL DEVELOPMENTAL CYCLE

Myxobacteria, with the exception of newly identified marine forms, are aerobic soil or terrestrial bacteria that prefer relatively neutral or nonextreme growth environments. These bacteria are intriguing for (1) their ability to form metabolically inactive, environmentally resistant endospore-like forms, called **myxospores**, organized into sometimes elaborate and beautiful macroscopic structures called **fruiting bodies**, (2) their social behavior; and (3) their predatory nature. The social behaviors of these bacteria are important aspects of both their vegetative growth and developmental cycle. In particular, the cell-to-cell interactions, the cell signaling, and the gliding motility that mediate such social behaviors make these organisms especially fascinating. Most of the basic knowledge of the life cycle of myxobacteria has come from the study of *Myxococcus xanthus*, and, thus, this organism has become the prototype myxobacterium. Therefore, unless indicated otherwise, our discussion of the myxobacterial life cycle comes from studies on *M. xanthus*.

Life Cycle of Myxobacteria

The life cycle of *M. xanthus* is illustrated schematically in Figure 19-3. As with endospore-forming bacteria, the life cycle consists of a vegetative cycle, during which growth and cell division occurs, and a developmental cycle, during which myxospores and fruiting bodies are formed.

The vegetative cycle is itself very interesting. In nature or in the laboratory, when supplied with sufficient nutrients and favorable growth conditions, the bacteria can grow with generation times as low as 3.5 hours. Even in the vegetative state, an important aspect of myxobacterial growth involves cell-to-cell interactions that create localized high cell densities. Myxobacteria, as a group, also excrete a wide assortment of hydrolytic enzymes, allowing them to grow on a variety of macromolecules including proteins, simple and complex polysaccharides, peptidoglycans, and nucleic acids. In

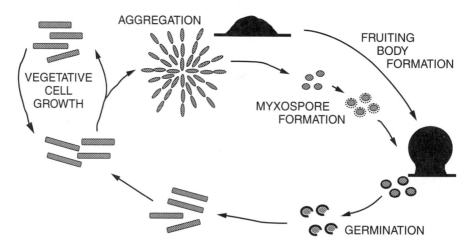

Fig. 19-3. Schematic representation of *Myxococcus xanthus* life cycle. Structures are not drawn to scale. Fruiting bodies can be a few hundredths of a millimeter in diameter and vegetative cells are 5 to 7 by 0.7 μm. (*Source*: Modified from Dworkin, 2000. In Y. V. Brun and L. J. Shimkets (eds.), *Prokaryotic Development*, ASM Press: Washington, DC.)

addition, many myxobacteria produce a broad spectrum of antibiotics. Thus, like their closest relatives, the bdellovibrios, myxobacteria are capable of killing and lysing a number of other bacteria, fungi, and protozoa. In fact, myxobacteria are thought to prey on other living cells, and this activity, coupled with their tendency to "socialize" (staying close together in groups and maintaining a localized high cell density), has caused researchers to call this behavior the "**wolf-pack effect**." Studies with *M. xanthus* have shown that cells, although capable of individual movement (**adventurous** or **A-motility**), stay together and move in swarms (**social** or **S-motility**). This type of behavior is believed to optimize feeding on macromolecules by increasing the local concentration of hydrolytic enzymes.

Vegetative cells will continue to divide until nutrient downshift or limitation is sensed. At this point, cells can either enter a stationary-phase condition of very slow growth, or if a solid surface is present and a high cell density is achieved, cells will enter into the developmental stage. Entry into either a stationary phase or the developmental stage is a biologically costly process for myxobacteria, since both are accompanied by a considerable loss in viable cells. Thus, the choice to enter into the developmental stage is not automatic. It is only when the three conditions of a decline in available nutrients are met, a solid surface is present (to allow for **gliding motility**), and there is high cell density that cells will enter into the developmental mode. In a process not completely understood but involving cell-signaling mechanisms and cell-to-cell interactions, cells begin to aggregate, forming **aggregation centers**. Interestingly, aggregation coincides with an extensive loss of viable cells in the population due to lysis in a process called **developmental autolysis**. The surviving cells go on to form myxospores and fruiting bodies. When conditions become favorable once again, the myxospores can germinate to produce metabolically active, highly motile, vegetative cells.

Aggregation and Fruiting Body Formation

Aggregation. Nutrient limitation in the presence of a solid surface and at a high cell density triggers the first steps in the developmental cycle of *M. xanthus*. Unlike the responses of endospore-forming and non-spore-forming bacteria to nutrient depletion, the initial stage of the developmental cycle of myxobacteria involves an energy-intensive aggregation. The initiation and progression of aggregation involves both intracellular signals in individual cells and intercellular signals. The intracellular signal is believed to be mediated by *relA*-dependent (p)ppGpp accumulation, since *M. xanthus* relies on primarily an amino acid–based diet, and depletion of one or more amino acids will trigger a stringent response.

Two additional factors are proposed to play a role in determining entry into the developmental cycle: the products of the *socE* and *csgA* genes. The *socE* gene product is unknown and does not exhibit homology to any other proteins in the databases. The *csgA* gene encodes two products, a 17 KDa and a 25 kDa protein, which function as the C-signal (see below). The *socE* gene is highly expressed in vegetative cells and becomes depleted in nongrowing starved cells as a result of inhibition by ppGpp or the stringent response. It is the depletion of the *socE* product that is believed to be important in SocE's role in regulating the induction of the developmental cycle.

In contrast, the *csgA* gene is poorly expressed in vegetative cells and is induced some 5-fold during development. It is partially repressed by the *socE* product and positively regulated by ppGpp and interestingly another intercellular signal encoded by the *bsgA* gene, which encodes the B-signal (see below). The C-signal functions throughout the developmental cycle, but early on, when at low concentrations, it is needed for growth arrest and directional motility for aggregation. Thus, it is the induction of the stringent response and the balance between SocE and CsgA that appear to be important in the decision to enter into development.

One of the effects of inducing a *relA*-dependent stringent response is the production of the earliest intercellular signal, the A-signal. **A-signal** appears to be a mixture of both amino acids and peptides that are generated by proteases secreted by the myxobacteria. Generation of the A-signal involves at least three genes: *asgA* (*asg*, meaning A-signal gene), *asgB*, and *asgC*. Other signals include (1) the **B-signal**, which has not been identified as of yet but is dependent on the *bsgA* gene product; (2) the **C-signal**, which is a 17 KDa and a 25 kDa protein encoded by the *csgA* gene; (3) the **D-signal**, the least understood of the extracellular signals, which is believed to be a mixture of fatty acids and is dependent on the *dsgA* gene; and (4) the **E-signal**, which appears to be long branched-chain fatty acids and requires the *esgA* and *esgB* gene products. Response to, and for some the production of, these extracellular signals (A-, B-, C-, D- and E-signals) requires high cell densities. High cell densities are essential for the effective exchange and response to cell signals as well as the necessary direct cell-to-cell interactions.

Direct cell-to-cell interactions involve the production of cell surface appendages called **fibrils**. Fibrils are filamentous structures up to 50 μm long and are composed of polysaccharides and closely associated **integral fibril proteins (IFPs)**. Fibrils are important for cell-to-cell cohesion and the characteristic social behavior associated with the developmental cycle. This was confirmed genetically by demonstrating that strains lacking fibrils, *dsp* mutants, are also defective in several social functions. In particular, fibrils are proposed to function in the normal cell-to-cell positioning required for efficient intercellular signaling. Part of this function appears to lie in their ability to allow individual cells to perceive adjacent cells and analyze cell density. Important to this ability is a fibril-associated ADP-ribosylation activity.

The requirement of a solid surface for the induction of development in *M. xanthus* reflects the need for gliding motility (e.g., S-motility) in myxobacteria aggregation and fruiting body formation. As with other bacteria that exhibit gliding motility, the underlying mechanism is not completely understood. This motility is central to formation of the aggregation centers and cellular alignment within the developing fruiting body. Although the mechanism of S-motility is not well understood, it requires the ability to produce polar type IV pili. Mutants in pilus biosynthesis, *pil* mutants, as well as in the S-motility genes, *sgl* and *tgl* mutants, exhibit defects in fruiting body formation, further supporting the need for gliding or S-motility in the developmental cycle.

S-motility is not a random process. In both vegetative swarming and developmental aggregation, cells respond chemotactically, moving toward attractants (nutrients) and away from repellents. One such attractant is phosphatidylethanolamine (PE), both dilauroyl and dioleoyl PE. At least two signal transduction systems play a role in developmental aggregation: the **Frz** and **Dif systems**. Both are composed of components showing homology to the Che systems of enteric bacteria. Each acts at different stages of development. The Dif system functions early in the aggregation process, since *dif* (defect in fruiting) mutants fail to aggregate beyond the early stages of fruiting body formation. The Frz system functions in both vegetative swarming activity and at later stages in fruiting body formation. Mutants in the *frz* genes exhibit as a "frizzy" phenotype and form defective aggregates but not discrete mounds. Both of these systems are proposed to sense and respond to developmental signaling molecules and perhaps other extracellular factors to control developmental aggregation/motility.

Fruiting Body Formation. The induction of the developmental cycle ultimately leads to the formation of a myxospore-filled fruiting body in *M. xanthus* and other myxobacteria. The size and complexity of fruiting bodies ranges considerably among the myxobacteria (Figure 19-4) from very simple sacs of myxospores (*Nanocystis exedans*) to elaborate branched tree–like structures (*Stigmatella* or *Chondromyces*). Why there is such a difference in fruiting body morphology is unclear but may reflect the habitats that the particular genera or species typically populates and the effect on dispersal of myxospores into the environment.

As mentioned, fruiting body formation is initiated by the presence of three conditions: nutrient limitation, solid surface, and high cell density. These conditions lead to accumulation of the intracellular signal **(p)ppGpp**, which in turn regulates the transcription of genes necessary for the production of the first developmental signal, A-signal. Based on evidence to date, the generation of these two signals is required for the initiation of development and fruiting body formation. Within a few hours following the onset of starvation, cells begin to aggregate, forming aggregation centers. Microscopic examination shows cells moving in spiral patterns within stacked monolayers, eventually forming a hemispheric mound. In the outer densely packed portions are vegetative cells moving in a single direction, clockwise or counterclockwise, around the fruiting body periphery. A less densely packed inner core contains nonmotile myxospores. Based on experiments using *lacZ* fusions to several

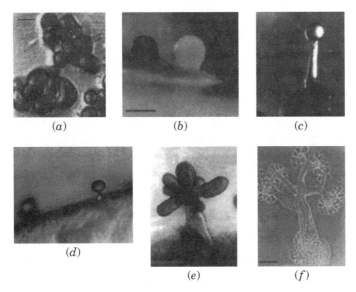

(a)   (b)   (c)

(d)

(e)   (f)

Fig. 19-4. Examples of fruiting bodies produced by the myxobacteria. Fruiting bodies of (*a*) *Cystobacter fuscus* (bar, 100 μm); (*b*) *Myxococcus fulvus* (bar, 100 μm); (*c*) *Myxococcus stiptatis* (fruiting body about 170 μm high); (*d*) *Mellitangium* sp. strain Hp (fruiting body about 40 μm high); (*e*) *Stigmatella aurantiaca* (bar, 50 μm); and (*f*) *Chondromyces crocatus* (bar, 100 μm). (*Source*: Modified from Dworkin, 2000. In Y. V. Brun and L. J. Shimkets (eds.), *Prokaryotic Development*, ASM Press: Washington, DC.)

developmental genes, it is proposed that C-signal, produced in the outer regions by the circling cells, induces myxospore formation, and these cell forms ultimately get moved into the inner domains by the movement of the undifferentiated cells in the outer areas of the fruiting body.

The myxospores of *M. xanthus* have been studied in the most detail and their characteristics have been extrapolated to those of other genera/species. Myxospores are metabolically quiescent and exhibit significant resistance to high temperatures, UV irradiation, and desiccation, especially compared to vegetative cells. Although generally not as resistant as endospores, they are sufficiently environmentally resistant to survive the extremes they would typically encounter in their natural milieu.

Genetics of *Myxococcus xanthus* Development

Some of the genes involved in *M. xanthus* development are listed in Tables 19-3, 19-4, and 19-5.

Genes Involved in Signaling. As mentioned, the primary carbon energy source of myxobacteria is amino acids. The limitation of any amino acid induces development and fruiting body formation. It is well known in enteric bacteria that amino acid starvation induces the stringent response. Thus, it is not surprising that studies showed that *relA*-dependent (p)ppGpp accumulation is an early intracellular signal essential to development and that ***relA*** mutants are defective in myxospore and fruiting body formation. Two additional early signals are dependent on the ***socE*** and ***csgA*** genes. The ***socE* gene** is an essential gene and does not exhibit homology to anything in the various

TABLE 19-3. A Partial List of Genes Involved in *Myxococcus xanthus* Development: Intracellular and Intercellular Signaling

Gene	Proposed Function of Product or Developmental Phenotype of Mutants
Intracellular Signaling	
relA	(p)ppGpp synthetase I; defective in myxospores and fruiting body formation
socE	Essential gene; negative regulator of development and *relA*-dependent accumulation of (p)ppGpp
csgA	C-signal gene encodes 17 and 25 kDa proteins; 25 kDa protein is a short-chain alcohol dehydrogenase
Intercellular Signaling	
asgA	A-signal gene; putative response regulator-kinase
asgB	A-signal gene; putative transcription factor (helix-turn-helix protein)
asgC	A-signal gene; σ^{70} subunit of RNA polymerase
bsgA	B-signal gene; 90.4 kDa putative ATP-dependent protease (Lon homologue)
csgA	C-signal gene encodes 17 and 25 kDa proteins; 25 kDa protein is a short-chain alcohol dehydrogenase
dsgA	D-signal gene; homologue of *E. coli* translation initiation factor (IF)-3
esgA	E-signal gene; E1 decarboxylase of branched-chain keto-acid dehydrogenase (with *esgB* product)
esgB	E-signal gene; E1 decarboxylase of branched-chain keto-acid dehydrogenase (with *esgA* product)

TABLE 19-4. A Partial List of Genes Involved in *Myxococcus xanthus* Development: Aggregation and/or S-motility

Gene	Proposed Function of Product or Developmental Phenotype of Mutants
dsp	Function unknown; mutants lack fibrils
ifp-20	IFP-20; fibril protein; mutants defective in side-to-side cohesion
pilA	Major pilus subunit, prepilin
pilB	Type II secretion with PilC
pilC	Type II secretion with PilB
pilG	ABC transporter system with PilH and PilI
pilH	ABC transporter system with PilG and PilI
pilI	ABC transporter system with Pilg and PilH
pilR	Transcriptional regulator of *pil* operon; part of two-component sensor-kinase-regulator system with PilS
pilS	Histidine-protein kinase; part of two-component sensor-kinase-regulator system with PilR; negatively regulates PilR activity
pilT	Pilus retraction; mutants defective in S-motility but form pili
sglA(pilQ)	Secretin involved in macromolecule transport
sglK	DnaK (HSP70) homologue
tgl	Mutants defective in pilus biogenesis, S-motility, and development
wzm	Lipopolysaccharide (LPS) biosynthesis
wzt	Lipopolysaccharide (LPS) biosynthesis
wbgA	Lipopolysaccharide (LPS) biosynthesis
mrpA	Histidine-protein kinase; mutants defective in sporulation
mrpB	NtrC-like response regulator
mrpC	cAMP receptor protein (CRP)-like transcriptional regulator
rpoN	σ^{54} subunit of RNA polymerase

databases. Studies indicate that SocE acts as a negative regulator of development and myxospore formation, since loss of SocE leads to growth arrest and myxospore formation. SocE depletion in the cell promotes the *relA*-dependent accumulation of (p)ppGpp. The **csgA gene** is discussed below, but it also is involved in the early signaling events required for growth stoppage and entry into the developmental cycle (Table 19-3).

Five intercellular signals, designated A, B, C, D, and E, have been identified as mediators of the aggregation and fruiting body formation (Table 19-3). The earliest acting signal involved in fruiting body and myxospore formation is the **A-signal**, a mixture of both amino acids and peptides produced by proteases secreted by the myxobacteria. Three genes, *asgA asgB*, and *asgC*, are required for A-signal generation. The **asgA gene** product is proposed to be a response regulator-kinase, since analysis of its amino acid sequence demonstrates homology to the transmitter domains of histidine protein kinases and the receiver domain of response regulator. The **asgB gene** encodes a putative transcription factor based on the presence of a helix-turn-helix domain. The third gene, **asgC**, is a *rpoD* homologue encoding the σ^{70} subunit of RNA polymerase.

The B-signal acts early in development but has yet to be identified. It is dependent on the **bsgA gene** product, a 90.4 kDa protein that appears to be an ATP-dependent protease homologous to Lon protease.

The most extensively characterized of the intercellular signals is the C-signal. It is encoded by the **csgA gene**, which encodes for both a 17 and a 25 kDa protein.

TABLE 19-5. A Partial List of Genes Involved in *Myxococcus xanthus* Development: Chemotaxis Signal Transduction Systems

Gene	Proposed Function of Product or Developmental Phenotype of Mutants
Dif System	
DifA	Methyl-accepting chemotaxis protein (MCP) Tar homologue; receptor for Dif system
difB	Unknown function; no homologue
difC	CheW homologue; signal transduction from DifA to DifE
difD	CheY homologue; response regulator
difE	CheA homologue; histidine-protein kinase that phosphorylates DifD
Frz System	
frzA	CheW homologue; signal transduction from FrzCD to FrzE
frzB	Unknown function; no homologue; interacts with FrzCD
frzCD	Methyl-accepting chemotaxis protein (MCP) Tar homologue; receptor for Frz system
frzE	CheA homologue (N-terminus); histidine-protein kinase
	CheY homologue (C-terminus); response regulator
frzF	CheR homologue (C-terminus); methyltransferase
frzG	CheB homologue; methylesterase
frzS	No homologue; putative regulator of S-motility
frzZ	CheY homologue (N-terminus); CheY homologue (C-terminus); putative response regulator not central part of Frz system
abcA	ABC transporter/exporter; may interact with FrzZ; mutants exhibit "frizzy" phenotype in *pilQ1* background
rpoE1	Extracytoplasmic function (ECF) σ factor; may interact with FrzZ; mutants are defective in vegetative swarming and developmental aggregation

Both function as the C-signal and are cell surface associated. The larger 25 kDa protein is proposed to be a short-chain alcohol dehydrogenase. *csgA* mutants are unable to aggregate or sporulate and fail to express several developmental genes. At low concentrations, C-signal functions in the growth arrest required for initiating the developmental cycle. It is positively regulated by (p)ppGpp and positively regulates *relA*. At higher concentrations and at high cell densities, C-signal is essential for directed motility during aggregation and fruiting body formation as well as differentiation into myxospores.

The D-signal is one of the most puzzling of the extracellular signals. The D-signal itself is believed to be a mixture of fatty acids, but it is dependent on the ***dsgA*** **gene** whose product is a homologue of the *E. coli* initiation factor (IF)-3. Hence the confusing nature of this signal, since it is difficult to hypothesize a reasonable relationship between the effect of a *dsgA* mutation on development and the ability to rescue the *dsgA* mutant phenotype with a mixture of fatty acids.

The E-signal is dependent on two genes: ***esgA*** and ***esgB***. The E-signal itself is proposed to be long branched-chain fatty acids. The *esgA* and *esgB* gene products appear to be an E1 decarboxylase of the branched-chain keto-acid dehydrogenase; thus, they may play a role in generating the E-signal from fatty acids in the environment.

Genes Involved in Aggregation and/or S-motility. In addition to the intracellular and intercellular signals, a number of other gene functions and structures are required for aggregation and S-motility (Table 19-4). These include **fibrils** that are essential to cell-to-cell interactions and many social activities. The production of fibrils is dependent on at least two genes: *dsp* and *ifp-20*. The function of the *dsp* gene product is unclear, but *dsp* mutants lack fibrils. The *ifp-20* gene product is one of the fibril proteins. Mutants unable to produce IFP-20 are defective in the side-to-side cohesion important for maximal signal exchange and thus are defective in cell-to-cell interactions required for development.

As mentioned, the production of type IV pili is required for gliding motility and thus social aggregation and fruiting body formation. The genes essential for pilus biosynthesis are organized into an operon, ***pilBTCSRAGHID***. The ***pilA* gene** encodes the major pilus subunit, prepilin. The ***pilB*** and ***pilC* genes** are proposed to encode proteins involved in type II secretion. The ***pilG***, ***pilH***, and ***pilI* genes** appear to make up an ABC transporter system. The ***pilR*** and ***pilS* genes** are proposed to make up a two-component sensor kinase-regulator signal transduction system, with PilS functioning as a histidine-protein kinase that negatively regulates PilR, a transcriptional regulator of the *pil* operon. The ***pilT* gene** encodes a protein that functions in pilus retraction that is essential in S-motility. With the exception of *pilT* mutants, mutants in any of the *pil* genes do not produce pili; *pilT* mutants produce pili but are defective in S-motility.

Mutations in the ***sgl*** and ***tgl* genes** also result in defects in S-motility and fruiting body formation. The ***sglA* gene** encodes a secretin involved in macromolecular transport — in particular pilus production — and was renamed *pilQ*. A spontaneous mutation, *sglA1 (pilQ1)*, is found to be in a number of laboratory *M. xanthus* strains, which allows for dispersed growth in liquid cultures but does not affect development. Another gene, ***sglK***, has been found to encode a DnaK (HSP70) homologue required for S-motility and cell-to-cell cohesion but is not essential for growth. The ***tgl* gene** encodes a product needed for pilus biogenesis, which accounts for its defect in S-motility and fruiting body formation.

Genes Involved in Chemotaxis Signal Transduction Systems. As mentioned, gliding motility is directed by and dependent on at least two signal transduction systems, Dif and Frz, which function at different stages in the developmental process.

The ***dif* genes** encode a putative signal transduction system that functions early in the aggregation process (Table 19-5). The *dif* genes are organized into an operon, ***difABCDE***. *difA* or *difE* mutants are defective in fruiting body formation. The deficiency appears to lie in the regulation of S-motility-associated gliding behavior, supporting a role for the Dif system as a signal transduction system. Further evidence for this role is the finding that several of the *dif* gene products encode homologues of the Che system of enteric bacteria. The **DifA protein** is a homologue of the methyl-accepting chemotaxis protein (MCP) Tar and thus is proposed to be a receptor for the Dif system. The ***difB* gene product** does not exhibit homology to any protein in the databases and thus far has not been assigned a function. The **DifC protein** is a homologue of the CheW protein and thus likely interacts with both the DifA protein and the CheA histidine-kinase protein homologue, the **DifE protein**. The ***difD* gene product, DifD protein**, is a CheY homologue and thus is proposed to function as a response regulator in the Dif system that is regulated by the DifE histidine-kinase. Thus far, no CheB or CheR homologues in the Dif system have been identified.

The Frz system is encoded by genes (Table 19-5) located in the same region on the *M. xanthus* chromosome and organized into an operon, *frzABCDEGF*, and as separate genes, *frzZ* and *frzS*. The *frz* genes are essential for vegetative swarming, aggregation, and formation of the fruiting body. The designation *frz* comes from the frizzy appearance of the aborted mounds seen in *frz* mutants. Similar to the Dif system, many of the *frz* genes encode for Che system homologues. The *frzA* **gene product** is a CheW homologue that likely acts as an adapter protein, interacting with the MCP homologue **FrzCD protein** and the CheA histidine kinase homologue **FrzE protein**. Both the FrzCD and FrzE proteins exhibit unique features relative to their Dif and Che homologues. The FrzCD protein does not possess a membrane-spanning region and therefore probably remains cytoplasmic, suggesting that signal input must come from an as-yet unidentified membrane receptor transducer or is generated within the cytoplasm.

The FrzE protein appears to have two functions: its N-terminal end is homologous to the CheA protein, but its C terminus exhibits homology to the CheY response regulator. The FrzCD protein interacts with the **FrzB protein**; although it is not known why, since FrzB has no homologue and its function is unknown. The C terminus of the **FrzF protein** is a CheR homologue and is proposed to possess methyltransferase activity. The **FrzG protein** is a homologue of the CheB protein and possesses methylesterase activity. Thus, FrzF is proposed to methylate FrzCD while FrzG demethylates FrzCD.

The *frzS* gene encodes a new component of the Frz system. Mutants in *frzS* exhibit the frizzy phenotype and have only a slight increase in reversal frequency compared with wild-type strains but interestingly exhibit a major defect in vegetative swarming. This suggests that FrzS plays a role in social motility. The **FrzZ protein** possesses two distinct domains, both of which exhibit homology to the CheY protein, and is therefore proposed to be a response regulator. However, phenotypic data suggest that it is not a central part of the Frz system but rather a regulator of directed motility.

The FrzZ protein was found to interact with two other proteins encoded by the *abcA* and *rpoE1* **genes**, respectively. AbcA protein shows homology to members of the ABC transporter family. Analysis indicates that it may be an exporter important in the secretion of a molecule involved in developmental aggregation. *abcA* mutants exhibit a frizzy phenotype but only in a *sglA1(pilQ1)* background, making their role in development less clear. The RpoE1 protein is a member of the extracytoplasmic function (ECF) σ-factor family. The *rpoE1* gene is located downstream of the *frzZ* gene and upstream of the *frzS* gene. The *rpoE1* and *frzS* genes are apparently part of an operon with two other genes. Mutants of *rpoE1* do not exhibit the frizzy phenotype but show defects in vegetative swarming and aspects of developmental aggregation.

Other Genes. Mutations in three genes, *wzm*, *wzt*, and *wbgA*, involved in *M. xanthus* lipopolysaccharide (LPS) O-antigen biosynthesis, are also defective in S-motility and aggregation. Since these mutants produce normal type IV pili and fibrils, a role is indicated for LPS O-antigen in S-motility.

Recently, three new genes involved in aggregation and sporulation have been described: *mrpA*, *mrpB*, and *mrpC* (for *Myxococcus* regulatory protein). The *mrpA* gene appears to encode a histidine-kinase homologue and is upstream of, and forms an operon with, the *mrpB* gene, which encodes a NtrC-like response regulator. *mrpC* is transcribed separately and encodes for a cAMP receptor protein (CRP)-like transcriptional regulator. Mutants in both *mrpB* and *mrpC* are defective in aggregation and sporulation, while *mrpA* mutants are defective in sporulation only. All three genes are developmentally regulated, with MrpB being required for *mrpC* expression and MrpC autoregulating itself.

Another gene that is important for *M. xanthus* development and vegetative growth is ***rpoN***. The *rpoN* gene encodes a σ^{54} homologue. In *M. xanthus*, σ^{54} appears to be essential for growth but is also required for the expression of several developmental genes. In fact, the MrpB gene is proposed to be a σ^{54} activator involved in the expression of several σ^{54}-dependent developmental genes.

CAULOBACTER DIFFERENTIATION

Stalked bacteria are a broadly diverse collection of Gram-negative bacteria that are members of the α *Proteobacteria*. They are referred to as stalked bacteria because at some stage in their life cycle they all possess at least one extension from the cell surface called a stalk — also known as **prostheca**, appendages, or hypha.

Caulobacter crescentus is the best studied of the so-called stalked bacteria. It is a crescent-shaped bacillus populating aquatic environments. It is interesting and unique for a variety of reasons. For example, unlike most other differentiating bacteria, *C. crescentus* does not differentiate in response to nutritional stress or environmental cues. Instead, differentiation is a central part of the growth cycle of this bacterium, occurring during each cell division. As a part of this growth cycle, two very different cells are formed: one that is motile and incapable of replicating its DNA or dividing (swarmer cells) and the other that is nonmotile but capable of DNA replication and cell division (stalked cells). The generation of different cell forms during the life cycle of *C. crescentus* brings up intriguing questions: How does the dividing cell differentially produce structures at opposite poles prior to cell division? Why is one cell form able to reproduce whereas the other is not?

Life Cycle of *Caulobacter crescentus*

A schematic representation of the *C. crescentus* life cycle is presented in Figure 19-5. As for most of the stalked bacteria, each division cycle yields two very different cells: a swarmer cell and a stalked cell. The swarmer cell has a single polar flagellum and pili (at the flagellar pole) and is capable of chemotaxis. The main purpose of the swarmer cell is dissemination of the bacteria to new and different environs. This presumably enhances the probability of finding niches with sufficient nutrient supply and less competition for available nutrients. In contrast, the stalked cell is sessile — that is, typically attached to some surface via its stalk/holdfast structure. It is capable of replicating its DNA and dividing to form a swarmer cell and another stalked cell. Thus, unlike cell division in most cells, cell division in *Caulobacter* and most other stalked bacteria yields only one daughter cell capable of further reproduction.

The swarmer cell, although metabolically active and capable of chemotaxis, is unable to replicate its DNA and divide. After some time and in response to an as-yet unknown intracellular signal, the swarmer cell sheds its flagellum and pili. A stalk is produced at the same pole, replacing the flagellum while DNA replication is initiated. As stalk synthesis progresses, the holdfast organelle is formed at the tip. The holdfast functions directly in attachment of the stalked cell to surfaces. This newly formed stalked cell

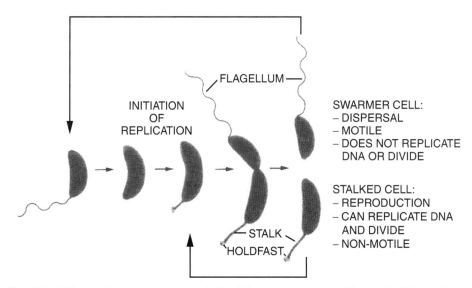

Fig. 19-5. Schematic representation of *Caulobacter crescentus* life cycle. Figure shows electron micrographs of *C. crescentus* cells at different stages of its cell cycle. The flagella are enhanced for illustration purposes. Also at the flagellated poles of swarmer cells but not visible would be pili. (*Source*: Modified from Brun and Janakiraman, 2000. In Y. V. Brun and L. J. Shimkets (eds.), *Prokaryotic Development*, ASM Press: Washington, DC.)

elongates and begins to synthesize a flagellum at the opposite pole, eventually yielding a **predivisional cell** with a flagellum at one pole and a stalk, with holdfast, at the other. The predivisional cell ultimately divides to yield a swarmer cell and a stalked cell. The swarmer cell is released into the environmental milieu to allow the search for more nutrients and less competition. The stalked cell, on the other hand, is capable of initiating a new round of cell division almost immediately.

The Stalk, the Holdfast, and the Flagellum: Structure, Genetics, and Regulation

The Stalk. The stalk of *C. crescentus* can be described as an extension of the cell body. The center of the stalk is cytoplasmic in origin and appears to be contiguous with the cytoplasm of the cell but devoid of ribosomes and DNA. The surface of the stalk is also continuous with the cell surface structure of the cell body. Along the stalk are **crossbands** of peptidoglycan and perhaps other material. These crossbands link the inner and outer membranes to provide rigidity to the stalk. Interestingly, the numbers of crossbands are believed to reflect the age of the cell. One crossband is formed during each cell cycle, and recent evidence indicates that the stalk grows linearly with each generation.

Stalk biosynthesis is localized to its base. The enzyme apparatus necessary for stalk biogenesis—at least the stalk peptidoglycan—is located there. Components of this machinery include the enzymes that catalyze peptidoglycan biosynthesis, collectively known as penicillin-binding proteins (PBPs). PBPs involved in stalk peptidoglycan biosynthesis appear to be somewhat different from those involved in peptidoglycan production in the cell proper. PBPX, PBPY, and PBP2 appear to be the major PBPs involved in stalk peptidoglycan biosynthesis.

Stalk biogenesis is an impressive undertaking for the cell. It not only requires the redirecting of cell surface biosynthesis to a specific site in the cell and in a perpendicular direction but, it also requires that it be initiated at the correct time in the cell cycle. In addition, the cell must control stalk elongation in response to extracellular phosphate levels.

Although *C. crescentus* differentiation is not responsive to extracellular phosphate concentrations, phosphate limitation does have an extraordinary effect on stalk biosynthesis. Phosphate starvation stimulates stalk biosynthesis, leading to stalks 15 to 30 times the length of stalks (1–2 μm) produced in excess phosphate conditions. Phosphate deprivation has been shown to stimulate stalk biosynthesis in other stalked bacteria as well, suggesting that this is a common environmental stress encountered by these typically aquatic bacteria. The increased stalk length during phosphate limitation can have many benefits. For example, elongating the stalk will increase the surface area of the cell, which allows for more efficient uptake and utilization of available phosphate sources as well as other nutrients. Not surprisingly, some aspects of stalk biosynthesis are under the control of the *pho* **regulon** of *C. crescentus*. Mutants of the high-affinity phosphate transporter system encoded by the *pstSCAB* **operon** exhibit a constitutive long stalk phenotype independent of phosphate concentration.

PstSCAB has also been shown in *E. coli* to regulate the Pho regulon by complexing with the sensor-kinase **PhoR**, preventing the phosphorylation of its cognate response regulator **PhoB**. Thus, mutants of the *pstSCAB* genes would not form the repression complex, allowing phospho-PhoR to activate PhoB; phospho-PhoB would then go on to activate, directly or indirectly, genes involved in stalk biosynthesis. This model is supported by findings that *phoB* mutants are defective in stalk elongation in response to phosphate limitation. The stalk biosynthetic genes regulated by PhoB have not been identified.

In addition to Pho regulation, stalk biosynthesis exhibits other levels of control. Mutants of the *rpoN* **gene**, (encodes the σ^{54} subunit of RNA polymerase), do not produce flagella or stalks, and have cell division defects. Mutants of, the histidine-protein kinase encoding, *pleC* **gene** do not produce stalks or pili and have nonfunctional flagella. Mutants of the *pleD* **gene**, which encodes the putative cognate response regulator, are defective in stalk elongation but appear to exhibit normal motility. Moreover, mutants of the global response regulator gene, *ctrA*, exhibit defects in stalk biosynthesis as well as flagellum biosynthesis, DNA replication, and cell division. Interestingly, the Pho regulation of stalk biosynthesis appears to be independent of, or can bypass, the cell cycle control of stalk biosynthesis, since phosphate limitation can induce stalk elongation in mutants that lack stalks during phosphate excess.

The Holdfast. As mentioned, the holdfast is located at the tip of the stalk and functions as an attachment organelle. It mediates the attachment of the stalked cell to a surface or other substrate. Chemical analysis of the holdfast indicates that it is a complex polysaccharide containing a number of acidic residues such as uronic acids. The synthesis of the holdfast is restricted to during differentiation of the swarmer cell to the stalked cell. It is first detected at the tip of the stalk of a predivisional cell and is not found in swarmer cells or at the flagellar pole of the predivisional cell.

The genes involved in holdfast synthesis are still not well characterized. However, a four-gene operon, *hfaABDC* **operon**, has been identified as being important in holdfast attachment to the cell. The roles of each of these gene are still unclear, but sequence analysis has revealed some clues. The *hfaA* gene product was proposed to exhibit homology to the C termini of pilus tip proteins of *E. coli* and *Serratia marcescens*. However, this was not confirmed by BLAST or PSI-BLAST (see Chapter 4) analysis of known proteins in the databases. The *hfaB* gene product exhibits partial homology to the CsgA protein of *E. coli*, which is involved in *curli* fiber (temperature-controlled thin, coiled fibrillar pili) formation. The *hfaC* gene product shows similarity to members of the ABC transporter family. The *hfaD* gene product possesses three potential membrane-spanning regions and is proposed to function in anchoring the holdfast to the membrane at the tip of the stalk. Mutants of *hfaA, hfaB*, and *hfaD* all show defects in holdfast attachment, exhibiting a high frequency of holdfast shedding. Interestingly, *hfaC* mutants do not show any defects in holdfast production or attachment.

The transcription of *hfaABD* shows cell cycle control. It is greatest in the swarmer compartment of the predivisional cell. However, the holdfast does not appear until after swarmer cell differentiation in the next cell cycle, so it is unclear why *hfaABD* transcription occurs preferentially in the swarmer cell compartment of the predivisional cell. Some researchers have suggested that it provides for the preloading of the swarmer cell with *hfa* mRNA and/or proteins so that it is ready for holdfast production during swarmer cell differentiation into the stalked predivisional cell.

The Flagellum. During each cell cycle, *C. crescentus* divides to generate a nonmotile stalked cell (see above) and a motile swarmer cell. This swarmer cell possesses a **single polar flagellum** formed at the pole opposite to the stalked pole in the predivisional cell. The flagellum of *C. crescentus* is very similar to the flagellum of Gram-negative bacteria described in Chapter 7. Like the flagellum of these bacteria, the biosynthesis of the *C. crescentus* flagellum involves some 50 genes encoding structural, export, and regulatory proteins. In addition, the flagellum is assembled in stages and as three subassemblies.

The first subassembly is known as the basal body, which transverses the inner and outer membranes. The basal body is composed of the C-ring/switch, MS-ring, E-ring, P-ring, L-ring, and distal and proximal rods. The basal body anchors the flagellum to the cell (MS-ring) and also acts as a rotor to turn the flagellum (C-ring/switch) during chemotaxis. Also, associated with the basal body is the flagellar (type III–like) export system. The second subassembly is the hook, which is extracellular and attached to the basal body via the distal rod. The third subassembly is the flagellar filament attached to the hook and acting as the propeller to move the swarmer cell through the environmental milieu.

The flagellum is put together in stages. The genes involved in flagellar regulation and synthesis in each stage are placed into four classes, I through IV. Flagellum synthesis is controlled by cues from the cell cycle. As a result of these cues, the *ctrA* gene (**class I gene**) gets turned on and activates the σ^{70}-RNA polymerase holoenzyme to transcribe the **class II genes**, which include genes that encode for the MS-ring (*fliF* gene) and the flagellar C-ring/switch (*fliG, fliM*, and *fliN* genes). The *fliF, fliG*, and *fliN* genes are found in an operon with two other genes: *flbD* and *flbE*. FlbD and FlbE are *trans*-acting regulators that negatively control class II gene expression and positively regulate class III gene expression. The *fliM* gene is in an unlinked operon with the *fliL* gene. The *fliL* gene product is needed for flagellar function but not for flagellum assembly. Other class II genes include those that encode the components needed for the export of the extracytoplasmic/external components of the flagellum. These are the *flhA, flhB, flhE, fliQ, fliR, fliI, fliJ, fliP*, and *fliO* genes. The flagellar export system is homologous to the inner membrane portion of the type III secretion systems of several bacterial pathogens. MS-ring, C-ring/switch, and export system assembly are required not only for the assembly of the remaining flagellum components but also for the expression of class III genes.

FlbD, FlbE, and the integration host factor (IHF) protein positively regulate the expression of **class III genes**. The class III genes are transcribed by an RNA polymerase holoenzyme containing the σ^{54} (the *rpoN* gene product). The next assembled structures are the proximal and distal rods along with the E-, P-, and L-rings of the basal body. The *flgB, flgC*, and *flgF* genes encode the proximal rod while the *flgG* gene encodes the distal rod. These genes compose an operon: *flgBCFG*. The formation of the rod is followed by the assembly of the P-ring (encoded by the *flgI* gene) and the L-ring (encoded by the *flgH* gene). The flagellum of *C. crescentus* possesses an additional outer ring called the E-ring, which is encoded by the *flaD* gene. The E-ring lies between the P-ring and the MS-ring within the periplasm.

Upon completion of the basal body, the next subassembly to be put together is the hook, which is also encoded by class III genes. The hook and hook-associated proteins are encoded by the *flgE* gene and the *flgK* and *flgL* genes, respectively. The FlgE, FlgK, and FlgL proteins are secreted through the flagellar export system. The completion of the basal body and hook is required for production of the flagellar filament by suppressing the activity of the *flbT* gene product, which is proposed to be a negative posttranscriptional regulator of flagellin synthesis. The flagellin genes are class IV genes.

Like the class III genes, FlbD, FlbE, and the integration host factor (IHF) protein positively regulate the expression of **class IV genes**, and the class IV genes are also transcribed by a σ^{54}–RNA polymerase holoenzyme. The flagellin protein secretion via the flagellar export system and assembly into the flagellar filament are the last stages in flagellum synthesis. In *C. crescentus*, six different flagellin genes are present in two separate gene clusters. The α cluster is composed of *fljL, fljK*, and *fljJ* genes, which encode a 27.5 kDa, 25 kDa, and 28.5 kDa flagellin protein, respectively. The unlinked β cluster possesses three more genes, *fljM, fljN*, and *fljO*, each of which encodes a 25 kDa flagellin protein. The flagellin proteins are assembled in specific order. The 28.5 kDa flagellin is added proximal to the hook followed by addition of the 27.5 kDa flagellin. The 28.5 kDa and 27.5 kDa flagellins, although relatively minor components, are required for flagellum biosynthesis. The 25 kDa flagellins are added next and extend the flagellar filament; thus, they represent the most abundant flagellar component.

Several other genes have been identified that are required for the synthesis and/or assembly of a normal hook and/or filament. The *fliK* gene product controls assembly of FlgE monomers into the hook structure. A number of genes are required for correct flagellin synthesis including *flmA (flaA), flmB, flmC, flmD (flaR), flmE (flaZ), flmG (flhA)*, and *flmH (flaG)*. The *flmABCD* gene products are homologues of polysaccharide biosynthetic genes. The *flmEFGH* gene products appear to be similar to tryptophan monooxygenase, an O-linked acetylglucosamine transferase, and an acetyl transferase of other organisms. Based on these observations, the *flmEFGH* operon is proposed to encode a methyltransferase. Mutants of these genes produce flagellin proteins of anomalous molecular masses, indicating that they are involved in posttranslational modification of the flagellin proteins.

Regulation and Checkpoints of the Cell Cycle of *C. crescentus*

The successive stages of *C. crescentus* development all require the completion of the previous stage. The differentiation of the swarmer cell into the stalked predivisional cell must occur prior to the formation of the stalked and flagellated predivisional cell, which inturn must precede separation into the individual swarmer and stalked cells. In each successive stage, certain events must occur inorder for the cycle to proceed (Figure 19-5).

The formation of polar structures (flagellum and stalk/holdfast) is connected to the cell division cycle. Initially the flagellum of the swarmer cell is lost and replaced by the stalk and holdfast. The opposite pole of this predivisional cell then yields the flagellum for what will become the swarmer cell of that developmental cycle. The next developmental cycle initiated from that swarmer cell replaces the flagellum with the stalk/holdfast at that pole. If DNA replication is inhibited, this results in the blocking of class II early flagellar biosynthetic gene transcription. Mutants able to replicate their DNA but blocked in some later stage of cell division exhibit defects in their progression through the developmental cycle. The initiation of cell division is especially key to beginning the programmed differentiation of the two daughter cells in each cell cycle.

As mentioned, the *pleC* gene product encodes a histidine-protein kinase important in the regulation of flagellar rotation and stalk formation. An additional phenotype of *pleC* mutants is that they are blocked at early stages of cell division and the developmental cycle. In this role, PleC is believed to phosphorylate the response regulator homologue, the *divK* gene product, DivK. Thus, PleC and DivK are proposed to make up a signal transduction system that couples flagellum rotation activation and stalk formation to cell division.

As in *E. coli* and other bacteria, the control of cell division is a very complex and highly regulated process requiring the functions of several proteins. In many ways, cell division in *C. crescentus* is very similar to the process of cell division described for *E. coli* in Chapter 17. However, in *C. crescentus* it must occur in such a way that it yields two dissimilar daughter cells.

The accumulation and location of the tubulin-like GTPase FtsZ is key to the initiation of cell division. The structure and function of FtsZ is highly conserved among bacteria. As it accumulates, FtsZ, polymerizes, forming a ring structure at the future site of cell division, which is associated with the inner membrane. It is the first protein to localize to the site of cell division and is necessary for the localization of other cell division proteins at this site. As a result of its critical importance, FtsZ is tightly controlled with the cell cycle. FtsZ is only found in the stalked cell following cell division; it is absent in the swarmer cell. As the swarmer cell differentiates, FtsZ begins to accumulate at about the same time as DNA replication initiates. It increases rapidly and peaks around the time that cell constrictions can be seen in the predivisional cell. At this point, FtsZ levels rapidly decrease as a result of proteolytic and transcriptional control.

Unlike in *E. coli, ftsZ* transcription in *C. crescentus* is uncoupled from *ftsQ* and *ftsA* transcription due to the presence of a strong transcriptional terminator between *ftsA* and *ftsZ*. The *ftsZ* gene is transcribed from a single promoter just downstream from the terminator. The *ftsZ* gene transcription is turned off in swarmer cells but increased following the initiation of DNA replication during swarmer cell differentiation. Following DNA replication and the initiation of cell division, *ftsZ* transcription declines. After cell division, it increases again in the stalked cell. The CtrA global regulator controls FtsZ levels during the cell cycle by repressing *ftsZ* transcription.

Thus, CtrA expression during the cell cycle is opposite that of FtsZ; it is high in the swarmer cell, declines (due to proteolysis) at the time of DNA replication during swarmer cell differentiation, and then reappears as cell division begins. CtrA also negatively regulates DNA replication by binding to the origin of replication. Thus, the control of CtrA levels coordinates the initiation of DNA replication and cell division during the cell cycle.

In addition to cell cycle transcriptional control, FtsZ also undergoes cell cycle–dependent proteolytic control. The rapid decline in FtsZ levels after the beginning of cell division can only result from the degradation of FtsZ. Experiments with *ftsZ* constitutive mutants indicate that the variation in and localization of FtsZ levels during the cell cycle is primarily controlled by differential FtsZ degradation. The half-life of FtsZ is about 80 minutes during DNA replication to the beginning of cell division, at which point the half-life decreases to about 10 to 20 minutes. This change in stability may be linked to its polymerization into the FtsZ ring at the site of cell division.

Interestingly, in *C. crescentus*, transcription of the *ftsQA* operon occurs for the most part at the same time that *ftsZ* transcription is repressed. This is because in contrast to its role as a negative regulator of *ftsZ*, CtrA is a positive regulator of *ftsQA* transcription. Thus, most of *ftsQA* transcription occurs later than *ftsZ* transcription, peaking at the end of DNA replication and the beginning of the cell division process. Evidence indicates that FtsA is necessary for the final stages of cytokinesis in *C. crescentus*. The sequential transcription of *ftsZ* and *ftsA* may help to coordinate the end of DNA replication with the initiation of cytokinesis. It should be noted, however, that additional regulators besides CtrA play a role in controlling *ftsZ* and *ftsQA* transcription during the cell cycle.

CtrA controls flagellum biosynthesis, stalk biosynthesis, DNA replication, and cell division. CtrA also regulates DNA methylation. It is therefore not surprising that CtrA, itself, is meticulously controlled throughout the cell cycle. It is controlled at the level of stability, exhibiting increases in degradation at key points. The protease involved in CtrA degradation is the ClpXP homologue of *C. crescentus*. However, ClpXP is present throughout the cell cycle, suggesting that another protein/factor is required to promote or prevent CtrA degradation by ClpXP.

In addition to control by degradation, CtrA also exhibits regulation by phosphorylation. The phosphorylation of CtrA activates it. The signals that control the phosphorylation of CtrA with the cell cycle are not known. Several histidine-protein kinases have been proposed to play a role in CtrA control. One is encoded by the *cckA* gene. The CckA kinase is a membrane-bound enzyme that is expressed throughout the cell cycle but exhibits polar localization early in the predivisional cell. Mutants of this gene exhibit phenotypes similar to *ctrA* mutants. Two other kinases have been implicated in CtrA phosphorylation: *divJ* and *divL* gene products. Both these enzymes have been found to phosphorylate CtrA in vitro. Their role in CtrA regulation in vivo is not as clear-cut. For example, DivJ preferentially phosphorylates DivK in the presence of CtrA. However, both *divJ* and *divL* mutants exhibit cell division defects, indicating some role in cell division.

Critical to the cell division process is the delivery of the bacterial chromosome to each of the daughter cells. In *C. crescentus* a single round chromosomal replication occurs per cell cycle; during cell division a copy of the chromosome is delivered to each daughter cell by chromosomal segregation. For this bacterium, chromosomal segregation is proposed to involve homologues of the **ParA** and **ParB** proteins. The ParB protein is known to bind to DNA sequences located within 80 kb of the *oriC* region, downstream of the *parAB* operon. ParA and ParB localize at the poles of the late predivisional cell, and this localization is dependent on DNA replication, at least for ParB. Based on these findings and results from overexpression and mutants, it is hypothesized that ParA and ParB function in chromosomal partitioning or segregation during cell division.

BIBLIOGRAPHY

Endospore Formation

Adler, E., I. Barak, and P. Stragier. 2001. *Bacillus subtilis* locus encoding a killer protein and its antidote. *J. Bacteriol.* **183:**3574–81.

Barák, I., P. Prepiak, and F. Schmeisser. 1998. MinCD proteins control the septation process during sporulation of *Bacillus subtilis. J. Bacteriol.* **180:**5327–33.

Bath, J., L. J. Wu, J. Errington, and J. C. Wang. 2000. Role of *Bacillus subtilis* SpoIIIE in DNA transport across the mother cell–prespore division septum. *Science* **290:**995–7.

Bauer, T., S. Little, A. G. Stöver, and A. Driks. 1999. Functional regions of the *Bacillus subtilis* spore coat morphogenetic protein CotE. *J. Bacteriol.* **181:**7043–51.

Buchanan, C. E., A. O. Henriques, and P. J. Piggot. 1994. Cell wall changes during bacterial endospore formation. In J. M. Ghuysen and R. Hakenbeck (eds.), *Bacterial Cell Wall.* Elsevier, Amsterdam, The Netherlands, pp. 167–86.

Burkholder, W. F., and A. D. Grossman. 2000. Regulation of the initiation of endospore formation in *Bacillus subtilis.* In V. V. Brun and L. J. Shimkets (eds.), *Prokaryotic Development.* American Society for Microbiology, Washington, DC, pp. 151–66.

Driks, A. 1999. *Bacillus subtilis* spore coat. *Microbiol. Mol. Biol. Rev.* **63:**1–20.

Errington, J. 1993. *Bacillus subtilis* sporulation: regulation of gene expression and control of morphogenesis. *Microbiol. Rev.* **57:**1–33.

Graumann, P. L., and R. Losick. 2001. Coupling of asymmetric division to polar placement of replication origin regions in *Bacillus subtilis.* **183:**4052–60.

Hauser, P. M., and J. Errington. 1995. Characterization of cell cycle events during the onset of sporulation in *Bacillus subtilis. J. Bacteriol.* **177:**3923–31.

Jin, S., P. A. Levin, K. Matsuno, A. D. Grossman, and A. L. Sonenshein. 1997. Deletion of the *Bacillus subtilis* isocitrate dehydrogenase gene causes a block at stage I of sporulation. *J. Bacteriol.* **179:**4725–32.

Ju, J., and W. G. Haldenwang. 1999. The "pro" sequence of the sporulation-specific σ transcription factor σ^E directs it to the mother cell side of the sporulation septum. *J. Bacteriol.* **181:**6171–75.

Khvorova, A., L. Zhang, M. L. Higgins, and P. J. Piggot. 1998. The *spoIIE* locus is involved in the SpoOA-dependent switch in the location of FtsZ rings in *Bacillus subtilis. J. Bacteriol.* **180:**1256–1260.

Khvorova, A., V. Chary, D. W. Hilbert, and P. J. Piggot. 2000. The chromosomal location of the *Bacillus subtilis* sporulation gene *spoIIR* is important for its function. *J. Bacteriol.* **182:**4425–29.

Levin, P. A., J. J. Shim, and A. D. Grossman. 1998. Effect of *minCD* on FtsZ ring position and polar septation in *Bacillus subtilis. J. Bacteriol.* **180:**6048–51.

Liu, J., and P. Zuber. 1998. A molecular switch controlling competence and motility: competence regulatory factors ComS, MecA, and ComK control sigmaD-dependent gene expression in *Bacillus subtilis. J. Bacteriol.* **180:**4243–51.

Lucet, I., R. Borriss, and M. D. Yudkin. 1999. Purification, kinetic properties, and intracellular concentration of SpoIIE, and integral membrane protein that regulates sporulation in *Bacillus subtilis*. *J. Bacteriol.* **181**:3242–5.

Matsuno, K., and A. I. Sonenshein. 1999. Role of SpoVG in asymmetric septation in *Bacillus subtilis*. *J. Bacteriol.* **181**:3392–01.

Meador-Parton, J., and D. L. Popham. 2000. Structural analysis of *Bacillus subtilis* spore peptidoglycan during sporulation. *J. Bacteriol.* **182**:4491–9.

Mitchell, C., P. W. Morris, and J. C. Vary. 1992. Identification of proteins phosphorylated by ATP during sporulation in *Bacillus subtilis*. *J. Bacteriol.* **174**:2474–7.

Nicholson, W. I., N. Munakata, G. Horneck, H. J. Melosh, and P. Setlow. 2000. Resistance of *Bacillus* endospores to extreme terrestrial and extraterrestrial environments. *Microbiol. Mol. Biol. Rev.* **64**:548–72.

Ozin, A. J., A. O. Henriques, H. Yi, and C. P. Moran, Jr. 2000. Morphogenetic proteins SpoVID and SafA form a complex during assembly of the *Bacillus subtilis* spore coat. *J. Bacteriol.* **182**:1828–33.

Paidhungat, M., B. Setlow, A. Driks, and P. Setlow. 2000. Characterization of spores of *Bacillus subtilis*, which lack dipicolinic acid. *J. Bacteriol.* **182**:5505–12.

Peres, A. R., A. Abanes-deMello, and K. Pogliano. 2000. SpoIIE localizes to active sites of septal biogenesis and spatially regulates septal thinning during engulfment in *Bacillus subtilis*. *J. Bacteriol.* **182**:1096–108.

Piggot, P. J., J. E. Bylund, and M. L. Higgins. 1994. Morphogenesis and gene expression during sporulation. In P. Piggot et al. (eds.), *Regulation of Bacterial Differentiation*, American Society for Microbiology, Washington, DC, pp. 113–37.

Popham, D. L., J. Meador-Parton, C. E. Costello, and P. Setlow. 1999. Spore peptidoglycan structure in a *cwlDdacB* double mutant of *Bacillus subtilis*. *J. Bacteriol.* **181**:6205–09.

Quisel, J. D., and A. D. Grossman. 2000. Control of sporulation gene expression in *Bacillus subtilis* by the chromosome partitioning proteins Soj (ParA) and SpoOJ (ParB). *J. Bacteriol.* **182**:3446–51.

Quisel, J. D., D. C. Lin, and A. D. Grossman. 1999. Control of development by altered localization of a transcription factor in *B. subtilis*. *Mol. Cell.* **4**:665–72.

Ross, M. A., and P. Setlow. 2000. The *Bacillus subtilis* Hbsu protein modifies the effects of α/β-type, small acid-soluble spore proteins on DNA. *J. Bacteriol.* **182**:1942–8.

Setlow, B., K. A. McGinnis, K. Ragkousi, and P. Setlow. 2000. Effects of major spore-specific DNA binding proteins on *Bacillus subtilis* sporulation and spore properties. *J. Bacteriol.* **182**:6906–12.

Slepecky, R. A. 1993. What is a Bacillus? In R. H. Doi and M. McGloughlin (eds.), *Biology of Bacilli: Application to Industry*. Butterworth-Heinemann, Boston, MA, pp. 1–21.

Stragier, P., and R. Losick. 1996. Molecular genetics of sporulation in *Bacillus subtilis*. *Annu. Rev. Genet.* **30**:297–341.

Takamatsu, H., T. Kodama, A. Imamura, et al. 2000. The *Bacillus subtilis yabG* gene is transcribed by SigK RNA polymerase during sporulation, and *yabG* mutant spores have altered coat protein composition. *J. Bacteriol.* **182**:1883–8.

Germination and Outgrowth of Endospores

Atrih, A., P. Zöllner, G. Allmaier, M. P. Williamson, and S. J. Foster. 1998. Peptidoglycan structural dynamics during germination of *Bacillus subtilis* 168 endospores. *J. Bacteriol.* **180**:4603–12.

Behravan, J., H. Chirakkal, A. Masson, and A. Moir. 2000. Mutations in the *gerP* locus of *Bacillus subtilis* and *Bacillus cereus* affect access of germinants to their targets in spores. *J. Bacteriol.* **182**:1987–94.

Clements, M. O., and A. Moir. 1998. Role of the *gerI* operon of *Bacillus cereus* 569 in the response of spores to germinants. *J. Bacteriol.* **180**:6729–35.

Hayes, C. S., and P. Setlow. 2001. An α/β-type, small, acid-soluble spore protein which has a very high affinity for DNA prevents outgrowth of *Bacillus subtilis* spores. *J. Bacteriol.* **183**:2662–6.

Ishikawa, S., K. Yamane, and J. Sekiguchi. 1998. Regulation and characterization of a newly deduced cell wall hydrolase gene (*cwlJ*) which affects germination of *Bacillus subtilis* spores. *J. Bacteriol.* **180**:1375–80.

Kodama, T., H. Takamatsu, K. Asai, et al. 1999. The *Bacillus subtilis yaaH* gene is transcribed by sigE RNA polymerase during sporulation, and its product is involved in germination of spores. *J. Bacteriol.* **181**:4584–91.

Moir, A., and D. A. Smith. 1990. The genetics of bacterial spore germination. *Annu. Rev. Microbiol.* **44**:531–53.

Moriyama, R., H. Fukuoka, S. Miyata, et al. 1999. Expression of a germination-specific amidase, SleB, of bacilli in the forespore compartment of sporulating cells and its localization on the exterior side of the cortex in dormant spores. *J. Bacteriol.* **181**:2373–78.

Paidhungat, M., and P. Setlow. 2000. Role of Geer proteins in nutrient and nonnutrient triggering of spore germination in *Bacillus subtilis*. *J. Bacteriol.* **182**:2513–9.

Ragkousi, K., A. E. Cowan, M. A. Ross, and P. Setlow. 2000. Analysis of nucleoid morphology during germination and outgrowth of spores of *Bacillus* species. *J. Bacteriol.* **182**:5556–62.

Shimamoto, S., R. Moriyama, K. Sugimoto, S. Miyata, and S. Makino. 2001. Partial characterization of an enzyme fraction with protease activity which converts the spore peptidoglycan hydrolase (SleC) precursor to an active enzyme during germination of *Clostridium perfringens* S40 spores and analysis of a gene cluster involved in the activity. *J. Bacteriol.* **183**:3742–51.

Takamatsu, H., T. Kodama, T. Nakayama, and K. Watabe. 1999. Characterization of the *yrbA* gene of *Bacillus subtilis*, involved in resistance and germination of spores. *J. Bacteriol.* **181**:4986–94.

Thackray, P. D., J. Behravan, T. W. Southworth, and A. Moir. 2001. GerN, an antiporter homologue important in germination of *Bacillus cereus* endospores. *J. Bacteriol.* **183**:476–82.

Myxobacterial Developmental Cycle

Dworkin, M. 1986. *Developmental Biology of the Bacteria.* Benjamin Cummings Publishing Co., Inc., Menlo Park, CA.

Dworkin, M. 1996. Recent advances in the social and developmental biology of myxobacteria. *Microbiol. Rev.* **60**:70–102.

Dworkin, M. 2000. Introduction to the Myxobacteria. In Y. V. Brun and L. J. Shimkets (eds.), *Prokaryotic Development.* American Society for Microbiology Press, Washington, DC, pp. 221–42.

Kaiser, D. 2000. Cell-interactive sensing of the environment. In Y. V. Brun and L. J. Shimkets (eds.), *Prokaryotic Development.* American Society for Microbiology Press, Washington, DC, pp. 263–76.

Kearns, D. B., and L. J. Shimkets. 2001. Lipid chemotaxis and signal transduction in *Myxococcus xanthus*. *Trends Microbiol.* **9**:126–9.

Reichenbach, H. 1993. Biology of the myxobacteria: ecology and taxonomy. In M. Dworkin and D. Kaiser (eds.), *Myxobacteria II.* American Society for Microbiology Press, Washington, DC, pp. 13–62.

Shimkets, L. J. 2000. Growth, sporulation, and other tough decisions. In Y. V. Brun and L. J. Shimkets (eds.), *Prokaryotic Development*. American Society for Microbiology Press, Washington, DC, pp. 277–84.

Sun, H., and W. Shi. 2001. Genetic studies of *mrp*, a locus essential for cellular aggregation and sporulation of *Myxococcus xanthus*. *J. Bacteriol.* **183:**4786–95.

Wall, D., S. S. Wu, and D. Kaiser. 1998. Contact stimulation of Tgl and type IV pili in *Myxococcus xanthus*. *J. Bacteriol.* **180:**759–61.

Ward, M. J., and D. R. Zusman. 2000. Developmental aggregation and fruiting body formation in the gliding bacterium *Myxococcus xanthus*. In Y. V. Brun and L. J. Shimkets (eds.), *Prokaryotic Development*. American Society for Microbiology Press, Washington, DC, pp. 243–62.

Caulobacter Differentiation

Brun, Y. V., and R. Janakiraman. 2000. The dimorphic life cycle of *Caulobacter* and stalked bacteria. In Y. V. Brun and L. J. Shimkets (eds.), *Prokaryotic Development*. American Society for Microbiology Press, Washington, DC, pp. 297–318.

Domian, I. J., K. C. Quon, and L. Shapiro. 1997. Cell-type specific phosphorylation and proteolysis of a transcriptional regulator controls the G1-to-S transition in a bacterial cell cycle. *Cell* **90:**415–24.

Gober, J. W., and J. C. England. 2000. Regulation of flagellum biosynthesis and motility in *Caulobacter*. In Y. V. Brun and L. J. Shimkets (eds.), *Prokaryotic Development*. American Society for Microbiology Press, Washington, DC, pp. 319–40.

Gober, J. W., and M. Marques. 1995. Regulation of cellular differentiation in *Caulobacter crescentus*. *Microbiol. Rev.* **59:**31–47.

Hung, D., H. McAdams, and L. Shapiro. 2000. Regulation of the *Caulobacter* cell cycle. In Y. V. Brun and L. J. Shimkets (eds.), *Prokaryotic Development*. American Society for Microbiology Press, Washington, DC, pp. 361–78.

Ohta, N., T. W. Grebe, and A. Newton. 2000. Signal transduction and cell cycle checkpoints in developmental regulation of *Caulobacter*. In Y. V. Brun and L. J. Shimkets (eds.), *Prokaryotic Development*. American Society for Microbiology Press, Washington, DC, pp. 341–60.

Quardokus, E. M., N. Din, and Y. V. Brun. 2001. Cell cycle and positional constraints on FtsZ localization and the initiation of cell division in *Caulobacter crescentus*. *Mol. Microbiol.* **39:**949–59.

Reisenauer, A., K. Quon, and L. Shapiro. 1999. The CtrA response regulator mediates temporal control of gene expression during the *Caulobacter* cell cycle. *J. Bacteriol.* **181:** 2430–9.

Sackett, M. J., A. J. Kelly, and Y. V. Brun. 1998. Ordered expression of *ftsQA* and *ftsZ* during the *Caulobacter crescentus* cell cycle. *Mol. Microbiol.* **28:**421–34.

CHAPTER 20

HOST–PARASITE INTERACTIONS

For many microbes, we must consider growth and physiology from the point of view of host–parasite interactions as well as natural microcosms. Indeed, for many microbes growth upon or within plant or animal hosts is a significant portion of their life cycle. Not surprisingly, many of the physiologic processes described in previous chapters have been adapted to growth within the host microenvironments. However, many novel mechanisms have also evolved that are unique to the host–parasite relationship. Akin to growth under nonhost conditions, a major goal of the microbe is to obtain nutrients to support its growth. This chapter highlights several physiologic mechanisms and strategies employed by microbes during their interactions with host organisms. Unfortunately for the host, many of the strategies that pathogens employ to obtain nutrients and avoid host defenses result in damage to host cells and tissues, leading to a disease state.

OVERVIEW OF HOST–PARASITE RELATIONSHIPS

The first step in host–parasite interactions is the encounter between the microbe and the host. This can occur through a variety of methods ranging from direct contact with an infected animal or contaminated fomite, inhalation of contaminated aerosolized droplets or particles, ingestion of a contaminated food or beverage, intimate sexual contact, arthropod vectors, or contaminated needles and syringes. Following contact with the microbe, one of two possible situations develops: the microbe will be lost/shed without any significant interaction with the host or the microbe may establish a relationship with the host. This relationship can be a coexistence that is harmless to the microbe and host (**commensalism**), a mutual benefit to both the host and the microbe (**symbiosis**), or a benefit to the microbe at the expense of the host (**parasitism**). There are other types of relationships that can be established, but we will not consider them at this time. The microbe (**parasite**) in commensalism is referred to as a **commensal** while the benefiting parasite in parasitism is referred to as a **pathogen**. The **host** in or on which the parasite lives (plants, animals, or humans) is unharmed/benefits in commensalism/symbiosis while in parasitism the host is harmed or damaged in some manner. In either scenario, the microbe must establish a relationship with the host or infect the host.

Infection of the host involves the production of structures and/or products that allow the microbe to adhere to and colonize a host surface. These include structures and products described in Chapter 7 such as **pili** or **fimbriae**. In parasitism, damage to the host may result from pathogen-encoded structures or products and/or the host's response to the infection. This damage results in the production of a set of signs and symptoms (e.g., inflammation, fever, sneezing, coughing, aches and pains) characteristic of a **disease state**.

Commensal organisms typically inhabit or infect host body surfaces, and by the nature of their relationship they have evolved so as to occupy and effectively compete within a specific host niche without causing damage to the host under normal circumstances. As such, commensal organisms have become an important asset to the host by supplying important nutrients, breaking down material making nutrients available to the host, and/or occupying space, making it difficult for potential pathogenic microbes to establish themselves within the host. This population of commensal microbes can produce a biofilm that covers many of the host surfaces and is generally referred to as **normal** or **resident flora** for the host.

What makes a commensal organism different from a pathogen under normal conditions? It may be related to the recent finding that commensal organisms can actively block the host's immune response (e.g., inflammation) to their presence. Commensals are proposed to block the degradation of IκB, which normally inhibits the function of NF-κB. NF-κB is a key transcription factor in host cells; in particular, it activates the transcription of tumor necrosis factor (TNF) and interleukin (IL)-8, which in turn promote an inflammatory response. Thus, commensals are proposed to actively suppress the host inflammatory response.

For a pathogen, the next step after colonization is the entry of the microbe or a microbial product(s) across an epidermal or epithelial cell surface to enter deeper well-vascularized (both lymph and blood circulation) tissue. With the exception of microbes transmitted by blood-sucking arthropod vectors, or contaminated needles or syringes, entry into the host is for many microbes the difference between life as a commensal organism, confined to the host's body surfaces, and life as a pathogen, able to infect and elicit damage to host tissue.

Entry into the host may or may not involve the physical entry of the microbe into deeper tissue but instead may be limited to the entry of a microbe-derived product such as an exotoxin. The physical entry of the microbe involves microbial products, which mediate the direct invasion of epithelial cells or specialized epithelial cells called M cells, such as invasins. Microbial exotoxins and factors mediating host cell entry are discussed later in this chapter.

The next phases of host–pathogen interactions generally involve multiplication to increase cell number and/or the level of one or more cell products along with local and/or systemic spread/dissemination. In order to multiply and disseminate, the pathogen must be able to acquire nutrients to support its growth. However, the host, and the normal microbial flora, goes to great lengths to limit the nutrients available to incoming pathogens in order to limit their growth and spread. Thus, many pathogens have evolved mechanisms that allow them to successfully compete with the host for nutrients. For example, many bacterial pathogens possess systems designed to bind and take up iron from the host environments. But since the host and normal flora also compete actively for available iron, some pathogens have evolved ways to release iron from the host cells. These include the production of exotoxins known as **hemolysins**, which lyse red blood cells, releasing the iron-rich hemoglobin. This is a good example of the strategies that some pathogens employ to obtain nutrients.

The pathogen's goal is to acquire nutrients to support its growth, but the consequence is often damage to the host. This damage and/or that caused by the immune response leads to the development of the signs and symptoms associated with infectious disease.

STRUCTURES AND FUNCTIONS INVOLVED IN HOST–PARASITE INTERACTIONS

Adherence/Colonization

The first step in host–parasite interactions is adherence to host surfaces. This is necessary for the colonization of the host and the ability to cause disease. Because it is a critical step in host–parasite interactions, microbial pathogens have evolved a variety of structures and mechanisms to promote their adherence to host cells. Many express multiple adherence factors that allow them to colonize several hosts and/or a variety of tissues. Adherence factors generally fall into one of two broad groups: **fimbriae** or **pili** and **afimbrial adhesins**.

Fimbriae (Pili). Some of the best-characterized adherence structures are the fimbriae or pili of Gram-negative bacteria such as *E. coli* and *Salmonella enterica*. As described in Chapter 7, fimbriae (meaning threads or fibers) or pili (meaning hair) are terms used to describe similar proteinaceous adherence appendages. Recently, researchers have proposed that the term *fimbriae* be used to describe structures involved in bacterial–host cell adherence and that the term *pili* be used to describe structures involved in bacteria-to-bacteria adherence such as sex pili involved in conjugation. However, since no consensus has been reached on this nomenclature, the two terms are used interchangeably in this discussion.

A complex multiprotein apparatus carries out pilus biogenesis. A common scheme for pilus formation (except type IV pili; see below) involves a **chaperone/usher pathway**. In this scheme, pilus subunits are exported to the periplasm where they interact with **chaperone** proteins that stabilize the pilus subunits, preventing them from prematurely interacting with each other during their time in the periplasm. The pilus subunit is then transferred to the **usher** protein complex, which forms an oligomeric channel in the outer membrane and translocates the subunits across the outer membrane. In this way, the subunits are assembled to form a relatively long thread-like structure possessing a **tip protein** complex that is adhesive and directly interacts with the target receptor. The pilus is anchored to the outer membrane by another protein complex. Type IV pili are formed using a pathway similar to type II secretion systems described below and in Chapter 2.

One classification format groups pili into types, I, II, III, and IV. However, not all fimbriae will fit into these groupings. **Type I pili** were first identified for their ability to agglutinate red blood cells. Interestingly, this binding is blocked by the addition of mannose, so it is referred to as **mannose-sensitive hemagglutination**

(MSHA). Mannose-sensitive type I pili are produced by a large variety of Gram-negative bacteria—for example, *E. coli*, *Salmonella enterica*, *Klebsiella* spp., and *Vibrio* spp. It was later found that some type I–like pili are not inhibited from binding by mannose and therefore mediate **mannose-resistant hemagglutination (MRHA)**. Type I pili, however, are synthesized through the chaperone/usher pathway.

Type II pili are thought to be mutants of type I pili, since they are similar to type I pili but are nonadhesive. **Type III pili** mediate MRHA but only when red blood cells are pretreated with tannic acid. **Type IV pili** were originally described for their ability to mediate MRHA and as being immunologically different from type I pili. However, it was later shown that they are assembled through a different pathway than other pili.

The classification based on ability to mediate MSHA or MRHA is becoming outdated in light of the ability to characterize pili at the molecular level. For example, many of the pili genes have been sequenced and therefore they can be compared/aligned to one another and grouped based on similarities at the DNA or deduced amino acid sequence level. More recently, pili have been named based on specific characteristics owned by them—for example, pyelonephritis-associated pili (Pap or P pili), bundle-forming pili (BFP), or toxin coregulated pili (TCP). Both BFP and TCP are also type IV pili, while P pili are type I–like (see below).

The **Pap** or **P pili** of **uropathogenic *E. coli* (UPEC)** are among the best-studied pili, particularly in terms of their biogenesis. Thus, they have become a paradigm for the process of pilus formation. P pili are essential adhesin factors for UPEC strains to colonize the urinary tract and cause pyelonephritis. P pili are similar to type I pili in terms of gene organization and analogous sequences, but type I pili are flexible structures whereas P pili are rigid rod-like structures. The P pili are encoded for by the *pap* operon. The major subunit of P pili is encoded by the *papA* gene. The initial PapA subunit is anchored into the outer membrane by the PapH protein.

At the tip of the pilus is the tip fibrillum, PapE protein, and the tip adhesin, PapG. The PapD protein acts as a chaperonin required for transport of the pilus subunits from the inner membrane to the outer membrane, where they are accepted by the PapC protein, which is proposed to function as an usher protein. Two additional proteins, PapF and PapK, play a role in tip fibrillum synthesis. The P pili attach to specific host cell receptors in the upper urinary tract via the tip adhesin protein PapG. PapG binds to α-D-galactopyranosyl (1-4)-β-D-galactopyranoside [Galα-(1-4)-Gal] moieties of glycolipids on cells lining the upper urinary tract.

Afimbrial Adhesins. A pathogen of the respiratory mucosa *Bordetella pertussis*, the causative agent of whooping cough in children, expresses a number of adhesin factors including pili and several afimbrial adhesins. All these adhesin factors contribute to the difficult task of adherence to the ciliated respiratory epithelium and colonization of the upper respiratory mucosa. Examples of the latter include **filamentous hemagglutination (FHA)**, **pertactin**, **pertussis toxin** (see below), and the **BrkA protein**. FHA is a high-molecular-weight protein sharing homology with the high-molecular-weight (HMW) afimbrial adhesins of *Haemophilus influenzae* and many eukaryotic cell-to-cell adhesins.

FHA possesses the **RGD (arginine-glycine-aspartate) motif** characteristic of the recognition site of adhesion molecules that bind to eukaryotic surface integrin proteins. FHA binds to receptors on the ciliated respiratory epithelium and the leukocyte integrin CR3; the latter triggers bacterial uptake into macrophages without activating NADPH oxidase and the respiratory burst. Furthermore, the S2 and S3 subunits of pertussis toxin share similarities with eukaryotic selectin proteins — that is, cell adhesion molecules present on leukocytes (L-selectins) and endothelium (E- and P-selectins) that bind to mucin-like cell adhesion molecules (CAMs). Both pertactin and BrkA also possess RGD motifs and appear to be involved in host cell adherence. Thus, these four adhesins provide good examples of how bacteria have evolved adherence structures that may mimic host adherence molecules to promote their interactions with host cells.

Another example of an afimbrial adhesin molecule is the **YadA protein** of *Yersinia enterocolitica*, which binds to cell-associated fibronectin. Similarly, *Streptococcus pyogenes* also produces at least two afimbrial adhesin structures that bind to cellular fibronectin: **Protein F** and **M protein/lipoteichoic acid (LTA) complexes**. The LTA of *Staphylococcus aureus* also binds to fibronectin on cells, mediating adherence for these bacteria.

Host Cell Specificity. The specific adhesin–host receptor interactions dictate the host tissues/cells that microbes can adhere to/colonize, thereby causing disease. For example, in enteric bacteria the production of different pili or fimbriae (or more precisely the tip complex of the pili) determines the regions of the intestinal tract that the pathogen will adhere to and colonize.

The extreme acidic pH achieved in the stomach normally kills most ingested bacteria. However, *Helicobacter pylori* is able to colonize the stomach. It binds preferentially to the gastric mucosa by binding to the Lewis[b] blood group antigen expressed on gastric mucosal epithelium; however, the adhesin has not been definitively identified. This close interaction with the gastric mucosal epithelium is beneficial, since it is believed that the immediate environment around the gastric mucosa is neutralized to prevent damage to the epithelial cells.

Passage out of the stomach into the intestine provides several new challenges for the pathogen. Strains of *Vibrio cholerae*, enteropathogenic *E. coli* (EPEC), and enterotoxigenic *E. coli* (ETEC) adhere to and colonize the mucosal epithelium of the duodenum and proximal jejunum. EPEC strains produce damage to the jejunal epithelium (**attaching and effacing or A/E lesion**) by forming microcolonies referred to as localized adherence (LA). LA requires the production of **bundle-forming pili (BFP)** encoded by genes located on the EAF plasmid. In addition, the adherence of *V. cholerae* to epithelium of the proximal small intestine requires the chromosomally encoded **toxin coregulated pilus (TCP)**. Interestingly, the colonization of ETEC strains of the proximal small intestine in humans requires plasmid-encoded fimbriae referred to as **colonization factor antigens** (CFAs); and, in piglets and calves, pili called **K88 pili**. This is a good example of pathogenic microbes expressing different adhesin factors to colonize different hosts.

Progression into the distal jejunum and ileum also requires the expression of certain adhesin factors for colonization by pathogens. *Salmonella enterica* serovar Typhimurium adherence to epithelial cells in the distal small intestine involves at least two pili systems: **long polar fimbriae (LPF)** and **plasmid-encoded fimbriae (PEF)**. LPF appear to bind to specialized epithelial cells, associated with Peyer's patches, called M cells, and PEF appear to bind to receptors on intestinal epithelium. LPF appear to be necessary for adherence associated with *Salmonella* invasion while PEF appear to be involved in colonization of the mucosal surface without invasion. Thus, a microbe can express different adhesin factors in the same host that result in different types of host–parasite interactions.

Colonization of the large intestine or colon requires that the incoming pathogen bind to colonic epithelium while competing with the normal resident flora for available space. Strains of enterohemorrhagic *E. coli* (EHEC) colonize the large intestine, causing A/E lesions via fimbriae encoded on a 60 MDa plasmid. Strains of enteroaggregative *E. coli* (EAEC) adhere to colonic mucosa in a distinct arrangement called aggregative adherence mediated by pili known as **aggregative adherence fimbriae (AAF)**.

Similar roles for afimbrial adhesins in binding to different host cells/tissues have been observed. FHA, pertactin, pertussis toxin, and BrkA of *B. pertussis* function in binding to receptors on ciliated respiratory epithelium, allowing this bacterium to cause disease in the upper respiratory tract; and, at least, FHA binds to the $\alpha\beta_2$ integrin CR3 on macrophages, which is believed to stimulate uptake of the bacterium into the macrophage without inducing the respiratory burst. This allows *B. pertussis* to evade an important host antimicrobial mechanism. Furthermore, strains of *S. pyogenes* produce Protein F and M protein/LTA that bind to cellular fibronectin, but protein F appears to be critical to the colonization of the nasopharyngeal epithelium, whereas M protein is key to binding to cells in the skin, suggesting additional factors may be involved.

Virulence Factor Secretion Systems

The majority of virulence factors are either surface-associated or secreted factors. Thus, systems used to export these factors are critical to the pathogenesis of many microbes. Interestingly, many of the secretion systems are conserved and highly homologous between species. In addition, many unrelated virulence factors can share basic transport systems. Furthermore, protein secretion in Gram-negative bacteria is more complex than that of Gram-positive bacteria due to the presence of an outer membrane.

Gram-Negative Bacterial Secretion Systems. To address some of these complexities, Gram-negative bacteria have evolved a number of different secretion systems designated types I through IV as well as other systems (described in detail in Chapter 2).

Type I secretion systems represent a family of structurally and functionally related protein complexes involved in the export of proteins, lacking the classic N-terminal signal sequence, through the inner and outer membrane to the exterior without a periplasmic intermediate. Thus, these are *sec*-independent pathways. Type I secretion pathways utilize three components to accomplish their task: an inner membrane ABC transporter protein (provides the energy for protein export), an outer membrane protein, and a periplasmic spanning protein that is anchored into the inner membrane and associated with the ABC transporter component.

The substrates for these secretion systems, although lacking an N-terminal signal peptide, possess a signal sequence within the C-terminal 60 amino acids. The prototype type I secretion system is involved in *E. coli* α-hemolysin secretion. However, several other virulence factors have been shown to exhibit type I secretion including: invasive adenylyl cyclase (CyaA) of *Bordetella pertussis*, leukotoxins (LktA) of *Pasteurella haemolytica*, alkaline protease (AprA) of *Pseudomonas aeruginosa*, LipA lipase and PtrA protease of *Serratia marcescens*, and PtrBC protease of *Erwinia chrysanthemi*.

Type II secretion systems are widely distributed among Gram-negative bacteria. This type of secretion involves two genetically and biochemically distinct translocation steps. The first step is mediated by the Sec pathway and requires an N-terminal signal sequence peptide. This step translocates the protein from the cytoplasm to the periplasm, where the signal peptide is cleaved and the protein folds into its secondary shape and perhaps tertiary assembly prior to being translocated across the outer membrane by the type II secretion system. Type II secretion systems are composed of at least 12 different gene products. Many plant and animal pathogens use this type of secretion for the export of exotoxins and hydrolytic enzymes important for their pathogenesis. The best-studied type II secretion system is the Pul system of *Klebsiella oxytoca* involved in pullulanase secretion; others include the Xcp system of *Pseudomonas aeruginosa*, which secretes elastase, exotoxin A, and phospholipase C; the Esp pathway involved in cholera toxin secretion in *Vibrio cholerae*; the Out system, which secretes pectic enzymes and cellulases in *Erwinia chrysanthemi*; and the Exe pathway implicated in amylase and protease secretion in *Aeromonas hydrophila*.

Type III secretion systems (TTSS) have relatively recently been described as major routes for export of virulence factors in human and animal pathogens. Export via TTSS occurs independently of the Sec system. However, many of the components of the TTSS apparatus appear to require the Sec pathway for their export. The TTSS apparatus is a multiprotein complex that spans the inner-to-outer membrane and extends from the surface forming a needle-like structure (Fig. 20-1).

Unlike other secretion systems, which can export proteins that are active in the extracellular milieu, TTSS appear to have specifically evolved for the direct translocation of proteins from the bacterial cytoplasm into the host cell cytoplasm. Thus, secretion via TTSS is believed to be regulated by bacterial adherence to the surface of target cells. Over the past few years, several plant, animal, and human pathogens have been shown to encode TTSS apparati. These TTSS and the proteins they secrete, for the most part, are encoded on specific regions of the bacteria's chromosome that have apparently been acquired via horizontal transfer (from other genera/species rather than parental ancestors).

The chromosomal loci that encode the TTSS and its secreted proteins are referred to as **pathogenicity islands**. Examples of pathogens (Table 20-1) that produce TTSS include (1) *Yersinia* species (*Y. pestis, Y. enterocolitica,* and *Y. pseudotuberculosis*), which secrete Yops proteins that inhibit host phagocyte activity among other functions; (2) *Salmonella enterica* serovars, which encode at least two different TTSS on two separate *Salmonella* pathogenicity islands (SPI-1 and SPI-2) and secrete proteins that promote invasion of nonphagocytic cells within the small intestine (SPI-1) and systemic infection and disease (SPI-2); (3) enteropathogenic (EPEC) and enterohemorrhagic (EHEC) *Escherichia coli*, rabbit-diarrheagenic *E. coli* (RDEC-1), and the rodent pathogen *Citrobacter rodentium*, which produce related TTSS apparati encoded on a pathogenicity island referred to as the **locus of enterocyte effacement (LEE)** and secrete proteins required for the generation of the attaching and effacement (A/E) lesion; (4) *Shigella* species, which secrete Ipa proteins required for invasion of nonphagocytic cells and induction of apoptosis in host cells; (5) *Pseudomonas aeruginosa*, which secretes various exoenzymes, such as exoenzyme S, needed for hematogenous dissemination and inhibition of host T cell functions; and (6) the plant pathogens *Pseudomonas syringae, P. solanacearum, Xanthomonas campestris,* and *Erwinia chrysanthemi*, which secrete proteins called harpins that are involved in generating plant tissue damage.

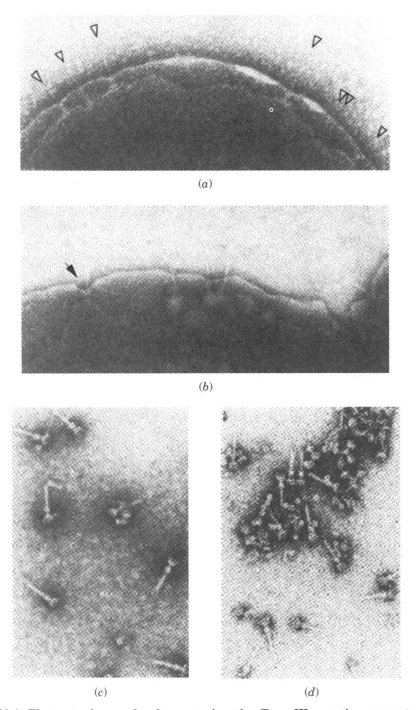

(a)

(b)

(c)   (d)

Fig. 20-1. Electron micrographs demonstrating the Type III secretion apparatus of *Salmonella enterica* serovar Typhimurium. (*a*) and (*b*) EM of osmotically shocked nonflagellated *S.* Typhimurium showing needle complexes spanning the inner-to-outer membranes and extending from the surface (open arrows). (*c*) and (*d*) EM of isolated needle complexes from an enriched fraction of a CsCl gradient. Excerpted from T. Kubori, et al. 1998. *Science* **280**:602–605, copyright (1998) American Association for the Advancement of Science.

TABLE 20-1. A Partial List of Components and Effector Proteins, and Their Functions, of Bacterial Type III Secretion Systems (TTSS)

Bacteria	Effector Protein	Function
Yersinia spp.	YopB/D	Pore-forming translocase
	YopK	Control of translocation
	YopN	Regulation of secretion
	YopE	Rho GTPase-activating protein causing F-actin filament disruption
	YopT	Causes F-actin filament disruption
	YopH	Protein tyrosine phosphatases that blocks phagocytosis and respiratory burst
	YpoJ/P	Bind to mitogen-activated protein kinase kinase (MAPKK) and IκB kinase β, preventing cytokine synthesis and promoting host cell apoptosis
	YopM	Binds to host thrombin
	YpkA	Ser/Thr protein kinase
	YscM	Gene regulation
Enteropathogenic *Escherichia coli* (EPEC)	Tir	Translocated intimin receptor needed for intimin-mediated adherence
	EspA	Activation of epithelial cell signal transduction
	EspB/D	Tyrosine phosphorylation of Tir; cell signaling
Salmonella enterica (SPI-1)	InvJ	Part of TTSS apparatus
	SspA[a]	Binds to F-actin, stabilizing F-actin
	SspB/C/D[a]	Effector protein translocation; binds to IL-1 converting enzyme (ICE or caspase I), promoting apoptosis
	SopB	Inositol phosphate phosphatase that promotes fluid secretion and inflammation
	SopE	Guanine-nucleotide exchange factor (GEF) and activates Rho GTPases, promotes membrane ruffling
	SptP	Tyrosine protein phosphatase; GTPase, activating protein Rac and Cdc42 (counteracts SopE effect)
	SspH1	Calf virulence
(SPI-2)	SseB	Component of TTSS surface translocon
	SpiC	Blocks endosomal trafficking
	SspH1/2	Calf virulence
Shigella flexneri	IpaA	Binds to vinculin and controls cell invasion
	IpaB	Binds ICE protease to promote apoptosis
	IpaB/C	Binds to $\alpha_2\beta_1$ integrins and CD44 to promote membrane ruffling
	IpaD	Regulates secretion
Pseudomonas syringae	AvrB	Promotes plant hypersensitivity response (HR)
	AvrPto	Binds to Pto kinase, resulting in induction of HR
	HopP	Promotes HR
	HrpA	TTSS pilus
	HrpW/Z	Effector protein translocation (?) and HR induction

[a] SspA, SspB, SspC and SspD are also known as SipA, SipB, SipC and SipD, respectively.

Type IV secretion systems are another mechanism of macromolecular export in Gram-negative bacteria. Type IV systems are a diverse family of secretion pathways that export an array of substrates from large nucleoprotein conjugation intermediates to both monomeric and polymeric protein complexes. Moreover, the cells targeted for secreted substrate delivery are also diverse, ranging from bacteria to fungi to plants to animals. The three defining type IV secretion systems are the VirB/D4 transfer system of *Agrobacterium tumefaciens* that exports T-DNA into susceptible plant cells, leading to plant tumors; the conjugal transfer system of conjugative IncN plasmid pKM101; and the pertussis toxin (see below) exporter Ptl of *Bordetella pertussis*. Additional type IV secretion systems identified to play a role in virulence include the Cag system of *Helicobacter pylori* required for export of the CagA protein, which causes rearrangements of the host cell cytoskeleton, and the Dot/Icm system of *Legionella pneumophila* which secretes an unidentified effector, allowing survival and growth in macrophages possibly by blocking phagosome-lysosome fusion. Further systems have been proposed to be encoded by *Brucella*, *Bartonella*, and *Rickettsia* spp. that are thought to aid in intracellular survival of these bacteria.

Other types of secretion systems have also been described that do not fit into the type I–IV definitions. **Autotransporters** are represented by a family of proteins that appear to mediate their own export through the outer membranes. This type of secretion utilizes the Sec pathway for export across the inner membrane, but all the information for translocation across the outer membrane is located in the protein itself. Thus, this represents an additional mechanism of export utilized by Gram-negative bacteria. The paradigm for this form of secretion is the IgA protease of *Neisseria gonorrhoeae*. Other members of this family include IgA protease, putative adherence and invasion proteins of *Haemophilus influenzae*; BrkA, pertactin, and other outer membrane proteins in *Bordetella pertussis*; a serine protease of *Serratia marcescens*; the vacuolating cytotoxin of *Helicobacter pylori*; SepA protein secreted by *Shigella flexneri*; and EspC protein secreted by EPEC strains.

An additional secretion mechanism has recently been described in Gram-negative bacteria. The **two-partner secretion (TPS) pathways** are composed of a secretion domain-possessing exoprotein (protein being transported) of the TpsA family and a channel-forming outer membrane transporter protein of the TpsB family. Interestingly, each TpsB transporter appears to be specific for the export of its cognate exoprotein and tends to be encoded by genes organized into an operon. Based on analysis of genome sequences, TPS pathways appear to be widespread among Gram-negative bacteria. Three TPS systems have been relatively well characterized: secretion of the filamentous hemagglutinin (FHA) of *Bordetella pertussis*, the ShlA Ca^{+2}-independent cytolysin (hemolysin) of *Serratia marcescens*, and the high-molecular-weight adhesins (HMW1 and HMW2) of *Haemophilus influenzae*. Other less-well-characterized TPS systems include systems for the secretion of the cytolysins of *Proteus mirabilis*, *Edwardsiella tarda*, and *Haemophilus ducreyi*.

Gram-Positive Bacterial Secretion Systems. Secretion of proteins in Gram-positive bacteria is generally simpler than in Gram-negative bacteria, primarily because of the absence of a second outer membrane. Surface proteins or secreted proteins all possess a classic signal sequence peptide. In addition, such proteins have a region of tandem repeats and a C-terminal region rich in proline and glycine residues and a conserved hexapeptide (LPXTGE). Examples of surface-associated or secreted proteins in Gram-positive bacteria include F, M, and G proteins of *Streptococcus pyogenes*, the P1 adhesin of *Streptococcus mutans*, internalin of *Listeria monocytogenes*, and the fibronectin-binding protein and protein A of *Staphylococcus aureus*.

Exotoxins

From the point of view of the microbial pathogen, its host is like a warehouse of food for which you need membership to gain access. The host and its normal resident flora are established and geared to efficiently obtain nutrients. The incoming pathogen must either outcompete the host/resident flora or develop ways to increase the availability of nutrients to support its growth. Thus, many pathogens either secrete, or put on their surface, enzymes that can degrade host constituents and/or lyse cells to release nutrients. For example, pathogens may secrete proteases/peptidases that degrade host proteins (e.g., elastase, collagenase, IgA protease) to smaller peptides or amino acids or secrete lipases that degrade lipids or phospholipids to their fatty acid and sugar moieties. The latter may result in cell lysis, since cell membranes are rich in (phospho)lipids. Therefore, the bacterium attempting to obtain nutrients damages the host through the breakdown of its constituents. It is not surprising then that many pathogens secrete one or more exotoxins, and these exotoxins play a key role in pathogenesis. Although a variety of microbes produce exotoxins, there are relatively few general structural or mechanistic schemes that have evolved for microbial exotoxins.

A-B Toxins. Many exotoxins, although they may differ in their activity or target cell affected, exhibit a common structural scheme. They possess one or more **binding** or **B subunits** that lack enzyme activity but bind to host receptors. These subunits are required for entry of the toxin into the target cell. An **activity** or **A subunit** possesses the enzyme activity or toxic activity that acts on the target cell. A-B toxins are synthesized by both Gram-negative and Gram-positive bacteria and target a variety of different host cell proteins and functions.

One group of A-B toxins possesses a common enzyme activity but differs in the protein or function it targets. This group utilizes the redox cofactor nicotinamide adenine dinucleotide (NAD) as a substrate, cleaving the bond between the nicotinamide (NAm) group and the ribose molecule to yield NAm and **adenosine diphosphate ribose (ADP-ribose or ADP-R)**.

$$\text{NAm-ribose-P-P-ribose-Ade} \longrightarrow \text{NAm} + \text{ribose-P-P-ribose-Ade}$$

$$\text{(NAD)} \qquad\qquad\qquad \text{(ADP-ribose or ADP-R)}$$

The ADP-R is then transferred to a target molecule (e.g., host protein), altering its structure and/or function and resulting in some physiologic effect. This activity is referred to ADP ribosylation activity, and the toxins possessing this activity are known as **ADP ribosyltransferases** or **ADP-ribosylating toxins**.

Examples of ADP-ribosylating toxins are diverse in terms of the bacteria that produce them and the protein targeted. The classic example of an ADP ribosylating toxin is the **diphtheria toxin (DT)** produced by *Corynebacterium diphtheriae*. DT and the identically acting **exotoxin A (ExoA)**, from *Pseudomonas aeruginosa*, both ADP-ribosylate the eukaryotic elongation factor (EF)-2, inhibiting protein synthesis and ultimately leading to cell death (Fig. 20-2). DT binds to the heparin-binding epidermal

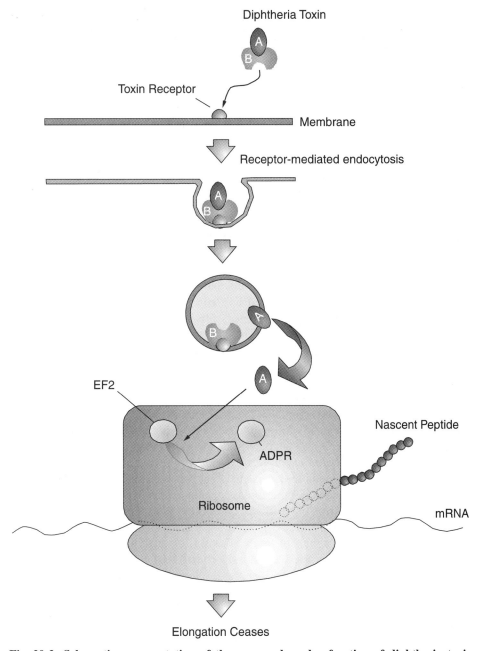

Fig. 20-2. Schematic representation of the proposed mode of action of diphtheria toxin (DTX). DTX binds to EGF receptor on host cell surface, enters the cell via receptor mediated endocytosis. The A subunit is then released from the endosome, targets the ribosome ADP-ribosylating EF-2 inactivating it. Inactivation of EF-2 stops elongation of the nascent polypeptide.

growth factor (EGF), entering the cell through an early endosome while ExoA binds to α_2-macroglobulin, entering the cell following retrograde transport to the endoplasmic reticulum (ER).

The **cholera toxin (CTX)** of *Vibrio cholerae* and **heat-labile enterotoxin (LT)** of enterotoxigenic *E. coli* (ETEC) are both ADP-ribosylating toxins that target the $G_{S\alpha}$ regulatory protein of adenylyl cyclase (AC), resulting in increased AC activity and elevated levels of cAMP (Fig. 20-3). Elevated cAMP levels in the enterocytes affected by CTX and LT cause increased Cl^- ion secretion and inhibition of Na^+ ion absorption, ultimately resulting in watery diarrhea in the patient. CTX is encoded by a bacteriophage (CTXϕ) while LT is plasmid-encoded. Both bind to surface gangliosides such as GM1 and enter the cell via retrograde transport to the ER, where the A subunit is activated.

Pertussis toxin (PT) of *Bordetella pertussis* is an ADP-ribosylating toxin that targets the $G_{i,o}$ regulatory protein of AC, preventing its inhibition and resulting in elevated levels of cAMP. In contrast, to CTX and LT, PT targets respiratory epithelium and results in many of the symptoms associated with whooping cough. PT is secreted via a type IV secretion system and binds to a surface receptor on respiratory epithelium via its B subunits, delivering the A subunit into the cell cytoplasm. The **C2 toxin** from *Clostridium botulinum* types C and D, **toxin A (ToxA)** and **toxin B (ToxB)** of *C. difficile*, and **iota toxin** of *C. perfringens* are ADP-ribosylating toxins that also possess binding/translocation activity. They enter the cell through endosomes and ADP-ribosylate G-actin monomers and prevent actin polymerization.

Other A-B toxins do not possess ADP-ribosylating activity but instead act through a variety of different mechanisms. The **Shiga toxin (STX)** of *Shigella* spp. and the **Shiga-like toxins (SLT)** of enterohemorrhagic *E. coli* (EHEC) strains both possess *N*-glycosidase activity that targets the 28S rRNA, damaging the ribosome, blocking protein synthesis, and eventually killing the cell. Following binding via its B subunits, the A subunit enters the cell through retrograde transport to the ER. **Invasive adenylyl cyclase toxin (ACT)** of *Bordetella pertussis* enters the cell directly, where in the cytoplasm it acts as a calmodulin-dependent adenylyl cyclase that raises cAMP levels in the cell, generally disrupting intracellular signaling pathways.

The plasmid-encoded tripartite **anthrax toxin** of *Bacillus anthracis* is composed of an edema factor (EF), lethal factor (LF), and protective antigen (PA). The PA subunit acts as the binding or B subunit for both EF and LF, which act independent of each other. The EF, like ACT, is a calmodulin-dependent adenylyl cyclase that produces similar effects on the cell. The LF is Zn^{+2}-metalloprotease, which cleaves host cell mitogen-activated protein kinase kinases (MAPKK), inhibiting cell proliferation and likely causing cell death.

The **cytolethal distending toxins (CDT)** of *Haemophilus ducreyi*, *Shigella* spp., *Campylobacter* spp., and several *E. coli* strains are A-B toxins that indirectly inhibit $p34^{cdc2}$ dephosphorylation, arresting the cell cycle in G2 and resulting in cytoplasmic distention and cell enlargement; the CDT of *C. jejuni* possesses deoxyribonuclease I–like activity that appears to be required for its action.

The **tetanus neurotoxin (TeNT)** of *Clostridium tetani* and **botulism neurotoxin (BoNT)** of *Clostridium botulinum* both are A-B toxins composed of a heavy chain (B subunit) and light chain (A subunit). The light chain possesses metalloprotease activity that cleave synaptosomal-associated proteins. Both TeNT and BoNT inhibit neuroexocytosis. However, TeNT acts on central inhibitory interneurons, preventing

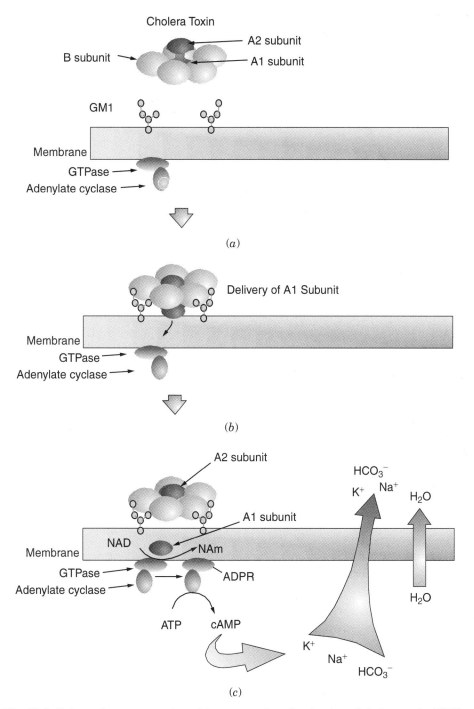

Fig. 20-3. Schematic representation of the proposed mode of action of cholera toxin (CTX).
(A) CTX binds via its B subunits to surface receptor, GM1. (B) The A1 subunit is then delivered
across the membrane; (C) A1 subunit ADP-ribosylates the $G_{s\alpha}$ (GTPase) subunit, locking
Adenylate cyclase in the "on" position leading to elevated levels of cAMP and excretion of
electrolytes and loss of water from the cell.

inhibitory neurotransmitter release, causing constant muscle stimulation and a spastic paralysis; BoNT acts on peripheral motor neurons, blocking acetylcholine (ACh) release and muscle stimulation, leading to a flaccid paralysis.

Surface-Acting Toxins. A number of bacteria secrete exotoxins that act at the surface of the target cell to cause some physiologic effect or cell lysis. Strains of both *Staphylococcus aureus* and *Streptococcus pyogenes* produce toxins that act as **superantigens**. They nonspecifically cross-bridge class II MHC molecules on antigen-presenting cells and the T cell (antigen) receptor (TCR), which results in the polyclonal activation of helper T cells and copious cytokine secretion. Superantigen toxins include TSST-1, SPE-A, SPE-C, SSA, SEA-SHE, and SpeB.

The **heat-stable enterotoxins (ST)** produced by strains of ETEC, *Yersinia enterocolitica*, *Vibrio cholerae*, and *Citrobacter freundii* bind to receptors on the cell surface to activate guanylyl cyclase activity, elevating intracellular cGMP levels and activating cGMP kinase II. The net effect is very similar to the effect of LT or CTX on enterocytes, inducing Cl^- ion secretion and blocking Na^+ ion absorption. The result in the patient is watery diarrhea.

Several bacteria produce exotoxins that form pores in the cell membranes of host cells, resulting in cell lysis. Examples of these **lysins** are **streptolysin O (SLO)** of *S. pyogenes*, which binds to cholesterol in the membrane and polymerizes to form large pores, disrupting cell membrane integrity; similarly, **perfringolysin** of *C. perfringens*, **pneumolysin** of *Streptococcus pneumoniae*, **listeriolysin O (LLO)** of *Listeria monocytogenes*, and **tetanolysin** of *C. tetani* also form large pores in the cell membrane, causing enormous electrolyte, metabolite, and protein loss with eventual cell death. Other pore-forming toxins generate smaller pores and include the **α-toxin** and **leukocidin** of *S. aureus* and **aerolysin** of *Aeromonas hydrophila*. Pore formation by these toxins leads to uncontrolled ion and solute flux and stimulates eicosanoids and endonuclease production, cytokine release, cell death, or apoptosis.

Other membrane-acting toxins include the (1) **δ-toxin** of *S. aureus* and *Staphylococcus epidermidis*, which binds nonspecifically to membranes where it self associates to form micelles and leads to osmotic cell lysis; (2) the **vacuolating cytotoxin (VacA)** of *Helicobacter pylori*, which is activated by a cellular protease and forms oligomers in the cell membrane and late endosome membrane, inhibiting endosomal trafficking and causing the formation of large vacuoles; and (3) the secreted phospholipases (PL) of several bacterial pathogens. Examples of PLs include the Zn^{+2}-dependent phospholipase C activity, possessing **α-toxin** of *C. perfringens* and **PLC B** of *L. monocytogenes*, and the phosphatidylinositol-hydrolysing toxin (PLC A) of *L. monocytogenes* that result in cell lysis/hemolysis.

"Injected" Toxins. Several of the effector proteins secreted directly into the cytoplasm of host cells by type III (described previously and in Chapter 2; Table 20-1) or type IV secretion systems alter cell physiology/function or result in cell death, and thus can be considered exotoxins. The **YpkA** and **YopO** of *Y. pseudotuberculosis* and *Y. enterocolitica* both are protein kinases, but their target proteins are unknown. In addition, **YopH** of *Y. pseudotuberculosis* is a protein tyrosine phosphatase that dephosphorylates the Crk-associated tyrosine kinase substrate Cas and inhibits its phagocytosis by host macrophages.

The **YopE** protein [a GTPase-activating protein (GAP) for Rho GTPases] and the **YopT** protein (modifies RhoA in an unknown manner, leading to its redistribution in the cell) of *Y. pseudotuberculosis* and *Y. enterocolitica* cause disruption of the actin cytoskeleton of the host cell. **Exoenzyme S or ExoS** of *Pseudomonas aeruginosa* has two functional domains: the N-terminal domain acts as a GAP on Rho GTPases, causing disruption of the actin cytoskeleton, while the C-terminal domain acts to ADP-ribosylate the Ras protein. The **SopE** protein of *Salmonella enterica* sv. Typhimurium has guanine-nucleotide exchange factor (GEF) activity and activates Rho GTPases, causing actin cytoskeleton reorganization that results in membrane ruffling (Fig. 20-4). The **SptP** protein is a GAP for Rac and Cdc42 and counteracts the effects of SopE.

The **YopP** and **YopJ** proteins of *Y. pseudotuberculosis* and *Y. enterocolitica* bind to IκB kinase β, ultimately inhibiting the eukaryotic transcription factor NF-κB, and bind to MAPKK, affecting signaling pathways to inhibit cytokine production. Additionally, YopJ as well as **SipB/C/D** of *Salmonella enterica* and **IpaB** of *Shigella* spp. bind to interleukin-1 (IL-1)-converting enzyme (ICE or caspase I) and stimulate apoptosis of the host cell.

In *Shigella flexneri*, the **IpaA** and **VirA** proteins bind to vinculin, control cell invasion, and later control cell-to-cell spread. IpaB, in addition to stimulating apoptosis, binds to $\alpha_5\beta_1$ integrin along with **IpaC/D** to cause membrane ruffling and to promote cell division. IpaC/D also stimulates phagosomal lysis to release the bacteria into the cytoplasm.

The **SopB** protein secreted by *S. enterica* serovar Dublin and *S. flexneri* is an inositol phosphate phosphatase, which ultimately results in increased Cl$^-$ ion secretion by enterocytes and diarrhea in the host. The **ExoY adenylyl cyclase** of *Pseudomonas aeruginosa* is a calmodulin-independent enzyme that results in increased cAMP levels in host cells.

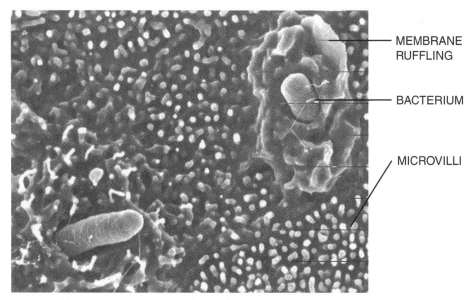

MEMBRANE RUFFLING

BACTERIUM

MICROVILLI

Fig. 20-4. Enhanced electron micrograph showing membrane ruffling and invasion of *Salmonella enterica* serovar Typhimuirum. Taken from Galán and Zhou. 2000. *Proc. Natl. Acad. Sci. USA* **97**:8754–8761. Copyright (2000) National Academy of Sciences, U.S.A.

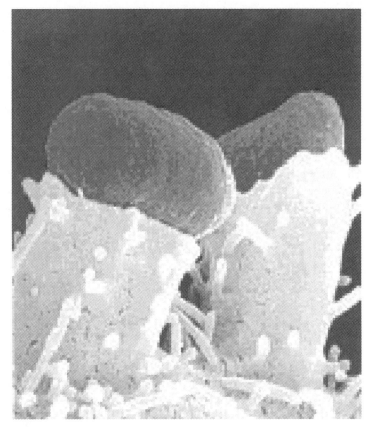

Fig. 20-5. Enhanced electron micrograph showing the formation of the "pedestal" under adhering enteropathogenic *Escherichia coli* (EPEC) cells. Taken from I. Rosenshine et al., 1996. *EMBO J* **15**:2613–2624.

EPEC strains secrete **translocated intimin receptor (Tir)**, into the host cell which functions as the receptor for the **intimin** adhesin that allows bacteria to attach, actin to be nucleated, and formation of a "pedestal" under the adhering bacterium (Fig. 20-5).

The **CagA** protein of *Helicobacter pylori* is secreted by a type IV secretion system. It is tyrosine phosphorylated by cellular kinases to stimulate pedestal formation in a similar manner to Tir.

Quorum Sensing

Although bacteria are generally thought of as single-cell organisms living solitary lives, it has become increasingly clear that this is not the case. In fact, many bacteria sense and respond to the presence of, and communicate with, other bacteria in their environment. These interactions are cell density dependent and may occur between individual cells of a single species and/or with other bacterial species. This phenomenon is referred to as quorum sensing (see Chapter 5). In general, each member in the population releases a signal molecule, called an autoinducer or pheromone, which can be sensed by neighboring cells. As the number of cells in the population increases, the concentration of autoinducer/pheromone increases. Once a threshold of autoinducer is reached, the bacteria respond by activating a transcriptional regulator protein that induces or represses a set of genes. This allows bacteria to gauge the cell density of their own population and/or other bacteria.

It is now apparent that many pathogens link quorum sensing to virulence factor secretion. This provides a mechanism for bacteria to achieve sufficient numbers in, and produce a successful infection of, the host before producing virulence factors that may elicit host defense systems. A possible rationale is that early in the infection sufficient nutrients are likely present to support growth of a small number of cells, but as cell numbers increase, nutrients become more limiting, so the bacteria produce virulence factors to cause the release of nutrients from host tissues/cells. Coincidently, by that point, they have reached sufficient numbers to withstand the host defenses. Studies of numerous pathogens have revealed that generally the mechanisms used by Gram-negative bacteria and Gram-positive bacteria are fundamentally different, particularly in the nature of the signal molecule employed.

Gram-Negative Bacterial Pathogens. Most quorum-sensing systems identified in Gram-negative bacteria use *N*-**acyl homoserine lactones** (AHL) as their autoinducer (AI) molecules. However, other signaling molecules have also been described that appear to be chemically different from AHL. These molecules then bind to and activate a transcriptional regulator, or R protein, which regulates the expression of various target genes. A partial list of Gram-negative bacterial autoinducer synthases and regulators along with their cognate autoinducers are presented in Table 20-2.

The paradigm system of quorum sensing is that of the squid parasite, *Vibrio fischeri*, which colonizes the light organ of squids (see Chapter 5). In this bacterium, AI is produced by the **LuxI protein** and binds to the **LuxR protein**; the LuxR–AI complex then transcriptionally activates genes necessary for bioluminescence as well as *luxI* itself. Thus, colonization of the squid light organ resulting in high cell density leads to sufficient AI production to induce the bioluminescence genes. Many Gram-negative bacterial pathogens utilize a similar quorum-sensing scheme.

The best-characterized quorum-sensing systems related to virulence are those of *Pseudomonas aeruginosa*. *P. aeruginosa* infects a wide range of organisms from invertebrates to plants to animals. In humans, it is a frequent opportunistic pathogen of patients who (1) are immunocompromised (e.g., cancer and AIDS patients), (2) are suffering from burns, (3) have indwelling medical devices, (4) are undergoing prolonged aggressive antimicrobial therapies or (5) are suffering from cystic fibrosis (CF). Its virulence reflects an extraordinary arsenal of cell surface–associated and secreted virulence factors. The expression of many of these virulence factors is found to be cell density dependent and under the control of at least two separate quorum sensing systems: *las* and *rhl*.

Virulence genes controlled by quorum sensing include *lasA* (staphylolytic protease), *lasB* (elastase), *apr* (alkaline protease), *toxA* (exotoxin A), and *xcp* (encode a type II secretion system). The *las* system is composed of the **LasR** transcriptional activator and **LasI** autoinducer synthase, which synthesizes the cognate AI for this system. A second quorum-sensing system, *rhl*, is composed of the **RhlR** transcriptional activator and **RhlI** autoinducer synthase, which produces the cognate AI for this system.

LasR and RhlR are LuxR homologues while LasI and RhlI are LuxI homologues. As in the *lux* system, LasR–AI and RhlR–AI complexes induce the expression of their cognate AI synthases as well as other genes. In addition to several virulence

TABLE 20-2. A Partial List of Quorum-Sensing Systems and Functions in Bacterial Pathogens

Bacteria	Regulator/AI Synthase	Signal Molecule[a]	Function
Vibrio fischeri	LuxR/LuxI	3-Oxo-C_6-HSL	Bioluminescence
Pseudomonas aeruginosa	LasR/LasI	3-Oxo-C_{12}-HSL	Virulence factor production; RhlR expression; biofilm formation
	RhlR/RhlI	C_4-HSL	Virulence factor production; rhamno-lipid synthesis; RpoS expression; secondary metabolites
Burkholderia cepacia	CepR/CepI	C_8-HSL	Virulence factor production
Escherichia coli (EHEC)	SdiA/?	Unknown	Cell division control; virulence factor production
Salmonella enterica serovar Typhimurium	SdiA/?	Unknown	Cell division control; virulence plasmid gene regulation
Erwinia carotovora	ExpR/ExpI	3-Oxo-C_6-HSL	Virulence factor production
	CarR/CarI	3-Oxo-C_6-HSL	Carbapenem production
Agrobacterium tumefaciens	TraR/TraI	3-Oxo-C_8-HSL	Ti plasmid conjugation
Aeromonas hydrophilia	AhyR/AhyI	C_4-HSL	Exoprotease production
Aeromonas salmonicida	AsaR/AsaI	C_4-HSL	Exoprotease production
Enterobacter agglomerans	EagR/EagI	3-Oxo-C_6-HSL	Unknown
Yersinia enterocolitica	YenR/YenI	C_6-HSL	Unknown
Yersinia pseudotuberculosis	YesR/YesI	C_8-HSL	Unknown

[a] HSL, homoserine lactone.

genes, RhlR–AI induces *rhlAB* (required for rhamnolipid biosynthesis), *rpoS* (the stress-/stationary-phase σ factor, σ^S), and the genes for the production of pyocyanin. *P. aeruginosa* appears to produce a number of minor AHL molecules, but the functions of these are not clear. A recent report indicates that *P. aeruginosa* induces another AI molecule that is chemically different from the AHL molecules. This AI appears to be a quinolone derivative, 2-heptyl-3-hydroxy-4-quinolone, known as PQS. The genes encoding the LuxI/LuxR homologues for this AI have yet to be identified.

However, PQS production is under LasR control, and RhlR is essential for its activity. In addition to controlling virulence gene expression, a direct role for quorum sensing in *P. aeruginosa* pathogenesis has been determined in the nematode, plant, and burned mouse models. Mutants deficient in quorum sensing or virulence factors controlled by quorum sensing are attenuated in all three models of *P. aeruginosa* virulence. Furthermore, transcripts for *lasR*, *lasA*, *lasB*, and *toxA* accumulate in *P. aeruginosa*, infecting the lungs of CF patients.

Quorum sensing has also been implicated in the related bacterium *Burkholderia cepacia*, which is an emerging pathogen in CF patients usually as a coinfecting agent with *P. aeruginosa*. *B. cepacia* possesses both a LuxR and LuxI homologue called **CepR** and **CepI**, respectively. It is not clear whether *B. cepacia* actually produces its cognate AHL in sufficient amounts to act as an AI in vivo. The same conditions that lead to the accumulation of *P. aeruginosa* AHLs result in a 1000-fold less amount of *B. cepacia* AHL. This, along with the findings that *P. aeruginosa* AHLs can increase the pathogenicity of *B. cepacia*, has led to the proposal that *B. cepacia* uses AIs synthesized by other bacteria to regulate aspects of its pathogenicity and thus profits from the energy expenditure made by other bacteria.

In this same vein, a similar scenario has been proposed for both *E. coli* and *Salmonella enterica* serovar Typhimurium. Both possess a LuxR homologue encoded by the ***sdiA* gene**. However, neither appears to possess a LuxI AI synthase homologue that produces a cognate AI molecule. Interestingly, they both can produce an AI molecule that induces the *Vibrio harveyi* luminescence genes, but the SdiA protein from both organisms does not respond to this AI. The SdiA protein of both organisms does respond to AI molecules from other bacterial species. Thus, it is proposed that pathogenic strains of *E. coli*, such as EHEC O157 : H7 strains, and *S.* Typhimurium can sense and respond to AHL produced by other bacteria in the vicinity via SdiA, which then goes on to regulate various target genes. Among the genes regulated by SdiA in both EHEC and *Salmonella* is the *ftsQAZ* operon. Virulence genes regulated by SdiA in EHEC include the genes encoding EspD and intimin, both of which are negatively regulated by SdiA; thus, SdiA appears to regulate colonization by O157 : H7 strains.

In *S.* Typhimurium, the SdiA–AI complex was determined to positively regulate four genes on the *S.* Typhimurium virulence plasmid that appear to compose an operon. These four *sdiA*-regulated genes are *srgA* (ORF8), *srgB* (ORF9), *rck*, and *srgC* (ORF11). The *srgA* gene encodes a *dsbA* homologue, suggesting that it may function as a disulfide bond isomerase involved in protein folding in the periplasm. The *srgB* gene encodes a putative secreted protein (possesses a lipoprotein signal sequence) that shows no significant homology to any gene in the databases. Rck is an outer membrane protein that confers resistance to complement killing (i.e., human serum resistance) and may also play a role in adherence to host cells. The *srgC* gene encodes a putative regulatory protein of the AraC family, although a function for SrgC has not been reported.

The plant pathogen *Erwinia carotovora* also employs quorum sensing to control virulence factors during its pathogenesis in plants. Key to this pathogenesis is the production of various plant tissue–degrading enzymes. The regulation of these exoenzymes is cell density dependent and controlled by the LuxR/LuxI homologues, **ExpR** and **ExpI**. Interestingly, while *expI* mutants are deficient in exoenzyme production, *expR* mutants do not exhibit any defects in exoenzyme production and ExpR overexpression decreases enzyme production, indicating that ExpR acts as a negative regulator. Additionally, *E. carotovora* possess a second quorum-sensing system involved in the production of the antibiotic carbapenem. This second quorum-sensing system involves the **CarR** and **CarI**, LuxR/LuxI homologues. In this system, CarR–AI positively regulates the carbapenem biosynthetic genes. The carbapenem would then inhibit any competing bacteria from participating in the banquet of nutrients released from the degradation of plant tissues. Thus, this bacterium links the regulation of plant tissue degradation and carbapenem production.

The etiologic agent of crown gall tumors in plants, *Agrobacterium tumefaciens*, uses quorum sensing to mediate conjugal transfer of its Ti plasmid. The Ti plasmid can be transferred to the nucleus of plant cells, where it induces tumor formation. The Ti plasmid is also a conjugative plasmid possessing *tra* genes capable of mediating its transfer via conjugation between bacterial cells. The **TraR** protein, a LuxR homologue, and an autoinducer synthase, **TraI**, positively control conjugal transfer of the Ti plasmid. The TraR–cognate AI complex positively regulates *tra* regulon expression, including *traI*. TraR exhibits regulation by either octopines or agrocinopines A and B through the regulatory proteins OccR (octopine-responsive activator) and AccR (agrocinopines-responsive repressor). *A. tumefaciens* can carry one of two types of Ti plasmids that direct the production of octopines or agrocinopines by the plant and their utilization by the bacteria: octopine-type Ti plasmids or nopaline-type Ti plasmids, respectively. Thus, accumulation of octopine results in OccR activation of *traR* expression, or accumulation of agrocinopines A and B relieves repression by AccR, leading to increased *traR* gene expression; then, once a cell density is achieved for sufficient AI accumulation to occur to activate TraR, the *tra* genes are expressed and conjugative transfer is stimulated.

Gram-Positive Bacterial Pathogens. The quorum-sensing systems of Gram-positive bacteria are fundamentally different from those of Gram-negative bacteria. They do not produce AHL and use sensor-kinase/regulator two-component signal transduction systems. Instead of AHL molecules, they produce peptide signals that interact with the sensor-kinase, which then phosphorylates the cognate regulator protein. Quorum sensing regulates a number of processes in Gram-positive bacteria such as conjugation in *Enterococcus faecalis* and competence in *Bacillus subtilis* and *Streptococcus pneumoniae*.

A good example of how quorum sensing is used to control virulence in Gram-positive bacteria is its role in regulating *Staphylococcus aureus* virulence. Like *P. aeruginosa*, *S. aureus* is the etiologic agent of a variety of diseases ranging from relatively minor skin lesions (folliculitis) to potentially fatal pneumonias or bacteremias. *S. aureus* also produces an impressive collection of virulence factors expressed in stages during infection. Early on, the bacterium synthesizes a number of surface-associated factors involved in adherence to host tissues (collagen and fibronectin-binding proteins) and host defense evasion (Protein A). Presumably these factors allow *S. aureus* to grow and multiply to reach a local high cell density. At this point, surface factor expression is downregulated and expression of a plethora of secreted products is increased, including several lipases and proteases, hyaluronidase, hemolysins, α-toxin, and enterotoxins.

This successive regulation of virulence factors is under the control of two regulatory loci: *agr* (accessory gene regulator) and *sar* (staphylococcal accessory gene regulator). The *agr* locus contains two divergently transcribed operons encoding two polycistronic mRNAs referred to as RNAII and RNAIII. The *agrBDCA* operon (RNAII) encodes **AgrC**, a signal transducer protein, and **AgrA**, a response regulator. AgrB and AgrD are involved in producing the quorum-sensing signal molecule, an octopeptide cleaved from AgrD and exported extracellularly by the AgrB protein. RNAIII encodes the δ-hemolysin and also functions as a regulatory RNA molecule. During quorum sensing,

AgrC binds to the peptide signal molecule, phosphorylates itself (autophosphorylation), and then phosphorylates AgrA. Phospho-AgrA (AgrA~P) then induces RNAIII expression, which goes on to stimulate the expression of numerous secreted products and to downregulate specific surface protein expression as well as increase the expression of *agrBDCA*. Additionally, the SarA protein is a positive regulator, inducing the expression of both RNAII and RNAIII.

PARADIGMS OF BACTERIAL PATHOGENESIS

Enteropathogenic *Escherichia coli*

Enteropathogenic *E. coli* (EPEC) strains cause enteritis not by invading the intestinal epithelial cell but by tightly adhering to the enterocyte surface and inducing cytoskeletal reorganization, producing attachment and effacement (A/E) lesions. EPEC strains bind to the apical surface of the enterocyte via EAF (EPEC adherence factor) plasmid-encoded bundle-forming pili (BFP). This initial adherence is referred to as localized adherence (LA) and is the first stage of EPEC pathogenesis.

LA is followed by induction of a number of signal transduction pathways including increases in intracellular inositol triphosphate (IP$_3$) and Ca^{+2} concentrations. This signal transduction induction is dependent on gene products encoded within the locus of enterocyte effacement (LEE) found on a 35 kb pathogenicity island on the EPEC chromosome. LEE encodes a type III secretion system (*sep/esc*; TTSS), the secreted proteins EspA-D, and the adhesin intimin and its secreted receptor Tir. These are all needed for host cell signal transduction that results in intimate adherence, which contributes to A/E lesion formation.

Translocation of Tir, and its subsequent tyrosine phosphorylation, and the EspA, EspB, and EspD proteins, requires the TTSS while the EspC protein is a member of the autotransporter family and mediates its own export. The interaction of intimin with phospho-Tir mediates tight intimate adherence of the bacterial cell to the host cell surface. The effect of intimate adherence and secreted proteins on the cell cytoskeleton (colocalization of cytoskeletal components) results in the formation of a pedestal structure (Fig. 20-5) characteristic of the A/E lesion. Similar A/E lesions are induced by enterohemorrhagic *E. coli* (EHEC) O157:H7, rabbit-diarrheal *E. coli* (RDEC-1), *Citrobacter rodentium*, and *H. pylori* strains.

Salmonella Enterica Serovars

S. Typhimurium, shortly after adherence to the apical surface of enterocytes, stimulates a striking response at the cell surface. It causes an outward extrusion of the surface at the site of adherence referred to as **membrane ruffling** and entry into the cell via a macropinocytosis process (Fig. 20-4). Furthermore, Ca^{+2} and IP$_3$ levels increase activating their cognate signal transduction pathways and actin filaments are reorganized. *Salmonella* invasion is rapid, with the bacteria being internalized in vacuoles within just a few minutes. Interestingly, the host cell surface structure returns to normal soon afterward in a process promoted by bacterial-encoded products. Genes encoding invasion factors are located in one of several pathogenicity islands identified on the *S. enterica* chromosome, referred to as SPI-1. SPI-1 encodes a TTSS and several secreted proteins. Several of these secreted proteins mediate actin reorganization and membrane ruffling by regulating host cell Rho GTPases or functioning as a IP_3 phosphatase.

Salmonella serovars are found in membrane-bound vacuoles within both non-phagocytic and phagocytic cells. These ***Salmonella*-containing vacuoles (SCV)** possess host cell markers that appear to be determined by some bacterial-encoded function(s). Interestingly, the presence of SCV membrane markers changes over time. Furthermore, the SCV escapes normal endocytic trafficking in both enterocytes and macrophages. Intracellular survival is dependent on genes located in a second pathogenicity island, SPI-2. SPI-2 encodes another TTSS as well as several secreted effector proteins. At least one of these, SpiC, inhibits endosomal trafficking. The SCV is larger than normal and is sometimes referred to as a spacious phagosome. It is proposed that *Salmonella* can actively prevent fusion of the phagosome with lysosomes; however, this appears to be incomplete, since there is evidence that they can and do survive within the phagolysosome. Clearly, though, the phagosome or phagolysosome that they occupy is different from the normal phagosome or phagolysosome. One key event that is critical for virulence of these bacteria is the acidification of the immediate milieu in the SCV. Another interesting aspect of *Salmonella* invasion is that entry into macrophages frequently induces apoptosis while entry into epithelial cells does not. This suggests that *Salmonella* enters epithelial cells and macrophages by two distinct mechanisms.

Essential to *Salmonella* pathogenesis is the entry into and passage through specialized epithelial cells called M cells with subsequent release of the bacteria into the submucosa. In the submucosa they typically enter macrophages where the bacteria can survive and multiply. Survival in macrophages is promoted by functions encoded on SPI2 and regulated by the PhoP/PhoQ two-component regulatory system. PhoP/PhoQ sense and respond to low Mg^{+2} ion concentrations, activating and repressing specific genes. Dissemination from the intestinal tract or the associated lymphoid tissue is influenced by the host and the serovar involved. In mice, *S.* Typhimurium can disseminate systemically to the spleen and liver, whereas in humans it remains localized within the intestines and does not typically spread beyond the regional lymph nodes.

Systemic spread is dependent on functions encoded on pathogenicity islands and the virulence plasmid. Specifically, four genes, *spvABCD* (*spv*, meaning *Salmonella* plasmid virulence), composing an operon on the virulence plasmid, are needed for systemic spread of *Salmonella* in appropriate hosts. The expression of this operon is dependent on an upstream regulatory gene, *spvR*, which is divergently transcribed from the *spvABCD* operon. Expression of *spvR*, and thus *spvABCD*, is induced by a specific set of environmental conditions found within macrophages. This regulation assures that the *spv* operon is expressed inside the correct host niche. Serovars causing enteritis in humans do not typically disseminate. However, entry of these organisms into the intestinal epithelium stimulates the epithelial cell to produce a number of chemokines (soluble proteins that act as neutrophil chemotaxins) that cause neutrophils to migrate to the area, transmigrate across the epithelial cell layer into the intestinal lumen, and contribute to inflammation and ulceration of the bowel wall.

Listeria Monocytogenes

L. monocytogenes is a Gram-positive bacterium that can actively invade both nonphagocytic and phagocytic cells. Its invasion is mediated by InlA, a member of the internalin family of bacterial invasion proteins. Like the Ipa proteins of *Shigella*, InlA mediates epithelial cell invasion at the basolateral rather than apical surface of the cell. Not coincidentally, the InlA receptor is E-cadherin, a cell-to-cell adhesin molecule expressed on the basolateral surface of epithelial cells. A related surface molecule called InlB mediates entry into hepatocytes. InlA functions in epithelial cell invasion but not hepatocytes, while the opposite is true for InlB. Upon internalization, the bacteria escape the vacuole through the action of **listeriolysin O (LLO)** (see this chapter under surface acting toxins) and perhaps additional phospholipase C activities. LLO is activated by low pH, which may be achieved in the initial vacuole in which it resides. Interestingly, only a relatively small percentage of internalized bacteria escape the vacuole into the cytoplasm; those that do not are killed.

Free in the cytoplasm, *L. monocytogenes* directs the condensation of actin at one pole of the bacterial cell and reorganizes it into a **polar tail** or **"comet" of polymerized actin** that moves the bacteria through the cytoplasm and into adjacent host cells. The listerial surface protein ActA localizes to one pole of the cell and mediates this actin reorganization and accumulation. Additional host cytoskeletal components and bacterial proteins also appear to be involved. This allows the bacteria to infect and spread within host tissue evading typical host immune responses. *Shigella flexneri* exhibits a similar pathogenic scheme. It also can escape the internalized vacuole, can polymerize actin to form comets, and can spread from cell to cell (Fig. 20-6).

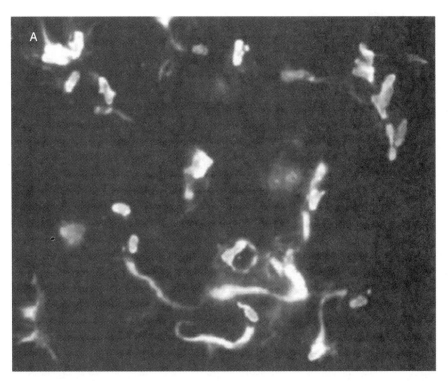

Fig. 20-6. Immunofluorescent photomicrograph showing actin tail or "comets" associated with individual *Shigella* bacterium. Courtesy of *Trends in Microbiology* (1996) Vol. 4, no. 6, p. 220.

Chlamydia spp

Unlike *Salmonella* and *Listeria*, which are **facultative intracellular pathogens**, chlamydiae are **obligate intracellular pathogens** and can only grow within living cells. They can be found in two cell forms: the **elementary body (EB)**, which can adhere to and invade host cells but cannot replicate itself, and the **reticulate body (RB)**, which is the intracellular actively growing cell form but is not infectious. Binding of the EB to the target cell results in the uptake of the cell into a vacuole. This vacuole remains nonacidic throughout the time spent inside the cell. It is atypical in that it lacks significant amounts of host markers in the vacuole membrane. This vacuole also avoids the normal endocytic and exocytic trafficking pathways and does not fuse to lysosomes. Within the vacuoles, the EB converts to the RB and begins to divide. RB multiplication leads to the formation of a cell-packed inclusion in the infected cell. Interestingly, sphingomyelin from the Golgi apparatus of the cell gets incorporated into the inclusion and eventually into the cell surface of the bacteria. The bacteria also appear to secrete proteins that become incorporated into the membrane of the bacterial-laden vacuole, which are proposed to direct vacuolar trafficking so as to promote intracellular growth of the RB. Late in the infection cycle, the RB cell form converts to the EB, which is released from the cell by cell lysis.

BIBLIOGRAPHY

Overview

Casadevall, A., and L.-A. Pirofski. 1999. Host–pathogen interactions: redefining the basic concepts of virulence and pathogenicity. *Infect. Immun.* **67**:3703–13.

Casadevall, A., and L.-A. Pirofski. 2000. Host–pathogen interactions: basic concepts of microbial commensalisms, colonization, infection and disease. *Infect. Immun.* **68**:6511–8.

Finlay, B. B., and S. Falkow. 1997. Common themes in microbial pathogenicity revisited. *Microbiol. Mol. Biol. Rev.* **61**:136–69.

Neish, A. S., A. T. Gewirtz, H. Zeng, A. N. Young, M. E. Hobert, V. Karmali, A. S. Rao, and J. L. Madara. 2000. Prokaryotic regulation of epithelial responses by inhibition of IκB-α ubiquitination. *Science* **289**:1560–3.

Xavier, R. J., and D. K. Podolsky. 2000. How to get along — friendly microbes in a hostile world. *Science* **289**:1483–4.

Adherence/Colonization

Baumler, A. J., R. M. Tsolis, and F. Heffron. 1996. The *lpf* fimbrial operon mediates adhesion of *Salmonella typhimurium* to murine Peyer's patches. *Proc. Natl. Acad. Sci. USA* **93**:279–83.

Edwards, R. A., and J. L. Puente. 1998. Fimbrial expression in enteric bacteria: a critical step in intestinal pathogenesis. *Trends Microbiol.* **6**:282–7.

Finlay, B. B., and S. Falkow. 1997. Common themes in microbial pathogenicity revisited. *Microbiol. Mol. Biol. Rev.* **61**:136–69.

Hanski, E., and M. Caparon. 1992. Protein F, a fibronectin-binding protein is an adhesin of the group A streptococcus, *Streptococcus pyogenes*. *Proc. Natl. Acad. Sci. USA* **89**:6172–6.

Sauer, F. G., K. Fütterer, J. S. Pickner, K. W. Dodson, S. J. Hultgren, and G. Waksman. 1999. Structural basis of chaperone function and pilus biogenesis. *Science* **285**:1058–66.

Schulze-Koops, H., H. Burkardt, J. Heeseman, T. Kirsch, B. Swoboda, C. Bull, S. Goodman, and F. Emmrich. 1993. Outer membrane protein YadA of enteropathogenic yersinae mediates specific binding to cellular but not plasma fibronectin. *Infect. Immun.* **61**:2513–9.

Virulence Factor Secretion Systems

Christie, P. J., and J. P. Vogel. 2000. Bacterial type IV secretion: conjugation systems adapted to deliver effector molecules to host cells. *Trends Microbiol.* **8**:354–60.

Fernandez, L. A., and V. de Lorenzo. 2001. Formation of disulphide bonds during secretion of proteins through the periplasmic-independent type I pathway. *Mol. Microbiol.* **40**:332–46.

Finlay, B. B., and S. Falkow. 1997. Common themes in microbial pathogenicity revisited. *Microbiol. Mol. Biol. Rev.* **61**:136–69.

Hueck, C. J. 1998. Type III protein secretion systems in bacterial pathogens of animals and plants. *Microbiol. Mol. Biol. Rev.* **62**:379–433.

Jacob-Dubuisson, F., C. Locht, and R. Antoine. 2001. Two-partner secretion in gram-negative bacteria: a thrifty, specific pathway for large virulence proteins. *Mol. Microbiol.* **40**:306–13.

Kubori, T., Y. Matsushima, D. Nakamura, J. Uralil, M. Lara-Tejero, A. Sukhan, J. Galán, and S.-I. Aizawa. 1998. Supramolecular structure of the *Salmonella typhimurium* type III protein secretion system. *Science* **280**:602–5.

Plano, G. V., J. B. Day, and F. Ferracci. 2001. Type III export: new uses for an old pathway. *Mol. Microbiol.* **40**:284–93.

Sandkvist, M. 2001. Biology of type II secretion. *Mol. Microbiol.* **40**:271–83.

Scherer, C. A., E. Cooper, and S. I. Miller. 2000. The *Salmonella* type III secretion translocon protein SspC is inserted into the epithelial cell plasma membrane upon infection. *Mol. Microbiol.* **37**:1133–45.

Warawa, J., B. B. Finlay, and B. Kenny. 1999. Type III secretion-dependent hemolytic activity of enteropathogenic *Escherichia coli*. *Infect. Immun.* **67**:5538–40.

Exotoxins

Covacci, A., J. L. Telford, G. Del Giudice, J. Parsonnet, and R. Rappuoli. 1999. *Helicobacter pylori* virulence and genetic geography. *Science* **284**:1328–33.

Devinney, R., O. Steel-Mortimer, and B. B. Finlay. 2000. Phosphatases and kinases delivered to the host cell by bacterial pathogens. *Trends Microbiol.* **8**:27–31.

Duesbery, N. S., and G. F. Van de Woude. 1999. Anthrax toxins. *Cell Mol. Life Sci.* **55**:1599–609.

Fu, Y., and J. E. Galán. 1999. A *Salmonella* protein antagonizes Rac-1 and Cdc42 to mediate host-cell recovery after bacterial invasion. *Nature* **401**:293–7.

Galán, J. E. and Y. Zhou. 2000. Striking a balance: modulation of actin cytoskeleton by *Salmonella*. *Proc. Natl. Acad. Sci. USA* **97**:8754–8761.

Hueck, C. J. 1998. Type III protein secretion systems in bacterial pathogens of animals and plants. *Microbiol. Mol. Biol. Rev.* **62**:379–433.

Johannes, L., and B. Goud. 1998. Surfing on a retrograde wave: how does Shiga toxin reach the endoplasmic reticulum? *Trends Cell Biol.* **8**:158–62.

Lara-Tejero, M., and J. E. Galán. 2000. A bacterial toxin that controls cell cycle progression as a deoxyribonuclease I–like protein. *Science* **290**:354–7.

O'Loughlin, E. V., and R. M. Robins-Browne. 2001. Effect of Shiga toxin and Shiga-like toxins on eukaryotic cells. *Microbes Infect.* **3**:493–507.

Orth, K., L. E. Palmer, Z. Q. Bao, S. Stewart, A. E. Rudolph, J. B. Bliska, and J. E. Dixon. 1999. Inhibition of the mitogen-activated protein kinase kinase superfamily by a *Yersinia* effector. *Science* **285**:1920–3.

Papageorgiou, A. C., and K. R. Acharya. Microbial superantigens: from structure to function. *Trends Microbiol.* **8**:369–75.

Pickett, C. L., and C. A. Whitehouse. 1999. The cytolethal distending toxin family. *Trends Microbiol.* **7**:292–7.

Rappouli, R., and M. Pizza. 2000. Bacterial toxins. In P. Cossart, P. Boquet, S. Nomark, and R. Rappuoli (eds.), *Cellular Microbiology.* American Society for Microbiology Press, Washington, DC, pp. 265–291.

Rosenshine, I., S. Ruschkowski, M. Stein, D. J. Reinscheid, S. D. Mills and B. B. Finlay. 1996. Apathogenic bacterium triggers epithelial signals to form a functional bacterial receptor that mediates actin pseudopod formation. *EMBO J.* **15**:2613–2624.

Sears, C. L., and J. B. Kaper. 1996. Enteric bacterial toxins: mechanisms of action and linkage to intestinal secretion. *Microbiol. Rev.* **60**:167–215.

Shatursky O, A. P. Heuck, L. A. Shepard, J. Rossjohn, M. W. Parker, A. E. Johnson, and R. K. Tweten. 1999. The mechanism of membrane insertion for a cholesterol-dependent cytolysin: a novel paradigm for pore-forming toxins. *Cell* **99**:293–9.

Zumbihl, R., M. Aepfelbacher, A. Andor, C. A. Jacobi, K. Ruckdeschel, B. Rouot, and J. Heesemann. 1999. The cytotoxin YopT of *Yersinia enterocolitica* induces modification and cellular redistribution of the small GTP-binding protein RhoA. *J. Biol. Chem.* **274**:29,289–93.

Quorum Sensing

Ahmer, B. M., J. van Reeuwijk, C. D. Timmers, P. J. Valentine, and F. Heffron. 1998. *Salmonella typhimurium* encodes an SdiA homolog, a putative quorum sensor of the LuxR family, that regulates genes on the virulence plasmid. *J. Bacteriol.* **180**:1185–93.

De Kievit, T. R., and B. H. Iglewski. 2000. Bacterial quorum-sensing in pathogenic relationships. *Infect. Immun.* **68**:4839–49.

Dunny, G. M., H. Hirt, and S. Erlandsen. 1999. Multiple roles for enterococcal sex pheromone peptides in conjugation, plasmid maintenance and pathogenesis. In R. England, G. Hobbs, N. Bainton, and D. M. Roberts (eds.), *Microbial Signaling and Communication.* University Press, Cambridge, U.K., pp. 117–38.

Dunny, G. M., and B. A. B. Leonard. 1997. Cell-cell communication in Gram-positive bacteria. *Annu. Rev. Microbiol.* **51**:527–64.

Fuqua, W. C., S. C. Winans, and E. P. Greenberg. 1996. Census and consensus in bacterial ecosystems: the LuxR-LuxI family of quorum-sensing transcriptional regulators. *Annu. Rev. Microbiol.* **50**:727–51.

Kanamaru, K., K. Kanamaru, I. Tatsuno, T. Tobe, and C. Sasakawa. 2000. SdiA, an *Escherichia coli* homologue of quorum-sensing regulators, controls the expression of virulence factor in entreohaemorrhagic *Escherichia coli* O157 : H7. *Mol. Microbiol.* **38**:805–16.

Michael, B., J. N. Smith, S. Swift, F. Heffron, and B. M. Ahmer. 2001. SdiA of *Salmonella enterica* is a LuxR homolog that detects mixed microbial communities. *J. Bacteriol.* **183**:5733–42.

Pesci, E. C., J. B. Milbank, J. P. Pearson, S. McKnight, A. Kende, E. P. Greenberg, and B. H. Iglewski. 1999. Quinolone signaling in the cell-to-cell communication systems of *Pseudomonas aeruginosa.* *Proc. Natl. Acad. Sci. USA* **96**:11229–34.

Pierson, L. S., III, D. W. Wood, and S. Beck von Bodman. 1999. Quorum-sensing in plant-associated bacteria. In G. M. Dunny and S. C. Winans (eds.), *Cell-Cell Signaling in Bacteria.* American Society for Microbiology Press, Washington, DC, pp. 101–16.

Surrette, M. G., and B. L. Bassler. 1999. Regulation of autoinducer production in *Salmonella typhimurium. Mol. Microbiol.* **31:**585–95.

Wu, H., Z. Song, M. Giskov, G. Doring, D. Worlitzsch, K. Mathee, J. Rygaard, and N. Høiby. 2001. *Pseudomonas aeruginosa* mutations in *lasI* and *rhlI* quorum-sensing systems result in milder chronic lung infection. *Microbiology* **147:** 1105–13.

Paradigms of Bacterial Pathogenesis

Buchmeier, N. A., and F. Heffron. 1991. Inhibition of macrophage phagosome-lysosome fusion by *Salmonella typhimurium. Infect. Immun.* **59:**2232–38.

Ernst, R. K., T. Guina, and S. I. Miller. 1999. How intracellular bacteria survive: surface modifications that promote resistance to host innate immune responses. *J. Infec. Dis.* **179(Suppl 2):**S326–30.

Garcia-del Portillo, F., and B. B. Finlay. 1995. Targeting of *Salmonella typhimurium* to vesicles containing lysosomal membrane glycoproteins bypasses compartments with mannose 6-phosphate receptors. *J. Cell Biol.* **129:**81–97.

Garcia-del Portillo, F., and B. B. Finlay. 1995. The varied lifestyles of intracellular pathogens within eukaryotic vacuolar compartments. *Trends Microbiol.* **3:**373–80.

Hacker, J., and J. B. Kaper. 2000. Pathogenicity islands and the evolution of microbes. *Annu. Rev. Microbiol.* **54:**641–79.

Hackstadt, T., M. A. Scidmore, and D. D. Rockey. 1995. Lipid metabolism in *Chlamydia trachomatis*-infected cells: directed trafficking of Golgi-derived sphingolipids to the chlamydial inclusion. *Proc. Natl. Acad. Sci. USA* **92:**877–4881.

Kenny, B., and B. B. Finlay. 1995. Protein secretion by enteropahtogenic *Escherichia coli* is essential for transducing signals to epithelial cells. *Proc. Natl. Acad. Sci. USA* **92:**7991–5.

Mengaud, J., H. Ohayon, P. Gounon, R. M. Mege, and P. Cossart. 1996. E-cadherin is the receptor for internalin, a surface protein required for entry of *L. monocytogenes* into epithelial cells. *Cell* **84:**923–32.

Monack, D. M., B. Raupach, A. E. Hromocky, and S. Falkow. 1996. *Salmonella typhimurium* invasion induces apoptosis in infected macrophages. *Proc. Natl. Acad. Sci. USA* **93:**9833–8.

Moulder, J. W. 1991. Interaction of chlamydiae and host cells in vitro. *Microbiol. Rev.* **55:**143–90.

Nataro, J. P., and J. B. Kaper. 1998. Diarrheagenic *Escherichia coli. Clin. Microbiol. Rev.* **11:**142–201.

Smith, G. A., D. A. Portnoy, and J. A. Theriot. 1995. Asymmetric distribution of the *Listeria monocytogenes* ActA protein is required and sufficient to direct actin-based motility. *Mol. Microbiol.* **17:**945–51.

INDEX